DIAGNOSTIC PROCEDURES FOR MYCOTIC AND PARASITIC INFECTIONS

Prepared under the auspices of the
Committee on Laboratory Standards and Practices
of the
American Public Health Association

7th EDITION

Diagnostic Procedures For Mycotic And Parasitic Infections

Coordinating Editor

Berttina B. Wentworth, PhD

Editorial Board

Marilyn S. Bartlett - Barbara E. Robinson, PhD - Ira F. Salkin, PhD

APHA

American Public Health Association

Seventh Edition

AMERICAN PUBLIC HEALTH ASSOCIATION, INC.
1015 Fifteenth Street, NW
Washington, DC 20005

William H. McBeath, MD, MPH, Executive Director

Library of Congress Cataloging-in-Publication Data

Diagnostic procedures for mycotic and parasitic infections.
Together with: Diagnostic procedures for bacterial infections, constitutes a rev. ed. of:
Diagnostic procedures for bacterial, mycotic, and parasitic infections. 6th ed. c1981.
Wentworth, Berttina B.
Diagnostic procedures for mycotic and parasitic infections.

Includes index.
1. Diagnostic mycology. 2. Diagnostic parasitology.
I. Title.
QR248.W46 1988 616.9′69075 88-16681
ISBN 0-87553-156-3

3M10/88

Printed and bound in the United States of America
Typography: Monotype Composition Company, Inc., Baltimore, Maryland
Set in: *Times Roman, Helvetica*
Text and Binding: Port City Press, Baltimore, Maryland
Cover Design: Donya Melanson Associates, Boston, Massachusetts

TABLE OF CONTENTS

PREFACE

Knowledge in the fields of diagnostic mycology and parasitology has expanded so greatly since publication of the 6th edition of "Diagnostic Procedures for Bacterial, Mycotic and Parasitic Infections" in 1981 that these infections now require a volume of their own as a companion to "Diagnostic Procedures for Bacterial Infections." Hitherto clinically insignificant fungal and parasitic organisms have emerged as important opportunistic pathogens for immunosuppressed patients and diagnoses of infections in patients with the acquired immune deficiency syndrome (AIDS) have brought these opportunists to worldwide attention and concern.

Consideration of these opportunistic fungi and parasites has lead to considerable expansion in almost all the chapters of this volume. There is no longer a separate chapter on opportunistic fungal agents, but they have been covered as appropriate in each of the chapters in the Mycology Section. Discussion of phaeohyphomycosis and mycetoma are now incorporated into the chapter on "Molds Involved in Subcutaneous Infections," which describes the phaeohyphomycoses and hyalohyphomycoses. The medical importance of the aerobic actinomycetes has led to inclusion of a new chapter dealing with these organisms.

The advent of new antifungal agents and new serologic tests for diagnosis of fungal diseases has required significant expansion of the chapters covering these topics. A glossary of terminology and a chapter on media and stains are new for the Mycology Section.

Expansion of tests for diagnosis of parasitic diseases has required a new, separate chapter for serology of parasitic infections and a very important new chapter on the care and use of the microscope has been added to the Parasitology Section. Although the microscope is the most important piece of equipment used by the mycologist and parasitologist, few laboratories give this essential component the care it deserves and needs. This new chapter has very practical and well described instructions for obtaining optimum results from microscope use. It will be important reading for bacteriologists also.

As a service to readers, we have provided trade and proprietary names as sources of laboratory equipment and supplies, but such information does not constitute endorsement by either the editors or the American Public Health Association. A list of the names and addresses of the manufacturers cited appears at the end of the book.

The editors wish to thank all the authors who contributed to this book for their excellent chapters and to thank the publications staff of APHA for their continuing support and encouragement.

We ask that readers who have questions concerning any material in this book please refer all inquiries to the Coordinating Editor, Berttina B. Wentworth, rather than to individual authors or editors. Questions can then be referred to the appropriate person.

Berttina B. Wentworth, Ira F. Salkin, Barbara E. Robinson and Marilyn S. Bartlett.

CONTRIBUTORS

Marilyn S. Bartlett, M.S.
Professor, Department of Pathology
Indiana University School of Medicine
Indianapolis, IN 46223

Donald L. Greer, Ph.D.
Director, Mycology and Mycobacteriology Laboratory
Louisiana State University Medical Center
New Orleans, LA 70112

John H. Haines, Ph.D.
Senior Scientist in Mycology
New York State Museum
Albany, NY 12230

Anne W. Howell, B.S., MT (ASCP)
Medical Technologist
Barnes Hospital
St. Louis, MO 63110

Geoffrey A. Land, Ph.D.
Director, Microbiology and Immunology
Methodist Medical Center
Dallas, TX 75265

Earl G. Long, Ph.D.
Parasitologist
Parasitic Disease Control
Centers for Disease Control
Atlanta, GA 30333

Frank L. Lyon, Ph.D.
Assistant Professor
Department of Microbiology and Immunology
Kirksville College of Osteopathic Medicine
Kirksville, MO 63502

Michael R. McGinnis, Ph.D.
Professor, Department of Pathology
University of Texas Medical Branch
Galveston, TX 77550

Dorothy Mae Melvin, Ph.D.
Division of Laboratory Training and Consultation
Centers for Disease Control
Atlanta, GA 30333

Thomas G. Mitchell, Ph.D.
Director, Clinical Mycology Laboratory
Duke University Medical Center
Durham, NC 27710

Stephen A. Moser, Ph.D.
Director, Clinical Microbiology and Serology
Washington University Medical Center
St. Louis, MO 63178

Anthony D. Oldham, B.Sc.
President
Scientific Instrument Center, Inc.
Westerville, OH 43081

Arvind A. Padhye, Ph.D.
Chief, Fungus Reference and Investigative Branch
Centers for Disease Control
Atlanta, GA 30333

David H. Pincus, M.S.
Senior Scientist
Analytab Products—Division of Sherwood Medical
Plainview, NY 11803

Harry D. Pratt, Ph.D.
Vector Borne Disease Control
Training Service
Centers for Disease Control
Atlanta, GA 30333

Michael G. Rinaldi, Ph.D.
Associate Professor
Department of Pathology and
Medical Technology
University of Texas Health Science
Center
San Antonio, TX 78284

Barbara E. Robinson, Ph.D.
Chief, Diagnostic Microbiology
Section
Michigan Department of Public
Health
Lansing, MI 48909

Stanley A. Rosenthal, Ph.D.
Associate Professor of
Experimental Dermatology
New York University Medical
Center
New York, NY 10016

Ira F. Salkin, Ph.D.
Director, Laboratories for
Mycology
New York State Department of
Health
Albany, NY 12201

Wiley A. Schell, M.S.
Clinical Mycology Laboratory
Duke University Medical Center
Durham, NC 27710

Margarita Silva-Hutner
Consultant in Mycology
Columbia-Presbyterian Medical
Center
New York, NY 10032

James W. Smith, M.D.
Professor, Department of Pathology
Indiana University School of
Medicine
Indianapolis, IN 46223

Joseph L. Staneck, Ph.D.
Professor, Department of Pathology
and Laboratory Medicine
University of Cincinnati Medical
Center
Cincinnati, OH 45267

Kenneth W. Walls, Ph.D.
Division of Parasitic Diseases
Centers for Disease Control
Atlanta, GA 30033

Irene Weitzman, Ph.D.
Chief, Mycology and
Mycobacteriology Laboratory
New York City Department of
Health
New York, NY 10016

CHAPTER 1

THE MEDICALLY IMPORTANT FUNGI: AN INTRODUCTION

John H. Haines, Ph.D. and Ira F. Salkin, Ph.D.

INTRODUCTION

The fungi, like the proverbial elephant experienced by the blind men, can be defined in different ways. To some, fungi are mushrooms and toadstools or considered edible as opposed to poisonous. To others, they are divided into yeasts and molds or pathogens and nonpathogens. The fungi are all of these and more. To obtain a better understanding of medically important fungi, it is best to consider human pathogens within the framework of a broad and diverse assemblage of organisms having an inextricable role in nature. In addition, one must also note that all major groups of fungi contain genera and species which affect humans in one way or another. However, it is important to keep in mind that less than 200 of the more than 60,000 known fungal species are pathogenic for humans and/or lower animals.[4]

The criteria used to differentiate the fungi from other organisms, as well as criteria used to distinguish the major groups within the fungi, are constantly changing as we increase our knowledge of mycology. The names and classification schemes used in this book represent only a point on this everchanging continuum.

The fungi are eukaryotic (possess an organized nucleus and associated subcellular structures), heterotrophic (require an exogenous carbon source), achlorophyllous (lack chlorophyll) microorganisms. They release extracellular enzymes to break down the substrates upon which they are growing, permitting absorption of required nutrients into their multinucleate body forms. The developing body may be unicellular, filamentous or both, and is generally surrounded by a rigid, protective cell wall. The cell wall is chiefly composed of chitin, or less frequently, of cellulose-like materials. Fungi may reproduce sexually, asexually or both, through the formation of unicellular or multicellular spores. The fungi may be saprobes, parasites or mutualists, but the boundaries among these categories are seldom firm and many fungi change their life styles as dictated by environmental conditions.

VEGETATIVE VERSUS REPRODUCTIVE

All fungi have an assimilative, growing or vegetative phase and a spore-bearing or reproductive phase. These stages in a fungal life cycle may occur within the same time and space and be morphologically identical or they may occur at different times, on different substrates and be morphologically dissimilar. For example, the vegetative and reproductive cells of *Saccharomyces cerevisiae* (Baker's yeast) are one and the same, but the formation of spores within the cells occurs only after vegetative development has ceased. Alternatively, although the vegetative and reproductive stages of mushroom-forming fungi may occur at the same time and on the same substrate, the vegetative stage is morphologically quite dissimilar from the spore-bearing phase.

The vegetative stage of most fungi (except of yeasts) is composed of filaments (hyphae) which interweave to form a mat (mycelium). The hyphae are divided by cross-walls (septa) into cell-like units. However, the septa are not complete and bear an opening through which cytoplasm, subcellular structures and even nuclei may migrate. In this sense the fungi are functionally unicellular and multinucleate with the entire mycelium acting as a single cell.

Although the vegetative phases of fungi are physiologically diverse due to their adaptation to a wide variety of nutrient substrates, they are, in general, morphologically indistinguishable from each other. Except in rare cases, it is impossible to identify a fungus on the basis of the appearance of its mycelium. The vegetative phase may be as short-lived as a few days or as long-lived as several decades. However, in all instances it precedes the reproductive stage.

The reproductive stage is not involved in the assimilation of nutrients from the substrate, but rather derives its entire nutrition from the vegetative phase. The reproductive phases of fungi, in contrast to the vegetative stages, exhibit great morphological diversity, with no two fungi being alike. Reproduction in fungi may be as simple as the formation of single spores at the tips of nonspecialized hyphae or as complex as the development of spores in or on highly modified hyphae within macroscopic structures composed of pseudotissues known as fruiting bodies, e.g., mushrooms. It is this great diversity in the mechanisms of spore formation, the morphology of the spores and the structures in which they may be formed which serves as the basis of fungal identification. However, fungi may cease spore formation when isolated and grown on artificial media. The subsequent chapters describe the media and techniques which have been developed to induce reproductive phases in culture which will permit identification of fungal pathogens in clinical laboratories.

Molds, Yeasts and Dimorphism

The term mold is one of convenience and has no taxonomic validity. It is used to describe virtually any fungus whose vegetative growth consists of an interwoven mat of hyphae. The term is applied to fungi as diverse as primitive water molds to the more advanced fungal forms which are pathogens of humans.

The term yeast when used in the taxonomic sense refers to a more or less homogenous group of fungi whose vegetative growth form is a single cell and which share biochemical, physiological and reproductive characteristics. Specifically, yeast refers to members of the Endomycetales of the Ascomycotina. In contrast, yeastlike is simply a morphologic definition and has no taxonomic status. It is applied to a rather heterogenous group of fungi which share one characteristic in common, i.e., the vegetative growth form is similar to a true yeast in that it consists of a single cell. Growth takes place by the enlargement and duplication of the single cell through either fission (simple cell division) or more commonly through a process called budding. For a more detailed description of both yeasts and yeastlike fungi, the reader should consult Chapter 6, "Human Infections Caused by Yeastlike Fungi."

Dimorphism refers to the ability of a very limited number of fungi to form both a moldlike and a yeastlike vegetative form. The term is used primarily in association with zoopathogens which are yeastlike (Y form) in the host and moldlike (M form) on artificial media at room temperature. The conditions favoring development of the Y form in most of these fungi are: 1) an incubation temperature of 37°C, 2) a relatively high CO_2 tension of 5-10%, 3) the presence of organic sulfur-containing nutrients and 4) high hexose carbohydrate concentrations in the growth medium.[2] Depending on the organism involved, if one or more of these criteria are not met, the yeast may convert to a predominance of the M form.

Dimorphism should not be confused with pleomorphism. The latter has long been misapplied in medical mycology to those fungi which lose their ability to form spores or conidia when maintained on artificial media. In contrast, pleomorphism should only be used to refer to the ability of a fungus to produce more than one conidia-forming state.

Anamorph versus Teleomorph

Although life cycles of the fungi are varied and complex, those of human pathogens, which are predominantly Deuteromycotina exhibit variations on a single theme. The vegetative mycelium grows throughout the nutrient substrate and with time, forms conidia which are capable of producing new, identical colonies. Conidia are spores borne on specialized hyphal branches (conidiophores) and are asexual, i.e., the nuclei which they contain have been formed by mitotic divisions. Consequently, the conidia are genetically identical. The mycelium with its conidia is the anamorphic state. It is haploid at all times.

The same mycelium that produces conidia may form a physical connection to another mycelium of a compatible mating strain, which ultimately leads to the sexual stage. The diploid cells that result from the physical connection are referred to as either asci or basidia. It is in or on these structures that the haploid sexual spores are formed through meiosis. The sexual stage of most of these fungi is borne within a larger, protective structure called a fruiting body. The mycelium with its fruiting bodies and sexual spores is the teleomorphic state.

The life cycle of a fungus is seldom equally divided between these two states and in many instances one state may be rare or missing altogether. In medically important fungi, it is the teleomorph state which is generally missing or only induced on artificial media. However, it is of theoretical and practical importance to identify the teleomorph since the majority of our knowledge of systematic relationships is based on the sexual state. Locating a fungus in the taxonomic framework is more than a mere academic exercise since information as to the organism's ecology, physiology, morphology and other pertinent characteristics can be obtained from this identification.

Taxonomy and Nomenclature[3]

Taxonomy is the study of the relationship of organisms, whereas nomenclature is the system used to apply names to these organisms. The application of names to fungi, as with other organisms, follows specific rules. The rules are used to provide stability in the names of organisms and to establish a hierarchical system of classification which reflects phylogenetic relationships. The rules for naming fungi are contained within the International Code of Botanical Nomenclature (ICBN), in which special provisions have been made to meet the natural idiosyncracies of the fungi.[11]

The ICBN stipulates that each fungus shall have one and only one correct name and that name shall be a binomial composed of the genus and species epithets. It is not surprising, given the morphologic and physiologic variability in fungi and the inherent problems in identifying these microorganisms, that a single fungus may have been given several names by different investigators. For such circumstances, the ICBN stipulates that the first genus and/or species name published for the fungus is the correct one. This rule of priority has a starting date of 1753, the date of publication of Linnaeus' major work on plant nomenclature. Thus, in selecting a name for a new genus and/or species one must be certain that it has not been used since 1753 as the name for any other fungus. Fortunately, virtually none of the names for medically important fungi date from this era.

Another provision of the ICBN requires that each species, subspecies and variety be represented by a single, permanently preserved, type specimen. Although this rule is crucial to most of plant and fungal taxonomy, it is less than ideal for medically important fungi because it further stipulates that the type may not be a living culture. A dried, dead specimen cannot provide the biochemical, physiological or developmental characteristics essential for the identification of zoopathogens. Alternatively, although such characters may be determined from a living culture, a culture has a greater probability of becoming nonsporulating, contaminated or simply lost. The most practical solution to this problem for newly described, medically important fungi would be to provide for each: 1) a living culture, 2) a permanent dried specimen as required by present ICBN rules and 3) 12 or more lyophilized cultures. All of these representatives of the new fungus could be stored in a national or international fungal collection.

If the rules contained in the ICBN are to insure stability and to prevent arbitrary nomenclatural changes, why then do the names of fungi change? The most common reason is that definitions of the genera are revised as new information becomes available. For example, *Mucor pusillus* was found to be sufficiently different from other species within the genus *Mucor* that it could no longer be accommodated. As a result, the new genus *Rhizomucor* was erected for it.

A unique problem in fungal taxonomy is the presence of a teleomorph and an anamorph within a single life cycle. For practical reasons, both states are given binomials and are treated as individual entities even though they represent different portions of the life cycle of a single species. The ICBN specifies that the correct name for the whole fungus, all stages, is the name applied to the teleomorph. The anamorph is grouped into an artificial or form-division, the Deuteromycotina. For example, the medically important anamorph *Cryptococcus neoformans* was linked within the last decade to its rarely seen teleomorph, *Filobasidiella*.[6] If one follows the ICBN, the name of this fungus must now be *Filobasidiella neoformans*. However, since only the anamorphic states of zoopathogens are isolated from clinical specimens (with very few exceptions) and since clinicians are most familiar with the names applied to the anamorphs, medical mycologists can continue to use the anamorph binomials, such as *Cryptococcus neoformans*.

CLASSIFICATION OF FUNGI

The fungi are arranged in a hierarchical classification scheme which includes divisions, classes, orders, families, genera, species and intermediate categories of all of the above. It is a useful framework upon which to organize our knowledge of these organisms. However, the higher and more inclusive categories, i.e., orders and above, seem to be rearranged rather frequently as new theories and information are presented in the literature. We discuss in this chapter a conservative, somewhat fragmentary classification scheme, as it includes only those fungi which are most often implicated in human diseases.

Subdivision Zygomycotina (Zygomycetes)

These molds are characterized by: rapid growth on standard nutrient media, broad, nonseptate hyphae and a peculiar, thick-walled, single-celled sexual structure, the zygospore.[9] The anamorph-teleomorph concept is not applied to members of this group. Although most commonly saprobes, members of several genera within this subdivision have been implicated as the etiologic agents of opportunistic infections in humans and lower animals. These include *Absidia, Apophysomyces, Basidiobolus, Conidiobolus, Cunninghamella, Mortierella, Rhizomucor* and *Rhizopus*.[8]

Subdivision Ascomycotina (Ascomycetes)

Meiospores (ascospores) formed inside specialized cells (asci) characterize the members of this group. Although many members have anamorph states in the deuteromycetes, by definition, ascomycetes are all teleomorphic forms whether or not they have associated anamorphs. Sexual reproduction usually includes a stage whose ploidy is neither exactly haploid nor exactly diploid. Rather one finds non-fused, compatible pairs of genetically dissimilar nuclei within each cell of a restricted hyphal system (dikaryon phase) inside a fruiting body. The fruiting bodies of members of the Ascomycotina tend to be macroscopic, with some as large as common mushrooms. However, the teleomorphs of the human pathogens which are members of this group tend to be more primitive forms which develop very small or no fruiting bodies (Table 1:1).

Order Endomycetales ("true" yeasts)

This group includes the "true" yeasts or those fungi which exhibit a single-celled, generally budding, vegetative growth form and which reproduce sexually through the formation of ascospores. Fungi which have a similar vegetative form but do not form ascospores are classified within the Deuteromycotina class Blastomycetes.

Table 1:1 Ascomycete Teleomorphs of Selected, Medically Important Anamorphs

Ascomycete Genus	Ascomycete Order	Pathogenic Anamorph	Disease
Ajellomyces	Onygenales	*Blastomyces*	Blastomycosis
Ajellomyces	Onygenales	*Histoplasma*	Histoplasmosis
Arthroderma	Onygenales	*Trichophyton*	Dermatophytosis
Arthroderma	Onygenales	*Microsporum*	Dermatophytosis
Eurotium	Eurotiales	*Aspergillus*	Aspergillosis
Kluyveromyces	Endomycetales	*Candida*	Candidiasis
Lodderomyces	Endomycetales	*Candida*	Candidiasis
Pichia	Endomycetales	*Candida*	Candidiasis
Piedraia	Myriangiales	Not known	Black piedra

Orders Eurotiales, Gymnoascales and Myriangiales

Members of these groups are more advanced ascomycetes in which asci are formed within minute fruiting bodies which are barely visible to the unaided eye. Included in these groups are the infrequently encountered teleomorph states of some of the most important human pathogens, e.g., *Histoplasma capsulatum* and *Blastomyces dermatitidis*.

Subdivision Basidiomycotina (Basidiomycetes)

The basidiomycetes contain the most conspicuous and familiar of all fungi, the mushrooms. The order Agaricales contains most of the edible and poisonous mushrooms and their relatives. Although they have been only rarely reported as associated with human infections, medical mycologists are frequently called upon for advice and assistance in cases involving mushroom poisonings. The teleomorph state of one of the best known fungal zoopathogens, *Cryptococcus neoformans*, is a member of the basidiomycetes. *F. neoformans* is a close relative of a rather large group of basidiomycete plant parasites, the Ustilaginales or "smut fungi".[6]

Subdivision Deuteromycotina (Deuteromycetes or Imperfects)

This group is composed of conidia-forming fungi. Most of the medically important fungi are classified within the Deuteromycetes. In contrast to the other subdivisions, this group is totally artificial and includes the anamorphs of ascomycetes and a few basidiomycetes. However, this subdivision is a practical necessity since only about 10-15% of all anamorphs have proven links with teleomorphs.[5] In addition, many of the remaining anamorphs may no longer form teleomorph states at all. There have been numerous attempts to structure this group into artificial classes, orders and families.[10] The most recent and possibly the most successful attempt at establishing a systematic framework is based upon the methods of conidium production. However, since this classification is in a state of ongoing development, it seems more useful to list the medically important anamorph genera and their methods of conidium production, as shown in Table 1:2. A complete discussion of the Deuteromycetes and their relationships may be found in a work by E. S. Luttrell.[7]

The following selected deuteromycete categories are a modification of the scheme described by Carmichael and co-workers[1] and are based on the vegetative growth forms and mechanisms of conidial production. The names which are used are descriptive and are not meant to indicate any taxonomic rank (Table 1:2).

Blastomycetes

Fungi with a budding, yeastlike vegetative growth form are included within this group. It contains the anamorphic states of true yeasts, yeastlike

Table 1:2 Medically Important Genera of Deuteromycetes

Genus	Conidia/Vegetative Type	Disease
Aspergillus	Phialidic hyphomycete	Aspergillosis
Blastomyces	Blastic hyphomycete	Blastomycosis
Candida	Blastomycete	Candidiasis
Cladosporium	Blastic hyphomycete	Chromoblastomycosis
Coccidioides	Arthric hyphomycete	Coccidioidomycosis
Cryptococcus	Blastomycete	Cryptococcosis
Epidermophyton	Thallic hyphomycete	Dermatophytosis
Exophiala	Annellidic hyphomycete	Phaeohyphomycosis
Fonsecaea	Phialidic hyphomycete	Chromoblastomycosis
Histoplasma	Blastic hyphomycete	Histoplasmosis
Madurella	Phialidic hyphomycete	Mycetoma
Malassezia	Blastomycete	Pityriasis
Microsporum	Thallic hyphomycete	Dermatophytosis
Paracoccidioides	Blastic hyphomycete	Paracoccidioidomycosis
Penicillium	Phialidic hyphomycete	Hyalohyphomycosis
Phaeoannellomyces	Annellidic hyphomycete	Tinea nigra
Phialophora	Phialidic hyphomycete	Phaeohyphomycosis
Scedosporium	Annellidic hyphomycete	Mycetoma
Sporothrix	Blastic hyphomycete	Sporotrichosis
Torulopsis	Blastomycete	Toruloposis
Trichophyton	Blastic hyphomycete	Dermatophytosis
Trichosporon	Blastomycete	White piedra
Wangiella	Phialidic hyphomycete	Phaeohyphomycosis
Xylohypha	Blastic hyphomycete	Phaeohyphomycosis

anamorphs of basidiomycetes and yeastlike forms with no known teleomorph. Blastoconidia, annelloconidia, phialoconidia and arthroconidia may be formed by blastomycetes, as well as true hyphae and pseudohyphae. Included in this group are such zoopathogens as *Blastoschizomyces, Candida, Cryptococcus, Rhodotorula, Torulopsis* and *Trichosporon*.

Phialidic Hyphomycetes

A vegetative growth form composed of true hyphae and conidia (phialoconidia) formed from a fixed growing point inside a specialized cell, the phialide, is characteristic of members of this group. The phialide does not elongate during the successive production of conidia. *Aspergillus, Paecilomyces, Penicillium* and *Phialophora* are all examples of potentially pathogenic genera which are members of this group.

Annellidic Hyphomycetes

A vegetative growth form consisting of a mycelium and conidia (annelloconidia) formed within a specialized cell which elongates with the successive formation of conidia is characteristic of genera included in this form-group. As each conidium is successively formed within the specialized cell it leaves a remnant of its cell wall, which in turn causes the elongation of the conidia-forming cell. *Exophiala, Scopulariopsis* and *Scedosporium* are all examples of annellidic hyphomycetes.

Thallic (Aleuric) Hyphomycetes

The vegetative growth form is composed of true hyphae. The conidia (aleuroconidia or hollothallic conidia) are formed solitarily at the ends of nonspecialized hyphae and are released by the fracture of a supporting cell. *Epidermophyton, Microsporum* and *Trichophyton* all form thallic conidia.

Arthric Hyphomycetes

The vegetative growth form is composed of a mycelium and the conidia (arthroconidia) are formed as a result of the disarticulation of the hyphal cells. *Coccidioides immitis* is probably the best known pathogenic fungus within this group.

Blastic Hyphomycetes

The vegetative growth form consists of true hyphae, and the conidia (blastoconidia) are formed as a result of the ballooning out of the tips of specialized, generally aerial hyphae. The initial and succeeding conidia may in turn form conidia by this ballooning process, resulting in chains, often branched, of conidia with the youngest conidium at the apex. *Blastomyces, Cladosporium* and *Histoplasma* are examples of zoopathogens within this group.

NATURAL RESERVOIRS OF PATHOGENIC FUNGI

It is important to keep in mind when working with pathogenic fungi that the human body is not the natural habitat of these infectious agents. Although a few dermatophytes are found almost exclusively in association with man, all of the remaining pathogenic fungi have other natural reservoirs. Some are found associated with lower animals, but the vast majority are soil-inhabiting organisms. The biology of nonpathogenic fungi is discussed by Hudson.[4] Humans are rarely-used hosts and the fungi have had to undergo considerable adaptation to survive and utilize this habitat. They are weak pathogens in general and unless directly inoculated through the external barrier of the skin, inhaled in high concentrations or gain entry to severely immunocompromised patients, they cannot cause serious or life-threatening infections.

REFERENCES

1. Carmichael JW, Kendrick WB, Conners IL, Sigler L. Genera of hyphomycetes. Edmonton: University of Alberta Press, 1980:366.
2. Cole GT, Kozawa Y. Dimorphism. In: Cole GT, Kendrick B, eds. Biology of conidial fungi; vol 1. New York: Academic Press, 1981:97-133.
3. Hennebert GL, Weresub LK. Terms for states and forms of fungi, their names and types. In: Kendrick B, ed. The whole fungus; vol 1. Ottawa: National Museum of Natural Sciences, National Museums of Canada, 1979:27-30.
4. Hudson, HJ. Fungal biology. London: Edward Arnold, 1986:298.
5. Kendrick B. The fifth kingdom. Waterloo: Mycologue Publications, 1985:363.
6. Kwon-Chung KJ. A new genus, *Filobasidiella*, the perfect state of *C. neoformans*. Mycologia 1973:67;1197-2000.
7. Luttrell ES. Deuteromycetes and their relationships. In: Kendrick B, ed. The whole fungus; vol 1. Ottawa: National Museum of Natural Sciences, National Museums of Canada, 1979:241-64.
8. McGinnis MR. Laboratory handbook of medical mycology. New York: Academic Press, 1980:661.
9. O'Donnel KL. Zygometes in culture. Athens, GA: Department of Botany, University of Georgia, 1979:257.
10. Subramanian CV. Hyphomycetes taxonomy and biology. London: Academic Press, 1983:502.
11. Voss EG, ed. International code of botanical nomenclature. In: Stafleu FA, ed. Regnum vegetabile; vol 111. Bohn: W. Junk, 1983:472.

CHAPTER 2

COLLECTION, TRANSPORT AND PROCESSING OF CLINICAL SPECIMENS

Barbara E. Robinson, Ph.D. and Arvind A. Padhye, Ph.D.

INTRODUCTION

Proper collection, transport and processing of clinical specimens are essential for the correct diagnosis of mycotic infections. Improperly collected specimens cannot be improved upon once received in the laboratory. Specimens must be collected from the site of infection by aseptic technique, since cultures inoculated with contaminated specimens become overgrown by contaminants which may obscure or even suppress development of pathogens. Often a tentative or even definitive identification may be achieved within minutes by serologic testing or by microscopic examination of a properly collected clinical specimen. Sufficient clinical material should be collected to allow for microscopic observation and culture of the infecting agent. In the event that only a single swab is used to obtain the clinical specimen, the culture should be set up before a slide preparation is made.

SAFETY

Because of the documented hazards associated with working with pathogenic fungi,[35] the mycology laboratory must have well defined safety procedures which protect laboratory personnel as well as the surrounding environment. It is the responsibility of everyone involved in the operation of the laboratory to be familiar with these procedures and to use them in the routine processing of specimens. The safety practices must address all areas of potential danger, including biological hazards related to the isolation and identification of fungi, disinfection of laboratory equipment and elimination of biological waste. A specimen or culture of filamentous fungi should only be manipulated in a biological safety cabinet regardless of whether or not a pathogenic organism is suspected. Two types of safety cabinets provide a suitable environment for working with most fungal cultures.[58] A Class I cabinet has an inward face velocity of at least 75 feet per minute. This will usually protect the laboratory personnel but may allow contamination of the specimen from the inward flow of air. A Class II cabinet is a vertical laminar flow cabinet and will protect both the worker and the specimen from contamination. Since the identity of a fungus is often known only after extensive work with the organism, a Class II cabinet should be used in the mycology laboratory.[27]

Personnel in the mycology laboratory should follow standard laboratory practices. Additionally, protective gloves and clothing should be worn by persons working with specimens or cultures, all infectious material must be decontaminated before being discarded, biohazard signs must be present and access to the laboratory should be controlled. Even while working in an approved safety cabinet, personnel should make a continuous, conscious effort to reduce aerosol production.

All laboratory surfaces which come in contact with potentially infectious specimens or cultures should be decontaminated on a daily basis. Several disinfectants are available which are effective against fungi, such as alcohol (ethyl, isopropyl, 70-95%), phenolic compounds (0.5-5.0%), iodophors (30-50 mg free iodine per liter) and chlorine compounds (50,000 ppm available chlorine). Disinfectant-filled bottles should always be available on the bench surface or within the biological safety cabinet when cultures or specimens are handled.

Employees should be trained not to panic if an accident occurs. Good laboratory practices dictate that laboratory personnel know how to handle a potentially hazardous situation. In the event of an accident, the chairperson of the safety committee or the officer responsible for safety at the institution must be notified immediately. If a minor spill happens, cover the exposed material with absorbent paper or pad and saturate with disinfectant. The spill should be left this way for approximately one hour before the area is cleaned. If a major accident occurs within the safety cabinet,[27] the spill should be covered and disinfected. The interior surfaces of the safety cabinet, including the drain pans, must also be decontaminated. All materials used to disinfect the safety cabinet, including those saturated with disinfectant, should be autoclaved before being discarded. The safety cabinet should be left on during the decontamination procedure to reduce the possibility of contamination of the surrounding environment. If a major accident occurs outside of a biological safety cabinet but in a room with negative air pressure, all personnel in the affected area should leave the room as quickly as possible and wait approximately 30 minutes for any aerosols to settle before instituting cleanup proceedings. A sign saying "Do Not Enter, Contaminated Area" should be placed on the door. While waiting, all contaminated clothing should be removed and placed in an autoclavable laundry bag for immediate decontamination. If the laboratory does not have negative air pressure, it is necessary to begin to clean the spill immediately. Personnel must wear protective clothing, masks and gloves to disinfect all surfaces and equipment. All materials used to disinfect the room must be autoclaved before disposal.

All inoculation of media, specimen manipulation and subculturing of fungal growth must be performed only in an approved biological safety cabinet. This will decrease the chance of contaminating the specimen as well as protect the laboratory worker and environment. Furthermore, a culture must never be opened to sniff the odor of the growing colony! The use of plastic coverslips instead of glass will provide less opportunity for accidental inoculation with shattered coverglasses.

DIRECT EXAMINATION

Although 2-3 weeks are often required before fungal growth is noted, direct microscopic examination of clinical specimens can very quickly reveal the presence of fungal elements and can aid in the selection of plating media. In addition, the knowledge that fungal elements are present in a specimen will allow the clinician to make an informed decision regarding the importance of fungi obtained from a culture. Frequently, direct examination provides the only evidence of fungal infection. Direct microscopy can be used for most liquid or solid specimens. It may be necessary to macerate tissues or to dilute viscous liquids in a digestant such as potassium hydroxide (KOH) or saline prior to observation. The slide preparation should be prepared by placing a drop of the specimen on a clean glass slide, adding a drop of diluent or stain, such as lactophenol cotton blue if necessary, and placing a coverslip over the preparation.

Several different types of microscopes can be used. The light or brightfield microscope is the instrument most commonly available in laboratories and clinics. The slide preparation is placed above the light source on this microscope. The condenser, which is between the specimen and the light source, prevents glare by focusing the beam of light. Since most nondematiaceous fungi have a low refractive index, it is necessary to use either reduced light for viewing or to apply a substance which will selectively stain the fungal elements.

The use of a phase contrast microscope alleviates the problem of visualizing fungi of low refractive index with a brightfield microscope. With a phase contrast microscope, light, which is deflected by the different refractive indices present in a specimen, is deflected again by a special phase contrast lens. Light which is in phase (traveling in the same direction) is seen as a bright area while light which is out of phase (traveling in different directions) is seen as a dark area.

Differential interference contrast microscopy (Nomarski optics) has elements of both phase contrast and brightfield microscopy. Nomarski microscopy provides a three-dimensional image of the specimen which gives an added degree of clarity. The image is produced by prisms which split the light beam prior to its transmittance through the specimen and prisms which combine the light beam at the objective lens. Specimens to be examined by Nomarski microscopy are prepared in the same manner as for an unstained brightfield examination under low light or for phase contrast microscopy.

Fluorescence microscopy can be used to visualize fungi in clinical specimens. A fluorochrome dye is added to the object to be examined. Absorption of ultraviolet light causes the molecules of dye to become excited and raised to a higher energy level. As the dye molecules return to their preexcitation stage, longer wavelength light is emitted as excess energy. This causes the stained object to be viewed as a bright subject against a dark background.

Unfortunately, the advantages afforded by a dissecting microscope are not widely appreciated in the clinical microbiology laboratory. The use of this microscope can provide valuable information about growth characteristics such as texture, structural formation and fruiting bodies. All of this information can aid in the identification of unknown fungi.[27]

Two types of specialized stains or mounting fluids can enhance the viewing of fungal elements. These two are (1) stains or specialized mounting fluids used with liquid or solid clinical material and (2) stains used on dried, fixed material. Formulae and preparation of these reagents are found in Chapter 11.

1. Potassium hydroxide (KOH) can be used with either liquid or solid specimens. It is most commonly prepared as a 10% aqueous solution. A drop of the solution is mixed with a small portion of the specimen on a clean glass slide and a coverglass is placed over the specimen. The specimen is heated gently or allowed to sit at room temperature until the alkali clears the specimen, i.e., has digested the protein-containing material. The fungal polysaccharide-containing cell wall is resistant to alkali digestion and can thus be easily distinguished from other cellular debris. KOH solution should not be stored in a glass bottle as minerals from the glass will be leached and will appear as flocculent material in the bottle.

2. Lactophenol cotton blue (LPCB) can be used as a mounting medium to aid brightfield microscopic visualization of fungi in clinical specimens or when examining tease-mounts or slide cultures. The same medium without cotton blue can be used for phase contrast microscopy. LPCB will turn most fungi a pale blue while the other components of the stain act as a mild digestant and decontaminant. A drop of LPCB can be added to a KOH wet mount to increase the visibility of the fungal elements.

3. An extremely rapid means for demonstrating the presence of fungi in clinical specimens is the use of Cellufluor white (CFW) (calcofluor, Polys-

ciences, Inc.), a brightening agent which has been shown[11] to preferentially bind to cellulose and chitin in the fungal cell wall. Its use avoids the inaccuracies associated with reading KOH preparations. In addition, it is more sensitive than a Gram stain and is more rapid than the periodic acid-Schiff (PAS) or Gomori methenamine silver (GMS) stains. CFW does not interfere with subsequent staining by PAS or GMS, should confirmation with a permanent stain be necessary. CFW has also been shown to provide a rapid and nonspecific method to demonstrate fungi in deparaffinized and rehydrated tissue sections.[13,33] In retrospective and prospective studies, Monheit et al[32] found that CFW could be added to the Papanicolaou stain without altering its diagnostic cytopathologic features. Their studies indicate that demonstration of fungi by a combination of light and fluorescent microscopy is far more effective than is examination of specimens by light microscopy alone.

A one-step aqueous stain technique is effective for rapidly screening tissues for fungi. CFW should be used as a 0.1% solution containing from 0.01% to 0.08% Evans blue as a counterstain. It can be added directly to a clinical specimen on a glass slide. If it is used in combination with KOH, the two solutions should be stored separately and one drop of each added separately to the clinical specimen on the slide since the solutions tend to precipitate if stored together. Any combination of filters producing blue-white wave length light (500 nm) suitable for fluorescein, such as a K530 excitation filter and a BG 12 barrier filter, can be used. Fungi will fluoresce light blue or apple-green depending on the light source and filters which are used.

4. Traditionally, mycologists have used India ink or nigrosin to demonstrate encapsulated fungi. Microscopic examination with India ink should never be relied upon as the sole means to determine the presence of encapsulated cells in spinal fluid because the test lacks sensitivity and specificity. The ink particles act as a negative stain of the background material. Encapsulated cells appear to have a halo as the polysaccharide capsule prevents uptake of the ink by the cells. To prepare an India ink specimen for examination, place a drop on a slide; add enough ink to the sediment or fluid specimen so that the preparation is brown, not black; and add a coverslip. Examine the specimen under both low and high power magnification with reduced light on a brightfield microscope.

5. Clinical material on a slide can be fixed and stained with a variety of procedures. The Gram stain can be used to determine the presence of fungi. Although Gram stain will adequately stain most of the yeastlike fungi, some, e.g., *Cryptococcus* spp., will appear atypical and mottled.[5] It is very easy to overlook hyphal elements as they will often not be stained by the Gram reaction. Since the Gram stain is so widely utilized in the diagnostic microbiology laboratory, the microbiologist should be aware of the limitations of this stain for observation of fungi in clinical materials.

6. A modified acidfast stain should be used if infection with *Nocardia*

spp. is suspected. The cell walls of these bacteria contain fatty acids which are similar to the mycolic acids in *Mycobacterium* spp., but the *Nocardia* acids contain fewer carbon atoms. The modified acidfast stain uses a 0.5%-1.0% aqueous acid with a shorter time for decolorization than that used with acid-alcohol. The pathogenic *Nocardia* spp. will be decolorized by acid-alcohol but are resistant to decolorization with the weaker acid solution.

7. Several stains generally thought to be reserved for histologic preparations can be used for the direct examination of mycological specimens. The physician usually has a high index of suspicion of a fungal infection before requesting a specialized stain. The Giemsa stain which is commonly used in hematology, can be used for detecting fungi, (e.g., *Histoplasma capsulatum*) in bone marrow aspirates and other tissues. The Giemsa stain should be used in conjunction with other fungal tissue stains as it will often not stain fungi other than *H. capsulatum*. Stains such asGMS, PAS and the mucicarmine stain can be used. Use of these is more involved and requires more reagents than the previously described stains; therefore, most clinical microbiology laboratories do not perform these stains on a routine basis. The differential reactions in the GMS and PAS stains rely upon acidification to aldehydes of adjacent hydroxyl groups in cellular polysaccharides.[7] In the GMS stain, the aldehydes form a methenamine silver nitrate complex and the fungal cell walls appear brown-black. In the PAS stain, the aldehydes form a purple-red reaction with the Schiff reagent and the fungi appear that color.

CULTURE

Media

A battery of appropriate culture media should be used since no single culture medium is satisfactory for the recovery of all fungal agents. The battery should generally include one medium containing cycloheximide (500 mg/L), which inhibits rapidly-growing, contaminating molds as well as such pathogens as *Cryptococcus neoformans, Pseudallescheria boydii, Scedosporium inflatum* and many of the opportunistic zoopathogens. A second medium should be enriched with 5-10% sheep blood. Antibacterial agents such as chloramphenicol (50 mg/L), or the combination of chloramphenicol and gentamicin (5 mg/L), or penicillin (20,000 IU/L) and streptomycin (40 mg/L) (which is less expensive), should be incorporated into one medium. Each medium should be inoculated with at least 0.5 mL of a liquid or homogenized specimen or enough minced tissue to provide optimum conditions for growth of fungal agents. Liquid specimens should be streaked across the surface of the agar to allow isolated colonies to grow: tissue should be embedded into the agar. The inoculated media should be incubated for at least 4 weeks at 25-30°C, observed daily for the first week and at weekly intervals thereafter. It is not necessary to incubate primary isolation cultures at 35-37°C, because most pathogens, including those that cause systemic infection, do not grow well at this elevated temperature.

When preparing media to be used for the isolation and cultivation of fungi, it is helpful to increase the agar content of the product to 1.7-2.0% instead of the commonly used 1.5%. Also, if petri plates (100 mm) are used, 25-35 mL of media should be used to fill each plate. Both of these measures will act to decrease dehydration of the medium during incubation. The pH of each medium must be tested following sterilization if prepared "in house" or upon receipt of commercial products. The pH values allowed by the commercial producers of media vary; therefore, the desired pH must be specified to the manufacturer. Frequently, the pH of the medium prevents isolation of fungi.

Traditionally, media used for the isolation and identification of fungi have been prepared in bottles or tubes, which has both advantages and disadvantages. It is extremely dangerous to use media in petri plates in the mycology laboratory. Although there is a definite advantage in using the large surface area of a petri dish, an aerosol is created every time the plate is opened. Therefore, all plated media should be sealed with oxygen-permeable tape or placed in a plastic bag after inoculation. The use of media in screwcapped tubes or bottles provides safety and also reduces dehydration. However, the decreased gas exchange and/or increased relative humidity that occur in sealed petri dishes as well as in screwcapped containers causes many laboratories to use cotton-plugged slants rather than screw-capped containers.

The following types of media are generally used for initial plating of clinical specimens for fungal culture. Formulae and preparation of these media may be found in Chapter 11.

1. Blood agar: Of all the isolation media that are used, sheep blood agar is one of the most important for it supports growth of many fastidious, dimorphic pathogenic fungi. The addition of antibacterial antibiotics will reduce possible contamination. If an actinomycotic infection is suspected or if direct examination reveals narrow, branched filaments typical of *Nocardia* or *Actinomyces* spp., cultures should be incubated both aerobically and anaerobically at 35-37°C, using at least one medium which does not contain antibacterial antibiotics. Sheep blood should be used since blood from humans may possess antifungal activity and thus inhibit the growth of some yeasts. Furthermore, sporulation may be inhibited on blood-containing media.

2. Brain heart infusion agar (BHIA): BHIA is an enriched medium used for isolation of *Actinomyces* spp. and for the primary recovery of dimorphic fungal pathogens. Addition of 5% defibrinated sheep blood often enhances the isolation of *H. capsulatum* and can be used to convert the mold to its yeast form.

3. Cystine heart hemoglobin agar (CHHA): CHHA is a highly enriched

nutritional medium. When fortified with antibacterial antibiotics, it can be used for the isolation of systemic dimorphic pathogens from clinical specimens. When prepared without antibiotics, it can be used for isolation of pathogenic actinomycetes.

4. Sabouraud glucose agar (SGA): SGA which contains 4% glucose at a pH of 5.6 is the medium most commonly used in clinical mycology. Because the high concentration of dextrose inhibits the sporulation of many fungi, the use of SGA as a primary recovery medium is enhanced by reduction of the dextrose concentration to 2% and by adjustment of the pH to neutral.[43] SGA supplemented with chloramphenicol and cycloheximide is especially suitable for isolation of dermatophytes. This selective SGA is available commercially as Mycobiotic Agar (Difco Laboratories) or Mycosel Agar (BBL Microbiology Systems).[9]

Bacterial contamination in fungal cultures can be controlled by the addition to the medium of penicillin (20,000 IU/L), streptomycin (40 mg/L) or other broad-spectrum antibiotics such as chloramphenicol (50 mg/L) and gentamicin (5 mg/L if used in combination with chloramphenicol or 50 mg/L if used alone). Cycloheximide (Actidione, Upjohn Company, 500 mg/L) will initially or totally inhibit the development of rapid-growing fungal contaminants.

5. Dermatophyte test medium (DTM): DTM was developed by Taplin, et al[53] as a field medium. Dermatophyte growth can be easily recognized because it changes the pH indicator (phenol red) from yellow to red. The medium is supplemented with antibacterial antibiotics to inhibit bacterial growth and cycloheximide to suppress fungal contaminants. However, Salkin[47] and Jacobs and Russell[17] have shown that the growth of nondermatophytic fungi such as *H. capsulatum* and *Blastomyces dermatitidis* can also cause a change in the pH indicator. It is therefore necessary to identify by microscopic examination and other pertinent tests all organisms that elicit a color change.

6. Sabhi agar: Sabhi medium was developed primarily for the isolation of *H. capsulatum*.[12] The addition of blood (5% of the total volume) to the medium increases the frequency of isolation of *H. capsulatum* from clinical material. Unlike BHIA, Sabhi is not useful in the conversion of *H. capsulatum* from its mold to yeast form. Sabhi is the medium of choice for isolation of fungal pathogens from spinal fluid by the membrance filter technique.[12]

7. Yeast extract phosphate agar with ammonium hydroxide (YEP): YEP medium was developed for isolation of *B. dermatitidis* and *H. capsulatum* from contaminated respiratory specimens.[50] This medium also supports the growth of *Coccidioides immitis* but is not satisfactory for the isolation of *C. neoformans* or *Aspergillus fumigatus*.[56]

The clinical specimen is streaked onto the surface of YEP in a tube or bottle. One drop of concentrated ammonium hydroxide (NH_4OH) is immediately added to the surface. Most bacteria are inhibited by the alkaline conditions produced by the ammonium hydroxide.[56] YEP can also be fortified with 50 mg/L of chloramphenicol.[12] Cycloheximide should not be included in the medium since it is inactivated by alkaline conditions. Mycologists using YEP should be careful not to incubate it in the same incubator as other media containing a pH indicator. The ammonium hydroxide vapor in the incubator can cause a false increase in the pH of the other incubating media.

8. Inhibitory mold agar (IMA): IMA medium was devised for the selective isolation of fungi from heavily contaminated specimens. IMA contains both gentamicin (5 mg/L) and chloramphenicol (125 mg/L) to prevent overgrowth of bacteria. This product should never be used as the sole medium for isolation of fungi because it is so highly inhibitory.

Specimens

Blood Specimens:

Fungemia, the presence of fungi in the blood, can represent a medical emergency and clinical microbiology laboratories must utilize procedures which will reliably detect the presence of fungi in specimens of blood. The previously reported unreliability of blood cultures to detect opportunistic fungal infections or fungal sepsis has been well documented.[60] However, as it has been estimated that more than 200,000 cases of septicemia occur annually[61] and that the number of cases of fungemia is expected to increase in proportion to an increase in the number of debilitated, immunocompromised patients,[30] it is essential that clinical microbiology laboratories be able to rapidly determine whether fungi are present in blood specimens. Since many improvements have been made in blood culturing systems, the detection of fungal sepsis is now more reliable.

The details of skin decontamination, venipuncture and the number and frequency of blood specimens necessary to document fungemia has been well described by Reller, Murray and MacLowry.[40] The most important factors for reliable isolation of fungi from blood by conventional techniques are the volume of blood sampled, the conditions of incubation and the culture medium.

The volume of blood cultured may be the most important variable affecting reliable recovery of fungi.[59] Ilstrup and Washington[15] and others[48,55] have demonstrated that the yield from culture of 10 mL or more of blood increases with increased volume. Since fungal sepsis may be intermittent or transient, 2-3 separately collected blood samples should be cultured.[40] All blood should be diluted 1:10 to 1:20 (vol/vol); therefore, if greater than 10 mL of blood is collected per venipuncture and the total volume of medium in the culture bottle is 45-90 mL, several blood culture bottles should be inoculated.

A number of different media have been used for recovery of fungi from blood. These include brain heart infusion, Columbia and trypticase soy broths, among others. Roberts and Washington[46] showed that a biphasic blood culture medium consisting of brain heart infusion broth/agar was significantly better for the isolation of fungi than any blood culture media designated for bacterial growth. This medium will also allow for growth of *N. asteroides*, a rare isolate from blood. Additionally, the time required for detection of positive cultures is decreased with biphasic medium.[3,22,44]

All conventional blood cultures for fungi should be incubated at 30°C for 30 days and should be checked for growth on days 1, 2 and 7 and weekly thereafter. This can be done macroscopically or microscopically. Often, yeast development will not affect the turbidity of liquid blood culture media, but the use of a biphasic blood culture medium will permit the detection of yeast growth as discrete colonies on the agar surface. All fungal blood cultures should be vented. Significantly higher rates of isolation of *Candida* spp. from vented bottles have been documented.[16,40,44]

Fungal development in non-biphasic blood cultures should be assessed for growth by subculturing a portion of the medium onto SGA. McCarthy and Senne showed that approximately the same number of positive blood cultures were detected by examination of a portion of the blood culture broth with an acridine orange stain as were detected by a 24 h subculture of the broth.[26] Because of its fluorescent nature, this stain is approximately 10-fold more sensitive than methylene blue or Gram stains and it is also more cost-effective to examine a portion of the culture microscopically than to perform a subculture.

Numerous studies have compared the rate of recovery of fungi with conventional blood culture broths to the rate with biphasic blood culture systems (Septi-Chek, Roche Laboratories), radiometric detection systems (Bactec 460, Johnston Laboratories), nonradiometric detection systems (Bactec NR 660, Johnston Laboratories) and lysis centrifugation (Isolator, DuPont deNemours). The lysis-centrifugation system may result in a higher contamination rate than that obtained with other systems; however, it also detects more fungi in less time than that required by conventional techniques.[2,4,6,23] The radiometric and nonradiometric detection systems also provide decreased time to detection of growth but the organism(s) must still be grown on a solid medium before a final identification can be made.[14,20,36,62] An added advantage of lysis centrifugation is that filamentous fungi are isolated better than with other systems.[4,34]

The lysis-centrifugation system consists of a double-stoppered, evacuated tube containing a red blood cell lysing agent and anticoagulants. Following venipuncture and centrifugation, the cellular debris is removed from the bottom of the tube and inoculated onto the surface of agar media. The media are incubated and inspected on a daily to weekly basis for fungal growth. In the radiometric system, specimens of blood are placed in broth containing ^{14}C-labeled substrates. Growth is detected by probes placed into the headspace of the broth bottle to detect radioactive carbon dioxide produced from utilization of ^{14}C-labeled substrates such as glucose. Growth of fungi in the nonradiometric system is detected by quantitation of the carbon dioxide in the headspace of the bottle with an infrared spectrophotometer.

Cerebrospinal fluid specimens:

Since very few fungal cells may be present in the cerebrospinal fluid (CSF) of patients with fungal meningitis, at least 3-4 mL of CSF should be collected in a sterile, tightly sealed or screwcapped tube containing 0.01 g sodium citrate per 5 mL of spinal fluid. Thorough decontamination of the skin before a lumbar puncture is essential for collecting CSF specimens. A CSF specimen without anticoagulants should also be submitted for serologic studies. The specimens should be transported to the laboratory as soon as possible to ensure a rapid diagnosis. However, if a delay in transport is unavoidable, the CSF need not be refrigerated since it serves as an ideal culture medium for growth of fungal pathogens at 25-30°C.

Filtration is the optimum means of detecting fungi in volumes of CSF greater than 2 mL. Filtering CSF through a 0.45 μm membrane deposits all cellular components onto the filter membrane and the filtrate can then be used for serological assays. The membrane is gently removed from the filter with a pair of forceps and placed onto the surface of an agar plate with the cellular contents in contact with the medium.[12] The position of the filter should be changed every other day for 1 week. When visible growth is evident, it is generally concentrated in the area where the filter has been placed. Extreme caution must be exercised when manipulating the filter, particularly when any filamentous growth is noted and all work must be performed in a Class II biological safety cabinet. If the patient is suspected of having coccidioidomycosis, this technique should not be used.

Centrifuging the spinal fluid at 2500 rpm for 15 min and removing the sediment for culturing is an alternative method for handling volumes greater than 2 mL. Although a direct microscopic examination could then be done on the sediment, this is generally unrewarding as the sensitivity of such examinations ranges from 26-52%.[28] The sediment is best cultured by resuspending it in a small amount of CSF and placing several drops on the surface of the medium. Desiccation is prevented by not spreading the CSF over the surface and the organisms are allowed to acclimate.

When there is less than 2 mL of the specimen, place drops of the fluid directly on the surface of the medium as previously described. An enriched medium without blood should be used for isolation from CSF and cycloheximide should be avoided since *C. neoformans* and other pathogenic fungi can be inhibited by this agent.

Corneal scrapings (mycotic keratitis):

Scrapings from corneal ulcers are collected by ophthalmologists by scraping areas of ulceration with the help of a Kimura spatula and a slit-lamp microscope. Such scrapings should be transported to the laboratory as quickly as possible in 2-3 drops of sterile saline.[38,63]

Mycotic keratitis is diagnosed by examining stained smears and cultures of the scrapings. Smears are stained by the CFW, Gram, Giemsa or GMS procedures; the last is the most reliable.[37]

Plates containing blood agar and SGA, both fortified with 50 mg/L of gentamicin sulfate, should be inoculated in a series of "C"-shaped cuts into the agar[38] and incubated at 25-28°C for culture of corneal scraping. The plates should be examined daily for fungal growth originating from the "C" cuts. Growth of fungi away from the "C" cuts should be considered airborne contaminants. Fungal growth appearing in the "C" cuts should be subcultured promptly to a fresh medium for identification.

Superficial mycoses and dermatophytosis specimens:

In superficial mycoses, the affected area is generally the cornified layers of epidermis and the suprafollicular portion of the hair. In dermatophytosis, keratinized epidermal tissue and its appendages (hair, nails) are affected.

The surface of the affected area should be cleansed with cotton gauze moistened with 70% ethyl alcohol or sterile water. As the lesions of tinea versicolor and hairs infected by certain species of *Microsporum* and *Trichophyton* fluoresce under Wood's light (ultraviolet radiation at 3660 A), this light is very useful in selecting the site from which specimens should be obtained.

Since dermatophytic skin lesions progress in a centrifugal manner, the peripheral area of the lesion is the site of active fungal growth and the center is the area of healing. Skin scrapings should be collected from the active peripheral areas with a sterile scalpel. The tops of newly formed vesicles can be clipped off and submitted for laboratory examination.

Superficially infected hairs (in white and black piedra) should be clipped off. At least 10-15 such hairs with nodules should be collected for laboratory studies. Patients suspected of having fungal scalp infection should be examined in a darkened room with Wood's light. Hairs infected by some *Microsporum* and *Trichophyton* species fluoresce a bright green to yellow-green under Wood's light and even minimal infection, in which only a few hairs are infected, can be detected with this instrument. The infected hairs are plucked with forceps. Such infected hairs are generally loose in the follicles and can be removed easily without causing pain to the patient. Sterile combs or brushes[24] can be used instead of forceps to obtain hair from the scalp or from animal fur. In patients where loss of hair has caused alopecia, the hairs that have broken off close to the scalp look like black dots. Such areas should be scraped with a scalpel.

Scrapings from infected nails should be collected with a sterile scalpel. To decrease contamination, the initial scrapings from the nail surface should be discarded and not submitted for laboratory study. If the patient has thickened or split nails, the nails should be clipped and the clippings submitted to the laboratory.

Skin, nail and hair specimens should be kept dry and sent to the laboratory wrapped in clean paper packets or envelopes, sandwiched between sterile glass slides or placed in screwcapped glass tubes. If the specimen is properly collected, there should not be a problem with overgrowth of contaminating bacteria or saprophytic fungi.

Epilated hair, minced skin or nail shavings should be examined in CFW with or without 10% KOH or in 10% KOH alone for the presence of dermatophyte hyphae and arthroconidia. Nail clippings may be ground in a mortar before direct examination. Skin scrapings, hair and nail clippings should be spread onto the surface of SGA containing chloramphenicol and cycloheximide. For specimens heavily contaminated with bacteria, use of DTM may be appropriate. All culture plates or slants should be incubated at 25-30°C for at least 4 weeks before being discarded as negative.

Peritoneal, pleural and synovial fluids:

These fluids are collected by needle aspiration in an aseptic manner and should be transported to the laboratory as quickly as possible. A minimum of 1.0 mL of specimen should be collected. If immediate processing is not possible, specimens can be stored at 4°C overnight before being cultured.

If a large amount of a specimen is available, it should be centrifuged (2000 rpm for 10-15 min) before being cultured. About 0.5 mL of the sediment or specimen should be spread onto the surface of at least 2 enriched media, one with blood and the other without blood. Alternately, body fluids can be processed by lysis centrifugation, which will improve the recovery of fungi in culture.[54] Acridine orange or CFW staining of the sediment will enhance the visualization of fungal elements better than will Gram staining.[10]

Genital:

Genital specimens are generally collected on a sterile swab and should be transported to the microbiology laboratory as soon as possible. Often the laboratory diagnosis can be made by microscopic examination of a wet-mount KOH preparation. Specimens which do not show budding yeasts and pseudohyphae characteristic of *Candida* spp. should be cultured on a nonselective agar (e.g., SGA), on a selective agar containing cycloheximide and an antibacterial agent and on CHHA which contains antibacterial agents.

Intravascular catheters:

Indwelling intravascular devices associated with local or systemic signs of infection may require removal and subsequent culturing. The catheter tip should be processed by rolling it over the surface of blood agar and CHHA.[25] The tip should then be placed into a tube of trypticase soy broth and both broth and agar plates incubated at 25-30°C for 3-4 weeks. Most growth will occur within the first 3 days.[18] Semiquantitative cultures yielding greater than 15 colonies are highly likely to be associated with septicemia or fungemia as opposed to those cultures yielding 14 or less colonies.

Lower respiratory tract and gastric secretions (sputum, bronchial washings, tracheal aspirates, gastric washings):

The collection of a sputum specimen should be supervised by personnel aware of its importance. Before the specimen is collected, the patient should be informed of the type of specimen required, the patient's teeth brushed and the mouth rinsed with a mouthwash or several changes of sterile water. Only specimens expectorated from deep within the lungs are useful. All specimens from the respiratory tract should be collected in sterile widemouth bottles or sputum cups. The entire first morning expectorate is the optimal specimen. The consistency and cellular makeup of sputum often help in determining the quality of a specimen. Respiratory secretions which contain large numbers of epithelial cells and multiple types of bacteria characteristic of the oropharynx are most likely produced from the upper respiratory tract and are saliva or significantly contaminated with saliva. Bronchial or gastric washings should be transported to the laboratory as soon as possible, either in the original containers in which they were collected or in sterile, screwcapped tubes or jars. Gastric specimens should be refrigerated and neutralized as soon as possible to avoid the possibly harmful effects on the fungi of a very low pH.

Specimens from the respiratory tract should be processed as soon as possible after they are received in the laboratory. The 2% NaOH sputum decontamination method used for mycobacteria is inappropriate for the recovery of fungi; however, N-acetyl-l-cysteine and dithiothreitol have been shown to be nontoxic for most fungi,[45] and the mucolytic activity of these agents allows for effective concentration of fungi in sputum specimens.[39] A drop of the digested and concentrated specimen can be placed on a clean glass slide and examined with the use of CFW or 10% KOH. When the dimorphic fungi such as *H. capsulatum, B. dermatitidis* or *C. immitis* are anticipated as etiologic agents, YEP medium with ammonium hydroxide may be used.[50,56] If *N. asteroides* is suspected as a cause of respiratory disorders, the use of paraffin baiting has been shown to provide enhanced recovery of the offending bacterium.[49]

Bronchoscopy specimens may be centrifuged before being cultured since there is often a moderate volume of the specimen. Mucus plugs should be separated from the remainder of the specimen and cultured. Viscous specimens may be mixed with broth and pipetted with a widebore pipette or glass tubing onto culture media. At least 0.5 mL of the specimen should be spread onto the surface of each of 3 media: 2 enriched media without blood, such as BHIA and Sabhi agar, and one medium containing blood. One of the 3 media selected should contain cycloheximide; however, one should also be free of cycloheximide so that pathogens such as *C. neoformans* and *P. boydii*, which are sensitive to cycloheximide, can be isolated. If culture slants are used, the tubes should be incubated in a horizontal or slanted position for 12-24 h to allow absorption of the specimen by the media, thereby preventing growth from occurring only at the bottom of the slant.

Upper respiratory tract specimens (ear, nose, nasopharynx and oral cavity):

Specimens are usually submitted on sterile swabs. Transportation to the laboratory should be rapid; however, overnight storage in a refrigerator is satisfactory.

Specimens should be streaked onto the surface of at least two enriched media. Preferably, one should be without blood enrichment and free of cycloheximide; the other medium should include blood and cycloheximide to inhibit growth of saprobic molds.

Subcutaneous and deep tissue specimens (sinus tracts, abscesses, ulcers, fistulas, biopsies and tissues):

Whenever possible, clinical material from subcutaneous tissues should be obtained by aspiration of a closed lesion to minimize contamination. The specimens should be collected in sterile screwcapped tubes. If a small volume of material is obtained, a few drops of sterile broth should be added to prevent drying of the specimen before it reaches the laboratory.

If the subcutaneous lesion is open, clinical material can be collected by curetting the sinus tract, abscess, ulcer or fistula to obtain material deep from the tract and tissue from the wall of the lesion. If cotton swabs are used, they should be moistened first with sterile water or saline and the specimen should be placed in a screwcapped tube and transported to the laboratory immediately.

In general, biopsied or autopsied tissue specimens should be divided aseptically by the surgeon for culture and histopathology. All specimens for mycologic study should be placed in sterile containers. Surgical specimens should be transported to the laboratory as soon as possible after collection. If immediate examination and culture is not possible, specimens should be stored at 4°C for no longer than 8-10 h and a few drops of sterile, nonbacteriostatic saline or sterile water added to the container to prevent desiccation of the specimen.

Biopsied tissue from eumycotic mycetomas should be examined under a dissecting microscope for the presence of granules. Visible granules should be picked up with dissecting needles and washed with several changes of saline containing antibacterial antibiotics before being cultured. The color and morphology of a granule should be noted, then it should be crushed between two glass slides and examined microscopically in KOH. Granules should be cultured on Sabouraud agar with and without antibiotics, both incubated at 30°C.

Tissue from subcutaneous lesions and biopsied specimens should be cut into small pieces with sterilized scissors and a pair of forceps, using aseptic technique, or processed with an automated tissue processing device (Stomacher, Tekmar). A portion of the homogenized or minced tissue is placed on a clean glass slide and examined microscopically with CFW, LPCB, Gram, acidfast or PAS stains. At least 3 media including one enriched with blood and one with cycloheximide should be used for culture. Approximately 0.5-1.0 mL of the homogenized material should be streaked onto the surface of the media. All culture plates are incubated for at least 4 weeks and periodically examined for growth. When dimorphic pathogenic fungi are suspected as etiologic agents, the incubation period should be extended to 6 weeks before cultures are reported as negative.

Urine and fecal specimens:

A urine specimen can be collected as a midstream clean-catch or as a catheterized specimen. The urine should be collected in sterile tubes or sterile widemouth, screwcapped jars. Fungal cultures of stool specimens should be discouraged; however, if culture is necessary, diarrheic stool specimens should be collected in sterile screwcapped jars or waxlined paper containers. If specimens cannot be cultured promptly, they should be stored at 4°C for no longer than 10-12 h. Urine and fecal specimens should not be kept at room temperature for long periods, since bacteria and yeasts will replicate rapidly.

The urine specimen should be centrifuged and a minimum quantity of 0.5 mL of the sediment should be cultured on appropriate media, especially when a dimorphic pathogen or *C. neoformans* is suspected. For quantitation of *Candida* spp. in urine, inhibitory mold agar, BHIA or Sabhi agar can be used.[43]

A fecal specimen can be stained with LPCB or mixed with a few drops of sterile water and examined as an unstained preparation. For isolation of *B. dermatitidis, H. capsulatum* and other fungi, fecal specimens or rectal swabs are cultured on SGA containing cycloheximide and chloramphenicol.

Serologic specimens:

Specimens for fungal serology must be taken by aseptic techniques and a minimum of 10 mL of blood should be collected. After the blood has clotted and the serum has separated, the serum should be removed aseptically and preserved by adding merthiolate (ethyl mercurithiosalicylic acid, sodium salt) to a final concentration of 1:10,000. A 1% stock solution of merthiolate can be prepared by dissolving 1.4 g of sodium borate and 1.0 g merthiolate in distilled water to obtain 100 mL of final solution. Then 0.01 mL of the stock solution per 1.0 mL of serum or other body fluid specimen is used. Spinal fluid specimens should also be collected aseptically when meningeal involvement is anticipated. If CSF is to be cultured, no preservative should be added.

The importance of rapid and presumptive diagnosis offered by serologic tests should not be underestimated. Diagnosis achieved by histology, isolation and identification of the etiologic agent is often time-consuming. In such situations, immunologic procedures should be used to obtain rapid and presumptive evidence of infection; immunologic reactions often provide the first clues of fungal infection. Serologic tests also provide information on the effects of therapy and sometimes narrow the choice of methods to use for isolating and identifying the causal agent.[21]

Environmental specimens (soil samples, bird, bat and pigeon droppings for H. capsulatum *and* C. neoformans*:*

Proper sampling procedures should be followed in collecting soil and bird or pigeon droppings.[1] The location of each sampling site in relation to the roosting sites must be noted and recorded for each specimen. Sterile spoons, tongue depressors or other tools should be used to collect 4 oz samples placed in plastic bags. Environmental samples often are infested with mites; therefore, all samples should be treated with a mitacide before handling in the laboratory.

H. capsulatum can be isolated from soil by injection of laboratory mice but this requires at least 6-8 weeks.[8] The isolation of *C. neoformans* from pigeon droppings is achieved by serially diluting (beginning at 1:5) the excrement samples in saline fortified with antibacterial antibiotics and streaking 0.1 mL of each dilution onto plates of birdseed agar or trypan blue agar.[42,57] Colonies of *C. neoformans* develop a characteristic brown or blue pigment on each medium respectively and are therefore easily recognized. The isolates are then subcultured, purified and their identity confirmed by biochemical analysis.

Proficiency test specimens:

Proficiency test specimens are usually received as lyophilized cultures and should only be manipulated in a biological safety cabinet. To rehydrate, aseptically add a small amount (0.5-1.0 mL) of sterile broth or water to the vial and gently mix. After allowing the mixture to rehydrate for approximately 30 min, inoculate the suspension to the normal battery of media and make slide preparations. Any remaining material should be held at 4°C until the culture is finalized and the results received. Guidelines for the identification of proficiency test unknowns have been delineated by McGinnis and Salkin.[29]

STOCK CULTURE COLLECTIONS

All laboratories performing mycological examinations should have the capability of maintaining a stock culture collection. Fungi deposited in the collection can be used for quality control, teaching purposes and preservation of unusual and reference isolates. It is essential that accurate records be kept of all fungi in the collection. The minimum data for each isolate should include the name, date, numerical designation and source of the isolate, dates of subculturing, storage medium and method of storage.

A number of methods have been described for the preservation of fungi.[27,31] The most common means include freezing at -70°C, freeze-drying, storage under mineral oil and storage in sterile water. Each method has advantages and disadvantages (Table 2:1). Isolates to be stored should be actively growing on potato dextrose agar or other similarly deficient media. Before storage, the tube or bottle containing the culture should be identified with the name of the organism, the culture number and the date. To subculture, place a portion of the fungus onto the agar and incubate at 25-30°C until growth occurs. The identity of all isolates should be reconfirmed by tease preparation or slide culture after storage.

Sending Specimens, Cultures or Sera to Other Laboratories

All specimens and etiologic agents should be packaged and shipped in accord with Public Health Service, United States Department of Transpor-

Table 2:1 Techniques for Preservation of Stock Cultures

Method	Technique	Advantages	Disadvantages
Freezing	Screwcapped culture at -70°C	Easy, rapid	If culture thaws, must be subcultured
Freeze-drying	Follow manufacturer's instructions	Viable for many years	Time-consuming, need special equipment
Mineral oil	Pour layer of sterile oil over fungus in screwcapped tube	Easy, rapid	Messy, need to leave room at top of tube
Water culture	Place conidia in water in sterile vial. Store at 25°C	Easy, rapid, viable for many years	Easy to contaminate

tation, United States Postal Service or Air Transport Association regulations, as applicable. Detailed instructions for transportation of clinical specimens and etiologic agents have been described by Richardson[41] and published by these regulatory agencies.

Control of Mites

Mites are small anthropods belonging to the order *Acarina*. Free-living species are destructive pests in the laboratory. In nature, these mites inhabit leaf litter and soil and are carried into the laboratory on clothing, shoes, paper cartons and new equipment as well as in or on clinical specimens.

Adult mites are about 0.2 mm long and cannot be detected with the naked eye. Their eggs, even though larger, are transparent and usually go undetected. Their small size can pose a serious threat to an entire culture collection. Mites are able to enter cultures plates and cotton-plugged tubes and move from one plate or tube to another, carrying with them bacteria and a variety of fungal contaminants.

The first step for avoiding mite infestation is to periodically clean desk tops with an acaricidal solution such as Kelthane, Tedion V-18, chlorocide or Actellic.[51] An ammonia solution is also acaricidal and is commonly available. Since solutions may be toxic or irritating to the skin, workers using them should wear plastic gloves. Camphor and paradichlorobenzene crystals (moth balls) are safe when used as a mite repellent and can be spread on shelves, in cabinets or as a thin layer in boxes or cans containing culture plates and tubes. However, these compounds can be fungistatic in high concentrations or after prolonged use and paradichlorobenzene may induce fungal mutations.

Sealing culture tubes with cigarette paper and a copper sulphate gelatin adhesive is recommended by the American Type Culture Collection as an effective method to prevent mite infestation.[19,52] As soon as a mite infestation is detected, immediate action should be taken. Infested cultures should be frozen at -20°C for 24 h, thawed and subcultured repeatedly until mite-free growth is obtained. The subcultures should then be sealed.[19] Addition of hexachlorocyclohexane (10 mg/L) to media used for fungal cultures will kill any mites which may be present and will not affect fungal growth.[27]

REFERENCES

1. Ajello W, Weeks RJ. Soil decontamination and other control measures. In: DiSalvo AF, ed. Occupational mycoses. Philadelphia: Lea & Febiger, 1983:229-38.
2. Bille J, Edson RS, Roberts GD. Clinical evaluation of the lysis-centrifugation blood culture system for the detection of fungemia and comparison with a conventional biphasic broth blood culture system. J Clin Microbiol 1984;19:126-8.
3. Bille J, Roberts GD, Washington JA II. Retrospective comparison of three blood culture

media for the recovery of yeasts from clinical specimens. Eur J Clin Microbiol 1983;2:22-5.

4. Bille J, Stockman L, Roberts GD, Horstmeier CD, Ilstrup DM. Evaluation of a lysis-centrifugation system for recovery of yeasts and filamentous fungi from blood. J Clin Microbiol 1983;18:469-71.
5. Bottone EJ. *Cryptococcus neoformans*: pitfalls in diagnosis through evaluation of Gram-stained smears of purulent exudates. J Clin Microbiol 1980;12:790-1.
6. Brannon P, Kiehn TE. Clinical comparison of lysis-centrifugation and radiometric resin systems for blood culture. J Clin Microbiol 1986;24:886-7.
7. Chandler FW, Kaplan W, Ajello L. Color atlas and text of the histopathology of mycotic diseases. 1st ed. Chicago: Year Book Medical Publications, 1980.
8. Emmons CW. Isolation of *Histoplasma capsulatum* from soil. Public Health Rep 1949;64:892-6.
9. Emmons CW, Binford CH, Utz JP, Kwon-Chung KJ. Medical mycology. 3rd ed. Philadelphia: Lea & Febiger, 1977:592.
10. Fessia S, Cocanour B, Ryan S. Microbiological examination of peritoneal dialysis fluid using a fluorescent acridine orange stain. Lab Med 1986;17:404-6.
11. Hageage GJ, Harrington BJ. Use of calcofluor white in clinical mycology. Lab Med 1984;15:109-12.
12. Haley LD, Callaway CS. Laboratory methods in medical mycology. Washington, DC: U.S. Government Printing Office, 1978; DHEW publication no. (CDC)79-8361.
13. Hollander H, Keilig W, Bauer J, Rothemund E. A reliable fluorescent stain for fungi in tissue and clinical specimens. Mycopathologia 1984;88:131-4.
14. Hopfer RL, Orengo A, Chesnut S, Wenglar M. Radiometric detection of yeasts in blood cultures of cancer patients. J Clin Microbiol 1980;12:329-31.
15. Ilstrup DM, Washington JA II. The importance of volume of blood cultured in the detection of bacteremia and fungemia. Diagn Microbiol Infect Dis 1983;1:107-10.
16. Ilstrup DM, Washington JA II. Effects of atmosphere of incubation on recovery of bacteria and yeasts from blood cultures in tryptic soy broth. Diagn Microbiol Infect Dis 1983;1:215-19.
17. Jacobs PH, Russell B. Dermatophyte test medium for systemic fungi. JAMA 1973;224:1649.
18. Jones PG, Hopfer RL, Elting L, Jackson JA, Fainstein V, Bodey GP. Semiquantitative cultures of intravascular catheters from cancer patients. Diagn Microbiol Infect Dis 1986;4:299-306.
19. Jong SC. Prevention and control of mite infestation in fungus cultures. ATCC Quarter Newsletter 1987;7:1,7.
20. Jungkind D. Millan J, Allen S, Dyke J, Hill E. Clinical comparison of a new automated infrared blood culture system with the Bactec 460 system. J Clin Microbiol 1986;23:262-6.
21. Kaufman L, Reiss E. Serodiagnosis of fungal diseases. In: Lennette EH, Balows A, Hausler WJ Jr, Shadomy HJ, eds. 4th ed. Manual of clinical microbiology. Washington, DC: American Society for Microbiology, 1985:924-44.
22. Kiehn TE, Capitolo C, Mayo JB, Armstrong D. Comparative recovery of fungi from biphasic and conventional blood culture media. J Clin Microbiol 1981;14:681-3.
23. Kiehn TE, Wong B, Edwards FF, Armstrong D. Comparative recovery of bacteria and yeasts from lysis-centrifugation and a conventional blood culture system. J Clin Microbiol 1983;18:300-4.
24. Mackenzie DWR. "Hairbrush diagnosis" in detection and eradication of nonfluorescent scalp ringworm. Br Med J 1963;2:363-5.
25. Maki DG, Weise CE, Sarafin HW. A semiquantitative culture method for identifying intravenous-catheter-related infection. N Engl J Med 1977;296:1305-9.
26. McCarthy LR, Senne JE. Evaluation of acridine orange stain for detection of microorganisms in blood cultures. J Clin Microbiol 1980;11:281-5.

27. McGinnis MR. Laboratory handbook of medical mycology. 1st ed. New York: Academic Press, 1980:661.
28. McGinnis MR. Detection of fungi in cerebrospinal fluid. Am J Med 1983;75(suppl 1B):129-38.
29. McGinnis MR, Salkin IF. Identification of molds commonly used in proficiency tests. Lab Med 1986;17:138-42.
30. Meunier-Carpentier F, Kiehn TE, Armstrong D. Fungemia in the immunocompromised host; changing patterns, antigenemia, high mortality. Am J Med 1981;71:363-70.
31. Miguens MP. Methods for maintaining stock cultures. Mykosen 1985;28:134-7.
32. Monheit JE, Brown G, Kott MM, Schmidt WA, Moore DG. Calcofluor white detection of fungi in cytopathology. Am J Clin Pathol 1986;85:222-5.
33. Monheit JE, Cowan DF, Moore DG. Rapid detection of fungi in tissues using calcofluor white and fluorescent microscopy. Arch Pathol Lab Med 1984;108:616-18.
34. Paya CV, Roberts GD, Cockerill FR III. Transient fungemia in acute pulmonary histoplasmosis: detection by new blood-culturing techniques. J Infect Dis 1987;156:313-15.
35. Pike RM, Laboratory-associated infections: incidence, fatalities, causes and prevention. Ann Rev Microbiol 1979;33:41-66.
36. Prevost E, Bannister E. Detection of yeast septicemia by biphasic and radiometric methods. J Clin Microbiol 1981;13:655-60.
37. Rebell G. *Fusarium* infections in human and veterinary medicine. In: Nelson PE, Toussoun TA, Cook RJ, eds. *Fusarium*: diseases, biology and taxonomy. University Park: Pennsylvania State University Press, 1981:210-20.
38. Rebell G, Forster RK. Fungi of keratomycosis. In: Lennette EH, Balows A, Hausler WJ Jr, Truant JP, eds. 3rd ed. Manual of clinical microbiology. Washington, DC: American Society for Microbiology 1981:553-61.
39. Reep BR, Kaplan W. The use of N-acetyl-l-cysteine and dithiothreitol to process sputa for mycological and fluorescent antibody examinations. Health Lab Sci 1972;9:118-24.
40. Reller LB, Murray PR, MacLowry JD. Blood cultures II. In: Washington JA II, ed. Cumitech 1A. Washington, DC: American Society for Microbiology, 1982:1-11.
41. Richardson JH. Transportation of clinical specimens and etiologic agents. In: Wentworth BB, Baselski VS, Doern GV, et al, eds. Diagnostic procedures for bacterial infections. 7th ed. Washington, DC: American Public Health Association, 1987:747-60.
42. Rippon JW. Medical mycology, the pathogenic fungi and the pathogenic actinomycetes. 2nd ed. Philadelphia: WB Saunders, 1982:842.
43. Roberts GD, Goodman NL, Land GA, Larsh HW, McGinnis MR. Detection and recovery of fungi in clinical specimens. In: Lennette EH, Balows A, Hausler WJ Jr, Shadomy HJ, eds. Manual of clinical microbiology. 4th ed. Washington, DC: American Society for Microbiology 1985:500-13.
44. Roberts GD, Horstmeier C, Hall M, Washington JA II. Recovery of yeast from vented blood culture bottles. J Clin Microbiol 1975;2:18-20.
45. Roberts GD, Karlson AG, DeYoung DR. Recovery of pathogenic fungi from clinical specimens submitted for mycobacteriological culture. J Clin Microbiol 1976;3:47-8.
46. Roberts GD, Washington JA II. Detection of fungi in blood cultures. J Clin Microbiol 1975;1:309-10.
47. Salkin IF. Dermatophyte test medium: evaluation with nondermatophyte pathogens. Appl Microbiol 1973;26:134-7.
48. Salventi JF, Davies TA, Randall EL, Whitaker S, Waters JR. Effect of blood dilution on recovery of organisms from clinical blood cultures in medium containing sodium polyanethol sulfonate. J Clin Microbiol 1979;9:248-52.
49. Singh M, Sandhu RS, Randhawa HS. Comparison of paraffin baiting and conventional culture techniques for isolation of *Nocardia asteroides* from sputum. J Clin Microbiol 1987;25:176-7.
50. Smith CD, Goodman NL. Improved culture method for the isolation of *Histoplasma*

capsulatum and *Blastomyces dermatitidis* from contaminated specimens. Am J Clin Pathol 1975;63:276-80.
51. Smith D, Onions AHS. The preservation and maintenance of living fungi. 1st ed. Kew, England: Commonwealth Mycological Institute, 1983:117.
52. Snyder WC, Hansen HN. Control of culture mites by cigarette paper barriers. Mycologia 1946;38:455-62.
53. Taplin D, Zaias N, Rebell G, Blank H. Isolation and recognition of dermatophytes on a new medium (DTM). Arch Dermatol 1969;99:203-9.
54. Taylor TC, Poole-Warren LA, Grundy RE. Increased microbial yield from continuous ambulatory peritoneal dialysis peritonitis effluent after chemical or physical disruption of phagocytes. J Clin Microbiol 1987;25:580-3.
55. Tenney JH, Reller LB, Mirrett S, Wang W-LL, Weinstein MP. Controlled evaluation of the volume of blood cultured in detection of bacteremia and fungemia. J Clin Microbiol 1982;15:558-61.
56. Thompson DE, Kaplan W, Phillips BJ. The effect of freezing and the influence of isolation medium on the recovery of pathogenic fungi from sputum. Mycopathologia 1977;61:105-9.
57. Vickers RM, McElligott JJ Jr, Rihs JD, Postic B. Medium containing trypan blue and antibiotics for the detection of *Cryptococcus neoformans* in clinical samples. Appl Microbiol 1974;27:38-42.
58. Wagner WM. What do we know about our biological safety cabinets? Clin Microbiol Newsl 1986;8:138-40.
59. Washington JA II. Conventional approaches to blood culture. In: Washington JA II, ed. The detection of septicemia. West Palm Beach, FL: CRC Press, 1978:41-87.
60. Washington JA II. Cultures for miscellaneous organisms. In: Washington JA II, ed. The detection of septicemia. West Palm Beach, FL: CRC Press, 1978:89-99.
61. Washington JA II, Ilstrup DM. Blood cultures: issues and controversies. Rev Infect Dis 1986;8:792-802.
62. Weinstein MP, Reller LB, Mirrett S, Stratton CW, Reimer LG, Wang W-LL. Controlled evaluation of the agar-slide and radiometric blood culture systems for the detection of bacteremia and fungemia. J Clin Microbiol 1986;23:221-5.
63. Wilson LA, Sexton RB. Laboratory aids in diagnosis. In: Duane T, ed. Clinical ophthalmology. Hagerstown, PA: Harper and Row, 1976;4:1-15.

CHAPTER 3

SUPERFICIAL AND CUTANEOUS INFECTIONS CAUSED BY MOLDS: DERMATOMYCOSES

Irene Weitzman, Ph.D., Stanley A. Rosenthal, Ph.D.
and Margarita Silva-Hutner, Ph.D.

INTRODUCTION

Dermatomycoses are fungal infections of man and lower animals. These infections are restricted to the skin and appendages because the causative fungi usually are unable to invade deeper tissues or organs. Based on host response, clinical similarity of lesions, relatedness of the etiologic fungi and frequency of occurrence, the dermatomycoses can be further subdivided into: (a) superficial, (b) dermatophytic and (c) opportunistic mycoses.

Superficial mycoses are those in which the causative fungi colonize the cornified layers of the epidermis and the suprafollicular portion of the hair. The diseases they produce involve changes in the pigment of the skin (i.e., pityriasis versicolor and tinea nigra) or the formation of nodules along the hair shaft distal to the follicle (i.e., black piedra and white piedra). The status of these fungi in the external environment has not been precisely determined. Pityriasis versicolor, caused by a yeast, will be discussed in Chapter 6, "Human Infections Caused by Yeastlike Fungi."

Dermatophytic cutaneous mycoses are caused by dermatophytes, a group of related fungi belonging to the Ascomycetes, family Gymnoascaceae, whose natural habitat is the soil or the skin and appendages of humans and animals. Dermatophytes can invade the cellular layers of the epidermis and sometimes the dermis.

Opportunistic dermatomycoses are cutaneous infections, often resembling those produced by dermatophytes, but are caused by fungi that are normally saprobes or plant pathogens. These fungi are typically of low virulence and their hosts are often debilitated or similarly compromised.

The first section of this chapter deals exclusively with the superficial dermatomycoses caused by molds. The dermatophytic and the opportunistic mycoses are discussed in succeeding sections.

SUPERFICIAL MYCOSES

The superficial mycoses are caused by molds usually of such low pathogenicity that they do not attack living cellular layers of the skin and nails. The infection is typically asymptomatic and gives the host little physical discomfort because cellular response from the host is usually absent. The patient often consults a physician for cosmetic reasons. The fungi in this group, the diseases they cause and their most recent or common synonyms are listed in Table 3:1.

TINEA NIGRA (KERATOMYCOSES NIGRICANS PALMARIS)

Definition

This disease is an asymptomatic, superficial, mycotic infection of the stratum corneum affecting mostly the palmar but also the plantar skin. It is characterized by the appearance of gray, light-brown or black non-scaly macules that gradually enlarge.

Historical Perspectives

The early reporting of this disease and its separation from pityriasis versicolor is not clear.[98] The first authentic description appears to have been by Cerqueira in Bahia, Brazil in 1891. He called this disease keratomycosis

Table 3:1 Molds Causing Superficial Mycoses

Fungus	Recent or Common Synonyms	Name of Disease
Phaeoannellomyces werneckii	*Exophiala werneckii* *Cladosporium werneckii* *Dematium werneckii* *Pullularia werneckii*	tinea (keratomycosis) nigra
Stenella araguata	*Cladosporium castellanii* *Cladosporium araguatum*	tinea nigra
Piedraia hortae	*Trichosporon hortai*	black piedra, tinea nodosa
Piedraia quintanilhai		black piedra in Central African primates
Trichosporon beigelii	*Trichosporon cutaneum*	white piedra, chignon disease

nigricans palmaris, but it was not until 1916 that his findings, along with eight other cases, were published by his son, Cerqueira-Pinto.[26] In 1921, Horta discovered the first case in Rio de Janeiro, isolated the fungus and called it *Cladosporium werneckii*.[59] The polymorphic nature of this fungus, with illustrations of three distinct synanamorphs and an apparently predominant yeast form, was provided in a publication by Langeron and Horta.[72] Von Arx[10] proposed a transfer to the hyphomycete (filamentous) genus *Exophiala* since *E. werneckii* produces annelloconidia. This was the generally accepted taxon until the recent publication by McGinnis et al.[81] who concluded that this fungus should be considered a blastomycete (yeast) rather than a hyphomycete since the annellidic yeast cells are the most stable distinctive anamorph. According to these authors, this fungus belongs to the black yeasts in the genus *Phaeoannellomyces*, family Phaeococcomycetaceae, Class Blastomycetes.

A dematiaceous (dark) hyphomycete in Venezuela has been implicated as another, but very rare, agent of tinea nigra. Borelli and Marcano[18] proposed the name *Cladosporium castellanii* for this fungus, which was later reported as a synonym of *Stenella araguata*.[80]

Etiologic Agents

The primary causative agent of tinea nigra is *Phaeoannellomyces (Exophiala* or *Cladosporium) werneckii*. *S. araguata* appears to be a very rare agent of this disease, limited so far to Venezuela.[18]

Clinical Manifestations

Tinea nigra is characterized by flat, sharply marginated, darkly pigmented, non-scaly patches on the skin, most commonly on the palms but also on the soles or dorsa of the feet. The color is usually mottled with deeper pigmentation at the advancing periphery. It has been likened to a stain produced on the skin by silver nitrate (Fig. 3:1).

Epidemiology and Public Health Significance

The disease is most common in tropical areas of Central and South America, Africa and Asia.[98] Between 1960-1970, it was reported to be prevalent along coastal areas of southeastern United States, particularly North Carolina.[121] Five cases were reported from Australia[60] and an autochthonous case in The Netherlands.[111] Most cases outside of endemic areas have been in individuals who had travelled to endemic areas.[63,98,132] The disease is probably acquired exogenously since organisms similar to *P. werneckii* have been reported as abundant in soil, humus, sewage and decaying vegetation.[98] *P. werneckii* was also obtained from sea water.[62]

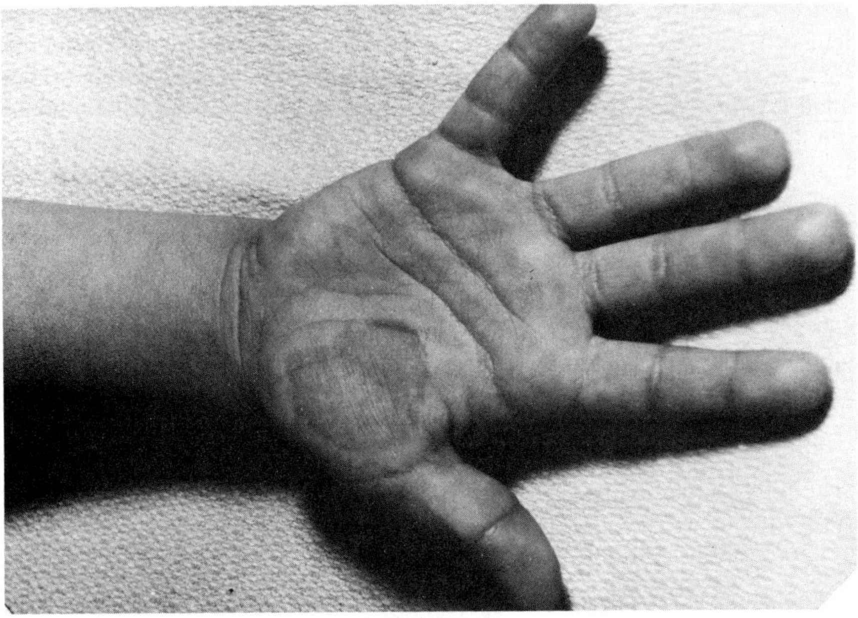

Figure 3:1. Lesion of tinea nigra on palmar surface of hand. Note dark pigmentation and absence of elevation of lesion. X1.

Familial spread has not been reported. All races and age groups are apparently susceptible, although most patients have been less than 19 years old, with females outnumbering males 3 to 1. [98] The incubation period is believed to be around 2-7 weeks,[41] although a lesion was reported 20 years after an experimental inoculation.[17]

The disease poses no threat to the patient's health and is only of cosmetic importance if left untreated. Once medical help is sought, however, it is important that the physician distinguish tinea nigra from malignant melanoma, which tinea most closely resembles,[132] or other dermatological conditions which it clinically simulates, such as junctional nevus, contact dermatitis, pigmentation of Addison's disease, post-inflammatory melanoses, melanoses from syphilis or pinta and staining due to chemicals, dyes and pigments.[60,98] Confusion of tinea nigra with malignant melanoma may result in surgical mutilation that could be prevented by a simple skin scraping examined in KOH.[132]

Collection, Transport and Processing of Specimens

Scales are collected by scraping the lesion. The scales should be examined in a KOH preparation and cultured promptly. Scales should be inoculated onto Sabouraud dextrose agar and Sabouraud dextrose agar with chloramphenicol and cycloheximide. See Chapter 2, "Collection, Transport and Processing of Specimens," for further details.

Identification of the Etiologic Agents

A presumptive diagnosis can usually be made clinically and confirmed by a KOH preparation of skin scrapings. Fungal elements are usually present in large numbers and are seen as light- brown, frequently branched, septate filaments 1.5-5 μm in diameter. Sinuous hyphal fragments and budding cells may also be seen (Fig. 3:2). The presence of pigmented filaments and septate cells with bipolar budding in the KOH preparation confirms the diagnosis of tinea nigra (Fig. 3:3).

In culture at 25°C on Sabouraud dextrose agar with or without antibiotics, *P. werneckii* grows slowly and appears initially as black, shiny, moist yeastlike colonies; occasionally they are essentially hyphal.[81] After two or

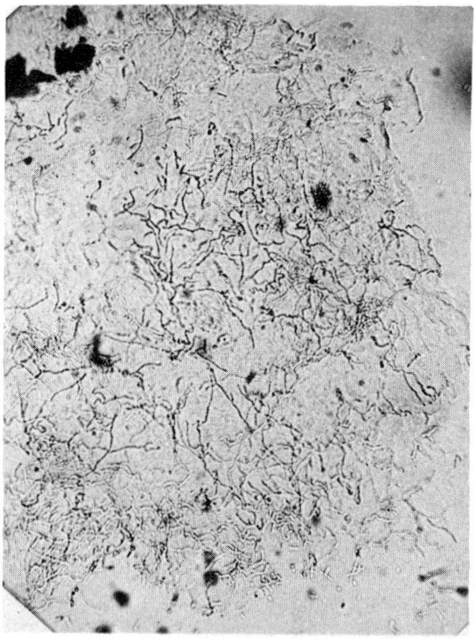

Figure 3:2. *Phaeoannellomyces (Exophiala, Cladosporium) werneckii* in KOH preparation of skin scrapings from lesion of tinea nigra. Note abundance of long, branching, dematiaceous (dark-colored) hyphae and small clusters of ellipsoid spores. X125.

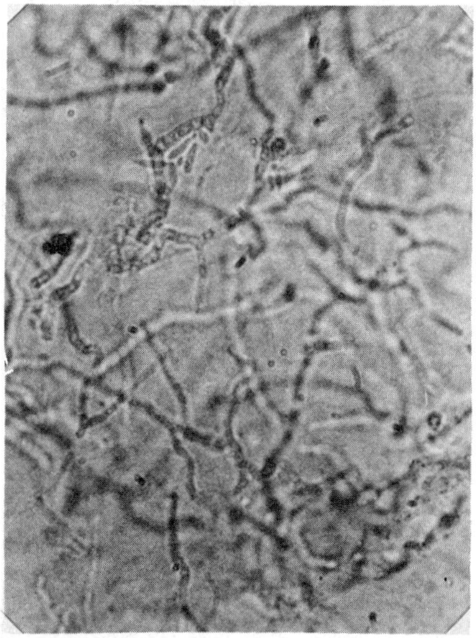

Figure 3:3. *P. werneckii* in KOH preparation of skin scrapings from lesion of tinea nigra: higher magnification, X560.

more weeks of additional incubation, velvety-gray to gray-black aerial mycelium appears over portions of the yeastlike growth and may eventually predominate.

The microscopic appearance correlates well with the gross appearance of the colony. The initial yeastlike colonies consist of one to predominantly two-celled, pale brown to deeply pigmented, cylindrical to spindle-shaped cells, some apparently budding, that are rounded at one end and taper towards the other with annellations at the tapered end (Fig. 3:4). As the colony becomes more mycelial, a corresponding change in microscopic morphology occurs, with development of some pseudohyphae and hyphae, that varies with the particular isolate.[81] The hyphae are usually olivaceous, thin or broad, often thick-walled and septate at frequent intervals. Along these hyphae, one to two-celled hyaline to olivaceous annelloconidia arise from intercalary annelides or from annelides integrated within the hyphae

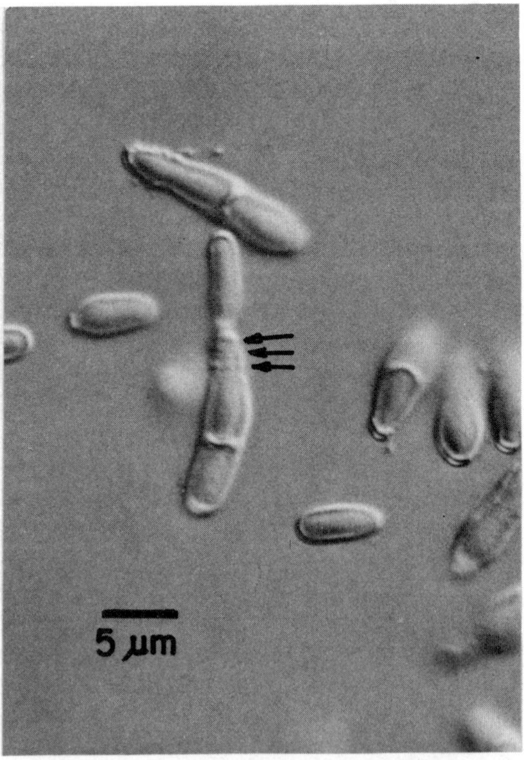

Figure 3:4. *P. werneckii* on potato dextrose agar. Note annellidic yeast cells (arrow). Differential interference contrast microscopy. Bar equals 5 µm. Reprinted with permission.[81]

(Fig. 3:5). The annelloconidia tend to accumulate in balls that eventually slide down along the sides of the hyphae. Seceded conidia may produce new conidia by budding, with each successive conidium resulting in annellations. Branching chains of conidia have been observed in older cultures of the mycelial form[24] and this type of sporulation was the basis for classification in the genus *Cladosporium*. However, it has been recommended that the naming of polymorphic fungi be based on the single most stable and distinctive anamorph[81], which in the case of *P. werneckii* is the annellidic yeast cell.[81]

According to Mok,[83] *P. werneckii* decomposes casein but not tyrosine, xanthine, hypoxanthine nor starch, does not liquefy gelatin and has a maximum growth temperature of 42°C. Kane and Albreish[62] differentiated *P. werneckii* from other dematiaceous fungi by its high tolerance to salt and its ability to grow luxuriantly and exhibit yeastlike growth in 15% NaCl.

BLACK PIEDRA

Definition

Black piedra is a fungal infection of the hair shaft characterized by the presence of discrete, hard and gritty, dark-brown to black nodules (Fig. 3:6) which adhere firmly to the hair of the scalp, less commonly to the beard or mustache, and rarely to the axillary or pubic hairs of humans.

Historical Perspectives

According to Rippon[98], black piedra was described by "Malgoi-Hoes in 1901", called *Trichosporon* spp. by Horta in 1911 and *Trichosporum hortai* by Brumpt in 1913. Fonseca and Area Leao renamed the fungus *Piedraia hortae* (as *hortai*) because of observation of the sexual stage and recognition of its affinity to the ascomycetes.

Etiologic Agents

The etiologic agent of black piedra in humans is *P. hortae*. *P. quintanilhai* has been reported to produce black piedra in Central African mammals.[120] Van Uden described *P. quintanilhai* as a new species on the basis of the absence of polar filaments on the ascospores.

Clinical Manifestations

The presence of dark, hard, gritty nodules (piedra in Spanish means stone) firmly adhering to the hair shaft is characteristic of this infection. Nodules

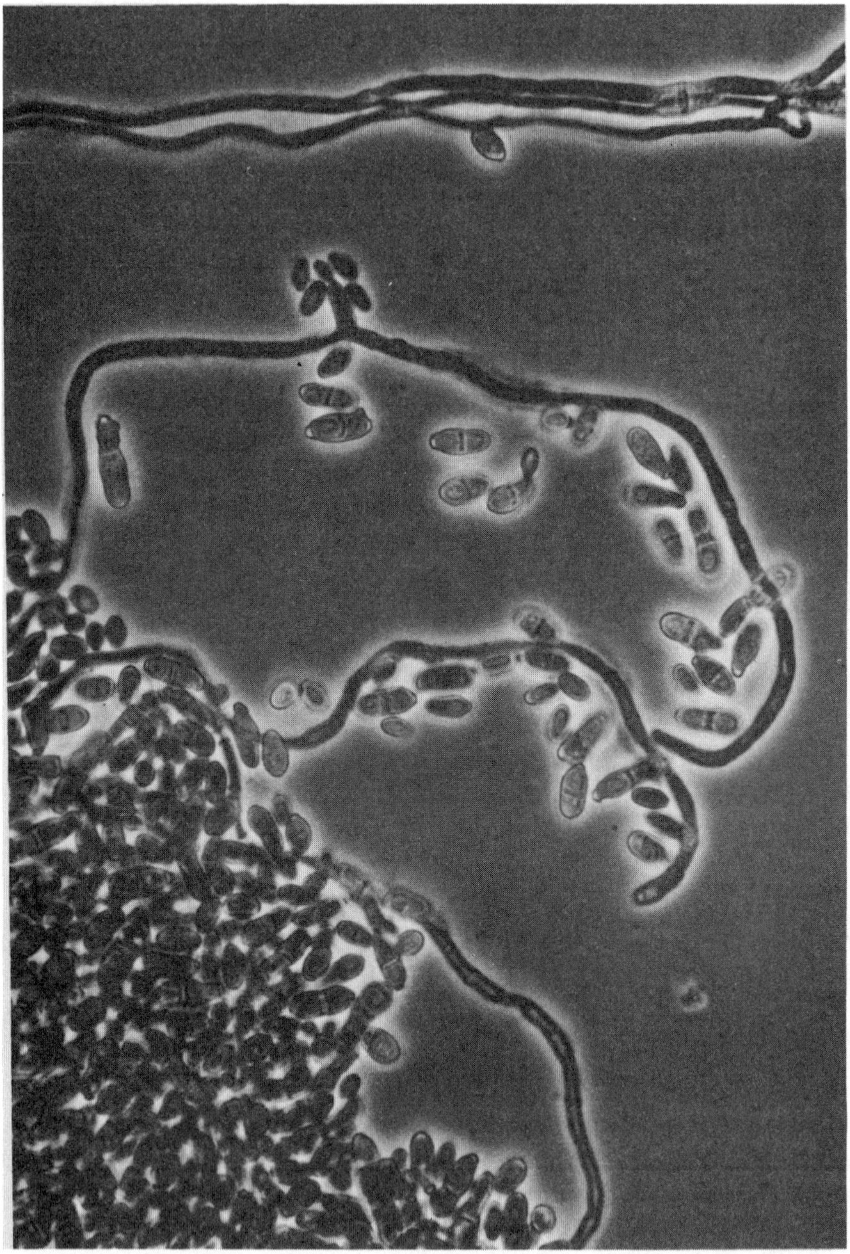

Figure 3:5. *P. werneckii* **after 2 weeks on potato dextrose agar. Note dematiaceous septate hyphae, intercalary annellide and bicellular annelloconidia. Phase contrast microscopy. X630. Reprinted with permission.[79]**

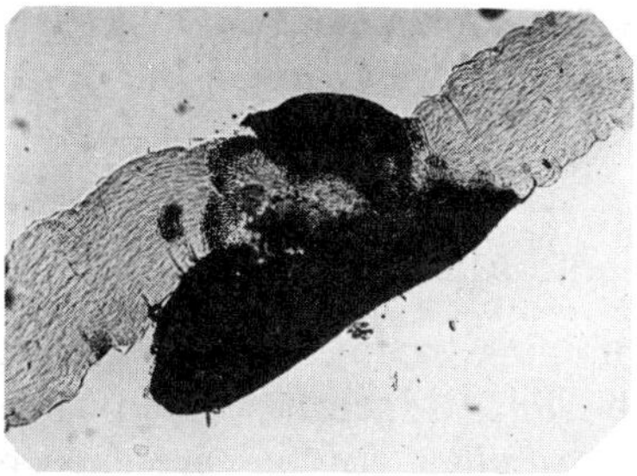

Figure 3:6. Black piedra nodule on hair from human scalp. Note fungal filaments along advancing border of nodule. KOH preparation. X100.

range in size from microscopic to 3 mm in diameter. In humans, the fungus grows on but not inside the hair shaft. Although the fungus penetrates and disrupts the cuticle while growing around it, deep penetration and complete disruption of the hair is not observed.[28] In contrast, black piedra in animals can result in severe pilar damage, destruction of the cortex and medulla with breaking of the hair at the site of the nodule.[67,120]

Epidemiology and Public Health Significance

Black piedra occurs in humid and tropical regions of South and Central America, Southeast Asia and Africa. It has been reported in humans in the Surinam, Venezuela, Paraguay, Argentina, Uraguay and Brazil, and from the southeast Asian countries of Indonesia, Malaysia, Thailand and Indochina.[40] Black piedra was observed in animals held in captivity for prolonged periods.[67,120] The source of infection is believed to be exogenous and of non-animal origin, probably an undetermined source in nature.[67,120] A combination of climatic conditions (abundant rainfall, high humidity and a temperature of 26°C) and the use of possibly contaminated plant oil on the hair probably results in the endemicity of this disease in Brazilian Indians in the Amazon area.[40] It has been suggested that infection may spread to neighboring hairs of an infected human by the dissemination of free ascospores through gaps in the nodule.[28] There are no reports of animal-to-animal transmission, but an epidemic reported by Carrion in 1938[23] and 1965[25] suggests that person-to-person transmission can occur. This situation involved 45 boys in a government orphanage in Puerto Rico who shared wet combs when "styling" their hair. In areas of high humidity it is possible for piedra to become a minor public health problem. Hygenic measures should be observed when an outbreak occurs.

Collection, Transport and Processing of Specimens

Hairs containing nodules are required for microscopic examination and culture. If culture is desired, hair should be inoculated promptly onto

Sabouraud agar with chloramphenicol. Cycloheximide has been used successfully to isolate *P. hortae*,[28,40] although the presence of cycloheximide has been reported to inhibit the growth of the fungus.[6]

Identification of Etiologic Agents

Hair fragments containing nodules are examined in a 10-20% KOH preparation. After the slide is gently heated and the nodule carefully squashed, the slide is examined for the presence of compact masses of septate, dark-brown to black hyphae on the surface of the hair, round to oval asci and hyaline, curved, fusiform, aseptate ascospores (Fig. 3:7) that bear one or more appendages. The nodule is composed of fungal pseudotissue (pseudoparenchyma) that is actually an ascostroma (harbors asci and ascospores in special cavities called locules).

Colonies on Sabouraud agar are slow-growing, heaped, dark-brown to black, glabrous or covered by short aerial mycelium. Hyphae are dematiaceous; conidia are absent. The teleomorph, not generally observed in culture, was reported by Takashio and Vanbreuseghem.[116]

Ascocarps, which are produced on hair, are variable in size and shape, black, rounded to elongate, composed of vertical rows of thick-walled, dark-brown polygonal cells and usually surround the hair shaft. Loculi are irregularly distributed, globose to ovate and open to the surface by an inconspicuous, ostiolar pore. Asci are ellipsoidal to obovate, 30-60 μm and contain 2-8 fusiform, curved, hyaline ascospores with a single appendage at one or both ends.[122] Piedraiaceae are now classified in the subclass Ascoloculomycetidae, order Dothideales.[122]

WHITE PIEDRA (CHIGNON DISEASE, PIEDRA COLOMBIANA, PIEDRA NOSTRAS)

Definition

White piedra has distinct parasitic morphology, appearing as soft, white, ivory or beige, oval nodules occurring along the hair shafts, mostly of the mustache and beard hairs, less frequently on the scalp, axillary or pubic hairs of humans. Lower animals may also be infected, especially the horse.[98]

Historical Perspectives

According to Carmo-Sousa[22], Kuchenmeister and Rabenhorst in 1867 considered the etiologic agent of chignon disease as an alga which they described and named *Pleurococcus beigelii*. The description was based on material received from Beigel who later showed (1869) that the agent of the

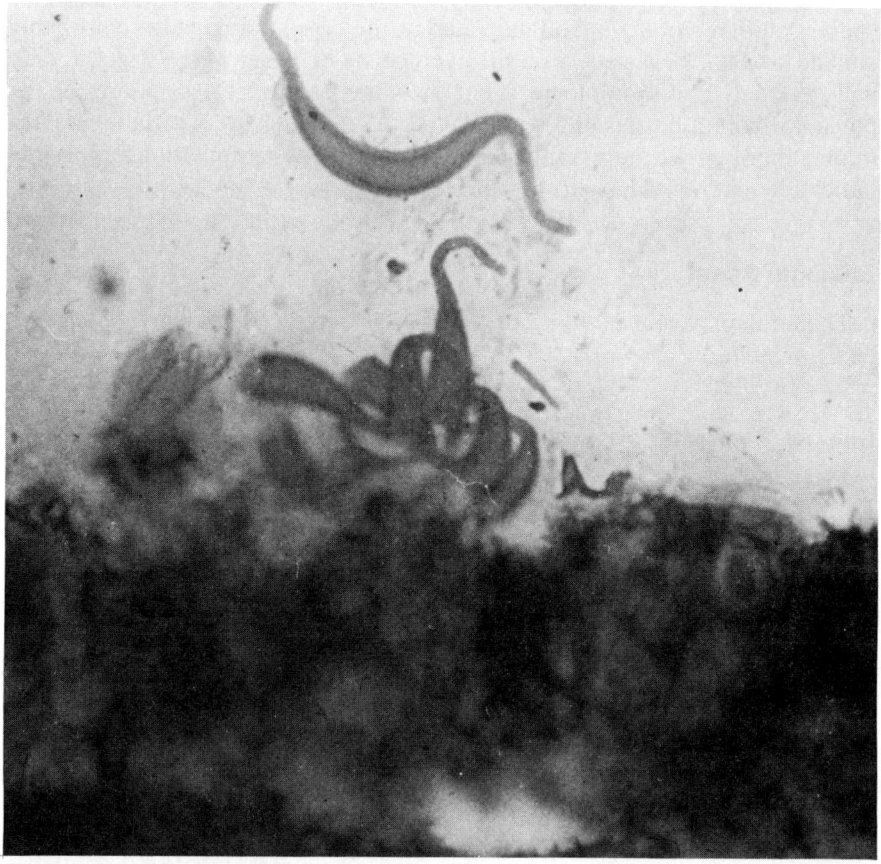

Figure 3:7. Crushed nodule of black piedra. Note aseptate, curved, fusiform ascospores with polar appendages. Periodic acid-Schiff stain. Approximately X1000.

disease was a fungus. Behrend (1890) isolated an organism from a mustache infection in Germany which he described and named *Trichosporon ovoides*. As a result of further studies, he concluded that the fungus in the nodules from chignon disease, from the mustache infection and from hairs from Colombian piedra were the same agent. Vuillemin (1902) observed that the fungus from nodules from a case of piedra in France appeared to be identical with *P. beigelii* and transferred it to the genus *Trichosporon* (-um); thus the binomial *T. beigelii* has priority. De Beurmann et al. (1909) described *Oidium cutaneum* which had been isolated from a skin wound; Ota (in 1926) brought *O. cutaneum* into the genus *Trichosporon* as *T. cutaneum*. Because there was no type material available for *T. beigelii* and the original description for the species was considered insufficient, Diddens and Lodder used the species epithet *cutaneum* and indicated as the type strain a culture probably studied by Ota.[70] However, other mycologists consider *beigelii* as the only valid epithet for the etiologic agent of white piedra.[6] Descriptions of the physiological properties of *T. beigelii (cutaneum)* indicate a variable pattern in maximum growth temperature and carbohydrate assimilation,[70] possibly indicating heterogenicity in this species.

Etiologic Agent

The etiologic agent of white piedra is considered by many mycologists to be *T. beigelii (cutaneum)*.[6]

Clinical Manifestations

White piedra is characterized by the presence of soft, white to beige accretions around the hair shafts of the moustache, beard, scalp, axillary and genital regions of the body of humans and the fur of lower animals. These nodules may occur at intervals or coalesce to form an extensive mass around the hair shaft. The hair may be weakened by the fungus and may break at these areas.

Epidemiology and Public Health Significance

White piedra occurs in temperate zones, sporadically in the United States and Europe and more commonly in South America, the Orient and in the temperate periphery of the black piedra geographic belt.[98] Although the source of infection is unknown, *T. beigelii* and similar organisms may be found in soil, air, sputum and on body surfaces.[98] White piedra does not constitute a public health problem but may be mistaken for nits of pediculosis capitis.[52]

Collection, Transport and Processing of Specimens

Hairs with nodules are required for the laboratory diagnosis of white piedra. See Chapter 2, "Collection, Transport and Processing of Specimens," for further details regarding these procedures.

Identification of the Etiologic Agent

Microscopic examination of squashed nodules in KOH will reveal oval to rectangular blastoconidia (budding cells), arthroconidia 2-4 μm in diameter and hyphae intermingled in either a mosaic arrangement or aligned perpendicular to the hair axis. Bacteria may surround the nodule as a zooglea.

T. beigelii grows rapidly in routine culture media. Colonies are creamy in consistency when young but soon become membranous and wrinkled, forming radial furrows and irregular folding with age. Aerial hyphae may develop along the edges of the colony, particularly upon subculture. The fungus usually grows well at 37°C but may be variable.[70] Microscopically, the presence of arthroconidia, blastoconidia and septate hyphae is diagnostic of the genus. Pseudomycelium with blastoconidia often occurs. Physiological studies are necessary for species identification and cultures must be pure for this purpose. *T. beigelii (cutaneum)* does not ferment carbohydrates; it assimilates dextrose, lactose, D-xylose and inositol. The following commonly used carbon compounds are assimilated by some isolates and not by others: galactose, sucrose, maltose, cellobiose, trehalose, raffinose, L-arabinose and erythritol. This organism does not assimilate potassium nitrate, is urease-positive and gives a positive reaction with Diazonium Blue B salt.[70]

DERMATOPHYTOSES

Definition

The dermatophytoses (ringworm) are fungal infections of the keratinized epidermal tissues (i.e., skin, hair, nails, feathers, fur and horns) of humans and animals, caused by closely related fungi whose asexual forms (anamorphs) belong to the genera *Microsporum, Trichophyton* and *Epidermophyton*. Both pathogenic and nonpathogenic species of these genera have an affinity for keratin and can digest this scleroprotein. These fungi may exist as soil saprophytes or as zoopathogens, some of which are exclusively pathogenic for man.

Historical Perspectives

Clinical recognition of the tineas (ringworm diseases) as distinct from other dermatologic lesions has existed since "the most remote times."[118] However, their fungal etiology was not discovered until 1837 when Remak described the scutula of favus as made up of mold filaments.[118] In 1839, Schoenlein demonstrated the connection between this disease and the plant kingdom.[118] The general acceptance of the fungal cause of the tineas, and indeed, the beginning of the science of medical mycology occurred between 1841-1844 when David Gruby presented his observations before the French Academy of Sciences.[69] Gruby described in detail the parasitic (though not cultural) morphology of the fungi in four distinct dermatophytoses: favus (*T. schoenleinii*), tinea barbae (*T. mentagrophytes*), the small-spored, ectothrix type of tinea capitis in children (*M. audouinii*) and the large-spored, endothrix type of tinea capitis (*T. tonsurans*). Using pure cultures, Raymond Sabouraud conducted precise studies on the dermatophytes then available, devised culture media appropriate for their differentiation and described the colonial and microscopic morphology of the various species and species groups. His publications, beginning in 1892, culminated in the monumental treatise "Les Teignes" (The Tineas) published in Paris in 1910.[107] Sabouraud's original "proof" medium was used by pioneer medical mycologists during the next three decades. It contained a special crude maltose and a special peptone (Chasaings') obtainable from Paris. This supply of maltose and peptone became inaccessible during World War II and substitutes were sought. After chemical analyses of the two substances, the crude maltose was replaced by chemically-pure dextrose, its chief and essential ingredient, and other suitable sources of peptone became commercially available in the United States and elsewhere. The result is Sabouraud dextrose-peptone agar (SDA), today's basic mycological medium. Among other milestones in the study of the dermatophytoses are: (1) the isolation of trichophytin by Plato and Neisser in 1902,[118] paving the way for the elucidation of the cause of dermatophytids and other allergic reactions to this group of fungi; and (2) the discovery of Margarot and Daveze in 1925[69] of the fluorescence of hairs infected by certain *Microsporum* species when these hairs are exposed to a Wood's light.

Although early attempts at classification of the dermatophytes were based on parasitic morphology, taxonomic proposals by Ota and Langeron in 1923,[87] Langeron and Milochevitch in 1930[73] and Emmons in 1934,[35] were based on morphology in culture. Emmons' classification system, expanded by Ajello[1,3] and others, forms the basis for today's nomenclature. Physiological characteristics and other tests have supplemented morphologic criteria for the differentiation of the dermatophytes.[50] The discovery of griseofulvin revolutionized the treatment of dermatophyte infections.[46]

Etiologic Agents

The etiologic agents of the dermatophytoses belong to the genera *Microsporum, Trichophyton* and *Epidermophyton*. Emmons' classification scheme is essentially based on the asexual spores of these three genera.[35] This scheme, updated by Ajello,[1,3] is the basis of the current classification of the anamorphic or asexual state of the dermatophytes.

Many of the dermatophytes reproduce sexually as well as asexually. The teleomorph (sexual state) of these species belong to the genus *Arthroderma* of the family Gymnoascaceae of the Ascomycota. Until recently the teleomorph of the genus *Microsporum* had been named *Nannizzia* and the name *Arthroderma* reserved for the teleomorph of the genus *Trichophyton*. However, a careful evaluation of the characteristics used to define *Nannizzia* and *Arthroderma* concluded that these characteristics represent a continuum and that these two taxa are congeneric.[129] Based on priority, the correct name is *Arthroderma*. The perfect states of the *Microsporum* and the *Trichophyton* species are listed in Table 3:2. The teleomorph for the *Epidermophyton* spp. has not yet been discovered. The majority of these fungi are heterothallic; that is, sexual reproduction requires the pairing of compatible mating types (referred to as + and - or as A and a) on a suitable medium. The best results are obtained on soil and hair, but special nonkeratinous agar media are suitable substrates for most species.[55,91]

Most of the *Microsporum* and *Trichophyton* species produce two kinds of conidia: large, multiseptate conidia (macroconidia) and small, aseptate conidia (microconidia). Some species, notably *T. terrestre*, produce conidia that are transitional between microconidia and macroconidia.

Differential features of *Microsporum, Trichophyton* and *Epidermophyton* are found in Tables 3:3 and 3:4

Microsporum:

The distinguishing feature of the genus is the roughened surface of the macroconidia, often described as verrucose, verruculose or echinulate. The roughened wall may be very prominent, as in *M. canis*, or difficult to discern

Table 3:2 Teleomorphs (Sexual States) of *Microsporum* and *Trichophyton* Species

Telemorph	*Anamorph*
Arthroderma borellii	*Microsporum amazonicum*
A. cajetanii	*M. cookei*
A. corniculatum	*M. boulardii*
A. fulvum	*M. fulvum complex*
A. grubyi	*M. vanbreuseghemii*
A. gypseum	*M. gypseum complex*
A. incurvatum	*M. gypseum complex*
A. obtusum	*M. nanum*
A. otae	*M. canis* var. *canis* var. *distortum*
A. persicolor	*M. persicolor*
A. racemosum	*M. racemosum*
A. benhamiae	*Trichophyton mentagrophytes* complex
A. ciferii	*T. georgiae*
A. flavescens	*T. flavescens*
A. gertlerii	*T. vanbreuseghemii*
A. gloriae	*T. gloriae*
A. insingulare	*T. terrestre* complex
A. lenticularum	*T. terrestre* complex
A. quadrifidum	*T. terrestre* complex
A. simii	*T. simii*
A. uncinatum	*T. ajelloi*
A. vanbreuseghemii	*T. mentagrophytes* complex

Table 3:3 Characteristic Features of Pathogenic *Microsporum* and *Epidermophyton* Species

Species	Colony on Sabouraud Dextrose Agar	Microscopic Morphology	Differential Features
Microsporum			
M. audounii	Grayish-white to buff, flat, spreading, velvety; light salmon pink to light reddish-brown on reverse	Usually no conidia; apiculate, terminal chlamydospores are the only characteristic features; pectinate hyphae may be present (Fig. 3:8)	Poor growth and brownish discoloration on rice grains; sporulating strains have irregular fusiform and elongated macroconidia with few septa at irregular intervals (Fig. 3:9)
var. *rivalieri**	Grayish-white; folded; pale apricot on reverse; more rapid growth than *M. audounii*	Inflated, bulbous, pectinate	Pathogenic for guinea pigs; macroconida produced on soil and hair by some isolates
var. *langeronii**	Rose to tan with radial grooves	May have more microconidia and macroconidia than *M. audouinii*	Pathogenic for guinea pigs
M. canis			
var. *canis*	White to pale buff, fluffy or velvety with radial grooves; yellow-orange to orange-brown or colorless on reverse	Numerous fusiform macroconidia with thick walls, up to 15 septa; 18–125 μm × 5–25 μm, with asymetric knobbed apex (Fig. 3:10); few microconidia	Good growth and sporulation on rice grains
var. *distortum*	White to pale buff; usually with radial grooves; colorless to yellowish on reverse	Macroconidia are distorted, bizzare in shape, 20–60 μm x 7–27 μm; microconidia abundant (Fig. 3:11)	
M. cookei	Yellowish to reddish tan; flat; powdery, granular or downy; reverse dark purple-red	Macroconidia numerous, some resembling *M. gypseum* but most have thicker walls (1–5 μm); microconidia abundant	
M. equinum	White to pale salmon; some folding; velvety to finely powdery; buff to salmon on reverse	Macroconidia, infrequent on Sabouraud agar; are elliptical to fusiform, 18–60 μm × 5–13 μm, 2 to 4-celled; thick walls	Macroconidia stimulated by Niger seed-medium 8 agar[65]; does not perforate hair in vitro

M. ferrugineum	Yellowish to rust colored; folded; waxy; very slow grower	Usually no conidia; numerous chlamydospores irregular hyphae and long, straight, coarse hyphae with prominent septa ("bamboo hyphae")	Some isolates produce spindle-shaped, multicellular thick-walled macroconidia on freezing agar or on diluted SDA[119]; light yellow colonies on Lowenstein-Jensen medium differentiates *M. ferrugineum* from dark, reddish-brown colonies of *T. soudanense*
M. gallinae	White, tinged with pink, slightly folded; downy; raspberry-red, diffusing pigment on reverse	Macroconidia fairly abundant, blunt tipped, 6–8 μm x 15–50 μm, 5 to 6-celled; cell walls usually smooth but may be echinulate; pyriform microconidia	Conidiation stimulated by growth on media containing yeast extract
M. gypseum complex (*M. gypseum* and *M. fulvum*)	Pale buff, rosy buff to light cinnamon, white border; flat, powdery, granular to floccose; buff to reddish-brown on reverse	Abundant macroconidia, 25–60 μm × 7.5–15 μm; fusiform to cylindrical up to 6 septa, thin walled; microconidia moderately abundant (Fig. 3:12)	
M. fulvum	Usually floccose	Macroconidia more cylindrical but with tapering ends	Teleomorph *A. fulvum*
M. gypseum	Usually coarse to finely granular	Macroconidia ellipsoidal to fusiform	Teleomorph *A. incurvatum, A gypseum*
M. nanum	Cream to downy; reddish brown on reverse	Abundant obovate to clavate macroconidia 10.5–30 μm × 6.5–13 μm, usually 2-celled; microconidia few to moderate	Grows more slowly than members of *M. gypseum* complex; also must differentiate from *Trichothecium roseum*.
M. persicolor	Yellowish-buff becoming peach to pink; flat; powdery to downy; reddish brown on reverse	Abundant microconidia, spherical to pyriform (few clavate), stalked, borne mostly in grape-like clusters but also singly along the sides of hyphae; thin-walled macroconidia, often produces spirals	Differentiated from *T. mentagrophytes* by echinulations of macroconidia which are more evident on cereal agar or soil and hair; pink to wine-rose colonies on cereal agar or sugar-free peptone agar[88]; teleomorph = *A. persicolor*

Table 3:3 Characteristic Features of Pathogenic *Microsporum* and *Epidermophyton* Species (*Continued*)

M. praecox	Cream to yellowish-tan; folded powdery; pale yellow to orange on reverse	Numerous long fusiform macroconidia, some with apical appendages, 40–90 μm × 7–17 μm, 2–8 septa, thin-walled; microconidia absent.	
M. racemosum	White, cream or buff; flat; finely granular; reverse dark purple-red	Macroconidia abundant, fusiform to ellipsoidal, 41–77 μm × 9 μm, 3–8 septa, moderately thick walls; numerous microconidia, mostly stalked and produced in grape-like clusters	
M. vanbreuseghemii	Pink to deep rose, light buff or yellowish; flat; coarsely granular to downy; reverse cream to pale yellow	Abundant cylindrofusiform macroconidia, 43.8–87.5 μm, thick walled, up to 12 septa; numerous pyriform to obovate microconidia borne singly along sides of hyphae	
Epidermophyton *E. floccosum*	Yellowish-olive to olive green; flat to radially folded; suede-like; brownish-orange on reverse; white tufts common on surface of older cultures (pleomorphism)	Abundant, broad clavate, smooth-walled macroconidia, 20–40 μm × 6–8 μm, single or in clusters, 0–4 septa; chlamydospores common in older cultures (Fig. 3:13)	No microconidia

*Considered by some as a variety, by others as a distinct species.

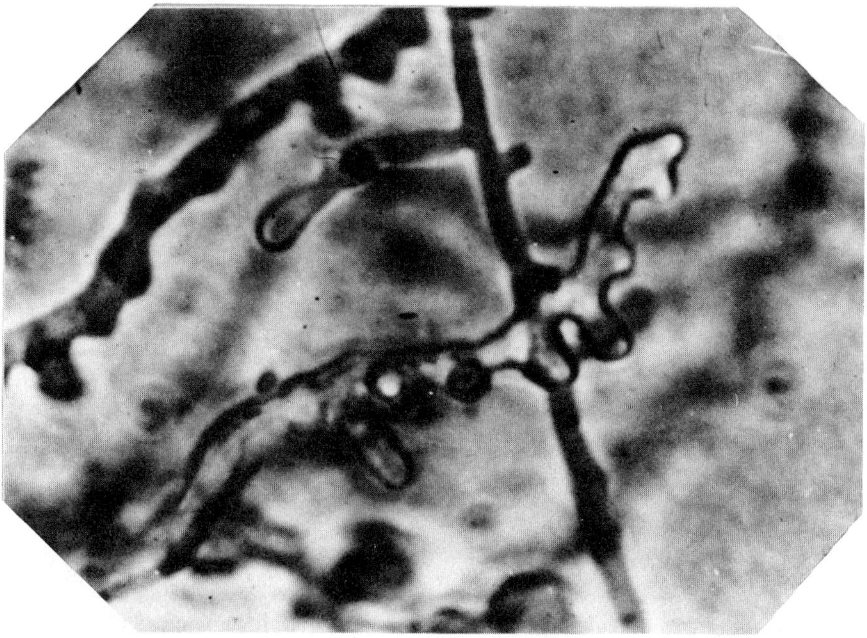

Figure 3:8. Pectinate organs (curved hyphae with many short side branches). These structures are abundantly produced on Sabouraud agar by cultures of *M. audouinii* and aid in its identification. Lactophenol preparation. X1000.

without an oil immersion lens, as in *M. persicolor.* Scanning electron microscopic studies have revealed that these are vesicles rather than spines, suggesting the term vesiculate rather than echinulate.[93]

Microsporum species usually sporulate profusely and produce numerous macroconidia and smaller numbers of microconidia. *M. audouinii* and *M. ferrugineum* are notable exceptions since sporulation is usually rare or absent unless induced on special media. Macroconidia are usually borne singly on the conidiophores, are multiseptate (1-15 septa), typically spindle-shaped (fusiform) but may range from obovate to cylindrofusiform, are thin to thick-walled and range from 60-160 μm long and from 6-25 μm wide. The *Microsporum* species are listed in Table 3:5.

Trichophyton:

Macroconidia are smooth-walled, borne singly or in clusters on the conidiophores and may be rare, abundant or absent, depending upon the species, the isolate and the culture medium. Macroconidia are usually multiseptate (1-12 septa), variable in shape (clavate, cylindrical, cylindro-fusiform), thin to thick-walled and range from 8-86 μm long by 4-14 μm

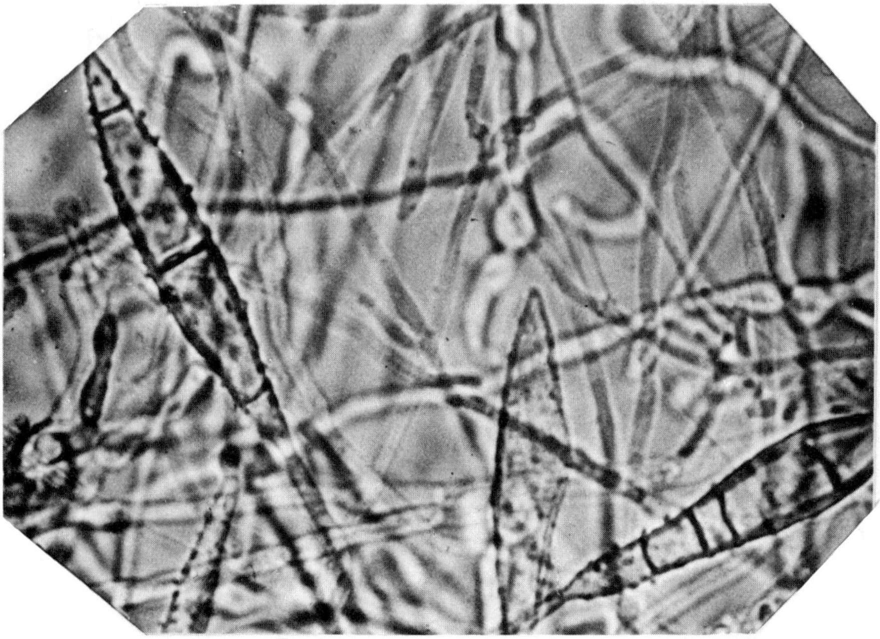

Figure 3:9. Macroconidia of *M. audouinii* (not a regular finding) on honey agar plus yeast extract. Note rough-walled, fusiform macroconidia with septa at irregular intervals and with thickened outer walls. Lactophenol preparation. X1000.

wide.[90] Microconidia are smooth-walled, globose, pyriform (pear-shaped) or clavate (club-shaped), and may be borne singly along the sides of the hyphae (*en thyrse*), or in grapelike clusters (*en grappe*). Transitional forms between micro- and macroconidia are characteristic of some species or varieties. The *Trichophyton* species are listed in Table 3:6.

Tables 3:2, 3:3 and 3:4 list *M. gypseum, T. mentagrophytes* and *T. terrestre* as complexes rather than as single species. These complexes were revealed when mating studies among the asexual species led to the discovery of more than one teleomorph per anamorph. For example, *M. gypseum* was found to be the anamorph of three teleomorphs, *Arthroderma (N.) gypseum, A. (N.) incurvatum* and *A. (N.) fulvum*.[112] Since it is difficult for the routine laboratory to differentiate between the anamorphs we may simply refer to the asexual state (including *M. fulvum* which was formerly considered to be synonymous with *M. gypseum*) as the *M. gypseum* complex. Similarly, a complex of species was recognized for the anamorph, *T. mentagrophytes* (teleomorphs *A. benhamiae* and *A. vanbreuseghemii*) and for *T. terrestre* (teleomorphs *A. quadrifidum, A. lenticularum* and *A. insingulare*).

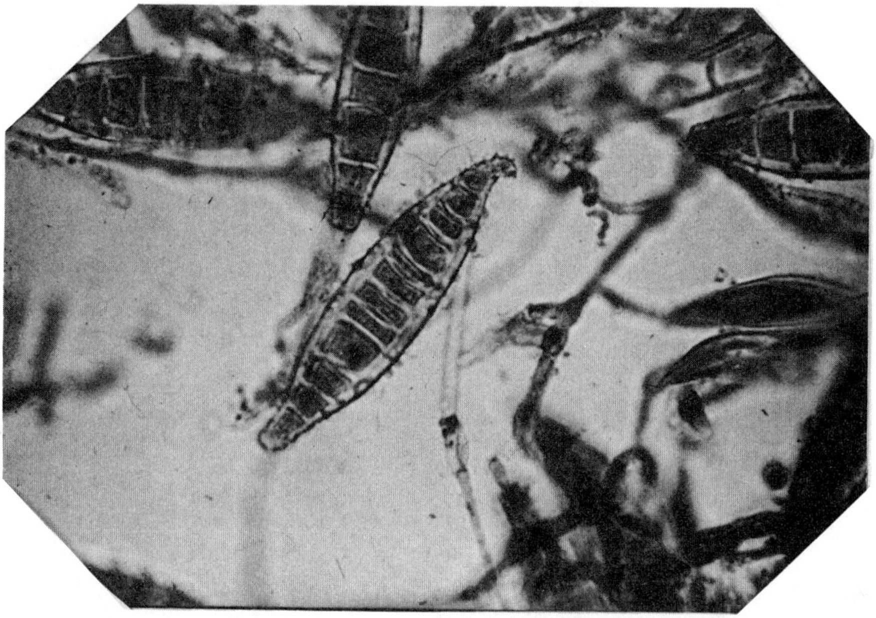

Figure 3:10. Macroconidia of *M. canis* on Sabouraud agar. Note knob-like ends of fusiform macroconidia with rough walls thickest at level of medial septa. These occur at regular intervals unlike those of *M. audouinii* (Figure 3:9). Lactophenol-blue preparation accounts for darker staining of cytoplasm. X1000. Compare with Figure 3:11.

Epidermophyton:

Macroconidia are smooth-walled, borne singly or in clusters, obovate to broadly clavate, have 0-4 septa and range in size from 20-40 μm by 6-12 μm. Cultures produce no microconidia. Until recently, the genus was monotypic; *E. floccosum* was the only species. However, a new saprobic species, *E. stockdaleae*, has been reported.[94]

Mating reactions may be helpful in identifying a species complex or an atypical isolate. These mating reactions should be done at reference laboratories or by individuals engaged in research in this field, since specialized media, an extensive collection of stock cultures of each species and isolates of each mating type are required.

Certain saprobic or rarely pathogenic fungi (e.g., *T. terrestre*) can resemble some pathogenic species. Many of these fungi also have one or more of the properties of dermatophytes, such as the ability to turn Dermatophyte Test Medium (DTM) red, the production of perforating organs in vitro, urease activity and the ability to grow on media containing cycloheximide. Table

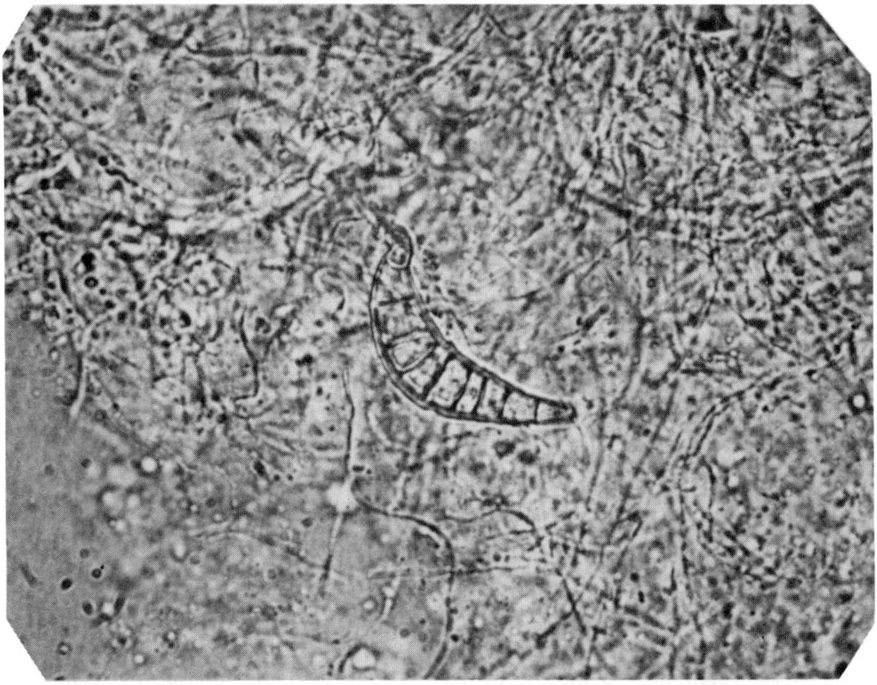

Figure 3:11. Macroconidium of *M. canis* var. *distortum* on Sabouraud agar. Note characteristic boomerang (bent) shape of the macroconidium. Lactophenol preparation. X600.

3:7 lists some of these fungi, the dermatophytes for which they may be mistaken and some differentiating characteristics.

Clinical Manifestations

Brief descriptions of the various clinical manifestations follow and descriptions of hair infections are summarized in Table 3:8.

The clinical manifestations of dermatophytic infections (tineas, ringworm) may range from mild to severe depending on the virulence of the infecting agent, the anatomical location of the lesion(s) and host factors such as age, sex and immune status. A single fungal species can infect many anatomical locations and therefore produce various types of lesions, and conversely, it is possible for different species to produce clinically identical lesions.

Traditionally, the diseases caused by dermatophytes have been named as they present to the observer; i.e., according to their anatomical location: (a) tinea capitis (scalp); (b) tinea barbae (beard); (c) tinea corporis (face and

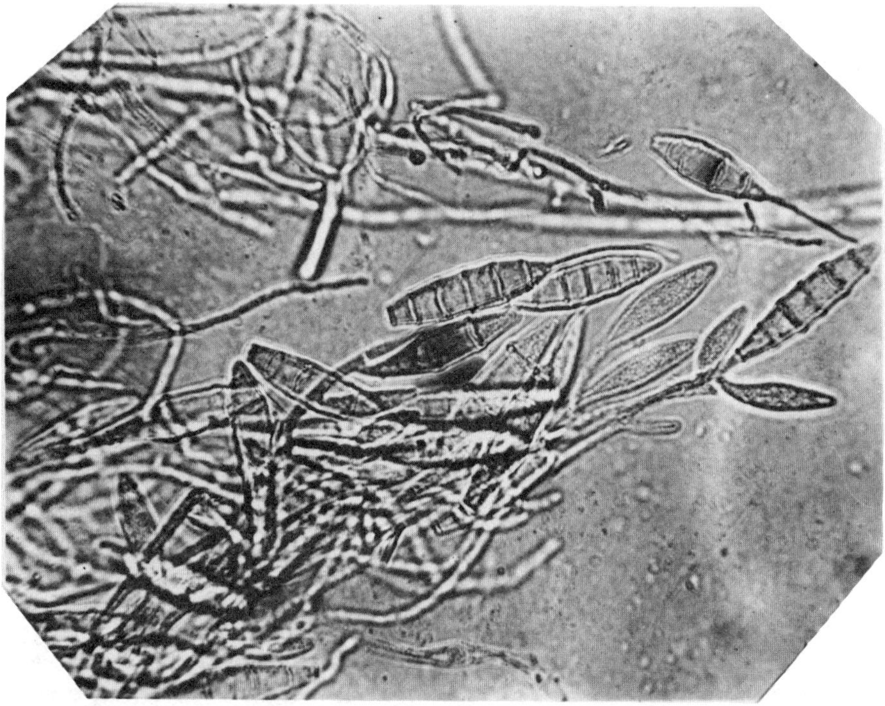

Figure 3:12. Macroconidia of *M. gypseum* on Sabouraud agar. Note shorter, blunter shape of the macroconidia, fewer numbers of septa and uniformity of thickness of macroconidial wall. Compare with *M. canis* (Figure 3:10). Lactophenol preparation. X450.

trunk); (d) tinea axillaris (armpits); (e) tinea cruris (groin); (f) tinea pedis (feet); (g) tinea manuum (hands); and (h) tinea unguium (nails).

Terms describing the morphology of the lesions have also been coined: (a) tinea circinata for infections showing discrete circular lesions with a raised border; (b) tinea imbricata for lesions showing concentric layers of scales whose borders overlap like tiles on a roof (imbricate pattern); (c) kerion celsi for raised, boggy, inflammatory lesions with a smooth, pink-to-red surface and pustules; (d) Majocchi's granuloma for lesions involving deeper tissues around hair follicles; and (e) favus (or tinea favosa) for infections producing a honeycomb pattern of saucer-like depressions called scutula (tiny shields).

Tinea capitis:

This infection is characterized by one or more areas of alopecia (bald spots) caused by the breaking off of hairs invaded by the fungus (Figs. 3:21,

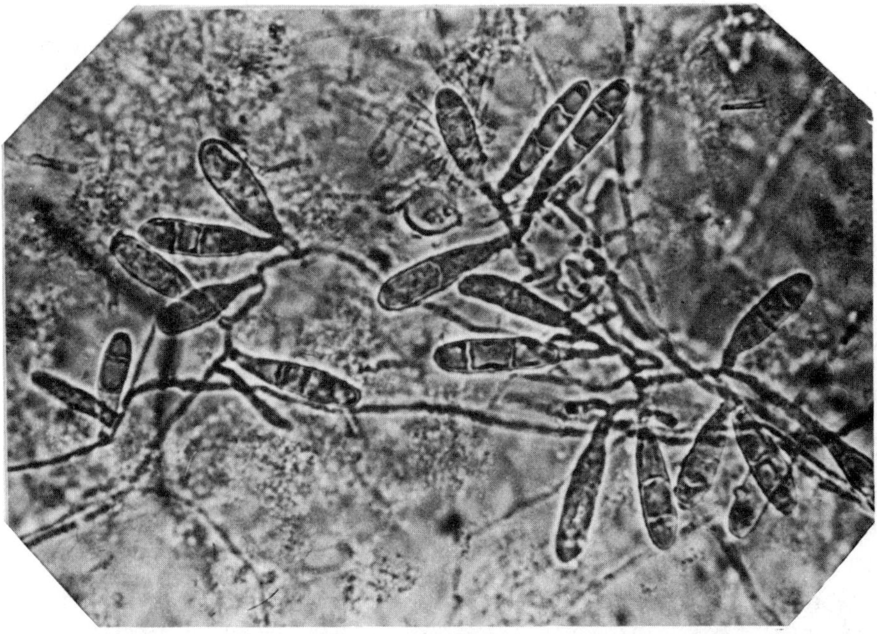

Figure 3:13. Macroconidia of *Epidermophyton floccosum* on Sabouraud agar. Note clusters of clavate (club-shaped) macroconidia with 1-3 septa, smooth walls and rounded distal ends. Lactophenol-blue preparation. X450.

3:22). The skin surface is also invaded by hyphae. This disease occurs mostly in children, but certain fungi, i.e., *T. tonsurans* and *T. schoenleinii*, produce lesions that persist into adulthood. The clinical picture is influenced by the infecting fungus. The *Microsporum* species, *T. mentagrophytes* and *T. verrucosum* produce arthroconidia outside the hair so that infected hairs are covered with a grayish sheath (ectothrix infection, Fig. 3:23). Hairs break off 2-5 mm above skin level. When seen under Wood's light (filtered ultraviolet light, peak wavelength 365 nm), *Microsporum*-infected hairs may emit a greenish fluorescence.

In *T. tonsurans* and *T. violaceum* infections, the arthroconidia are produced inside the hair (i.e., endothrix infection, Fig. 3:24). The hairs break off flush with the scalp, and subsurface hairs coil into minute knots resembling dots under the scaly skin surface. These infections in patients with dark hair are called "black dot tinea." Certain patients respond to fungal infection of the scalp with an acute inflammatory reaction at the site of infection, which becomes raised, smooth-surfaced, pink-to-red, boggy and pustular (kerion, Fig. 3:22). Kerions are usually produced in response to zoophilic fungi such as *M. canis, T. mentagrophytes* var. *mentagrophytes*

Table 3:4 Characteristic Features of Pathogenic *Trichophyton* Species

Species	Colony on Sabouraud Dextrose Agar	Microscopic Morphology	Differential Features
T. ajelloi	Cream or orange-tan, flat, powdery; reverse blackish-purple, sometimes non-pigmented	Microconidia rare, macroconidia numerous, cylindrical to fusiform, thick-walled, multiseptate, 5 to 12-celled	
T. concentricum	Buff to beige; elevated and convoluted; glabrous to velvety; no undersurface color; slow growing	Micro- and macroconidia usually absent; chlamydospores may be present	50% of isolates are stimulated by thiamine, others are autotrophic
T. equinum	Cream colored; flat; fluffy; reverse yellow becoming reddish-brown	Pyriform or spherical microconidia; macroconidia rare, similar to those of *T. mentagrophytes*	Requires nicotinic acid; an autotrophic variety has been described[110]
T. gourvilii	Pink to red; heaped up; convoluted; glabrous; a brown pigment may diffuse into medium	Typical *Trichophyton*-type macroconidia and microconidia usually found	
T. megninii	Pink to rose; radially folded; suede-like; reverse wine-red	Pyriform to clavate microconidia; macroconidia rare, similar to those of *T. rubrum*.	Requires l-histidine.
T. mentagrophytes	Cream, tan, or pink; flat; powdery to granular or fluffy; reverse light tan, yellow, red or reddish-brown; sometimes producing a diffusible melanoid pigment	Round to pyriform microconidia in clusters or singly along the hyphae; clavate macroconidia present in some strains; coiled hyphae (spirals) are usually seen (Figs. 3:14, 3:15)	Urease positive; perforates hair in vitro; grows at 37°C (Fig. 3:16)
T. raubitschekii	Buff; raised center with radial grooves; velvety to granular; reverse blood-red	Microconidia clavate, globose or subspherical, sessile or on short stalks along unbranched hyphae; macroconidia abundant, thin and elongate with blunt ends, 5 to 9-celled.	Urease positive; in vitro hair perforation test negative; brown pigmentation on casein-dextrose agar
T. rubrum	White; fluffy (seldom powdery); reverse wine-red, sometimes yellow, orange or a diffusible melanoid pigment	Pyriform microconidia usually along unbranched hyphae; macroconidia absent to rare, thin, cylindrical-to-clavate in granular cultures (Figs. 3:17, 3:18)	Urease test usually negative; in vitro hair perforation test negative; red undersurface pigment best on cornmeal dextrose agar

Table 3:4 Characteristic Features of Pathogenic *Trichophyton* Species (*Continued*)

T. schoenleinii	White to tan; elevated and convoluted; glabrous or waxy, becoming velvety on subculture; reverse lacking pigment; slow growing	Micro- and macroconidia rarely seen; chlamydospores often numerous; hyphal tips often show "nail head" morphohology and branch to form antler-like structures (favic chandeliers)	Autotrophic for vitamins which differentiates it from *T. verrucosum*
T. simii	Pale to buff; flat or slightly convoluted; powdery; reverse straw to salmon	Numerous macroconidia, some fragment or develop swellings resembling chlamydospores; pyriform microconidia and spirals may be found	
T. soudanense	Yellow-orange (like dried apricots); flat with convolutions; suede-like texture; fringed (eyelash) periphery; reverse yellow to orange-yellow; slow growing	Pyriform microconidia are rare; macroconidia very rare; reflex branching characteristic	Reflex branching, growth stimulated at 37°C
T. terrestre	White, pale yellow or red; flat; granular to downy; reverse pale yellow, yellowish-brown or red	Microconidia clavate to pyriform, single or clustered, short, intermediate or fully extended to attain the size of macroconidia; these are clavate to cylindrical, thin-walled, 2 to 6-celled	Usually no growth at 37°C
T. tonsurans	Color varies with isolate (yellow, cream, white, pink, brown, gray, etc.); convoluted usually, sometimes flat; velvety to powdery; reverse dark brown to mahogany-red; slow growing	Clavate to elongate microconidia, some swollen into balloon forms and some 2-celled, attached to branched conidiophore by a short stalk; macroconidia rare (Figs. 3:19, 3:20)	Stimulated by thiamine
T. verrucosum	White to yellowish-tan; heaped, flat or convoluted; glabrous or downy; no reverse pigment; slow growing	Usually no micro- or macroconidia; chlamydospores usually numerous and in chains	All strains requre thiamine; most require inositol as well; growth stimulated at 37°C
var. *album*	White; heaped and folded; glabrous to downy	Usually no micro- or macroconidia; chlamydospores usually numerous and in chains	All strains require thiamine; most require inositol as well; growth stimulated at 37°C

var. *discoides*	Cream to tan; flat-discoid; velvety	Usually no micro- or macroconidia; chlamydospores usually numerous and in chains	All strains require thiamine; most require inositol as well; growth stimulated at 37°C
var. *ochraceum*	Yellow-ochre; convoluted; glabrous	Usually no micro- or macroconidia; chlamydospores usually numerous and in chains	All strains require thiamine; most require inositol as well; growth stimulated at 37°C
T. violaceum	Violet or lavender; heaped and convoluted; glabrous or velvety; purple undersurface; slow growing	Micro- and macroconidia usually lacking; chlamydospores may be found	Growth and sporulation stimulated by thiamine
T. yaoundei	Initially buff, turning to chocolate-brown; convoluted; glabrous, reverse brown, diffusible; slow growing	Pyriform microconidia are rare; macroconidia not found; chlamydospores seen.	

Figure 3:14. Microconidia and macroconidia of *T. mentagrophytes* var. *mentagrophytes* on Sabouraud agar. Note clusters of detached spherical microconidia, thin-walled fusiform macroconidia and loosely spiraled hyphae (see also Figure 3:15). Lactophenol blue preparation. X600.

and *T. verrucosum* or to geophilic fungi such as *M. gypseum*, although other dermatophytes may elicit them on occasion.

Tinea barbae:

This infection occurs in two clinical forms: dry and suppurative. In dry lesions (sycosis barbae), the beard involvement resembles the gray patches of alopecia described for tinea capitis. The hairs break off a few millimeters above the skin surface, leaving a straight hair stump surrounded by a sheath of arthroconidia. The most common agent of dry tinea barbae in the United States is *T. mentagrophytes*, although beard infections produced by *T. rubrum* have been reported.[27] When the *T. mentagrophytes* infection is of animal origin, kerions are produced on the bearded area. In Europe (particularly Portugal, France and Belgium), dry tinea barbae is produced by *T. megninii*.[85,106]

Suppurative tinea barbae is most often seen among dairy farmers, who contract *T. verrucosum* infections from cattle. The kerions are raised, erythematous, pustular and boggy. Infected hairs are broken off a few

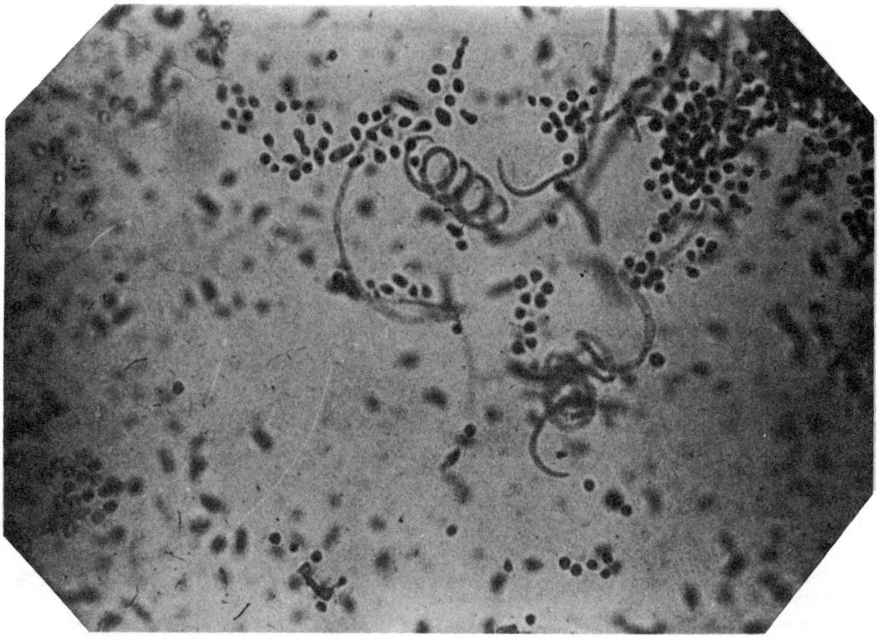

Figure 3:15. Tighter coils produced by hyphae of *T. mentagrophytes* are also seen by this variety on Sabouraud agar. Lactophenol-blue preparation. X450.

millimeters above the skin but are also loosened by the inflammatory process and often fall out spontaneously.

Tinea corporis:

This term usually refers to dermatophyte infections of the glabrous skin, excluding nails or intertriginous areas. As in tinea capitis, the lesions of tinea corporis show a certain degree of variation depending on the patient's age and on the causative fungus. In children, face lesions commonly accompany fungal infections of the scalp, particularly those produced by *M. canis* or *M. audouinii* (Fig. 3:25). Acute lesions of this sort are circinate, i.e., circular, with a raised, active border. Similar lesions may also appear on the wrists, neck and throat area (but rarely on the face) of the mother or other young adults in contact with an infected child. These lesions are seldom seen in the elderly, who are more likely to suffer from chronic tinea corporis.

In contrast to acute tinea circinata, chronic dermatophyte lesions of the glabrous skin are more often found throughout the trunk and extremities, occasionally invading the face and are frequently associated with tinea pedis and tinea manuum. On a worldwide basis, *T. rubrum* is the most common

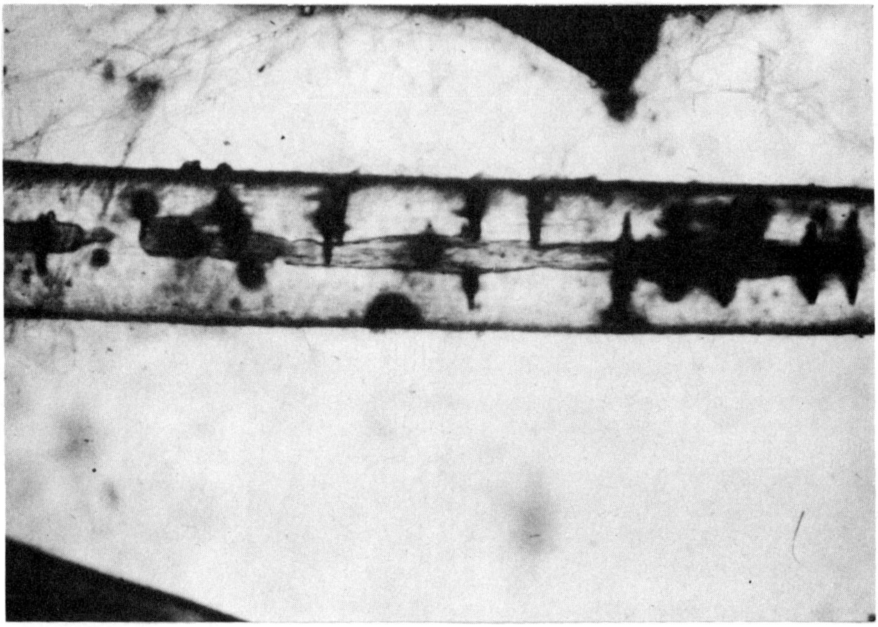

Figure 3:16. Wedge-shaped hair perforations produced by *T. mentagrophytes* during standard in vitro test (see text for details). X600.

agent of this type of generalized, chronic infection. In localized geographic areas, other species, such as *T. tonsurans, T. schoenleinii* (Fig. 3:26) and *T. concentricum* may predominate. The clinical picture of *T. tonsurans* lesions is less generalized, showing local spreading areas of induration and a granulomatous reaction. Lesions caused by *T. schoenleinii* occur on the glabrous skin and scalp of both children and adults. These lesions are characteristically funnel- or saucer-shaped depressions (scutula) centered around hair follicles. They may occur in clusters covering extensive areas.

Tinea imbricata, caused by *T. concentricum*, is a distinct type of tinea corporis in which the lesions appear as concentric circles of imbricate scales (overhanging scales, like tiles on a roof). The lesions cover the trunk, face and extremeties.

Tinea unguium:

The toe- or fingernail undergoes one or more of the following changes as a result of the infection: is discolored (yellow or white), loses hardness and is easily crumbled, thickens, becomes grooved, separates from the nail bed, and/or accumulates subungual debris (Figs. 3:27, 3:28). The most recalcitrant type of tinea unguium is caused by *T. rubrum*, which is also the most

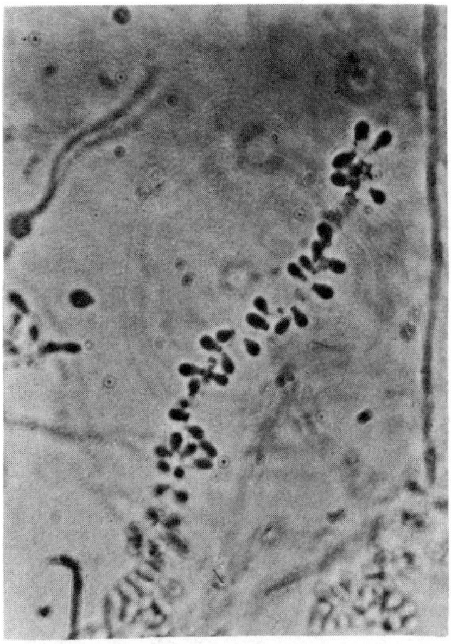

Figure 3:17. Microconidia of *T. rubrum* on Sabouraud agar. Note pyriform (pear) shaped conidia arranged *en thyrse*, i.e., along sides of unbranched hyphae which characteristically shrivel and become inconspicuous as the conidia mature. Lactophenolblue preparation. X1000.

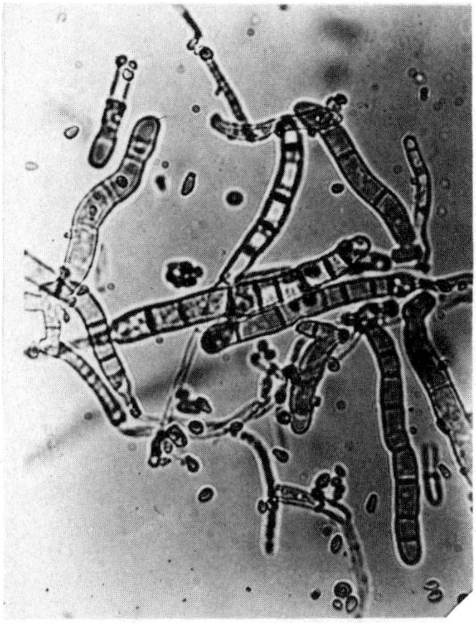

Figure 3:18. Pencil-shaped and curved macroconidia characteristic of *T. rubrum* (granular variety). X1000.

frequent cause of spongy, hyperkeratotic, deeply invaded and indurated nails. Chalky, white, superficial discolorations of nails are produced most often by *T. mentagrophytes*. *E. floccosum* and *T. tonsurans* nail infections are less frequent.

In addition to dermatophytes, other fungi such as *Candida, Aspergillus, Scopulariopsis, Fusarium* and *Acremonium* can infect nails, causing a clinically similar condition, called onychomycosis (see section on opportunistic dermatomycosis).

Tinea cruris:

Lesions in the groin area are characterized by sharply demarcated, raised edges on which scales and vesicles may appear. *T. rubrum* and *E. floccosum* are the most common agents. These lesions, which are red or tawny brown,

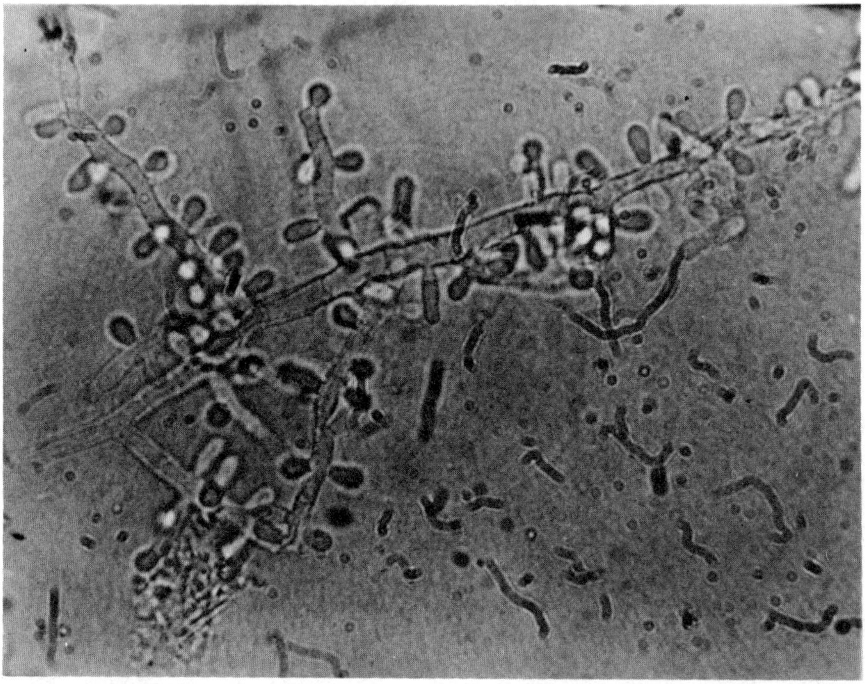

Figure 3:19. Microconidia of *T. tonsurans*. Note right-angled branching and "pine-tree" shape of conidiophore whose main axis is thickened, also note clavate shape and firm attachment of each conidium with septum at its base. X1000.

must be differentiated from clinically-similar lesions caused by *Candida* and from erythrasma.

Tinea pedis:

The disease varies widely in its clinical manifestations. The most common is the interdigital (intertriginous) type, with maceration, peeling, itching and painful fissuring (Fig. 3:29). The spaces between the fourth and fifth toes are most frequently attacked. The soles may have hyperkeratotic patches or blisters (Fig. 3:30). The most common agents of tinea pedis are *T. mentagrophytes* and *T. rubrum*. The former frequently causes inflammatory, vesicular lesions, while *T. rubrum* more often causes hyperkeratotic, chronic lesions. *E. floccosum* is third in frequency.

Tinea manuum:

This condition appears most frequently as a unilateral, diffuse hyperkeratosis of the palm and fingers. Other clinical manifestations are exfoliation,

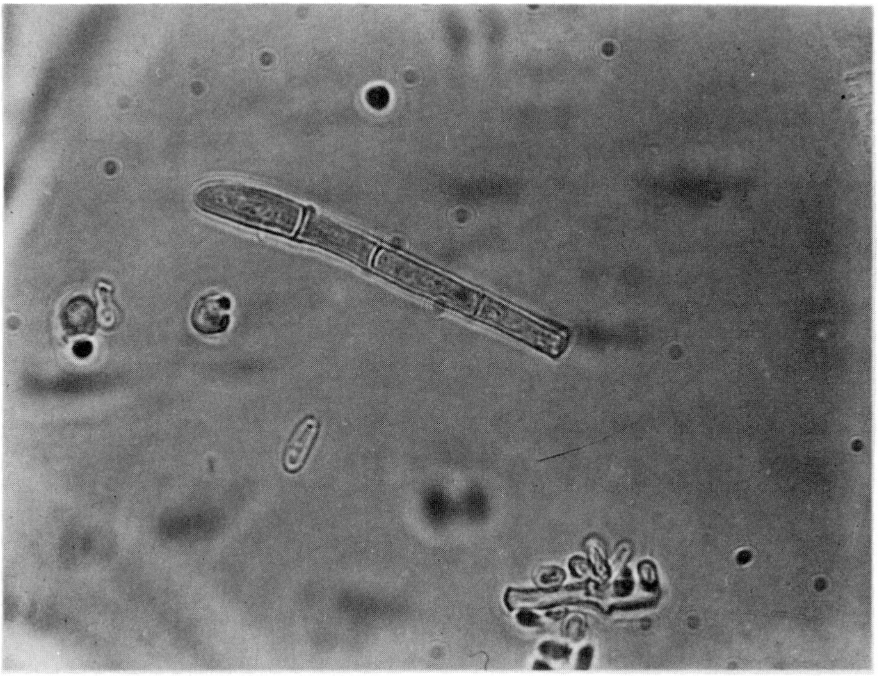

Figure 3:20. Clavate (club shaped) macroconidium of *T. tonsurans* (not a regular finding). X1000.

Table 3:5 *Microsporum* Species

M. amazonicum	*M. gallinae*
M. audouinii	*M. gypseum**
M. boullardii	*M. nanum*
M. canis var. *canis*	*M. persicolor*
M. canis var. *distortum*	*M. praecox*
M. cookei	*M. racemosum*
M. equinum	*M. ripariae*
M. ferrugineum	*M. vanbreuseghemii*
*M. fulvum**	

**M. gypseum* complex.

papules and follicular patches. The most frequent agent is *T. rubrum*; less frequent is *T. mentagrophytes*. Tinea manuum must be differentiated from ''dermatophytid'' reactions, which appear on the fingers or palms and are an allergic response to a dermatophyte infection on another part of the body.

Table 3:6 *Trichophyton* Species

T. ajelloi	*T. raubitschekii*
T. concentricum	*T. rubrum*
T. equinum	*T. schoenleinii*
T. flavescens	*T. simii*
T. georgiae	*T. soundanense*
T. gourvilii	*T. terrestre* complex
T. mariatii	*T. tonsurans*
T. megninii	*T. vanbreuseghemii*
T. mentagrophytes complex	*T. verrucosum*
T. phaseoliforme	*T. violaceum*
	T. yaoundei

Table 3:7 Some Saprobic Fungi Mistaken for Dermatophytes

Saprobe	Dermatophyte	Saprobe Differentiated from Dermatophyte by:
Chrysosporium tropicum	*T. mentagrophytes*	Larger microconidia 3–4 × 6–7 μm may be slightly roughened, pyriform to clavate, with a broadly truncate base
C. keratinophilum	*M. nanum*	Aseptate conidia, typically smooth (may be slightly roughened); colony granular
Trichothecium roseum	*M. nanum*	Smooth conidial walls, conidia are produced in short chains at apex of simple conidiophore and remain as a group if undisturbed (slide culture recommended)
Trichophyton terrestre	*T. mentagrophytes*	Transitional forms intermediate between microconidia and macroconidia; species studied show poor or no growth at 37°C;[90] teleomorph = *A. quadrifidum, A. lenticalarum, A. insingulare*
Fusarium species	*T. rubrum*	Crescent- or boat-shaped macroconidia; microconidia characteristically in short chains or clusters

Epidemiology and Public Health Significance

Prevalence:

Since ringworm or other superficial fungal infections are not reported to public health departments, the true number of such infections are not known. On the basis of sales of antidermatophyte medications, the prevalence of infection appears considerable.[2] A clinical and laboratory survey[61] of over 20,000 persons in the United States has given some data on the approximate prevalence of fungal infections. These data show that fungal infections constitute the second largest group of skin conditions in the general population (81 per 1,000), second only to diseases of the sebaceous glands

Table 3:8 General Characteristics of Hair Infections by Dermatophytes in Humans

Disease Name and Variations	Site	Type of Infection	Most Prevalent in:	Usually Caused by:
A. Tinea capitis	Scalp			
1. Gray-patch-ringworm		Bald scaly spot with straight hair stumps	Children	*M. canis*, *M. audouinii*
2. Black-dot-tinea		Bald, scaly spot with coiled hair stumps (black dots)	Children and adults	*T. tonsurans*, *T. violaceum*
3. Kerion celsi		Raised, inflammatory and pustular bald area	Children	*M. canis*, *M. gypseum*, *T. mentagrophytes*, *T. verrucosum*, *T. violaceum*
B. Favus	Scalp and glabrous skin	Honeycombed clusters of saucer-shaped depressions; mousy odor; permanent hair loss	Children and adults	*T. schoenleinii*
C. Tinea barbae	Bearded area			
1. Dry type		Scaly with straight hair stumps	Adult men	*T. megninii*, *T. mentagrophytes*, *T. rubrum*
2. Suppurating type		Inflammatory; pustular	Dairy farmers	*T. verrucosum*

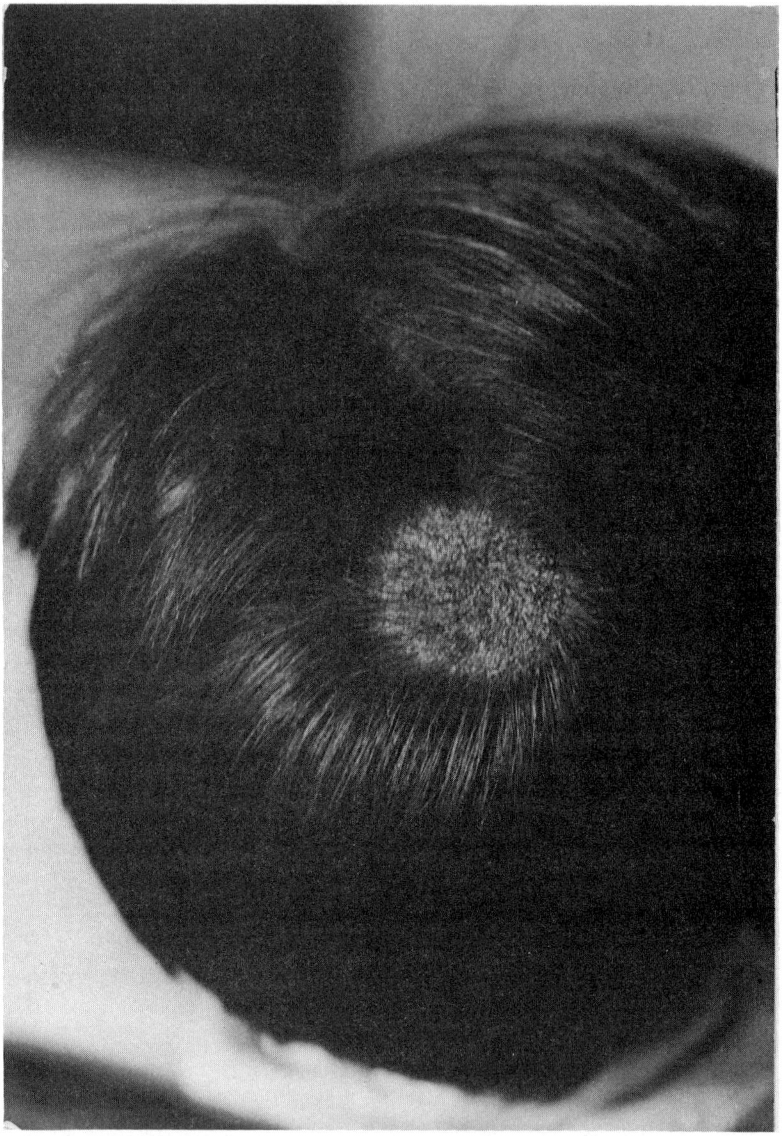

Figure 3:21. Non-inflammatory tinea capitis in a child infected with *Microsporum canis*. Note circular area of alopecia with scaly surface and remaining hair stumps, each showing a grayish sheath (ectothrix arthroconidia). See also Figures 3:22 and 3:23.

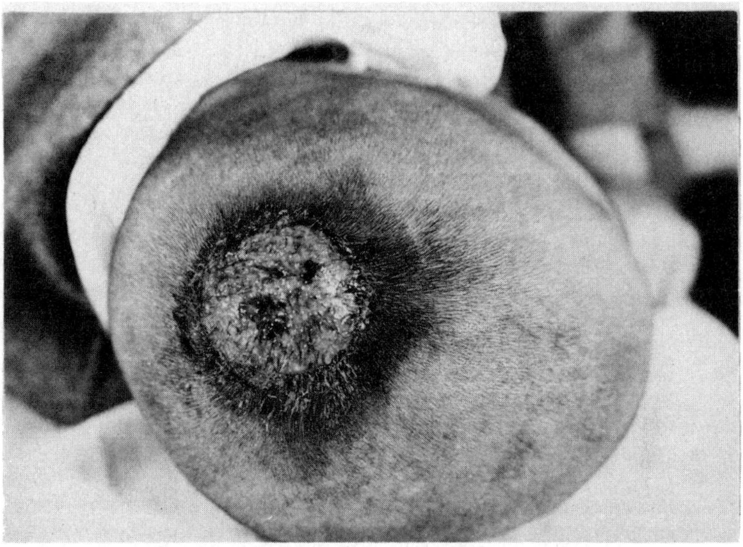

Figure 3:22. Inflammatory (kerion type) lesion also produced by *M. canis* in another child.

(85 per 1,000). This means about 15.7 million persons in the United States have at least one fungal infection. Table 3:9 lists the estimated prevalence of the most common superficial fungal diseases. In addition, the table shows males more frequently infected than females in all categories, with a ratio as high as 14:1 in the case of tinea cruris.

Ecology:

The dermatophytes and related keratinophilic species may be grouped into three categories based on host preference and natural habitat (Table 3:10). Anthropophilic species infect humans and rarely infect lower animals, zoophilic species primarily infect lower animals and geophilic species are essentially soil organisms. The major importance of these groupings is to aid in determining the source of an infection, for example, an infection in a human by *M. canis* usually implicates a cat or less likely a dog, as the culprit.

Anthropophilic dermatophytes may be found in all three anamorphs, e.g., *M. audouinii, T. tonsurans* and *E. floccosum*. They infect humans almost exclusively; rare animal infections have been traced to sustained close human contact.[68,71] There is no evidence of a saprobic niche in nature. Infection in humans is usually attributed to direct contact or indirectly to

Figure 3:23. Photomicrograph of hair with ectothrix type of infection by *M. canis*. Note mosaic of spherical arthrospores surrounding hair and extending as a sheath beyond its perimeter. KOH preparation. X1000.

fomites (combs, brushes, backs of chairs, bed linens, towels, etc.) or to aerosols which may carry infected scales and hairs to others.[42]

Zoophilic dermatophytes belong to the genera *Microsporum* and *Trichophyton*. Commonly recognized zoophilic fungi include *M. canis, T. mentagrophytes* var. *mentagrophytes* and *T. verrucosum*. They primarily attack lower animals. However, human infections do occur either through direct contact with infected animals or indirectly by contact with fomites. Arthroconidia of *T. verrucosum* may remain viable on fence posts, doors, wooden pens, etc., for years.[99] Dermatophytes that naturally infect domestic and wild animals are listed in Table 3:11. A more detailed survey of animal hosts infected by *M. canis* and *T. mentagrophytes* is reported elsewhere.[99]

Geophilic dermatophytes and closely related species include members of the genera *Microsporum* and *Trichophyton*, as well as *Epidermophyton stockdaleae*. These fungi live in soil which is often seeded with keratinaceous material acting as an enrichment medium. Exposure to the soil is the main source of infection for humans and lower animals, although direct and indirect contact with infected humans and animals are also modes of transmission.

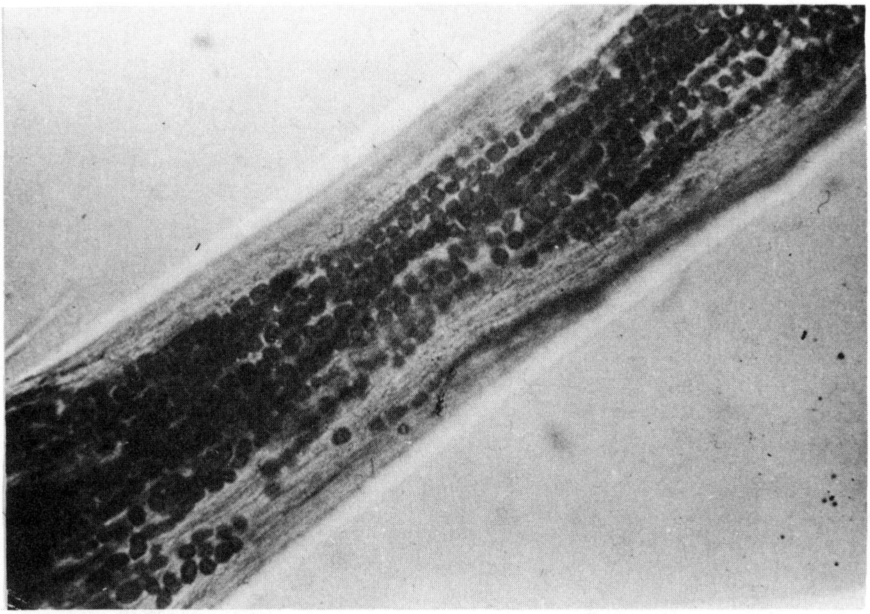

Figure 3:24. Photomicrograph of hair with endothrix type of infection produced by *Trichophyton violaceum*. Note internal position of arthroconidia. Polychrome-blue stain. X600.

Members of the *M. gypseum* complex are representative geophilic dermatophytes. *M. nanum*, included here in this category, is considered by others as zoophilic because it primarily infects pigs[12,98] and is isolated from pig yards. However, the presence of typical macroconidia in pig yards suggests its saprobic existence in soil, especially since macroconidia are not formed in vivo.[7] *M. praecox*, reported to cause human infections,[32,101,130] is considered geophilic since it has been isolated from sites associated with horses and typical macroconidia were found in dust collected from stables.[32] Infections in horses have not been reported. *M. persicolor*, a rare human pathogen, is considered zoophilic because it is frequently associated with voles.[38] However, others consider it geophilic because it has been isolated from soil.[6]

Contagion and Geographic Distribution of Anamorphs

Tinea capitis:

This highly contagious disease spreads rapidly within a family or school community. The great epidemic of the eastern United States caused by *M.*

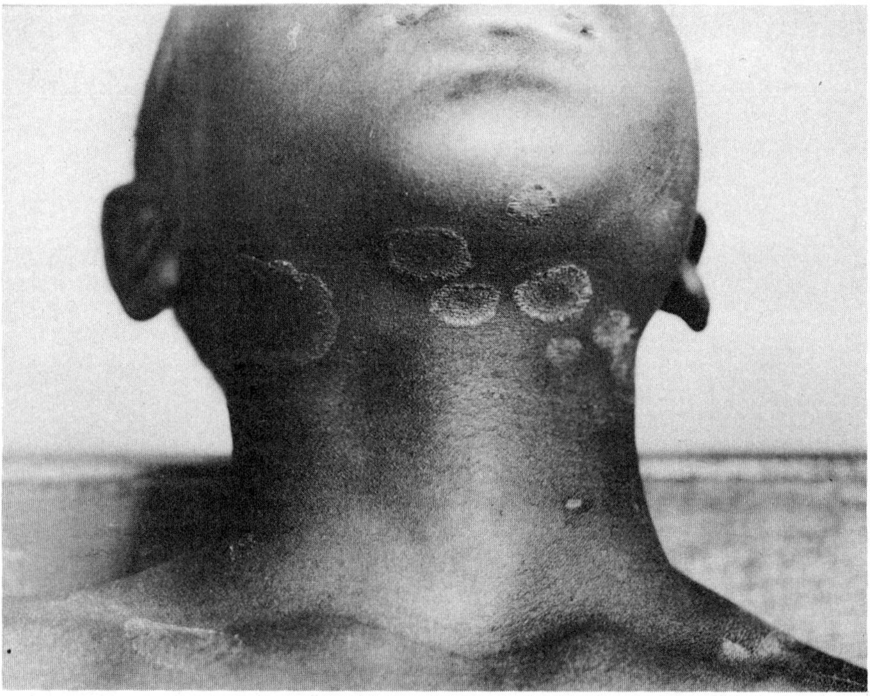

Figure 3:25. Tinea corporis in a child infected with *M. audouini*. Note abrupt, raised, scaly and vesicular border of the lesions. This is usually contracted from a sibling with tinea capitis.

audouinii during the 1940s is an example of the extent of contagion and spread of tinea capitis. Since *M. audouinii* is an anthropophilic fungus, the person-to-person chain has to be broken to terminate an epidemic.

In recent years, *T. tonsurans* has replaced *M. audouinii* as the principal cause of ringworm of the scalp in the United States and in Canada. This fungal species is believed to have been brought into the United States from Mexico, Central and South America and the Caribbean area. In western Europe, parts of South America, Australia and New Zealand, *M. canis* is frequently contracted from cats and to a lesser degree from dogs, monkeys and other furry pets. *T. violaceum*, an anthropophilic fungus, is the dominant agent of tinea capitis in eastern Europe.[99]

Tinea pedis:

The epidemiology of tinea pedis is still a controversial subject. Some believe that contact with dermatophytes incites clinical disease. Others feel that host factors, such as immunologic status, are the determinants of

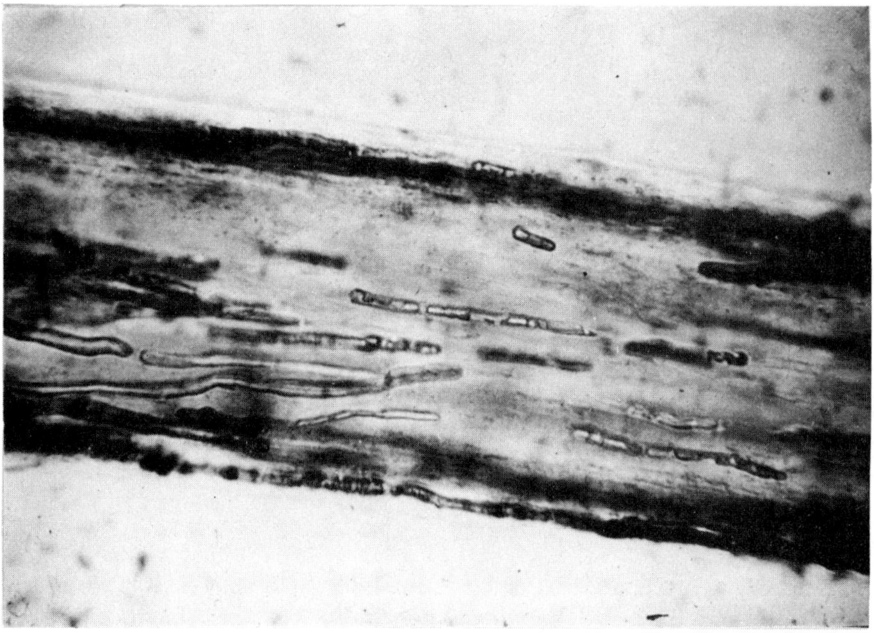

Figure 3:26. Photomicrograph of hair from case of favus (*T. schoenleinii* infection). Note internal septate hyphae and "air tunnels," i.e., digested portions of hair replaced by trapped air. KOH preparation. X600.

infection and that exogenous exposure to dermatophytes is relatively unimportant. There is evidence to support both points of view.

The following facts support the theory that exogenous exposure to dermatophytes is important in transmitting tinea pedis:

a. Viable dermatophytes are readily shed from the feet of persons with tinea pedis. These fungi have been recovered from areas where people walk barefoot (i.e., the floors of shower rooms) and from towels, shoes and socks.[5,44,45,51]

b. Epidemiologic data suggest person-to-person transmission. The incidence of tinea pedis increased threefold when communal showers were installed in a coal mine for use by the workers.[48] Other infections by this same species tend to occur within families of patients with *T. rubrum* infections.[76] However, when extreme precautions are taken to prevent contact between a patient with tinea pedis and other members of his household, no cross-infection occurs.[36]

c. Strains of dermatophytes isolated within families are similar. The growth characteristics of the isolates of *T. rubrum* isolated from various persons within a family are often similar.[36]

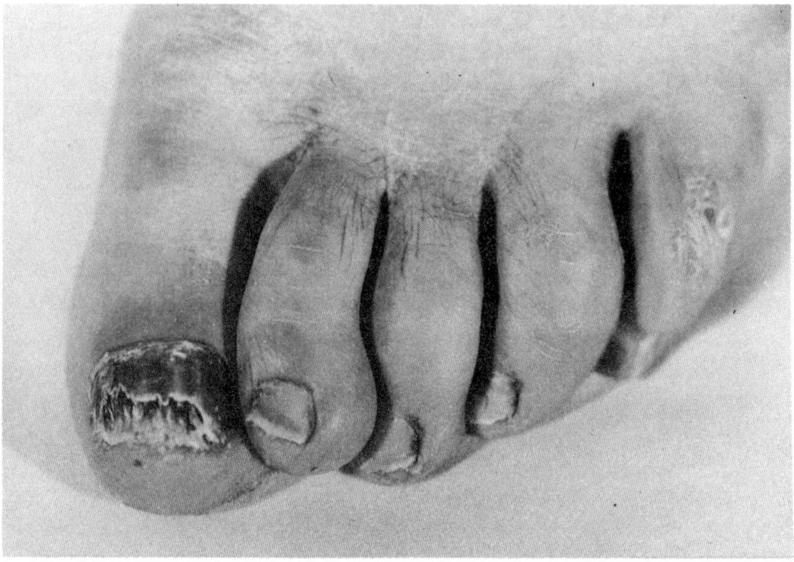

Figure 3:27. Onychomycosis of large toe nail. Note smooth surface of nail with thickened, darkened, subungual debris.

The following facts support the theory that exogenous exposure to dermatophytes is not important in transmitting tinea pedis:

a. Experimental infection has failed. The fungus-free feet of volunteers were placed in foot baths containing high concentrations of viable dermatophytes, either from cultural material or in scales directly from the feet of patients with tinea pedis. Clinical fungal infections of the feet were not induced in any of 68 subjects thus exposed.[14] Only when the skin of the feet was deliberately blistered with cantharidin before exposure could "takes" be induced.[13]

b. Person-to-person transmission has rarely been demonstrated. Contrary to the epidemiologic data supporting the importance of direct contact in transmitting tinea pedis, other similar studies support the opposite view. A survey of 88 dermatologists in the United States on the contagiousness of tinea pedis and tinea cruris among their patients revealed only 4 instances out of surely hundreds of thousands of contacts where family transmission was strongly suspected.[114]

Hopkins et al.[58] examined groups of soldiers at an army base during World War II and found that species of dermatophytes occurred in about the same ratio in all groups examined. If contagion were important in the spread of tinea pedis, one would expect a particular species to predominate within an individual group of soldiers. This has not been the case, except for one instance involving *T. mentagrophytes* infection (*A. benhamiae*,

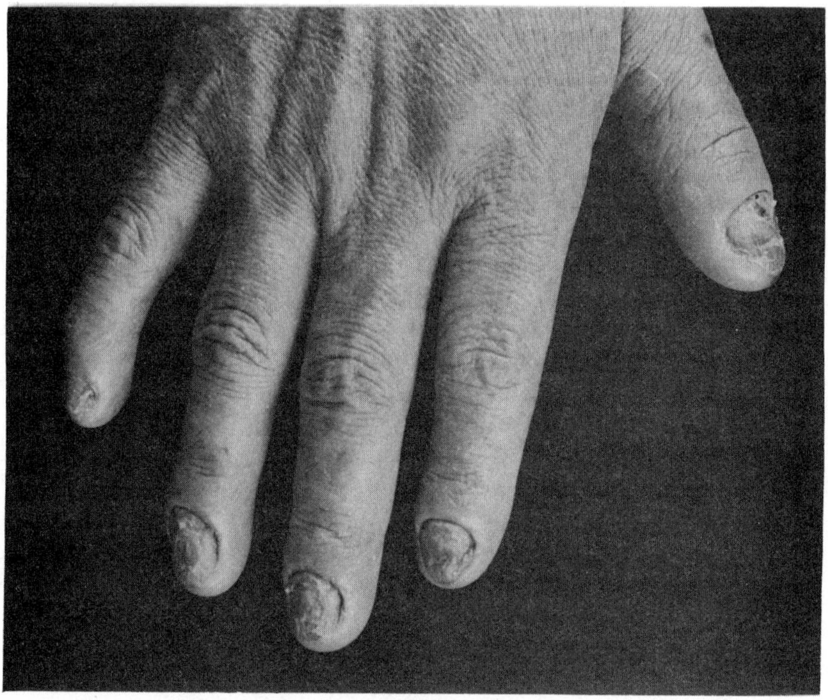

Figure 3:28. Onychomycosis of five fingernails infected with *T. rubrum*.

mating type (a) or (-)) among American soldiers in the Mekong Delta in Vietnam.[8] In this instance, however, an exogenous source of infection was implicated, rather than human-to-human transmission. Host factors such as poor hygiene, excess moisture and occlusive clothing, particularly combat boots, facilitated the epidemic.

c. Dermatophytes are found on clinically normal feet. Most surveys indicate that dermatophytes can be isolated from feet which do not have clinically apparent infection, with frequency of isolation ranging from 1-40%.[103]

Tinea cruris:

Person-to-person transmission of tinea cruris apparently occurs rarely,[114] although Beare, Gentles and Mackenzie[15] have shown that the disease can be transmitted by sharing towels and clothing.

Tinea corporis:

There is little evidence that tinea corporis is transmissible, except for the fact that acute infections are frequently associated with episodes of tinea

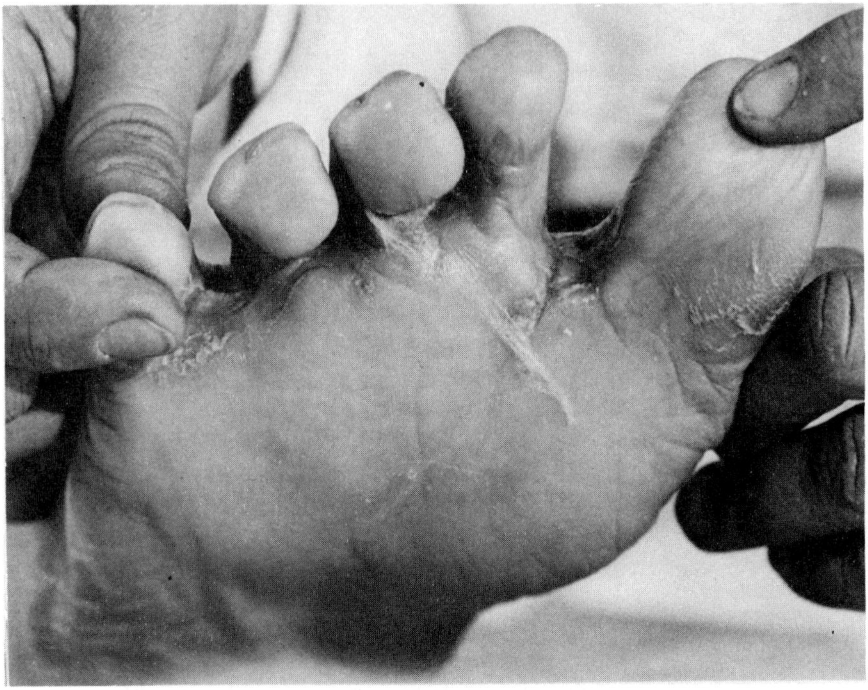

Figure 3:29. Tinea pedis. Non-inflammatory, scaly, desquamating lesions of interdigital and subdigital areas seen in chronic cases.

capitis in the immediate contacts of a patient with glabrous skin lesions. Chronic tinea corporis is not considered contagious but as evidence of the special susceptibility of a given patient.

Tinea unguium:

There is little evidence of contagion in tinea unguium, particularly since some patients continue for many years with only one, two or three involved toe or fingernails without extension to any adjacent nails.

Geographic Distribution of the Teleomorphs

Identification of the teleomorphic state of a dermatophyte and of its mating type is important in assessing its epidemiologic and evolutionary status. Mating studies revealed that the inflammatory infections manifested by troops in southeast Asia were due to *A. benhamiae* (a) or (-) mating type (anamorph, *T. mentagrophytes*), which was acquired there rather than carried from the United States where (A) or (+) mating type predominates.[8,98]

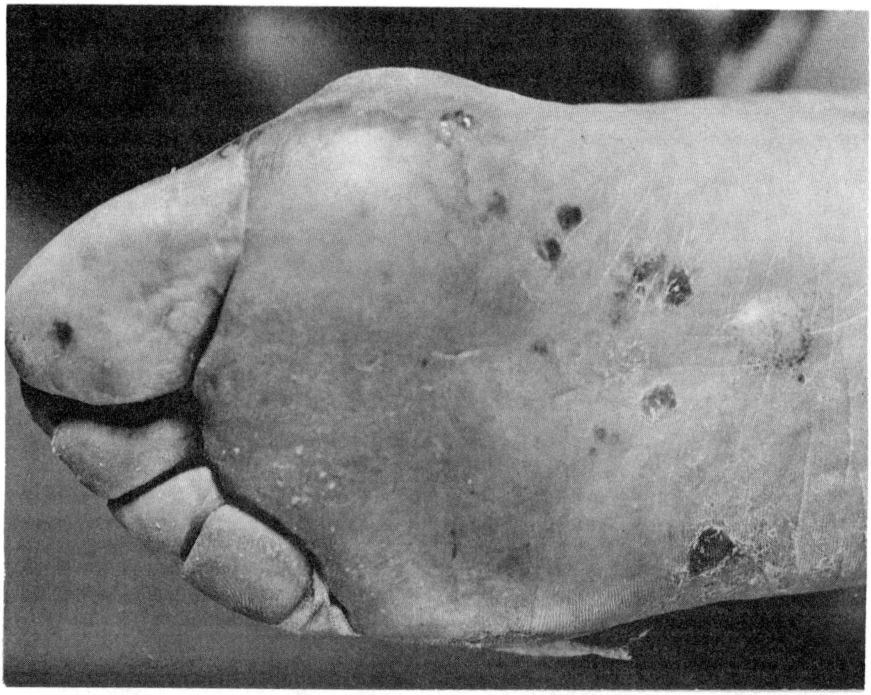

Figure 3:30. Blister formation in acute tinea pedis caused by *T. mentagrophytes*.

Table 3:9 Prevalence of Superficial Fungal Infections in the United States[61]

Infection	Rate per 1,000 Persons	Total Number of cases (in millions)	Male to Female Ratio
Tinea pedis	38.7	7.5	6:1
Tinea unguium	21.8	4.2	2:1
Tinea versicolor	8.4	1.6	2:1
Tinea cruris	6.7	1.3	14:1
Other tineas	5.5	1.1	
Totals	81.1	15.7	4:1

Although *A. benhamiae* is prevalent in southeast Asia, it is absent in Japan where *A. vanbreuseghemii* (the other teleomorph of *T. mentagrophytes*) exists primarily as (+) mating type.[57] Both *A. benhamiae* and *A. vanbreuseghemii* are found in Europe and in Africa.[31,92,115] To date, only *A. benhamiae* has been isolated from the United States and Canada.[89]

Sexually reactive isolates of *A. (N.) otae*, the teleomorph of *M. canis*, was reported only as the (-) mating type in a collection of 198 cultures from

Table 3:10 Grouping of the Dermatophytes and Related Keratinophilic Fungi on the Basis of Natural Habitat and Host Preference

Anthropophilic	Zoophilic	Geophilic
M. audounii	*M. canis* var. *canis*	*M. amazonicum*
M. ferrugineum	*M. canis* var. *distortum*	*M. boullardii*
M. concentricum	*M. gallinae*	*M. cookei*
T. gourvilli	*M. persicolor*	*M. fulvum*
T. megninii	*T. equinum*	*M. gypseum*
T. mentagrophytes	*T. mentagrophytes*	*M. nanum*
var. interdigitale	var. *mentagrophytes*	*M. persicolor*
T. rubrum	var. *guinckeanum*	*M. praecox*
T. schoenleinii	var. *erinacei*	*M. racemosum*
T. soudanense	*T. simii*	*M. ripariae*
T. tonsurans	*T. verrucosum*	*M. vanbreuseghemii*
T. violaceum		*T. ajelloi*
T. yaoundei		*T. georgiae*
E. floccosum		*T. gloriae*
		T. phaseoliforme
		T. simii
		T. terrestre complex
		T. vanbreuseghemii
		E. stockdaleae

12 countries in North and South America, Europe and Africa.[131] This mating type was also predominant in Japan.[55] Only four isolates have been found as (+) mating type, all isolated in Japan.[117]

In contrast to the unequal distribution of mating types of sexually-reproducing zoophilic species, geophilic species from clinical isolations of *A. (N.) gypsea* and *A. (N.) incurvata* (anamorph, *M. gypseum*) in the United States and Japan are represented by both mating types.[56,128]

Most of the teleomorphic species of the dermatophytes and related fungi are geophilic, a few are zoophilic (Tables 3:2 and 3:10) and only one is anthropophilic. Studies of the mating type of anthropophilic dermatophytes using Stockdale's method[113] have revealed that most species exist predominantly as a single mating type. It appears that the evolution from geophilic species to anthropophilic species resulted in a loss of sexual fertility and unequal distribution of mating types or in the loss of the other mating type.[125]

Collection, Transport and Processing of Specimens

Please refer to Chapter 2, "Collection, Transport and Processing of Specimens," in which these topics for all fungal diseases will be discussed.

Identification of the Etiologic Agents

Direct microscopic examination of specimens (KOH examination):

The methods for preparing specimens for microscopic examination for fungi have been presented in Chapter 2. The microscopic findings will vary,

Table 3:11 Dermatophytes Causing Natural infections in Domestic and Wild Animals[98]

Animal	Dermatophytes
1. Cat	
Frequent:	*M. canis,* M. gypseum, T. mentagrophytes*
Rare:	*M. canis* var. *distortum, M. gallinae, M. vanbreuseghemii, T. rubrum, T. schoenleinii, T. verrucosum*
2. Dog	
Frequent:	*M. canis,*, M. gypseum, T. mentagrophytes*
Rare:	*M. audouinii, M. canis* var. *distortum, M. cookei, M. gallinae, M. nanum, M. persicolor, M. vanbreuseghemii, T. equinum, T. rubrum, T. simii, T. verrucosum, T. violaceum, E. floccosum*
3. Horse	
Frequent:	*M. gypseum, T. equinum,* T. mentagrophytes, T. verrucosum*
Rare:	*M. canis* var. *distortum, M. equinum*
4. Cattle	
Frequent:	*T. mentagrophytes, T. verrucosum**
Rare:	*M. canis, M. gypseum*
5. Pig	
Frequent:	*M. nanum,* T. mentagrophytes*
Rare:	*M. canis, T. verrucosum*
6. Monkeys	
Frequent:	*M. canis,* M. gypseum, T. mentagrophytes,* T. simii** (India)
Rare:	*M. audouinii, M. canis* var. *distortum, M. cookei*
7. Fowl	
Frequent:	*M. gallinae,* T. simii*
8. Rodents	
Frequent:	*M. canis, M. gypseum, M. persicolor** (voles), *T. mentagrophytes**
Rare:	*M. gallinae, M. vanbreuseghemii*

*Most common etiologic agent(s)

depending on the type of clinical specimen examined. All types of specimens should be scanned first for fungi under low power magnification (about 100X); suggestive structures should then be confirmed under higher magnification (about 430X).

Hair: Two main types of hair invasion are seen by direct examination of infected hairs:

1. Ectothrix infection, where the fungus appears as a mosaic of spherical arthroconidia, surrounding the hair as a stocking or sheath:

a) Arthroconidia 2-3 μm, suggest infection with *M. canis, M. audouinii* or *M. ferrugineum* (Fig. 3:23);

b) Arthroconidia 3-5 μm that retain a linear arrangement suggest infection by the *T. mentagrophytes* complex ("ectothrix microides");

c) Arthroconidia 5-10 μm ("megaspore ectothrix") suggest infection with *T. verrucosum* (if numerous), the *M. gypseum* complex or *M. nanum* (if sparse).

Table 3:12 Geographic Distribution of the Dermatophytes[98,99]

Species	Geographic Area
Epidermophyton	
E. floccosum	Cosmopolitan
Microsporum	
**Microsporum audounii*	United States, Canada, Great Britain, Western Europe, Northeastern and Central Africa
M. canis	Scandinavia, England, United States
M. ferrugineum	Japan, China, Russia, Central Africa
M. gypseum complex	Cosmopolitan
Trichophyton	
T. concentricum	South Pacific, Central and South America
T. gourvillii	Northeastern Africa
T. megninii	Portugal, Spain, France, Belgium, The Netherlands, Sardinia, Italy
T. mentagrophytes	Cosmopolitan
T. rubrum	Cosmopolitan
T. schoenleinii	Eastern Canada, Greenland, Iceland, all Europe, Asia Minor (Near East) Mediterranean Countries
T. simii	India
T. soudanense	Eastern and Central Africa
T. tonsurans	Caribbean, Central America, North America (Mexico and United States), Northeastern South America (Columbia, Venezuela, Ecuador, Peru)
T. verrucosum	Cosmopolitan
T. violaceum	Spain, Portugal, Southern France, Mediterranean Basin, Northwestern Africa, Eastern Europe, all of Asia (except Siberia and Southeast), Southeastern Australia
T. yaoundei	Central and Southeastern Africa

*M. audouinii has largely disappeared from developed countries.

2. Endothrix infection, where the fungus appears as arthroconidia within the hair:

a) Spherical arthroconidia, 4-8 μm, in a mosaic pattern suggest infection with *T. tonsurans* or *T. violaceum* (Fig. 3:24);

b) Long cylindrical arthroconidia in a linear arrangement and air tunnels suggest infection with *T. schoenleinii* (favus, Fig. 3:26).

Glabrous skin: Fungal elements appear as hyaline, septate, branching hyphae of varying length, 2-4 μm in diameter, sometimes fragmenting into arthroconidia (Figs. 3:31, 3:32). These arthroconidia are cylindrical at first, spherical after disarticulation from the hyphae. It is important to differentiate between fungal hyphae and artifacts such as "mosaic fungus," cotton fibers, collagen fibers, etc.[74]

Nails: The microscopic picture is the same in nail preparations as in preparations of glabrous skin, except that chains of beadlike, spherical

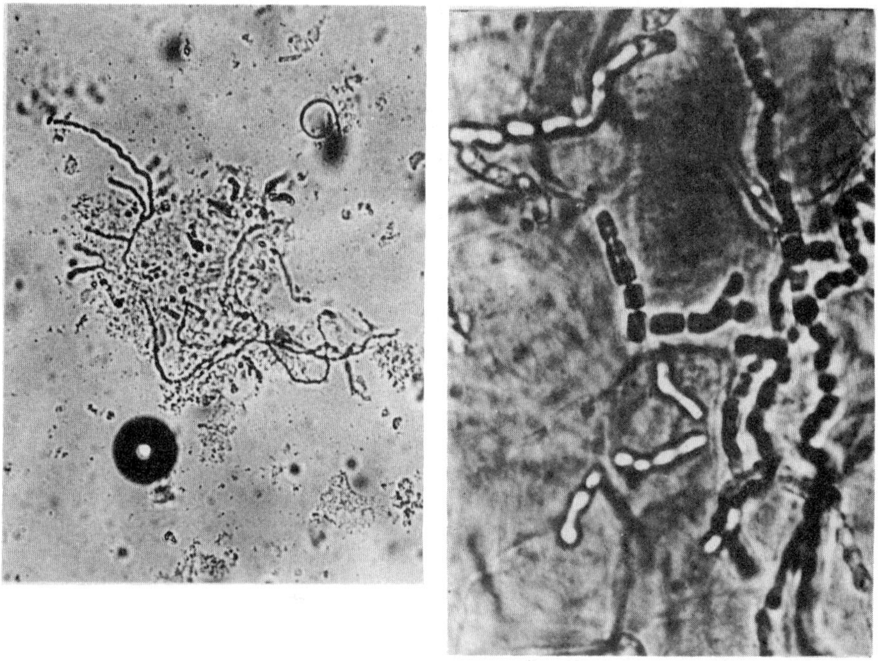

Figure 3:31. Photomicrograph of dermatophytic hyphae in KOH preparation of skin scrapings. Note branching, septate hyphae. X150.

arthroconidia, as well as long, thin, hyaline hyphae are more often observed in nails.

Selection and Inoculation of Primary Culture Media

For many years the standard medium used in medical mycology was Sabouraud dextrose (glucose) agar which is a modification of the agar originally formulated by R. Sabouraud. It is essentially a peptone-glucose agar with pH of 5.6. Many variations in its formulation have been proposed. Although this medium is an excellent medium for the growth of most dermatophytes, it is also an excellent growth medium for saprobic fungi and for many bacteria, as well. Since these latter, unwanted microorganisms often overgrow the isolation medium, they may interfere with the isolation of the dermatophytes. Sabouraud dextrose agar has been improved by the addition of antibacterials (especially chloramphenicol) to inhibit bacterial growth and cycloheximide to inhibit the development of most saprobic fungi. The addition of these two antibiotics to Sabouraud dextrose agar makes it selective and increases its effectiveness as an isolation medium.

Another formulation is Dermatophyte Test Medium (DTM) which, in addition to antibacterials and cycloheximide, also contains phenol red, a pH indicator. When dermatophytes grow on this medium they turn it alkaline so that the phenol red changes from yellow to red. However, DTM has its limitations; fungi other than dermatophytes can change the color of the medium from yellow to red.[82,108]

Two media should therefore be used for the primary isolation of dermatophytes: one with antibiotics and the other without antibiotics (Sabouraud dextrose agar). Both media should be used since certain yeasts and filamentous fungi causing cutaneous infections are inhibited by cycloheximide.

Culture of skin, hair and nails:

In general, small fragments of the specimen should be inoculated into 4-6 sites on the surface of the primary isolation media. Care should be taken that the inoculum is in direct contact with the medium. Slight embedding of the specimen will insure such contact. The inoculated sites should be well separated from each other to facilitate subculture of colonies of dermatophytes and to diminish the chance of overgrowth by contaminants.

Cultures are usually incubated within 23-30°C, but should also be incubated at 37°C if *T. verrucosum* is suspected (material from cattle or dairy farmers). It is important that the cultures have access to air during incubation. All tubes of media should either be cotton-plugged or have caps loosened after inoculation, if screwcapped. If containers are tightly capped, dermatophyte growth may be impeded.

Cultures should be examined for growth after one week. If growth is present, make subcultures from the periphery of the colony to minimize contamination. In our experience, cultures that are negative after 2 weeks of incubation usually remain negative on prolonged incubation. However, if *T. verrucosum* or another slow-growing dermatophyte is suspected, the cultures may be held for up to 4 weeks. Some mycologists prefer to retain cultures of microscopically-positive specimens for the longer period before discarding them as negative.

Identification Procedures

Identification of species of *Microsporum* and *Trichophyton* is often based on colony characteristics and on the morphology of the macroconidia. These criteria alone may be insufficient for specific identification in the genus *Trichophyton*, since colonial and microscopic morphology vary and some isolates fail to sporulate when subcultured. Also, many dermatophytes, particularly on Sabouraud dextrose agar, develop white, fluffy tufts on the surface of the colony which may eventually overgrow the original culture, a condition often called pleomorphism. Subcultures of these tufts show loss in normal pigmentation and sporulation. This phenomenon may result from mutation[126] or chromosomal aberration[34] and occurs most frequently in the *M. gypseum* complex, *M. canis, E. floccosum* and the *T. mentagrophytes* complex.

Physiological tests are often required in conjunction with morphology to identify species correctly. Such tests include: perforation of hair in vitro, special nutritional requirements, comparison of growth at room temperature and at 37°C and pigment production on special media.

Table 3:13 describes a sequence of procedures that may be necessary to arrive at a definitive identification of a dermatophyte isolate. The techniques for performing these tests are described. Further aids to identification are found in the illustrations and in Tables 3:3 and 3:4.

Colony characteristics of dermatophytes:

The following sets of characteristics are important in observing the gross appearance of a colony of a dermatophyte. Very often an isolate can be speciated with near certainty by observing characteristic gross features, but it must be remembered that variation occurs quite frequently.

a) Color of surface and reverse. Colloquial names for colors (i.e., beige, brick-red, etc.) are often used to described colonies, although interpretations may vary with the observer. Color standards have been developed to attempt uniformity in color dcsignation.[96]

b) Topography. This term refers to a colony's contours, margin, elevation, folding, etc.

Table 3:13 Sequence of Procedures for Identification of Dermatophytes in Pure Culture

1. Observe colony for the development of pigment both on the surface and reverse, for characteristic topography, texture and rate of growth. Consult identification charts (Table 3:3 and 3:4) to arrive at a presumptive identification.
2. Prepare teased mounts and search for characteristic structures (Figs. 3:8–3:20; Tables 3:3 and 3:4). If this search is inconclusive, proceed to step 3.
3. Prepare and examine slide cultures for characteristic structures. If the findings are still inconclusive, proceed to step 4.
4. Perform as many of the following physiological tests as necessary (see text for details).
   A. Hair perforation
   B. Urease production
   C. Nutritional requirements (if a *Trichophyton* is suspected)
   D. Growth on rice grains (if a *Microsporum* is suspected)
   (Note: Teased mounts prepared from media for C and D, often show structures that were absent on routine media).
   E. Effect of incubation at 37°C on growth
   F. Sexual reproduction using tester strains (reference laboratories)
5. Arrive at final identification.

c) Texture. This term describes the appearance of the surface of the colony, using such subjective terms as powdery, granular, waxy, felt-like, downy or velvety.

d) Rate of growth. Determined by measuring colony diameters after a given period of incubation. Growth rates range from slow (1 mm/week) to fast (10 mm/week).

Microscopic morphology:

Observation of the morphology and arrangement of the conidia is essential for species identification. This micromorphology can be observed in either culture mounts (teased mounts), plastic tape mounts and/or slide cultures.

a) Teased mounts. Use stiff needles to tease apart a small fragment of the aerial mycelium in a drop or two of a mounting medium on a slide. The mounting medium is usually lactophenol-cotton blue solution or colorless lactophenol. Examine microscopically after adding a coverslip.

b) Plastic tape mounts. Adequate microscopic preparations may also be made from petri dish cultures of heavily sporulating fungi by using transparent, pressure-sensitive tape, such as No. 800 acetate-backed tape (Minnesota Mining and Manufacturing Co.).[105] A flag of tape is fastened to a wooden applicator stick and the sticky surface touched to the surface of the thallus (colony) just proximal to the advancing periphery. The tape is then pressed sticky side down on a slide with a drop of lactophenol cotton blue stain and examined under the microscope directly. The tape selectively removes branches of hyphae bearing conidia (since these lie at the surface of the colony) and avoids the taxonomically valueless vegetative mycelium.

c) Slide cultures. A slide culture is often necessary to avoid disturbing aerial hyphae and developing conidia so that conidial ontogeny (manner of origin and development) and spore arrangement can be observed. Only two of several slide culture techniques will be described. Each of the techniques requires a sterile culture chamber, consisting of a sterile microscope slide, coverslip and petri dish. Since a humid atmosphere must be provided, a bent glass rod ("V" shaped or "U" shaped) is added to the bottom of the petri dish to support the glass slide above the sterile water in the bottom which provides humid conditions. Sterile plastic petri dishes having three to four chambers may be substituted for plain petri dishes with glass supports. Culture media for slide cultures should be poor in nutrients; this stimulates sporulation and inhibits vegetative mycelium. Cornmeal dextrose agar and potato dextrose agar are suitable.

Method One: Lay the glass slide across a piece of bent glass tubing. Then with a pipette dispense enough agar to cover an area equivalent to a 22 x 40 mm coverslip onto the slide. After the agar solidifies, inoculate the fungus in two parallel streaks onto the surface of the agar. Add sterile distilled water to the bottom of the petri dish and incubate the preparation until the fungus reaches the proper state of development. When this is attained, dry the agar by leaving the petri dish cover slightly ajar. After the agar dries, add a few drops of absolute alcohol to the area of growth to wet it down, followed by mounting fluid and a coverslip.

Method Two: This technique includes the same preparation of the slide as in Method One. However, with this technique, two preparations may be obtained. Cut a 10 mm square of agar from a solidified agar plate, place it on a slide, and inoculate the four edges of the agar square. Alternatively, a round slab of agar may be cut from a poured plate by using the rim of a test tube ("cookie-cutter" technique). Place a sterile coverslip over the inoculated agar slab. Add sterile water to the bottom of the petri dish to provide moisture.

When the culture is ready to be examined, remove the coverslip and wet down the growth with a few drops of alcohol. Place the coverslip growth-side down on a slide containing a drop of mounting fluid. Make a second preparation from the slide by removing the block of agar, wetting down the growth adherent to the slide, adding the mounting fluid and a coverslip.

Examination of one or more of the above slide preparations may disclose the characteristic microscopic morphology of the species. When such an examination does not provide sufficient information for species identification, more tests are necessary to determine certain physiological characteristics.

Physiological tests:

a) Hair perforation.[4] Several fragments of human hair[109] (many prefer children's hair) are cut into lengths of about 1 cm and autoclaved at 120°C for 10 min. Add about 25 mL of sterile distilled water, two or three drops of sterile 10% yeast extract and several fragments of human hair to a 100 mm diameter petri dish. Inoculate the hairs with the organism under study (from young cultures on Sabouraud agar) and incubate at 25°C (room temperature) for up to 4 weeks.

After 7-10 days of incubation, aseptically remove a few hairs covered with fungal filaments. Mount them in lactophenol cotton blue or other mounting medium and examine under low power. Perforations appear as wedge-shaped erosions occurring at irregular intervals along the hair shaft (Fig. 3:16). The presence of these perforations constitutes a postive test.

If perforations are not seen after incubation for 10 days, the plates should be reincubated and reexamined at intervals up to 28 days before being discarded as negative. Hyphae growing on the surface of the hair, but with no perforations, constitutes a negative test.

Quality control: *T. mentagrophytes*, expected result = perforations formed within 28 days. *T. rubrum*, expected result = perforations not formed at the time the test is terminated. Table 3:14 lists the in vitro hair reactions of some dermatophytes. This test is most useful for differentiating *T. mentagrophytes* from *T. rubrum*.

b) Urease formation.[104] Either Christensen urea agar (urea agar base + 1.5-2.0% agar) or urea broth can be used. The broth is more reliable than the agar for the demonstration of urease.[64]

Inoculate the medium with the fungus under study and incubate for 7 days at 25°C. Screwcaps should be slightly loosened to allow adequate aeration. The appearance of a purple-red color in the medium indicates a positive reaction; no color change or a slight color change to orange or pale pink indicates a negative reaction. The test should be terminated after 7 days' incubation.

Quality control: *T. mentagrophytes*, expected result = positive reaction within 7 days. *T. rubrum*, expected result = negative after 7 days. This test is most useful in differentiating *T. mentagrophytes* from *T. rubrum*. Kane et al.[66] have reported that *T. raubitschekii*, which closely resembles *T. rubrum*, is urease-positive.

c) Special nutritional requirements. Certain species of *Trichophyton* have either absolute or partial growth requirements for specific vitamins or amino acids (Table 3:15).[16,49,102] Based on this information, Georg and Camp[50] devised an identification scheme to differentiate these species from other dermatophytes that resemble them morphologically. The media used in this test are available commercially as "Trichophyton agars 1-7" (Difco Laboratories, Detroit, MI). Although the details for the preparation of these media are described elsewhere, their general composition is summarized below:

Table 3:14 In Vitro Hair Perforation Test*[4]

Perforations formed	Perforations not formed
M. canis	*E. floccosum*
M. canis var. *distortum*	*M. audounii*
M. cookei	*M. equinum*
M. fulvum	*M. ferrugineum*
M. gypseum	*M. praecox*
M. nanum	*T. concentricum*
M. persicolor	*T. equinum*
M. racemosum	*T. gourvilii*
M. vanbreuseghemii	*T. megninii*
T. mentagrophytes	*T. rubrum*
T. terrestre	*T. verrucosum*
T. simii	*T. violaceum*
T. tonsurans	*T. yaoundei*
var. sulfureum	
subvar. perforans	

*The hair perforation test is most useful in distinguishing between strains of *T. rubrum* and *T. mentagrophytes*.

Table 3:15 Growth Response of Certain Dermatophyte Species on Trichophyton Agars 1–7[50]

Species	Trichophyton Agar No.						
	1	2	3	4	5	6	7
T. verrucosum (84% of strains)	0*	±	4+	0			
T. verrocosum (16% of strains)	0	0	4+	4+			
T. schoenleinii	4+	4+	4+	4+			
T. concentricum (50% of strains)	4+	4+	4+	4+			
T. concentricum (50% of strains)	2+	2+	4+	4+			
T. tonsurans	±			4+			
T. violaceum	±			4+			
T. mentagrophytes	4+			4+	4+		
T. rubrum	4+			4+			
T. megninii						0	4+
M. gallinae						4+	4+
T. equinum	0				4+		

*0 = No growth
± = Trace of growth
2+ = Moderate growth
4+ = Good growth

Medium 1: Casein basal medium (vitamin-free)
Medium 2: Casein basal medium plus inositol
Medium 3: Casein basal medium plus thiamin and inositol
Medium 4: Casein basal medium plus thiamin
Medium 5: Casein basal medium plus nicotinic acid
Medium 6: NH_4NO_3 basal medium
Medium 7: NH_4NO_3 basal medium plus histidine

The results obtained when certain dermatophytes are cultured on these media are summarized in Table 3:15. It is important to use small (pin head-sized) inocula to avoid carrying over nutrients that would invalidate this test. Incubate at 25°C until positive controls show good growth.

Quality Control: Every time a nutritional test is performed, appropriate controls should be included to verify the proper reaction of the media. When an unidentified dermatophyte is inoculated onto Trichophyton agars 1-4, the controls, usually *T. rubrum* or *T. mentagrophytes* and *T. tonsurans* should be also. *T. rubrum* or *T. mentagrophytes* should grow equally well on 1-4, whereas *T. tonsurans* will show only trace growth on 1 and 2 and good growth on 3 and 4. When testing for inositol, inositol-requiring *T. verrucosum* should be a control; for nicotinic acid, *T. equinum*, and for histidine, *T. megninii*.

d) Growth on rice grains. The inability of *M. audouinii* to grow well on sterile, cooked rice grains is a useful clue in the identification of this poorly sporulating species.[30] The medium is prepared by mixing 1 part raw rice grains to 3 parts of water in a flask or test tube and subsequently "cooking" by autoclaving. The surface of the rice is inoculated with the suspect isolate and the rice incubated at 25°C for up to 2 weeks.

Quality Control: Inoculate rice with *M. audouinii*, expected results = slight brownish discoloration of the medium and a trace of growth; and with either *M. canis* or *M. gypseum*, expected result = vigorous growth and sporulation resembling that obtained on Sabouraud dextrose agar.

e) Effect of incubation at 37°C on growth. Place small inocula of about equal size on the surface of two slants of Sabouraud agar. Incubate one tube at room temperature (25°C) and the other at 37°C. After 1-2 weeks' incubation, compare amount of growth (particularly colony diameter).

Quality Control: *T. mentagrophytes*, expected result = equal growth at both temperatures; *T. terrestre*, expected result = good growth at 25°C, no growth to poor growth at 37°C.

f) Production of teleomorph. Certain dermatophytes may defy classification by routine procedures, either because they are so atypical of the species or are so similar to each other that they form a complex of anamorphs (the complexes of *T. terrestre, T mentagrophytes* and *M. gypseum*), but as a rule it is not necessary for diagnostic purposes to determine the teleomorph; identification of the anamorph is sufficient.

Occasionally, it may be difficult to differentiate a saprobic species from a pathogen (distinguishing *T. terrestre* from *T. mentagrophytes* is a case in point). In such an instance, definitive identification could be based on obtaining the teleomorph. Since most of the teleomorphic species are heterothallic, the formation of gymnothecia with asci and ascospores requires pairing (mating, crossing) the unknown isolate with tester strains comprising the two mating types (designated as + and - or as A and a) of a species. This mating must be done on a suitable culture medium,[55,91] and may be done by cutting cubes of growth from agar cultures of the two isolates to be crossed and placing them side by side on the appropriate medium or by preparing conidial suspensions.[127] This procedure is best left to reference laboratories.

Epilogue

Although we have outlined a rather extensive series of identification procedures, many will not be necessary to identify most isolates. The great majority of dermatophytes can be identified by the first two or three steps (gross and microscopic examination of isolates) outlined in Table 3:13. The remaining dermatophytes may require one or more of the physiological tests for definitive identification.

OPPORTUNISTIC DERMATOMYCOSES

Definition

The opportunistic dermatomycoses are cutaneous infections sometimes resembling those produced by dermatophytes, but are caused by a wide variety of fungi that are normally saprobes or plant pathogens. These fungi are usually of low virulence and their hosts are often debilitated or similarly compromised.

Etiologic Agents

Since the opportunistic fungi are normally saprobes or plant pathogens, their isolation in culture is not, in itself, proof of pathogenicity. A nondermatophytic fungus that is isolated repeatedly from a patient, especially in large numbers, may be suspected of playing a potentially pathogenic role if one or more of the following criteria are met: (a) observation on direct microscopic examination of an adequate number of fungal elements whose morphology is compatible with the organism isolated; (b) observation of a histopathologic reaction in response to the fungal invasion; (c) repeated failure to isolate a dermatophyte on selective media; and (d) disappearance of the fungus with clinical cure.

Nondermatophytic filamentous fungi reported as probable etiologic agents of cutaneous infection are listed in Table 3:16. This list is selective since the literature is replete with reports of single or a few isolations of other opportunistic fungi. More frequent isolations have been reported from abnormal nails[124,134,135] but the significance of these isolations has not been precisely established.

Table 3:16 Cutaneous Infections by Non-Dermatophytic Fungi

Site	Causal Organisms
1. Nails[21,37,47,78,84,86,95,97,123,124,133,134,135]	*Acremonium (Cephalosporium) roseo-griseum*
	Alternaria spp.
	Aspergillus spp. A. candidus, A. flavus, A. fumigatus, A. glaucus, A. niger, A. restrictus, A. sydowii, A. terreus, A. ustus, A. versicolor
	Fusarium oxysporum
	Hendersonula toruloidea
	Scytalidium anamorph
	Pyrenochaeta unguis-hominis
	Scopulariopsis brevicaulis
	Scytalidium hyalinum
2. Skin[9,11,19,20,21,29,33,43,47,53,54,75,77,79,84,95,100]	*Alternaria* spp. (*A. alternata, A. tenuis,* etc.)
	Aphanoascus fulvescens
	Aspergillus flavus, A. fumigatus, A. niger
	Fusarium moniliforme, F. oxysporum, F. solani
	Hendersonula toruloidea
	Scytalidium anamorph
	Rhizopus rhizopodiformis
	Scytalidium hyalinum

Some of these fungi, notably *Alternaria* spp.,[19] *Aphanoascus fulvescens*,[77,100] *Hendersonula toruloidea* and its *Scytalidium* anamorph and *Scytalidium hyalinum*,[54,84] may cause skin and nail infections clinically indistinguishable from a dermatophyte infection. *Aspergillus flavus, A. fumigatus* and *A. niger* were reported as etiologic agents of primary cutaneous aspergillosis at the sites of intravenous cannulas or in the vicinity where arm boards were taped to the extremities of leukemic children.[53] A similar type of infection was reported for a dematiaceous fungus, *Bipolaris (Drechslera) spicifera*.[39] Nosocomial outbreaks of mucormycosis of the skin caused by *Rhizopus rhizopodiformis* resulted from the use of contaminated adhesive tape at the site of surgical wounds.[43]

Clinical Manifestations

Most of the nondermatophytic fungi infecting the nail cause distal subungual onychomycosis.[134] Fungi (other than the dermatophyte, *T. mentagrophytes*) most frequently producing leukonychia mycotica (white superficial onychomycosis) include *Acremonium roseo-griseum, Aspergillus terreus* and *Fusarium oxysporum*.[133,134] *F. oxysporum* may be the most frequent cause of this infection after *T. mentagrophytes*.[97]

Infections by *H. toruloidea* and *S. hyalinum* usually resemble a dry *T. rubrum* infection with clinical features limited to the hands, feet and nails.[54,84] However, distinctive aspects of these infections include the absence of dorsal infection on the feet, lateral and distal invasion of the nail with extensive onycholysis and paronychia of the fingers.[54] However, toenail infections cannot be distinguished from those caused by a dermatophyte.[54]

Primary cutaneous aspergillosis in children with leukemia who are neutropenic and require intravenous therapy usually begins as erythematous papules or plaques and progresses through a hemorrhagic bullous stage to a purpuric ulcer with a central necrotic eschar.[53]

Epidemiology and Public Health Significance

Although most cases of infection due to *H. toruloidea* have been diagnosed in the United Kingdom and France, the patients were mainly immigrants from the tropics and subtropics, i.e., the Indian subcontinent, the West Indies, the Near East and West Africa.[84] Patients with infections due to *S. hyalinum* were generally from the West Indies, West Africa and Guyana.[84] Most patients were between 20 to 49 years old. A survey conducted in West Tobago to determine the prevalence of infection in the general population due to *H. toruloidea* and *S. hyalinum* indicated that infections caused by fungi are common in this area but generally asymptomatic.[9] *H. toruloidea* is a plant pathogen isolated from a variety of hosts in the tropics, subtropics,

Near East and parts of the United States; *S. hyalinum* has not been isolated from the environment.[54,84]

Collection, Transport and Processing of Specimens

H. toruloidea has been reported to remain viable in dry scales for six months.[20] Both *H. toruloidea* and *S. hyalinum* and most of the opportunistic fungi listed in Table 3:16 are sensitive to cycloheximide.[21,47,54]

Identification of Etiologic Agents

Potassium hydroxide (KOH) preparations of skin scrapings and nails from patients with infections due to *H. toruloidea* and *S. hyalinum* often resemble infections caused by dermatophytes.[21,47] However, these may often be differentiated from dermatophytes by a more sinuous appearance, variable width and in many specimens, broader hyphae having a double-contoured appearance.[84] In most KOH preparations, the hyphae of *S. hyalinum* cannot be distinguished from those of *H. toruloidea* except in the rare cases when the latter produces pigmented hyphae.[84]

Most isolates of *H. toruloidea* grow rapidly. Colonies are described as dark mouse-gray on the surface with a brownish-gray to black reverse. Coiled hyphae and branched dark hyphae which break up into cylindrical or rounded chains of one to two-celled, thick-walled arthroconidia (*Scytalidium* anamorph) are characteristic features.[20] Cultures may differ, however, in growth rate and in abundance of aerial hyphae and arthroconidia.

Pycnidia are usually absent but when present are sparse, solitary or in groups, stromatic with long necks. Pycnidial walls are several layers thick and the outermost walls are thick and dark. Pycnidiospores are ellipsoid to ovoid, initially uniseptate and hyaline, becoming brown and three-celled when mature.[95]

S. hyalinum colonies are white and cottony. Hyphae are pale or colorless, septate, single or in bundles and fragment to form cylindric, ellipsoid or rounded chains of arthroconidia which are one to two-celled, sometimes slightly roughened. *S. hyalinum* may resemble *Geotrichum* but differs by its irregular production of arthrocondia that are often septate.[21]

REFERENCES

1. Ajello L. A taxonomic review of the dermatophytes and related species. Sabouraudia 1968;6:147-59.
2. Ajello L. The medical mycological iceberg. In: Proceedings of the international symposium of the mycoses. P.A.H.O. Scientific Publication no. 205, 1970:3-12.
3. Ajello L. Taxonomy of the dermatophytes. A review of their perfect and imperfect states.

In: Iwata K, ed. Recent advances in medical and veterinary mycology. Tokyo: University of Tokyo Press, 1977:289-97.
4. Ajello L, Georg LK. In vitro hair cultures for distinguishing between atypical isolates of *Trichophyton mentagrophytes* and *Trichophyton rubrum*. Mycopathologia 1957;8:3-17.
5. Ajello L, Getz ME. Recovery of dermatophytes from shoes and shower stalls. J Invest Dermatol 1954;22:17-24.
6. Ajello L, Padhye AA. Dermatophytes and the agents of superficial mycoses. In: Lennette EH, Balows A, Hausler WJ Jr, Shadomy HJ, eds. Manual of clinical microbiology. 4th ed. Washington, DC: American Society for Microbiology, 1985:514-25.
7. Ajello L, Varsavsky E, Ginther OJ, Bubash G. The natural history of *Microsporum nanum*. Mycologia 1964;56:873-84.
8. Allen AM, Taplin D. Epidemic *Trichophyton mentagrophytes* infections in servicemen. JAMA 1973;226:864-7.
9. Allison VY, Hay RJ, Campbell CK. *Hendersonula toruloidea* and *Scytalidium hyalinum* infections in Tobago. Br J Dermatol 1984;111:371-2.
10. Arx JA von. Genera of fungi sporulating in pure culture. Lehre, Germany: Cramer, 1970:180.
11. Austwick PKC. *Fusarium* infections in man and animals. In: Moss MO, Smith JE, eds. The applied mycology of *Fusarium*. Symposium of the British Mycological Society London 1982. London: Cambridge University Press, 1984:129-40.
12. Babel DE, Rogers AL. Dermatophytes: their contribution to infectious disease in North America. Clin Microbiol Newsletter 1983;5:81-5.
13. Baer RL, Rosenthal SA. The biology of fungous infections of the feet. JAMA 1966;197:1017-20.
14. Baer RL, Rosenthal SA, Litt JZ, Rogachefsky H. Experimental investigations on the mechanism producing acute dermatophytosis of the feet. JAMA 1956;160:184-90.
15. Beare JM, Gentles JC, Mackenzie DWR. Mycology. In: Rook A, Wilkonson DS, Ebling FJG, eds. Textbook of dermatology. Oxford: Blackwell Scientific, 1972:694-805.
16. Benham RW. Nutritional studies of the dermatophytes - effect on growth and morphology with special reference to the production of macroconidia. Trans NY Acad Sci 1953;15:102-6.
17. Blank H. Tinea nigra: a twenty-year incubation period? J Am Acad Dermatol 1979:1:49-51.
18. Borelli D, Marcano C. *Cladosporium castellanii* nova species agente de tinea nigra. Castellania 1973;1:151-4.
19. Botticher WW. *Alternaria* as a possible human pathogen. Sabouraudia 1966;4:256-8.
20. Campbell CK. Studies on *Hendersonula toruloidea* isolated from human skin and nail. Sabouraudia 1974;12:150-6.
21. Campbell CK, Mulder JL. Skin and nail infections by *Scytalidium hyalinum* sp. nov. Sabouraudia 1977;15:161-6.
22. Carmo-Sousa L Do. *Trichosporon* Behrend. In: Lodder J, ed. The yeasts, a taxonomic study. Amsterdam-London: North-Holland, 1971;1301-52.
23. Carrion AL. In: Bachman GW, ed. Report of the director of the school of tropical medicine. New York: Columbia University Press, 1938;25.
24. Carrion AL. Yeastlike dematiaceous fungi infecting the human skin. Arch Dermatol Syphilol 1965;61:996-1009.
25. Carrion AL. Dermatomycoses in Puerto Rico. Arch Dermatol 1965;91:431-8.
26. Cerqueira-Pinto AGC. Keratomycose nigricans palmar (Dissertation). Tese, Faculdade de Medicina da Bahia, Brasil, 1916. 69 p.
27. Champion RH, Whittle CH. Sycosis barbae due to *Trichophyton rubrum*. Trans St. John's Hosp Dermatol Soc 1960;44:114.
28. Chong KC, Adams BA, Soo-Hoo TS. Morphology of *Piedra hortai*. Sabouraudia 1975;12:157-60.

29. Collins MS, Rinaldi MG. Cutaneous infection in man caused by *Fusarium moniliforme*. Sabouraudia 1977;15:151-60.
30. Conant NF. Studies on the genus *Microsporum*. I. Cultural studies. Arch Dermatol Syphilol 1936;33:665-83.
31. Contet-Audonneau N, Percebois G. Mating types of strains of the "*mentagrophytes*" complex in relation to clinical lesions. In: Vanbreuseghem R, DeVroey C, eds. Sexuality and pathogenicity of fungi. Proceedings of the Third International Colloquium on Medical Mycology. Paris: Masson, 1981:160-63.
32. DeVroey C, Wuytack-Raes C, Fossoul F. Isolation of saprophytic *Microsporum praecox* Rivalier from sites associated with horses. Sabouraudia 1983;21:255-7.
33. Dijk E van, van den Berg WHHW, Landwehr AJ. *Fusarium solani* infection of a hypertensive leg ulcer in a diabetic. Mykosen 1980;23:603-6.
34. El-Ani AS. The cytogenics of the conidium in *Microsporum gypseum* and of pleomorphism and the dual phenomenon in fungi. Mycologia 1968;60:999-1015.
35. Emmons CW. Dermatophytes. Natural grouping based on the form of the spores and accessory organs. Arch Dermatol Syphilol 1934;30:337-62.
36. English MP. *Trichophyton rubrum* infections in families. Br Med J 1957;1:744-6.
37. English MP. Infection of the finger-nail by *Pyrenochaeta unguis-hominis*. Br J Dermatol 1980;103:91-3.
38. English MP, Southern HN. *Trichophyton persicolor* infection in a population of small wild animals. Sabouraudia 1967;5:302-9.
39. Estes SA, Merz WG, Maxwell LG. Primary cutaneous phaeohyphomycosis caused by *Drechslera spicifera*. Arch Dermatol 1977;113:813-15.
40. Fischman O. Black piedra among Brazilian indians. Rev Inst Med Trop Sao Paulo 1973;15:103-6.
41. Fischman O, Soares EC, Alchorne MMA, Baptista G, Camargo ZP. Tinea nigra contracted in Spain. Bol. Micologico 1983;1:68-70.
42. Friedman L, Derbes VJ, Hodges EP, Sinski JT. The isolation of dermatophytes from the air. J. Invest Dermatol 1960;35:3-5.
43. Gartenberg G, Bottone EJ, Keusch GT, Weitzman I. Hospital-acquired mucormycosis (*Rhizopus rhizopodiformis*) of skin and subcutaneous tissue. N Engl J Med 1978;299:1115-18.
44. Gentles JC. The isolation of dermatophytes from the floors of communal bathing places. J Clin Pathol 1956;9:374-7.
45. Gentles JC. Athletes foot fungi on floors of communal bathing places. Br Med J 1957;1:746-8.
46. Gentles JC. Experimental ringworm in guinea pigs; oral treatment with griseofulvin. Nature 1958;182:476-7.
47. Gentles JC, Evans EGV. Infection of the feet and nails with *Hendersonula toruloidea*. Sabouraudia 1970;8:72-5.
48. Gentles JC, Holmes JG. Foot ringworm in coal miners. Br J Ind Med 1957;14:22-9.
49. Georg LK. Dermatophytosis, new methods in classification. Atlanta: US Dept Health, Education and Welfare, Centers for Disease Control, 1957.
50. Georg LK, Camp LB. Routine nutritional tests for the identification of dermatophytes. J Bacteriol 1957;74:113-21.
51. Gip L. Estimation of incidence of dermatophytes on floor areas after barefoot walking with washed and unwashed feet. Acta Derm Venereol 1967;47:89-93.
52. Gold I, Sommer B, Urson S, Schewach-Millet M. White piedra: a frequently misdiagnosed infection of hair. Int J Dermatol 1984;23:621-3.
53. Grossman ME, Fithian EC, Behrens C, Bissinger J, Fracaro M, Neu HC. Primary cutaneous aspergillosis in six leukemic children. J Am Acad Dermatol 1985;12:313-18.
54. Hay RJ, Moore MK. Clinical features of superficial fungal infections caused by *Hendersonula toruloidea* and *Scytalidium hyalinum*. Br J Dermatol 1984;110:677-83.

55. Hironaga M, Nozaki K, Watanabe S. Ascocarp production by *Nannizzia otae* on keratinous and non-keratinous agar media and mating behavior of *N. otae* and 123 Japanese isolates of *Microsporum canis*. Mycopathologia 1980;72:135-41.
56. Hironaga M, Tanaka S, Watanabe S. Distribution of mating types among clinical isolates of the *Microsporum gypseum* complex. Mycopathologia 1982;77:31-5.
57. Hironaga M, Watanabe S. Mating behavior of 334 Japanese isolates of *Trichophyton mentagrophytes* in relation to their ecological status. Mycologia 1980;72:1159-70.
58. Hopkins JG, Hillegas AB, Ledin RB, Rebell G, Camp E. Dermatophytosis at an infantry post. J Invest Dermatol 1947;8:291-316.
59. Horta P. Sobre um caso de tinha preta e um novo cogumelo (*Cladosporium werneckii*). Rev Med Cirug Brazil 1921;29:269-74.
60. Isaacs F, Reiss-Levy E. Tinea nigra plantaris: A case report. Aust J Dermatol 1980;21:13-15.
61. Johnson MLT, Roberts J. Prevalence of dermatological disease among persons 1-74 years of age: United States. Washington, DC: Advance data from vital and health statistics; USDHEW No. 4, 1977:1-6.
62. Kane J, Albreish S. Differentiation of *Phaeoannellomyces werneckii* and other dematiaceous fungi of medical importance on various concentrations of sodium chloride. (Abstract). Annual meeting of The American Society for Microbiology. Washington, DC: American Society for Microbiology, 1986:86.
63. Kane J, Birkett B, Fischer JB. Tinea nigra infection in Canada. Sabouraudia 1976;14:327-30.
64. Kane J, Fischer JB. The differentiation of *Trichophyton rubrum* and *T. mentagrophytes* by use of Christensen's urea broth. Can J Microbiol 1971;17:911-13.
65. Kane J, Padhye AA, Ajello L. *Microsporum equinum* in North America. J Clin Microbiol 1982;16:943-7.
66. Kane J, Salkin IF, Weitzman I, Smitka C. *Trichophyton raubitschekii*, sp. nov. Mycotaxon 1981;3:259-66.
67. Kaplan W. The occurrence of black piedra in primate pelts. Trop Geogr Med 1959;11:115-26.
68. Kaplan W, Gump RH. Ringworm in a dog caused by *Trichophyton rubrum*. Vet Med 1958;53:139-42.
69. Keddie FM. Medical mycology 1841-1870. In: Poynter FN, ed. Medicine and science in the 1860's. London: Wellcome Institute of the History of Medicine; Publication New Series; vol 16, 1968:137-49.
70. Kreger-Van Rij NJW. *Trichosporon* Behrend. In: Kreger-Van Rij, ed. The yeasts, a taxonomic study. 3rd ed. Amsterdam: Elsevier, 1984:933-1062.
71. Kushida T, Watanabe S. Canine ringworm caused by *Trichophyton rubrum*: probable transmission from man to animal. Sabouraudia 1975;13:30-2.
72. Langeron M, Horta P. Note complementaire sul le *Cladosporium werneckii*: Horta 1921. Bull Soc Pathol Exot Filiales 1922;15:381-3.
73. Langeron M, Milochevitch S. Morphologie des dermatophytes sur milieux naturels et milieux a base de polysaccharides. Essai de classification (deuxieme memoire). Ann de Parasitol 1930;8:465-508.
74. Lewis GM, Hopper ME. An introduction to medical mycology. Chicago: Year Book Publishers, 1939:208-10.
75. Male O, Pehamberger H. Die Kutane alternariose-fallberichte und literaturubersicht. Mykosen 1984;28:278-305.
76. Many H, Derbes VJ, Friedman L. *Trichophyton rubrum*: exposure and infection within household groups. Arch Dermatol 1960;82:226-9.
77. Marin G, Campos R. Dermatofitosis por *Aphanoascus fulvescens*. Sabouraudia 1984;22:311-14.

78. McAleer R. Fungal infections of the nails in Western Australia. Mycopathologia 1981;73:115-20.
79. McGinnis MR. Dematiaceous fungi. In: Lennette EH, Balows A, Hausler WR Jr, Shadomy HJ, eds. Manual of clinical microbiology. 4th ed. Washington, DC: American Society for Microbiology, 1985:561-74.
80. McGinnis MR, Padhye AA. *Cladosporium castellanii* is a synonym of *Stenella araguata*. Mycotaxon 1978;7:415-18.
81. McGinnis MR, Schell W, Carson J. *Phaeoannellomyces* and the *Phaeococcomycetaceae*, a new dematiaceous blastomycete taxa. Sabouraudia 1985;23:179-88.
82. Merz WG, Berger CL, Silva-Hutner M. Media with pH indicators for the isolation of dermatophytes. Arch Dermatol 1970;102:545-7.
83. Mok WY. Nature and identification of *Exophiala werneckii*. J Clin Microbiol 1982;16:976-8.
84. Moore MK. *Hendersonula toruloidea* and *Scytalidium hyalinum* infections in London, England. J Med Vet Mycol 1986;24:219-30.
85. Muijs D. *Trichophyton rosaceum*. Ned Tijdschr Geneeskd 1916;2:1985-92.
86. Onsberg P. *Scopulariopsis brevicaulis* in nails. Dermatologia 1980;161:259-64.
87. Ota M, Langeron M. Nouvelle classification des dermatophytes. Ann de Parasitol 1923;1:305-36.
88. Padhye AA, Blank F, Koblenzer PJ, Spatz S, Ajello L. *Microsporum persicolor* infection in the United States. Arch Dermatol 1973;108:561-2.
89. Padhye AA, Carmichael JW. Mating behavior of *Trichophyton mentagrophytes* varieties paired with *Arthroderma benhamiae* mating types. Sabouraudia 1969;7:178-81.
90. Padhye AA, Carmichael JW. The genus *Arthroderma* Berkeley. Can J Bot 1971;49:1525-40.
91. Padhye AA, Sekohn AS, Carmichael JW. Ascocarp production by *Arthroderma* and *Nannizzia* species on keratinous and non-keratinous media. Sabouraudia 1973;11: 109-14.
92. Pereiro Miguens M. Relation between the mating type of strains of the "*mentagrophytes*" complex and the clinical lesions. In: Vanbreuseghem R, DeVroey C, eds. Sexuality and pathogenicity of fungi. Proceedings of the Third International Colloquium on Medical Mycology. Paris: Masson, 1981:164-9.
93. Pier AC, Rhoades KR, Hayes TL, Gallagher J. Scanning electron microscopy of selected dermatophytes of veterinary importance. Am J Vet Res 1972;33:607-13.
94. Prochacki H, Engelhardt-Zasada L. *Epidermophyton stockdaleae* sp. nov. Mycopathologia 1974;54:341-5.
95. Punithalingam E. Sphaeropsidales in culture from humans. Nova Hedwiga 1979;31:119-58.
96. Rayner RW. A mycological colour chart. Commonwealth Mycological Institute, Kew, Surrey, England, 1970.
97. Rebell G. *Fusarium* infections in human and veterinary medicine. In: Nelson RE, Toussoun TA, Cook RJ, eds. *Fusarium*: disease, biology, taxonomy. University Park: Pennsylvania State University Press, 1981:210-20.
98. Rippon JW. Medical mycology; the pathogenic fungi and the pathogenic actinomycetes. 2nd ed. Philadelphia: W.B. Saunders, 1982:140-8.
99. Rippon JW. The changing epidemiology and emerging patterns of dermatophyte species. In: McGinnis MR, ed. Current topics in medical mycology. New York: Springer-Verlag, 1985:209-34.
100. Rippon JW, Lee FC, McMillen S. Dermatophyte infection caused by *Aphanoascus fulvescens*. Arch Dermatol 1970;102:552-5.
101. Rivalier E. Description de *Sabouraudites praecox* nova species suivie de remarques sur le genre *Sabouraudites*. Ann Inst Pasteur 1954;86:276-84.

102. Robbins WJ. Growth requirements of dermatophytes. Ann NY Acad Sci 1950;50:1357-61.
103. Rosenthal SA. The epidemiology of tinea pedis. In: Robinson HM Jr, ed. The diagnosis and treatment of fungal infections. Springfield, IL: CC Thomas, 1974:515-26.
104. Rosenthal SA, Sokolsky H. Enzymatic studies with pathogenic fungi. Dermatol Int 1965;4:72-9.
105. Roth F Jr. Microscopic preparations from cultures to show aleuriospores. In: Rebell G, Taplin D, eds. Dermatophytes. Their recognition and identification. rev. ed. Coral Gables, FL: University of Miami Press, 1970:114.
106. Sabouraud R. Contribution a l'etude de la trichophytie humaine. Les trichophyton animaux sur l'homme: Trichophyties pilaires de la barbe. Ann Dermatol Syphiligr 1893;4:814-35.
107. Sabouraud R. Les teignes. Paris: Masson, 1910.
108. Salkin IF. Dermatophyte test medium: evaluation with nondermatophytic pathogens. Appl Microbiol 1973;26:134-7.
109. Salkin IF, Hollick GE, Hurd NJ, Kemna ME. Evaluation of human hair sources for the in vitro hair perforation test. J Clin Microbiol 1985;22:1048-9.
110. Smith JMB, Jolly RD, Georg LK, Connole MD. *Trichophyton equinum* var. *autorophicum*; its characteristics and geographic distribution. Sabouraudia 1968;6:296-304.
111. Soei KI. An autochthonous case of tinea nigra in the Netherlands. Ned Tijdschr Geneeskd 1984;128:2347-8.
112. Stockdale PM. The *Microsporum gypseum* complex (*Nannizzia incurvata* Stockd., *N. gypsea* (Nann.) comb. nov., *N. fulva* sp. nov). Sabouraudia 1963;3:114-26.
113. Stockdale PM. Sexual stimulation between *Arthroderma simii* Stockd., MacKenzie and Austwick and related species. Sabouraudia 1968;6:176-81.
114. Sulzberger MB, Baer RL, Hecht R. Common fungous infections of the feet and groin. Negligible role of exposure in causing attacks. Arch Dermatol 1942;45:670-5.
115. Takashio M. Observations on African and European strains of *Arthroderma benhamiae*. Int J Dermatol 1974;13:94-101.
116. Takashio M, Vanbreuseghem R. Production of ascospores by *Piedraia hortai* in vitro. Mycologia 1971;63:612-18.
117. Takatori K, Hasegawa A. Mating experiment of *Microsporum canis* and *M. equinum* isolated from animal with *Nannizzia otae*. Mycopathologia 1985;90:59-63.
118. Vanbreuseghem R. Mycoses of man and animals. Springfield: CC Thomas, 1958:71-4.
119. Vanbreuseghem R, DeVroey C, Takashio M. Production of macroconidia by *Microsporum ferrugineum*, Ota 1922. Sabouraudia 1970;7:252-6.
120. Van Uden N, Barros-Machado A, Castelo Branco R. On black piedra in Central African mammals caused by the ascomycete *Piedraia quintanilhai* nov. spec. Rev Brasiliera Porgugesa Biol Geral 1963;3:271-6.
121. Van Velsor H, Singletary H. Tinea nigra palmaris. Arch Dermatol 1964;90:50-61.
122. Vries GA de. Ascomycetes: *Eurotiales, Sphaeriales* and *Dothideales*. In: Howard DH, ed. Fungi pathogenic for humans and animals, Part A. Biology. New York: Marcel Dekker, 1983:81-111.
123. Wadhwani K, Srivastava AK. Some cases of onychomycosis from North India in different working environments. Mycopathologia 1985;92:149-50.
124. Walshe MM, English MP. Fungi in nails. Br J Dermatol 1966;78:198-207.
125. Watanabe S, Hironaga M. Differences or similarities of the clinical lesions produced by "+" and "-" type members of the "*mentagrophytes*" complex in Japan. In: Vanbreuseghem R, DeVroey C, eds. Sexuality and pathogenicity of fungi. Proceedings of the Third International Colloquium on Medical Mycology. Paris: Masson, 1981:83-96.
126. Weitzman I. Variation in *Microsporum gypseum*. I. A genetic study of pleomorphism. Sabouraudia 1964;3:195-204.

127. Weitzman I, Allderdice PW, Silva-Hutner M, Miller OJ. Meiosis in *Arthroderma benhamiae* (= *Trichophyton mentagrophytes*). Sabouraudia 1968;6:232-7.
128. Weitzman I, Gordon MA, Rosenthal SA. Determination of the perfect state, mating type and elastase activity in clinical isolates of the *Microsporum gypseum* complex. J Invest Dematol 1971;57:278-82.
129. Weitzman I, McGinnis MR, Padhye AA, Ajello L. The genus *Arthroderma* and its later synonym *Nannizzia*. Mycotaxon 1986;25:505-18.
130. Weitzman I, McMillen S. Isolation in the United States of a culture resembling *M. praecox*. Mycopathologia 1980;70:181-6.
131. Weitzman I, Padhye AA. Mating behavior of *Nannizzia otae* (= *Microsporum canis*). Mycopathologia 1978;64:17-22.
132. Yaffee HS. Tinea nigra palmaris resembling malignant melanoma. N Engl J Med 1970;283:1112.
133. Zaias N. Superficial white onychomycosis. Sabouraudia 1966;5:99-103.
134. Zaias N. Onychomycosis. Arch Dermatol 1972;105:263-74.
135. Zaias N, Oertel I, Elliott DR. Fungi in toe nails. J Invest Dermatol 1969;53:140-2.

CHAPTER 4

MOLDS INVOLVED IN SUBCUTANEOUS INFECTIONS

Wiley A. Schell, M.S. and Michael R. McGinnis, Ph.D.

INTRODUCTION

Subcutaneous mycoses are a diverse group of infections caused by a wide variety of fungi. Most of these fungi are saprobes in soil or on woody plant detritus. Some can be isolated from water, others are plant pathogens and still others are parasites of animals. Though these fungi are usually introduced into subcutaneous or cutaneous tissue via traumatic inoculation, any breach in the cutaneous barrier (such as an indwelling dialysis catheter) can provide a portal of entry with subsequent infection. In some specific syndromes, infection is thought to begin following inhalation of conidia. Subcutaneous mycoses are chronic and usually localized, though dissemination to other organs can occur, resulting in systemic infection.

Until fairly recently, discussions of subcutaneous mycoses were limited primarily to eumycotic mycetoma, sporotrichosis and chromoblastomycosis. These infections still account for most cases of subcutaneous mycoses. During the past 15 years, however, many fungi which previously were not recognized as pathogens have been shown to cause opportunistic subcutaneous infections. These diseases, along with other previously unclassified mycoses, are grouped into two broad categories: phaeohyphomycosis and hyalohyphomycosis. Without doubt, the realm of subcutaneous infections has become one of the most rapidly changing aspects of the mycoses.

SPOROTRICHOSIS

Despite geographical differences, sporotrichosis is probably the most frequently encountered subcutaneous mycosis in the world. Since the mycoses are not regarded as reportable diseases, comparative data are unavailable.[9] Sporotrichosis is a varied disease caused by the single species

Sporothrix schenckii. S. schenckii is known to cause eye infections, sinusitis and pulmonary infection.[253,263] Even a vocal cord infection has been documented in one case[5] and implicated in another.[280] Its typical clinical presentation, however, is that of a chronic cutaneous and subcutaneous infection. Although extremities are most frequently involved, Fukushiro found that 92% of 529 cases in children involved the face.[96]

The first report that documented the fungal etiology of sporotrichosis was published in 1898.[265] By 1932, less than 200 cases had been reported in the United States.[133] The increase in recognition of the disease in subsequent years has been dramatic.[211,253] The Japanese experience is typical: 13 cases were reported through 1945, but 2,248 cases have been seen since then.[96] The largest single reported outbreak of sporotrichosis occurred during 1941-1944 when 2,825 cases were seen in gold miners of the African Transvaal region.[281] Outbreaks continue to occur occasionally, along with individual cases.

Clinical Manifestations

In subcutaneous sporotrichosis, lesions begin as firm nodules which gradually develop into softened ulcers. The lesions may remain fixed, with satellite lesions arising, or they may progress to form a series of ulcers which develop and spread along the lymphatic system (Fig. 4:1). Muscu-

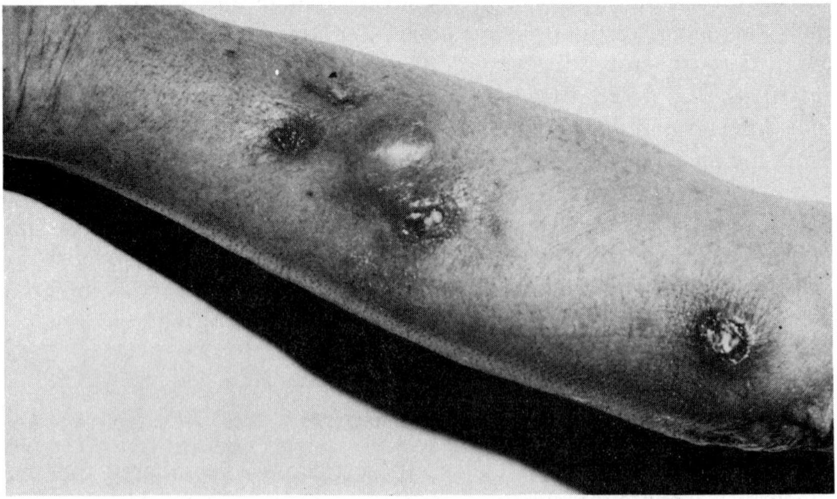

Figure 4:1. Sporotrichosis. Subcutaneous nodules and ulcerating lesions developing up the forearm via the lymphatic system.

loskeletal involvement[18,19,51,165] and dissemination[207,273,280] which may involve the central nervous system[92] are rare manifestations.

Pulmonary sporotrichosis has been infrequently documented,[92,238,241,255] but it may be more common than previously realized.[169,253] This form of the disease follows inhalation of conidia. The infection is insidious and sometimes asymptomatic. Symptoms may include low grade fever, weight loss, night sweats, productive cough with or without hemoptysis, and pleuritic pain.[238,241,255] Chest films may show adenopathy, nodules or cavitation, usually involving the upper lobes, or diffuse involvement similar to tuberculosis.[60,61,238,253,255]

Epidemiology and Public Health Significance

S. schenckii is widespread in nature and has been isolated from a wide variety of plant materials. Infections are most often associated with splinters, abrasions from woody material, thorn pricks (often from rose bushes), sphagnum moss and grasses such as hay. *S. schenckii* has also been isolated from two commercial brands of potting soil being screened for *Aspergillus* species.[148] Sporotrichosis is most often associated with florists, farmers, gardeners, nursery and forestry workers and others whose occupation or avocation brings them in frequent contact with plants or soil. Also at risk in regions where manual labor is common, especially in rural areas of developing countries, are farm workers who do not regularly wear shoes or who wear open sandals. It is well established that these individuals have increased exposure to the fungus and have a greater probability of developing sporotrichosis.[90,129,169] Preventive measures include prudent use of gloves, clothing and shoes to protect extremities.

Noncutaneous infection has been traced to dusty conditions existing in a sphagnum moss packing plant.[5] Improving ventilation and wearing respiratory masks is advisable under such conditions. *S. schenckii*, found as a persistent contaminant in a nursery barn, was successfully controlled with a phenol-base disinfectant.[186] Treating contaminated lumber with fungicides has been effective in eliminating outbreaks of sporotrichosis associated with skin abrasions from mine timbers.[281] Not surprisingly, *S. schenckii* has also been found as a contaminant of food. It has been isolated from frankfurters[7] and culinary mushrooms,[147] but there is no evidence thus far of food as a source of infection. *S. schenckii* was also found to be a contaminant of intravenous fluid.[179] Although a patient was infused with 100 mL of the contaminated fluid, no evidence of sporotrichosis was detected.

Recently, *S. schenckii* was found during a routine examination of floor and wall surfaces of an indoor swimming pool complex.[283] The fungus was isolated in moderate numbers from several swabs of the floor adjacent to a shower room. There were no plants or wooden material in the building. Mycological screening of such facilities has normally been intended to detect and aid in control of dermatophytes and certain yeasts. Personnel in such settings should be alert to the possible presence of *S. schenckii* as well.

Early experiments suggested that a skin break is not necessary to establish cutaneous infection.[281] A 1945 report[279] that suggested person-to-person transmission, however, lacked supportive investigation, and an account of infections acquired from handling contaminated wound dressings has not been substantiated.[90]

Though uncommon, cases of animal to human transmission have been known for many years. A recent series of cases suggests that veterinarians and their staffs may be at increased risk of contracting sporotrichosis from infected animals. In 1980, an entire family developed sporotrichosis, evidently from their infected house cat.[246] A veterinary technician and a veterinary student also contracted sporotrichosis soon after handling the animal. None recalled being scratched or bitten by the cat. Other similar cases have been reported.[72,223,261,266] Sporotrichosis has also been seen in a number of other animals including horse, chimpanzee, donkey, mule, cattle, dog, fowl, camel, rat, mouse, dolphin, fox, boar and armadillo.[144] Veterinarians and their assistants should be cautious when treating cats and other animals that are suffering from ulcers or abscesses.

A three-year study conducted in Guatemala followed 53 cases of sporotrichosis among people living in the Lake Ayarza District.[183] Among these 53 patients, 6 infections were attributed to trauma by woody material, 24 individuals recalled trauma associated with handling fish taken from the lake and 23 had no recollection of contact with fish but were users of the lake.

Reports of differences in male to female ratios among sporotrichosis patients are conflicting. These numbers may well reflect the sexual ratio among various occupations, and thus correlate with frequency of exposure to the fungus rather than physiologic predilection for infection based upon sex.

Identification of Etiologic Agent

Microscopic examination of clinical specimens is often unproductive since fungal cells are present in very small numbers. Fluorescent stain preparations using calcofluor white-potassium hydroxide,[113] and tissue sections stained with the periodic acid-Schiff (PAS) or Gomori methenamine silver (GMS) methods are far more sensitive than traditional KOH preparations.[286] In cases where biopsies are obtained but cultures are not requested, examination of tissue sections stained with specific *S. schenckii* fluorescent antibody (FA) conjugate can quickly provide a specific identification.[145] Unfortunately, FA reagents are maintained only at a few reference centers (such as the Centers for Disease Control) and are not commercially available. In tissue, *S. schenckii* appears sparsely as isolated cells, 3-5 μm in diameter, which are usually round in shape and commonly do not show budding (Fig. 4:2). "Cigar-shaped" buds[81] are rarely present in human infections. Asteroid bodies, though frequently seen in cases from Africa[281] and Japan,[96] are uncommon in the United States.

Isolates of *S. schenckii* can usually be detected within 3-5 days of incubation at 30°C. Colonies are initially cream-colored, without aerial hyphae, smooth and moist. The colony gradually turns dark brown, often in irregular patches, and aerial hyphae develop which cause the colony to become velvety (or, more rarely, lanose). There is, however, much variation among isolates and even between subcultures of the same isolate. Some will remain smooth with little or no aerial hyphae, while others will remain white to cream-colored with little or no brown pigmentation. Microscopically, conidiophores are elongate and compactly sympodial, each with a swollen apex bearing 1-celled, obovate (egg-shaped), hyaline to sub-hyaline conidia on denticles in "flower-like" clusters (Fig. 4:3). Most isolates also show conidia borne singly along the hyphae (Fig. 4:4). These are usually subglobose but may be triangular in some isolates (Fig.4:5). They may be colorless, but often are dark brown and are chiefly responsible for the dark color of the colony. Because its conidia are easily dislodged, a teased preparation of the mold may not be revealing. Slide culture, "squash prep"

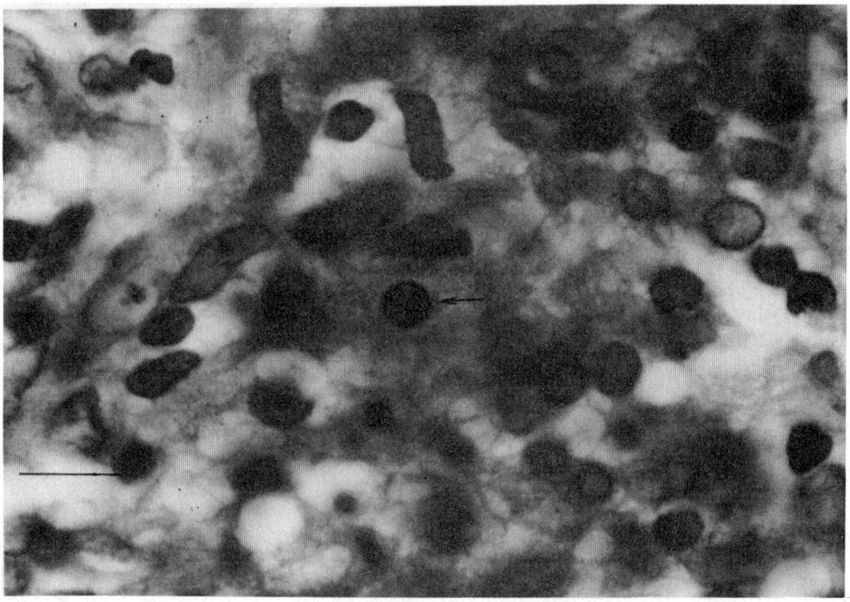

Figure 4:2. Solitary round cells (arrow), usually without buds, characterize *Sporothrix schenckii* in tissue. Subcutaneous biopsy, PAS stain. Bar = 10 μm.

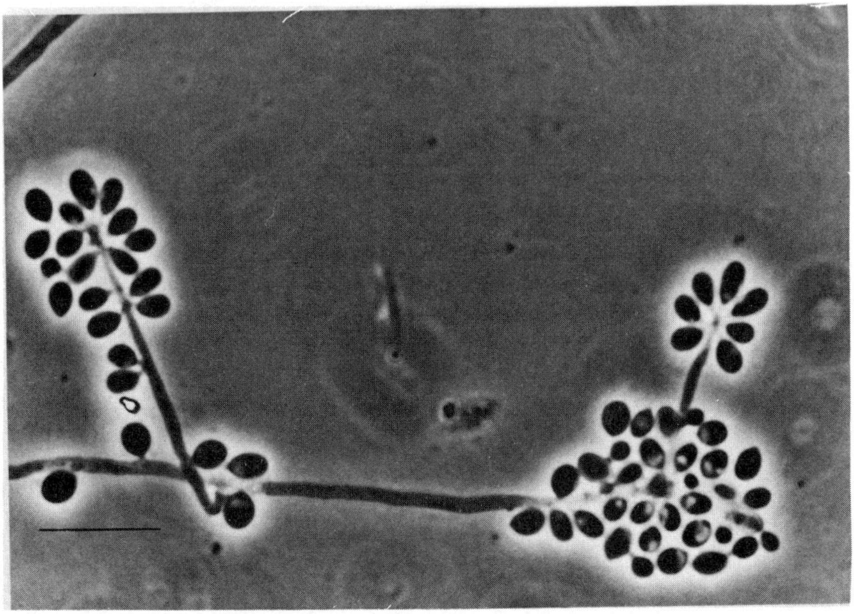

Figure 4:3. Sympodial conidiogenesis in *S. schenckii* sometimes results in a flower-like appearance. Bar = 10 μm.

or scotch tape techniques are needed to study conidial ontogeny and arrangement.

Since there are many species of *Sporothrix* as well as other genera that resemble *Sporothrix*,[124,126,127] it is necessary to demonstrate that a suspected isolate of *S. schenckii* will grow as a stable yeast form at 35-37°C. Of the morphologically similar fungi, *Sporothrix cyanescens* may exhibit a limited and unstable yeast form.[128] When cultured at 36-37°C on brain heart infusion agar or other enriched media, most isolates of *S. schenckii* will quickly show partial or complete conversion to the stable yeast form characterized by elongate "cigar-shaped" blastoconidia (Fig. 4:6). Occasional isolates are more difficult to convert and may require extended incubation (up to three weeks) and/or multiple subcultures.[166]

Significance of Laboratory Findings

Recovery of *S. schenckii* from lesions of patients with cutaneous, lymphocutaneous or deeper infections is diagnostic of sporotrichosis. An isolate from a respiratory specimen is also significant, but should be interpreted

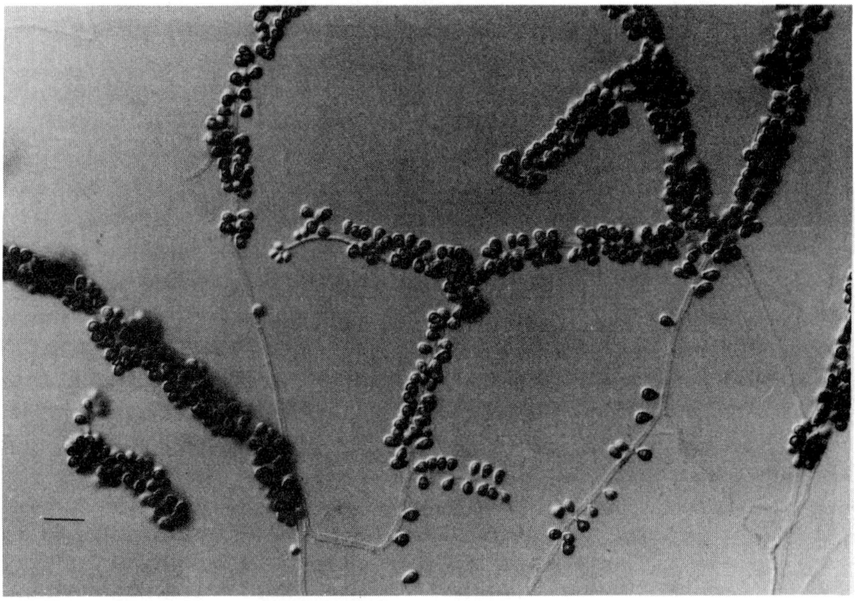

Figure 4:4. The "sleeve arrangement" of lateral, individually borne conidia is seen to a greater or lesser degree in most isolates of *S. schenckii*. These conidia may be pale but more often are brown in color. Bar = 10 μm.

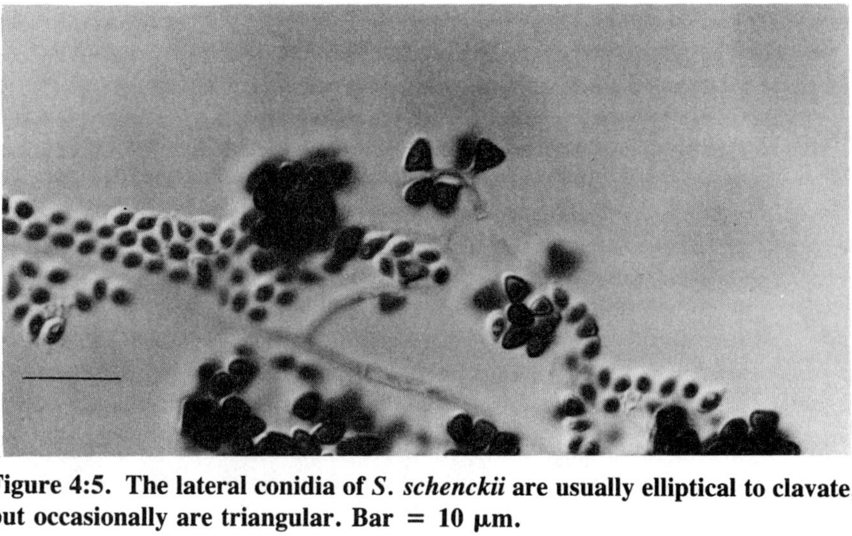

Figure 4:5. The lateral conidia of *S. schenckii* are usually elliptical to clavate, but occasionally are triangular. Bar = 10 μm.

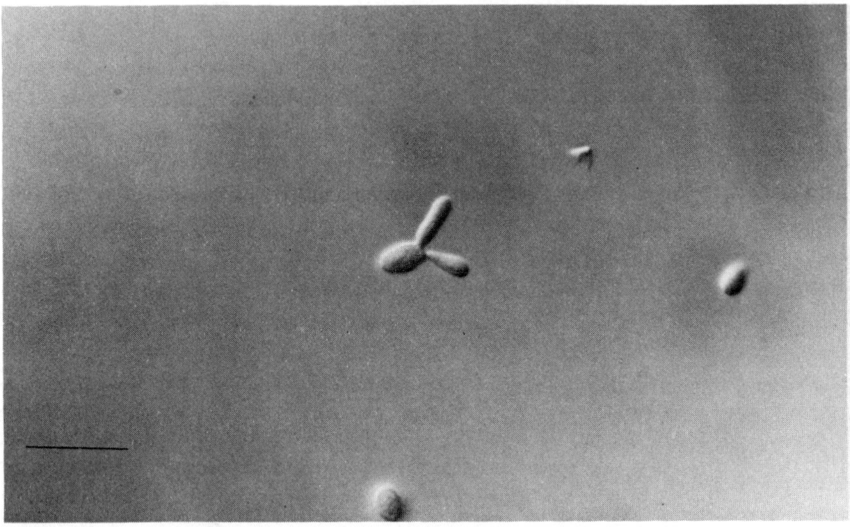

Figure 4:6. ***S. schenckii*** **can be converted in vitro to its yeast form which is characterized by cylindrical "cigar-shaped" buds. Bar = 10 μm.**

cautiously by the physician, since it is presently not clear whether subclinical, acute pulmonary sporotrichosis can occur and resolve untreated.[238,240,241]

Since *S. schenckii* is a conidial fungus that is widespread in nature, it is possible to recover it as a "contaminant" from cultures of clinical specimens. Practical experience, however, has shown that this is unlikely.

MYCETOMA

Mycetoma ("fungus tumor") is a chronic subcutaneous infection characterized by tumorous swelling of the affected area, the development of draining sinuses and the presence of "grains" (granules) which discharge from the sinuses. As the infection progresses, there is extensive destruction of soft tissue and ultimately, in most cases, of bone. Published reports have shown that at least 16 species of fungi may cause mycetoma. These fungi are saprobes in nature, found in soil and on dead plant material. Infection begins after the fungus is inoculated by trauma into cutaneous or subcutaneous tissue. Thus, any part of the body which is exposed to abrasions or wounds can become infected. It is not surprising that on a worldwide basis the feet are the most commonly involved part of the body.

Mycetoma is an ancient disease. References to infections that were probably mycetomas exist in ancient Sanscrit writings of the Hindus. Several other early descriptions have been noted.[133] In an 1845 medical report from

India, Godfrey summarized 3 cases of "disease of the foot not hitherto described," all of which appear to have been mycetomas. In one amputated foot, he first observed the presence of characteristic grains which he described as a black mass "much resembling fragments of coal."[103] It was not until 1860 that a fungal etiology of mycetoma was revealed by Vandyke Carter, a surgeon of the British India Army in Bombay.[48] In 1874, Carter published a comprehensive treatise on mycetoma including its history, symptoms, pathology, epidemiology, treatment and etiology.[49] Since that time it has been learned that mycetomas may have a bacterial (actinomycotic mycetoma) or a fungal (eumycotic mycetoma) etiology. This chapter will address only fungal mycetoma.

Clinical Manifestations

Since the infection is insidious and usually painless, detection of disease generally follows the initial wound by a period of several months to a year or longer. Thus, observations of the earliest stages of the disease are few. As the fungus grows there is fibrosis and a granulomatous reaction that normally results in a firm subcutaneous swelling (Fig. 4:7). The limb becomes progressively swollen and deformed, reflecting extensive soft tissue involvement. In most cases, there is eventually bone destruction. Sinus tracts

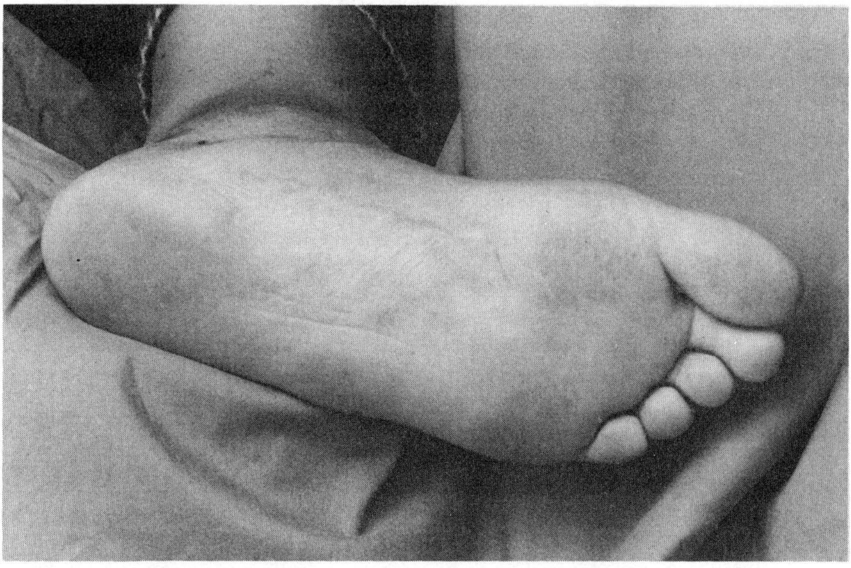

Figure 4:7. Mycetoma: ankle area and arch showing generalized subcutaneous swelling typical of early stages of infection. *Pseudallescheria boydii* infection of two years' duration.

typically develop within 3-12 months (Fig. 4:8). A serous to seropurulent fluid, containing fungal grains (Fig. 4:9) which vary in size from microscopic to 2 mm in diameter, is discharged through the tracts. The grains are irregular in shape and are white to yellowish, or brown to black depending on the species of fungus (Table 4:1). At least 2 genera (*Acremonium* and *Fusarium*) can produce either pale or dark granules.[98,115,234,289] Grains are deposited onto a patient's bandages, and they can be easily retrieved for microscopic and cultural studies with the aid of a stereo (dissecting) microscope.

Mycetomas are localized and chronic, but gradually extend via sinus tracts. Sinuses heal quickly with scarring (Fig. 4:8) even as new channels develop. Lymphatic involvement is rare. Though mycetoma is not life-threatening and the patient otherwise remains basically healthy, the disease frequently does not respond to chemotherapy and becomes incapacitating, leaving amputation as a final resort.

Epidemiology and Public Health Significance

The epidemiology is not entirely understood but clearly mycetoma is worldwide in occurrence (Table 4:2). The fungi involved are saprobes,

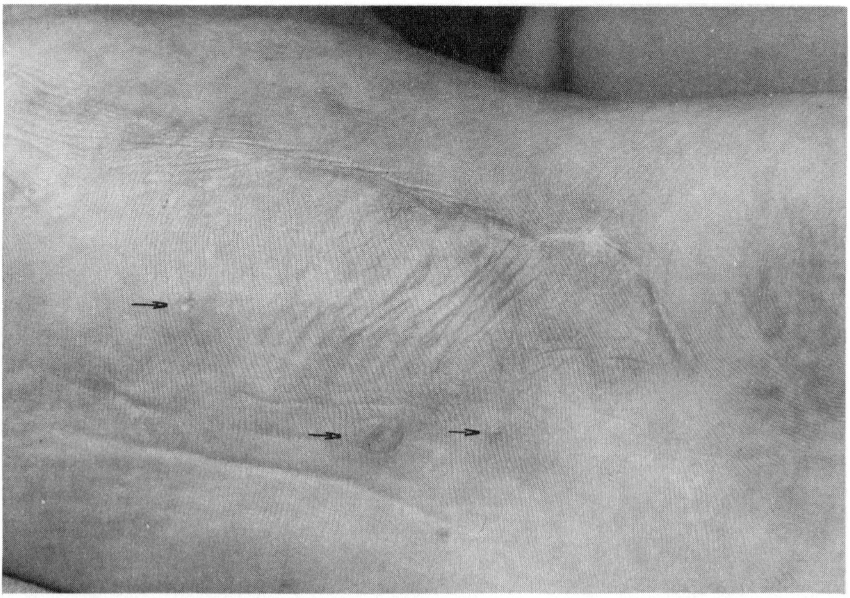

Figure 4:8. Mycetoma: inactive sinuses (arrows) that have healed, with scarring (same patient as in Fig. 4:7). Scar from previous surgical management is visible at top.

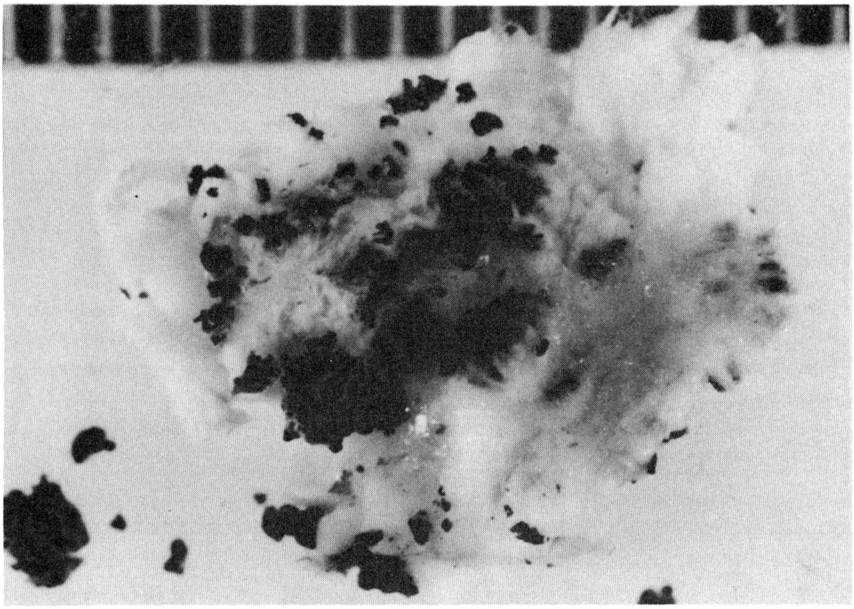

Figure 4:9. ***Madurella mycetomatis*** **granules (dark brown) in subcutaneous biopsy. Scale at top indicates millimeters.**

Table 4:1. Granule Characteristics of Etiologic Agents of Mycetoma

Granule	Fungi*
Pale (white to yellowish, soft	*Pseudallescheria boydii, Acremonium kiliense, A. falciforme, A. recifei, Aspergillus nidulans, Fusarium* spp.
Black to brown, firm to soft	*Leptosphaeria senegalensis, Exophiala jeanselmei, Curvularia lunata, C. geniculata, Pyrenochaeta romeroi, Acremonium kiliense, Leptosphaeria tompkinsii, Bipolaris spicifera, Fusarium solani*
Black, hard to brittle	*Madurella mycetomatis, Madurella grisea* (young granules may be softer)

* Listed in descending order of worldwide frequency for each granule type.

widely distributed in nature, but differing in their frequency among geographic regions. In North America, for example, *Pseudallescheria boydii* is the common cause of mycetoma, while in upper Africa and the Indian subcontinent, *Madurella mycetomatis* is by far the most common agent, even though both fungi are present in each region (Table 4:2).

Table 4:2. Geographic Occurrence of Mycetoma

Fungus*	Distribution†	Number of Reports‡	Key Ref.
Acremonium falciforme	North America, South America, West Indies, Senegal, Europe	9	45,47 87,115
A. kiliense	India, South America, West Indies	6	115,164 212
A. recifei	India, South America	6	162,163
Acremonium spp.	Thailand, South America, Europe	3	115
Aspergillus nidulans	Indian sub-continent, Africa, Europe	8	138,151
Bipolaris spicifera§	North America (in animals)	3	38
Curvularia lunata	Africa, Pakistan	7	151,176
C. geniculata	North America (in animals)	2	38,253
Exophiala jeanselmei	Indian sub-continent, North America, Thailand Europe	16	216,278 242
Fusarium spp.	Indian sub-continent, Jamaica, Thailand	6	119,289
Leptosphaeria senegalensis	Africa, India	117	23,160 178,247
L. tompkinsii	Africa	4	253
Madurella grisea	South America, Indian sub-continent, North America, Africa, West Indies, Central America	61	42,174 178,270
M. mycetomatis	Africa, Indian sub-continent, South America, North America, Europe, Indonesia	>1,000	1,65,110, 173,178 235
Pyrenochaeta romeroi	West Africa, South America	4	30,247
Pseudallescheria boydii	Indian sub-continent, North America, South America, Africa, West Indies, Europe, Fiji Islands	70	25,35, 110,119 151,178, 278

* Fungi with fewer than 3 reports are excluded.
† Listed in decreasing order of frequency.
‡ This survey, while representative of the distribution of reported fungal mycetoma, is not intended to be exhaustive.
§ Previously called *Drechslera spicifera*.

Poorly protected feet of people in rural areas are most often affected, but hand, leg or arm involvement is also seen regularly. Other sites may be affected depending on local habits. Among field workers in India, Singh[278] reported: 1) scalp infections in people who regularly transported goods by carrying them upon their heads; 2) infections of the perineum due to working in the fields from a squatting position while wearing the customary loose-fitting dhoti (loin cloth) and to cleansing of the anal area with soil after defecation; and 3) infection of the middle ear arising from use of field straw to remove wax from the ear canal. Joshi et al[138] traced *Aspergillus nidulans* mycetoma of the neck and upper back to the patient's practice of carrying goods upon his bare shoulders and upper back. Mycetomas are not contagious, but they do represent a significant and persistent threat to public health. Until more effective antifungal chemotherapy can be developed, the problem of mycetomas can only be dealt with by encouraging greater use of footwear and by striving for earlier recognition of infections, combined with aggressive surgical management and currently available chemotherapy.

Identification of Etiologic Agents

When mycetoma is suspected, granules collected from dressings or draining sinuses and examined microscopically will quickly reveal whether the infection is of an actinomycotic or fungal nature. Active sinuses may not be present at all times in a patient and biopsy may be necessary to establish the diagnosis. Color and consistency of the grains and consideration of the patient's geographic origin will suggest the likely etiologic agent(s) (Tables 4:1, 4:2). The granules consist of masses of hyphae sometimes so tightly packed that only peripheral filaments can be clearly distinguished as fungal elements (Figs. 4:10-4:13). In cases (mainly involving *P. boydii*) where minced and unstained biopsied tissue is being examined without

Figure 4:10. KOH preparation of *M. mycetomatis* granule. Note that the fungal nature of the material is difficult to discern. Bar = 10 μm.

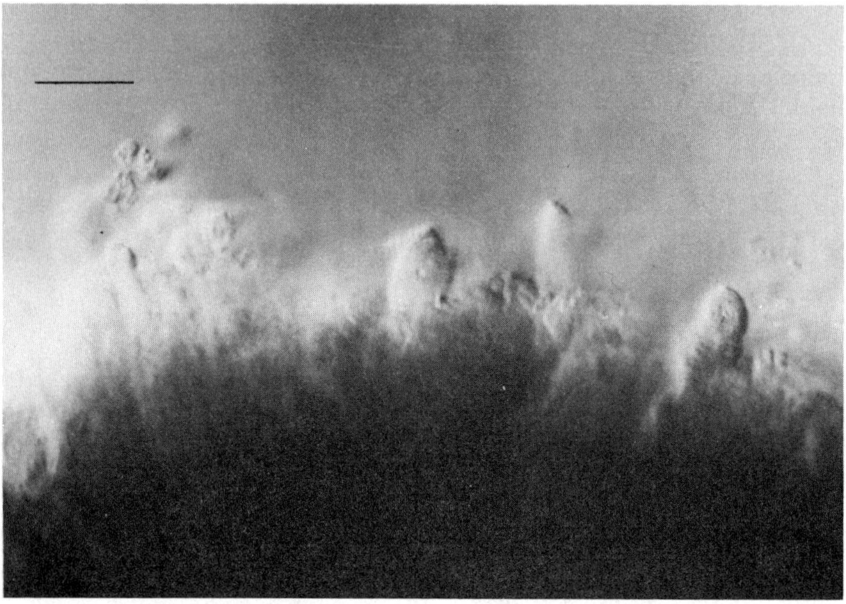

Figure 4:11. Higher power examination of *M. mycetomatis* granule reveals hyphal tips at the margin. Bar = 10 μm.

suspicion of mycetoma, it is possible that these masses will not be recognized as fungal granules. Examining stained material is a more sensitive procedure, but time-consuming. Calcofluor white[113] (also known as cellufluor) is an alternative to traditional staining that is both sensitive and as easy to prepare as the KOH preparation. However, calcofluor does not reliably detect fungal elements that are extremely dark, such as granules of *Madurella* and sclerotic cells of chromoblastomycosis. Prior to culturing, grains should be washed with sterile saline or distilled water to help remove other microorganisms. If microscopic examination clearly shows the granules to be fungal, the wash solution should contain penicillin and streptomycin or other antibacterial agents to help suppress bacterial overgrowth of cultures.

Rarely, in abnormal hosts, dermatophytes may form subcutaneous mycelial aggregates which resemble the granules of true mycetoma. However, such infections differ somewhat on histopathological grounds[13] and cause neither tumefaction nor sinus formation.

The most common agent of mycetoma in North America is *P. boydii (Scedosporium apiospermum)*. Identification is not difficult, but if necessary, can be confirmed by sending the isolate to a major reference center (such as the Centers for Disease Control) for testing with the exoantigen technique. Fluorescent antibody (FA) conjugated stains will specifically identify *P.*

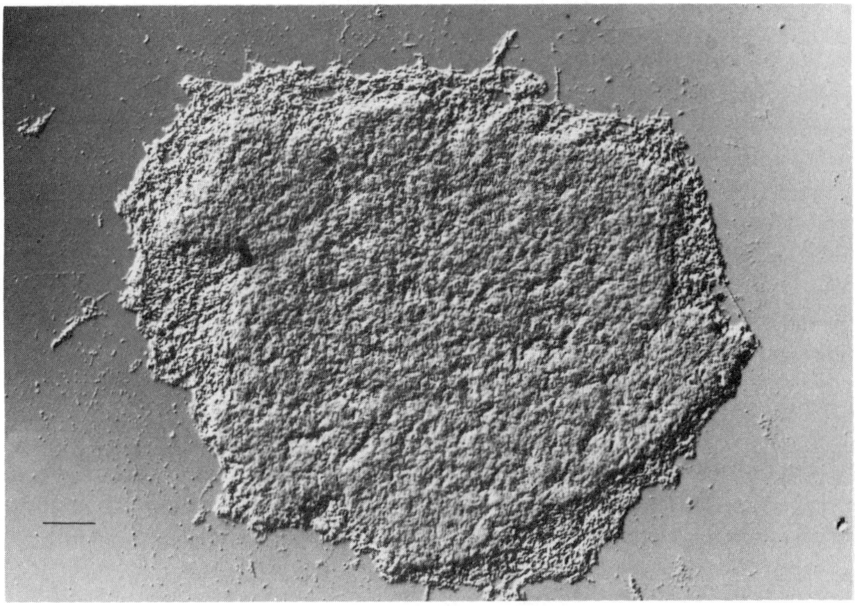

Figure 4:12. Granule of *P. boydii*. Bar = 100 μm.

boydii in histologic sections, independent of culture results. *P. boydii* is a common fungus of manure, soil and water and is regularly seen as a "contaminant" in cultures of respiratory specimens (particularly in patients with chronic lung disease, such as cystic fibrosis), and other nonsterile specimens. Isolates are frequently recovered from early cultures of dirty wounds. Although these isolates may not always cause infection, the patients should be monitored extremely closely, since infections with *P. boydii (S. apiospermum)* are certainly among the most difficult to cure. Determining the significance of an isolate and establishing a correct diagnosis requires careful communication between the physician and the laboratorian.

Isolation of a sterile dematiaceous mold from a lesion that is clinically consistent with mycetoma is significant. Since *Curvularia* may not sporulate readily on many routine media, V-8 juice agar may be required for inducing conidiation. *Madurella grisea* isolates are sterile, and less than 50% of *M. mycetomatis* isolates ever form conidia in culture. A presumptive identification of a *Madurella* species can be based upon clinical symptoms, grain characteristics, temperature requirements and assimilation tests for sucrose and lactose. *Leptosphaeria* has no conidia and appears sterile, but prolonged incubation (> 2 months) on nutritionally poor media allows formation of ascocarps and characteristic ascospores.

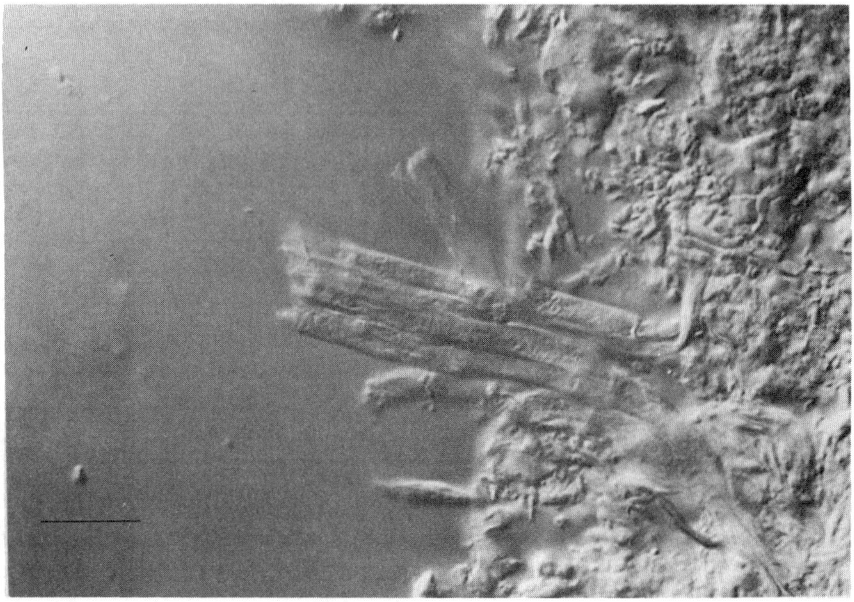

Figure 4:13. Septate, hyaline hyphae are discernible at the edge of crushed granule of *P. boydii*. Bar = 10 μm.

Acremonium kiliense:

Growth rate is moderate. Colonies are pinkish-orange, glabrous, shiny, usually with fasciculate aerial hyphae developing. Conidia are broadly elliptical to cylindrical, occasionally allantoid, 3-5.5 x 1-2 μm, accumulating in balls (Fig. 4:14). Phialides are long and tapering; collarettes are absent or vestigial (Fig. 4:15). Chlamydospores are readily formed. Agar may stain brown.[68]

A. falciforme (after Halde et al):[115]

Differs from *A. kiliense* by colonies becoming more downy, grayish-brown to grayish-violet; reverse is violet to purple. Conidiophores are long and repeatedly septate. Conidia are 0 to 1 septate, 7-8.5 x 2.7-3.2 μm.

A. recifei:

Colonies are white, developing pink to buff tinges, with brown reverse. Conidiophores are branched; phialides are long and tapering. Conidia are curved, tapered toward the base, 0 to 1 septate, 4-6 x 1.5-2 μm.[163]

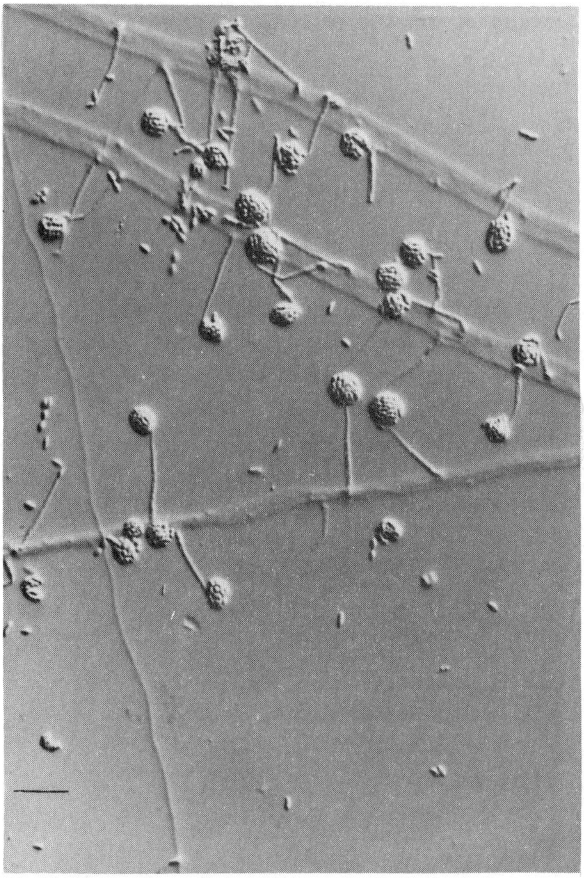

Figure 4:14. ***Acremonium kiliense*** **conidia accumulate in tight, wet balls at the apices of phialides. Bar = 20 μm.**

A. nidulans (after Raper and Fennell):[245]

Colonies are initially white, becoming dark green from conidiation, or buff to yellow from cleistothecia (*Emericella nidulans*) formation. Reverse is purplish-red, becoming dark with age. Conidiophores are brown. Vesicles are 8-10 μm in diameter. Conidial heads are biseriate, with phialides oriented upward, resulting in columnar heads. Conidia are green, globose, echinulate, 3-3.5 μm. Cleistothecia (teleomorph) are 100-200 μm, reddish-brown, accompanied by Hülle cells; asci are evanescent; ascospores are purplish-red, lenticular, 3.8-4.5 x 3.5-4 μm with 2 parallel equatorial rings.[55]

Leptosphaeria senegalensis:

Colonies are rapidly growing, downy, initially white, later greying, becoming beige-brown in the center. Margin is submerged, brown to black.

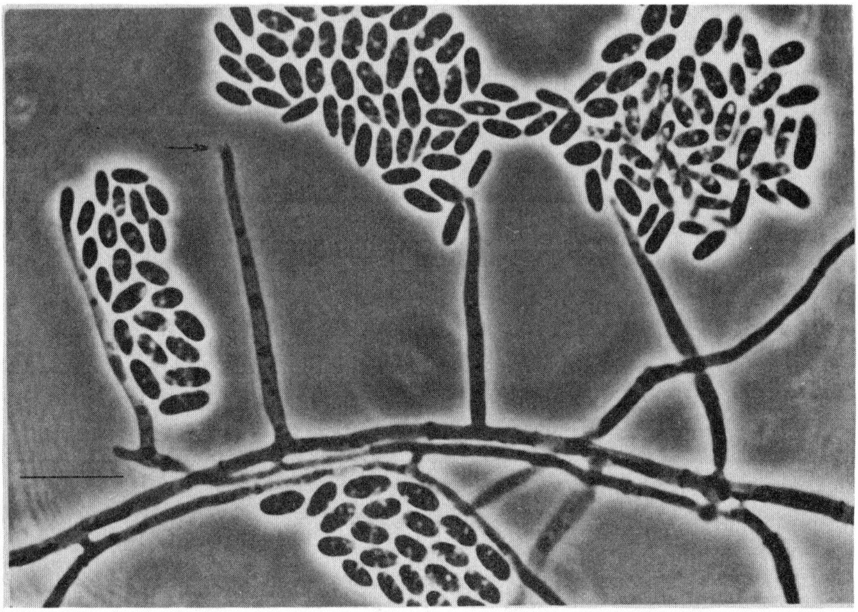

Figure 4:15. Phialides of *A. kiliense* are hyaline, long and gradually tapering (awl-shaped) and may or may not have a septum near the base. A vestigial collarette may be present (arrow). Bar = 10 μm.

Some isolates exhibit a rose tint. Upon subculture, reverse is black, with agar showing a rose tint and brown diffusible pigment. After 2 or more months on potato carrot agar, corn meal agar or similar deficient media, dark perithecia develop that are 100-300 μm in diameter, without neck or ostiole. Asci are 80-110 x 20 μm, clavate, bitunicate, containing 8 ascospores. Ascospores are usually 5-celled (up to 9 cells), 23-30 x 8-10 μm, hyaline to pale brown, fusiform. A turbinate, gelatinous sheath around each ascospore is seen when suspended in water and India ink.[78]

L. tompkinsii:

Differs from *L. senegalensis* mainly by slightly larger (32-45 x 8-11 μm) ascospores that predominantly have 7-8 cells. Ascospore sheaths are not turbinate.[77]

Madurella grisea:

Colonies are slow-growing, dark brown with grayish tints, reverse dark brown; velvety, folded and heaped, without diffusing pigment. Conidia are not present (other than two reports of pycnidia forming in isolates that may actually have been *Pyrenochaeta romeroi*).[31,184] Growth is better at 30°C than 37°C. Sucrose is assimilated; lactose assimilation is variable.[26,175,270]

M. mycetomatis:

Colonies are slow-growing, cream-colored, glabrous, folded, tough; with age becoming velvety, dark brown, staining the agar with a diffusing brown pigment. Growth is enhanced at 37°C; lactose is assimilated but sucrose is not. Colonies are sterile on routine media; conidiation is present in about 50% of stains grown on nutrient-poor media. Conidia are subglobose to pyriform, 3-5 μm, with truncate basal scar, occurring in balls, rarely in fragile chains. Phialides are variable, usually 9-11 μm long, tapering, often with collarette; but range from long (15 μm) and tubular to short (3 μm) and integrated. Occasionally 2 or 3 phialides may be borne upon a single branch. Large vesicles, terminal and intercalary, are often present. Sclerotia may be present.[1,65,174,189]

Pseudallescheria boydii (anamorph *S. apiospermum*):

Colonies are very rapid-growing, floccose, white, becoming grey with conidiation from the *S. apiospermum* anamorph, and usually becoming darker brown with age. Conidia are egg-shaped to clavate, truncate at the tapered end, subhyaline, grey to pale brown in mass. Unusual isolates have dark brown to black colonies and conidia and should be compared with *Scedosporium inflatum*. Conidia are produced singly (and remain attached to the conidiogenous cell) or multiply from annellides. Annellides are often characterized by swollen ring(s) (Figs. 4:16, 4:17), otherwise scars are not apparent. Occasional isolates also have a *Graphium* spp. synanamorph characterized by bundles of hyphae and annellidic conidiogenous cells fused into long (up to 1 mm) stalks (called synnemata) (Figs. 4:18, 4:19). Cleistothecia of the teleomorph (*P. boydii*) appear in most isolates after two or more weeks incubation, as small black dots, forming particularly within the agar or the interstices between the agar and the vessel. Cleistothecia are 100-300 μm, without a preformed opening (Fig. 4:20), splitting to release ascospores. Ascospores are pale to mid-brown, football-shaped, 4-5 x 7-9 μm (Fig. 4:21). Asci are evanescent and will not be visible in routine preparations.[35,121,194]

CHROMOBLASTOMYCOSIS

Chromoblastomycosis is usually a chronic, localized infection of cutaneous and subcutaneous tissue. It is characterized by darkly pigmented, thick-walled, muriform fungal cells known as sclerotic bodies in infected tissue. The disease is cosmopolitan, but occurs most frequently in individuals living

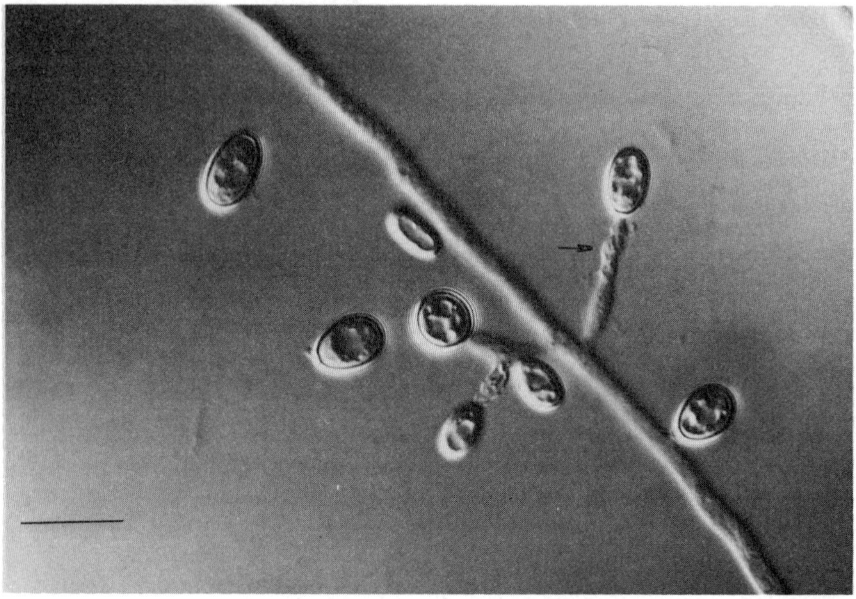

Figure 4:16. Slender annellide with swollen rings (arrow) and egg-shaped conidia with flat basal scar, typical of *Scedosporium apiospermum*. Bar = 10 μm.

in tropical and subtropical regions. Infection follows traumatic implantation of the etiologic agent. Fungi that cause chromoblastomycosis have been isolated from soil, plant material and wood.[67,100,131,227]

In 1914, Rudolph[257] described from Brazil the first cases of chromoblastomycosis. A year later Medlar[199] and Lane[167] both reported the first case in North America and were also the first to isolate an etiologic agent. Following these reports and others, the disease became known by a number of different names.[190] The name chromoblastomycosis (chromo = pigmented; blasto = budding) was proposed in 1922 by Terra et al.[288] Moore and Almeida,[206] however, felt that the term chromoblastomycosis was inappropriate and misleading because the etiologic agents do not form blastoconidia (buds) in tissue. Accordingly, in 1935 they proposed that chromoblastomycosis be renamed chromomycosis. Eventually, several more genera and species of dematiaceous fungi were reported as pathogens of humans. Unfortunately, infections caused by these fungi were also described as chromoblastomycosis even though they differed in several respects from the original concept of Terra et al. To eliminate the confusion surrounding the application of the terms chromoblastomycosis and chromomycosis, Ajello et al[11] proposed and McGinnis[190] subsequently defined the name

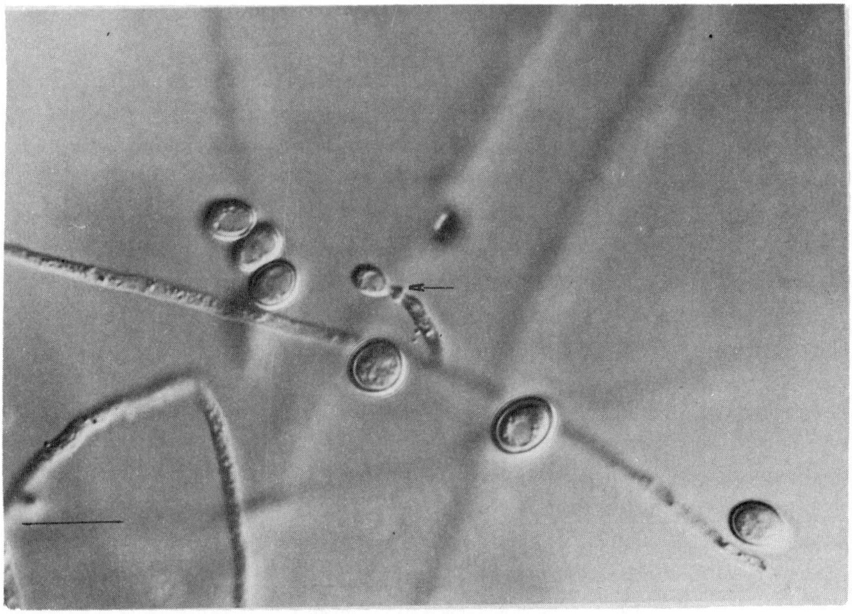

Figure 4:17. *S. apiospermum* with markedly swollen annellation (arrow) on a conidiogenous cell. These are present in at least some of the annellides in most isolates of *Scedosporium* species. Bar = 10 μm.

phaeohyphomycosis for those infections involving dematiaceous fungi that are clinically, mycologically and pathologically distinct from chromoblastomycosis sensu Terra et al.

The etiological agents of chromoblastomycosis are *Fonsecaea pedrosoi, F. compacta, Cladosporium carrionii, Phialophora verrucosa* and *Rhinocladiella aquaspersa*. Although the most common cause of chromoblastomycosis worldwide is *F. pedrosoi, P. verrucosa* is the most common in North America, and *C. carrionii* is the usual agent in Australia and South Africa.[253] Cases due to *R. aquaspersa* and *F. compacta* are extremely rare.[32,196,264] Why the predominant agent of chromoblastomycosis varies by region is not entirely understood; there is no evidence that these fungi are geographically restricted in nature. In Japan for example, Fukushiro[95] found that 86% of chromoblastomycosis infections were caused by *F. pedrosoi* and 1.6% by *P. verrucosa*, though Iwatsu et al[131] were able to obtain from natural substrates in Japan only 1 isolate of *F. pedrosoi* but 17 isolates of *P. verrucosa*.

Clinical Manifestations

Lesions of chromoblastomycosis usually begin as small, scaly, pink papules following traumatic inoculation with the fungus.[190] The disease

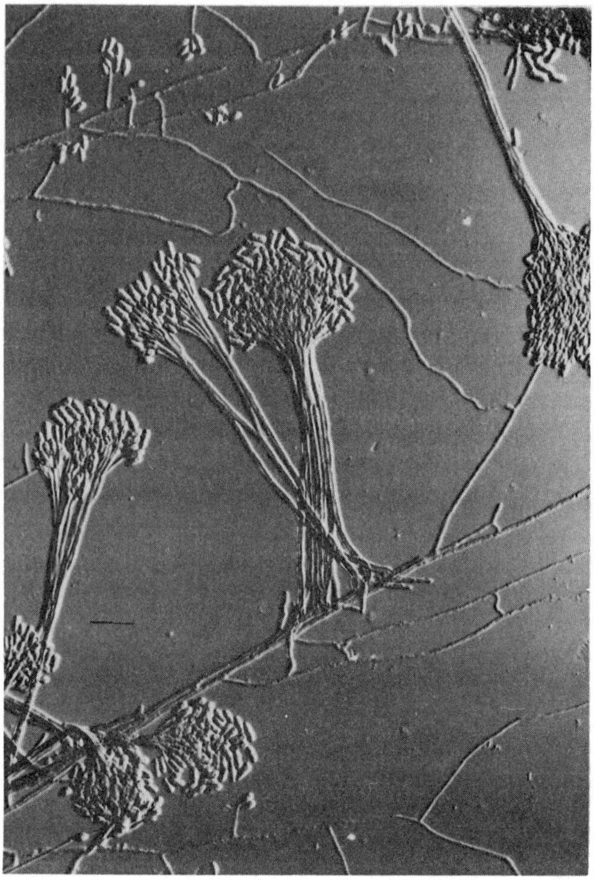

Figure 4:18. ***Graphium*** **anamorph of** ***S. apiospermum (Scedosporium*** **anamorph can be seen in left upper corner). Bar = 20 μm.**

develops most frequently on the foot or leg, though any exposed part of the body is vulnerable.[132,215] Papules may clinically resemble subcutaneous phaeohyphomycosis as they gradually increase in size to form superficial nodules. Enlarging nodules may progress to form raised irregular plaques, often becoming scaly and occasionally cracked or warty (Fig. 4:22). Large papillomatous growths may eventually develop. The lesions are painless but itching is a frequent complaint.[95] Although hematogenous spread is seen rarely,[95] satellite lesions may arise either from autoinoculation due to scratching the lesions or by spread of the fungus through the superficial lymphatics.[46] Chromoblastomycosis is a chronic, progressive and recalcitrant disease that may eventually become incapacitating. Complications include secondary bacterial infections and lymphostasis.

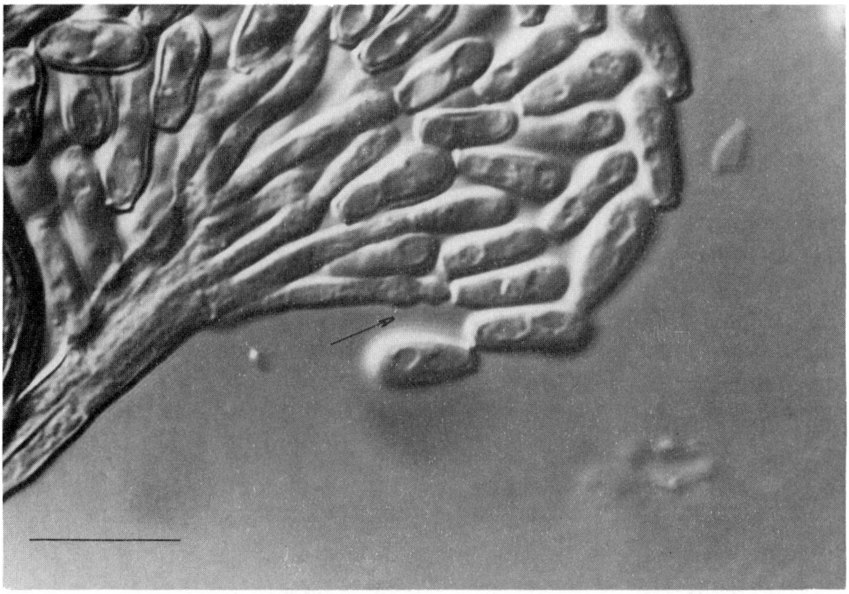

Figure 4:19. Annellides of *Graphium* spp. Note the two swollen rings (arrow) on one of the conidiogenous cells. *Phialographium* spp. differs from *Graphium* spp. by having phialides instead of annellides. Bar = 10 μm.

Histopathologically, chromoblastomycosis exhibits hyperkeratotic pseudoepitheliomatous hyperplasia with keratolytic microabscess formation in the epidermis. This response is not specific for chromoblastomycosis and may be seen in other mycoses such as blastomycosis, coccidioidomycosis and sporotrichosis. Transepithelial elimination expels damaged tissue and fungal cells through the epidermis as the lesions heal.[104] This process may result in the appearance of "black dots" on the lesion surface, which can be selectively examined for the detection of sclerotic cells. Dematiaceous, thick-walled, muriform sclerotic cells, 5-13 μm in diameter (Fig. 4:23), are the characteristic tissue form produced by all 5 agents of chromoblastomycosis. These are found within microabscesses or giant cells.[95] They do not reproduce by budding but rather by fission along the septa. Since it is not possible to determine the etiologic agent using histopathology alone, cultural studies are required.

Epidemiology and Public Health Significance

Chromoblastomycosis has been reported throughout the world. Approximately 80% of the cases studied and reported by Carrión[46] occurred in

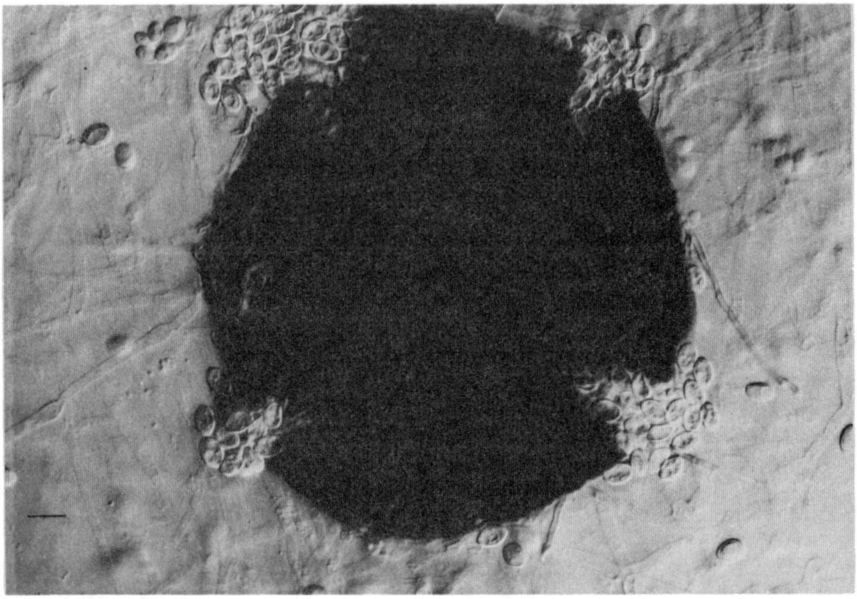

Figure 4:20. ***Pseudallescheria boydii*** **ascocarp (cleistothecium) splitting to release ascospores. Conidia from the** ***S. apiospermum*** **anamorph, visible in the background, are distinguished by their flat basal scars.**

patients living in tropical and subtropical regions. Most patients lived in rural areas and were involved in farming. There are conflicting reports concerning male to female patient ratios.[14,44,46] In contrast to several studies which concluded that most infections occur in males, Campin and Schareyj[44] found that 15 of their 34 patients were female. These differences may reflect sexual distribution among occupations in various regions, and possibly correlate with frequency of exposure to the fungus rather than physiologic predisposition to infection based upon sex.

Most patients are Caucasian, between 30 and 50 years old.[14,46,168] Blacks are the second most frequently infected race. Chromoblastomycosis has been reported in dogs, horses, toads and frogs, although many of these reports have not been well documented. The public health importance of chromoblastomycosis stems mainly from an increasing worldwide awareness of the disease. In Japan, for example, Fukushiro found 4 case reports prior to 1955. Since then, there have been 250 reports.[95] In addition, infections in immunosuppressed patients may be an emerging facet, and chromoblastomycosis has recently been seen in renal transplant patients.[301] In another case,[218] infection which developed in a patient undergoing unrelated treatment with a topical corticosteroid, resolved spontaneously when the steroid was discontinued.

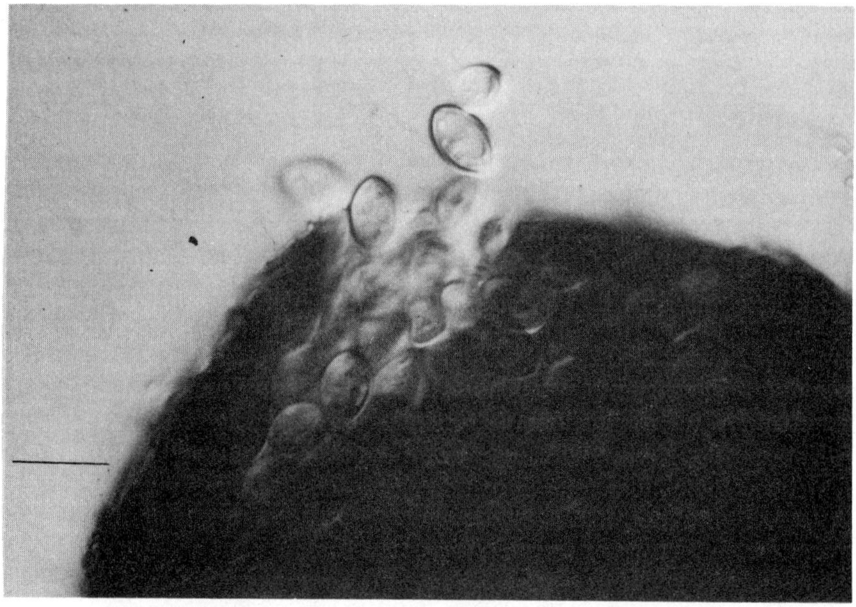

Figure 4:21. ***P. boydii.*** **Ascospores escaping from ruptured cleistothecium. In contrast to the conidia, the ascospores taper toward both ends.**

Identification of Etiologic Agents

Skin scrapings or biopsy specimens from the active border of lesions are examined in 10-20% potassium hydroxide for detection of sclerotic bodies (Fig. 4:23). In crusts from lesions, the fungi may instead appear as long, septate, dematiaceous hyphae.[46,95,168] "Black dots," which are often present on the surface of verrucous lesions should be examined for sclerotic bodies.[314] Unpublished observations have shown that the fluorescent stain, calcofluor white, is not reliable for detecting sclerotic cells. Effectiveness of other fluorescent stains is not known. Thus, when chromoblastomycosis is suspected, traditional methods of microscopic examination must be used.

All of the agents responsible for chromoblastomycosis are polymorphic, meaning they have the ability to form more than one anamorph in vitro. This may cause difficulties in their identification. The most practical approach to identifying a polymorphic fungus is to choose the most distinctive, conspicuous and stable anamorph as the basis of the name. An accompanying synanamorph which may be present can be referred to by a separate genus-level name (see "Phaeohyphomycosis: identification of etiologic agents").

Cladosporium carrionii:

Colonies are olivaceous-black, slow-growing and velvety in texture. Erect, lateral conidiophores end abruptly and bear outwardly spreading, acropetal

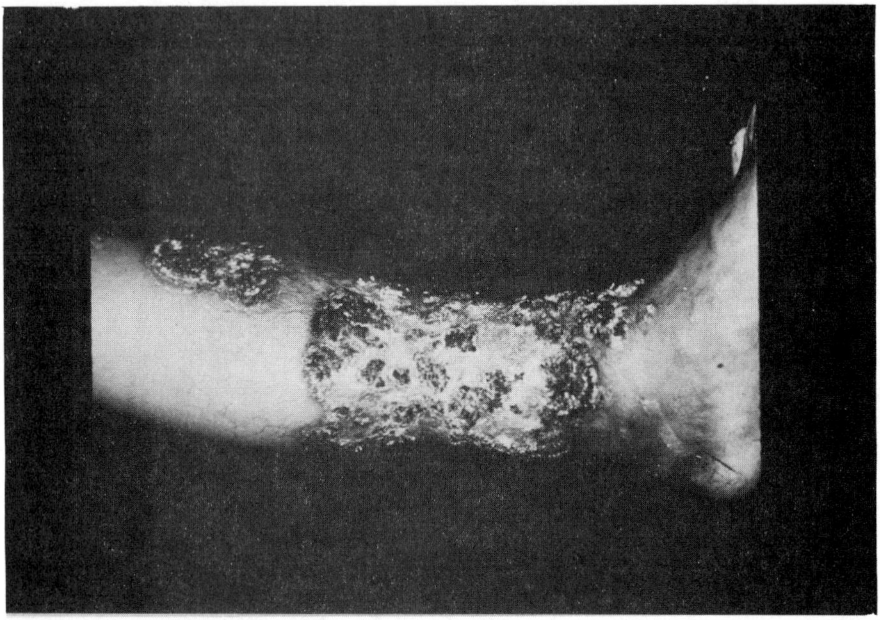

Figure 4:22. Chromoblastomycosis. Verrucous and scaly lesions, with scarring.

chains of blastoconidia (Fig. 4:24). *C. carrionii* differs from common saprobic species of *Cladosporium* by its conidial chains which branch only sparsely. Conidia are football-shaped, are more uniform in size and shape than with other species and are more tenaciously attached to each other. In contrast to most saprobic species, *C. carrionii* will liquify gelatin and grow at 37°C. A *Phialophora* synanamorph may also be formed in some strains when grown on enriched media such as lactrimel agar.[123,291]

Fonsecaea pedrosoi:

Colonies are olivaceous-black, slow-growing and velvety in texture. Conidiogenous cells are cylindrical, usually with an irregularly swollen apex that is studded with denticles where conidia were attached. Conidia are broadly clavate with a flat basal scar. Each conidium usually gives rise to 3 or more conidia in a repeating process that ultimately forms a complexly branched head of conidia (Fig. 4:25). Most conidia remain tenaciously attached even when the conidial apparatus is disrupted. Hyphae and conidia are subhyaline to brown. A *Rhinocladiella* synanamorph is usually present (Fig. 4:26). In rare isolates, a *Phialophora* synanamorph may also be present.[196]

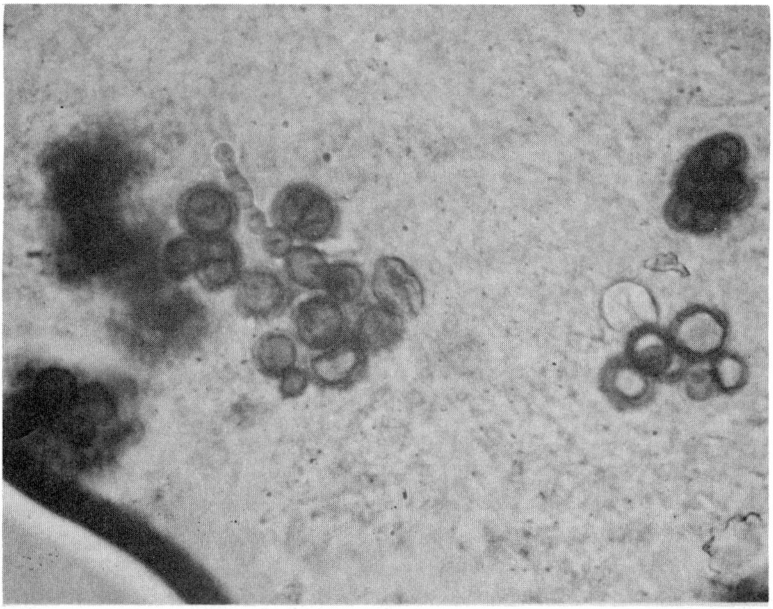

Figure 4:23. Brown, thick-walled, muriform sclerotic cells characterize chromoblastomycosis. Punch biopsy of leg, in 20% KOH.

F. compacta:

F. compacta differs from *F. pedrosoi* by conidia that are subglobose and have broad points of attachment (Fig. 4:27), conidial heads that are extremely compact and slow growing colonies that are more compact at maturity.[196]

Phialophora verrucosa:

Colonies are olivaceous-black, slow-growing and velvety in texture. Conidiogenous cells are phialides, vase-like in shape, with prominent, flaring collarettes which may be shallow or deep (Fig. 4:28). Conidia are broadly elliptical, bilaterally symmetrical, without basal scar, accumulating in balls at the tips of phialides.[57,268]

R. aquaspersa:

Colonies are olivaceous-black, velvety and slow-growing. Conidiophores are unbranched, erect, straight and tapering slightly; the conidiogenous zone is elongated due to sympodial proliferation and scarred where conidia were attached (Fig. 4:29). Conidia are elliptical to obovate, 0-(rarely 1)-septate, smooth, mid-brown, with dark basal scar. A *Wangiella* synanamorph may be present (Fig. 4:30), as may an *Exophiala* synanamorph.[264]

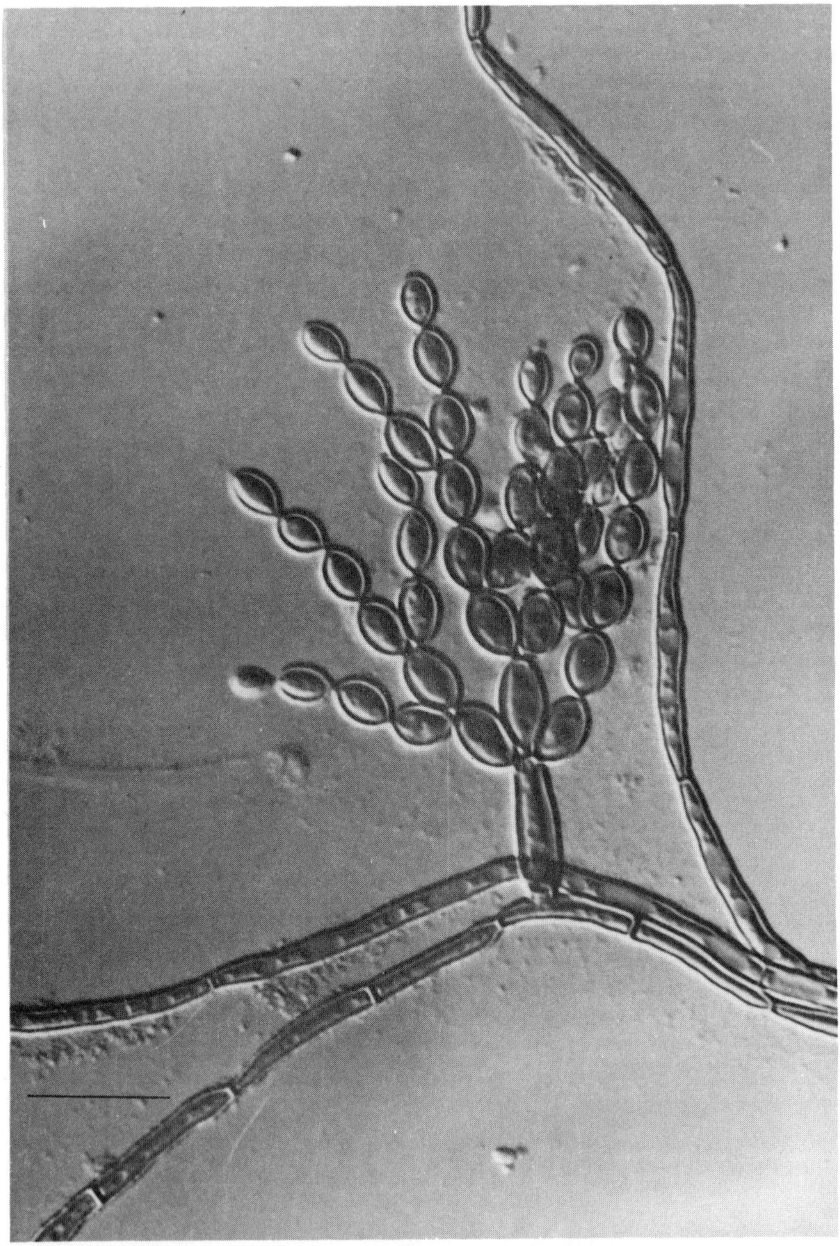

Figure 4:24. ***Cladosporium carrionii*** **is distinguished by short, erect conidiophores, sparsely branching chains of conidia, and conidia that are generally uniform in shape and size. Bar = 10 μm.**

Figure 4:25. ***Fonsecaea pedrosoi.*** **Conidia are formed in highly branched chains, usually of only 2 to 3 levels, which do not disarticulate readily (in contrast to *Cladosporium* species). Bar = 10 μm.**

PHAEOHYPHOMYCOSIS

The name phaeohyphomycosis is derived from the Greek roots phaio (dark), hyphae (web) and mykes (fungus). Phaeohyphomycosis is a collective disease name for fungal infections characterized by dematiaceous septate fungal elements growing within host tissue. The fungus may appear as regularly shaped, short to elongate hyphal elements; as distorted and swollen yeast-like elements; or a mixture of both. Phaeohyphomycosis was first described in 1907 and attributed to the fungus now known as *Exophiala jeanselmei*. A far greater awareness of the disease has developed over the last 80 years, and it is now attributed to more than 70 dematiaceous fungi.[10] The most common are species of *Exophiala, Wangiella, Bipolaris, Alternaria, Phialophora*[190,300] and *Curvularia*. Other fungi are involved less often (Table 4:3).

Some medical mycologists prefer the single term "chromomycosis" for all subcutaneous infections due to dematiaceous fungi, including phaeohyphomycosis and chromoblastomycosis. Phaeohyphomycosis, however, is a distinct disease entity and should be addressed separately from chromoblastomycosis. The two diseases have different etiologies, clinical expressions

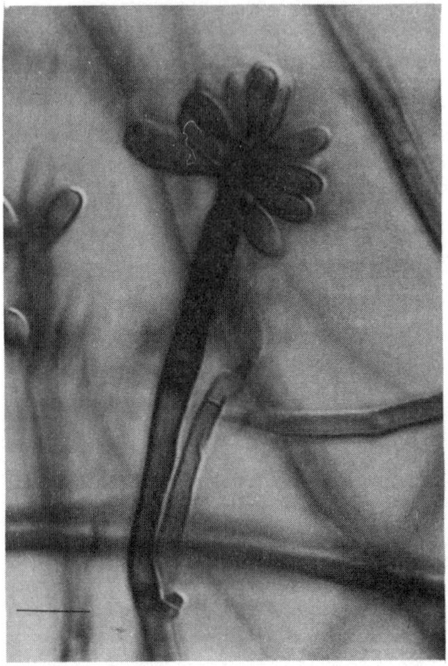

Figure 4:26. ***Rhinocladiella*** **anamorph of** ***F. pedrosoi*****, characterized by sympodial conidiogenesis. Each conidium is borne upon a denticle. Bar = 10 μm.**

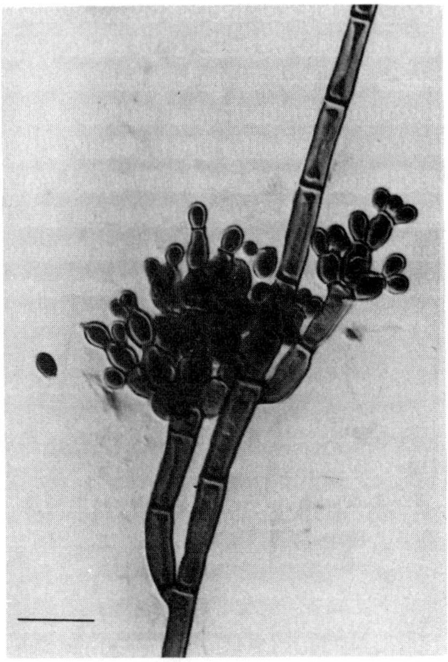

Figure 4:27. ***Fonsecaea compacta*** **differs from** ***F. pedrosoi*** **by having a more compact conidial arrangement and broader points of attachment between conidia. Bar = 20 μm.**

and tissue forms. Thus, the distinction between them is a useful one and they should not be merged under a common name.

Clinical Manifestations

The fungi which cause phaeohyphomycosis are associated with soil, wood and decaying plant material throughout the world.[67] The infection originates when the fungus is carried through a break in the epidermis, though most patients do not specifically recall any preceding injury to the affected area. Exposed extremities are the commonly involved sites though infections may originate at any body surface. In a few cases, disseminated infection was suspected to have begun in the lungs following inhalation of conidia,[4,91,205] but to date this remains an unproven possibility.

Symptoms of phaeohyphomycosis may be preceded by a long period of quiescence. Manifestations vary widely depending upon the fungus and the anatomic location, and definitive diagnosis rests upon demonstration of

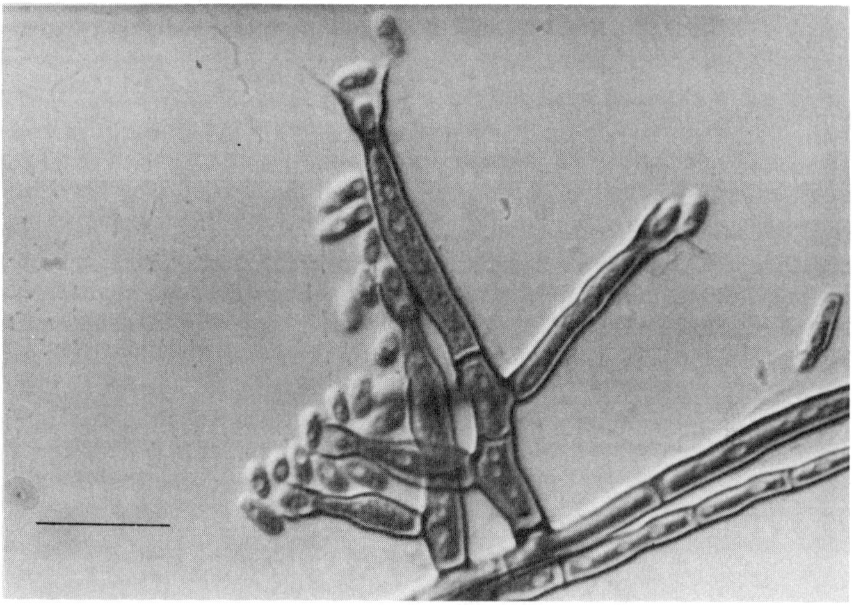

Figure 4:28. ***Phialophora verrucosa*** **is characterized by flask-shaped phialides with prominent, thick-walled, funnel-shaped collarettes. Bar = 10 μm.**

dematiaceous hyphal elements in a specimen by direct microscopic examination.

The most benign type of phaeohyphomycosis is exemplified by "black palm" (tinea nigra, Chapter 3, "Superficial and Cutaneous Infections Caused by Molds: Dermatomycoses.") in which the infection is extremely superficial, causing purely cosmetic problems. In contrast, the most grave type involves cerebral infection by the highly neurotropic fungus *Xylohypha bantiana*, which is almost always fatal.[269]

It is not yet certain whether cerebral phaeohyphomycosis caused by *X. bantiana* originates as a pulmonary or subcutaneous infection.[81,262,275,306] In one case initiated by a penetrating injury,[154] the identity of the etiologic agent as *X. bantiana* was not clearly established. The fungus may instead have been an isolate of *Wangiella dermatitidis*, which is also associated with cerebral infection.[95,116,221,274,292] However, there have been 3 confirmed reports of subcutaneous phaeohyphomycosis due to *X. bantiana*,[16,81,222] presumably initiated through traumatic inoculation. Since there are no well-documented cases of cerebral infection accompanied by subcutaneous lesions, it must be assumed that at least some cerebral infections may have a pulmonary origin.

By far the most common form of phaeohyphomycosis is chronic, localized,

Figure 4:29. ***Rhinocladiella aquaspersa.*** **Note the well developed, dark, thick-walled, denticulate, sympodial conidiogenous cell. Bar = 10 μm.**

subcutaneous infection, presenting in either of two ways. The subcutaneous cyst type develops as a result of granulomatous reaction followed by formation of a cavitary abscess. The epidermis remains unaffected. It may easily be mistaken for a ganglion, inclusion cyst or foreign body granuloma. The cyst wall is composed of a granulomatous inner layer and an outer layer of fibrous, connective tissue. The cyst, typically fluctuant, may be solitary or there may be an aggregation of cysts (Figs. 4:31, 4:32).[88,130,204,313] The cyst cavity is filled with pus and dematiaceous hyphal elements (Figs. 4:33, 4:34). Occasionally, an inoculation splinter can be found within the cyst.[130,293,315] In cases where cysts do not develop, the lesions may be painless, firm, solitary nodules without epidermal involvement (Fig. 4:35),[202] draining abscesses or ulcers,[195] or raised crusty or verrucose plaques (Fig. 4:36).[120,276]

In most cases of phaeohyphomycosis, the infection remains localized, but osteomyelitis,[12,309,316] arthritis[140,170] or dissemination[97,230,282] may occur.

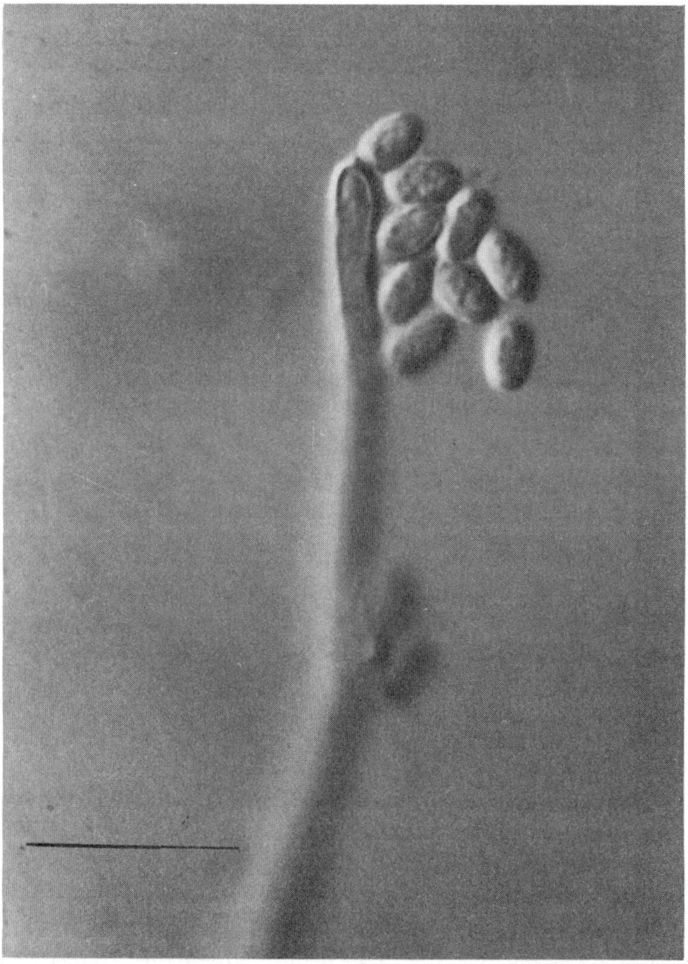

Figure 4:30. ***Wangiella*** **anamorph of** ***R. aquaspersa*****, characterized by phialides that lack collarettes. Bar = 10 μm.**

Dissemination is seen most often with *W. dermatitidis*,[95,221,243] *Bipolaris spicifera*[4,208] and *Curvularia*.[237,251]

Phaeohyphomycotic sinusitis begins when conidia are inhaled into the nasal passages, and for reasons which are unclear, germinate and grow into a tightly packed mass of hyphae. Most patients have been found to be immunocompetent. The large majority of cases are caused by species of *Curvularia* and *Bipolaris*.[4,117,231,251,256,312] Less often, fungi such as *Scedosporium (P. boydii)*,[39,56,252,253] which are normally regarded as hyaline (see "Hyalohyphomycosis") but which may become plainly dematiaceous with

Table 4:3. Significance of Laboratory Findings: Less Common Subcutaneous Pathogens*

Fungus	Disease†	Documented Sites of Infection	Key Ref.
Acremonium falciforme	m	subcutaneous	45,47,87, 115
A. kiliense	h,m	subcutaneous, bone, joints, pleural fluid	36,47,69, 87,115 302
A. recifei	m	subcutaneous	162,163
Aspergillus nidulans	m	subcutaneous	138,151
Exophiala spinifera	p	subcutaneous, bone	230,232, 282
E. moniliae	p	subcutaneous	181,198
Curvularia geniculata‡	m,p	subcutaneous, aortic valve	38,251
C. lunata‡	m,p	subcutaneous, lymph node, pleural fluid, lung, sinuses	151,176 251
Exserohilum rostrata‡	p	valve prosthesis, subcutaneous, sinus, bone	4,12 195,203
Lecythophora spp.	p	subcutaneous, endocarditis, peritonitis	8,250
Phialophora parasitica	p	subcutaneous, knee joint	11,140, 170,316
P. richardsiae	p	subcutaneous, bone, joint	63,209, 290,309
Pyrenochaeta spp.	m	subcutaneous	30,247
Rhinocladiella spp.	c,p	subcutaneous	32,254,264
Scedosporium inflatum	h	bone, joint, subcutaneous	259
Schizophyllum commune	h	subcutaneous, sinus, CSF	22,149 253

* The rarest pathogens (≤ 2 reports each) and cornea infections are excluded.
† c = chromoblastomycosis; h = hyalohyphomycosis; m = mycetoma; p = phaeohyphomycosis.
‡ Some of these infections became disseminated.

age, and *Exserohilum*[4,231] may be involved. Symptoms include congestion, pain, discharge, polyps and unresponsiveness to antibacterial therapy. Sinusitis is normally limited to colonization, but invasion of bone has been documented.[251,312]

Epidemiology and Public Health

The epidemiology of phaeohyphomycosis has important implications for modern medicine. With the exception of cases of keratitis,[263] sinusitis and cerebral phaeohyphomycosis, many if not most patients are immunocompromised or have some other underlying health problem. Nosocomial infections have also been documented. At least two cases of cutaneous

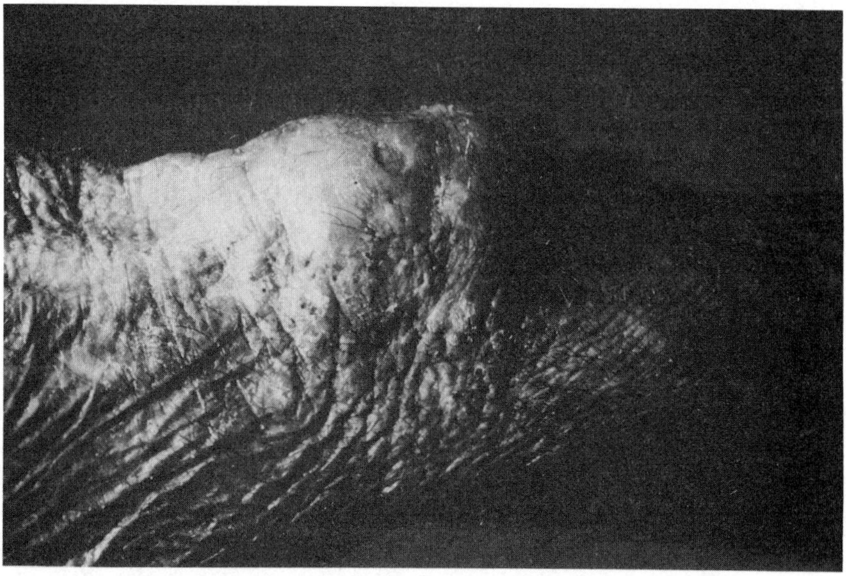

Figure 4:31. Phaeohyphomycosis of the wrist. This infection with *Exophiala jeanselmei* developed as multiple cysts. The epidermis was not effected by the fungus.

phaeohyphomycosis developed after skin was traumatized by adhesive tape.[86,203] Also noteworthy are the findings that several agents of phaeohyphomycosis are readily isolated from water, including bath water, waste water[220] and the reservoir of a humidifier.[219]

Chronic sinusitis due to dematiaceous fungi appears to be an underrecognized but increasingly reported phenomenon. In addition, the widespread use of soft contact lenses and the recent advent of extended-wear contact lenses may also foster a new syndrome of infection by dematiaceous fungi.[307] Continuous ambulatory peritoneal dialysis, a relatively new approach to dialysis, has been accompanied by several cases of peritonitis due to dematiaceous fungi.[4,66,228,256] Phaeohyphomycosis as a risk of intravenous drug use has been reported, e.g., Vartian et al[297] recently described endocarditis with subsequent spinal infection due to *W. dermatitidis* in a heroin addict.

Phaeohyphomycosis is clearly one of the most rapidly changing of the mycoses—the list of etiologic agents has grown steadily and now exceeds 70 fungi. Thus, we should expect a rising incidence of phaeohyphomycosis in the future. Clinicians and laboratorians need to develop an awareness of phaeohyphomycosis and become familiar with its etiologic agents.

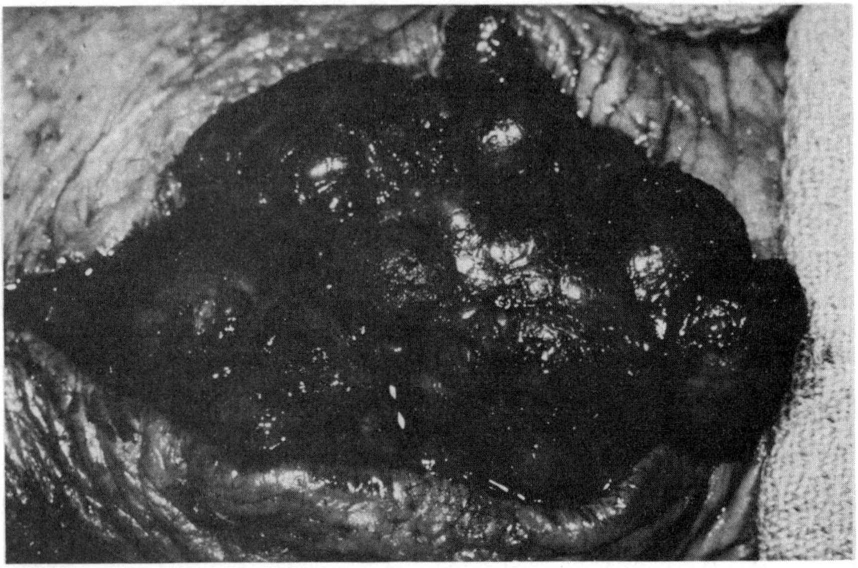

Figure 4:32. Surgical excision of the cysts shown in Fig. 4:31.

Identification of Etiologic Agents

Whenever fungal elements are seen during microscopic examination of any specimen, whether a subcutaneous one or not, they should be reviewed with brightfield microscopy to determine whether the cell walls have any degree of brown pigmentation. To do this reliably, it is imperative that the microscope illumination be adjusted correctly. This is easily accomplished using the Koehler method[2] (see Chapter 18, "Care and Use of the Microscope"). Phase contrast optics or the use of a fluorescent stain (e.g., calcofluor white) will not reveal the natural pigment of the fungus.

In vivo, the most common agent of phaeohyphomycosis, *E. jeanselmei*, is always dematiaceous to some degree. In contrast, other agents are less obviously pigmented and some such as *Curvularia, Bipolaris* and *Alternaria* are usually colorless. These could arguably be considered agents of hyalohyphomycosis, but since these fungi are plainly dematiaceous in vitro, and occasionally so in vivo, the infections they cause are included under phaeohyphomycosis. A careful search of the specimen may be required before the dematiaceous nature of the hyphae becomes apparent. It is important when reviewing histological sections to look at hematoxylin and eosin stained material, and if necessary, unstained sections. The shape of fungal elements is variable (Figs. 4:32, 4:33). The hyphae often are irregularly swollen and may appear to bud. They may even be confused with the yeast

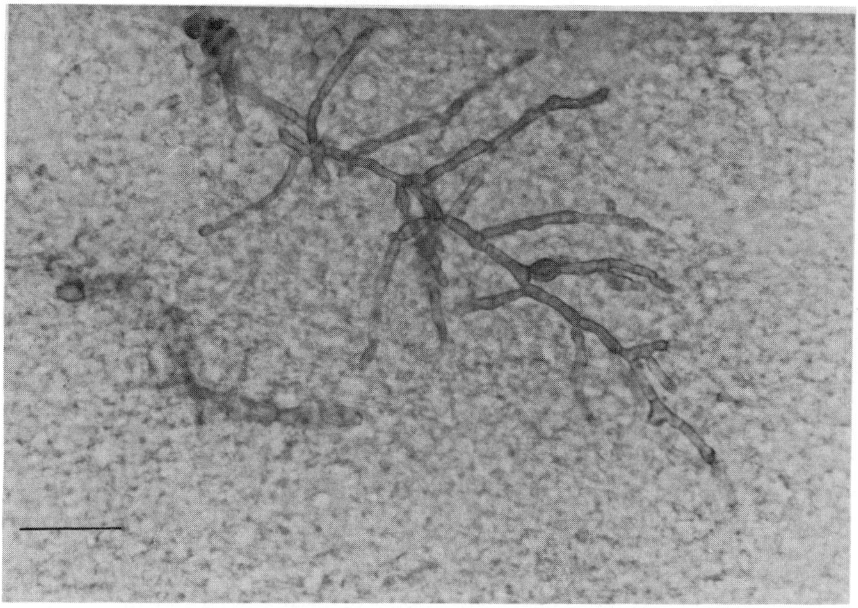

Figure 4:33. KOH preparation of aspirate from cyst shown in Fig. 4:31 revealed brown hyphae typical of phaeohyphomycosis. Bar = 10 μm.

and pseudohyphal forms seen in *Candida* if their dematiaceous nature is not detected.

The in vitro identification process for certain dematiaceous fungi is difficult and controversial. This is due to several factors: 1) the International Code of Botanical Nomenclature which governs fungal nomenclature and influences taxonomic decisions in mycology is written in a way that allows for, rather than discourages, contrasting opinions concerning what name to apply to a fungal anamorph. There is no ruling body to impose a binding decision upon disagreeing taxonomists; 2) Most of these fungi do not have a sexual phase in their life cycles, so mating studies that might resolve questions of relatedness are not possible. Therefore, identification of a mold rests mainly on morphological observations of asexual structures; 3) Many of these fungi are polymorphic, i.e., they have more than one conidial morphology or they have more than one mode of conidiogensis (or both). Since each of these synanamorphs of a fungus may be referred to by a separate name, the nomenclature may become quite confusing to many users.

An early approach to naming polymorphic fungi based the name (i.e., the identification) on whatever combination of anamorphs happened to be present in an isolate. Thus, various combinations of anamorphs which might

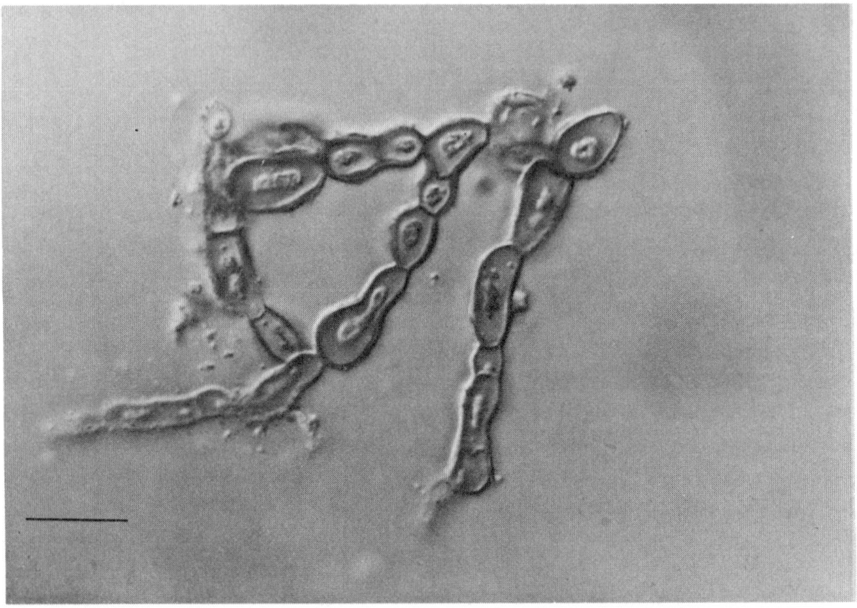

Figure 4:34. Moniliform dematiaceous hyphae, sometimes with budding forms, is also typical of phaeohyphomycosis. KOH preparation (same specimen as Fig. 4:33). Bar = 10 μm.

be present in separate isolates of the same fungal species could result in different identifications. Such an approach is obviously unwieldy and a simpler, more practical scheme is to base fungal names on the single anamorph that is judged to be the most distinctive, conspicuous and stable.[196] Any accompanying synanamorph(s) can then be referred to using a genus-level name. For example, an isolate of *F. pedrosoi* may exhibit a *Phialophora* synanamorph as well.

Recent research into alternative approaches to fungal identification has been promising. The exoantigen technique has been used experimentally to distinguish *X. bantiana, F. pedrosoi, P. verrucosa, W. dermatitidis, E. jeanselmei, C. carrionii* and other *Cladosporium* species.[84,85,122] Unfortunately, these are currently limited in usefulness, due to non-availability of commercial reagents.

Nitrate assimilation is an extremely helpful and very rapid test for distinguishing *W. dermatitidis* from species of *Exophiala*.[260,284] Temperature tolerance tests,[233] and to a lesser degree, gelatin liquefaction tests[106,130,182,229] are also helpful. Gene sequencing is a new approach that may ultimately resolve many of the taxonomic issues concerning the dematiaceous fungi.

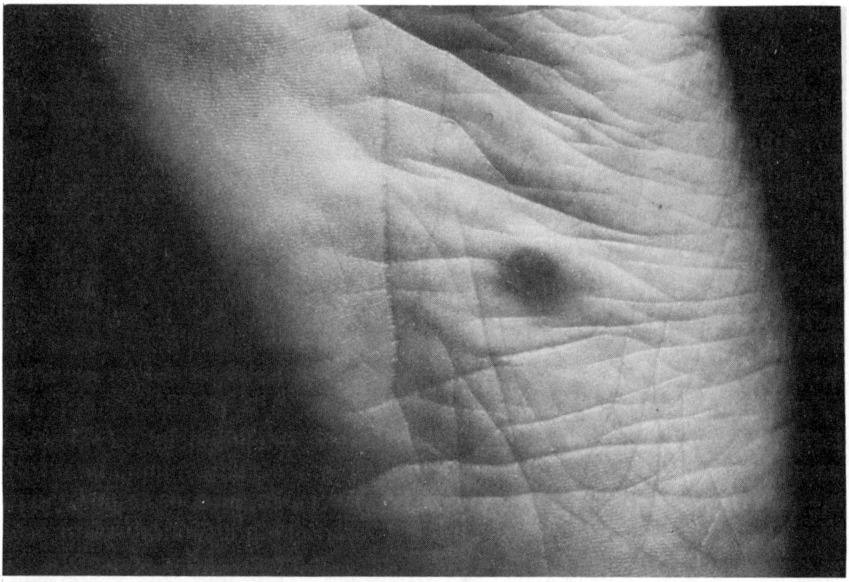

Figure 4:35. Early stage of phaeohyphomycosis. Violaceous, subcutaneous nodule, non-cyst type (*E. jeanselmei*).

Cultural and microscopic characteristics vary greatly according to individual fungi. *Curvularia* and *Bipolaris*, for example, are rapidly growing molds which are hyaline at first, but soon become mid to dark brown. *E. jeanselmei* is slower growing and usually appears first in its yeast form as a brown and pasty colony. This yeastlike form, designated as the *Phaeoanellomyces* anamorph by McGinnis and Schell,[197] reproduces by an annellidic mechanism (Fig. 4:38). With continued incubation, the colony progressively develops hyphae and takes on the appearance of a mold as the *Exophiala* anamorph (also annellidic) becomes predominant (Fig. 4:39). Thus, it is essential to study an unfamiliar isolate carefully, examining younger as well as older colonial and microscopic characteristics. Many microscopic fields should be examined, making note of the extreme as well as the predominant features. Teased preparations as well as slide cultures should be examined. Only in such a way can reliable conclusions be reached.

Bipolaris spicifera (after Ellis):[80]

Colonies are rapidly growing, floccose, grey, becoming dark brown. Conidiogenous cells are sympodial, geniculate. Conidia are broadly elliptical to cylindrical, 20-40 x 9-14 μg, consistantly 3 pseudoseptate, with a very small hyaline to pale zone just above the hilum (Fig. 4:37). The hilum is scarcely protuberant which helps distinguish it from *Exserohilum*.[195]

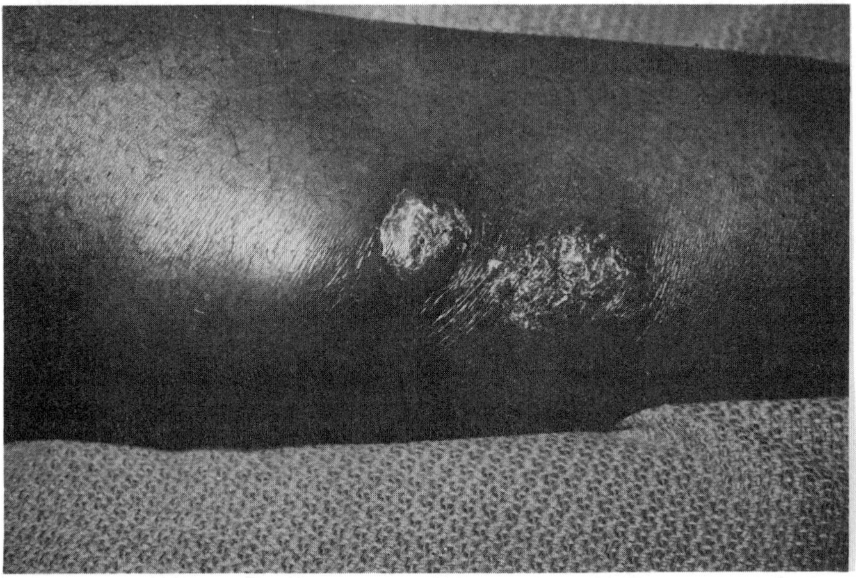

Figure 4:36. Phaeohyphomycosis involving epidermis as well as subcutaneous tissue (*E. jeanselmei*).

Curvularia lunata (after Ellis):[79]

Colonies are rapidly growing, floccose, hyaline at first, becoming mid-brown to blackish. Conidiophores are septate, lateral and terminal, pale brown to brown, sympodial in a geniculate fashion. Conidia are 18-32 (mean 23.5) x 8-16 (mean 12.5) μm, mostly 4-celled; septa sometimes of unequal thickness and darkness. Penultimate cell is curved, and is larger and darker than others; end cells are pale (Fig. 4:38).

C. geniculata (after Ellis):[79]

Differs from *C. lunata* in that colonies are grey to brownish-black; conidia are mostly 5-celled, longer and narrower, 18-37 (mean 28) x 8-14 (mean 10.3) μm.

Exophiala jeanselmei:

Colonies are dark brown to olivaceous-black, slow growing, initially yeastlike, forming conidia from yeastlike annellides (the *Phaeoannellomyces* synanamorph) (Fig. 4:39).[197] Colonies become velvety as hyphae develop. Hyphal annellides are lateral, often arising at nearly right angles, characterized by an irregular, elongating tip consisting of accumulated conidial

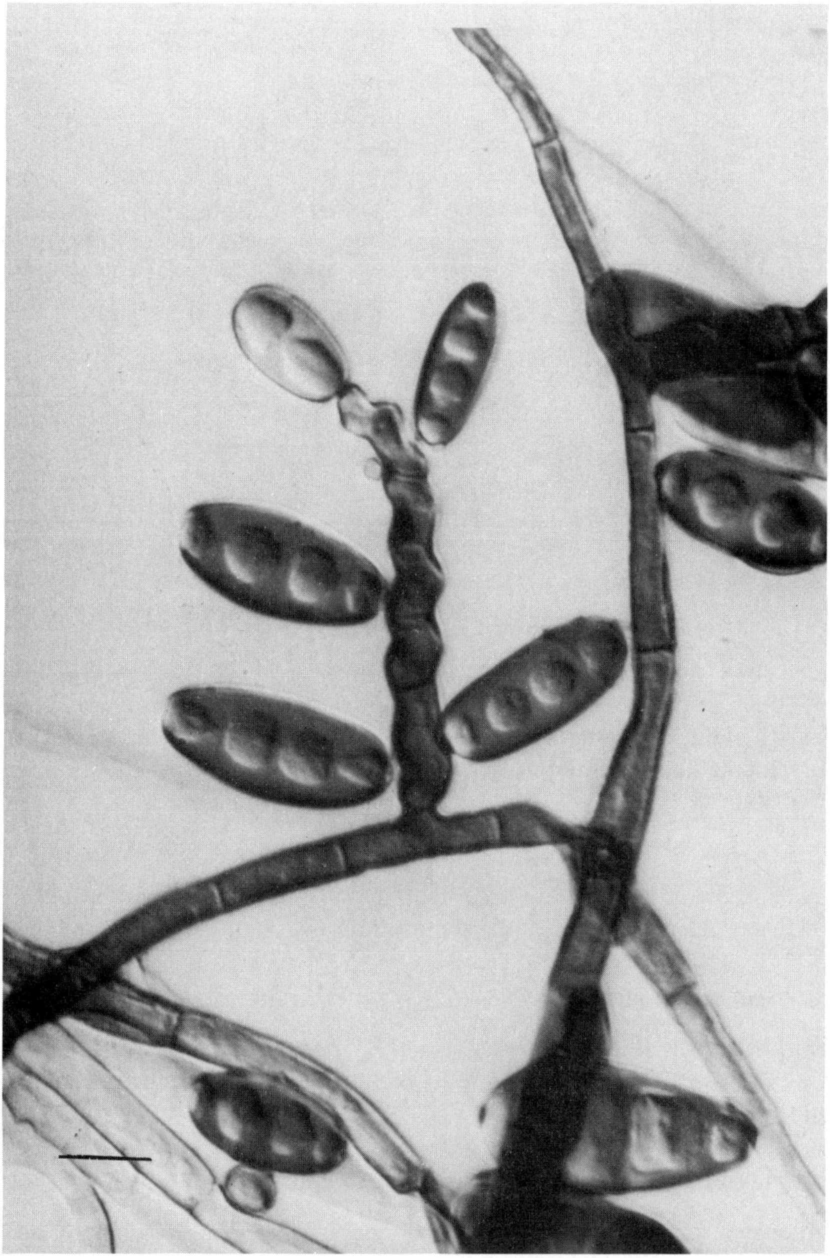

Figure 4:37. ***Bipolaris spicifera*****. Conidiogenous cell is sympodial, geniculate and dematiaceous. Conidia are pseudoseptate. Bar = 10 μm.**

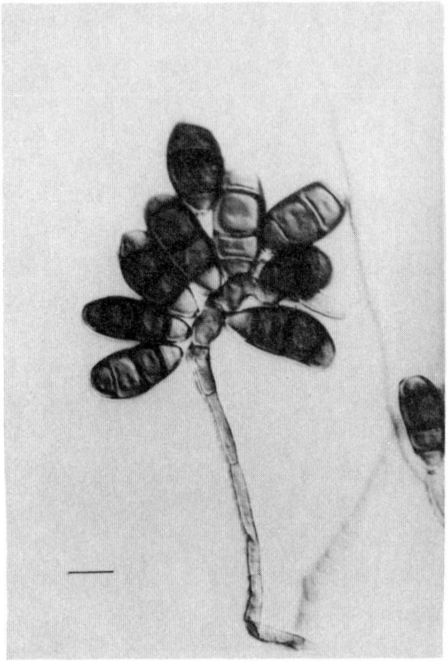

Figure 4:38. Conidiogenous cell is sympodial, geniculate and dematiaceous. Conidia have true septa; end cells are pale and penultimate cell is curved. Bar = 10 μm.

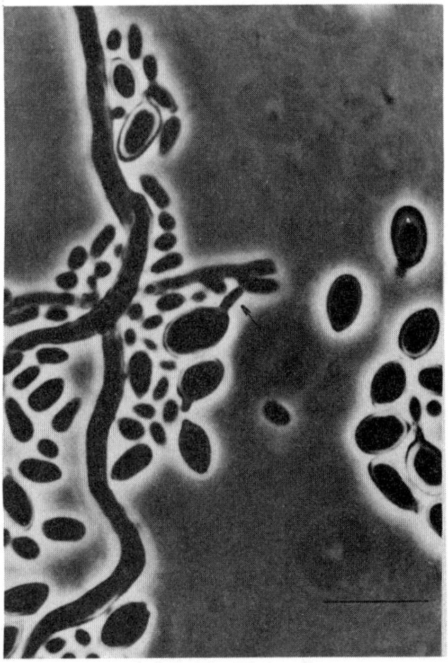

Figure 4:39. *Phaeoannellomyces* (yeast) anamorph of *E. jeanselmei*. The buildup of annellations (arrow) is apparent on several of the yeastlike cells. Bar = 10 μm.

scars (Fig. 4:40). Conidia are broadly elliptical, 3-5 μm, without basal scar. Growth occurs up to 37°C. Potassium nitrate is assimilated.[188]

E. jeanselmei var. *lecani-cornii*:

This organism is distinguished from *E. jeanselmei* var. *jeanselmei* by having a predominance of highly reduced annellidic loci which arise directly from the hyphae (Fig. 4:41).[125]

Phialophora parasitica:

Colonies are cream-colored when young, but a greyish-brown color soon develops in patches until the entire colony finally becomes brown. The margin may remain cream-colored. The reverse of the colony is light in color. Individual hyphae are mostly colorless. Phialides are quite variable in length and are long and tapering (awl-shaped). In some strains, phialides often arise from hyphal coils (Fig. 4:42). Collarettes are small and incon-

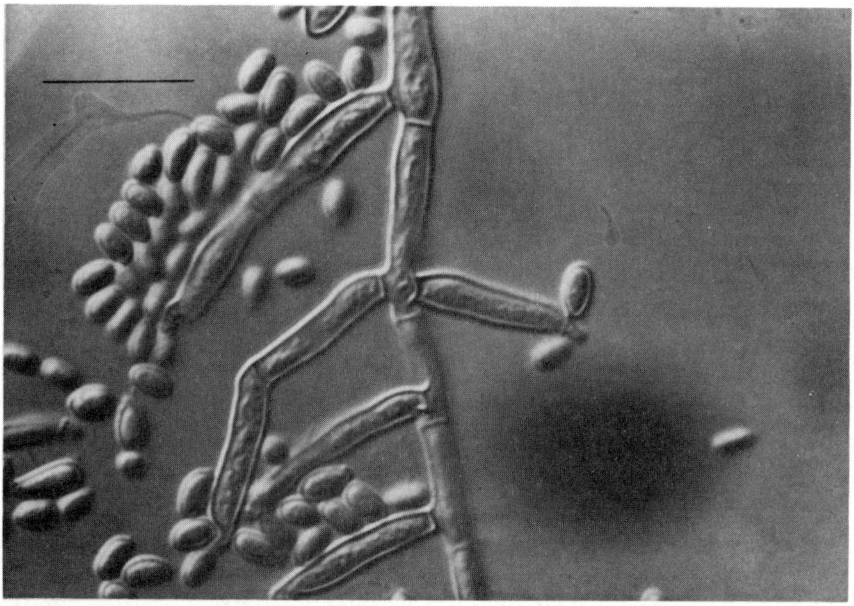

Figure 4:40. ***E. jeanselmei*. Several annellides are shown. Though individual scars are usually not discernible, their accumulation is manifest as a narrow stub, irregular in outline, at the tip of the annellide. Bar = 10 μm.**

spicuous. Long conidiophores (phialides with basal cell), which are often encrusted at the base (Fig. 4:43) are diagnostic for the fungus. Conidia are elliptical to cylindrical and often are curved. A limited yeast form which may be present is characterized by a long tubular extension from which conidia arise by a mechanism which has not yet been elucidated. Teased preparations as well as slide cultures are usually required to demonstrate all characteristics. Physiologic reactions are not helpful.[11,118]

P. richardsiae:

Colonies are light brown, becoming darker brown with age. Phialides are of two kinds: cylindric and tapering, with an inconspicuous collarette, giving rise to hyaline, ellipsoidal to slightly curved conidia; or with thick, conspicuous, flattened "saucer-shaped" collarette, giving rise to globose to subglobose conidia (Fig. 4:44). The phialides with distinctive saucer-shaped collarettes and the round conidia, are the basis for identifying this species.[57]

Wangiella. dermatitidis:

Colonies are dark brown to olivaceous-black, slow-growing, initially yeastlike, sometimes highly viscous; occasional isolates, particularly from

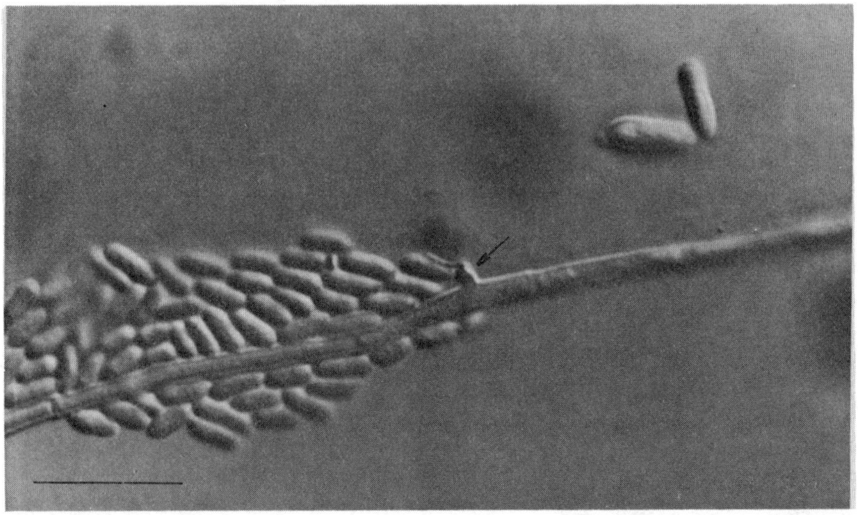

Figure 4:41. *E. jeanselmei* var *lecani-cornii* is characterized by the general lack of erect, lateral annellides. Instead, vegetative cells of the hypha themselves begin to function as annellides and the accumulation of scars arises from the main hyphal axis (arrow). Bar = 10 μm.

Japan[95] may be granular and have been described as mulberry-like in appearance. The yeastlike anamorph forms conidia by holoblastic multilateral budding (Fig. 4:45). Isolates become more or less velvety depending on the amount of aerial hyphae developed by the particular isolate. The identification of *W. dermatitidis* rests upon conidia arising from simple phialides or polyphialides (Fig. 4:46) which form laterally or terminally on the hyphae. The phialides do not have collarettes. Conidia are broadly elliptical, averaging 2.7-3.8 μm, and do not have a basal scar. Growth occurs at 40°C except for rare isolates and gelatin is not liquified.[182] Potassium nitrate is not assimilated.[260] Annellides may be seen if an *Exophiala* anamorph is present. Occasional isolates will also exhibit a *Phialophora* anamorph. These accompanying synanamorphs are irrelevant to an identification of *Wangiella* spp.[182,187]

X. bantiana:

Colonies are olivaceous-black and floccose; conidia occur in long, infrequently branched chains which do not disarticulate easily. Conidia are not well differentiated from vegetative hyphal cells (Fig. 4:47) and hila, if present, are inconspicuous. Conidia are light brown and usually average 3 x 6.4 μm.[193] Isolates grow at temperatures as high as 42°C. Since infection may possibly be acquired by inhalation, patients need have no alteration of

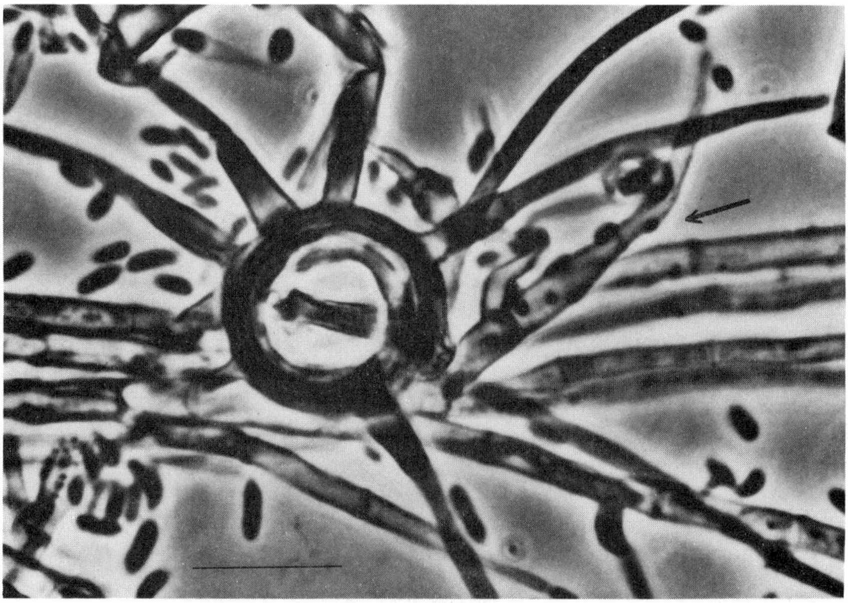

Figure 4:42. ***Phialophora parasitica*. Phialides often arise from hyphal coils. Encrustations are usually present at the base of at least some of the conidiophores (arrow). Bar = 10 μm.**

the immune response and infections are almost always fatal, *X. bantiana* cultures should be manipulated only within a biological safety cabinet.[192,193]

HYALOHYPHOMYCOSIS

In addition to reports of the more common subcutaneous infections caused by dematiaceous fungi, an increasing number of reports concern opportunistic molds which are not dematiaceous. These fungi grow in tissue as hyaline hyphal elements that are septate, branched or unbranched, and occasionally toruloid (Figs. 4:48, 4:49). Recently,[191] the term hyalohyphomycosis (hyalo: hyaline; hypho: hyphal) was proposed to refer to this heterogeneous group of infections in a concise and unambiguous manner. Hyalohyphomycosis was conceived as a complementary term to phaeohyphomycosis and, although it is not intended to replace such well established names as aspergillosis, infections caused by species of *Aspergillus* are readily accommodated by the umbrella term hyalohyphomycosis.

Hyalohyphomycosis usually involves subcutaneous soft tissue. Less common manifestations include dermal infection,[24,58] osteomyelitis,[34,36,64] arthritis,[302] fungemia,[52,153] endocarditis,[21,114,185,200,295] endophthalmitis,[172,201]

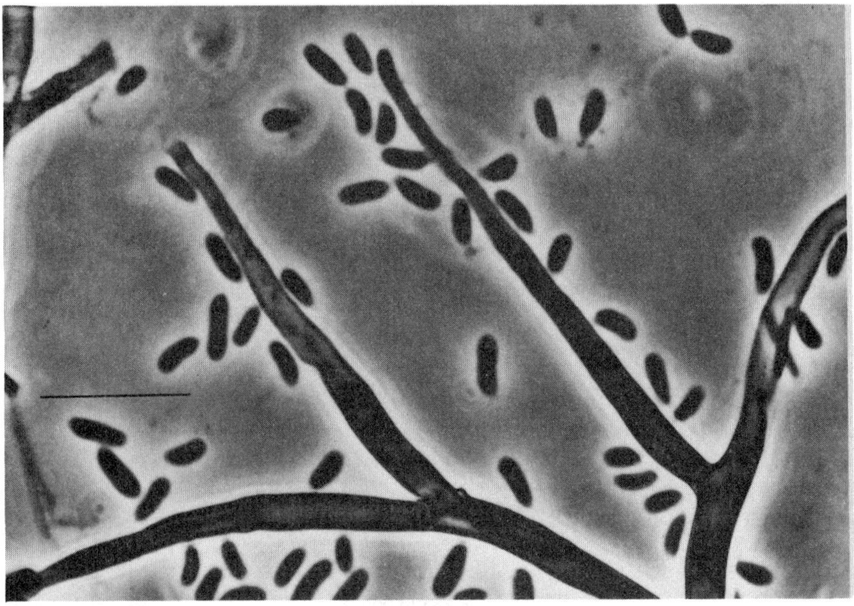

Figure 4:43. ***P. parasitica.*** **Long, tapering conidiophores (phialide with supporting cell), thin-walled and slightly flaring collarette and curved conidia are typical of this species. Bar = 10 μm.**

cerebral infection,[3,15,99,285,310] peritonitis,[76,150] and laryngeal infection.[89,152,244] As with other subcutaneous mycoses, infection begins when trauma (or surgery) allows the fungus a portal of entry. The fungi most frequently involved are species of *Aspergillus*, particularly of the *A. fumigatus, A. flavus* and *A. terreus* groups; species of *Fusarium*, particularly *F. solani* and *F. oxysporum; P. boydii* (i.e., infections that have not progressed to mycetoma formation), *S. apiospermum* and *S. inflatum*; and species of *Paecilomyces*, particularly *P. lilacinus* and *P. variotii*. Other fungi which are involved less frequently in hyalohyphomycosis are described in Table 4:3.

Clinical Manifestations

Symptoms of hyalohyphomycosis vary and are similar to those of phaeohyphomycosis. In subcutaneous infections, the affected area most often presents as a pustular nodule or papule with reddish to purplish coloration. Other cases are characterized by abscesses that ulcerate and may persist for many months or even years.[20,311] A hyalohyphomycotic abscess may also mimic a lipoma and hemorrhagic bullae which proceed to

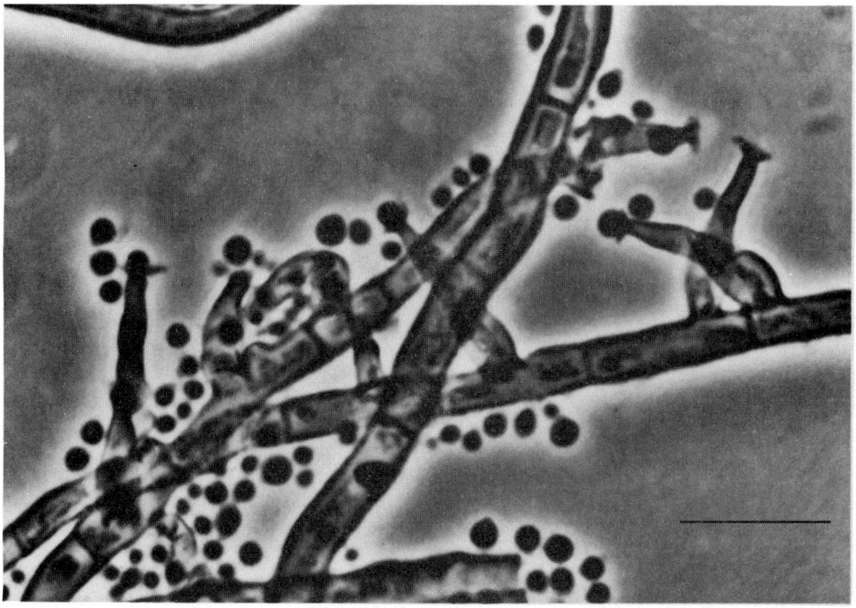

Figure 4:44. ***Phialophora richardsiae*** **is characterized by phialides with flattened, saucer-shaped collarettes and globose conidia. Bar = 10 μm.**

ulcerate may also be seen.[111] Rarely, hyalohyphomycosis can be disseminated.[3,112,180,213] As with phaeohyphomycosis, agents of hyalohyphomycosis cannot be reliably distinguished on clinical manifestations. Recently, in a diabetic patient, *Fusarium* infection of the nasal septum mimicked rhinocerebral zygomycosis. Culture results and careful review of the histologic sections correctly established the diagnosis.[296]

Epidemiology and Public Health Significance

With the exception of non-mycetoma infections due to *P. boydii*,[171,253] and infections acquired as surgical complications, nearly all patients who develop hyalohyphomycosis are, or are suspected to be, immunocompromised. *Fusarium* is rapidly becoming a leading pathogenic mold in such patients. Subcutaneous *Fusarium* infections have been seen in burn,[3,304] bone marrow transplant,[139,213] renal transplant,[311] cancer,[17,52,153,180,239,298] aplastic anemia,[112] diabetic[236,296] and other patients.[20,24,305]

Of particular interest is the rate of *Fusarium* infections among dialysis patients. In their review of 88 patients, Eisenberg et al[76] found that 7% developed *Fusarium* peritonitis and that *Fusarium* was the mold most commonly involved in peritonitis. Peritoneal dialysis is the treatment of

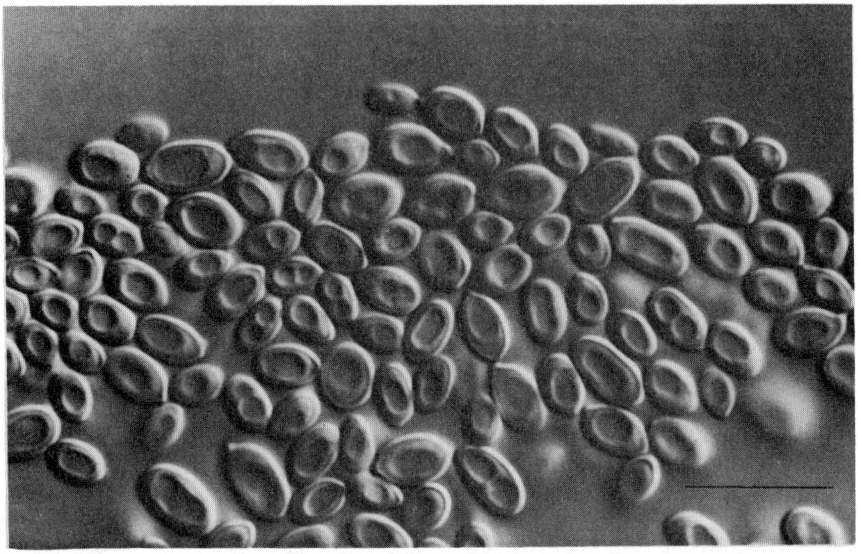

Figure 4:45. Yeast anamorph (*Phaeococcomyces*) of *Wangiella dermatitidis*. Note that conidia are formed by holoblastic budding rather than by annellides as in the yeast anamorph (*Phaeoannellomyces*) of *E. jeanselmei* (Fig. 4:39).

choice for end-stage renal disease and continuous ambulatory peritoneal dialysis is a widely used technique. Since the number of dialysis patients is increasing, the number of infections due to *Fusarium* and other molds will inevitably increase as well.

Another newly emerging pathogen is *S. inflatum*.[177] Salkin et al. recently reviewed 17 isolates, 10 of which were from subcutaneous infection in humans. At least 7 of these involved bone or joint infections.[259]

Leukemics,[59,71,105,111,277] transplant patients[43] and neonates[108,248] are also at increased risk of hyalohyphomycosis due to *Aspergillus* spp.[249] Only rarely have healthy people acquired subcutaneous infections with *Aspergillus* or *Fusarium* spp. In some cases, the patients have the usual history of penetrating trauma,[34,64,102,200,253] but in others the portal of entry is unknown.[137,146,172,214]

Since the fungi which cause hyalohyphomycosis are ubiquitous, the main epidemiological concern is to eliminate or reduce the presence of these fungi in areas where immunocompromised patients are treated. In addition to routine precautions against airborne fungi, care must be taken with supplies used in the treatment of these patients. Grossman et al.[111] recently traced the source of an outbreak of primary cutaneous aspergillosis in 6 leukemic children to contaminated adhesive tape and arm board covers from a storeroom which had suffered water damage to the suspended ceiling.

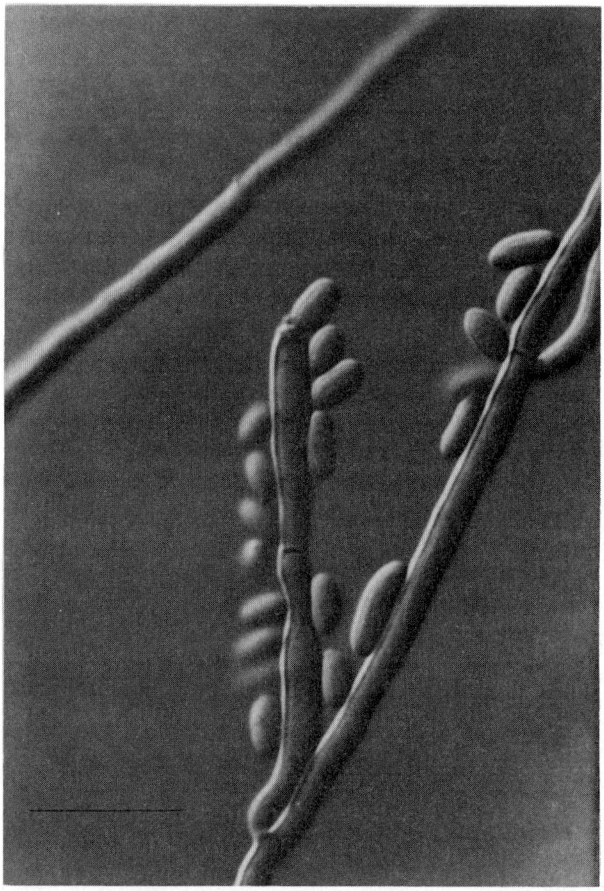

Figure 4:46. *W. dermatitidis*. Conidiogenesis is via phialides that do not have collarettes. Other anamorphs which may be present include *Exophiala* and, more rarely, *Phialophora*. Bar = 10 μm.

All 6 patients developed lesions at their venipuncture sites or where arm boards had been taped. The problem was eliminated by removal of contaminated products, disinfection of the storeroom and use of impermeable containers and disposable arm boards. In a similar case, cutaneous aspergillosis developed in a neonate after using Elastoplast to secure a chest tube.[108] Contaminated supplies were also the source of infection in a series of at least 12 cases of endophthalmitis caused by *P. lilacinus*.[201,224] Eleven of the infections were traced to a single contaminated lot of intraocular lens implants. Six of the infected eyes required enucleation and serious loss of vision resulted in the others. *P. lilacinus* infection has also been seen with corneal transplants, keratitis[107] and subcutaneous infections.[134] In addition,

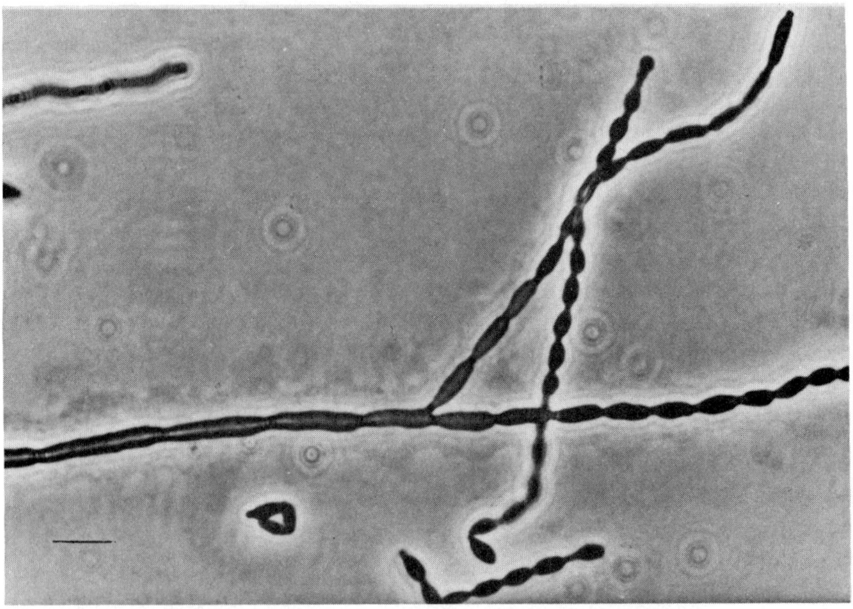

Figure 4:47. ***Xylohypha bantiana*****. Erect conidiophores are lacking and conidia are not abruptly demarcated from vegetative cells. Conidial chains branch infrequently and in most isolates do not disarticulate easily. Bar = 10 μm.**

P. variotii infections have accompanied the use of prosthetics in cardiac surgery.[114,185,295]

Identification of Etiologic Agents

Fungi that cause hyalohyphomycosis appear in infected tissue as colorless, septate, branched or unbranched hyphae, often relatively uniform in diameter, but which may be swollen or toruloid (Figs. 4:48, 4:49). The calcofluor white-KOH technique has been shown to be much more sensitive for microscopic detection of fungi than traditional methods.[286] Several other optical brighteners have also been screened and may eventually prove to be superior to calcofluor.[109,161] Special stains and brighteners increase the probability of detecting fungi, but to determine whether the hyphae are hyaline or dematiaceous, unstained material must be examined using brightfield optics. Since optical brighteners induce fluorescence without staining, one microscope can accomplish both goals. Once the fungal elements are detected by fluorescence, the ultraviolet light source is blocked and the incandescent lamp turned on. Before examining the hyphae for

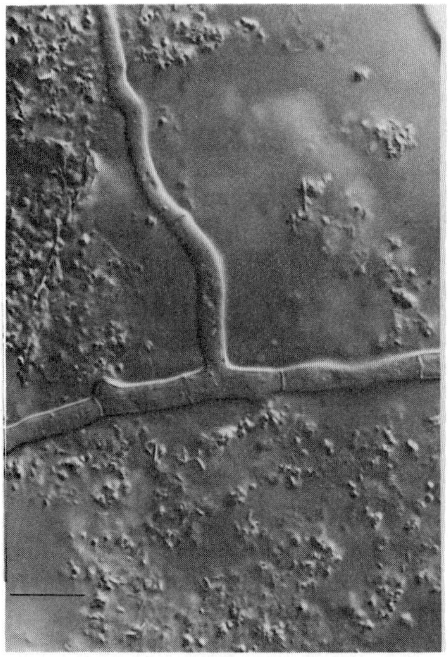

Figure 4:48. KOH preparation of dialysis fluid, showing hyaline, septate hyphae characteristic of hyalohyphomycosis. (*Fusarium solani*). Bar = 10 μm.

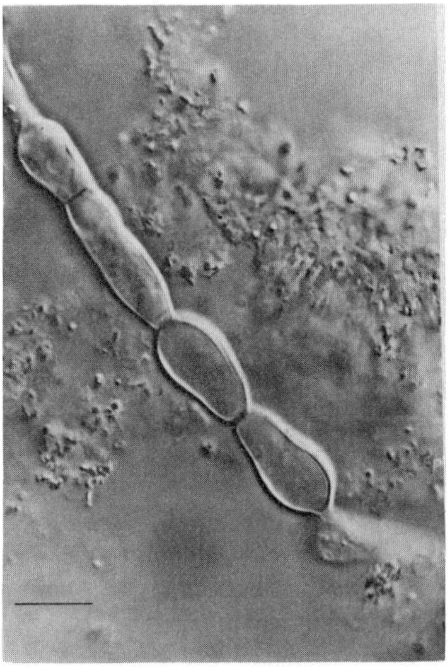

Figure 4:49. Same preparation as Fig. 4:48, showing moniliform hyphae. Budding forms may also be present in hyalohyphomycosis, depending on the fungus involved. Bar = 10 μm.

pigmentation, it is crucial to ensure that the microscope illumination has been properly adjusted using the Koehler technique[2] (see Chapter 18).

Identification at the genus level of the fungi which most often cause hyalohyphomycosis is relatively easy. Species identifications, however, can be difficult. Raper and Fennell[245] designated clusters of similar species within the genus *Aspergillus* as groups, such as the *A. flavus* group and *A. fumigatus* group. The *A. terreus* group consists of a single species which is usually easy to identify. *Aspergillus* differential medium[33,258] is helpful for isolates of the *A. flavus* group that are difficult to identify on morphologic grounds. Sterile isolates of *Aspergillus* are occasionally encountered. The identity of these at the group level can be established by use of the exoantigen technique.[272] In cases where hyphae are seen in biopsied tissue but cultures are negative or were not done, *Aspergillus* can be identified by fluorescent antibody staining.[50] Key characteristics of the most important groups follow, based upon Raper and Fennell.[245]

Aspergillus flavus group:

Consists of 14 species, as expanded by Christensen.[54] Conidial heads are uniseriate, biseriate or both; globose or radiate (sometimes columnar). Conidiophores are hyaline, often roughened, especially at upper portion. Conidia are globose to subglobose, 3.5-5.5 μm, usually roughened. Colonies and conidia are usually yellowish-green. In biseriate heads the metulae and phialides completely cover, and arise perpendicular to, the vesicle.

A. fumigatus group:

Consists of 11 species. Conidial heads are uniseriate. Phialides are not present on the lower portion of the vesicle. Phialides that arise from the sides of the vesicle are clearly curved in an upward direction, resulting in columnar chains of conidia. Conidiophores are hyaline (occasionally dark) and smooth. Conidia are globose to subglobose; very finely and inconspicuously echinulate; mostly 2.5-3.0 μm. Colonies and conidia are blue-green. Isolates are thermophilic, growing at ≥45°C.

A. terreus:

Conidial heads are biseriate, developing extremely tight columns with age; metulae and phialides are long and slender, oriented in an upward direction; conidiophores are hyaline, smooth. Conidia are globose to subglobose, smooth, much smaller (1.8-2.4 μm) than other common species. Colonies are bright yellow when young, becoming tan to cinnamon. Most isolates bear large (up to 7 μm), globose conidia attached singly to vegetative hyphae or occurring in clusters; they are best seen by preparing squash mounts of hyphae that are growing through the agar.

Fusarium oxysporum (after Booth):[29]

Colonies are thinly cottony, rapidly growing, white with salmon tones; blue to purplish tinges are usually present; colonies often develop moist aggregations of phialides and conidia (sporodochia) that are cream to tan in color. Microconidia are elliptical, sometimes curved, 5-12 x 2.2-3.5 μm; macroconidia are 3-5 septate, with a pointed apical cell and the distinctive "foot cell" (Fig. 4:50) that is characteristic of the genus. Phialides are lageniform, and have a short, inconspicuous, thin-walled collarette (Fig. 4:51). Often the microconidia are formed first with the macroconidia developing later. Chlamydospores are usually numerous, in pairs or short chains, smooth or rough-walled.[29,217]

F. solani:

Differs from *F. oxysporum* in that phialides which form microconidia are often elongate; microconidia are larger than in *F. oxysporum* and may

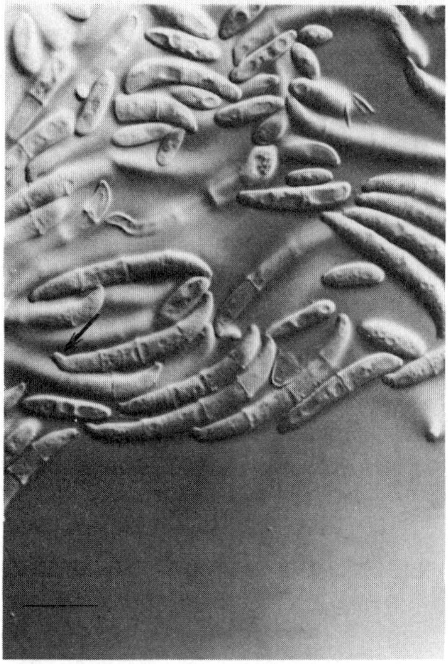

Figure 4:50. ***Fusarium*** **spp. Multicellular macroconidia with foot cells (arrow) characteristic of the genus. Unicellular microconidia are also shown. Bar = 10 μm.**

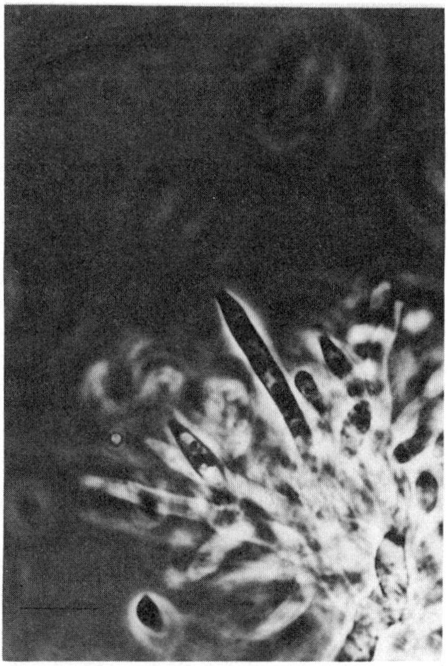

Figure 4:51. Phialides with inconspicuous, thin-walled collarette, typical for species of ***Fusarium*****. Bar = 10 μm.**

become one septate; macroconidia have a more blunt apex, and the foot cell is often indistinct.[29,217]

Scedosporium inflatum:

Differs from *S. apiospermum* in that its annellides are swollen at the base (Fig. 4:52), and often occur in clusters; colonies are usually more darkly pigmented than those of *S. apiospermum*; a teleomorph is unknown. *Scopulariopsis brumptii* substantially resembles *S. inflatum* but can be distinguished by its conidia which develop in dry chains.[177,259]

SUBCUTANEOUS ZYGOMYCOSIS

Zygomycosis due to fungi of the class Zygomycetes, order Mucorales (e.g., *Rhizopus*), occurs primarily in compromised hosts and is a well-known phenomenon of worldwide proportions. In contrast, a much rarer kind of zygomycosis caused by *Conidiobolus* and *Basidiobolus* (members of the

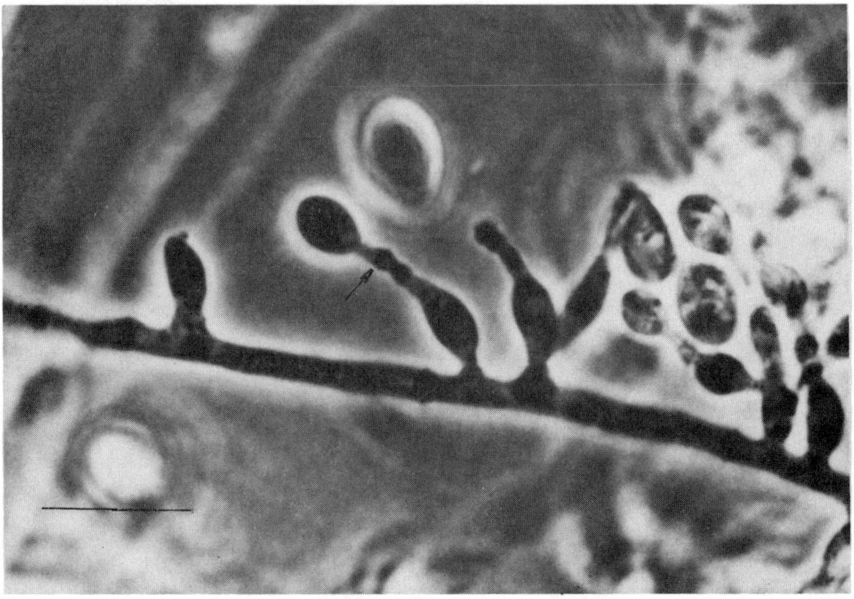

Figure 4:52. ***Scedosporium inflatum*** **has swollen annellations (arrow) typical of the genus, but is distinguished by having annellides that are noticeably bulged ("inflated") in the middle. Bar = 10 μm.**

class Zygomycetes, order Entomophthorales) occurs in otherwise healthy people and is seen almost exclusively in tropical climates. These rarer cases of zygomycosis fall into two distinct clinical syndromes. Infections caused by *Basidiobolus* normally begin when the fungus is inoculated by trauma into cutaneous or subcutaneous tissue—thus, most cases involve an extremity or the trunk. Infections with *Conidiobolus*, however, involve mainly the nasal regions with spread to adjacent facial areas and are presumed to begin following the inhalation of spores. In both circumstances, the infection is chronic, spreads slowly and remains more or less localized.

Although *B. ranarum* was first described in 1886[75] and *C. coronatus* in 1897,[82] infections in humans were not attributed to these fungi until 1956[136] and 1965,[37] respectively. A second species of *Conidiobolus, C. incongruus*, has also been shown to be pathogenic.[41,74,94,101] Infections caused by these fungi have been variously called subcutaneous phycomycosis, basidiobolomycosis, entomophthoramycosis, entomophthoromycosis, rhinophycomycosis and rhinoentomophthoromycosis. The umbrella term "zygomycosis" includes all infections caused by fungi belonging to the class Zygomycetes, whether they are classified in the order Mucorales or Entomophthorales. Referring to an infection as "subcutaneous zygomycosis due to *B. ranarum*", for example, lends coherence to the nomenclature of mycoses and precludes the need for multiple disease names based upon lower taxonomic ranks.

Clinical Manifestations

In most instances of subcutaneous zygomycosis due to *B. ranarum*, the early manifestation is a subcutaneous nodule that is painless, firm, well demarcated and movable. The nodule is not attached to the skin or underlying tissue and can often be lifted at its edges.[40] It gradually increases in size and may reach 1-2 (up to 4) cm in thickness. The diameter of the lesion may be several centimeters or it may extend as a subcutaneous plaque over large areas of the body.[27] According to Joe and Eng,[135] the nodule may not grow evenly—one portion may enlarge while another may regress—and growth of the lesion may be periodic. The overlying skin is initially unaffected but eventually may become discolored in red, blue or dark hues. Lymph nodes are very rarely involved.[141] The time lapse between onset of symptoms and the seeking of medical attention normally ranges from 1 to 12 months. Symptoms may mimic those of osteomyelitis,[27] malignancy[28,40,159] or worm infection.[136] Spontaneous resolution has been noted in several patients.[40,83,136]

In addition to these typical presentations, there have been several very unusual cases of *B. ranarum* zygomycosis. Infections involving muscle tissue have developed after intramuscular injection.[142,143] In one patient, a small subcutaneous nodule quickly developed within 6 months into a large plaque which covered the right buttock, part of the right thigh, the lower abdomen and the pubic region. Enlargement of the right inguinal lymph nodes was also noted.

In another unusual case, the patient complained of a painless depression in his palate, which was followed a few weeks later by nasal congestion and facial numbness. Upon admission, perforation of the hard palate was noted and roentgenograms revealed infection of the maxillary sinus and partial destruction of the orbit.[73]

Of particular interest are 2 reports of gastrointestinal infection. A 4 year-old boy suffering from abdominal pain, fever, sweating and diarrhea was found to have a large (15.5 x 7.5 x 3 cm), hard mass which involved the stomach, colon and hepatic lobes.[6] The patient later died of surgical complications. A 69-year old diabetic man (later shown to be anergic) complaining of abdominal pain, fever, nausea, vomiting and constipation was found to have a firm abdominal mass that involved the ileum, cecum and colon.[267] Two weeks after resection of the mass, more surgery was needed to remove a mass adjacent to the duodenum. Two weeks later, the patient died. Aguiar et al.[6] identified 4 additional cases of gastrointestinal infection in the literature. The origin of these infections is not known but in light of the natural occurrence of *B. ranarum* colonization in the digestive tract of amphibians and reptiles, one could speculate that ingestion of the fungus may have led to colonization and subsequent infection.

Zygomycosis due to *C. coronatus* is significantly different from infections caused by *B. ranarum*. In particular, the former begin in the nasal mucosa and are confined to the rhinofacial areas. Symptoms include nasal congestion and obstruction, swelling (which is usually painless) and nosebleed in some cases. The swelling is progressive and may eventually cause hideous facial hypertrophy and deformity. Symptoms may be confused with those of osteomyelitis,[37] nonspecific chronic inflammation[271] and carcinoma.

In one very unusual case, a 20 year-old female presented with a subcutaneous mass of the breast, low grade fever, dramatic weight loss and cough with occasional hemoptysis.[41] *C. incongruus* was recovered from culture[94] and histopathological study of skin and subcutaneous biopsies, as well as autopsy specimens from esophagus, jejunum, lung, lymph nodes, liver and mediastinum, revealed hyphae consistent with *Conidiobolus* spp.

The only other report of zygomycosis due to *C. incongruus* concerned a 15-month old male who survived a mediastinal mass that involved the pericardium and main stem bronchus.[74,101,158]

Epidemiology

The genus *Basidiobolus* was created in 1886 to accommodate fungi that were discovered to exist as parasites within the digestive tract of amphibians and reptiles.[75] The fungus occurs as spherical cells within the gut, and after excretion can survive for several months under favorable conditions.[303] The fungus can be isolated easily from fresh excrement,[226] but verified reports of isolation from soil[62] or plant debris[70] are extremely rare.

A 1966 review by Klokke et al.[159] found 71 cases of zygomycosis due to *B. ranarum*. By 1980, the total reported was nearly 200 cases.[299] The disease is found almost exclusively in tropical and subtropical regions, particularly of the African continent. *Basidiobolus* infection in temperate climates is known, but is exceedingly rare.[73] It is seen predominantly in children rather than in adults, and at a somewhat higher frequency in males than females.[40,210] The particularly high incidence of buttocks and thigh involvement has led to speculation that the use of contaminated leaves for cleaning after defecation may often be responsible for traumatic inoculation of the fungus.[210]

Nearly 100 cases of subcutaneous zygomycosis due to *Conidiobolus* species have been reported.[53,225,271] It has not been conclusively demonstrated, but all evidence suggests, that inhalation of spores is the avenue of infection. Most of the infections were in tropical and subtropical regions, and males were involved more often than females. In contrast to infections with *B. ranarum*, however, children were infected far less often than adults.[225,271] Though infections are predominantly from tropical or subtropical regions, *C. coronatus* can be readily isolated from damp soil, leaf litter and parasitized insects of temperate areas.[157,294,303] *C. coronatus* is also known as the cause of chronic nasal granuloma in horses.[82]

Identification of Etiologic Agents

Species of *Basidiobolus* and *Conidiobolus* are identical in appearance within infected tissue. Potassium hydroxide preparations reveal hyaline, sparsely septate, thin-walled hyphae, irregularly 8-22 μm in diameter. Histopathologic examination reveals two other very striking characteristics that distinguish zygomycosis due to members of the Entomophthorales from members of the Mucorales. First, as seen with the hematoxylin and eosin stain, there is an invariably intense eosinophilic layer of granular material enveloping the hyphae in a sleeve-like fashion (the Splendore-Hoeppli phenomenon). This material is also periodic acid-Schiff stain (PAS) positive. Second, there is no invasion of blood vessels, as is the strong tendency in zygomycosis due to fungi of the order Mucorales.

B. ranarum:

Colony descriptions have been consistent. The fungus grows faster at 30°C than 37°C, and becomes visible in 2-3 days as greyish to pale yellow colonies that are smooth, flat, radially folded and waxy in appearance.[6,27,28,40,73,83,136,159,299] Occasional isolates may show tan to light brown coloration. Spores (Fig. 4:53) are formed singly upon phototropic sporophores that burst and forcibly discharge the spores (ballistospores). Either another sporophore or vegetative hyphae may develop directly from the discharged spore. A second kind of morphologically distinct spore (Fig. 4:54) also develops directly from the spent ballistospores. It arises singly upon a sporophore, has an adhesive tip and is passively released. *B. ranarum* is homothallic and within 1-2 weeks will form zygospores (Fig. 4:55) that are characterized by 2 beak-like appendages (gametangia) which remain attached to the spore.[155,303] Fresh isolates of *Basidiobolus* and *Conidiobolus* may stop sporulating in a short period of time.[157] In such circumstances, Fromentin[93] has shown that biochemical tests can be useful in separating the 2 genera. The taxonomy of *Basidiobolus* species is disputed;[81,93,156,157,189,308] for the purposes of this treatment, all pathogenic isolates are considered to be *B. ranarum*.

C. coronatus:

Colony characteristics are essentially the same as in *B. ranarum*. However, ballistospores are more obviously apparent as the culture vessel becomes powdered with discharged spores. Soon, satellite colonies appear. Several kinds of spores may be present. Primary (i.e., the first to be formed) ballistospores develop from the vegetative hyphae (Fig. 4:56). Each of these

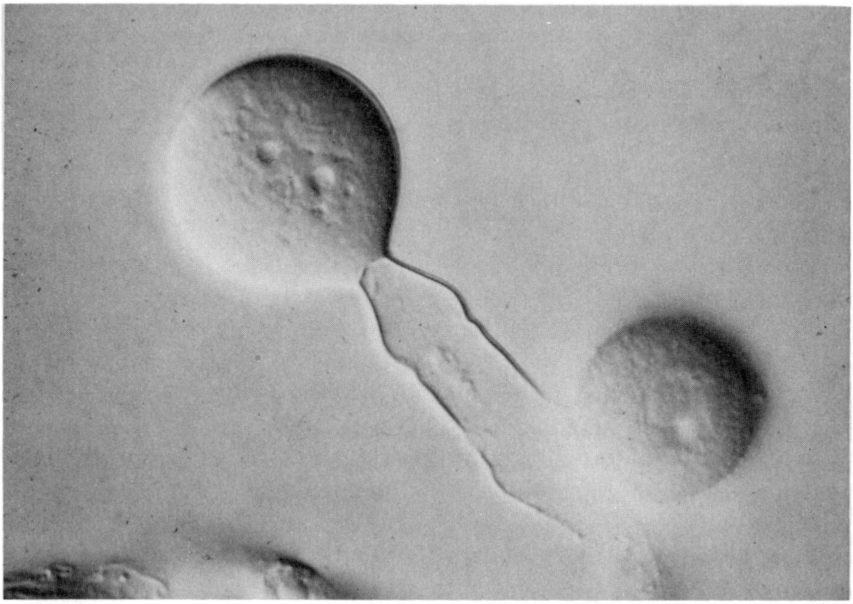

Figure 4:53. ***Basidiobolus*** **spp. ballistospore on sporophore.**

discharged spores may then directly give rise to a sporophore that bears a morphologically similar ballistospore (these are called replicative spores; Fig. 4:57). In addition, any of the spores may synchronously form multiple sporophores that each bear a single, smaller ballistospore (these are called multiplicative spores or microspores). Any of the foregoing spores may produce many hairlike appendages that are 3-15 μm long (and would then be referred to as villose spores). Villose spores (Fig. 4:58) are found only in *C. coronatus*,[157] but not all isolates of this species will readily form them.[157,225,271,287,294] Zygospores are not known to occur in this species.

C. incongruus:

This species differs from *C. coronatus* by the presence of zygospores and the absence of villose spores.[157]

Significance of Laboratory Findings

Of the fungi discussed in this chapter, virtually all are known to occur in association with living or decaying plant parts or soil, and most if not all of them propagate by aerial dispersal of spores. Thus, any of these fungi could potentially be encountered as contaminants in cultures of clinical specimens. For many of the better known subcutaneous pathogens, however, experience

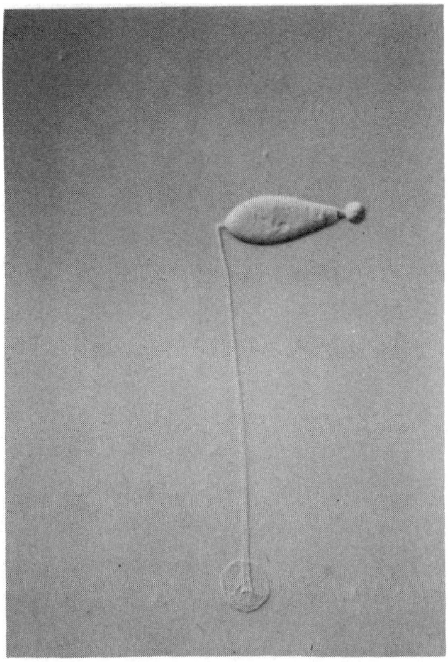

Figure 4:54. ***Basidiobolus ranarum*** **adhesive-tipped spore, passively released.**

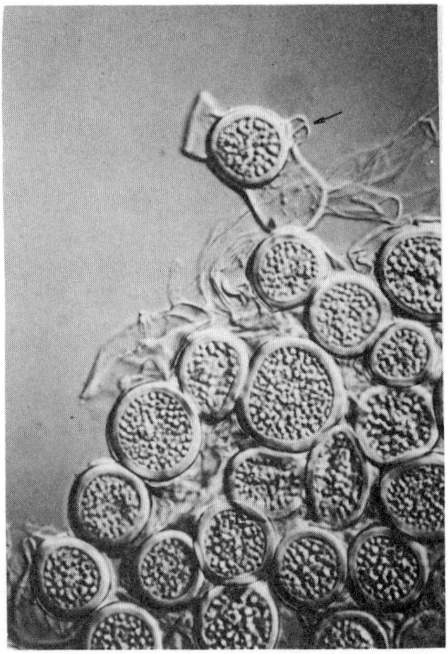

Figure 4:55. ***B. ranarum*** **zygospores with characteristic "beaks" (arrow).**

has shown that this is a remote possibility. When these fungi are isolated and correctly identified (as with *S. schenckii, P, verrucosa, F. pedrosoi, M. mycetomatis* and *M. grisea*), there is seldom any confusion concerning the significance of the isolate. In contrast, with the more recently emerged opportunistic pathogens, as well as some of the rarer pathogens, the significance of laboratory findings may be much less certain to the laboratorian and physician and must therefore be evaluated cautiously.

Table 4:3 lists fungi that can be encountered both as rare subcutaneous pathogens and as contaminants. Their isolation from culture of clinical specimens does not necessarily establish a diagnosis. There are several criteria which should be considered before ascribing significance to any of these fungi. Most important is the finding by microscopic examination of fungal elements within the clinical specimen. These must be morphologically consistent with the fungus that was isolated in culture. Repeated isolations are also suggestive of infection. Fungal colonies should be evaluated, when possible, to confirm that they are growing from the inoculum streak. Finally, other observations, such as failure to isolate any other pathogenic organism and failure of the infection to respond to antibacterial therapy may be supportive of a diagnosis of fungal infection.

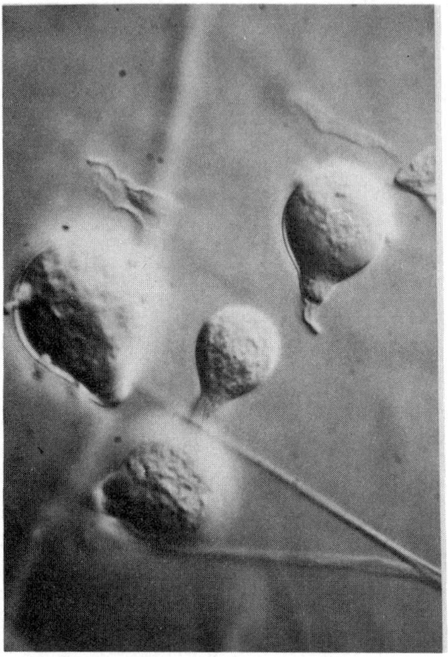

Figure 4:56. ***Conidiobolus coronatus*** **ballistospores.**

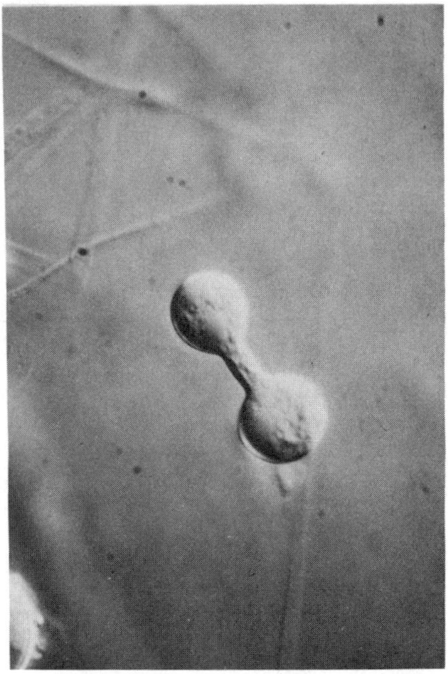

Figure 4:57. ***C. coronatus*** **replicative spore arising directly from a previously discharged ballistospore.**

REFERENCES

1. Abbot P. Mycetoma in the Sudan. Trans R Soc Trop Med Hyg 1956;50:11-24.
2. Abramowitz M. Microscope basics and beyond. Lake Success: Olympus Corp., 1985:19-21.
3. Abramowski CR, Quinn D, Bradford WD, Conant NF. Systemic infection by *Fusarium* in a burned child. J Pediatr 1974;84:561-4.
4. Adam RD, Paquin ML, Petersen EA, et al. Phaeohyphomycosis caused by the fungal genera *Bipolaris* and *Exserohilum*: a report of 9 cases and review of the literature. Medicine 1986;65:203-17.
5. Agger WA, Seager GM. Granulomas of the vocal cord caused by *Sporothrix schenckii*. Laryngoscope 1985;95:595-6.
6. Aguiar E de, Moraes WC, Londero AT. Gastrointestinal entomophthoromycosis caused by *Basidiobolus haptosporus*. Mycopathologia 1980;72:101-5.
7. Ahearn DG, Kaplan W. Occurrence of *Sporotrichum schenckii* on a cold-stored meat product. Am J Epidemiol 1969;80:116-24.
8. Ahmad S, Johnson RJ, Hillier S, Shelton WR, Rinaldi MG. Fungal peritonitis caused by *Lecythophora mutabilis*. J Clin Microbiol 1985;22:182-6.
9. Ajello L. The medical mycological iceberg. PAHO Scientific Publication no. 205, 1970:3-10.
10. Ajello L. Hyalohyphomycosis and phaeohyphomycosis: two global disease entities of public health importance. Eur J Epidemiol 1986;2:243-51.

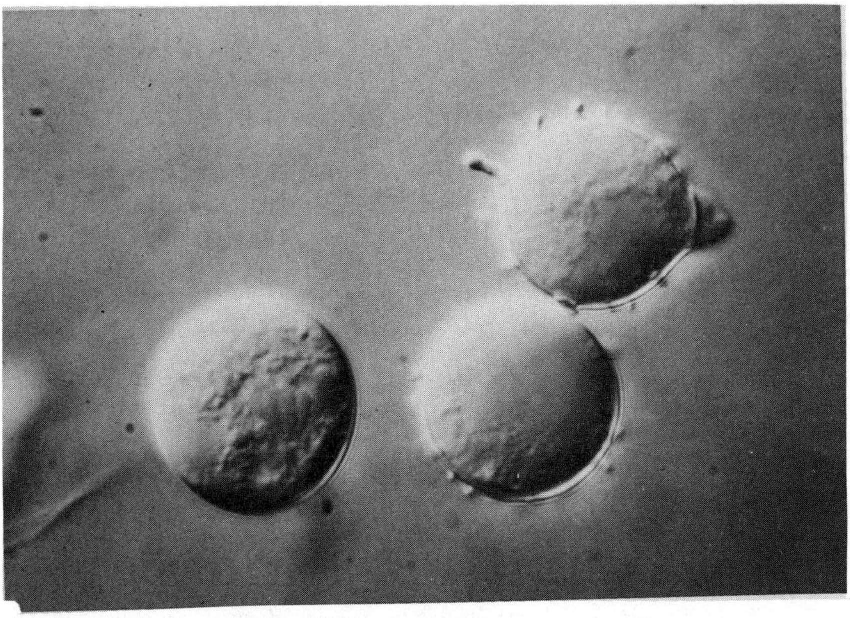

Figure 4:58. ***C. coronatus*** **villose spores.**

11. Ajello L, Georg LK, Steigbigel RT, et al. A case of phaeohyphomycosis caused by a new species of *Phialophora*. Mycologia 1974;66:490-8.
12. Ajello L, Iger M, Wybel R, Vigil FJ. *Drechslera rostrata* as an agent of phaeohyphomycosis. Mycologia 1980;72:1094-1102.
13. Ajello L, Kaplan W, Chandler FW. Dermatophyte mycetomas: fact or fiction? Proceedings of the Fifth International Conference on the Mycoses. Superficial, cutaneous and subcutaneous infections. PAHO Scientific Publication no. 396, 1980:135-40.
14. Al-Doory Y. Chromomycosis. Missoula, MT: Mountain Press Publishing, 1972:9-50.
15. Alsip SG, Cobbs CG. *Pseudallescheria boydii* infection of the central nervous system in a cardiac transplant patient. South Med J 1986;79:383-4.
16. Amma SM, Paniker CKJ, Iype PT, Rangaswamy S. Phaeohyphomycosis caused by *Cladosporium bantianum* in Kerala (India). Sabouraudia 1979;17:419-23.
17. Anaissie E, Kantarjian H, Jones P. *Fusarium*: a newly recognized fungal pathogen in immunosuppressed patients. Cancer 1986;57:2141-5.
18. Arenas R, Latapi F. Sporotrichose generalisee. Bull Soc Pathol Exot 1984;77:385-91.
19. Atdjian M, Granda JL, Ingberg HO, Kaplan BL. Systemic sporotrichosis polytenosynovitis with median and ulnar nerve entrapment. JAMA 1980;243:1841-2.
20. Attapattu MC, Anandakrishnan C. Extensive subcutaneous hyphomycosis caused by *Fusarium oxysporum*. Sabouraudia: J Med Vet Mycol 1986;24:105-11.
21. Barst RJ, Prince AS, Neu HC. *Aspergillus* endocarditis in children: case report and review of the literature. Pediatrics 1981;68:73-8.
22. Batista AC, Maia JA, Singer R. Basidioneuromycosis of man. Soc Biol Pernambuco Anals 1955;13:52-60.
23. Baylet J, Camain R, Segretain G. Identification des agents des maduromycoses de Senegal et de la Mauritanie. Description dúne espece nouvelle. Bull Soc Pathol Exot 1959; 52:448-77.

24. Benjamin RP, Callaway JL, Conant NF. Facial granulomas associated with *Fusarium* infection. Arch Dermatol 1970;101:598-600.
25. Bezes H. L'aspect chirurgical des mycetomes a Dakar. A propos dúne statistique personelle de 60 observations (I). J Chir(Paris) 1961;82:13-32.
26. Biasoli MS, Alvarez DP, Bracalenti BC de. Study of a *Madurella grisea* strain isolated from a foot mycetoma. Mycopathologia 1986;94:117-21.
27. Bittencourt AL, Londero AT, Araujo M, Mendonca N, Bastos J. Occurrence of subcutaneous zygomycosis caused by *Basidiobolus haptosporus* in Brazil. Mycopathologia 1979;68:101-4.
28. Bittencourt AL, Serra G, Sadigursky M, Araujo M, Compos M, Sampaio L. Subcutaneous zygomycosis caused by *Basidiobolus haptosporus*: presentation of a case mimicking Burkitt's lymphoma. Am J Trop Med Hyg 1982;31:370-3.
29. Booth C. The genus *Fusarium*. Kew, Surrey, England: Commonwealth Mycological Institute, 1971.
30. Borelli D. *Pyrenochaeta romeroi* n. sp. Sep Rev Dermatol Venez 1959;1:2.
31. Borelli D. Medios caseros para micologia. Arch Venez Med Trop Parasitol Med 1962;4:301-10.
32. Borelli D. *Acrotheca aquaspersa* nova species agente de cromomicosis. Acta Cient Venezolana 1972;23:193-6.
33. Bothast RJ, Fennell DI. A medium for rapid identification and enumeration of *Aspergillus flavus* and related organisms. Mycologia 1974;66:365-9.
34. Bourgignon RL, Walsh AF, Flynn JC, Baro C, Spinos E. *Fusarium* species osteomyelitis. J Bone Joint Surg 1976;58:722-3.
35. Boyd MF, Crutchfield ED. A contribution to the study of mycetoma in North America. Am J Trop Med 1921;1:215-89.
36. Brabender W, Ketcherside J, Hodges GR, Rengachary S, Barnes WG. *Acremonium kiliense* osteomyelitis of the calvarium. Neurosurg 1985;16:554-6.
37. Bras G, Gordon CC, Emmons CW, Prendegast KM, Sugar M. A case of phycomycosis observed in Jamaica; infection with *Entomophthora coronata*. Am J Trop Med Hyg 1965;14:141-5.
38. Brodey RS, Schryver HF, Deubler MJ, Kaplan W, Ajello L. Mycetoma in a dog. J Am Vet Med Assoc 1967;151:442-51.
39. Bryan CS, DiSalvo AF, Kaufman L, Kaplan W, Brill AH, Abbott DC. *Petriellidium boydii* infection of the sphenoid sinus. Am J Clin Pathol 1980;74:846-51.
40. Burkitt DP, Wilson AMM, Jelliffe DB. Subcutaneous phycomycosis: a review of 31 cases seen in Uganda. Br Med J 1964;1:1669-72.
41. Busapakum R, Youngchaiyud U, Sriumpal S, Segretain G, Fromentin H. Disseminated infection with *Conidiobolus incongruus*. Sabouraudia 1983;21:323-30.
42. Butz WC, Ajello A. Black grain mycetoma; a case due to *Madurella grisea*. Arch Dermatol 1971;104:197-201.
43. Byrd BF, Weiner MH, McGee ZA. *Aspergillus* spinal epidural abscess. JAMA 1982;248:3138-9.
44. Campins H, Scharyj M. Chromoblastomicosis comentarios sobre 24 casos, con estudio clinico, histologico, y micologico. Gac Med Caracas 1953;61:127-51.
45. Carrión AL. Estudio micologico de un caso de micetoma por *Cephalosporium* en Puerto Rico. Mycopathologia 1939;2:165-70.
46. Carrión AL. Chromoblastomycosis. Ann NY Acad Sci 1950;50:1255-82.
47. Carrión AL. *Cephalosporium falciforme* sp. nov., a new etiologic agent of maduromycosis. Mycologia 1951;43:522-3.
48. Carter HV. On mycetoma. Br For Med Chir Rev (London) 1863;32:198-203.
49. Carter HV. On mycetoma or the fungus disease of India. London: J and A Churchhill, 1874.
50. Chandler FW, Kaplan W, Ajello L. Color atlas and text of the histopathology of mycotic diseases. Chicago: Year Book Medical Publishers, 1980:23-5.

51. Chang AC, Destouet JM, Murphy WA. Musculoskeletal sporotrichosis. Skel Rad 1984; 12:23-8.
52. Chaulk CP, Smith PW, Feagler JR, Verdirame J, Commers JR. Fungemia due to *Fusarium solani* in an immunocompromised child. Pediatr Infect Dis 1986;5:363-6.
53. Chauvin JL, Drohet E, Dupont B. Nouveau cas de rhino-entomophthoromycose. Ann Oto-Laryngol 1982;99:563-8.
54. Christensen M. A synoptic key and evaluation of species in the *Aspergillus flavus* group. Mycologia 1981;73:1056-84.
55. Christensen M, States JS. *Aspergillus nidulans* group: *Aspergillus navahoensis* and a revised synoptic key. Mycologia 1982;74:226-35.
56. Chu FWK, Sooy CD. *Pseudallescheria boydii*: a new pathogen in AIDS. Mycol Observer 1986;6:1-7.
57. Cole GT, Kendrick B. Taxonomic studies of *Phialophora*. Mycologia 1973;65:661-88.
58. Collins MS, Rinaldi MG. Cutaneous infection in man caused by *Fusarium moniliforme*. Sabouraudia 1977;15:151-60.
59. Colman MF. Invasive *Aspergillus* of the head and neck. Laryngoscope 1985;95:898-9.
60. Comstock C, Woolson AH. Roentgenology of sporotrichosis. Am J Roentgenol Radium Ther Nucl Med 1975;125:651-5.
61. Conant NF, Smith DT, Baker RD, Callaway JL. Manual of clinical mycology. 3rd ed. Philadelphia: WB Saunders, 1971:417-57.
62. Coremans-Pelseneer J. Biologie des champignons du genre *Basidiobolus* Eidam 1886 saprophytisme et pouvoir pathogene. Act Zool Pathol 1974;60:7-143.
63. Corrado ML, Weitzman I, Stanek A, Goetz R, Agyare E. Subcutaneous infection with *Phialophora richardsiae* and its susceptibility to 5-fluorocytosine, amphotericin B and miconazole. Sabouraudia 1980;18:97-104.
64. Corrall CJ, Merz WG, Rekedal K, Hughes WT. *Aspergillus* osteomyelitis in an immunocompetent adolescent: a case report and review of the literature. Pediatrics 1982;70: 455-61.
65. Destombes P, Andre M, Segretain G, Mariat F, Camain R, Nazimoff O. Contribution a l'etude des mycetomes en Afrique Francaise. Bull Soc Pathol Exot 1958;51:815-75.
66. DeVault GA Jr, Brown ST III, King JW, Fowler M, Oberle A. Tenckhoff catheter obstruction resulting from invasion by *Curvularia lunata* in the absence of peritonitis. Am J Kidney Dis 1985;6:124-7.
67. Dixon DM, Shadomy HJ, Shadomy S. Dematiaceous fungal pathogens isolated from nature. Mycopathologia 1980;70:153-61.
68. Domsch KH, Gams W, Anderson T. Compendium of soil fungi. London: Academic Press, 1980:22-3.
69. Douglas R, Simpson SE.*Cephalosporium* in pleural fluid. Am Rev Tuber 1943;48:237-40.
70. Drechsler C. A *Basidiobolus* producing elongated secondary conidia with adhesive beaks. Bull Torrey Bot Club 1947;74:403-13.
71. Dreizen S, Bodey G, McCredie KB, Keating MJ. Orofacial aspergillosis in acute leukemia. Oral Surg Oral Med Oral Pathol 1985;59:499-504.
72. Dunstan RW, Langham RF, Reimann KA, Wakenell PS. Feline sporotrichosis: a report of five cases with transmission to humans. J Am Acad Dermatol 1986;15:37-45.
73. Dworzack DL, Pollock AS, Hodges GR, Barnes WG, Ajello L, Padhye A. Zygomycosis of the maxillary sinus and palate caused by *Basidiobolus haptosporus*. Arch Intern Med 1978;138:1274-6.
74. Eckert HL, Khoury GH, Pore RS, Gilbert EF, Gaskell JR. Deep entomophthora phycomycotic infection reported for the first time in the United States. Chest 1972;61:392-4.
75. Eidam E. *Basidiobolus* eine neue gattung der Entomophthoraceen. Beit Biol Pflanz 1886;4:181-251.
76. Eisenberg ES, Leviton I, Soeiro R. Fungal peritonitis in patients receiving peritoneal dialysis: experience with 11 patients and review of the literature. Rev Infect Dis 1986; 8:309-21.

77. El-Ani AS. A new species of *Leptosphaeria*, an etiologic agent of mycetoma. Mycologia 1966;58:406-11.
78. El-Ani AS, Gordon MA. The ascospore sheath and taxonomy of *Leptosphaeria senegalensis*. Mycologia 1965;57:275-8.
79. Ellis MB. Dematiaceous hyphomycetes. VII: *Curvularia, Brachysporium*, etc. Mycological Paper no. 106. Kew, Surrey, England: Commonwealth Agricultural Bureau, 1966.
80. Ellis MB. Dematiaceous hyphomycetes. Kew, Surrey, England: Commonwealth Agricultural Bureau, 1971:416-17.
81. Emmons CW, Binford CH, Utz JP, Kwon-Chung KJ. Medical mycology. 3rd ed. Philadelphia: Lea and Febiger, 1977:472.
82. Emmons CW, Bridges CH. *Entomophthora coronata*, the etiologic agent of a phycomycosis of horses. Mycologia 1961;53:307-12.
83. Emmons CW, Joe L, Eng NT, Pohan A, Kertopati S, Meulen A van der. *Basidiobolus* and *Cercospora* from human infections. Mycologia 1957;49:1-10.
84. Espinel-Ingroff A, Shadomy S, Dixon D, Goldson P. Exoantigen test for *Cladosporium bantianum, Fonsecaea pedrosoi* and *Phialophora verrucosa*. J Clin Microbiol 1986;23:305-10.
85. Espinel-Ingroff A, Shadomy S, Kerkering TM, Shadomy HJ. Exoantigen test for differentiation of *Exophiala jeanselmei* and *Wangiella dermatitidis* isolates from other dematiaceous fungi. J Clin Microbiol 1984;20:23-7.
86. Estes SA, Merz WG, Maxwell LG. Primary cutaneous phaeohyphomycosis caused by *Drechslera spicifera*. Arch Dermatol 1977;113:813-15.
87. Etta van JL, Peterson LR, Gerding DN. *Acremonium falciforme (Cephalosporium falciforme)* mycetoma in a renal transplant patient. Arch Dermatol 1983;119:707-8.
88. Fathizadeh A, Rippon JW, Rosenfeld SI, Fretzin DF, Lorincz AL. Pheomycotic cyst in an immunosuppressed host. J Am Acad Dermatol 1981;5:423-7.
89. Ferlito A. Primary aspergillosis of the larynx. J Laryngol Otol 1974;88:1257-63.
90. Foerster HR. Sporotrichosis. Am J Med Sci 1924;167:55-76.
91. Friedman AD, Campos JM, Rorke LB, Bruce DA, Arbeter AM. Fatal recurrent *Curvularia* brain abscess. J Pediatr 1981;99:413-15.
92. Friedman SJ, Doyle JA. Extracutaneous sporotrichosis. Int J Dermatol 1983;22:171-6.
93. Fromentin H. Enzymatic characterization with the API ZYM system of Entomophthorales potentially pathogenic to man. Curr Microbiol 1982;7:315-18.
94. Fromentin H, Segretain GL, Segretain GM. Identification d'un *Conidiobolus incongruus* agent dúne mycose profonde en Thailande. Bull Soc Fran Mycol Med 1981;10:77-80.
95. Fukushiro R. Chromomycosis in Japan. Int J Dermatol 1983:22:221-9.
96. Fukushiro R. Epidemiology and ecology of sporotrichosis in Japan. Zentral Bakteriol Hyg A 1984;257:228-33.
97. Fuste FJ, Ajello L, Threlkeld R, Henry JE Jr. *Drechslera hawaiiensis*: causative agent of a fatal fungal meningo-encephalitis. Sabouraudia 1973;11:59-63.
98. Gaind ML, Padhye AA, Thirumalachar MJ. Madura foot in India caused by *Cephalosporium infestans* sp. nov. Sabouraudia 1962;1:230-3.
99. Gari M, Fruit J, Rousseaux P, et al. *Scedosporium (Monosporium) apiospermum*: multiple brain abscesses. J Med Vet Mycol 1985;23:371-6.
100. Gezuele E, Mackinnon JE, Conti-Diaz IA. The frequent isolation of *Phialophora verrucosa* from natural sources. Sabouraudia 1972;10:266-73.
101. Gilbert EF, Khoury GH, Pore RS. Histopathological identification of *Entomophthora* phycomycosis. Arch Pathol 1970;90:583-7.
102. Glotzbach RE. *Aspergillus terreus* infection of pseudoaneurysm of aortofemoral vascular graft with contiguous vertebral osteomyelitis. Am J Clin Pathol 1982;77;224-7.
103. Godfrey J. Disease of the foot not hitherto described. Lancet 1846;1:593-4.
104. Goette DK, Robertson D. Transepithelial elimination in chromomycosis. Arch Dermatol 1984;120:400-1.
105. Goldberg B, Eversmann WW, Eitzen EM. Invasive aspergillosis of the hand. J Hand Surg 1982;7:38-42.

106. Gonzalez MS, Alfonso B, Seckinger D, Padhye AA. Subcutaneous phaeohyphomycosis caused by *Cladosporium devriesii*, sp. nov. Sabouraudia: J Med Vet Mycol 1984;22:427-32.
107. Gordon MA, Norton SW. Corneal transplant infection by *Paecilomyces lilacinus*. J Med Vet Mycol 1985;23:295-301.
108. Granstein RD, First LR, Sober AJ. Primary cutaneous aspergillosis in a premature neonate. Br J Dermatol 1980;103:681-4.
109. Green LK, Moore DG. Fluorescent compounds that nonspecifically stain fungi. Lab Med 1987;18:456-8.
110. Green WO Jr, Adams TE. Mycetoma in the United States; a review and report of seven additional cases. Am J Clin Pathol 1964;42:75-91.
111. Grossman ME, Fithian EC, Behrens C, Bissinger J, Fracaro M, Neu H. Primary cutaneous aspergillosis in six leukemic children. J Am Acad Dermatol 1985;12:313-18.
112. Gutmann L, Chou SM, Pore RS. Fusariosis, myasthenic syndrome and aplastic anemia. Neurology 1975;22:922-6.
113. Hageage GJ, Harrington BJ. Use of calcofluor white in clinical mycology. Lab Med 1984;15:109-12.
114. Haldane EV, MacDonald JL, Gittens WO, Yuce K, Rooyen van CE. Prosthetic valvular endocarditis due to the fungus *Paecilomyces*. Can Med Assoc J 1974;111:963-8.
115. Halde C, Padhye AA, Haley LD, Rinaldi MG, Kay D, Leeper R. *Acremonium falciforme* as a cause of mycetoma in California. Sabouraudia 1976;14:319-26.
116. Harada S, Tokishi U, Kusunoki T. Systemic chromomycosis. J Dermatol 1976;3:13-17.
117. Harpster WH, Gonzalez C, Opal SM. Pansinusitis caused by the fungus *Drechslera*. Otolaryngol Head Neck Surg 1985;93:683-5.
118. Hawksworth DL, Gibson IAS, Gams W. *Phialophora parasitica* associated with disease conditions in various trees. Trans Brit Mycol Soc 1976;66:427-31.
119. Hay RJ, Mackenzie WR. Mycetoma (madura foot) in the United Kingdom—a survey of forty-four cases. Clin Exp Dermatol 1983;8:553-62.
120. Hernanz Del Palacio A, Conde-Zurita JM, Reyes Pecharroman S, Rodriquez Noriega A. A case of *Alternaria alternata* (Fr.) Keissler infection of the knee. Clin Exp Dermatol 1983;8:641-6.
121. Hironaga M, Watanabe S. Annellated conidiogenous cells in *Petriellidium boydii (Scedosporium apiospermum)*. Sabouraudia 1980;18:261-8.
122. Honbo S, Padhye AA, Ajello L. The relationship of *Cladosporium carrionii* to *Cladophialophora ajelloi*. J Med Vet Mycol 1984;22:209-18.
123. Honbo S, Standard PG, Padhye AA, Ajello A, Kaufman L. Antigenic relationships among *Cladosporium* species of medical importance. J Med Vet Mycol 1984;22:301-10.
124. Hoog GS de. The Genera *Blastobotrys, Sporothrix, Calcarisporium* and *Calcarisporiella* gen. nov. Stud Mycol No. 7. Centraalbureau voor Schimmelcultures. Baarn, 1974:8,12-66.
125. Hoog GS de. The black yeasts and allied Hyphomycetes. Stud Mycol No. 15. Centraalbureau voor Schimmelcultures. Baarn, 1977:1-128.
126. Hoog GS de, Constantinescu O. A new species of *Sporothrix* from calf skin. Antonie van Leeuwenhoek 1981;47:367-70.
127. Hoog GS de, Rantio-Lehtimaki AH, Smith MT: *Blastobotrys, Sporothrix* and *Trichosporiella*: generic delimitation, new species, and a *Stephanoascus* teleomorph. Antonie van Leeuwenhoek 1985;51:79-109.
128. Hoog GS de, Vries GA de. Two new species of *Sporothrix* and their relation to *Blastobotrys nivea*. Antonie van Leeuwenhoek 1973;39:515-20.
129. Itoh M, Okamoto S, Kariya H. Survey of 200 cases of sporotrichosis. Dermatologica 1986;172:209-13.
130. Iwatsu T, Miyagi M. Phaeomycotic cyst: a case with a lesion containing a wooden splinter. Arch Dermatol 1984;120:1209-11.
131. Iwatsu T, Miyagi M, Okamoto S. Isolation of *Phialophora verrucosa* and *Fonsecaea pedrosoi* from nature in Japan. Mycopathologia 1981;75:149-58.
132. Iwatsu T, Tokano M, Okamoto S. Auricular chromomycosis. Arch Dermatol 1983;119:88-9.

133. Jacobson HP. Fungous diseases: a clinico-mycological text. Springfield, IL: CC Thomas, 1932:121-48.
134. Jade KB, Lyons MF, Gnann JW. *Paecilomyces lilacinus* cellulitis in an immunocompromised patient. Arch Dermatol 1986;122:1169-70.
135. Joe LK, Eng N-IT. Subcutaneous phycomycosis: a new disease found in Indonesia. Ann NY Acad Sci 1960;89:4-16.
136. Joe LK, Eng N-IT, Pohan A, Meulen H van der. *Basidiobolus ranarum* as a cause of subcutaneous mycosis in Indonesia. Arch Dermatol 1956;74:378-83.
137. Jordan JM, Waters K, Caldwell DS. *Aspergillus flavus*: an unusual cause of chest wall inflammation in an immunocompetent host. J Rheumatol 1986;13:660-2.
138. Joshi KR, Mathur DR, Sharma JC, Vyas MCR, Sanghvi A. Mycetoma caused by *Aspergillus nidulans* in India. J Trop Med Hyg 1985;88:41-4.
139. June CH, Beatty PG, Shulman HM, Rinaldi MG. Disseminated *Fusarium moniliforme* infection after allogenic marrow transplantation. South Med J 1986;79:511-35.
140. Kaell AT, Weitzman I. Acute monoarticular arthritis due to *Phialophora parasitica*. Am J Med 1983;74:519-22.
141. Kamalam A, Thambiah AS. Basidiobolomycosis with lymph node involvement. Sabouraudia 1975;13:44-8.
142. Kamalam A, Thambiah AS. Basidiobolomycosis following injection. Mykosen 1982;25: 512-16.
143. Kamalam A, Thambiah AS. Muscle invasion by *Basidiobolus haptosporus*. J Med Vet Mycology 1984;22:273-7.
144. Kaplan W, Broderson JR, Pacific JN. Spontaneous systemic sporotrichosis in nine-banded armadillos (*Dasypus novemcinctus*). Sabouraudia 1982;20:289-94.
145. Kaplan W, Ochoa AG. Application of the fluorescent antibody technique to the rapid diagnosis of sporotrichosis. J Lab Clin Med 1963;62:835-41.
146. Karayannopolous SI, Stylianea A. *Aspergillus flavus* in prostatic fluid. Br J Urol 1981;53:192.
147. Kazanas N, Jackson G. *Sporothrix schenckii* isolated from edible black fungus mushrooms. J Food Protect 1983;46:714-16.
148. Kenyon EM, Russell LH, McMurray DN. Isolation of *Sporothrix schenckii* from potting soil. Mycopathologia 1984;87:128.
149. Kern MA, Uecker FA. Maxillary sinus infection caused by the homobasidiomycetous fungus *Schizophyllum commune*. J Clin Microbiol 1986;23:1001-5.
150. Kerr CM, Perfect JR, Craven PC, et al. Fungal peritonitis in patients on continuous ambulatory peritoneal dialysis. Ann Intern Med 1983;99:334-7.
151. Khan KA, Khan AF, Masih M, Farooqi AH, Ansari AM. Clinical and pathological findings of mycetoma with special reference to the etiology. Asian Med J 1984;27:250-7.
152. Kheir SM, Flint A, Moss JA. Primary aspergillosis of the larynx simulating carcinoma. Hum Pathol 1983;14:184-6.
153. Kiehn TE, Nelson PE, Bernard EM, Edwards FF, Koziner B, Armstrong D. Catheter-associated fungemia caused by *Fusarium chlamydosporum* in a patient with lymphocytic lymphoma. J Clin Microbiol 1985;21:501-4.
154. Kim RC, Hodge CJ, Lamberson HV, Weiner LB. Traumatic intracerebral implantation of *Cladosporium trichoides*. Neurology 1981;31:1145-8.
155. King DS. *Basidiobolus haptosporus*. In: Fuller MS, ed. Lower fungi in the laboratory. Athens: Department of Botany, University of Georgia, 1978:153, 156.
156. King DS. Systematics of fungi causing entomophthoromycosis. Mycologia 1979;71:731-45.
157. King DS. Entomophthorales. In: Howard DH, ed. Fungi pathogenic for humans and animals: part A. Biology. New York: Marcel Dekker, 1983:63-72.
158. King DS, Jong SC. Identity of the etiological agent of the first deep entomophthoraceous infection of man in the United States. Mycologia 1976;68:181-3.
159. Klokke AH, Job CK, Warlow PFM. Subcutaneous phycomycosis in India. Trop Geog Med 1966;18:20-5.

160. Klokke AH, Swamidasan G, Anguli R, Verghese A. The causal agents of mycetoma in south India. Trans R Soc Trop Med Hyg 1968;62:509-16.
161. Koch HH, Pimsler M. Evaluation of Uvitex 2B: a nonspecific fluorescent stain for detecting and identifying fungi and algae in tissue. Lab Med 1987;18:603-6.
162. Koshi G, Kurian T, Mathai R, Abraham J, Joseph LBM, Selvapandian AJ. Diverse etiologic agents from mycetoma cases. Ind J Med Res 1981;74:539-43.
163. Koshi G, Padhye AA, Ajello L, Chandler FW. *Acremonium recifei* as an agent of mycetoma in India. Am J Trop Med Hyg 1979;28:692-6.
164. Koshi G, Victor N, Chacko J. Causal agents in mycetoma of the foot in southern India. Sabouraudia 1972;10:14-18.
165. Kumar R, Smissen van der E, Jorizzo J. Systemic sporotrichosis with osteomyelitis. J Can Assoc Radiol 1984;35:83-4.
166. Kwon-Chung KJ. Comparison of isolates of *Sporothrix schenckii* obtained from fixed cutaneous lesions with isolates from other types of lesions. J Infect Dis 1979;139:424-31.
167. Lane CG. A cutaneous lesion caused by a new fungus (*Phialophora verrucosa*). J Cut Dis 1915;33:840-6.
168. Lavalle P. Chromoblastomycosis in Mexico. PAHO Scientific Publication no. 396, 1980:235-47.
169. Lavalle P, Mariat F. Sporotricosis. Bull Inst Pasteur 1983;81:295-322.
170. Lavarde V, Bedrossian J, Bievre de C, Vacher C. Un cas de phaeomycose a *Phialophora parasitica* chez un transplante deuxieme observation mondiale. Bull Soc Fran Mycol Med 1982;11:273-6.
171. Lazarus HS, Myers JP, Brocker RJ. Post-craniotomy wound infection caused by *Pseudallescheria boydii*. J Neurosurg 1986;64:153-4.
172. Leiberman TW, Ferry AP, Bottone EJ. *Fusarium solani* endophthalmitis without primary corneal involvement. Am J Ophthalmol 1979;88:764-7.
173. Lynch JB. Mycetoma in the Sudan. Ann R Coll Surg Engl 1964;35:319-40.
174. Mackinnon HE, Ferrada LV, Montemayor L. Investigaciones sobre las maduromicosis y sus agentes. Anales Facultad Medicina 1949;34:231-300.
175. Mackinnon JE, Ferrada-Urzua LV, Montemayor L. *Madurella grisea* n. sp.: a new species of fungus producing the black variety of maduromycosis in South America. Mycopathologia 1949;4:384-92.
176. Magoub ES. Mycetomas caused by *Curvularia lunata, Madurella grisea, Aspergillus nidulans* and *Nocardia brasiliensis* in Sudan. Sabouraudia 1973;11:179-82.
177. Malloch D, Salkin IF. A new species of *Scedosporium* associated with osteomyelitis in humans. Mycotaxon 1984;21:247-55.
178. Mariat F. Sur la distribution geographique et la repartition des agents de mycetomes. Bull Soc Pathol Exot 1963;56:35-45.
179. Matlow AG, Goldman CB, Muklow MG, Kane J. Contamination of intravenous fluid with *Sporothrix schenckii*. J Infect 1985;10:169-71.
180. Matsuda T, Matsumoto T. Disseminated hyalohyphomycosis in a leukemic patient. Arch Dermatol 1986;122:1171-5.
181. Matsumoto T, Nishimoto K, Kimura K, Padhye AA, Ajello A, McGinnis MR. Phaeohyphomycosis caused by *Exophiala moniliae*. J Med Vet Mycol 1984;22:17-26.
182. Matsumoto T, Padhye AA, Ajello A, Standard PG. Critical review of human isolates of *Wangiella dermatitidis*. Mycologia 1984;76:232-49.
183. Mayorga R, Caceres A, Concepcion T, et al. Etude dúne zone déndemie sporotrichosique au Guatemala. Sabouraudia 1978;16:185-98.
184. Mayorga R, Close de Leon JE. Sur une myctome Guatemalteque a grains noir. Sabouraudia 1966;4:210-14.
185. McClellan JR, Hamilton JD, Alexander JA, Wolfe WG, Reed JB. *Paecilomyces varioti* (sic) endocarditis on a prosthetic aortic valve. J Thorac Cardiovasc Surg 1976;71:472-5.
186. McDonough ES, Lewis AL, Meister M. *Sporothrix (Sporotrichum) schenckii* in a nursery barn containing sphagnum. Public Health Rep 1970;85:579-86.

187. McGinnis MR. *Wangiella*, a new genus to accommodate *Hormiscium dermatitidis*. Mycotaxon 1977;5:353-63.
188. McGinnis MR. Taxonomy of *Exophiala jeanselmei* (Langeron) McGinnis and Padhye. Mycopathologia 1979;65:79-87.
189. McGinnis MR. Recent developments and changes in medical mycology. Ann Rev Microbiol 1980;34:109-35.
190. McGinnis MR. Chromoblastomycosis and phaeohyphomycosis: new concepts, diagnosis and mycology. Am Acad Dermatol 1983;8:1-16.
191. McGinnis MR, Ajello L, Schell WA. Mycotic diseases: a proposed nomenclature. Int J Dermatol 1985;24:9-15.
192. McGinnis MR, Borelli D. *Cladosporium bantianum* and its synonym *Cladosporium trichoides*. Mycotaxon 1981;13:127-36.
193. McGinnis MR, Borelli D, Padhye AA, Ajello L. Reclassification of *Cladosporium bantianum* in the genus *Xylohypha*. J Clin Microbiol 1986;23:1148-51.
194. McGinnis MR, Padhye AA, Ajello L. *Pseudallescheria* Negroni et Fischer, 1943, and its later synonym *Petriellidium* Malloch, 1970. Mycotaxon 1982;14:94-102.
195. McGinnis MR, Rinaldi MG, Winn RE. Emerging agents of phaeohyphomycosis: pathogenic species of *Bipolaris* and *Exserohilum*. J Clin Microbiol 1986;24:250-9.
196. McGinnis MR, Schell WA. The genus *Fonsecaea* and its relationship to the genera *Cladosporium, Phialophora, Ramichloridium* and *Rhinocladiella*. PAHO Scientific Publication no. 396, 1980:215-24.
197. McGinnis MR, Schell WA, Carson J. *Phaeoannellomyces* and the Phaeococcomycetaceae, new dematiaceous blastomycete taxa. J Med Vet Mycol 1985;23:179-88.
198. McGinnis MR, Sorrel DF, Miller RL, Kaminski GW. Subcutaneous phaeohyphomycosis caused by *Exophiala moniliae*. Mycopathologia 1981;73:69-72.
199. Medlar EM. A cutaneous infection caused by a new fungus, *Phialophora verrucosa*, with a study of the fungus. J Med Res 1915;32:507-21.
200. Mershon JC, Samuelson DR, Laymon TE. Left ventricular "fibrous body" aneurysm caused by *Aspergillus* endocarditis. Am J Cardiol 1968;22:281-5.
201. Miller GR, Rebell G, Magoon RC, Kulvin SM, Forster RK. Intravitreal antimycotic therapy and the cure of mycotic endophthalmitis caused by a *Paecilomyces lilacinus* contaminated pseudophakos. Opthal Surg 1978;9:54-63.
202. Mitchell AJ, Solomon AR, Beneke ES, Anderson TF. Subcutaneous alternariosis. J Am Acad Dermatol 1983;8:673-6.
203. Moneymaker CS, Shenep JL, Person TA, Field ML, Jenkins JJ. Primary cutaneous phaeohyphomycosis due to *Exserohilum rostratum (Drechslera rostrata)* in a child with leukemia. Pediatr Infect Dis 1986;5:380-2.
204. Monroe PW, Floyd WE. Chromohyphomycosis of the hand due to *Exophiala jeanselmei (Phialophora jeanselmei, Phialophora gougerotii)*—case report and review. J Hand Surg 1981;6:370-3.
205. Monte de la SM, Hutchins GM. Disseminated *Curvularia* infection. Arch Pathol Lab Med 1985;109:872-4.
206. Moore M,, Almeida F. Etiologic agents of chromomycosis (chromoblastomycosis of Terra, Torres, Fonseca and Leao, 1922) of North and South America. Rev Biol Hyg 1935;6:94-7.
207. Morgan MA, Cockerill FR III, Cortese DA, Roberts GD. Disseminated sporotrichosis with *Sporothrix schenckii* fungemia. Diagn Microbiol Infect Dis 1984;2:151-5.
208. Morton SJ, Midthun K, Merz WG. Granulomatous encephalitis caused by *Bipolaris hawaiiensis*. Arch Pathol Lab Med 1986;110:1183-5.
209. Moskowitz LB, Cleary TJ, McGinnis MR, Thomson CB. *Phialophora richardsiae* in a lesion appearing as a giant cell tumor of the tendon sheath. Arch Pathol Lab Med 1983;107:374-6.

210. Mugerwa JW. Subcutaneous phycomycosis in Uganda. Br J Dermatol 1976;94: 539-44.
211. Muir DB, Pritchard RC. *Sporothrix schenckii*— incidence in the Sydney region. Aust J Dermatol 1984;25:27-8.
212. Murray IG, Holt HD. Is *Cephalosporium acremonium* capable of producing maduromycosis? Mycopathol Mycol Appl 1963;22:336-8.
213. Mutton KJ, Lucas TJ, Harkness JL. Disseminated *Fusarium* infection. Med J Aust 1980;2:624-5.
214. Myers JT, Dunn AD. *Aspergillus* infection of the hand. JAMA 1930;95:794-6.
215. Nakamura T, Grant JA, Threlkeld R, Wible L. Primary chromoblastomycosis of the nasal septum. Am J Clin Pathol 1972;58:365-70.
216. Neilsen HS Jr, Conant NF, Weinberg T, Reback JF. Report of a mycetoma due to *Phialophora jeanselmei* and undescribed characteristics of the fungus. Sabouraudia 1968;6:330-3.
217. Nelson PE, Toussoun TA, Marasas WFO. *Fusarium* species: an illustrated manual for identification. University Park: Pennsylvania State University Press, 1983.
218. Nishimoto K, Yoshimura S, Honma K. Chromomycosis spontaneously healed. Int J Dermatol 1984;23:408-10.
219. Nishimura K, Miyagi M. Studies on a saprophyte of *Exophiala dermatitidis* isolated from a humidifier. Mycopathologia 1982;77:173-81.
220. Nishimura K, Miyaji M, Taguchi H, Tanaka R. Fungus flora in water in bathtubs and in sludge of drainpipes of bathrooms: (1) frequent isolation of the genus *Exophiala*. [Abstract] IXth International Congress, International Society for Human and Animal Mycology, 1985:3-48.
221. Nishitani H, Nishitani K, Numaguchi Y, Honbo S. Cerebral chromomycosis. J Comput Assist Tomog 1982;6:624-6.
222. Nsanzumuhire H, Vollum D, Poltera AA. Chromomycosis due to *Cladosporium trichoides* treated with 5-fluorocytosine. Am J Clin Pathol 1974;61:257-63.
223. Nusbaum BP, Gulbas N, Horwitz SN. Sporotrichosis acquired from a cat. J Am Acad Dermatol 1983;8:386-91.
224. O'Day DN. Fungal endophthalmitis caused by *Paecilomyces lilacinus* after intraocular lens implantation. Am J Ophthalmol 1977;83:131-2.
225. Okafor BC, Gugnami HC. Nasal entomophthoromycosis in Nigerian Igbos. Trop Geog Med 1983;35:53-7.
226. Okafor JI, Testrake D, Mushinsky HR, Yangco BG. A *Basidiobolus* sp. and its association with reptiles and amphibians in southern Florida. J Med Vet Mycology 1984;22: 47-51.
227. Okeke CN, Gugnami HC. Studies of pathogenic dematiaceous fungi: isolation from natural sources. Mycopathologia 1986;94:19-25.
228. O'Sullivan FX, Bradly RS, Lynch JM, et al. Peritonitis due to *Drechslera spicifera* complicating continuous ambulatory peritoneal dialysis. Ann Intern Med 1981;94:213-14.
229. Padhye AA. Comparative study of *Phialophora jeanselmei* and *P. gougerotii* by morphological, biochemical and immunological methods. PAHO Scientific Publication no. 356, 1978:60-5.
230. Padhye AA, Ajello L, Chandler FW, et al. Phaeohyphomycosis in El Salvador caused by *Exophiala spinifera*. Am J Trop Med Hyg 1983;32:799-803.
231. Padhye AA, Ajello L, Weiden MA, Steinbronn KK. Phaeohyphomycosis of the nasal sinuses caused by a new species of *Exserohilum*. J Clin Microbiol 1986;24;245-9.
232. Padhye AA, Kaplan W, Neuman MA, Case P, Radcliffe GN. Subcutaneous phaeohyphomycosis caused by *Exophiala spinifera*. J Med Vet Mycol 1984;22:493-500.
233. Padhye AA, McGinnis MR, Ajello A. Thermotolerance of *Wangiella dermatitidis*. J Clin Microbiol 1978;8:424-6.

234. Padhye AA, Sukapure RS, Thirumalachar MJ. *Cephalosporium madurae* n. sp., cause of Madura foot in India. Mycopathol Mycol Appl 1962;16:315-22.
235. Pankovich AM, Auerbach BJ, Metzger WI, Barreta T. Development of maduromycosis (*Madurella mycetomi*) after nailing of a closed tibial fracture. Clin Orth Rel Res 1981;154:220-2.
236. Parent D, Depre G, Pelseneer-Coremans J, Depierreux M, Schoutens-Serruys E, Heenen M. Arthrite a *Fusarium* chez un patient burundais. Bull Soc Fran Mycol Med 1985; 14:201-4.
237. Pierce NF, Millan JC, Bender BS, Curtis JL. Disseminated *Curvularia* infection. Arch Pathol Lab Med 1986;110:959-61.
238. Pluss JL, Opal SM. Pulmonary sporotrichosis: review of treatment and outcome. Medicine 1986;65:143-53.
239. Poirot JL, Lamporte JP, Gueho E. Mycose profunde a *Fusarium*. La Press Med 1985;14:2300-1.
240. Powell KE, Taylor A, Phillips BJ, et al. Cutaneous sporotrichosis in forestry workers. JAMA 1978;240:232-5.
241. Pueringer RJ, Iber C, Deike M, Davies S. Spontaneous remission of extensive pulmonary sporotrichosis. Ann Intern Med 1986;104:36-7.
242. Pupaibul K, Sindhuphak W, Chindamporn A. Mycetoma of the hand caused by *Phialophora jeanselmei*. Mykosen 1982;25:321-30.
243. Rajam RV, Kandhari KC, Thirumalachar MJ. Chromoblastomycosis caused by a rare yeastlike dematiaceous fungus. Mycopathologia 1958;9:5-19.
244. Rao PB. Aspergillosis of larynx. J Laryngol Otol 1969;83:377-9.
245. Raper KB, Fennell DI. The genus *Aspergillus*. Malabar: RE Krieger, 1965:242,361-7, 567-74.
246. Read SI, Sperling LC. Feline sporotrichosis. Arch Dermatol 1982;18:429-31.
247. Rey M, Baylet R, Camain R. Donnees actuelles sur les mycetomes: a propos de 214 cas africains. Ann Dermatol 1962;89:511-27.
248. Rhine WD, Arvin AM, Stevenson DK. Neonatal aspergillosis: a case report and review of the literature. Clin Pediatr 1986;25:400-3.
249. Rinaldi MG. Invasive aspergillosis. Rev Infect Dis 1983;5:1061-77.
250. Rinaldi MG, McCoy EL, Winn DF. Gluteal abscess caused by *Phialophora hoffmannii* and review of this organism in human mycoses. J Clin Microbiol 1982;16:181-5.
251. Rinaldi MG, Phillips P, Schwartz JG, et al. Human *Curvularia* infections: report of five cases and review of the literature. Diagn Microbiol Infect Dis 1987;6:27-39.
252. Rippon JW. Clinical spectrum of petriellidosis: mycetoma to systemic opportunist. PAHO Scientific Publication no. 396, 1980:276-95.
253. Rippon JW. Medical mycology: the pathogenic fungi and the pathogenic actinomycetes. 2nd ed. Philadelphia: WB Saunders, 1982.
254. Rippon JW, Arnow PM, Larson RA, Zang KL. Golden tongue syndrome caused by *Ramichloridium schulzeri*. Arch Dermatol 1985;121:892-4.
255. Rohatgi PK. Pulmonary sporotrichosis. South Med J 1980;73;1611-17.
256. Rolston KVI, Hopfer RL, Larson DL. Infections caused by *Drechslera* species: case report and review of the literature. Rev Infect Dis 1985;7:525-9.
257. Rudolph M. Uber die brasilianische "Figueira." Arch Schiffs Tropen-Hygen 1914;18:498.
258. Salkin IF, Gordon MA. Evaluation of *Aspergillus* differential medium. J Clin Microbiol 1975;2:74-5.
259. Salkin IF, McGinnis MR, Dykstra MJ, Rinaldi MG. *Scedosporium inflatum*, an emerging pathogen. J Clin Microbiol 1988;26:498-503.
260. Pincus DH, Salkin IF, Hurd NJ, Levy IL, Kemna MA. Modification of potassium nitrate assimilation test for identification of clinically important yeasts. J Clin Microbiol 1988;26: 366-8.
261. Samorodin CS, Sing B. Ketoconazole-treated sporotrichosis in a veterinarian. Cutis 1984;33:487-8.

262. Sandhyamani S, Bhatia R, Mohatpatra LN, Roy S. Cerebral cladosporiosis. Surg Neurol 1981;15:431-4.
263. Schell WA. Oculomycoses caused by dematiaceous fungi. Washington, DC: PAHO Publication no. 479, 1986:105-9.
264. Schell WA, McGinnis MR, Borelli D. *Rhinocladiella aquaspersa*, a new combination for *Acrotheca aquaspersa*. Mycotaxon 1983;17:341-8.
265. Schenk BR. On refractory subcutaneous abscesses caused by a fungus possibly related to the sporotricha. John Hopkins Hosp Bull 1898;9:286-91.
266. Schiappacasse RH, Colville JM, Wong PK, Markowitz A. Sporotrichosis associated with an infected cat. Cutis 1985;35:268-70.
267. Schmidt JH, Howard RJ, Chen JL, Pierson K. First culture-proven gastrointestinal entomophthoromycosis in the United States: a case report and review of the literature. Mycopathologia 1986;95:101-4.
268. Schol-Schwarz MB. Revision of the genus *Phialophora* (Moniliales). Persoonia 1970; 6:59-94.
269. Seaworth BJ, Kwon-Chung KJ, Hamilton JD, Perfect JR. Brain abscess caused by a variety of *Cladosporium trichoides*. J Clin Pathol 1983;79:747-52.
270. Segretain G, Destombes P. Recherche sur les mycetomes a *Madurella grisea* et *Pyrenochaeta romeroi*. Sabouraudia 1969;7:51-61.
271. Segura JJ, Gonzalez K, Berrocal J, Marin G. Rhinoentomophthoromycosis: report of the first two cases observed in Costa Rica (Central America) and review of the literature. Am J Trop Med Hyg 1981;30:1078-84.
272. Sekhon AS, Standard PG, Kaufman L, Garg AK, Cifuentes P. Grouping of *Aspergillus* species with exoantigens. Diagn Immunol 1986;4:112-16.
273. Selman SH, Hampel N. Systemic sporotrichosis: diagnosis through biopsy of epididymal mass. Urology 1982;20:620-1.
274. Shimazono Y, Kiminori I, Torii H, Otsuka R. Brain abscess due to *Hormodendrum dermatitidis* (Kano) Conant, 1953. Folia Psych Neurol Jap 1963;17:80-96.
275. Shimosaka S, Waga S. Cerebral chromoblastomycosis complicated by meningitis and multiple fungal aneurysms after resection of a granuloma. J Neurosurg 1983;59:158-61.
276. Sindhuphak W, MacDonald E, Head E, Hudson RD. *Exophiala jeanselmei* infection in a postrenal transplant patient. J Am Acad Dermatol 1985;13:877-81.
277. Singer MI, House JL. Invasive mycotic infections in the immunocompromised child. Otolaryngol Head Neck Surg 1979;87:32-4.
278. Singh H. Mycetoma in India. Ind J Surg 1979;41:577-97.
279. Smith LM. Sporotrichosis: report of four clinically atypical cases. South Med J 1945; 38:505-9.
280. Smith PW, Loomis GW, Luckasen JL, Osterholm RK. Disseminated cutaneous sporotrichosis: three illustrative cases. Arch Dermatol 1981;117:143-4.
281. Sporotrichosis infection on mines of the Witwatersrand. A symposium. Helm MAF, Berman C. The clinical, therapeutic and epidemiological features. Johannesburg: Transvaal Chamber of Mines, 1947:59-67.
282. Spraker MK. Rare infection responds to itraconazole therapy. Mycol Observer 1985;5:3.
283. Staib F, Grosse G. Isolation of *Sporothrix schenckii* from the floor of an indoor swimming pool. Zentral Bakteriol Hyg I Abt Orig B 1983;177:499-506.
284. Steadham JE, Geis PA, Simmank JL. Use of carbohydrate and nitrate assimilations in the identification of fungi. Diagn Microbiol Infect Dis 1986;5:71-5.
285. Steinberg GK, Britt RH, Enzmann DR, Finlay JL, Arvin AM. *Fusarium* brain abscess. J Neurosurg 1983;56:598-601.
286. Sutphin JE, Robinson NM, Wilhemus KR, Osato MS. Improved detection of oculomycoses using induced fluorescence with Cellufluor. Ophthalmology 1986;93:416-17.
287. Taylor GD, Sekhon AS, Tyrrell DLJ, Goldsand G. Rhinofacial zygomycosis caused by *Conidiobolus coronatus*: a case report including in vitro sensitivity to antimycotic agents. Am J Trop Med Hyg 1987;36:398-401.

288. Terra F, Torres M, da Fonseca O, et al. Novo typo de dermatite verrucosa mycose por *Acrotheca* com associacao de leishmaniosa. Brasil Med 1922;2:263-8.
289. Thianprasit M, Sivayathorn A. Black dot mycetoma. Mykosen 1984;27:219-26.
290. Torstrick RF, Harrison K, Heckman JD, Johnson JE. Chronic bursitis caused by *Phialophora richardsiae*. J Bone Joint Surg 1979;61:772-4.
291. Trejos A. *Cladosporium carrionii* n. sp. and the problem of Cladosporia isolated from chromoblastomycosis. Rev Biol Trop 1954;2:75-112.
292. Tsai CY, Lu YC, Wang LT, Hsu TL, Sung JL. Systemic chromoblastomycosis due to *Hormodendrum dermatitidis* (Kano) Conant. Report of the first cases in Taiwan. Am J Clin Pathol 1966;46:103-14.
293. Tschen JA, Knox JM, McGavran MH, Duncan C. Chromomycosis: the association of fungal elements and wood splinters. Arch Dermatol 1984;120:107-8.
294. Tucker BE, Benjamin RK. *Conidiobolus coronatus*. In: Fuller MS, ed. Lower fungi in the laboratory. Athens: Department of Botany, University of Georgia, 1978:157-8.
295. Uys CJ, Don PA, Schrire V, Barnard CN. Endocarditis following cardiac surgery due to the fungus *Paecilomyces*. S Afr Med J 1963;21:1276-80.
296. Valenstein P, Schell WA. Primary intranasal *Fusarium* infection: potential for confusion with rhinocerebral zygomycosis. Arch Pathol Lab Med 1986;110:751-4.
297. Vartian CV, Shales DM, Padhye AA, Ajello L. *Wangiella dermatitidis* endocarditis in an intravenous drug user. Am J Med 1985;78:703-7.
298. Villarroya JCP, Fernandez-Guerrero ML, Moran V, et al. *Fusarium oxysporum* infection. J Infect 1982;5:307-10.
299. Vismer HF, Beer HA de, Dreyer L. Subcutaneous phycomycosis caused by *Basidiobolus haptosporus* (Drechsler, 1947). S Afr Med J 1980;58:644.
300. Viviani MA, Tortorano AM, Laria G, Giannetti A, Bignotti G. Two new cases of cutaneous alternariosis with a review of the literature. Mycopathologia 1986;96:3-12.
301. Wackym PA, Gray GF, Richie RE, Gregg CR. Cutaneous chromomycosis in renal transplant patients. Arch Intern Med 1985;145:1036-7.
302. Ward HP, Martin WJ, Ivins JC, Weed LA. *Cephalosporium* arthritis. Proc Mayo Clin 1961;36:337-43.
303. Webster J. Introduction to fungi. Cambridge: Cambridge University Press, 1970:143-50.
304. Wheeler MS, McGinnis MR, Schell WS, Walker DH. *Fusarium* infections in burned patients. Am J Clin Pathol 1981;75:304-11.
305. Willemsen MJ, Coninck AL de, Cormans-Pelseneer JE, Marichal-Pipeleers MA, Roseeuw DI. Parasitic invasion of *Fusarium oxysporum* in an arterial ulcer in an otherwise healthy patient. Mykosen 1986;29:248-52.
306. Wilson E. Cerebral abscess caused by *Cladosporium bantianum*. Case report. Pathology 1982;14:91-6.
307. Wilson LA, Ahern DG. Association of fungi with extended-wear soft contact lenses. Am J Ophthalmol 1986;101:434-6.
308. Yangco BG, Nettlow A, Okafor JI, Park J, Strake D Te. Comparative antigenic studies of species of *Basidiobolus* and other medically important fungi. J Clin Microbiol 1986;23:679-82.
309. Yangco BG, TeStrake D, Okafor J. *Phialophora richardsiae* isolated from infected human bone: morphological, physiological and antifungal susceptibility studies. Mycopathologica 1984:86:103-11.
310. Yoo D, Lee WHS, Kwon-Chung KJ. Brain abscess due to *Pseudallescheria boydii* associated with primary non-Hodgkin's lymphoma of the central nervous system: a case report and literature review. Rev Infect Dis 1985;7:272-7.
311. Young CN, Meyers AM. Opportunistic fungal infection by *Fusarium oxysporum* in a renal transplant patient. Sabouraudia 1979;17:219-23.

312. Young CN, Swart JG, Ackerman D, Davidge-Pitts K. Nasal obstruction and bone erosion caused by *Drechslera hawaiiensis*. J Laryngol Otol 1978;92:137-43.
313. Zackheim HS, Halde C, Goodman RS, Marchasin S, Bunke HJ Jr. Phaeohyphomycotic cyst of the skin caused by *Exophiala jeanselmei*. J Am Acad Dermatol 1985;12:207-12.
314. Zaias N, Rebell G. A simple and accurate diagnostic method in chromoblastomycosis. Arch Dermatol 1973;108:545-6.
315. Ziefer A, Connor DH. Phaeomycotic cyst: a clinicopathologic study of twenty-five patients. Am J Trop Med Hyg 1980;28:901-11.
316. Ziza JM, Dupont B, Boissonnas A, et al. Osteoarthrites a champignons noirs (dematies). Ann Med Interne 1985;136:393-7.

CHAPTER 5

SYSTEMIC MYCOSES

Stephen A. Moser, Ph.D., Frank L. Lyon, Ph.D. and Donald L. Greer, Ph.D.

INTRODUCTION

Systemic mycoses are caused by several genera of fungi; some are dimorphic, i.e., grow as a mold in environmental conditions and either as a yeast (*Blastomyces dermatitidis, Histoplasma capsulatum, Paracoccidioides braziliensis, Sporothrix schenkii*) or spherule in vivo, while others exist solely as a mold or as a yeast in both environments. Besides the "classic" systemic mycoses, opportunistic infections of humans caused by other molds are included in this chapter. Members of the genus *Aspergillus* and the zygomycetes are discussed, with limited comment on some rare agents causing disease with systemic manifestations. Many fungi other than those discussed in this chapter are capable of causing systemic disease. The list of isolated instances of diseases caused by fungi previously unknown as human or animal pathogens is extensive and continues to expand.

Systemic mycoses are similar in that they are acquired by inhalation of infectious particles, most probably conidia, produced by fungi normally resident in soil or other niches in the environment. Although cryptococcosis and candidiasis sometimes are manifest as systemic diseases, they are covered in Chapter 6, "Human Infections Caused by Yeastlike Fungi," and sporotrichosis, primarily a cutaneous disease with uncommon systemic manifestations, is discussed in Chapter 4, "Molds Involved in Subcutaneous Infections."

The primary infection in blastomycosis, coccidioidomycosis, histoplasmosis and paracoccidioidomycosis is thought to occur in the lower respiratory tract. Most healthy individuals have a remarkable capacity to handle the primary infection and the resulting disease is usually asymptomatic or minimally symptomatic with spontaneous healing. Progression to serious, potentially fatal systemic disease occurs as a very uncommon event.

However, when it does occur, any organ may become involved, with different fungal species exhibiting differing predilections. Individuals with diminished immunologic functions (e.g., the result of malignancies, prolonged treatment with corticosteroids, antineoplastic agents or other immunosuppressants) are more likely to develop the more serious systemic manifestations rather than the benign disease seen in the general population.

The systemic mycoses are part of the diagnostic differential in acute and chronic pneumonias, solitary pulmonary lesions and chronic extrapulmonary infectious processes. The diseases produced by these fungi are clinically and radiologically indistinguishable from each other or from tuberculosis, various granulomas, lymphomas or other malignancies. Therefore, the laboratory plays an important role in the diagnosis of these infections.

ASPERGILLOSIS

The group of diseases known as aspergillosis includes toxicosis (aflatoxin), allergic reactions, infections localized to the skin, external ear, nasal sinuses and the orbit of the eye, pulmonary and disseminated multi-organ disease caused by a member of the genus *Aspergillus*. Pulmonary aspergillosis is the most common form and includes allergic, transient colonization, with or without formation of an aspergilloma, and invasive tissue disease. Invasive disease has been recognized recently as a frequent opportunistic infection of immunocompromised patients, especially those with hematological malignancies and neutropenia.

Clinical Manifestations

Aspergillosis is usually acquired by inhalation of airborne spores, initially resulting in pulmonary disease. Detailed descriptions of the clinical manifestations can be found in several excellent textbooks[18,203] and review articles.[68,150,202,254,269] Therefore, only a brief summary of the most common syndromes follows, with subsequent discussion limited to the role of the clinical laboratory in diagnosing infectious processes due to *Aspergillus* spp.

Allergic bronchopulmonary aspergillosis is an asthmatic condition of atopic subjects manifest by eosinophilia and episodes of wheezing and coughing. A tenacious mucopurulent sputum and bronchial mucus plugs are characteristic features. Although the etiological agent can be isolated from these rubbery plugs in the sputum,[13,70,86] there is no invasion of tissue and the pathology is due to the host's immune response to *Aspergillus* antigens.

Colonizing aspergillosis (aspergilloma) usually develops in an ectatic bronchus, rarely as a primary disease,[195] or more often in pre-existing cavities due to tuberculosis, histoplasmosis or sarcoidosis.[248,253] Aspergillomas, which appear radiographically as solitary lesions, usually remain localized. However, some patients become hemoptysic as the result of erosion and rupture of large blood vessels adjacent to an aspergilloma. If patients with an aspergilloma are immunosuppressed, the process may then progress to tissue invasion.[107]

Invasive aspergillosis usually appears clinically as a pulmonary process with mycelia extending into tissue. Invasion subsequent to external otitis has been reported in an immunosuppressed patient,[186] illustrating the unexpected in a compromised host. As tissue is invaded by fungal hyphae there is necrosis, thrombosis and the possibility that infected emboli may disseminate the fungus.[271] Cutaneous and subcutaneous lesions are usually the result of hematogenous dissemination from a primary pulmonary infection.[259] Primary cutaneous lesions are rare but have been reported in pediatric patients as a result of longterm use of intravenous arm boards and nonsterile gauze.[158] The presence of immunosuppression, abrasion, moisture and fungal spores in the gauze were thought to be responsible for these cases.

Although most patients with invasive aspergillosis have immunosuppression and neutropenia as predisposing factors, a few cases of invasive disease have been reported in patients without these or any other identifiable predisposing factor.[119] Whether these cases were due to an undefined immune defect or to fungal strains with different pathogenic potential is not known.

Patients with the acquired immune deficiency syndrome (AIDS) contract aspergillosis[21] but there does not appear to be a correlation between aspergillosis and human immunodeficiency virus type 1 (HIV-1) infection. Therefore, aspergillosis is not part of the current case definition for AIDS.[35]

Epidemiology and Public Health Significance

Aspergillus spp. are ubiquitous and are frequent causes of plant and animal disease. Aspergillosis in the normal human population has limited public health significance. In contrast, up to 50% of invasive fungal infections documented at autopsy in patients with acute leukemia are due to *Aspergillus* spp.[113] Clinical and experimental evidence suggests that antineoplastic corticosteroid therapy and neutropenia predispose to an increased susceptibility. Patients at risk of opportunistic infections should be maintained in environments free from fungal conidia. Several clusters of disseminated aspergillosis have occurred during periods of hospital renovation.[112,149,230,239] If renovation or any similar activity causes an increase in the number of fungal particles in the air, the environment of patients at risk (e.g., renal transplant patients) should be monitored aerobiologically to insure that unnecessary exposure to infectious agents does not occur.

Collection, Transport and Processing of Specimens

Due to the type of patients with aspergillosis, most specimens are collected in hospitals and quickly transported to a laboratory. Delay in processing specimens should be short, otherwise concentrated antibiotic solutions (i.e., gentamicin at a final concentration of 50 μg/mL) should be added to prevent an increase in the bacterial population. Specimens suitable for culture include sputum, bronchoalveolar lavage or tissue biopsy from either the lungs or from the edge of cutaneous lesions. Unfortunately, venous blood cultures have not proved helpful in the diagnosis of disseminated aspergillosis or endocarditis.[180] One exception is a report of the recovery of *A. fumigatus* from blood of a patient with invasive external otitis.[186]

The aspergilli, like most opportunistic molds, are inhibited by cycloheximide, but chloramphenicol or gentamicin can be used to control bacterial contamination. The quantity of viable hyphae in clinical specimens may be small and not well disbursed, so media should be heavily inoculated and duplicate cultures on several media are often necessary for successful isolation. When agar plates are used, streak to obtain isolated colonies. Detailed methods for isolation of aspergilli are discussed in Chapter 2, "Collection, Transport and Processing of Specimens."

Identification of Etiologic Agents

A 10% KOH mount can be used for detection of hyphae in sputum or in the bronchus plugs associated with allergic disease. More recently, a dye specific for complex polysaccharides has been used to facilitate direct microscopic examination (see Chapter 2). Direct examination along with histopathologic examination of tissues produces the best evidence of infection by showing invasion of tissue. On direct examination, aspergilli appear as branching (ca. 45° angle), hyaline, septate hyphae (3-4 μm wide) with parallel, refractile walls, which cannot be differentiated from hyphae of other hyaline molds (except from those of the aseptate zygomycetes).

Tissue sections stained with hematoxylin and eosin (H and E) to assess tissue response, which in aspergillosis is typically an acute necrotic and pyogenic reaction, show poorly stained, repeated dichotomous branching fungal hyphae. Since other hyaline molds, e.g., *Pseudoallescheria, Fusarium* and *Acremonium*, have the same morphology, isolation of the fungus is required to establish its identity. Identification is important since aspergilli are susceptible to amphotericin B whereas other fungi (e.g., Pseudoallescheria) are resistant. Although rare, abortive conidiophores and scattered conidia can be seen in sections of tissues derived from an area exposed to air, such as a pulmonary cavity. When this occurs, it may be possible to tentatively identify the fungus as a member of the genus *Aspergillus*.[147] Although not generally available, there are both polyclonal and monoclonal antibodies which can be used to assist diagnosis when fixed tissue is the only material available.[187]

Since aspergilli encountered in the clinical laboratory grow rapidly at either ambient temperature or at 37°C, colonies are grossly visible within

2-4 days and conidial formation usually occurs within 7-10 days' incubation. Identification of aspergilli is based on both gross and microscopic morphology and pigment production. Occasionally, nonsporulating isolates are obtained from pulmonary lesions and are only identified by an exoantigen test.[124]

Keys using texture and color of colonies to differentiate *Aspergillus* are based on growth on Czapek-Dox agar at 25°C. All species of *Aspergillus* produce an anamorphic (asexual) stage characterized by erect unbranched conidiophores with swollen apices (vesicles). Depending on the species, a portion or the entire surface of the vesicle is covered by flask-shaped phialides (sterigmata). In some species, an intervening row of cells (metulae) support the phialides. Vesicles having metulae and phialides are called biseriate, whereas the term uniseriate is used to identify vesicles without metulae and with only a single row of phialides. Conidia are formed in unbranched chains from the tips of the phialides. The entire asexual reproductive structure is referred to as the conidial head. Cleistothecia (telemorph or sexual structures) and Hülle cells (thick-walled submerged asexual cells) (Fig.5:1) can aid in identification, if present.

Raper and Fennell list 132 *Aspergillus* species arranged in 18 groups in "The Genus *Aspergillus*." More recent estimates are closer to 200 species.[202] The aspergilli are widely distributed in the environment, are found in soil, on decaying vegetation and as agents of plant and animal disease. Seventeen

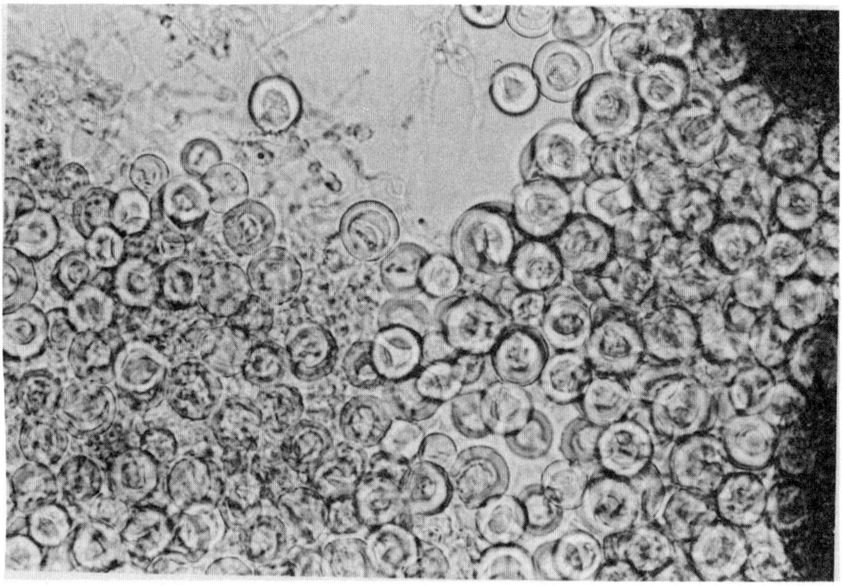

Figure 5:1. Lactophenol mount of thick-walled Hülle cells from *A. nidulans*, 450X.

species are reported to cause disease in humans.[161,202] *A. fumigatus* is the species most often isolated from clinical material and correlated with infection.[137,270] Other species that cause human disease include *A. flavus, A. niger* and *A. terreus: A. versicolor* and *A. nidulans* are found less often.

A brief description of the groups containing the most frequently encountered species are presented. Clinical laboratories with limited experience in mycology should not attempt to identify aspergilli beyond the group level. A more complete identification to the species level should be requested from a reference laboratory. For a more thorough taxonomic study of the aspergilli, refer to the classic monograph by Raper and Fennell.[194]

A. fumigatus Group (Fig.5:2)

Macroscopic growth appears first as a small white colony which develops green to grey-green or slate-grey pigment with age. Conidiophores are smooth-walled with a dome-shaped vesicle. The conidial heads are typically columnar and compact with uniseriate phialides covering only the upper half of the vesicles. Conidia are usually round and slightly rough. Emmons et al.[68] noted that *A. fumigatus* is one of the few aspergilli capable of growth at ≥45°C, a feature which can be used as an aid in identification.

A. flavus Group (Fig.5:3)

Colonies have a rough granular texture becoming yellow-green to green with age. Individual conidial heads are almost large enough to be visible to

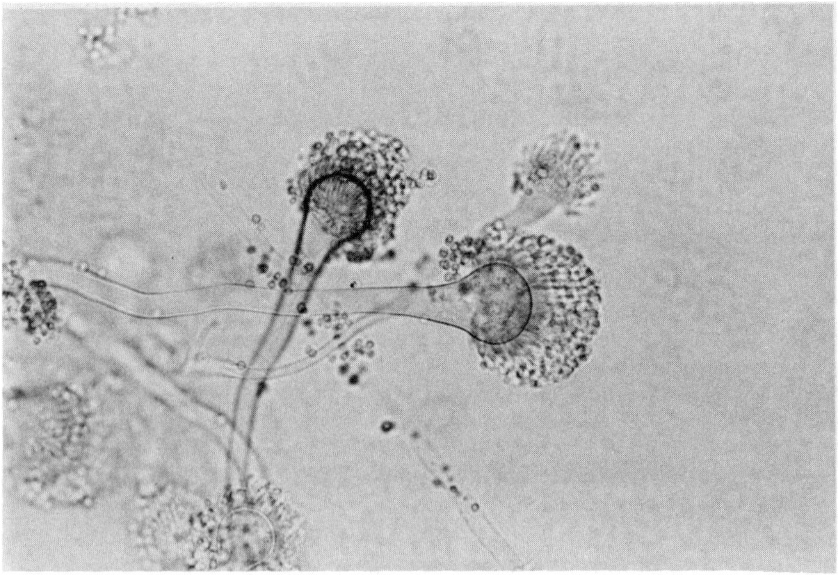

Figure 5:2. Lactophenol mount of *A. fumigatus* conidiophore, 450X.

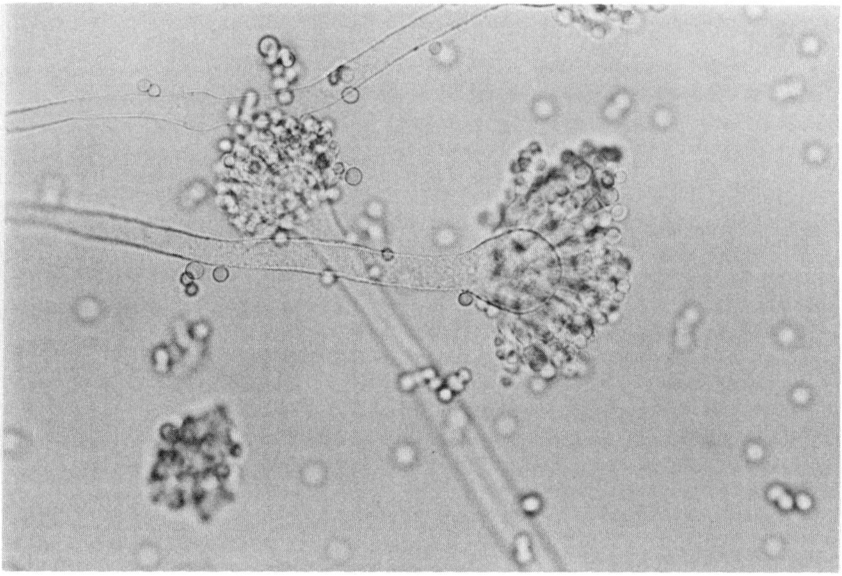

Figure 5:3. Lactophenol mount of *A. flavus* conidiophore, 450X.

the naked eye. Conidophores are rough-walled, ending in round to spherical vesicles. Phialides with (biseriate) or without (uniseriate) metulae cover the entire surface of the vesicle. With age the conidial heads split into radiate columns. Conidia are spherical and may have smooth or rough walls.

A. niger Group (Fig.5:4)

Colonies begin as scant, white, filamentous growth which rapidly produce large black conidial heads. The conidial heads are visible without magnification and impart a rough texture to colonies. The long conidiophores are smooth-walled and terminate in a spherical vesicle. The phialides are arranged on metulae covering the entire surface of the vesicle. Conidia are thick-walled, very rough and darkly pigmented. This is the only species of aspergilli with carbon-black conidia.

A. terreus Group (Fig.5:5)

The colony rapidly becomes cinnamon-brown to dark brown with age and has a fine powdery texture. The conidiophore is short, smooth-walled and colorless. The vesicle is dome-like to pyriform with metulae and phialides covering only the upper half. The conidial head is columnar, compact and long. Conidia are spherical to slightly oval and smooth.

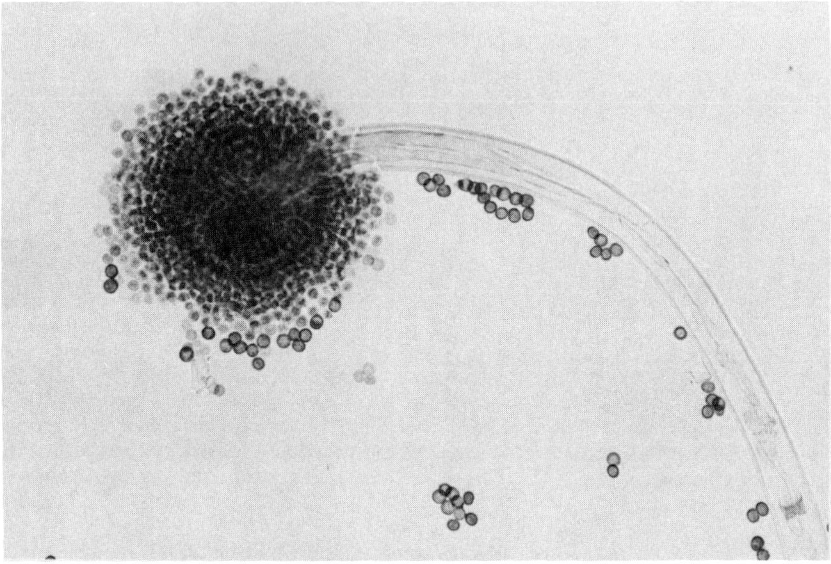

Figure 5:4. Lactophenol mount of *A. niger* conidiophore, 100X.

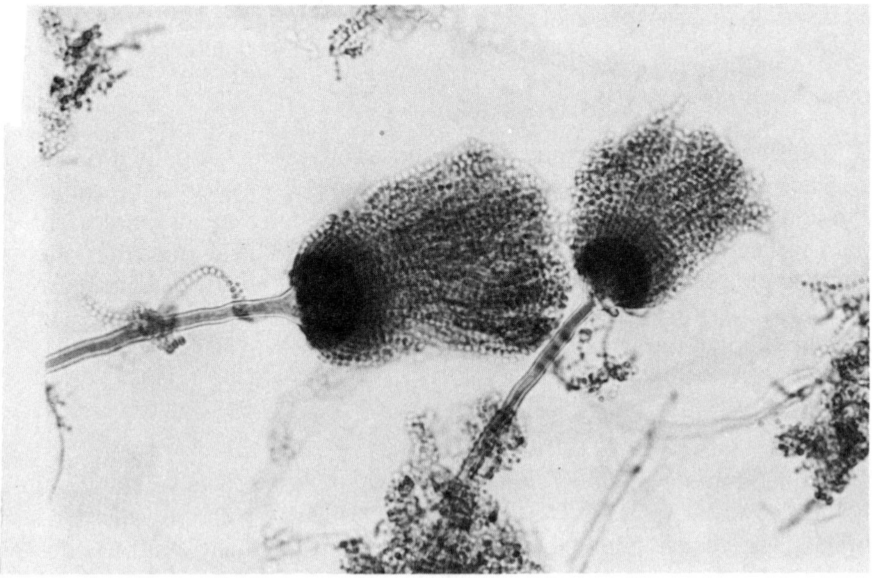

Figure 5:5. Lactophenol mount of *A. terreus* conidiophore, 450X.

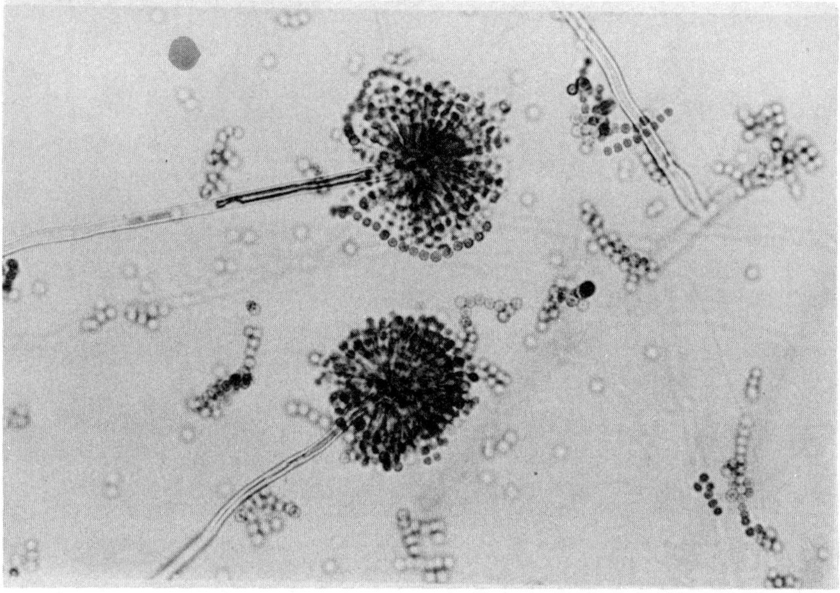

Figure 5:6. Lactophenol mount of *A. versicolor* conidiophore, 450X.

A. versicolor Group (Fig.5:6)

The species within this group vary in macroscopic appearance. The colonies may be white, yellow, orange, tan or yellow-green. Identification is based primarily upon the microscopic morphology. The conidiophores are smooth-walled, colorless and produce an elongate funnel-shaped vesicle. The phialides are mounted on metulae and produce chains of round conidia with rough walls. Conidial heads are radiate and some species produce Hülle cells.

A. flavipes Group (Fig.5:7)

Colonies are brown to buff colored and may resemble colonies of *A. terreus*. The conidiophores are thick-walled, smooth and light brown in color. The vesicle is oval to elongate and covered with sterigmata in a biseriate pattern. Conidial heads are loosely columnar, composed of spherical conidia. Hülle cells are present and irregular in morphology.

A. nidulans Group (Fig.5:8)

Colonies are multicolored, with green conidial heads, tan or yellow cleistothecia and a reddish pigment on the reverse side. The conidiophores

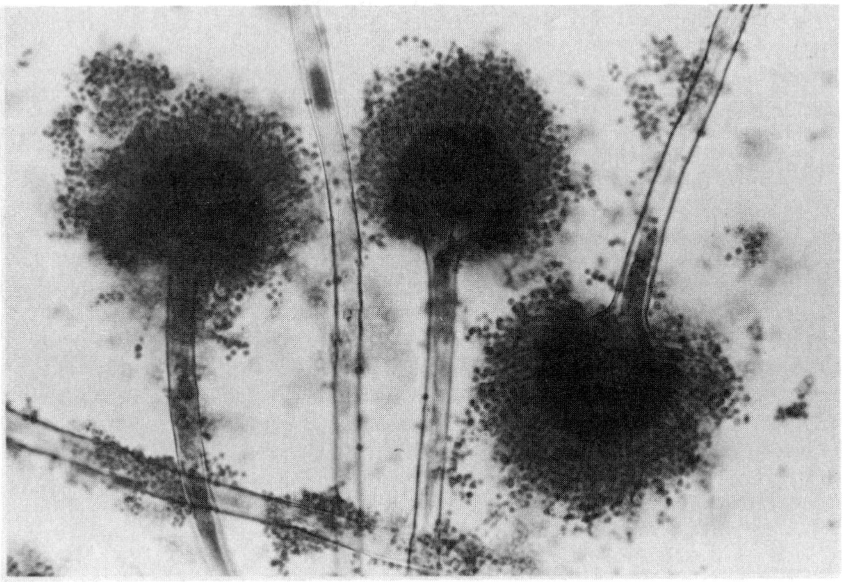

Figure 5:7. Lactophenol cotton blue mount of *A. flavipes* conidiophore, 450X.

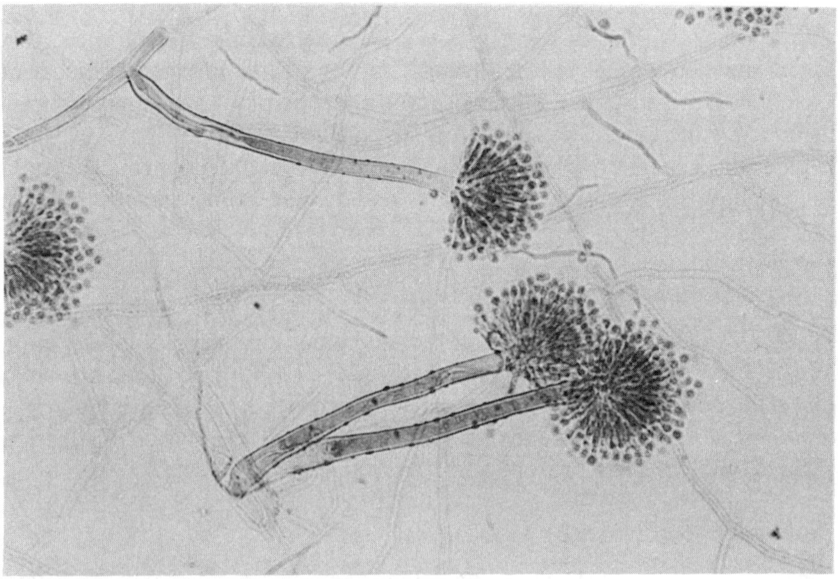

Figure 5:8. Lactophenol mount of *A. nidulans* conidiophore, 450X.

are brown, smooth-walled and terminate in a hemispherical vesicle covered with metulae and phialides (biseriate). Conidia are small, spherical and slightly roughened. Spherical, thick-walled Hülle cells are present and usually associated with the cleistothecia.

SIGNIFICANCE OF LABORATORY FINDINGS

Growth of *Aspergillus* spp. on culture media can be indicative of environmental contamination, colonization or invasive disease. Without histologic evidence, caution is required in making a diagnosis of aspergillosis based on the results of culture alone. However, since tissue is not always obtainable, direct microscopic findings and culture results should be communicated to the physician who must make a decision based upon both laboratory results and clinical findings.

Many laboratories feel the need to suppress reporting of "contaminated" cultures to minimize confusion from the reporting of extraneous isolates. However, there are no criteria by which the laboratory can discriminate between significant fungal isolates and those that are not. A report of the isolation of any fungus, including *Aspergillus*, is noteworthy in a patient population predisposed to opportunistic infections. Therefore, in the absence of a documented quality control problem associated with fungal contamination of media, all fungal isolates should be identified to at least the genus level and the results reported.

Bronchoalveolar lavage in patients with documented pulmonary aspergillosis has been used to compare the demonstration of morphologically compatible hyphae with growth in culture as measures for diagnosis.[114] An initial centrifugation to concentrate the specimen about 30-fold was followed by examination of stained cytocentrifuge slides and culture of the sediment. The presence of hyphae on direct examination, without consideration of growth in culture, had a 53% sensitivity and a 97% specificity for the diagnosis of invasive aspergillosis. Fungal cultures were positive in only 23% of patients and a combination of culture and microscopic results yielded an overall sensitivity of 58% and a specificity of 92%.

BLASTOMYCOSIS

Blastomycosis is a primary pulmonary infection of humans and a variety of other animals,[96,148] most notably dogs.[76] Infection may disseminate from an initial pulmonary focus to virtually any organ. For many years this organism was believed to produce two distinct diseases, a cutaneous form with the skin as the portal of entry and a pulmonary form when infectious particles were introduced into the lungs. It was not until the work of Schwarz and Baum[220] in 1951 that the concept of a single portal of entry in blastomycosis, i.e., the lungs, was accepted and the cutaneous form recognized as the result of dissemination from a primary pulmonary focus.

The etiologic agent of blastomycosis, *Blastomyces dermatitidis*, is dimorphic, growing as a large budding yeastlike form either in host tissue or at 35-37°C in vitro and as a typical moldlike form at lower temperatures, i.e., 25-30°C. The perfect stage, *Ajellomyces dermatitidis*, first described by McDonough and Lewis,[160] is an ascomycete with characteristics typical of the members of the family Gymnoascaceae. Both the yeastlike form[95] and conidia of the moldlike form[264] have been found capable of initiating infection and disease in experimental animals following inhalation. The precise nature of the infective particle(s) in human disease has not been defined, although it is generally assumed that the moldlike form exists in nature and that inhalation of aerosolized conidia alone or in combination with other small mold particles initiates pulmonary infection.[76]

Clinical Manifestations

Patients with blastomycosis may present with one of the following clinical forms: (1)self-limited primary or acute pulmonary disease which may resolve without antifungal therapy;[15,196,197,211] (2)chronic progressive pulmonary infection with or without involvement of extrapulmonary organs (e.g., skin,[19,221] bone,[16,171,219] urogenital,[60,106,175] central nervous system (CNS)[207] and other organs;[59,151,164] and (3)systemic disease without evidence of pulmonary involvement.[14]

Workers have postulated from observations of histoplasmosis and coccidioidomycosis that following airborne exposure to *B. dermatitidis* the majority of patients develop an influenza-like syndrome. Information obtained from recent point-source epidemics of blastomycosis[38,129] clearly demonstrates that a self-limited, acute pulmonary disease, similar to that found in histoplasmosis and coccidioidomycosis, is produced by *B. dermatitidis*. Data from these outbreaks, as well as observations following accidental exposure of laboratory workers[15] have shown that acute pulmonary disease develops after a mean incubation period of approximately 45 days (range = 21-106).[38,129,250] In contrast to cases in outbreaks, sporadic cases with self-limited acute infection are probably not seen by physicians and the diagnosis of blastomycosis is neither suspected nor documented.[197,211] Therefore, except for those in the few epidemics seen, patients that seek consultation and are diagnosed as having blastomycosis are more typically those with chronic nonresolving pulmonary or disseminated disease.

Chronic pulmonary blastomycosis may be difficult to diagnose by physical findings[210] and chest x-rays[92] or to differentiate from more common chronic lung diseases such as carcinoma or tuberculosis. This has led to surgical intervention in patients in whom the diagnosis of blastomycosis might have been made by careful scrutiny of cytologic specimens found negative for malignancy[208] and/or by performing serologic testing for the systemic mycoses.[265] It should be noted that, even if a mycosis is diagnosed, it is also possible for a malignancy and/or another infectious process to be present in the lungs at the same time.[25,174]

Disseminated blastomycosis is the result of extrapulmonary hematogenous spread of *B. dermatitidis* to one or more organ systems. Dissemination may occur concurrently with progressive disease of the lung or after spontaneous resolution of the pulmonary infection. *B. dermatitidis* usually disseminates to skin, bone, joints and the prostate gland,[221] but rarely to the gastrointestinal tract or central nervous system. However, in chronic meningitis, the diagnosis of blastomycosis must be considered, since untreated CNS infection may result in morbidity and mortality.[133]

Cutaneous lesions are so commonly observed in blastomycosis that it was initially believed to be a primary cutaneous infection.[80] But the cutaneous lesions are almost always secondary to hematogenous spread from a primary pulmonary focus[220] and they may develop months or even years after the primary infection. The lesions usually begin as small pustules which undergo central necrosis as they spread peripherally. A lesion enlarges to form an irregularly-shaped, crusted ulcer surrounded by raised erythematous margins. Lesions occur most frequently on the face, hands, feet, legs and ankles, but may occur anywhere on the skin.

Inoculation of *B. dermatitidis* directly into the skin, i.e., primary cutaneous blastomycosis, has been reported in workers in microbiology laboratories,[141,142] anatomic pathologists[143] and following a dog bite.[82] The resultant infection, which can be confused with sporotrichosis, takes the form of a mild, localized, chronic cutaneous lesion involving regional lymph nodes, which usually does not require systemic antifungal therapy. However, a reported accidental inoculation of an immunosuppressed patient[30] resulted in disseminated disease requiring antifungal therapy.

Endogenous reactivation has been described for blastomycosis,[144] but it is uncertain whether reactivation occurs because of the breakdown of pre-existing quiescent foci, as happens in histoplasmosis[102] and coccidioidomycosis.[94] Pulmonary and disseminated blastomycosis have been reported in immunosuppressed patients,[10,196,221] but AIDS has not been identified as a risk factor for disseminated blastomycosis as it has for histoplasmosis and coccidioidomycosis. Also, there is no distinct clinical syndrome associated with blastomycosis in immunocompromised patients[196] as has been reported for these other two mycoses, but blastomycosis in immunosuppressed patients results in the same range of disease seen in patients without immunologic defects.

Epidemiology and Public Health Significance

The endemic area for blastomycosis in the United States overlaps that for histoplasmosis. The distribution of human and canine cases in the United

States between 1885 and 1968 suggests that the endemic area for blastomycosis is even broader than that for histoplasmosis, extending farther eastward and northward.[76]

Blastomycosis occurs predominantly in North America, within the Mississippi River basin, the southeastern states and along the St. Lawrence seaway.[76] Confirmed cases have also been reported worldwide, notably in Africa,[9] India[8,193] and in the Middle East.[90,135] Although the African and American isolates are antigenically similar, attempts to mate them have resulted only in the production of sterile cleistothecia.[159,240]

Little is known about the incidence of primary pulmonary blastomycosis, due to the infrequent occurrence of point-source outbreaks. Suitable skin test antigens and serologic tests, which played an important role in defining the endemic areas for histoplasmosis and coccidioidomycosis, are not available for blastomycosis.

Delineation of the endemic area also has been difficult because of failure to define the natural habitat of *B. dermatitidis*. In turn, sparse knowledge of case distribution and circumstances of infection has probably impeded efforts to isolate *B. dermatitidis* from sources in nature. There have been only 3 published reports of isolation from soil[12,51,52] and 1 from pigeon manure.[214] There have been only 5 documented outbreaks in the United States; 1 in North Carolina,[232] 1 in Minnesota,[250] 1 in Illinois,[127] and 2 in Wisconsin.[38,129] Although attempts were made, *B. dermatitidis* was not cultured from the environment in 4 of these outbreaks. However, *B. dermatitidis* was recovered from a beaver dam in northern Wisconsin[129] in association with one large point-source outbreak of blastomycosis. The required reporting of blastomycosis in the State of Wisconsin resulted in a timely epidemiologic investigation and in the recovery of the organism from the environment. If blastomycosis were a reportable disease everywhere, our understanding of the epidemiology of blastomycosis might be greatly expanded.

Collection, Transport and Processing of Specimens

Handling and processing of specimens for the diagnosis of blastomycosis do not differ significantly from procedures used to diagnose other systemic fungal diseases (see Chapter 2, "Collection, Transport and Processing of Clinical Specimens"). Although *B. dermatitidis* is not as difficult to recover from sputum as is *H. capsulatum*, the etiologic agent is rarely known beforehand; therefore, all clinical specimens should be handled promptly. If a delay in processing is unavoidable, the specimen should preferably be stored at room temperature. Respiratory disease can be diagnosed by microscopic examination and culture of sputum. Multiple sputum specimen (a minimum of 3 and as many as 6) collected on consecutive days or at 2-3 day intervals are required for optimal recovery of *B. dermatitidis*. If examination of sputum does not yield a diagnosis, then a more aggressive approach using culture and microscopic examination of bronchoalveolar lavage fluid, post lavage sputum, transbronchial tissue biopsy, post-biopsy sputum or lung and lymph node specimens obtained at thoracotomy is necessary.

Selection of nutrient media for inoculation with clinical specimens depends upon the source of the material (see Chapter 2). Incubate primary cultures at 30°C rather than 25°C to induce more rapid growth and detection of fungi. This may not be practical in some laboratories since a special incubator is required. Laboratories lacking a 30°C incubator, or seeing yeastlike organisms on direct smears, should supplement their room temperature cultures by inoculating media capable of supporting the growth of the yeast phase and incubating them at 37°C in 5% CO_2. Kelley's agar is excellent for the growth of *B. dermatitidis* yeast phase and should be used in the primary setup when direct microscopic examination shows cells compatible with *B. dermatitidis*. It should be noted that cycloheximide inhibits the yeast form of *B. dermatitidis*.

In addition to sputum, the clinical specimens most likely to yield *B. dermatitidis* include pus, tissue biopsy from involved organs such as lung, prostatic fluid, urine and joint fluids. The usefulness of culturing prostatic fluid takes on added significance because as many as 11.5% of cases have been reported to have prostatic involvement.[221] Blood is unlikely to yield *B. dermatitidis*; the first and only report of isolation from blood was in 1987.[176]

Since *B. dermatitidis* grows readily on common laboratory media, its failure to grow in culture is due to improper handling or processing of specimens. The effects of lidocaine use in bronchoscopy procedures,[244,245] delay in specimen processing or improper storage conditions before culture[246] can result in failure to grow the organism.

Identification of Etiologic Agent

Direct microscopic examination of materials such as sputum or pus for *B. dermatitidis* yeasts is an important adjunct to culture in the laboratory diagnosis of blastomycosis. These yeasts are typically large (8-15 μm in diameter), have a thick refractile wall ("double-contoured") and lack a capsule. Single buds attached to the mother cell by a broad base (4-5 μm wide) often remain attached until they have reached a size equal to that of the mother cell (Fig.5:9). Occasionally, chains of connected yeast (Fig.5:10) and small yeast forms (2 μm in diameter) are observed.[221] Clinical mycologists have traditionally examined materials such as sputum, bronchial lavage fluid or purulent exudates with unstained wet mounts to which potassium hydroxide (KOH) was added to lyse interfering host cells, e.g., polymorphonuclear leukocytes (PMN). The sensitivity of this procedure depends

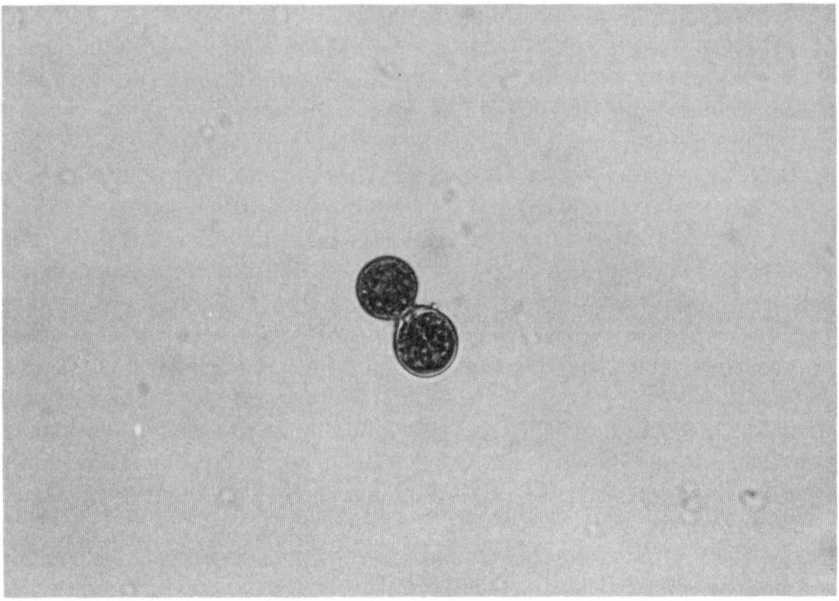

Figure 5:9. Lactophenol cotton blue mount of *B. dermatitidis* yeasts.

upon the number of organisms present and the expertise of the microscopist. Potassium hydroxide treatment of purulent material often results in the formation of artifacts which may resemble fungal cells to the inexperienced observer. Therefore, preparations both with and without KOH should be examined. The yeast phase of *B. dermatitidis* is more refractile than the artifacts and under optimal illumination often has a visible internal structure (e.g., the inner aspect of the thick cell wall), while non-fungal objects appear homogeneous and translucent (Fig.5:10). The use of phase contrast microscopy[205] and fluorescence with calcofluor staining[169,170] facilitates the observation of fungal elements in wet preparations. Alternately, dried, fixed smears stained with the periodic acid Schiff stain (PAS) readily reveal large, broad-based, budding yeasts of *B. dermatitidis*.

Due to the infrequent occurrence of the systemic mycoses compared to tuberculosis or pulmonary malignancies, specimens may not be submitted to the laboratory specifically for the detection of fungal pathogens. Therefore, cytotechnologists using the Papanicolaou stain to examine sputum for malignant cells should be trained to differentiate between blastoconidia of *Candida* spp., which are common in these specimens, and the larger thick-walled yeasts of *B. dermatitidis* (Figs.5:11, 5:12). Specimens submitted for bacteriologic culture and Gram stain probably will not reveal *B. dermatitidis* yeasts microscopically, since these forms do not retain the dyes used in the Gram stain. However, when initial cultures do not identify a bacterial

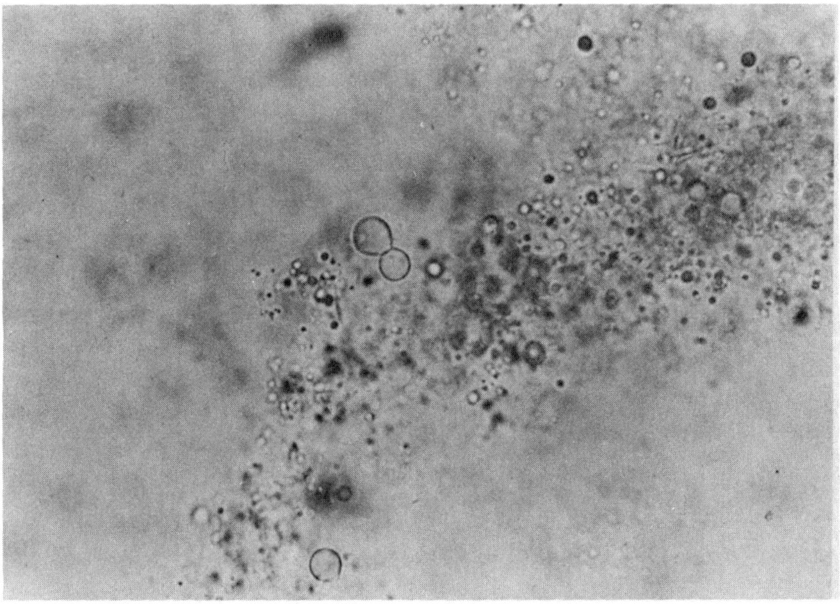

Figure 5:10. Potassium hydroxide preparation of pus from osteomyelitis lesion showing broad-based budding yeast (reproduced with permission of the Clinical Microbiology Newsletter).

etiology and the differential diagnosis is broadened to include a fungal infection, the original Gram stained smear can be decolorized, stained with PAS and reexamined for fungal cells.[227] There have been reports of the detection of hyphal forms in clinical material but such forms appear to be rare.[10,40,93]

In tissue specimens, the yeast forms are seen by examination of sections stained with PAS, Gridley's or Gomori methenamine silver (GMS) stains. The organisms are found in lesions characterized by suppuration and epithelioid cell granulomatous reactions with giant cells. Although not common, small *B. dermatitidis* yeast forms have been observed in tissue and cannot be differentiated from *H. capsulatum* yeasts by morphologic criteria alone.[203] *B. dermatitidis* yeasts in tissue are similar in size and therefore resemble *Cryptococcus neoformans*, but usually they can be separated by the mucicarmine stain which stains the carbohydrate capsule of *C. neoformans*. There are infrequent infections with a *C. neoformans* strain which does not produce a visible capsule and are thus more difficult to differentiate. Specific fluorescent antibody tests are useful in the differentiation of yeast forms in formalin-fixed and paraffin embedded tissue. The assays are currently available only at a few large reference laboratories, including the Centers for Disease Control (CDC).[118]

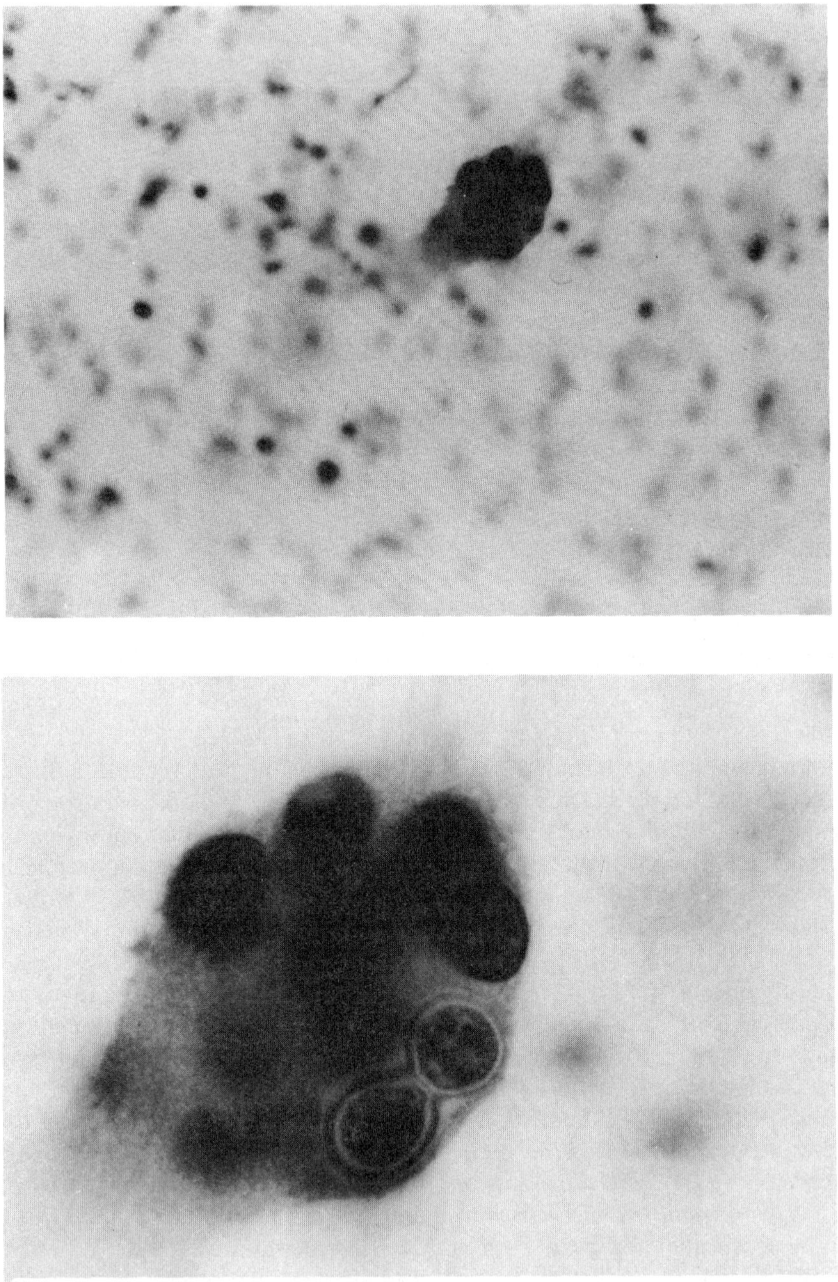

Figure 5:11. **Papanicolaou stained sputum specimen: (a) low power view of a giant cell; (b) higher magnification revealing the presence of a budding yeast consistent with *B. dermatitidis* (kindly provided by James E. Williams).**

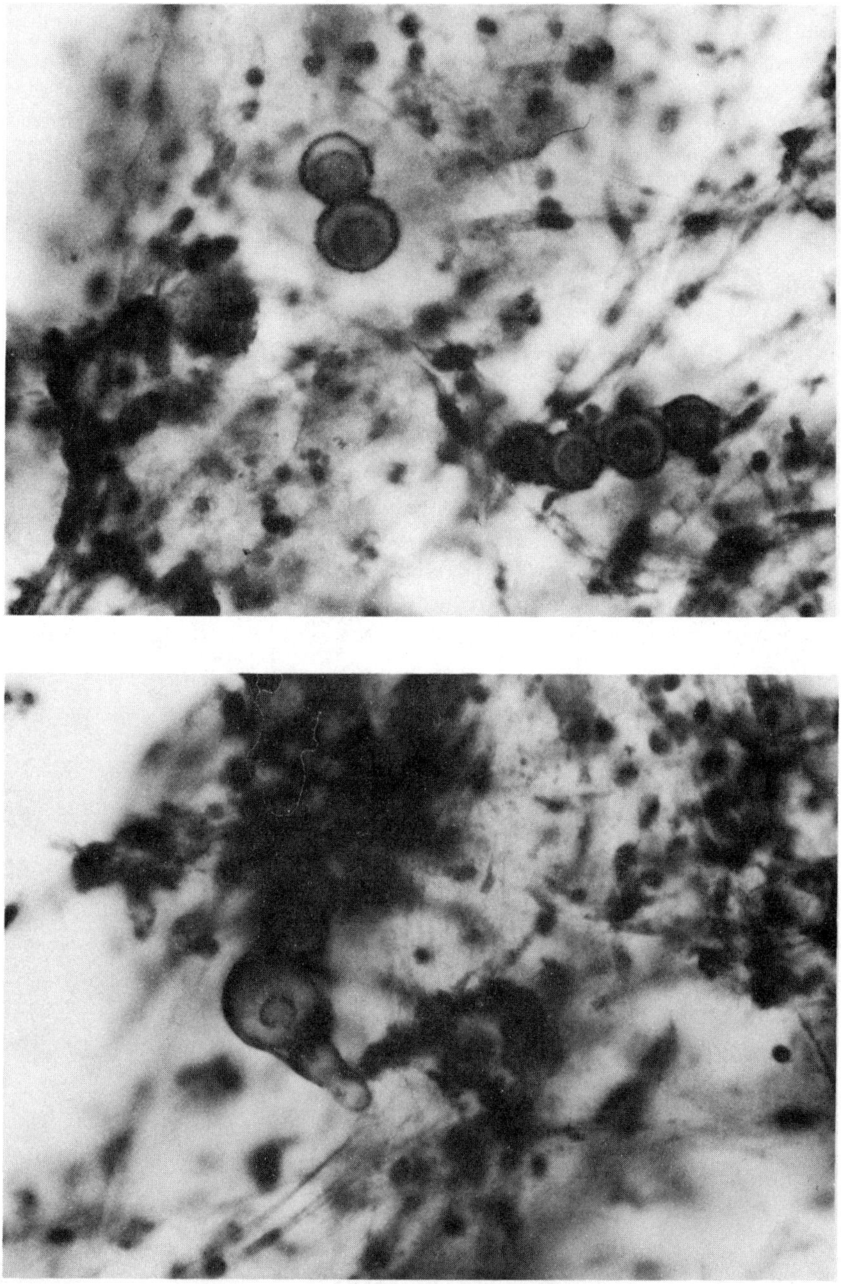

Figure 5:12. Papanicolaou stained sputum specimen: (a) typical broad-based budding cell and a chain of yeasts; (b) aberrant forms (kindly provided by James E. Williams).

Depending upon the source, the use of antifungal therapy, presence of competing microbial flora and the number of yeast cells present in the original specimen, incubation of cultures at 30°C will result in the appearance of the mold form in 1-6 weeks (rarely growth may be evident after 3-4 days). Colonies vary from white to tan in color, glabrous to cottony in texture and may or may not have abundant aerial mycelia. A "typical" colony appears as aerial growth which is moist and closely adherent to the agar surface, forming coremia which appear as protruding spicules (Fig.5:13). However, the appearance of the mold colony is not diagnostic and can resemble that of other fungi, notably *H. capsulatum* or saprobic molds, e.g., *Chrysosporium* spp. Globose, smooth-walled conidia which average 6 μm (range 4-10 μm) in diameter, are produced at the ends of slender conidiophores (Fig.5:14). In other isolates, conidia may be absent or sparse and it may be necessary to inoculate several different media (e.g., soil extract agar, Pine-Drouhet mold agar) before conidia are detected.

The definitive identification of *B. dermatitidis* is based on the thermal-dependent interconversion of mycelial and yeast forms of the fungus. To convert to the yeast form, inoculate a portion of the mold culture onto a slant of conversion media (e.g., Kelley's agar) and incubate at 37°C in 5% CO_2 in air. Moisture is important in the conversion process and ca. 0.5 mL of a similar liquid medium or sterile water should be added to each slant

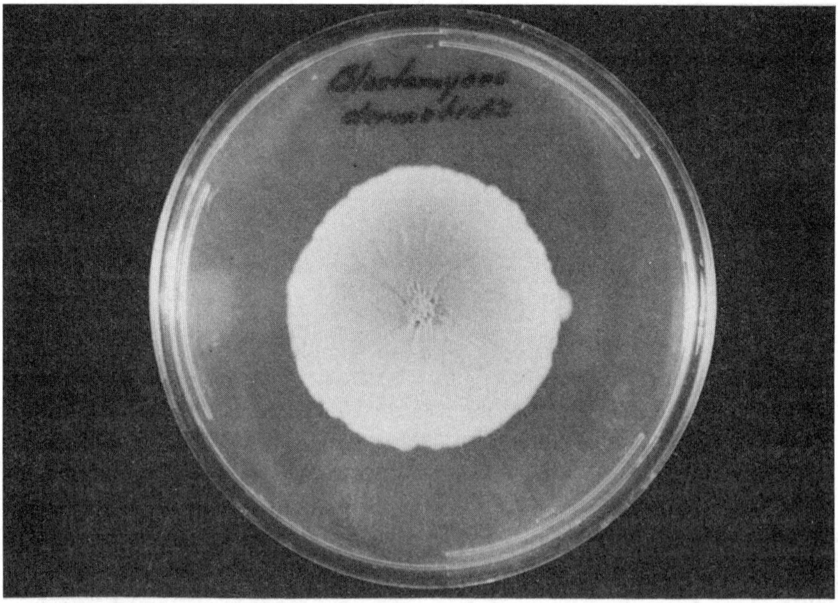

Figure 5:13. Colony of *B. dermatitidis* mold on Sabouraud dextrose agar showing coremia in the center.

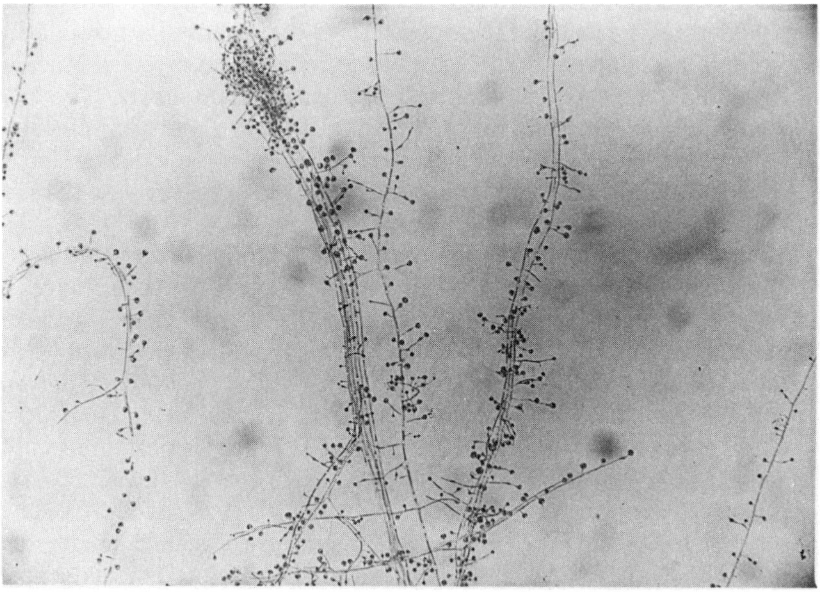

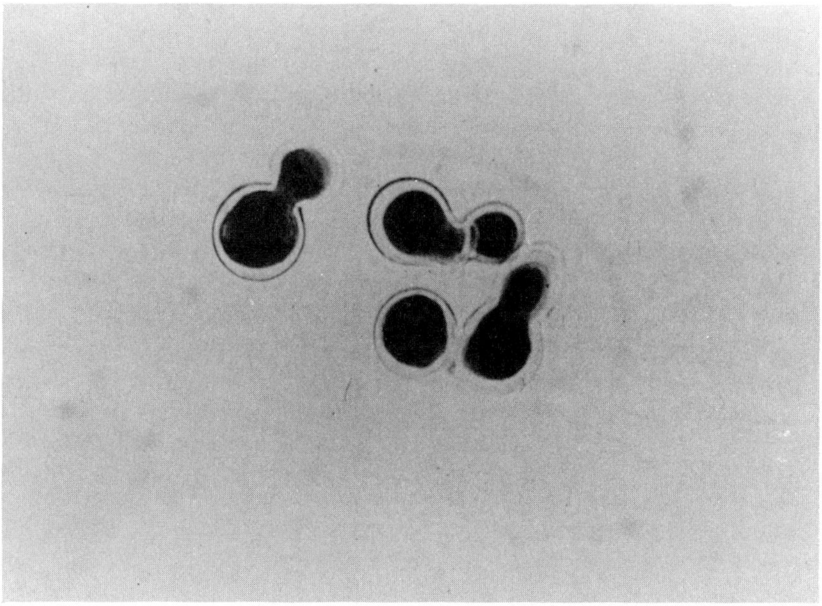

Figure 5:14. (a) Lactophenol cotton blue mount of a slide culture of *B. dermatitidis* mold phase showing abundant microconidia; (b) lactophenol cotton blue mount of *B. dermatitidis* yeast converted in vitro.

prior to inoculation. After a few days' incubation, the growth will begin to appear glabrous, typically at the periphery of the colony. Portions of these areas should be sampled, taking care not to remove the mycelial inoculum, and examined microscopically for broad-based, budding yeasts. If no yeasts are seen, transfer the glabrous material to a fresh slant and reincubate. Although several such transfers may be required to achieve conversion, the appearance of broad-based, budding yeasts with residual hyphae (Fig.5:15) is sufficient evidence to identify *B. dermatitidis*.

An occasional isolate of *B. dermatitidis* may not produce typical colonies, sporulate or readily convert to the yeast phase.[222] In such instances, an exoantigen test has been used to identify *B. dermatitidis*[223] and other medically important fungi. Exoantigen test reagents are available from several commercial sources. The technique requires growth of the mold form on an agar slant for 1-4 weeks. The slant is then flooded with 10 mL of 0.02% (w/v) aqueous merthiolate and held at room temperature for 24 h. The fluid is filter-sterilized with an 0.45 μm filter, concentrated 50-fold in an Amicon B-15 concentrator (Amicon Division, W. R. Grace and Company) and assayed for specific antigens by immunodiffusion against antisera to *B. dermatitidis, C. immitis* and *H. capsulatum* antigens. The presence of precipitin bands identical to the reference A precipitin band is considered specific for *B. dermatitidis*. Bands of nonidentity or partial identity with

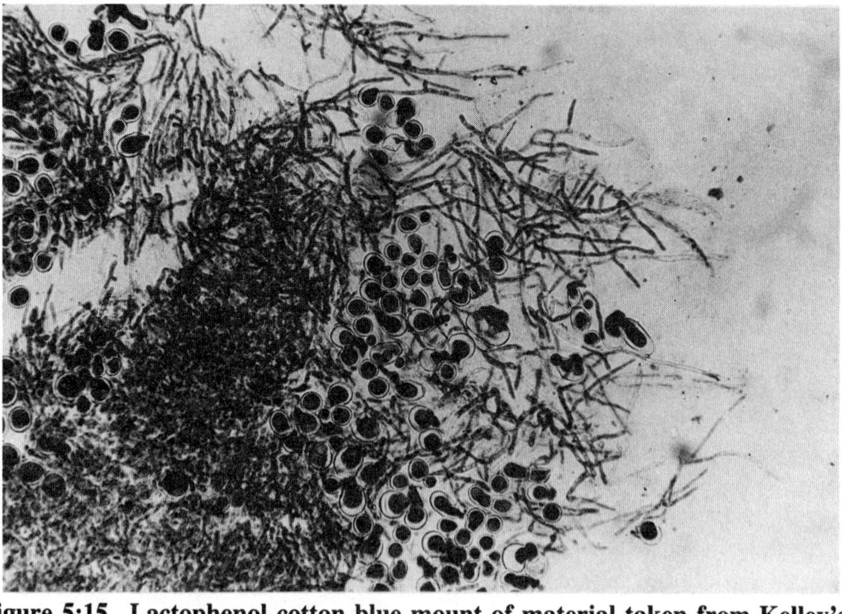

Figure 5:15. Lactophenol cotton blue mount of material taken from Kelley's agar incubated at 37°C for 7 days.

reference precipitin bands are ignored. Depending upon the modification of the method used, macro- versus microimmunodiffusion and the source of the reagents, a significant percentage (32%) of slant cultures of *B. dermatitidis* do not have detectable A antigen in their extract concentrates.[53] However, A antigen was detected when these isolates were retested by growth in brain heart infusion (BHI) broth shake-flasks.

Although the original description did not specify filter-sterilization of extracts prior to testing, it is clear that 0.02% aqueous merthiolate treatment is not adequate to sterilize the extract.[236] Therefore, all extracts must be prepared and filter-sterilized while working in a biohazard cabinet prior to concentration and testing.

Antimicrobial Susceptibility and Resistance

Since its introduction, amphotericin B has remained the drug of choice for life-threatening forms of blastomycosis. Rates of clinical response to this drug range from 66-93%[57] and depend upon the form of the disease and the immunologic status of the patient. There are no documented instances of in vitro resistance of *B. dermatitidis* to amphotericin B, yet treatment failures do occur and the side effects of therapy make an alternate antifungal drug desirable. Hydroxystilbamidine has been replaced by the imidazoles for therapy of mild blastomycosis. Two imidazoles, miconazole administered intravenously and ketoconazole administered orally, are used. Ketoconazole now has a definite role in the treatment of immunocompetent patients without life-threatening blastomycosis.[24,57]

Evaluation of Laboratory Findings

The isolation of *B. dermatitidis* from any human source is diagnostic for blastomycosis. Since identification may require weeks, the demonstration of typical yeast forms in clinical material is sufficient evidence to begin antifungal therapy. Therefore, preliminary findings must be immediately communicated to the physician. Because a morphologic identification is only presumptive and culture results may be pending, supportive evidence should be sought. Serologic testing should be done and cytologic and histopathologic material reexamined.

Commercially available serologic procedures include complement fixation (CF) and double immunodiffusion (DID). The CF test for blastomycosis is often negative and of little diagnostic value. The fungal reagents developed to date are crossreactive with antigens in sera from patients with other mycoses, especially histoplasmosis. Often sera from patients with culture-proven blastomycosis will react with one or both *H. capsulatum* antigens. The main problem, however, is that many patient with culturally proven blastomycosis may not have detectable levels of antibody. This is especially true when the disease process is not extensive. Among the newer serologic procedures is an immunodiffusion test that detects 80% of proven cases of blastomycosis seen in hospitals and to date is 100% specific.[122,128,265] An enzyme-linked immunosorbent assay (ELISA) test measuring antibody to the A antigen is promising for screening sera, which if positive, should be confirmed by DID.[128] Therefore, utilization of the DID rather than the CF for diagnosis is a useful and productive change (see Chapter 8, "Serodiagnosis of Human Fungal Infections" for a further discussion of serology).

COCCIDIOIDOMYCOSIS

Coccidioidomycosis was first recognized as a chronic progressive disease with cutaneous lesions[190] and thought to be caused by a protozoan. In 1900, Ophüls established that *Coccidioides immitis* is a fungus[182] and in 1905 confirmed its dimorphic nature.[181] The teleomorph of *C. immitis* has not been observed and therefore, its precise taxonomic position is unknown. Its anamorphic growth features have led some taxonomists to speculate that *C. immitis* is a zygomycete. Endospores are formed asexually within a sack (sporangium) either in vivo or in vitro under specific environmental conditions. *C. immitis* grows as a mold in the environment and in the laboratory at 25-37°C. The tissue form, the spherule, which develops from arthroconidia is unicellular (nonfilamentous) and develops in vitro only in special media (e.g., Converse).

As with other systemic mycoses, coccidioidomycosis was initially considered to be uniformly fatal. However, Dickson's careful studies in 1938 revealed that an acute self-limited form was much more common than the progressive disseminated form.[55] Military operations associated with World War II in the southwestern states, i.e., California and Arizona, resulted in a rapid expansion in knowledge of coccidioidomycosis.

The development of a sensitive and specific skin test reagent from the mold form (coccidioidin) and diagnostic serologic tests, i.e., tube precipitin and CF using coccidioidin, allowed investigators to establish the following important features of the disease: (1) coccidioidin skin test-positive individuals are not susceptible to reinfection; (2) coccidioidin skin test-negative persons entering an endemic area are susceptible to pulmonary infection; (3) primary pulmonary disease is usually self-limiting and results in conversion from negative to positive skin test status; (4) development of disseminated disease is more frequent in non-Caucasians; (5) endemic areas are restricted, with the highest incidence in Arizona; and (6) the incidence of disease varies in a seasonal pattern with the highest rates in late summer.

Clinical Manifestations

There are various ways to categorize the spectrum of coccidioidomycosis, but in general it includes asymptomatic primary pulmonary infection, symptomatic primary pulmonary infection, progressive chronic pulmonary disease and disseminated disease. These categories are not sharply delineated and often merge to form a continuous spectrum. Regardless of anatomic site, the histopathological response varies from granulomatous or suppurative to a mixture of both.

It has been estimated that 60% of those infected with *C. immitis* remain asymptomatic with skin test conversion as the only evidence of exposure.[231] The remaining 40% present a broad range of symptoms and severity of disease. The majority of symptomatic primary pulmonary infections are characterized by cough, fever, chest pain and recovery in 2-3 weeks without sequelae. However, approximately 5-10% of patients develop residual pulmonary disease.[58]

In addition to pulmonary involvement, there may be allergic manifestations in primary disease, including toxic erythema, erythema nodosum and erythema multiforme. Toxic erythema is a generalized rash which usually occurs 2-3 days after exposure to the fungus and before the development of a skin test reaction. It is reported most often in children and young adults and may be confused with other diseases such as scarlet fever. The syndrome known as "Valley Fever" is manifest as erythema nodosum or erythema multiforme, with or without arthralgias. Erythema nodosum lesions are typically restricted to the lower extremities, appear as red painful nodules under the skin and develop in the same time frame as the delayed hypersensitivity response to coccidioidin. Erythema nodosum is equated with a strong host immune response and a low probability of developing disseminated disease.[58]

Patients with persistent coccidioidomycosis lasting 6-8 weeks or longer present with chronic progressive pneumonia, miliary disease, cavities or nodules. Chronic pulmonary disease is characterized by prolonged symptoms, including cough, fever, weight loss, chest pain and a persistent CF titer, but without evidence of dissemination.[212] Miliary disease often is associated with a fulminant course and dissemination, especially in the immunosuppressed patient; serology is positive and fungemia has been reported.[58] Cavitation in symptomatic pulmonary disease is observed in fewer than 10% of patients (primarily Caucasians).[58] Pulmonary nodules have been shown to harbor viable *C. immitis*[204] as long as 15 years following primary disease[41] and thus may serve as a source for disseminated disease when the host is immunosuppressed.

Dissemination via hematogenous spread may occur in primary acute disease, persistent progressive pulmonary disease or in the immunosuppressed.[7,153] Disseminated disease occurs in only a small number of individuals, i.e., children less than 5 years old, adult males over 50 years of age, women in the 3rd trimester of pregnancy,[32] non-white males[110] and immunosuppressed individuals.[54] Dissemination from the primary pulmonary focus may result in disease involving any organ, including skin, bone, joints, liver, genitourinary and central nervous systems.[58] Since dissemination to the CNS is associated with a high rate of morbidity and mortality, patients at risk should be evaluated for CNS involvement.[58] The gastrointestinal tract is usually spared, as is not the case in histoplasmosis and paracoccidioidomycosis. Cutaneous lesions are a common manifestation of disseminated coccidioidomycosis but, as in other systemic mycoses, primary cutaneous disease is only associated with laboratory accidents.[234]

Epidemiology and Public Heath Significance

Coccidioidomycosis is geographically restricted to the semi-arid regions of the Western Hemisphere, including parts of Mexico, Guatemala, Honduras, Venezuela, Paraguay, Colombia and Argentina. In the United States, infection is associated with areas of southern California (e.g., San Joaquin Valley), southern Arizona, Nevada, New Mexico, parts of Utah and southwest Texas.

Maddy first recognized the similarity of the endemic areas to the Lower Sonoran life zone, a biologic zone characterized by hot summers (26-32°C mean temperature) and few winter freezes (4-12°C mean temperature), a low altitude and an alkaline soil.[155] It is of interest to note that *C. immitis*, at least in the laboratory, is capable of growth at 40°C, over a pH range of 3.5-9.0 and at high salt concentrations. In short, it has taken advantage of an ecologic niche with few microbial soil competitors. Outside of this environment, *C. immitis* appears less able to compete with other soil organisms and thus the endemic area does not enlarge even though the organism has been transported by wind or other means to areas where such optimal soil conditions do not exist.[184]

The fungus presumably grows following the rainy season when the soil is optimally moist. It then forms arthroconidia which are easily detached and disseminated by the wind as the soil dries. Since these weather conditions are seasonal, the disease is likewise seasonal to some extent. However, arthroconidia, once formed, are very hardy and will survive to remain infectious. Primary coccidioidomycosis occurring outside a recognized endemic area is thought to result from infection by spores carried on transported materials, as by California cotton to textile mills in the southeastern United States.[79]

Coccidioidomycosis is a significant public health issue in endemic areas. In some highly endemic areas, such as the lower half of the San Joaquin Valley and southern Arizona, 70% or more of the inhabitants become infected, with 40% of those infected expected to develop symptoms. There were 442, 500, 375 and 1095 new cases reported in California during 1975, 1976, 1977 and 1978, respectively.[183] The dramatic increase in cases in 1978 was the result of massive dust storms which spread *C. immitis* arthroconidia over large areas of California, including those in which the disease was not endemic.[184] Besides the introduction of arthroconidia into nonimmune populations, susceptible individuals move into endemic areas. A study of the impact of coccidioidomycosis on University of Arizona students from nonendemic areas found an annual number of cases of 25-55 (172 total) in a population with only 34% skin tests reactivity to either coccidioidin or spherulin.[126]

Knowledge of acquired resistance to second infection following recovery from a self-limited infection encourages a continued search for an effective means of artificial immunization. This vaccine would not be used to prevent coccidioidomycosis in the general population, but to modify the course of a naturally acquired infection and to reduce the incidence of the more severe progressive and disseminated forms in high risk populations (e.g., pretransplant patients). A vaccine of killed spherules has been tried on a limited scale, but with minimal success.[266]

Collection, Transport and Processing of Specimens

In primary asymptomatic or symptomatic pulmonary coccidioidomycosis with influenza-like symptoms, a diagnosis usually depends upon skin test or serologic conversion. Cough may be nonproductive and the mild course of disease does not warrant invasive measures.

The specimens collected depend upon the area of localization. Sputum is the most common specimen submitted to the laboratory from patients with persistent or progressive pulmonary disease. Regardless of the microscopic findings, all specimens are cultured (see Chapter 2). Unless there is microscopic evidence of *C. immitis* in 2 or more sputa, a minimum of 3 and not more than 6 specimens should be submitted. Bronchial aspirates are submitted when sputum cannot be produced nor induced. When bronchial lavage or sputa prove negative, cultures of lymph nodes and lung biopsy are appropriate. Empyema fluid, when present, or cavitary aspirates are also useful in diagnosing pulmonary disease. Pus is obtained from a closed abscess by percutaneous aspiration. Skin or mucous membrane lesions frequently accompany progressive coccidioidomycosis and can be curetted or biopsied for microscopy and culture.

In suspected CNS involvement, large volumes (ca.10-15 mL) of cerebrospinal fluid (CSF) are concentrated by centrifugation or filtration before culture and direct microscopy. *C. immitis* is not often isolated from blood unless the host is receiving high-dose steroid therapy[54] or is otherwise immunosuppressed.[7,27] Coccidioidouria occurs in as many as 25% of patients;[48] therefore, the sediment from 2000 x g centrifugation of approximately 200 mL of urine should be cultured on media which will inhibit the growth of bacterial contaminants (see Chapter 2).

Serologic testing is a useful adjunct to culture when coccidioidomycosis is suspected. The detection of antibodies can help to make a diagnosis;

antibody titers have prognostic value (a rising titer indicating possible disseminated disease) and are useful in monitoring the efficacy of chemotherapy (a falling titer indicating improvement). Antigenemia in coccidioidomycosis has been detected with radioimmunoassay[77] and ELISA[258] tests. However, neither of these assays are commercially available.

Identification of Etiologic Agents

C. immitis grows in tissues or body fluids in the form of spherical bodies called spherules, which range from 5 to 200 μm in diameter, but usually reach a diameter of only 30-60 μm. At maturity, the spherules have a thick wall (up to 2 μm) and are filled with globular or irregularly shaped spores called endospores that range from 2-5 μm in diameter (Fig.5:16). A mature spherule may contain anywhere from a few to several hundred endospores, either dispersed peripherally or throughout the spherule. After sporulation is complete, the wall of the spherule ruptures, releasing the endospores. These liberated spores gradually enlarge and develop into mature endosporulating spherules.

Mature spherules are recognized microscopically by their thick refractile walls and the presence of endospores (Fig.5:17). However, large mononu-

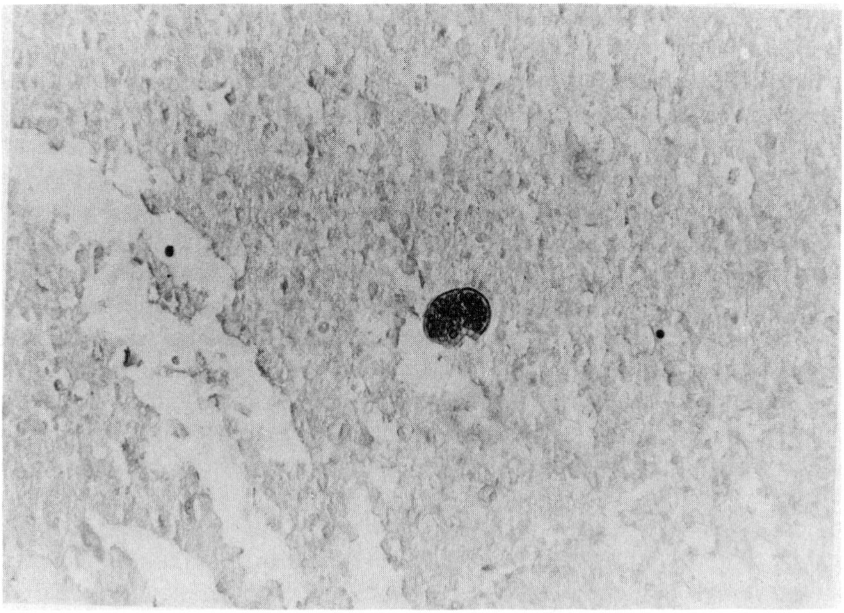

Figure 5:16. Gomori methenamine silver stain of a mature *C. immitis* spherule from a gluteal abscess.

clear cells filled with granules resemble mature spherules filled with endospores. The granules, unlike endospores, will not germinate and this can be determined by incubating a microscopic preparation (without KOH) for a few hours at 30°C in a moist chamber (to prevent dehydration). If the cell is an endospore, hyphae will protrude in all directions. Endospores recently released from a parent cell are virtually impossible to differentiate from host cell constituents, other debris or small yeasts in specimens from contaminated sources. If the sample is purulent, a drop of KOH is added and the preparation gently heated to digest host cells. Since KOH creates artifacts which may be confused with free endospores and thereby obfuscates the interpretation of the preparation, each specimen must also be examined without KOH.

Diagnosis of coccidioidomycosis by direct microscopic examination is contingent upon finding the fungus in its reproductive stage, viz., mature endosporulating spherules. A presumptive diagnosis of a systemic mycosis, albeit unidentified as to etiologic agent, can be made by the finding of an immature spherule. Hyphae or arthroconidia, rather than spherules, or a mixture of forms have been seen in sputum from patients with old cavitary lesions[192] or in abscess drainage.[267]

The nature of the container for primary isolation agar medium is a point of debate (see Chapter 2). Because of the highly infectious nature of the arthroconidia produced in culture, only culture tubes at least 25 x 150 mm with slanted agar surfaces should be used. Use of petri dishes, especially in geographic areas with a high incidence of coccidioidomycosis, is not recommended. Many isolates of *C. immitis* sporulate prolifically in culture and these highly infectious conidia are easily aerosolized. Therefore, all culture manipulations must be done in an appropriate biohazard cabinet. To minimize the hazard, transfer of cultures or preparation of tease mounts for microscopic examination should be attempted while growth is still young and sporulation limited.

Colonies of *C. immitis* usually appear by the 3rd to 5th day of incubation. Early growth is moist, slightly convex, and grayish in color, with a slightly granular surface. Characteristically, *C. immitis* forms abundant, rapidly growing, white aerial mycelium but there is considerable variation among isolates.[103] Some isolates remain glabrous, producing little aerial mycelium, especially in the central portion. Because of the rapidity of growth, routine bacteriological cultures incubated at 37°C can grow *C. immitis*. Therefore, technologists must be alert to the potential of the presence of the mold form on media used for culture of bacterial agents.

Microscopic examination of aerial growth from young cultures usually reveals the formation of rectangular or barrel-shaped arthroconidia (2-10 μm in diameter) that alternate with smaller empty cells (Fig.5:18). The empty cells rupture easily to free the arthroconidia, often leaving remnants of cells on the ends of the arthroconidia. However, the microscopic

(a)

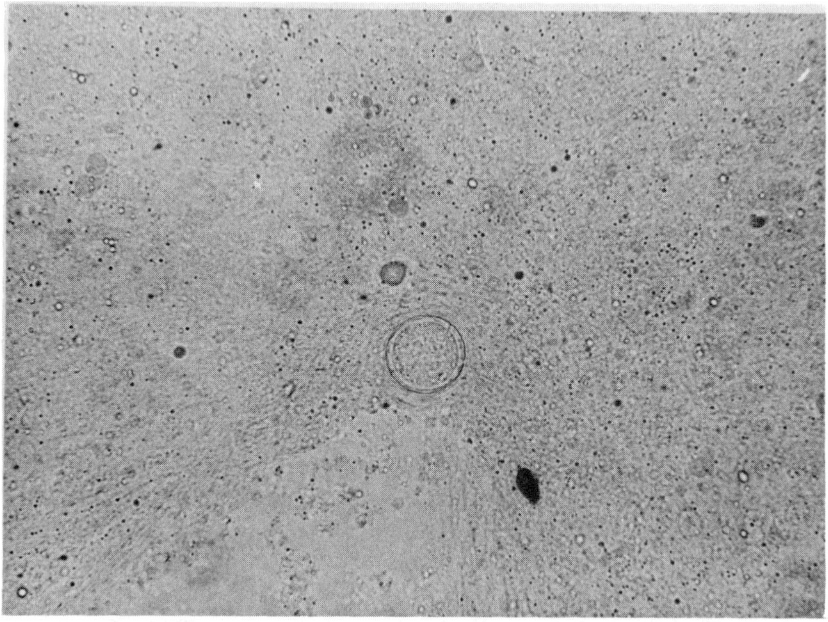

(b)

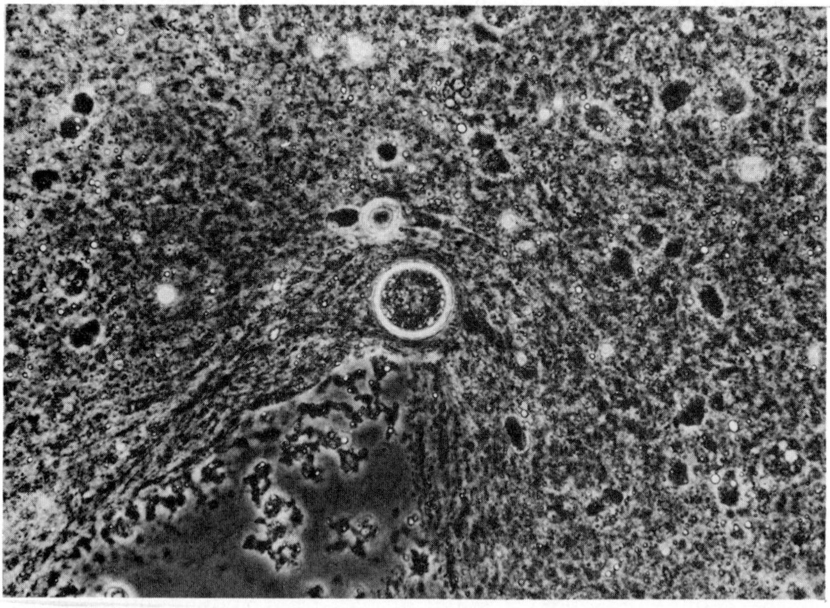

Figure 5:17. Potassium hydroxide preparations of pus from a gluteal abscess: (a) brightfield examination of an immature spherule of *C. immitis*; (b) same field with phase contrast; (c) brightfield examination of a ruptured mature spherule of *C. immitis*; (d) same field with phase contrast.

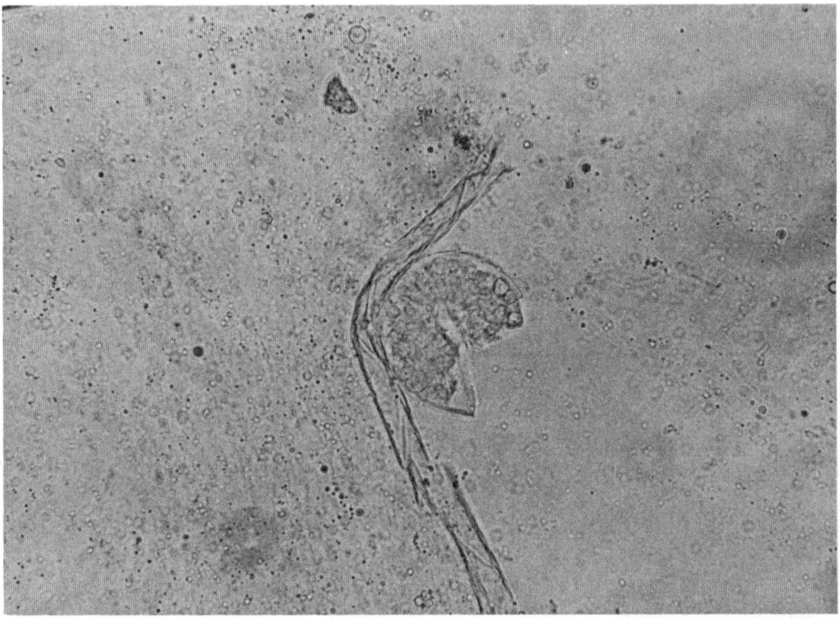

(c)

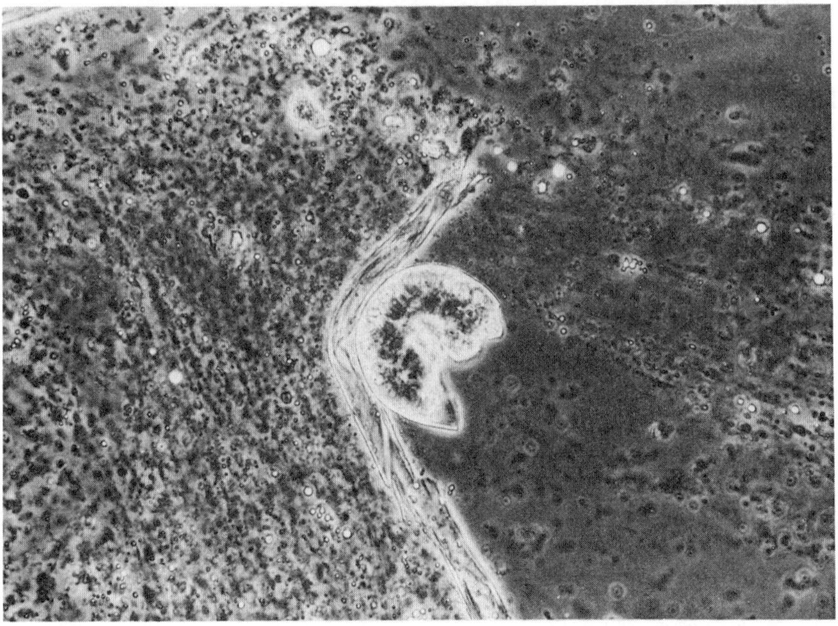

(d)

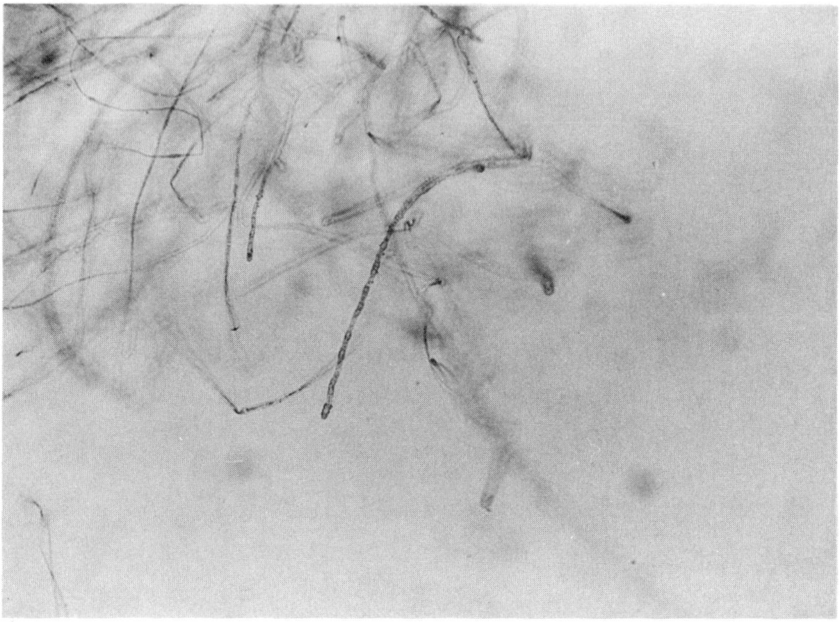

Figure 5:18. Lactophenol mount of the mold phase of *C. immitis* showing arthroconidia.

appearance varies in that arthroconidia may be thin-walled, elongated and lacking the alternate empty cells, or round rather than rectangular. Furthermore, contaminating saprobic fungi, e.g., *Oidiodendron* spp., *Auxarthron* spp. and *Malbranchea* spp., can be morphologically similar to typical *C. immitis*.[67,229]

Conversion of hyphal forms to endosporulating spherules is necessary for *C. immitis*. This can be accomplished by animal inoculation or incubation in a special medium under special conditions. Prepare a suspension of conidia and hyphae by flooding the surface of an agar culture with physiological saline containing a wetting agent such as Tween 80 (0.2-0.5%). Rub the surface growth gently with a stiff inoculating wire. Perform all these hazardous procedures within a biohazard cabinet. Transfer the suspension to a rubber-stoppered "vaccine" bottle for easy withdrawal with a needle and syringe. Inoculate a guinea pig intratesticularly with about 0.1 mL of suspension. If the fungus is *C. immitis*, a marked orchitis will develop within 2 weeks, when fluid can be withdrawn from the testes with a sterile syringe and needle for microscopic examination. If orchitis does not develop, the animal should be held for 2 more weeks and then sacrificed to examine the testes. If mice are used, several animals are inoculated intraperitoneally with no more than 1 mL of suspension per animal. Sacrifice one mouse at the end of a week and examine any exudate or lesion microscopically. If no evidence of coccidioidomycosis is found in the first mouse, repeat the procedure with the remaining mice at 2 and 4 weeks. Depending on the virulence of the isolate, the infectious process will be confined to the peritoneal cavity or spread to the viscera, especially the lungs. The finding of mature spherules containing endospores verifies the identification of *C. immitis*.

The in vitro conversion of the mold forms of *C. immitis* to spherules has been described as an alternate to animal inoculation. It is known that some isolates can be induced in vitro to form endosporulating spherules from arthroconidia by careful control of growth conditions. The procedure has been adapted for utilization in the clinical laboratory[242] and requires the preparation of slide cultures on modified Converse medium incubated in 20% CO_2 in air at 40°C. Since the manipulation of the fungus for inoculation of conversion medium presents considerable hazard, especially in laboratories unaccustomed to working with *C. immitis*, both animal inoculation and the in vitro demonstration of spherule formation should only be attempted by experienced personnel in reference laboratories. An exoantigen test (see Blastomycosis section for technique) offers a safer method for the identification of *C. immitis* cultures.[53,123] None of the fungi that morphologically resemble the mold phase of *C. immitis*, e.g., *Auxarthron* or *Malbranchea*, produced extracts that formed precipitin bands of identity with the reference system.[104] The antigen present in extracts of *C. immitis* appears to be the heat stable (HS) form.[42] Exoantigen testing is highly sensitive and specific, justifying a presumptive identification of *C. immitis*.

Antimicrobial Susceptibility and Resistance

As for other systemic mycoses, amphotericin B remains the most effective antifungal agent available for treatment of coccidioidomycosis. In contrast to the successful use of ketoconazole in certain forms of histoplasmosis and blastomycosis, the utility of this drug in the treatment of coccidioidomycosis is limited, since *C. immitis*, unlike *B. dermatitidis* and *H. capsulatum*, has shown in vitro resistance (Minimal Inhibitory Concentration $\geq$100 μg/mL) to the drug. However, no correlation between in vivo and in vitro data could be made.[225]

Evaluation of Laboratory Findings

Isolation of *C. immitis* from a patient without disease has never been reported. This fungus does not colonize the respiratory tract nor occur as an airborne contaminant in the clinical laboratory. Although the latter situation is always a possibility, current information suggests the possibility to be remote. Isolation of *C. immitis* from a clinical specimen, therefore, is positive confirmation of a diagnosis of coccidioidomycosis.

Diagnosis of primary coccidioidomycosis usually must be made by serology because most patients are unable to produce sputum. Fortunately, the antibody produced in this stage of the disease (precipitin) is highly specific, does not persist for any significant period of time beyond recovery and is not detectable in healthy individuals. Therefore, a positive tube precipitin test is acceptable evidence of acute coccidioidomycosis. A persistent serum CF titer ≥1:16 also strongly suggests the presence of disseminated disease, with a rise in titer associated with a poor prognosis. A declining titer is associated with a favorable response to chemotherapy. A positive CF test on CSF is the most sensitive indicator of CNS infection (91-96%), followed by culture (33-76%) and direct microscopic examination (8%).

HISTOPLASMOSIS

Histoplasmosis resembles the other systemic mycoses in that it has a primary pulmonary component, yet differs by disseminating to the reticuloendothelial system. The disease was first recognized in Panama in 1906 by Darling.[43] Utilizing histologic techniques, Darling recognized an intracellular microorganism, which was microscopically similar to the already described protozoan parasite, *Leishmania*, in tissue from a patient who had died of a disease similar to tuberculosis. In 1912, da Rocha-Lima reexamined Darling's original tissue sections and suggested that the organism was not a protozoan but more closely resembled the yeast seen in equine epizootic lymphangitis.[44] It was not until 1934 that reports appeared which unequivocally proved that the etiologic agent of histoplasmosis was a dimorphic fungus.[37,49]

The primary cause of histoplasmosis worldwide is *Histoplasma capsulatum* Darling var. *capsulatum*, but another variety, viz., *H. capsulatum* Darling var. *duboisii* (Vanbreuseghem) Ciferri is recognized as the agent of "African" histoplasmosis. A third fungus originally classified within the genus as a separate species, *H. farciminosum* (Rivolta) Ciferri and Redaelli, but now considered to represent a third variety, i.e., *H. capsulatum* var. *farciminosum*,[256] is the etiologic agent of epizootic lymphangitis of horses.[224] All three varieties are intracellular parasites of the reticuloendothelial system. In addition, they all exhibit dimorphism, i.e., they grow as a budding yeast in tissue or on specific enriched artificial media at 37°C, while in soil or on artificial media incubated at room temperature, they grow as a mold. *H. capsulatum* is classified as a deuteromycete, but its teleomorph is an ascomycete, *Ajellomyces capsulatus*.[136] The teleomorph of *B. dermatitidis*, i.e., *A. dermatitidis*, belongs to the same genus.[162]

H. capsulatum var. *duboisii* is the causative agent of "large form" African histoplasmosis. The term "large form" is derived from the presence of yeast cells which are characteristically much larger (8-15 μm diameter) in tissue than those of *H. capsulatum* var. *capsulatum* (2-4 μm). The former organism is considered a variety rather than a separate species since isolates of *H. capsulatum* var. *duboisii* can be mated with isolates of *H. capsulatum* var. *capsulatum* to form a teleomorph identical to *A. capsulatus*.[138]

Clinical Manifestations

It is generally accepted that histoplasmosis results from inhalation of components of the mycelial phase from an environmental source. Because of their small size (2-4 μm), microconidia can be deposited in the alveoli and therefore are the likely infectious particle.[74] The early events in the lung which determine the outcome of the host parasite interaction are unknown; however, it is reasonable to assume that the alveolar macrophage plays an important role. Once in the lung, the fungus converts to the yeast phase, which is preferentially localized in monocytic phagocytes, the predominant cells in the host inflammatory response. Infected foci progress to necrosis and eventually encapsulation of the site takes place. Calcification of the original foci and lymph node involvement are common.[153] Hematogenous dissemination, most commonly to the spleen, liver,[179] adrenal glands,[36] gastrointestinal tract,[91] skin, mucous membranes,[166] bone marrow[83] and less often the central nervous system,[152,163] joints[111] and genitourinary tract,[72,120] may occur even when the pulmonary infection is unnoticed. Vänek and Schwarz[252] found the occurrence of healed pulmonary complexes attributable to histoplasmosis was roughly equal to the frequency of histoplasmin skin test reactors in an endemic area. Further, 70% of autopsies of individuals dying of causes other than histoplasmosis had spleen and/or liver calcifications, representing the residual of self-limiting, unrecognized, disseminated histoplasmosis.

The clinical manifestations of disease can be broadly categorized into three forms: (1)primary histoplasmosis which can manifest as either acute or minimally symptomatic pulmonary disease; (2) chronic pulmonary forms and (3) progressive disseminated disease which can involve virtually any organ in the host.

The high frequency of positive histoplasmin skin tests in endemic areas coupled with the infrequency of overt clinical disease implies that most primary pulmonary histoplasmosis is asymptomatic or mild.[83] However, primary pulmonary disease, especially in epidemics, may present as moderate to severe symptomatic acute disease requiring medical intervention.[218] Residual foci of infection remain following primary disease and may contain viable cells of the etiologic agent,[45] as in coccidioidomycosis[204] and tuberculosis.[268] Only a small fraction of those infected develop chronic pulmonary or chronic progressive disease. Cavitary disease is found only in patients with coexisting chronic obstructive pulmonary disease[83] and is presumed to be the result of host immunologic responses rather than damage from the infection. Histoplasmomas, which can be either solitary or multiple solid lesions in the lungs, are often mistaken for carcinoma of the lung, with the definitive diagnosis obtained at thoracotomy.

Certain factors are implicated as predisposing individuals to disseminated histoplasmosis. In infants, considered immunologically immature, the disease is fulminant with a high rate of mortality; yeast cells of *H. capsulatum* are frequently seen in peripheral blood smears.[84] Therapeutic immunosuppression may result in reactivation of dormant *H. capsulatum*.[45] Therefore, kidney[46] and bone marrow[255] transplant recipients and patients with malignancies[237] are at increased risk of developing disseminated histoplasmosis. Although not originally associated with the acquired immune deficiency syndrome (AIDS), disseminated histoplasmosis is reported with increasing frequency in these patients.[156,261] The course of histoplasmosis in adults with AIDS more closely resembles the fulminant disease of infants rather than the disease observed in other immunosuppressed adults.[262] In addition, histoplasmosis in adults with AIDS is characterized by diverse cutaneous manifestations[98,115] and fungemia,[156] which are not observed as often in disseminated disease in adults without AIDS. Therefore, the disseminated form of histoplasmosis represents an opportunistic infection in a host with impaired cellular immunity,[28,156] and it is likely that there will be an increased incidence of disseminated disease as AIDS spreads to the areas endemic for histoplasmosis.[22] The CDC have revised the case description for AIDS to include disseminated histoplasmosis when it occurs in a patient with serologic or virologic evidence of HIV-1 (HTLV-III/LAV) infection.[34]

African, or large form histoplasmosis, is characterized by cutaneous lesions that may extend into subcutaneous tissue and eventually bone.[247] Although pulmonary manifestations, with or without calcification, are rare, lesions have been described in lung, as well as in spleen, liver and lymph nodes.[39] The reason the pathogenesis of this infection differs from that of classical histoplasmosis is not known.

Epidemiology and Public Health Significance

Portions of the United States and other countries have areas of high endemicity, generally located in tropical or subtropical climates, including most of the great river valleys of the world with the possible exception of the Nile River valley. Important endemic areas outside the tropics are the Ohio and Mississippi River valleys in the central United States. Part of the endemic area for histoplasmosis in the United States coincides with that for blastomycosis; however, unlike blastomycosis, histoplasmosis is absent in most southeastern states.

Although it is difficult to assess the public health importance of histoplasmosis, the results of skin test surveys suggest that infection with *H. capsulatum* var. *capsulatum* is widespread. It has been estimated that 40 million people in the United States have been infected[75] and that there are 200,000 new infections per year.[73] Between 80-90% of adults in endemic areas are skin test positive. It is impossible to determine the extent of clinical disease from these infections because the majority are mild and go undiagnosed. Those cases which are diagnosed are either severe sporadic ones or acute pulmonary disease associated with point-source epidemics. Outbreaks are relatively common and have been reported in association with a variety of outdoor activities.[173,189,206,216]

Histoplasmosis is optionally reported to the CDC and in 1984 a total of 357 cases were recorded.[33] Alabama had the greatest number, 81, followed by Missouri with 74, Illinois with 26 and Minnesota with 25 cases. However, since reporting is not mandatory, these data may not accurately reflect the actual distribution and total number of cases.

The natural reservoir for *H. capsulatum* var. *capsulatum* is the soil, particularly soil enriched with bat or bird feces[66], in areas with an average temperature of 22-29°C, an average relative humidity of 67-87% and annual rainfall of 1000 mm or more.[71,73] Birds themselves do not seem to become infected. Bats, on the other hand, are infected and may develop lesions containing viable yeast cells.[130] Dilution of bird excrement with soil and subsequent conversion to humus is important because *H. capsulatum* var. *capsulatum* will not grow on fresh feces. It is believed that a high nitrogen content of the droppings is associated with the presence of the fungus. When infested soil is disturbed, aerosols containing spores are generated and the inhalation of fungal elements of the appropriate size results in infection of a susceptible host.

Roosting places for chickens, blackbirds and starlings are often implicated as the point-sources of infection in both sporadic and epidemic disease. Contaminated bat caves can serve as an important point-source of *H. capsulatum* in regions where the general climate and soil conditions are unsuitable for the survival and growth of the fungus.[5] In a rural environment, most point-source infections and epidemics have occurred from soil enriched with chicken manure, while in an urban environment, infections are usually associated with starling or blackbird roosts.[213]

It is of interest to note that roughly equal numbers of plus and minus strains of *H. capsulatum*, which is heterothallic, are recovered from environmental sources. However, in a study of a large number of isolates obtained during two epidemics in Indianapolis, Indiana, the minus strain predominated in patients with disseminated disease.[139] This, in conjunction with an earlier report indicating that the minus strains convert more readily to the yeast phase, at least in vivo, than the plus strains, implies more variability in the pathogenicity of different *H. capsulatum* strains than previously described.

There are presently no generally accepted preventive measures, although attempts have been made to eradicate *H. capsulatum* from soil foci implicated in outbreaks.[257]

Collection, Transport and Processing of Specimen

The specimen most likely to yield diagnostic material depends upon the site of clinical disease. Regardless of the source, any clinical specimen suspected of containing *H. capsulatum* must be handled promptly, probably more so than specimens for any of the other systemic pathogenic fungi,[134] and should never be frozen.[246] Storage at room temperature is preferable when a delay in processing is unavoidable. Respiratory disease can be diagnosed by the examination of sputum and other respiratory specimens (see section on blastomycosis and Chapter 2).

Patients with disseminated disease are most readily identified by culturing peripheral blood, bone marrow, tissues from involved organs (e.g., skin) or urine if there is kidney or prostatic involvement. Culturing of venous blood can be a useful aid in diagnosis since hematogenous spread occurs with virtually every respiratory infection. If necessary, storage of blood in the lysis centrifugation system collection tube at 25°C for as long as 9 h has been shown not to significantly alter the number of cultures yielding *H. capsulatum*.[238] In acute disseminated histoplasmosis, yeast cells sometimes are detected in the peripheral circulation.[84,97] The presence of calcified lesions in the spleen and other organs indicates that such cells are also in the circulation, albeit in small numbers and perhaps only intermittently, thus going undetected, even in primary histoplasmosis[218] and in chronic cavitary disease.[233] However, microscopic examination of peripheral blood buffy coat in primary disease is generally not productive. But in disseminated disease in immunosuppressed patients, i.e., AIDS patients, organisms have been seen on both peripheral blood smears and buffy coat preparations and all technicians should be aware of this possibility. If the disease is widely disseminated, there may be mucocutaneous lesions or lesions in the kidney, in which case, kidney biopsy and urine should be cultured.

Specimens obtained from normally sterile sites are also placed onto enriched medium without inhibitory substances and incubated at 37°C, in addition to media incubated at 30°C. The tissue forms of most of the dimorphic fungal pathogens are inhibited by cycloheximide in selective media when incubated at 37°C. All cultures, especially those incubated at 37°C or any held longer than 2 weeks, should be closed in some way to prevent drying, e.g., petri plates place in nylon bags (M and Q Packaging).

Identification of Etiologic Agents

The tissue reaction in histoplasmosis resembles that of tuberculosis and cavitation is more frequent than in other systemic mycoses. Lesions in tissue are granulomatous with epithelioid cells, occasional giant cells and macrophages containing ingested yeasts.

Unstained preparations such as those with KOH are not useful in the diagnosis of histoplasmosis. Preparations from the peripheral blood or bone marrow, urine sediment (2,000 x g for 15 min), impression smears of biopsies or scrapings of mucosal lesions are stained with a polychrome methylene blue-eosin stain, such as Wright or Giemsa, and examined microscopically. Small intracellular yeasts, 2-4 μm in diameter, are present in monocytes, macrophages and rarely in polymorphonuclear leukocytes. Yeast forms characteristically appear as round bodies with a central stained area and a surrounding halo-like capsule. Although the presence of this capsular appearance gave rise to the species name, the yeast is not, in fact, encapsulated and the "halo" is actually an artifact created by shrinkage of the cytoplasm. An intracellular parasite which must be considered in a differential diagnosis is *Leishmania donovani.* When stained preparations are examined at 1000X magnification, *L. donovani* cells will be found to contain a kinetoplast, i.e., an accessory body located to one side of the nucleus, consisting of two portions united by a fibril. *H. capsulatum* yeast cells do not contain a similar structure. Furthermore, staining with PAS to show the presence of a cell wall will differentiate fungal and protozoan parasites in tissue.

Paraffin or plastic imbedded tissue specimens are prepared for light microscopy and stained with H and E (Fig.5:19) and GMS or PAS. PAS will stain both the cell wall and the internal structures of viable organisms; GMS will stain both viable and nonviable organisms.

Since it is sometimes difficult to obtain clinical material containing sufficient viable cells to recover the fungus in culture, diagnosis may rest upon histologic examination. This is also true when a fungal etiology was not recognized and tissue was not cultured. There are pitfalls in attempting the diagnosis of histoplasmosis by the use of histopathologic techniques alone. For example, the yeasts in tissue of the extremely rare disseminated infections caused by *Penicillium marneffei*[56] mimic the appearance of *H. capsulatum. P. marneffei* is a dimorphic fungus growing as a small (1-8 μm diameter) yeastlike cell in tissue and in culture at 37°C and as a mycelial form at 25°C.[50] The organism is present in histiocytes and is morphologically similar to *H. capsulatum*, but extracellular *P. marneffei* are distinguished by the formation of two-celled, short hyphae. *H. capsulatum* tends to grow as hyphae in intravascular infections, especially endocarditis.[105] Because of the size and staining of the opportunistic yeast *Torulopsis glabrata*, seen

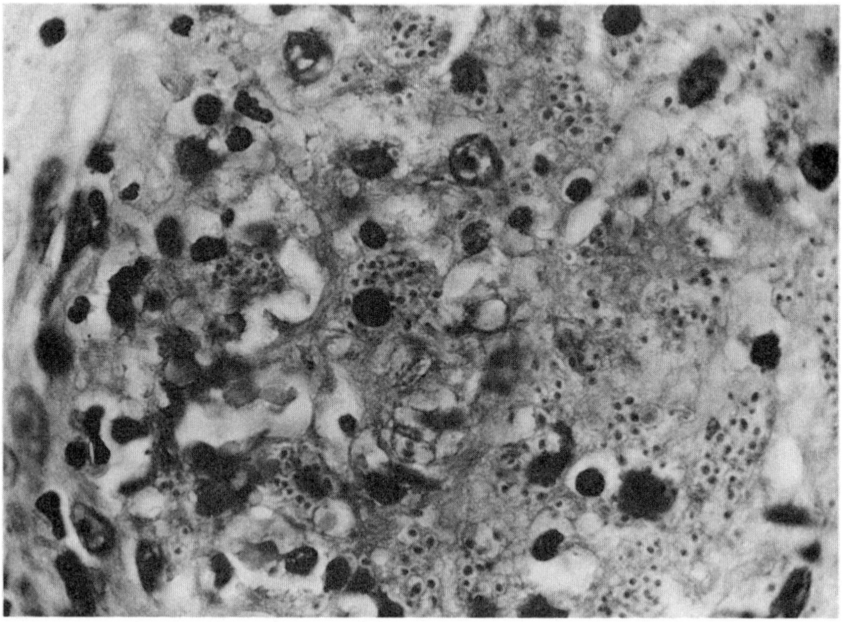

Figure 5:19. High power view of kidney tissue stained with hematoxylin and eosin revealing histiocytes with numerous yeasts compatible with *H. capsulatum* in their cytoplasm (material kindly provided by George S. Kobayashi).

especially in infections of immunosuppressed patients, it can be confused with *H. capsulatum*. Therefore, attempts should be made to confirm all histologic findings by culture.

Fluorescent antibody techniques have been applied to the identification of pathogenic fungi in tissue.[85,121] Kaplan[117] has presented an excellent review of the development of such techniques. Such methods are complicated by the presence of many crossreacting antibodies in antisera and absorption of the antisera must be done carefully. Moreover, the staining properties of cells with antisera seem to be dependent upon the age of the cell, with young cells staining well but older cells staining poorly. However, if no attempt was made to isolate the etiologic agent and fixed, paraffin embedded tissue sections are the only materials available, then fluorescent antibody techniques may be helpful in resolving the identity of the fungus. Tests are available in several of the larger public health laboratories in the United States.

The yeast cells of *H. capsulatum* var. *duboisii* are often found within macrophages and giant cells and exhibit similar staining properties to those of *H. capsulatum* var. *capsulatum*, but are distinguished by their larger size

(8-15 μm diameter). It is equally important to differentiate *H. capsulatum* var. *duboisii* from *B. dermatitidis*, which also causes disease in Africa. The main morphologic difference, not readily apparent, is that *H. capsulatum* var. *duboisii* buds by a narrow attachment to the parent cell as opposed to the broad-base budding of *B. dermatitidis*.

Colonies of *H. capsulatum* appear on laboratory media after 1-2 weeks' incubation. Cultures from patients who have had microscopic findings of yeasts compatible with *H. capsulatum* or a positive serology should be kept for 12 weeks before being considered negative. *H. capsulatum* grows in the mold form on media incubated at 30°C and as the yeast form on special media incubated at 37°C. The mold form begins as a white, cottony growth which may remain white and cottony or becomes powdery with a buff to tan coloration as it grows concentrically. Sometimes the mycelial mat appears to crack with age. The thallus consists of septate hyphae with conidiophores bearing macro- or microconidia. Macroconidia (8-14 μm) are the most distinctive morphologic feature of the mold form. They are tuberculate and the tubercle-like appendages on the surface are visualized even at low magnification (Fig.5:20). Smooth-walled macroconidia also occur and a single isolate may produce both types of macroconidia (Fig.5:20).

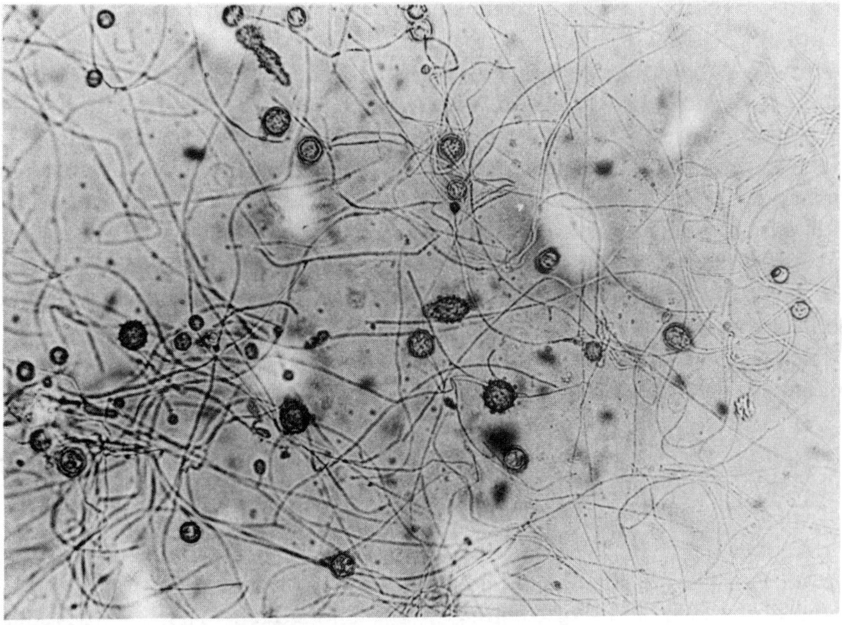

Figure 5:20. Tape preparation mounted in lactophenol cotton blue showing the various mold phase conidia of *H. capsulatum*.

Microconidia (2-4 μm diameter), generally regarded as the infectious particles, have smooth walls and vary in shape from round to pyriform (Fig.5:20). The attachment and arrangement of the microconidia on the hyphae may vary but typically they are attached to the lateral side of the hyphae by short conidiophores. Although variability in colonial morphology is not as common as it is for *C. immitis*, rare isolates produce red pigment or will not sporulate.[172] Therefore, atypical clinical isolates from appropriate specimens of symptomatic patients should not be considered contaminants if no sure identification can be made.

To make an unequivocal identification of *H. capsulatum*, the mold form must convert to the yeast state. This is done by adding approximately 0.5 mL of BHI broth to a tube with a slant of enriched medium (e.g., Pine-Drouhet yeast agar or glucose-cysteine-blood agar), inoculating a portion of the mycelium to the surface above the fluid level and incubating the tube at 37°C in 5% CO_2 in air. Peripheral areas of the inoculum should begin to look pasty after 4-7 days and portions of these areas should be transferred to fresh medium. Most mycelial cultures will convert to the yeast form after about 4 such weekly transfers. Rarely, however, will the conversion be complete and usually there are small (2-4 μm), thin-walled yeast cells, some budding singly, admixed with hyphal forms. These are easily differentiated from the thick-walled, yeastlike cells of *B. dermatitidis*. If the first growth observed on media incubated originally at 37°C is the yeast form, it is converted to the mycelial form simply by changing the temperature of incubation to 30°C.

Animal inoculation can be used as an aid in the identification of an isolate which does not convert to the yeast form in vitro. Mycelial growth is homogenized in saline with a tissue grinder in a biohazard cabinet and then injected intraperitoneally into mice. Mice are autopsied after 2-4 weeks and smears of the cut surfaces of the spleens or of any gross lesions are stained with Wright or Giemsa stains. Histologic sections and cultures of the spleens and lungs may be made. This potentially hazardous undertaking requires that both the inoculation area and the animal care facilities conform to biohazard safety level 2.[201]

The exoantigen test (see section on blastomycosis) is a valuable alternative to animal inoculation. The procedure provides a presumptive identification of *H. capsulatum* var. *capsulatum, H. capsulatum* var. *duboisii* or *H. capsulatum* var. *farciminosum* when extracts of cultures form precipitin bands of identity with reference H and/or M antigens.

Arthroderma tuberculatum, Chrysosporium spp., *Sepedonium* spp. and other fungi which produce tuberculate conidia that resemble those of *H. capsulatum*[125] are distinguished by their inability to be converted to a yeast phase or to form lines of identity in the exoantigen tests.[223,235] Mycelial cultures of *H. capsulatum* that do not form tuberculate macroconidia are indistinguishable from *B. dermatitidis*, but their yeast phases are different.

The mycelial form of *H. capsulatum* var. *duboisii* is indistinguishable from that of *H. capsulatum* var. *capsulatum*. Although the yeast cells of the former are larger, they are often mixed with small cells indistinguishable from those of *H. capsulatum* var. *capsulatum*.

Antimicrobial Susceptibility and Resistance

At present, the most effective antifungal agent for the treatment of histoplasmosis is amphotericin B. It is a toxic agent not to be used in a casual manner since many individuals with acute pulmonary histoplasmosis will recover spontaneously with only bed rest and symptomatic treatment. Therefore, amphotericin B is reserved primarily for immunosuppressed patients or those with progressive and chronic cavitary disease. Ketoconazole has been used successfully to treat the milder forms of histoplasmosis.[57]

Evaluation of Laboratory Findings

A definitive diagnosis of histoplasmosis depends upon isolation and identification of the causative agent. A presumptive diagnosis can be made by the demonstration of typical intracellular yeast forms in histologic sections and/or on the basis of serologic tests. Serology is frequently the only way of diagnosing acute primary disease. However, failure to detect antibody does not rule out infection. The presence of antibody should stimulate an effort to obtain material for culture.

PARACOCCIDIOIDOMYCOSIS

Paracoccidioidomycosis is a chronic, progressive respiratory disease with a wide variety of clinical manifestations. The primary tissue response is a mixed suppurative and granulomatous reaction. As a result of animal experimentation, human autopsy studies and case histories, it has become clear that the primary portal of entry is the lung. All other manifestations result from hematogenous spread.[88] As would be expected, the most prominent feature of the disease is primary lung involvement accompanied by secondary lesions of the oral and nasal mucosa. The main difficulty in defining the evolution of the disease is the long interval (months to years) between initial contact with the fungus and the appearance of clinically recognizable disease.

Paracoccidioides brasiliensis is the sole etiologic agent of paracoccidioidomycosis. This agent is a thermal dimorphic fungus occurring in tissue or on enriched media at 37°C as a yeast and at 25-30°C as a mold. The teleomorph has not been described.

Clinical Manifestations

Paracoccidioidomycosis is acquired through inhalation of viable fungal elements from the environment. As in other systemic mycotic diseases, initial infection is asymptomatic and may never develop beyond this stage. However, the infection may progress to symptomatic pulmonary or systemic disease in some individuals.[31,81,167,241]

The most common sites of diseases are pulmonary, gastrointestinal and mucocutaneous. Patients with pulmonary disease present with a variety of clinical syndromes ranging from mild "flu-like" distress to chronic progressive respiratory failure identical to pulmonary tuberculosis. Gastrointestinal involvement and prominent cervical lymphadenopathy are clinical features that distinguish this disease from blastomycosis. The cutaneous and mucocutaneous lesions are typically shallow, crusted granulomas similar to lesions of cutaneous tuberculosis, leishmaniasis and yaws.

Epidemiology and Public Health Significance

P. brasiliensis has been isolated from soil on at least two occasions.[47] However, the exact ecologic niche of *P. brasiliensis* is unknown. The disease is not found in domestic animals and reports of its isolation from bats[87] are not confirmed. Natural disease has been reported in a green monkey[109] and in armadillos.[177] In humans, the disease is found predominantly in adult males, but both males and females from rural areas are infected. Paracoccidioidomycosis is not communicable.

This is the most important systemic fungal disorder in South America. Skin test antigens have been used to define endemic areas. Disease is confined primarily to the subtropical forest areas of the continent with a few cases reported from similar ecological areas in most countries of Central America and Mexico. It was initially named the Brazilian disease because of its prevalence in that country. Brazil continues to have the highest number of reported cases, followed by Venezuela and Colombia.[88] There are 13 documented cases reported in the United States, all originating in Latin America.[3,200]

This is an important public health problem in Latin America, as illustrated by MacKinnon's literature survey of nearly 5,000 cases, most of which came from Brazil.[154] More recent reviews continue to show that this disease is the most frequently diagnosed respiratory mycosis in Latin America. The impact of this disease on morbidity and mortality in Latin America is indicated by the estimated annual incidence of clinical cases of 1-3 per 100,000 population.[88,200] The chronic course, frequent disability and lengthy treatment involved further emphasize the importance of this disease. Paracoccidioidomycosis is often confused with tuberculosis, which results in inappropriate therapy until the appearance of secondary manifestations after systemic spread.

Collection, Transport and Processing of Specimens

Sputum, tissue scrapings and biopsy material from cutaneous ulcers or lungs are the most common specimens examined. *P. brasiliensis* usually is not recovered from blood, but serum should be obtained for serology. Processing of specimens does not differ from that described for the other systemic mycoses (see Chapter 2).

Identification of the Etiologic Agent

Microscopic examination of unstained clinical material is especially valuable for the diagnosis of paracoccidioidomycosis. The demonstration of typical large yeast forms (30 μm in diameter is not uncommon) with multiple buds attached over the entire circumference, often by narrow "stalks," is pathognomonic. Mother cells and attached small buds may give the appearance of a "pilot's wheel" (Fig.5:21). The budding cells are slightly oval and have a thinner wall than the mother cell. Older cells may appear empty and collapsed and have thicker walls, but they seldom attain the thickness characteristic of *B. dermatitidis*.

Biopsy specimens can be fixed and stained by standard histologic tech-

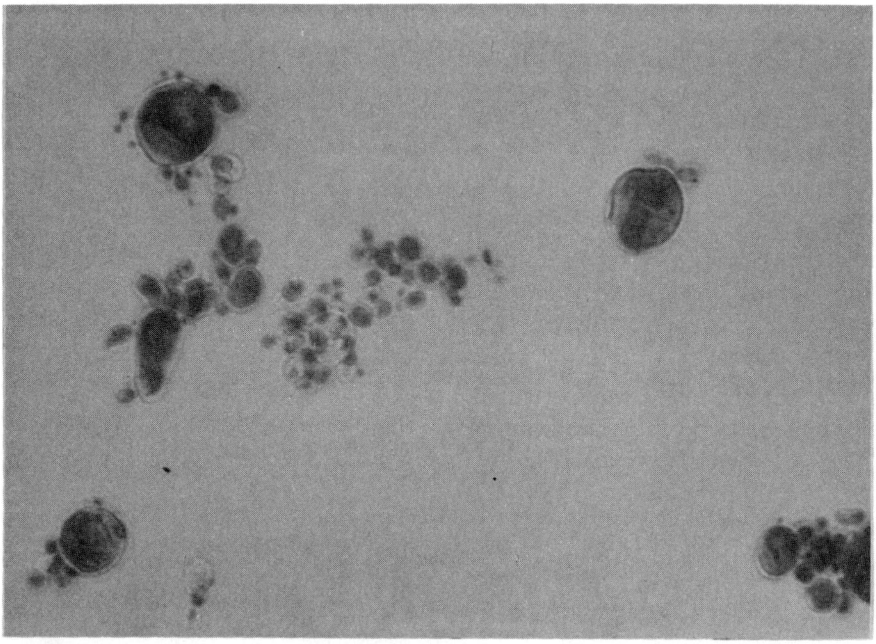

Figure 5:21. High power view of budding *P. brasiliensis* yeasts, 400X (kindly provided by Barbara E. Robinson).

niques. Typical budding forms can be seen in sections stained with H and E but are better visualized with special stains such as PAS and GMS (Fig.5:22).

P. brasiliensis grows on mycological media routinely used in diagnostic laboratories, e.g., Sabouraud dextrose agar or Sabouraud dextrose agar with chloramphenicol and cycloheximide. Specimens from normally sterile sources can be inoculated directly onto blood agar or other enriched media incubated at 30°C and 37°C. However, specimens from contaminated sites should be inoculated onto selective media incubated initially at 25-30°C.

P. brasiliensis, like the other dimorphic fungi, is identified on the basis of its morphology and colony characteristics when incubated at both 25-30°C and at 37°C. At room temperature, colonies are slow-growing, with dense, short aerial hyphae, initially white in color, but which may turn brown with age. Conidial formation in culture is rare, but when conidia are present, they are indistinguishable from those produced by *B. dermatitidis*. Restrepo found that more isolates sporulated on yeast extract medium than on Sabouraud agar, producing arthroconidia as well as blastoconidia. Primary isolates and recent stock cultures sporulated better than old stock cultures.[198]

To confirm the identification of *P. brasiliensis*, the mycelial form must

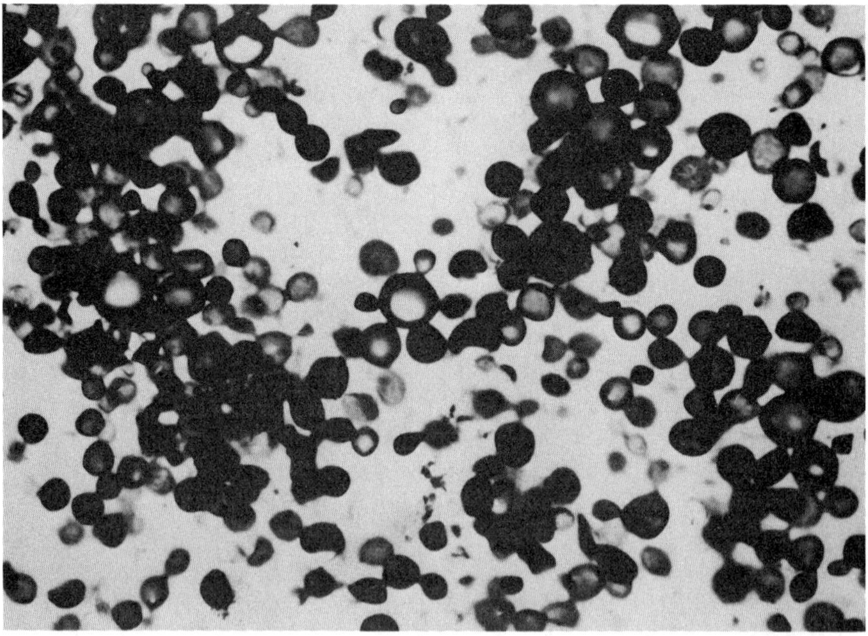

Figure 5:22. Photograph of *P. brasiliensis* in tissue, 400X (kindly provided by Michael R. McGinnis).

be converted to the yeast form or visa versa. The procedures are similar to those for *H. capsulatum* and *B. dermatitidis*. The slow-growing mycelial colony is transferred to either fresh blood agar, brain heart infusion agar with blood or Kelley's medium incubated at 37°C. Growth may be observed in 5-8 days. The transfer of a portion of the colony to fresh media aids in the conversion to the yeast stage. The typical large yeastlike cells with multiple buds must be seen to confirm the diagnosis. Since the mycelial form of *P. brasiliensis* is the slowest growing of the fungi causing systemic disease, laboratories sometimes have difficulty isolating it. Therefore, considerable reliance is placed upon the primary isolation of *P. brasiliensis* at 37°C rather than at room temperature.

Although serological tests are valuable adjuncts to the diagnosis, commercial reagents are not available and the tests are only performed in reference laboratories such as CDC. There are no commercially available skin test antigens for detecting sensitivity to *P. brasiliensis*. Antigens prepared from culture filtrates or polysaccharide fractions are being used in mycology centers throughout Latin America.[199]

Significance of Laboratory Findings

The isolation and identification of *P. brasiliensis* from a clinical specimen is required to establish a diagnosis. Evidence of disease may be shown by the demonstration of typical "pilot wheel" yeast forms in clinical specimens or by the detection of precipitin reactions with patients' sera.

ZYGOMYCOSIS

Zygomycosis (phycomycosis, mucormycosis) is an opportunistic fungal infection caused by several genera within the class Zygomycetes. The disease is usually associated with rhinocerebral infections, but primary infections can also involve the lungs, gastrointestinal tract, skin and rarely, following dissemination, other organ systems. Although the rhinocerebral disease is most common today, it was not until 1943 that a second case of rhinocerebral zygomycosis was reported.[89]

Although Alexopoulos and Mims[6] divide the class Zygomycetes into 6 orders, only Mucorales and Entomophthorales contain pathogenic organisms. McGinnis[161] recognizes the following species of Mucorales as agents of disease: *Absidia corymbifera, Cunninghamella bertholletiae, Mucor ramosissimus, M. rouxianus, Rhizomucor pusillus, Rhizopus arrhizus, R. microsporus, R. oryzae, R. rhizopodiformis* and *Saksenaea vasiformis*.[161] More recently, *Syncephalastrum* spp.,[116] *Cokeromyces recurvatus*[11] and *Apophysomyces elegans*[263] have been added to the list. Species in the order

Entomophthorales pathogenic for humans include *Basidiobolus haptosporus*, *Conidiobolus coronatus* and *C. incongruus*.[203,243]

Clinical Manifestations

Zygomycosis caused by a member of the order Mucorales is acute and rapidly fatal, most often associated with species of *Rhizopus*. The disease occurs mainly in 3 forms: rhinocerebral, pulmonary and intestinal. Rhinocerebral zygomycosis is most often seen in uncontrolled diabetes mellitus with acidosis. Disease begins in the nose or nasopharynx, extends into the nasal sinuses and orbits or directly into the meninges and brain by way of septic thromboses. The pulmonary form may resemble pneumonia, abscess or infarct and is associated with leukemia or lymphoma. It is a frequently incidental observation at autopsy. Lesions involving the gastrointestinal tract are uncommon and presumably arise by ingestion of the organisms.[2,185]

Cutaneous zygomycosis can arise in patients with burns and may disseminate.[29] Primary cutaneous infection has been reported following use of surgical dressings[23,78,228] and in trauma.[4,61,108,116,263] Other recently reported clinical manifestations include myocardial infarct due to disseminated zygomycosis[17,251] and chronic cystitis due to *Cokeromyces recurvatus*.[11] This was the first report of human disease due to a species of *Cokeromyces*.

Once predisposing conditions permit initiation of infection, this group of fungi have a remarkable capacity to invade blood vessels, seemingly without restriction. Thromboses and infarcts are characteristic features regardless of the organ involved. Small arteries can be seen filled with hyphae that penetrate into adjacent tissue and cut off the blood supply to the organ. The mechanism of this vascular damage has not been determined.

Zygomycosis due to species of *Basidiobolus* and *Conidiobolus* within the order Entomophthorales cause subcutaneous infections involving the trunk and extremities. Unlike species of the Mucorales, Entomophthorales species seldom cause life-threatening infections and can cause disease in previously healthy individuals. There have been reports of visceral infections due to this group of organisms.[217]

Several review articles[20,69,145,157] and textbooks[68,165,203] contain details of the clinical spectrum of zygomycotic diseases. A comprehensive bibliography of zygomycosis was published in 1979 by Ader and Dodd.[1]

Epidemiology and Public Health Significance

Like many other opportunistic pathogens, the zygomycetes are abundant in the environment. Exposure to these fungi often occurs as the organisms are known to be transient residents of the human integument. Disease due to the Mucorales is found throughout the world in patients with predisposing conditions. Zygomycosis due to the Entomophthorales occurs very infrequently in the United States, is primarily found in Africa and less frequently in India and southeast Asia. It is generally agreed that the incidence of zygomycosis has increased.[146] The increase is accredited to advances in modern medicine that provide life-saving intervention without curing predisposing disease, which allows opportunistically pathogenic organisms to multiply in host tissues.

Collection, Transport and Processing of Specimens

All too frequently, specimens received by the laboratory from zygomycotic infections are obtained at autopsy. Antemortem specimens from rhinocerebral zygomycosis include scrapings of the mucosal linings of the nares and sinuses or necrotic discharge material from facial eschars or the orbit of the eye. Respiratory specimens such as sputum or bronchial washings are submitted when pulmonary zygomycosis is suspected. Biopsies taken at the edge of necrotic areas are the most satisfactory specimens for isolation of the etiologic agent from burn patients. Localized cutaneous lesions are biopsied at the leading edge in subcutaneous zygomycosis.

The processing of these specimens does not differ from those for other fungal diseases described in Chapter 2. Zygomycetes are sensitive to cycloheximide and media containing this antibiotic should not be used for isolation. Clinical laboratories should immediately examine all specimens suspected of harboring zygomycetes and report the findings to the physician. The direct examination can provide a presumptive diagnosis so that treatment can begin immediately for these life-threatening diseases.

Identification of Etiologic Agents

Although direct microscopic examination of clinical material may be diagnostic of a zygomycete, the specimen must be cultured to identify the fungus. The specimen can be treated with KOH to improve visualization of fungal elements. The presence of aseptate hyphae in such preparations is suggestive of a zygomycete. Branching of the hyphae may be sparse and is usually at a 90° angle. Hyphae often have bizarre morphologic features such as bulging irregular walls, clubbing and granular cytoplasm. Care should be taken not to confuse aseptate hyphae with filaments of nerve fibers when examining brain tissue. Failure to observe any fungal elements does not necessarily indicate the absence of fungal disease.

Since fungal elements are often difficult to find in H and E stained histologic sections, special stains such as PAS and GMS should also be used. Often only fragments and cross-sections of hyphae are seen, so that GMS is the preferred stain. Care must be taken not to confuse the hyphae of zygomycetes with the narrow, septate, dichotomous hyphae of the aspergilli.

Zygomycete colonies are visible on agar media within 48-72 h when incubated at room temperature. Cultures should be observed daily or there will be confluent growth over the entire agar surface due to their rapid growth. Isolates from the order Mucorales produce abundant aerial mycelium, often filling the container with grey to light tan growth. Microscopic examination shows aseptate hyphae. Streaming of the protoplasm, implying lack of septation, can be seen when the specimen is mounted in water or saline. Isolates of the order Entomophthorales produce flat and folded colonies and hyphae with occasional septa.

Though all zygomycetes reproduce sexually, they are usually identified by the microscopic morphology of the asexual stage which is more likely to appear on laboratory media. The sporangiospores are produced in a large sac-like structure called a sporangium, borne on the tip of a supporting hypha, the sporangiophore. In some species, the sporangiosphore forms a protrusion (columnella) into the sporangium. The presence and morphology of the columnella are useful features in differentiation among some species. Species of *Rhizopus* and *Absidia* also produce rhizoids (root-like structures) that serve for identification. A hypha that connects two groups of rhizoids is called a stolon. There are no special biochemical reactions or nutritional requirements that aid in the identification of the pathogenic zygomycetes.

Descriptions of the most important pathogenic genera of the zygomycetes follows. For most laboratory work, identification to the genus level is adequate. However, references to keys for speciation will be mentioned for each genus. Even with these resources, however, identification is difficult and presumed pathogens should be confirmed by a recognized expert in the field of fungal taxonomy.

Order Mucorales

Rhizopus spp.:

The voluminous aerial hyphae are tenacious and resemble "steel wool." Solitary sporangiophores are long, unbranched and arise from the region of rhizoids, i.e., points where the stolons touch the substrate (Fig. 5:23). The sporangia are spherical and the columnella is hemispherical. The sporangiospores may be either smooth or rough.

Absidia spp.:

The aerial mycelia resemble *Rhizopus* spp. but are not as tenacious. The sporangiophores are branched and seen in groups of 3-5 arising from the internode of the stolons and not at the point of origin of the rhizoid. The rhizoids seen at the node are primitive and not as branched as those of *Rhizopus* spp. Sporangia, supported by an apophysis, are pear-shaped and

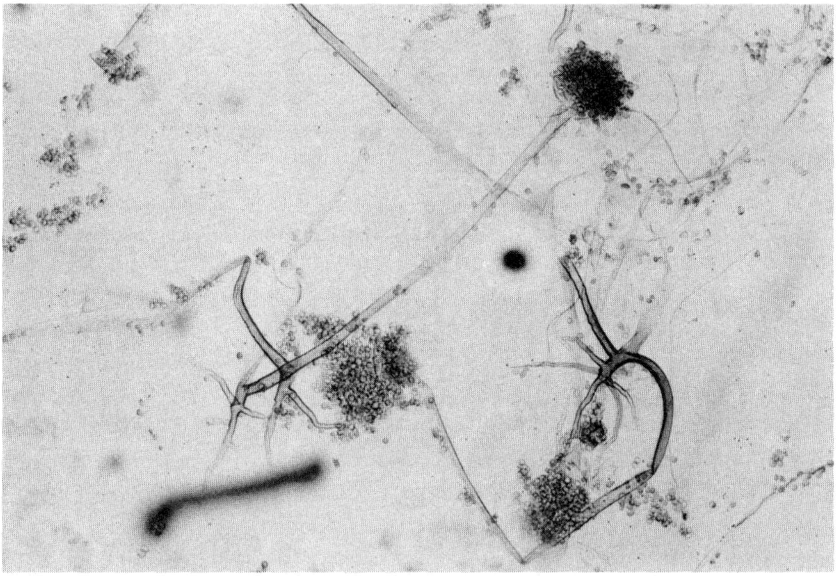

Figure 5:23. Lactophenol mount of *Rhizopus* spp. with rhizoids present at the base of the sporangiophore, 450X.

filled with round to ovoid sporangiospores.[63,64,99,100,101,178] The genus *Apophysomyces* contains the single species *A. elegans* which is similar to *Absidia*. The former differs in having a more pronounced apophysis that is funnel to bell-shaped and a "foot cell" similar to that found in *Aspergillus* at the base of the sporangiophore.[168]

Mucor spp.:

The aerial mycelia have a soft cottony texture. Rhizoids are not present and the sporangia are spherical with a columnella. Sporangiospores are spherical and smooth. There is some question whether any of the species of *Mucor* are actually associated with disease.[203] One species that was associated with disease[132] has been transferred to the genus *Rhizomucor*.[215] *Rhizomucor (Mucor) pusillus* resembles *Mucor* spp. Identifying features are the presence of vestigial rhizoids and the development of stolons in *Rhizomucor*. The sporangiophores are highly branched which distinguishes them from those of *Rhizopus* spp. Also, by definition, *Rhizomucor* can grow at 50°C.

Some other genera of Mucorales are agents of human zygomycosis. *Saksenaea vasiformis* (Fig.5:24) is strikingly characterized by sporangia shaped like long-necked vases.[4,61,65,188,191,249] Oval to spherical sporangiospores exude from the mouth of the sporangium. Rhizoids are present at the

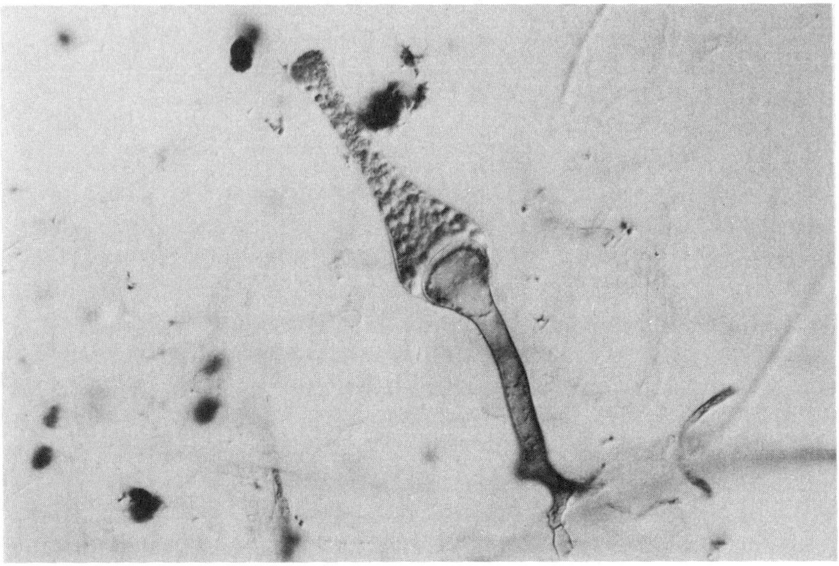

Figure 5:24. Lactophenol mount of *Saksenaea vasiformis* showing flask-shaped sporangium containing numerous sporangiospores, 450X.

base of the sporangiophore. Since this species rarely produces sporangia on laboratory media, sporulation should be induced by growing the *S. vasiformis* on cornmeal-sucrose-yeast extract agar for several days and then placing a small block of agar covered with growth on the surface of 1% water agar in a petri dish. After incubation, abundant sporulation will occur.[62]

Cunninghamella[26,131,140,209] is one of the genera of Mucorales that forms monosporous sporangia (sporangiola). The sporangiola are found crowded onto the surface of the swollen tip (vesicle) of a sporangiophore (Fig.5:25). The gross colony resembles that of *Mucor* spp. The sporangiophores are long, erect and unbranched. The taxonomy of this group has been reviewed by Weitzman and Crist.[260] *Syncephalastrum* spp. produce fast-growing cottony colonies with highly branched hyphae. The sporangiophore terminates in a vesicle upon which are formed cylindrical sporangia encircling the vesicle (Fig.5:26). Each tubular sporangium (merosporangium) contains a single row of sporangiospores. This asexual spore arrangement can be confused with that of the aspergilli.

Cokeromyces recurvatus is a dimorphic fungus. It was first described in 1949[226] and has only recently been associated with human disease.[11] At 30°C this organism is mycelial, but when incubated at 38°C only the yeast form is evident. The yeast phase has been found in clinical specimens. *C.*

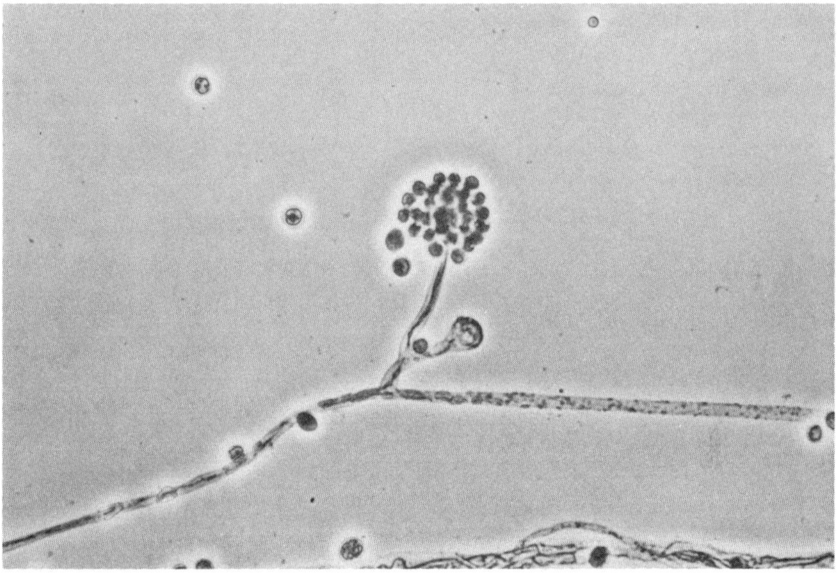

Figure 5:25. Lactophenol mount of *Cunninghamella* spp. showing sporangiola attached to the sporangiophore, 450X.

recurvatus produces numerous sporangiola on elongated recurved stalks arising from the terminal vesicle of the sporangiophores (Fig.5:27). Twelve to 20 spores are produced in each sporangiola. No true sporangia are produced.

Order Entomophthorales

The two recognized genera in the order Entomophthorales, *Basidiobolus* and *Conidiobolus*, both cause human disease and are similar in their gross colonial appearance. These zygomycetes grow in 3-4 days as flat, wrinkled colonies that adhere closely to the surface of the culture medium. The surface of the colony may have a short "bloom" of aerial hyphae, consisting of single-spored sporangiola that resemble conidia. These sporangiola are forcibly discharged from their sporangiophores. The hyphae of species in this order are septate. *Basidiobolus haptosporus* is the only species known to cause human disease. Large numbers of chlamydospores and zygospores are seen in culture. The uninucleate sporangiola are globose and contain a fragment of the broken sporangiophore from which they were forcibly discharged. The zygospores appear as round, thick-walled, smooth structures having two protuberances or "beaks" on their surface. This zygospore structure is the principal identifying feature for this genus.

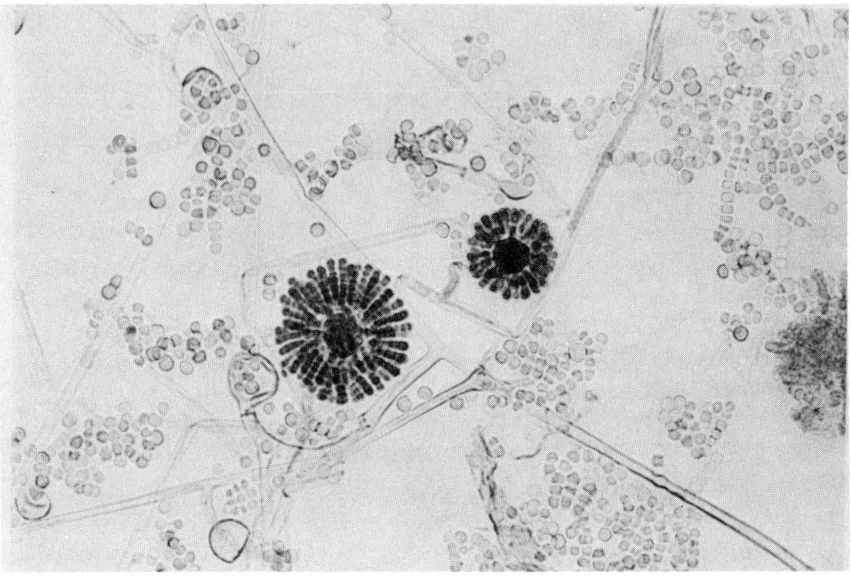

Figure 5:26. Lactophenol mount of *Syncephalastrum* spp. showing tubular sporangium (merosporangium) surrounding the vesicle. A single row of sporangiospores can be seen within each merosporagium, 450X.

Conidiobolus spp. resemble *Basidiobolus* in gross culture, but the former produce greater numbers of forcibly expelled sporangiola. The hyphae have fewer septa, but readily produce chlamydospores. Sporangiola are globose, smooth and show a prominent papilla where forcibly discharged from the sporangiophore (Fig.5:28). There is no hyphal fragment attached. Zygospores are not obvious. Some sporangiola produce secondary spores by multiple replication, giving a corona effect. Older sporangiola may show long, thin filaments (hairy spores).

Significance of Laboratory Findings

Diagnosis cannot be made by culture alone, especially if the specimen is from a contaminated source. Even when the specimen is from an ordinarily sterile body fluid or tissue, there is always the possibility that the fungus is a contaminant introduced during processing or examination of plates. The best diagnostic evidence of zygomycosis is observation of typical invasive forms in histopathologic sections with confirmation by isolation. Unfortunately, cultures may be negative even when aseptate hyphae are seen in tissue. Although diagnosis often is based upon the observation of hyphae

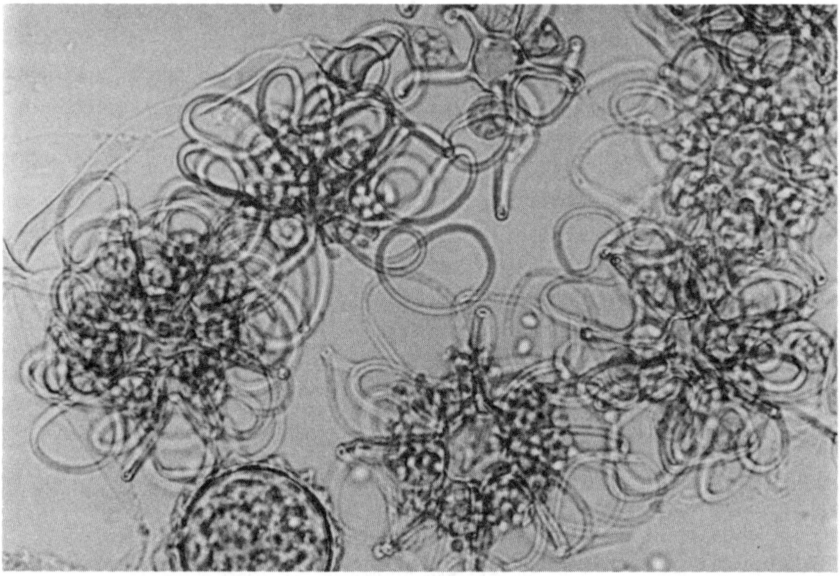

Figure 5:27. Lactophenol mount of *Cokeromyces recurvatus* showing the production of numerous sporangioles on elongated recurved stalks arising from the vesicle of a sporangiophore, 450X.

in sections, the typical morphology may not always be evident and the differential must be broadened to include aspergillosis versus zygomycosis.

REFERENCES

1. Ader PL, Dodd JK. Mucormycosis and entomophthoromycosis. A bibliography. Mycopathologia 1979;68:67-99.
2. Agha FP, Lee HH, Boland CR, Bradley SF. Mucormycoma of the colon: early diagnosis and successful management. AJR 1985;145:739-41.
3. Agia GA, Hurst DJ, Rogers WA. Paracoccidioidomycosis presenting as a cavitating pulmonary mass. Chest 1980;78:650-3.
4. Ajello L, Dean DF, Irwin RS. The zygomycete *Saksenaea vasiformis* as a pathogen of humans with a critical review of the etiology of zygomycosis. Mycologia 1976;68:52-62.
5. Ajello L, Kuttin ES, Beemer AM, Kaplan W, Padhye A. Occurrence of *Histoplasma capsulatum*. Darling 1906, in Israel, with a review of the current status of histoplasmosis in the Middle East. Am J Trop Med Hyg 1977;26:140-7.
6. Alexopoulos CJ, Mims CW. Introductory mycology. 3rd ed. New York: John Wiley, 1979.
7. Ampel NM, Ryan KJ, Carry PJ, Wieden MA, Schifman RB. Fungemia due to *Coccidioides immitis*. An analysis of 16 episodes in 15 patients and a review of the literature. Medicine 1986;65:312-21.
8. Andleigh HS. Blastomycosis in India: report of a case. Indian J Med Sci 1951;5:59.

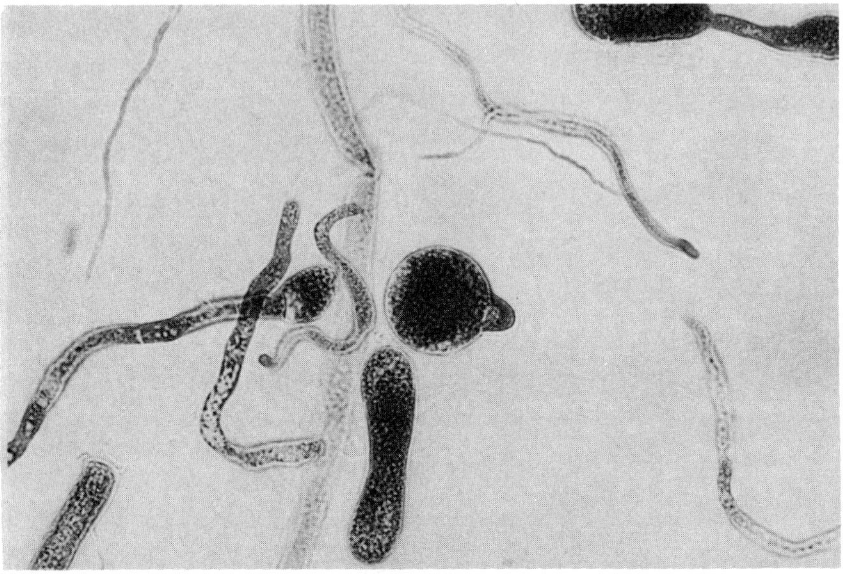

Figure 5:28. Lactophenol cotton blue mount of *Conidiobolus coronatus* showing the papilla on the globose, smooth spore.

9. Anjorin FI, Kazmi R, Malu AO, Lawande RV, Fakunle YM. A case of blastomycosis from Zaria Nigeria. Trans R Soc Trop Med Hyg 1984;78:577-80.
10. Atkinson JB, McCurley TL. Pulmonary blastomycosis: filamentous forms in an immunocompromised patient with fulminating respiratory failure. Hum Pathol 1983;14:186-8.
11. Axelrod P, Kwon-Chung KJ, Frawley P, Harvey R. Chronic cystitis due to *Cokeromyces recurvatus*: a case report. J Infect Dis 1987;155:1062-4.
12. Bakerspigel A, Kane J, Schaus D. Isolation of *Blastomyces dermatitidis* from an earthen floor in southwestern Ontario, Canada. J Clin Microbiol 1986;24:890-1.
13. Bardana EJ. Pulmonary aspergillosis. In: Al-Doory Y, Wagner GE, eds. Aspergillosis. Springfield, IL: CC Thomas, 1985:43-78.
14. Baughman RP, Kim CK, Bullock WE. Comparative diagnostic efficacy of bronchoalveolar lavage transbronchial biopsy and open-lung biopsy in experimental pulmonary histoplasmosis [Letter]. J Infect Dis 1986;153:376-7.
15. Baum GL, Lerner PI. Primary pulmonary blastomycosis: a laboratory-acquired infection. Ann Intern Med 1970;73:263-5.
16. Bayer AS, Scott VJ, Guze LB. Fungal arthritis. IV. Blastomycotic arthritis. Semin Arthritis Rheum 1979;9:145-51.
17. Benbow EW, McMahon RFT. Myocardial infarction caused by cardiac disease in disseminated zygomycosis. J Clin Pathol 1987;40:70-4.
18. Bennett JE. *Aspergillus* species. In: Mandell GL, Douglas RG Jr, Bennett JE, eds. Principles and practice of infectious diseases. New York: John Wiley, 1985:1447-51.
19. Bergstrom VW, Nugent G, Snider MC. Blastomycosis: report of a case with involvement of the skin and bones. Arch Dermatol Syphol 1937;36:70-6.
20. Bigby TD, Serota ML, Tierney LM, Matthay MA. Clinical spectrum of pulmonary mucormycosis. Chest 1986;89:435-9.

21. Blaser MJ, Cohn DL. Opportunistic infections in patients with AIDS: clues to the epidemiology of AIDS and the relative virulence of pathogens. Rev Infect Dis 1986;8:21-30.
22. Bonner JR, Alexander WJ, Dismukes WE, et al. Disseminated histoplasmosis in patients with the acquired immune deficiency syndrome. Arch Intern Med 1984;144:2178-81.
23. Bottone EF, Weitzman I, Hanna BA. *Rhizopus rhizopodiformis*: emerging etiological agent of mucormycosis. J Clin Microbiol 1979;9:530-50.
24. Bradsher RA, Rice DC, Abernathy RS. Ketoconazole therapy for endemic blastomycosis. Ann Intern Med 1985;103:872-9.
25. Brandsberg JW, Tosh FE, Furcolow ML. Concurrent infection with *Histoplasma capsulatum* and *Blastomyces dermatitidis*. N Engl J Med 1964;270:874-7.
26. Brennan RO, Crain BJ, Proctor AM, Durack MB. *Cunninghamella*: a newly recognized cause of rhinocerebral mucormycosis. Am J Clin Pathol 1983;80:98-102.
27. Brewer JH, Parrott CL, Rimland D. Disseminated coccidioidomycosis in a heart transplant recipient. Sabouraudia 1982;20:261-5.
28. Brivet F, Roulot D, Naveau S, et al. The acquired immunodeficiency syndrome: histoplasmosis. Ann Intern Med 1986;104:447-8.
29. Bruck HM, Nash G, Foley FD, Pruitt BA Jr. Opportunistic fungal infection of the burn wound with Phycomycetes and *Aspergillus*: a clinical-pathogenic review. Arch Surg 1971;102:476-82.
30. Butka BJ, Bennett SR, Johnson AC. Disseminated inoculation blastomycosis in a renal transplant recipient. Am Rev Respir Dis 1984;130:1180-3.
31. Castañeda OJ, Alarcön GS, Garcia MT, Lumbreras H. *Paracoccidioides brasiliensis* arthritis. Report of a case and review of the literature. J Rheumatol 1985;12:356-8.
32. Catanzaro A. Pulmonary mycosis in pregnant women. Chest 1984;86:14-18.
33. Centers for Disease Control. Annual summary 1984: Reported morbidity and mortality in the United States. Morbid Mortal Weekly Rep 1986;32:80.
34. Centers for Disease Control. Revision of the case definition of acquired immunodeficiency syndrome for national reporting-United States. Morbid Mortal Weekly Rep 1985;34:373-5.
35. Centers for Disease Control. Human immunodeficiency virus (HIV) infection codes: official authorized addendum ICD-9-CM. Effective January 1, 1988. Morbid Mortal Weekly Rep 1987;36:1-24.
36. Chooi LMF, Fine DP, Muchmore HG, Carter MD, Wilson DA, Tisdal RG. Bilaterally enlarged adrenal glands, an important clue to disseminated histoplasmosis. J Okla State Med Assoc 1985;78:301-3.
37. Ciferri R, Redaelli P. *Histoplasma capsulatum* Darling, the agent of "histoplasmosis:" systemic position and characteristics. J Trop Med Hyg 1934;37:278-80.
38. Cockerill FR, Roberts GD, Rosenblatt JE, Utz JP, Utz DC. Epidemic of pulmonary blastomycosis. (Namekagon fever) in Wisconsin canoeists. Chest 1984;86:688-92.
39. Cockshott WP, Lucas AO. Histoplasmosis duboisii. Quart J Med 1964;33:223-38.
40. Collins DN, Edwards MR. Filamentous forms of *Blastomyces dermatitidis* in mouse lung: light and electron microscopy. Sabouraudia 1970;7:237-40.
41. Cox AJ, Smith CE. Arrested pulmonary coccidioidomycosis granuloma. Arch Pathol 1939;27:717-34.
42. Cox RA, Britt LA. Isolation and identification of an exoantigen specific for *Coccidioides immitis*. Infect Immun 1986;52:138-43.
43. Darling ST. A protozoan general infection producing pseudotubercles in the lungs and focal necrosis in the liver, spleen and lymph nodes. JAMA 1906;46:1283-5.
44. DaRocha-Lima H. Beitrag zur kenntnis der Blastomykosen Lymphangitis epizottica und Histoplasmosia. Zentralbl Bakteriol 1913;67:233-49.
45. Davies SF, Khan M, Sarosi GA. Disseminated histoplasmosis in immunologically suppressed patients. Occurrence in a nonendemic area. Am J Med 1978;64:94-100.
46. Davies SF, Sarosi GA, Peterson PK, et al. Disseminated histoplasmosis in renal transplant recipients. Am J Surg 1979;137:686-91.

47. DeAlbornoz M. Isolation of *Paracoccidioides brasiliensis* from rural soil in Venezuela. Sabouraudia 1971;9:248-53.
48. DeFelice R, Wieden MA, Galgiani JN. The incidence and implications of coccidioidouria. Am Rev Respir Dis 1982;125:49-52.
49. DeMonbreun WA. The cultivation and cultural characteristics of Darling's *Histoplasma capsulatum*. Am J Trop Med 1934;14:93-125.
50. Deng Z, Connor DH. Progressive disseminated penicilliosis caused by *Penicillium marneffei*. Report of eight cases and differentiation of the causative organism from *Histoplasma capsulatum*. Am J Clin Pathol 1985;84:323-7.
51. Denton JF, DiSalvo AF. Isolation of *Blastomyces dermatitidis* from natural sites at Augusta, Georgia. Am J Trop Med Hyg 1964;13:716-22.
52. Denton JF, McDonough ES, Ajello J, Ausherman RJ. Isolation of *Blastomyces dermatitidis* from the soil. Science 1961;133:1126-7.
53. Denys GA, Newman MA, Standard PG. Evaluation of a commercial exoantigen test system for the rapid identification of systemic fungal pathogens. Am J Clin Pathol 1983;79:379-81.
54. Deresinski SC, Stevens DA. Coccidioidomycosis in compromised hosts. Experience at Stanford University Hospital. Medicine 1974;54:377-95.
55. Dickson EC, Gifford MA. *Coccidioides* infection (coccidioidomycosis): the primary type of infection. Arch Intern Med 1938;62:853-71.
56. DiSalvo AF, Fickling AM, Ajello L. Infection caused by *Penicillium marnefferi*: description of first natural infection in man. Am J Clin Pathol 1973;60:259-63.
57. Dismukes W, Karam G, Bowles C, et al. Treatment of blastomycosis and histoplasmosis with ketoconazole. Results of a prospective randomized clinical trial. National Institute of Allergy and Infectious Diseases Mycoses Study Group. Ann Intern Med 1985;103: 861-72.
58. Drutz DJ, Catanzaro A. Coccidioidomycosis Part II. Am Rev Respir Dis 1978;117: 727-71.
59. Dubuisson RL, Jones TB. Splenic abscess due to blastomycosis: scintigraphic, sonographic and CT evaluation. Am J Roentgenol 1983;140:66-8.
60. Dyer ML, Young TL, Kattine AA, Wilson DD. Blastomycosis in a papanicolaou smear. Acta Cytol 1983;27:285-7.
61. Ellis DH, Kaminski GW. Laboratory identification of *Saksenaea vasiformis*: a rare cause of zygomycosis in Australia. Sabouraudia 1985;23:137-40.
62. Ellis JJ, Ajello L. An unusual source for *Apophysomyces elegans* and a method for stimulating sporulation of *Saksenaea vasiformis*. Mycologia 1982;74:144-5.
63. Ellis JJ, Hesseltine CW. The genus *Absidia*: globosed-spored species. Mycologia 1965;57: 222-35.
64. Ellis JJ, Hesseltine CW. Species of *Absidia* with ovoid sporangiospores II. Sabouraudia 1966;5:59-77.
65. Ellis JJ, Hesseltine CW. Two new families of Mucorales. Mycologia 1974;66:87-95.
66. Emmons CW. Isolation of *Histoplasma capsulatum* from the soil. Public Health Rep 1949;64:892-6.
67. Emmons CW. Fungi which resemble *Coccidioides immitis*. In: Ajello L, ed. Coccidioidomycosis, proceedings of second coccidioidomycosis symposium. Phoenix: University of Arizona Press 1967:333-7.
68. Emmons CW, Binford CH, Utz JP, Kwon-Chung KJ. Medical mycology. 3rd ed. Philadelphia: Lea and Febiger, 1977.
69. Espinel-Ingroff A, Oakley LA, Kerkering TM. Opportunistic zygomycotic infections. A literature review. Mycopathologia 1987;97:33-41.
70. Fink JN. Allergic bronchopulmonary aspergillosis. Chest 1985;87(suppl):81-4.
71. Fonseca JC. Analisis estadistica y ecologio-epidemiologica de la sensibilidad a la histoplasmina en Colombia 1950-1968. Antioquia Medica 1971;21:109-54.

72. Frangos DN, Nyberg LM. Genitourinary fungal infections. South Med J 1986;79: 455-9.
73. Furcolow ML. Recent studies on the epidemiology of histoplasmosis. Ann NY Acad Sci 1958;72:127-64.
74. Furcolow ML. Airborne histoplasmosis. Bacteriol Rev 1961;25:301-9.
75. Furcolow ML. Environmental aspects of histoplasmosis. Arch Environ Health 1965; 10:4-10.
76. Furcolow ML, Chick EW, Busey JF, Menges RW. Prevalence and incidence studies of human and canine blastomycosis. I. Cases in the United States 1885-1968. Am Rev Respir Dis 1970;102:60-7.
77. Galgiani JN, Dugger KO, Ito JI, Wieden MA. Antigenemia in primary coccidioidomycosis. Am J Trop Med Hyg 1984;33:645-9.
78. Gartenberg G, Bottone EJ, Keusch GT, Weitzman I. Hospital acquired mucormycosis (*Rhizopus rhizopodiformis*) of skin and subcutaneous tissue. N Engl J Med 1978;299: 1115-18.
79. Gehlbach SH, Hamilton JD, Conant NF. Coccidioidomycosis, an occupational disease in cotton mill workers. Arch Intern Med 1973;131:254-5.
80. Gilchrist TC, Stokes WR. A case of pseudo-lupus vulgaris caused by a blastomyces. J Exp Med 1898;3:53-83.
81. Giraldo R, Restrepo A, Gutierrez F, et al. Pathogenesis of paracoccidioidomycosis: a model based on the study of 46 patients. Mycopathol Mycol Appl 1976;58:63-70.
82. Gnann JW Jr, Bressler GS, Bodet CA III, Avent CK. Human blastomycosis after a dog bite. Ann Intern Med 1983;98:48-9.
83. Goodwin RA Jr, DesPrez RM. Histoplasmosis. Am Rev Respir Dis 1978;117:929-56.
84. Goodwin RA Jr, Shapiro JL, Thurman GH, Thurman SS, DesPrez RM. Disseminated histoplasmosis: clinical and pathologic correlations. Medicine 1980;59:1-33.
85. Gordon MA. Fluorescent staining of *Histoplasma capsulatum*. J Bacteriol 1959;77:678-81.
86. Greenberger PA, Patterson R. Diagnosis and management of allergic bronchopulmonary aspergillosis. Ann Allergy 1986;56:444-8.
87. Greer DL, Bolanos B. Role of bats in the ecology of *Paracoccidioides brasiliensis*. The survival of *Paracoccidioides brasiliensis* in the intestinal tract of frugivorous bat *Artibes lituratus*. Sabouraudia 1977;15:273-9.
88. Greer DL, Restrepo A. The epidemiology of paracoccidioidomycosis. In: Al-Doory Y, ed. The epidemiology of human mycotic diseases. Springfield, IL: CC Thomas, 1975: 117-41.
89. Gregory JE, Golden A, Haymaker W. Mucormycosis of the central nervous system. A report of three cases. Bull Johns Hopkins Hosp 1943;73:403-19.
90. Hadasan FM, Jarrah T, Nassar V. The association of adenocarcinoma of the lung and blastomycosis from an unusual geographical location. Br J Dis Chest 1978;72:242-6.
91. Haggerty CM, Britton MC, Dorman JM, Marzoni FA. Gastrointestinal histoplasmosis in suspected acquired immunodeficiency syndrome. Western J Med 1985;143:244-6.
92. Halvorsen RA Jr, Duncan JD, Merton DF, Gallis HA, Putman CE. Pulmonary blastomycosis: radiologic manifestations. Radiology 1984;150:1-5.
93. Hardin HF, Scott DI. Blastomycosis. Occurrence of filamentous forms in vivo. Am J Clin Pathol 1974;62:104-6.
94. Harvey RP, Pappagianis D, Cochran J, Stevens DA. Otomycosis due to coccidioidomycosis. Arch Intern Med 1978;138:1434-5.
95. Harvey RP, Schmid ES, Carrington CC, Stevens DA. Mouse model of pulmonary blastomycosis: utility, simplicity and quantitative parameters. Am Rev Respir Dis 1978;117:695-703.
96. Hatkin JM, Phillips WE Jr, Utroska WR. Two cases of feline blastomycosis. J Am Anim Hosp Assoc 1975;15:217-20.

97. Henochowitz S, Sahovic E, Pistole M, Rodriques M, Macher A. Histoplasmosis diagnosed on peripheral blood smear from a patient with AIDS. JAMA 1985;253:3148.
98. Hernandez DE, Morgenstern J, Weiss E, et al. Cutaneous lesions of disseminated histoplasmosis in a Haitian man with the acquired immunodeficiency syndrome. Int J Dermatol 1986;25:117-18.
99. Hesseltine CW, Ellis JJ. Notes on Mucorales, especially *Absidia*. Mycologia 1961;53:406-26.
100. Hesseltine CW, Ellis JJ. The genus *Absidia: Gongronella* and cylindrical-spored species of *Absidia*. Mycologia 1964;56:568-601.
101. Hesseltine CW, Ellis JJ. Species of *Absidia* with ovoid sporangiospores. I. Mycologia 1966;58:761-85.
102. Holmberg K, Meyer RD. Fungal infections in patients with AIDS and AIDS-related complex. Scand J Infect Dis 1986;18:179-92.
103. Huppert M, Sun SH. Overview of mycology, and the mycology of *Coccidioides immitis*. In: Stevens DA, ed. Coccidioidomycosis, a text. New York: Plenum Medical Book, 1980:21-46.
104. Huppert M, Sun SH, Rice EH. Specificity of exoantigens for identifying cultures of *Coccidioides immitis*. J Clin Microbiol 1978;8:346-8.
105. Hutton JP, Durham JB, Miller DP, Everett ED. Hyphal forms of *Histoplasma capsulatum*. A common manifestation of intravascular infections. Arch Pathol Lab Med 1985;109: 330-2.
106. Inoshita T, Youngberg GA, Boelen LJ, Langston J. Blastomycosis presenting with prostatic involvement: report of 2 cases and review of the literature. J Urol 1983;130: 160-2.
107. Jewkes J, Kay PH, Paneth M, Citron KM. Pulmonary aspergilloma: analysis of prognosis in relation to haemoptysis and survey of treatment. Thorax 1983;38:572-8.
108. Johnson PC, Satterwhite TK, Monheit JE, Parks D. Primary cutaneous mucormycosis in trauma patients. J Trauma 1987;27:437-41.
109. Johnson WD, Lang CM. Paracoccidioidomycosis (South American blastomycosis) in a squirrel monkey. Vet Pathol 1977;19:568-71.
110. Johnson WM. Racial factors in coccidioidomycosis: mortality experience in Arizona. Ariz Med 1982;39:18-24.
111. Jones PG, Rolston K, Hopfer RL. Septic arthritis due to *Histoplasma capsulatum* in a leukemic patient. Ann Rheum Dis 1985;44:128-9.
112. Joseph JM. Aspergillosis in composting. In: DiSalvo AF, ed. Occupational mycoses. Philadelphia: Lea and Febiger, 1983:123-42.
113. Joshi JH, Schimpff SC. Infections in patients with lymphoma and myeloma. In: Mandell GL, Douglas RG Jr, Bennett JE, eds. Principles and practice of infectious diseases. New York: John Wiley, 1985:1658-62.
114. Kahn FW, Jones JM, England DM. The role of bronchoalveolar lavage in the diagnosis of invasive pulmonary aspergillosis. Am J Clin Pathol 1986;86:518-23.
115. Kalter DC, Tschen JA, Klima M. Maculopapular rash in a patient with acquired immunodeficiency syndrome. Disseminated histoplasmosis in acquired immunodeficiency syndrome (AIDS). Arch Dermatol 1985;121:1455-9.
116. Kamalam A, Thamblah AS. Cutaneous infection by *Syncephalastrum*. Sabouraudia 1980;18:19-20.
117. Kaplan W. Application of the fluorescent antibody technique to the diagnosis and study of histoplasmosis. In: Ajello L, Chick EW, Furcolow ML, eds. Histoplasmosis, proceedings of the second national conference. Springfield, IL: CC Thomas 1971:327-40.
118. Kaplan W, Kaufman L. Specific fluorescent antiglobulins for the detection and identification of *Blastomyces dermatitidis* yeast-phase cells. Mycopathol Mycol Appl 1963;19: 173-80.
119. Karam GH, Griffin FM Jr. Invasive pulmonary aspergillosis in nonimmunocompromised, nonneutropenic hosts. Rev Infect Dis 1986;8:357-63.

120. Kauffman CA, Slama TG, Wheat LJ. *Histoplasma capsulatum* epididymitis. J Urol 1981;125:434-5.
121. Kaufmann L, Kaplan W. Preparation of a fluorescent antibody specific for the yeast phase of *Histoplasma capsulatum*. J Bacteriol 1961;82:729-35.
122. Kaufman L, McLaughlin DW, Clark MJ, Blumer S. Specific immunodiffusion test for blastomycosis. Appl Microbiol 1973;26:244-7.
123. Kaufman L, Standard P. Immunoidentification of cultures of fungi pathogenic to man. Curr Microbiol 1978;1:135-40.
124. Kaufman L, Standard P, Padhye AA. Exoantigen tests for the immunoidentification of fungal cultures. Mycopathologia 1983;82:3-12.
125. Keddie F, Shadomy J, Barfanti M. Brief report on the isolation of *Arthroderma tuberculatum* from a human source. Mycopathol Mycol Appl 1963;20:1-2.
126. Kerrick SS, Lundergan LL, Galgiani JN. Coccidioidomycosis at a university health service. Am Rev Respir Dis 1985;131:100-2.
127. Kitchen MS, Reiber CD, Eastin GB. An urban epidemic of North American blastomycosis. Am Rev Respir Dis 1977;115:1063-6.
128. Klein BS, Kuritsky JN, Chappell WA, et al. Comparison of the enzyme immunoassay, immunodiffusion and complement fixation tests in detecting antibody in human serum to the A antigen of *Blastomyces dermatitidis*. Am Rev Respir Dis 1986;133:144-8.
129. Klein BS, Vergeront JM, Weeks RJ, et al. Isolation of *Blastomyces dermatitidis* in soil associated with a large outbreak of blastomycosis in Wisconsin. N Engl J Med 1986;314: 529-34.
130. Klite PD, Diercks FH. *Histoplasma capsulatum* in fecal contents and organs of bats in the canal zone. Am J Trop Med Hyg 1965;14:433-9.
131. Kolbeck PC, Makhoul RG, Bollinger RR, Sanfilippo F. Widely disseminated *Cunninghamella* mucormycosis in an adult renal transplant patient: case report and review of the literature. Am J Clin Pathol 1985;83:747-53.
132. Kramer BS, Hernandez AD, Reddick RL, Levine AS. Cutaneous infarction. Manifestation of disseminated mucormycosis. Arch Dermatol 1977;113:1075-6.
133. Kravitz GR, Davies SF, Eckman MR, Sarosi GA. Chronic blastomycotic meningitis. Am J Med 1981;71:501-5.
134. Kurnug JM. The isolation of *Histoplasma capsulatum* from sputum. Am Rev Tuberc 1952;66:578-87.
135. Kuttin ES, Beemer AM, Levij J, Ajello L, Kaplan W. Occurrence of *Blastomyces dermatitidis* in Israel. Am J Trop Med Hyg 1978;27:1203-5.
136. Kwon-Chung KJ. Sexual state of *Histoplasma capsulatum*. Science 1971;175:326.
137. Kwon-Chung KJ. A new pathogenic species of *Aspergillus* in the *Aspergillus fumigatus* series. Mycologia 1975;67:770-5.
138. Kwon-Chung KJ. Perfect state (*Emmonsiella capsulata*) of the fungus causing large-form African histoplasmosis. Mycologia 1975;67:980-90.
139. Kwon-Chung KJ, Bartlett MS, Wheat LJ. Distribution of the two mating types among *Histoplasma capsulatum* isolates obtained from an urban histoplasmosis outbreak. Sabouraudia 1984;22:155-7.
140. Kwon-Chung KJ, Young RC, Orland M. Pulmonary mucormycosis caused by *Cunninghamella elegans* in a patient with chronic myelogenous leukemia. Am J Clin Pathol 1975;64:544-8.
141. Landay ME, Schwarz J. Primary cutaneous blastomycosis. Arch Dermatol 1971;104: 408-11.
142. Larsh HW, Schwarz J. Accidental inoculation. Blastomycosis. Cutis 1977;19:334-7.
143. Larson DM, Eckman MR, Alger RL, Goldschmidt VG. Primary cutaneous (inoculation) blastomycosis: an occupational hazard of pathologists. Am J Clin Pathol 1983;79:253-5.
144. Laskey W, Sarosi GA. Endogenous activation in blastomycosis. Ann Intern Med 1978;88:50-2.
145. Lehrer RI. Mucormycosis (review). Ann Intern Med 1980;93:93-108.

146. Lehrer RI, Howard DH, Sypherd PS, Edwards JE, Sega GP, Winston DJ. Mucormycosis. Ann Intern Med 1980;93:93-108.
147. Lemos LB, Jensen AB. Pathology of aspergillosis. In: Al-Doory Y, Wagner GE, eds. Aspergillosis. Springfield: CC Thomas, 1985:156-95.
148. Lenhard A. Blastomycosis in a ferret. J Am Vet Med Assoc 1985;186:70-2.
149. Lentino JR, Rosenkranz MA, Michaels JA, Kurup VP, Rose HD, Rytel MW. Nosocomial aspergillosis: a retrospective review of airborne disease secondary to road construction and contaminated air conditioners. Am J Epidemiol 1982;116:430-7.
150. Levitz SM, Diamond RD. Changing patterns of aspergillosis infections. In: Stollerman GH, Harrington WJ, Lamont JT, Leonard JJ, Siperstein MD, eds. Advances in internal medicine; vol 30. Chicago: Year Book Medical Publishers, 1984:153-74.
151. Long ER. Specific necrosis in the infectious granulomas: demonstration in the testis in experimental tuberculosis, *Bacillus abortus* infection and blastomycosis. JAMA 1926;18: 1441-5.
152. Lyons RW, Andriole VT. Fungal infections of the CNS. Neurol Clin 1986;4:159-70.
153. MacDonald N, Steinhoff MC, Powell KR. Review of coccidioidomycosis in immunocompromised children. Am J Dis Child 1981;135:553-6.
154. Mackinnon JE. On the importance of South American blastomycosis. Mycopathol Mycol Appl 1970;41:187-93.
155. Maddy KT. The geographic distribution of *Coccidioides immitis* and possible ecologic implications. Arizona Med 1958;15:178-88.
156. Mandell W, Goldberg DM, Neu HC. Histoplasmosis in patients with the acquired immune deficiency syndrome. Am J Med 1986;81:974-8.
157. Marchevsky AM, Bottone EJ, Geller SA, Giger DK. The changing spectrum of disease, etiology and diagnosis of mucormycosis. Human Pathol 1980;11:457-64.
158. McCarty JM, Flam MS, Pullen G, Jones R, Kassel SH. Outbreak of primary cutaneous aspergillosis related to intravenous arm boards. J Pediatr 1986;108:721-4.
159. McDonough ES. Blastomycosis—epidemiology and biology of its etiologic agent *Ajellomyces dermatitidis*. Mycopathol Mycol Appl 1970;41:195-201.
160. McDonough ES, Lewis AL. *Blastomyces dermatitidis*: production of the sexual stage. Science 1967;156:528-9.
161. McGinnis MR. Laboratory handbook of medical mycology. New York: Academic Press, 1980.
162. McGinnis MR, Katz B. *Ajellomyces* and its synonym *Emmonsiella*. Mycotaxon 1979;8: 157-64.
163. McGregor JA, Kleinschmidt-DeMasters BK, Ogle J. Meningoencephalitis caused by *Histoplasma capsulatum* complicating pregnancy. Am J Obstet Gynecol 1986;154: 925-31.
164. McKenzie R, Khakoo R. Blastomycosis of the esophagus presenting with gastrointestinal bleeding. Gastroenterology 1985;88:1271-3.
165. Meyer RD. Agents of mucormycosis and related species. In: Mandell GL, Douglas RG Jr, Bennett JE, eds. Principles and practice of infectious diseases. New York: John Wiley, 1985:1452-6.
166. Meyer RD. Cutaneous and mucosal manifestations of the deep mycotic infections. Acta Derm Venereol (Stockh) 1986;121:57-72.
167. Minguetti G, Madalozzo LE. Paracoccidioidal granulomatosis of the brain. Arch Neurol 1983;40:100-2.
168. Misra PC, Srivastava KJ, Lata K. *Apophysomyces*, a new genus of the Mucorales. Mycotaxon 1979;8:377-82.
169. Monheit JG, Brown G, Kott MM, Schmidt WA, Moore DG. Calcofluor white detection of fungi in cytopathology. Am J Clin Pathol 1986;85:222-5.
170. Monheit JE, Cowan DF, Moore DG. Rapid detection of fungi in tissues using calcofluor white and fluorescence microscopy. Arch Pathol Lab Med 1984;108:616-18.
171. Moore RM, Green NE. Blastomycosis of bone. J Bone Joint Surg 1982;64:1097-1101.

172. Morris PR, Terreni AA, DiSalvo AF. Red-pigmented *Histoplasma capsulatum* - an unusual variant. J Med Vet Mycol 1986;24:231-3.
173. Morse DL, Gordon MA, Matte T, Eadie G. An outbreak of histoplasmosis in a prison. Am J Epidemiol 1985;122:253-61.
174. Moser SA, Friedman L, Varraux A. An atypical isolate of *Cryptococcus neoformans* cultured from sputum of a patient with pulmonary cancer and blastomycosis. J Clin Microbiol 1978;7:316-18.
175. Murray JJ, Clark CA, Lands RH, Heim CR, Burnett LS. Reactivation blastomycosis presenting as a tuboovarian abscess. Obstet Gynecol 1984;64:828-30.
176. Musial CE, Wilson WR, Sinkeldam IR, Roberts GD. Recovery of *Blastomyces dermatitidis* from blood of a patient with disseminated blastomycosis. J Clin Microbiol 1987;25: 1421-3.
177. Naiff RD, Ferreira LCL, Barrett TV, Naiff MF, Arias JR. Paracoccidioidomycose enzootica em tatus (Dasypus novemcinctus) no estaco do Para. Rev Inst Med Trop Sao Paulo 1986;28:19-27.
178. Nottebrock H, Scholer HJ, Wall M. Taxonomy and identification of mucormycosis-causing fungi. I. Synonymity of *Absidia ramosa* with *A. corymbifera*. Sabouraudia 1974;12:64-74.
179. Okudaira M, Straub M, Schwartz J. The etiology of discrete splenic and hepatic calcifications in an endemic area of histoplasmosis. Am J Pathol 1961;35:599-611.
180. Opal SM, Reller BL, Harrington G, Cannady P Jr. Aspergillus clavatus endocarditis involving a normal aortic valve following coronary artery surgery. Rev Infect Dis 1986;8:781-5.
181. Ophüls W. Further observations on a pathogenic mould formerly described as a protozoan. (*Coccidioides immitis, Coccidioides pyogenes*). J Exp Med 1905;6:443-85.
182. Ophüls W, Moffitt HC. A new pathogenic mould, (formerly described as a protozoan: *Coccidioides immitis*). Preliminary report. Phila Med J 1900;5:1471-2.
183. Pappagianis D. Epidemiology of coccidioidomycosis. In: Stevens DA, ed. Coccidioidomycosis, a text. New York: Plenum Medical Book Co., 1980:63-85.
184. Pappagianis D, Einstein H. Tempest from Tehachapi take toll. West J Med 1977;129:527-30.
185. Parra R, Arnau E, Julia A, Lopez A, Nadal A, Allende E. Survival after intestinal mucormycosis in acute myelogenous leukemia. Cancer 1986;58:2717-19.
186. Petrak RM, Pottage JC Jr, Levin S. Invasive external otitis caused by *Aspergillus fumigatus* in an immunocompromised patient. [Letter]. J Infect Dis 1985;151:196.
187. Phillips P, Weiner MH, Invasive aspergillosis diagnosed by immunohistochemistry with monoclonal and polyclonal reagents. Hum Pathol 1987;18:1015-24.
188. Pierce PF, Wood MB, Roberts GD, Fitzgerald RH Jr, Robertson C, Edson RS. *Saksenaea vasiformis* osteomyelitis. J Clin Microbiol 1987;25:933-5.
189. Pladson TR, Stiles MA, Kuritsky JN. Pulmonary histoplasmosis. A possible risk in people who cut decayed wood. Chest 1984;86:435-8.
190. Posadas A. Un neuvo caso de micosis fungoidea con psorospermias. Circulo Med Argent 1892;5:585-97.
191. Pritchard RC, Muir DB, Archer KH, Beith JM. Subcutaneous zygomycosis due to *Saksenaea vasiformis* in an infant. Med J Australia 1986;145:630-1.
192. Puckett F. Hyphae of *Coccidioides immitis* in tissues of the human host. Am Rev Tuberc 1954;70:320-7.
193. Randhawa HS, Khan ZU, Gaur SN. *Blastomyces dermatitidis* in India: first report of its isolation from clinical material. Sabouraudia 1983;21:215-21.
194. Raper KB, Fennell DI. The genus *Aspergillus*. Baltimore: Williams and Wilkins, 1965.
195. Raz R, Ephros M, Or R, Polacheck I. Primary pulmonary aspergilloma: case report and review of the literature. Israel J Med Sci 1986;22:400-3.
196. Recht LD, Davies SF, Eckman MR, Sarosi GA. Blastomycosis in immunosuppressed patients. Am Rev Respir Dis 1982;125:359-62.

197. Recht LD, Philips JR, Eckman MR, Sarosi GA. Self-limited blastomycosis: a report of thirteen cases. Am Rev Respir Dis 1979;120:1109-12.
198. Restrepo A. A reappraisal of the microscopical appearance of the mycelial phase of *Paracoccidioides brasiliensis*. Sabouraudia 1970;8:141-4.
199. Restrepo A, Cano LE, Tabares A. A comparison of mycelial filtrate - and yeast lysate - paracoccidioidin in patients with paracoccidioidomycosis. Mycopathologia 1983;84:49-54,166.
200. Restrepo A, Greer DL. Paracoccidioidomycosis. In: DiSalvo AF, ed. Occupational mycoses. Philadelphia: Lea and Febiger, 1983:43-64.
201. Richardson JH, Barkley EW, eds. Biosafety in microbiological and biomedical laboratories. Atlanta: US Dept Health and Human Services, Centers for Disease Control, 1984.
202. Rinaldi MG. Invasive aspergillosis. Rev Infect Dis 1983;5:1061-77.
203. Rippon JW. Medical mycology. 2nd ed. Philadelphia: WB Saunders, 1982:615-40.
204. Rivkin LM, Winn DF, Salyer JM. The surgical treament of pulmonary coccidioidomycosis. J Thorac Cardiovasc Surg 1961;42:401-12.
205. Roberts GD. Detection of fungi in clinical specimens by phase-contrast microscopy. J Clin Microbiol 1975;2:261-5.
206. Sacks JJ, Ajello L, Crockett LK. An outbreak and review of cave-associated histoplasmosis capsulati. J Med Vet Mycol 1986;24:313-27.
207. Salaki JS, Louria DB, Chmel H. Fungal and yeast infections of the central nervous system. Medicine 1984;63:108-32.
208. Sanders JS, Sarosi GA, Nollet DJ, Thompson JI. Exfoliative cytology in the rapid diagnosis of pulmonary blastomycosis. Chest 1977;72:193-6.
209. Sands JM, Macher AM, Ley TJ, Nienhuis AW. Disseminated infection caused by *Cunninghamella bertholletiae* in a patient with beta-thalassemia. Case report and review of the literature. Ann Intern Med 1985;102:59-63.
210. Sarosi GA, Davies SF. Blastomycosis. Am Rev Respir Dis 1979;120:911-38.
211. Sarosi GA, Hammerman KJ, Tosh FE, Kronenberg RS. Clinical features of acute pulmonary blastomycosis. N Engl J Med 1974;290:540-3.
212. Sarosi GA, Parker JD, Doto IL, Tosh FE. Chronic pulmonary coccidioidomycosis. N Engl J Med 1970;283:325-9.
213. Sarosi GA, Parker JD, Tosh FE. Histoplasmosis outbreaks: their patterns. In: Ajello L, Chick EW, Furcolow FL, eds. Histoplasmosis, proceedings of second national conference. Springfield: CC Thomas 1971:123-8.
214. Sarosi GA, Serstock DS. Isolation of *Blastomyces dermatitidis* from pigeon manure. Am Rev Respir Dis 1976;114:1179-83.
215. Schipper MAA. 1. On certain species of Mucor with a key to all accepted species. 2. On the genera *Rhizomucor* and *Parasitella*. Studies in mycology no. 17. Baarn, Netherland: Centralbureau voor Schimmelculture, 1978.
216. Schlech WF III, Wheat LJ, Ho JL, et al. Recurrent urban histoplasmosis, Indianapolis, Indiana 1980-1981. Am J Epidemiol 1983;118:301-12.
217. Schmidt JH, Howard RJ, Chen JL, Pierson KK. First culture-proven gastrointestinal entermophthoromycosis in the United States: a case report and review of the literature. Mycopathologia 1986;95:101-4.
218. Schwarz J. The pathogenesis of histoplasmosis. Trans NY Acad Sci 1958;20:541-8.
219. Schwarz J. What's new in mycotic bone and joint diseases? Pathol Res Pract 1984;178: 617-34.
220. Schwarz J, Baum GL. Blastomycosis. Am J Clin Pathol 1951;21:999-1029.
221. Schwarz J, Salfelder K. Blastomycosis. Curr Top Pathol 1977;65:165-200.
222. Sekhon AS, Jackson FL, Jacobs JH. Blastomycosis: report of the first case from Alberta, Canada. Mycopathologia 1982;79:65-9.
223. Sekhon AS, Standard PG, Kaufman L, Garg AK. Reliability of exoantigens for differentiating *Blastomyces dermatitidis* and *Histoplasma capsulatum* from *Chrysosporium* and *Geomyces* species. Diagn Microbiol Infect Dis 1986;4:215-21.

224. Selim SA, Soliman R, Osman K, Padhye AA, Ajello L. Studies on histoplasmosis farciminosi (epizootic lymphangitis) in Egypt. Isolation of *Histoplasma farciminosum* from cases of histoplasmosis farciminosi in horses and its morphological characteristics. Eur J Epidemiol 1985;1:84-9.
225. Shadomy S, White SC, Yu HP, Dismukes WE. Treatment of systemic mycoses with ketoconzaole: in vitro susceptibilities of clinical isolates of systemic and pathogenic fungi to ketoconazole. J Infect Dis 1985;152:1249-56.
226. Shanor L, Poitras AW, Benjamin RK. A new genus of the Choanephoraceae. Mycologia 1950;42:271-8.
227. Sheehan DC, Hrapchak BB. General staining considerations. Theory and practice of histopathology. 1980:133.
228. Sheldon DL, Johnson WC. Cutaneous mucormycosis. Two documented cases of suspected nosocomial cause. JAMA 1979;241:1032-4.
229. Sigler L, Carmichael JW. Taxonomy of *Malbranchea* and some other hyphomycetes with arthroconidia. Mycotaxon 1976;4:349-88.
230. Sinski JT. The epidemiology of aspergillosis. In: Al-Doory Y, Wagner GE, eds. Aspergillosis. Springfield: CC Thomas, 1985:25-42.
231. Smith CE, Whiting EG, Baker EE, Rosenberger HG, Beard RR, Saito MT. The use of coccidioidin. Am Rev Tuberc 1948;57:330-60.
232. Smith JR Jr, Harris JS, Conant NF, Smith DT. An epidemic of North American blastomycosis. JAMA 1951;158:641-5.
233. Snider HL, Winkler CF, Yam LT. Fungemia in chronic cavitary pulmonary histoplasmosis. J Infect Dis 1981;143:633.
234. Sorensen R, Cheu S. Accidental cutaneous coccidioidal infection in an immune person. Calif Med 1964;100:44-7.
235. Standard PG, Kaufman L. Specific immunological test for the rapid identification of members of the genus *Histoplasma*. J Clin Microbiol 1976;3:191-9.
236. Standard PG, Kaufman L. Safety considerations in handling exoantigen extracts from pathogenic fungi. J Clin Microbiol 1982;15:663-7.
237. Steinbrecher UP, Benaroya SH. Histoplasmosis and Lennert's lymphoma. Arch Intern Med 1979;139:596-7.
238. Stockman L, Roberts GD, Ilstrup DM. Effect of storage of the DuPont lysis-centrifugation system on recovery of bacteria and fungi in a prospective clinical trial. J Clin Microbiol 1984;19:283-5.
239. Streifel AJ, Lauer JL, Vesley D, Juni B, Rhame FS. *Aspergillus fumigatus* and other thermotolerant fungi generated by hospital building demolition. Appl Environ Microbiol 1983;46:375-8.
240. Sudman MS, Kaplan W. Antigenic relationship between American and African isolates of *Blastomyces dermatitidis* as determined by immunofluorescence. Appl Microbiol 1974;27:496-9.
241. Sugar AM, Restrepo A, Stevens DA. Paracoccidioidomycosis in the immunosuppressed host: report of a case and review of the literature. Am Rev Respir Dis 1984;129:340-2.
242. Sun HS, Huppert M, Vukovich KR. Rapid in vitro conversion and identification of *Coccidioides immitis*. J Clin Microbiol 1976;3:186-90.
243. Taylor GD, Sekhon AS, Tyrrell DLJ, Goldsand G. Rhinofacial zygomycosis caused by *Conidiobolus coronatus*: a case report including in vitro sensitivity to antimycotic agents. Am J Trop Med Hyg 1987;36:398-401.
244. Taylor MR, Boyce JM. Inhibition of *Blastomyces dermatitidis* by topical lidocaine. Chest 1983;84:431-2.
245. Taylor MR, Lawson LA, Lockard VG, Lockwood WR. Ultrastructural changes in *Blastomyces dermatitidis* after in vitro exposure to lidocaine. Mycopathologia 1984;88:173-80.
246. Thompson DW, Kaplan W, Phillips BJ. The effect of freezing and the influence of isolation medium on the recovery of pathogenic fungi from sputum. Mycopathologia 1977;61:105-9.

247. Thompson EM, Ellert J, Peters LL, Ajukiewicz A, Mabey D. *Histoplasma duboisii* infection of bone. Br J Radiol 1981;54:518-21.
248. Tomlinson FR, Sahn SA. Aspergilloma in sarcoid and tuberculosis. Chest 1987;92:505-8.
249. Torell J, Cooper BH, Helgeson NGP. Disseminated *Saksenaea vasiformis* infection. Am J Clin Pathol 1981;76:116-21.
250. Tosh FE, Hammerman KJ, Weeks RJ, Sarosi GA. A common source epidemic of North American blastomycosis. Am Rev Respir Dis 1974;109:525-9.
251. Tuder RM. Myocardial infarct in disseminated mucormycosis: case report with special emphasis on the pathogenic mechanisms. Mycopathologia 185;89:81-8.
252. Vanek J, Schwarz J. The gamut of histoplasmosis. Am J Med 1971;50:89-104.
253. Varkey B, Rose HD. Pulmonary aspergilloma: a rational approach to treatment. Am J Med 176;61:626-31.
254. Wahner HW, Hepper NGG, Anderson HA, Weed LA. Pulmonary aspergillosis. Ann Intern Med 1963;58:472-85.
255. Walsh TJ, Catchatourian R, Cohen H. Disseminated histoplasmosis complicating bone marrow transplantation. Am J Clin Pathol 1983;79:509-11.
256. Weeks RJ, Padhye AA, Ajello L. *Histoplasma capsulatum* variety *farciminosum*: a new combination for *Histoplasma farciminosum*. Mycologia 1985;77:964-70.
257. Weeks RJ, Tosh FE. Control of epidemic foci of *Histoplasma capsulatum*. In: Ajello L, Chick EW, Furcolow ML, eds. Histoplasmosis, proceedings of the second national conference. 1971:184-9.
258. Weiner MH. Antigenemia detected in human coccidioidomycosis. J Clin Microbiol 1983;18:136-42.
259. Weitzman I. Saprophytic molds as agents of cutaneous and subcutaneous infection in the immunocompromised host. Arch Dermatol 1986;122:1161-8.
260. Weitzman I, Crist MY. Studies with clinical isolates of *Cunninghamella* I. Mating behavior. Mycologia 1979;71:1024-33.
261. Wheat LJ, Slama TG, Zeckel ML. Histoplasmosis in the acquired immune deficiency syndrome. Am J Med 1985;78:203-10.
262. Wheat LJ, Small CB. Disseminated histoplasmosis in the acquired immune deficiency syndrome [Editorial]. Arch Intern Med 1984;144:2147-9.
263. Wieden MA, Steinbronn KK, Padhye AA, Ajello L, Chandler FW. Zygomycosis caused by *Apophysomyces elegans*. J Clin Microbiol 1985;22:522-6.
264. Williams JE, Moser SA. Chronic murine pulmonary blastomycosis induced by intratracheal inoculation of *Blastomyces dermatitidis* conidia. Am Rev Respir Dis 1987;135:17-25.
265. Williams JE, Murphy T, Standard PG, Phair JP. Serologic response in blastomycosis: diagnostic value of double immunodiffusion assay. Am Rev Respir Dis 1981;123:209-12.
266. Williams PL, Sable DL, Sorgen SP, et al. Immunologic responsiveness and safety associated with the *Coccidioides immitis* spherule vaccine in volunteers of white, black and Filipino ancestry. Am J Epidemiol 1984;119:591-602.
267. Wolf JE, Little JR, Pappagianis D, Kobayashi GS. Disseminated coccidioidomycosis in a patient with the acquired immune deficiency syndrome. Diagn Microbiol Infect Dis 1986;5:331-6.
268. Youmans GP. Tuberculosis. Philadelphia: WB Saunders, 1979.
269. Young RC, Bennett JE, Vogel CL, Carbone PP, De Vita VT. Aspergillosis. The spectrum of the disease in 98 patients. Medicine 1970;49:147-73.
270. Young RC, Jennings A, Bennett JE. Species identification of invasive aspergillosis in man. Am J Clin Pathol 1972;58:554-7.
271. Yu VL, Muder RR, Poorsattar A. Significance of isolation of *Aspergillus* from the respiratory tract in diagnosis of invasive pulmonary aspergillosis: results from a three-year prospective study. Am J Med 1986;81:249-54.

CHAPTER 6

HUMAN INFECTIONS CAUSED BY YEASTLIKE FUNGI

David H. Pincus, M.S. and Ira F. Salkin, Ph.D.

INTRODUCTION

This chapter will describe some of the pathogenic processes caused by yeasts and yeastlike fungi, as well as the methods commonly used in clinical laboratories for the identification of these zoopathogens. Some authors prefer to use the term yeast to refer to a specific, somewhat homogenous fungal group, the "true" yeasts, e.g., *Saccharomyces cerevisiae*. The major morphologic-defining characteristic of these fungi is a unicellular vegetative or growing stage. True yeasts also share biochemical, physiologic, reproductive and other characteristics. In contrast, the term yeastlike is used as a morphologic definition, i.e., to indicate fungi similar to true yeasts in their unicellular vegetative morphology. Yeastlike fungi represent a rather heterogenous assemblage of diverse organisms, including members of the Ascomycotina, Basidiomycotina and Deuteromycotina, as well as achlorophyllous algae, e.g., *Prototheca* (Fig. 6:1). For the sake of convenience and to avoid confusion, the term "yeast" will be used in this chapter to refer to all true yeasts, yeastlike fungi and algae routinely encountered in the clinical laboratory. Yeast colonies on standard nutrient agar media vary from smooth to wrinkled in topography, dull or dry to glossy or mucoid in appearance, soft or pasty to tough or leathery in texture and white, pink, orange, red, light tan, brown or black in color. Many taxa form pseudohyphae and/or true hyphae when grown on cornmeal + Tween 80 or cream of rice + Tween 80 media. They reproduce asexually (anamorph stage) through the formation of a diverse array of propagules collectively referred to as conidia, e.g., annelloconidia, arthroconidia, ballistoconidia, blastoconidia and phialoconidia.[43,55] Sexual reproduction (teleomorph stage) may be induced through the use of specific nutritional and environmental conditions[9,50,55] and results in the development of propagules called spores, i.e., ascospores or basidiospores.

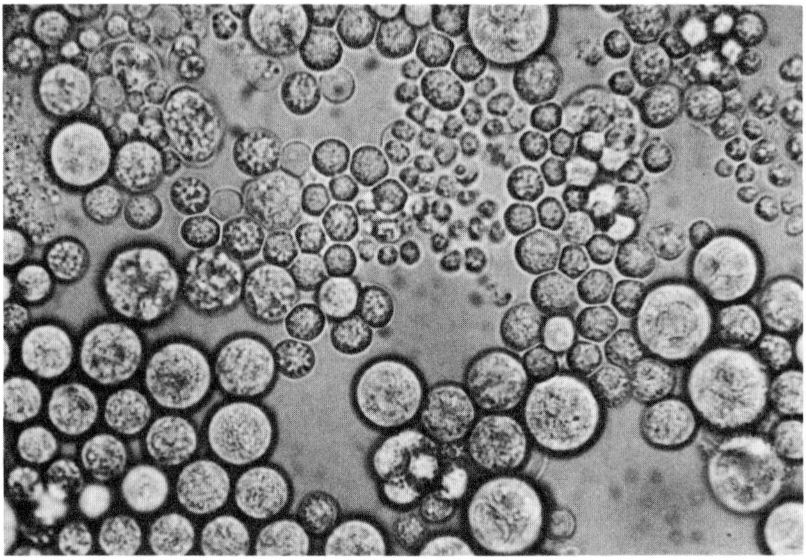

Figure 6:1. *Prototheca wickerhamii*, an achlorophyllous alga, produces sporangia and sporangiospores only (1000X).

Yeasts are ubiquitous in nature and have been isolated from a large and diverse number of substrates. They coexist with other microorganisms in a delicate balance as part of the normal microflora in humans and lower animals. Most yeast infections appear to be endogenous in origin and to have resulted from a disruption of this balance caused by iatrogenic or natural factors. Pregnancy, malnutrition, chronic alcoholism, diabetes mellitus, prolonged broad-spectrum antibacterial therapy, acquired immunodeficiency syndrome (AIDS) and malignant disorders such as lymphomas and leukemia may all provide the opportunity for yeasts to initiate infections (opportunistic diseases). In addition, the limited cell-mediated immunity found in neonates and the elderly or the suppression of cellular defenses through the use of corticosteroids, radiation therapy and cytotoxic drugs may also contribute to opportunistic yeast infections. Disruption of the primary defense barrier of the skin by parenteral drug abuse, indwelling catheters, hyperalimentation, trauma and burns may result in colonization or invasion by yeasts which are part of the microflora of the skin.[1,21,26,30,39,51,55,65,72,75,97]

Yeast infections range from the common and generally benign candidal vulvovaginitis and oral thrush to less frequently encountered but more serious diseases such as cryptococcal meningitis and candidal endocarditis. The less invasive forms of infection are bothersome to the patient and may require the use of topical antifungal agents. To reduce their relatively high morbidity and mortality, disseminated yeast infections require immediate diagnosis and prompt initiation of systemic therapy. Speciation of etiologic agents has become increasingly important with the emergence of new or normally saprobic species as agents of opportunistic systemic diseases, as well as the reported resistance of several taxa, e.g., *Candida lusitaniae*[31,33,59] or *Candida guilliermondii*,[23] to commonly used antifungal agents.

There are several serologic procedures currently available or in development for the rapid diagnosis of life-threatening yeast infections. The cryptococcal latex agglutination test provides a highly specific and sensitive means of detecting cryptococcal antigen in cerebrospinal fluid and serum.[11,41,57,94] Recent investigations have indicated that the detection of antigens, antibodies or fungal metabolites, e.g., D-arabinitol, mannose and mannan, in patients' sera may provide a rapid means of diagnosing systemic candidiasis.[6,22,60,61] Monoclonal antibodies have been shown to be useful in the identification of *Candida albicans* and may eventually be used in clinical laboratories as a diagnostic tool.[68]

In addition, several diagnostic media have been described to assist the laboratorian in the presumptive identification of yeast pathogens. For example, *Cryptococcus neoformans* may be rapidly identified on tween-oxgall-caffeic acid (TOC) agar,[27] bird/niger seed (*Guizotia abyssinica*) agar[20,55], and L-DOPA (3,4-dihydroxyphenylalanine)-containing media[16] by its formation of brown to black pigment. Several media have been described which stimulate germ tube production for the presumptive identification of *C. albicans*.[53] The rapid urease[76,96] and nitrate reductase[38,73] tests have been used for same-day identification of several clinically important yeast taxa.

Presumptive identification can also be accomplished through such routine practices as observation of colonial characteristics, e.g., pigment, texture and topography, or by microscopic examination of such morphologic features (Table 6:1) as pseudohyphae, true hyphae, chlamydospores, arthroconidia, ascospores, endospores, etc. on cornmeal + Tween 80 or cream of rice + Tween 80 agars.[10]

Another useful identification criterion overlooked frequently by laboratorians is the specimen source from which the yeast was isolated. Urinary tract infections are most often associated with *C. albicans* and *Torulopsis glabrata*. Fungal vulvovaginitis and oral thrush are almost always caused by *C. albicans*.[26,75] Transient catheter-associated fungemia is associated with a select group of taxa, e.g., *Candida parapsilosis* and *Malassezia furfur*.[39,51,70] A more detailed description of the specimen sources (Table 6:2) and clinical manifestations as well as the procedures used in the identification of yeast pathogens are presented in the following sections.

CLINICAL MANIFESTATIONS

Candidiasis

Candidiasis (candidosis, moniliasis) is quite probably the most protean disease of man, with virtually all tissue and organ systems subject to invasion

Table 6:1 Morphologic Characters of Selected Yeast Genera

	Morphologic characteristics								
Genus	BL*	PH	TH	AR	CL	AS	BS	SP	CP
Aureobasidium	+	−	+	+	−†	−	−	−	V‡
Blastoschizomyces	+	V	+	+	−	−	−	−	−
Candida	+	(+)	V	−	V	−	−	−	−
Cryptococcus	+	(−)	(−)	−	−	−	−	−	V
Geotrichum	−	−	+	+	−	−	−	−	−
Hansenula	+	V	V	−	−	(+)	−	−	−
Prototheca	−	−	−	−	−	−	−	+	−
Rhodotorula	+	(−)	(−)	−	(−)	−	−	−	+§
Saccharomyces	+	V	−	−	−	(+)	−	−	−
Sporobolomyces	+	V	V	−	−	−	+	−	+
Torulopsis	+	(−)	−	−	−	−	−	−	−
Trichosporon	+	V	+	+	−	−	−	−	−

*BL = blastoconidia
PH = pseudohyphae
TH = true hyphae
AR = arthroconidia
CL = chlamydospores
AS = ascospores
BS = ballistoconida
SP = sporangiospores (endospores)
CP = carotenoid pigment

\+ = present, − = absent, V = variable, (+) = usually present, (−) = usually absent or rudimentary if present.

† Thick-walled arthroconidia may resemble chlamydospores

‡ Pale pink pigment common in younger cultures; older cultures commonly develop olive-brown to black (dematiaceous) pigment.

§ Rare strains may not produce pigment.

by *Candida* species. *C. albicans*, the most common etiologic agent of candidiasis, is a normal constituent of the alimentary tract microflora and is commonly associated with several mucocutaneous sites on or in the human host. *Candida* vulvovaginitis in gravid females and oral candidiasis (thrush) in infants are relatively common, generally innocuous, clinical conditions. In fact, oral thrush in neonates may result from exposure to the etiologic agent during birth and the subsequent infection of the oral cavity prior to the establishment of normal bacterial flora. Cutaneous infections including intertriginous candidiasis, onychomycosis and paronychia may clinically resemble dermatophytoses. In contrast, *Candida* infections involving other mucocutaneous sites such as the alimentary tract and chronic mucocutaneous candidiasis are far more difficult to manage clinically and are often indicative of several underlying systemic disorders, e.g., leukemia, lymphoma or T-cell deficiencies. Systemic diseases such as urinary tract infections, meningitis, endocarditis, endophthalmitis and fungemia are usually associated with high morbidity and/or mortality. They are most often seen as opportunistic infections and the extent of their system involvement appears to be directly related to the severity of the underlying disorder. Although *C. albicans* is the predominant etiologic agent of all forms of candidiasis, other species of *Candida* may cause one or more of the diseases.

Table 6:2 Clinical Sources and References for Case Reports of Uncommon Yeast Infections

Isolates	Clinical Sources (44, 50, 59, 64, 89)	References
Aureobasidium pullulans	respiratory, skin, tissue, eye, nail	55,75,83
Blastoschizomyces capitatus	respiratory, skin, heart valve, blood	5,55
Candida ciferrii	nail, skin, ear	29
C. lipolytica	respiratory, genitourinary, eye, nail	55,75
C. lusitaniae	respiratory, genitourinary, skin, nail, feces, eye, ear, blood, oral	7,14,33,35,75
C. paratropicalis	respiratory, urine, blood, internal organs, skin, subcutaneous tissue	2,4,8
C. pseudotropicalis	respiratory, genitourinary, blood, nail, internal organs	55,63,75
C. rugosa	respiratory, skin, feces, blood	24,72,87
C. zeylanoides	respiratory, skin, feces, nail	55,75
Cryptococcus albidus	respiratory, genitourinary, skin, nail, feces, CSF	55,58,75
C. laurentii	respiratory, genitourinary, skin, CSF	
C. uniguttulatus	respiratory, nail, skin, CSF	52,55
Hansenula anomala	respiratory, skin, blood	55,75
Prototheca spp.*	skin, subcutaneous tissue, feces, olecranon bursa, respiratory, peritoneal	26,40,55,75
Rhodotorula spp.*	respiratory, skin, blood, CSF, genitourinary, internal organs, feces, eye, nail	26,55,75
Saccharomyces cerevisiae	respiratory, genitourinary, feces	55,75
Sporobolomyces salmonicolor	respiratory, skin, CSF	55
Torulopsis candida	respiratory, urine, skin, nail, bone	55,75,93

*Human isolates are almost exclusively *P. wickerhamii*. However, *P. zopfii* may occur rarely.[40] See Table 6:9 for differentiation of *Prototheca* spp. and *T. glabrata*. See Table 6:10 for differentiation of *Rhodotorula* spp.

C. tropicalis is associated with vulvovaginitis and onychomycosis, as well as bronchopulmonary, alimentary and systemic infections.[26,39,75] *C. parapsilosis* has been shown to cause endocarditis in parenteral drug abusers, otitis externa, paronychia and transient catheter-associated fungemia.[26,36,61,75] *C. guilliermondii* has been implicated in disseminated diseases and associated with infections similar to those caused by *C. parapsilosis*.[23,26,75] *C. stellatoidea* (a possible synonym of *C. albicans*)[79] may cause vulvovaginitis[9,75] and *C. krusei* has been shown to be the etiologic agent of vulvovaginitis and disseminated infections.[26,32,39,75] The *Candida* species which are infrequent agents of human disease are shown in Table 6:2 and the differential tests used in their identification are listed in Table 6:3.

Cryptococcosis

Although cryptococcosis (torulosis, European blastomycosis) is almost always caused by *C. neoformans*, other *Cryptococcus* spp. have been associated with the disease (Table 6:2). Table 6:4 lists physiological tests used to differentiate clinically significant species. Although reported as a commensal in immunocompetent individuals,[71] *C. neoformans* is most often found in nature in avian excreta, particularly pigeon droppings, and in soil contaminated with this material. The relative ease with which it is recovered from natural habitats lends support to the exogenous origin of cryptococcosis.[13] The antigenic properties of the conspicuous polysaccharide capsules which surround the cells permit the differentiation of *C. neoformans* into four serotypes, A, B, C and D. A and D serotypes are the most common cause of disease in temperate climates (including the United States with the exception of southern California). In contrast, B and C serotypes are recovered primarily from clinical cases in southern California, as well as from tropical and subtropical regions, especially in Asia.

The primary focus of infection is in the lungs and involvement is often asymptomatic. Extrapulmonary disease is frequently observed, especially in immunocompromised patients such as those with AIDS, Hodgkin's lymphoma and other neoplasms. Disseminated cryptococcosis most commonly involves the central nervous system, i.e., cryptococcal meningitis,[78] but the yeastlike agent has been recovered from such secondary sites of infection as skin, bone, urinary tract, prostate, testis, eye and heart. Although infections caused by serotypes B and C may be more refractory to therapy than those involving A and D serotypes,[82] all serotypes have been shown to be equally susceptible in vitro to several antifungal agents.[28]

Geotrichosis

Geotrichosis is a rare pulmonary infection caused by *Geotrichum candidum*. The etiologic role of *G. penicillatum (Trichosporon penicillatum)*, as

Table 6:3 Differentiation of Clinically Significant *Candida* Species

	Assimilation										Fermentation						
Species	Cel*	Gal	Ino	Lac	Mal	Raf	Rhm	Suc	Tre	Xyl	Dex	Gal	Mal	Suc	GTT	Ure	42C
C. albicans†	−	+	−	−	+	−	−	+	+	+	+	+/W	+	−‡	(+)	−	+
C. ciferrii	+	+	+	−	+	+	+	+	+	+	−	−	−	−	−	−	V
C. guilliermondii	+	+	−	−	+	+	V	+	+	+	+	+/W	−	+/W	−	−	V
C. krusei	−	−	−	−	−	−	−	−	−	−	+	−	−	−	−	V	+
C. lambica	−	−	−	−	−	−	−	−	−	+	+	−	−	−	−	V	−
C. lipolytica	V	V	−	−	−	−	−	−	−	−	−	−	−	−	−	V	−
C. lusitaniae	+	V	−	−	+	−	+	+	+	+	+	V	−‡	V	−	−	V
C. parapsilosis	−	+	−	−	+	−	−	+	+	+	+	V	−	−	−	−	V
C. paratropicalis	V	+	−	−	+	−	−	V§	+	+	+	+	−	+	−	−	+
C. pseudotropicalis	+	+	−	+	−	+	−	+	−	+	+	+	−	−	−	−	+
C. rugosa‖	−	+	−	−	−	−	−	−	−	+	−	−	−	−	−	−	V
C. stellatoidea	−	+	−	−	+	−	−	−	+	+	+	−	+	−	(+)	−	V
C. tropicalis	V	+	−	−	+	−	−	+	+	+	+	+	+	+	−	−	+
C. zeylanoides	−	V	−	−	−	−	−	−	+	−	−/W	−	−	−	−	−	−

*Cel = cellobiose
Gal = galactose
Ino = inositol
Lac = lactose
Mal = maltose

Raf = raffinose
Rhm = rhamnose
Suc = sucrose
Tre = trehalose
Xyl = xylose

Dex = dextrose (glucose)
GTT = germ tube test
Ure = urea hydrolysis
42C = growth at 42°C

+ = positive, − = negative, +/W = positive or weakly positive, −/W = negative or weakly positive, (+) = >90% strains positive, V = variable

†Rare isolates of *C. albicans* do not grow in defined assimilation media; these strains will exhibit a characteristic fermentation pattern and usually produce germ tubes and chlamydospores on appropriate media.[85]

‡Slight gas bubble may be formed.

§Sucrose assimilation usually delayed especially by auxanographic methods.

‖This species represents a complex of *C. pararugosa* and *C. rugosa*. Although xylose is positive and variable, respectively,[9] xylose positive strains appear to be more frequent clinical isolates. Arabinose assimilation (positive and negative, respectively) can be used for separation.

Table 6:4 Differentiation of Clinically Common *Cryptococcus* Species by Assimilation Tests

Species	Dulcitol	Lactose	Sucrose	KNO_3
C. albidus	V*	V	+	+
C. laurentii	V	+	+	–
C. neoformans	+	–	+	–
C. terreus	V	+/D	–	+
C. uniguttulatus	–	–	+	–

* V = variable, + = positive, – = negative, +/D = positive or delayed positive.

Table 6:5 Physiologic Characters of Clinically Important Arthroconidial Yeasts

Species	Galactose	Lactose*	Xylose
Blastoschizomyces capitatus	+/D†	–	–
Geotrichum species	+	–	+
Trichosporon beigelii	+	+	+

* Urea hydrolysis can be substituted for lactose since results are equivalent.
† + = positive, – = negative, V = variable, +/D = positive or delayed positive.

well as its taxonomic relationship to the genus *Geotrichum* remains unclear. Although the lung is the most common site of infection, bronchial, oral, gastrointestinal, conjunctival and cutaneous infections have been reported.[26,75] Diagnosis is complicated by the occurrence of *Geotrichum* as part of the normal flora of the mucous membranes (see Table 6:5 for salient characteristics of clinically important arthroconidia-forming yeasts).

Pityriasis Versicolor and Related Disorders

Pityriasis versicolor (tinea versicolor) is a superficial infection caused by *Malassezia furfur*. This yeast is unique in that it requires exogenous fatty acids or lipids for its in vitro growth (Tables 6:6 and 6:7). Although one may still find *Pityrosporum orbiculare* and *P. ovale* described as etiologic agents of this disease, these two taxa have been classified as *M. furfur*[31,80] by the rules of priority of the International Code of Botanical Nomenclature. Pityriasis versicolor is usually asymptomatic and primarily a cosmetic problem. However, *M. furfur* has also been described as the cause of folliculitis, dacryocystitis[75] and of fungemia in pediatric patients receiving hyperalimentation with lipid infusions.[51,70]

Torulopsosis

Torulopsosis is a systemic disease most commonly associated with the esophagus, vagina, kidneys and lungs.[26,75] Fungemia has also been reported,

Table 6:6 Key to the Identification of Medically Important Yeasts

1a	Germ tube test positive	2
b	Germ tube test negative	3
2a	Sucrose assimilated	*C. albicans*
b	Sucrose not assimilated	*C. stellatoidea*
3a	Pseudohyphae and true hyphae absent or rudimentary	4
b	Pseudohyphae or true hyphae present; arthroconidia absent; chlamydospores may be present	18
c	True hyphae with ballistoconidia present (satellite colonies appear in older cultures); pseudohyphae may be present; arthroconidia absent; urea hydrolyzed; KNO_3 assimilated; carotenoid pigment usually present	*Sp. salmonicolor*
d	True hyphae with arthroconidia present; blastoconidia absent	*Geotrichum* spp.
e	True hyphae, arthroconidia and blastoconidia present; pseudohyphae may be present	35
f	True hyphae, arthroconidia and large blastoconidia; olive-brown to black pigment formed in older cultures	*Aureobasidium* spp.
g	Sporangia with sporangiospores; hyphae absent	*Prototheca* spp. (Table 6:9)
4a	Urea hydrolyzed	5
b	Urea not hydrolyzed	10
5a	Inositol assimilated; carotenoid pigment usually absent	6
b	Inositol not assimilated; carotenoid pigment usually present	15
c	Inositol not assimilated; carotenoid pigment absent; waxy or crumbly colony; growth may be scant without addition of lipids or fatty acids to medium	*Malassezia* spp.
6a	KNO_3 assimilated	7
b	KNO_3 not assimilated	8
7a	Sucrose assimilated	*C. albidus*
b	Sucrose not assimilated	*C. terreus*
8a	Lactose assimilated; phenoloxidase negative	*C. laurentii*
b	Lactose not assimilated; phenoloxidase positive	*C. neoformans*
c	Lactose not assimilated; phenoloxidase negative or latent	9
9a	Dulcitol assimilated; 37°C growth usually positive	*C. neoformans*
b	Dulcitol not assimilated; 37°C growth usually negative	*C. uniguttulatus*
10a	Maltose and sucrose assimilated	11
b	Maltose and sucrose not assimilated	*T. glabrata* (syn. *C. glabrata*

Table 6:6 continued

11a	Cellobiose assimilated	12
b	Cellobiose not assimilated; ascospores usually present	*S. cerevisiae*
12a	KNO_3 assimilated; ascospores usually present	*H. anomala*
b	KNO_3 not assimilated	13
13a	Raffinose assimilated	14
b	Raffinose not assimilated	*C. lusitaniae*
14a	Blastoconidia mostly ovoid; 42°C growth variable; glucose fermented strongly	*C. guilliermondii*
b	Blastoconidia spheroidal; 42°C growth negative; glucose fermentation weak or negative	*T. candida* (syn. *C. famata*)
15a	KNO_3 assimilated	*R. glutinis*
b	KNO_3 not assimilated	16
16a	Maltose assimilated	*R. rubra* (syn. *R. mucilaginosa*
b	Maltose not assimilated	17
17a	Raffinose assimilated	*R. pilimanae*
b	Raffinose not assimilated	*R. minuta*
18a	Maltose or sucrose assimilated; lactose not assimilated	19
b	Maltose not assimilated; lactose and sucrose assimilated	*C. pseudotropicalis*
c	Maltose, sucrose and lactose not assimilated	31
19a	Chlamydospores present	20
b	Chlamydospores absent	21
20a	Sucrose assimilated	*C. albicans*
b	Sucrose not assimilated	*C. stellatoidea*
21a	Inositol assimilated	*C. ciferrii*
b	Inositol not assimilated	22
22a	Raffinose assimilated	23
b	Raffinose not assimilated	27
23a	KNO_3 assimilated	24
b	KNO_3 not assimilated	25

24a	Urea hydrolyzed	*R. glutinis*
b	Urea not hydrolyzed; ascospores usually present	*H. anomala*
25a	Urea hydrolyzed	*R. rubra* (syn. *R. mucilaginosa*)
b	Urea not hydrolyzed	26
26a	Cellobiose assimilated	*C. guilliermondii*
b	Cellobiose not assimilated; ascospores usually present	*S. cerevisiae*
27a	Rhamnose assimilated	*C. lusitaniae*
b	Rhamnose not assimilated	28
28a	Galactose, maltose and sucrose fermented	*C. tropicalis*
b	Maltose fermented; sucrose not fermented	29
c	Maltose and sucrose not fermented or latent and weakly fermented	*C. parapsilosis*
29a	Sucrose rapidly assimilated	*C. albicans*
b	Sucrose latently assimilated	*C. paratropicalis*
c	Sucrose not assimilated	30
30a	Galactose fermented	*C. paratropicalis*
b	Galactose not fermented	*C. stellatoidea*
31a	Trehalose assimilated; true hyphae absent; carotenoid pigment absent	*C. zeylanoides*
b	*Trehalose assimilated; true hyphae with indistinct phialides present; carotenoid-like pink pigment present*	*Lecythophora hoffmannii*[74] (syn. *Phialophora hoffmannii*)
c	Trehalose not assimilated	32
32a	Xylose assimilated	33
b	Xylose not assimilated	34
33a	Galactose assimilated	*C. rugosa*
b	Galactose not assimilated	*C. lambica*
34a	42°C growth positive; true hyphae absent	*C. krusei*
b	42°C growth negative; true hyphae present	*C. lipolytica*
35a	Lactose assimilated	36
b	Lactose not assimilated	*B. capitatus* (syn. *T. capitatum*)
36a	KNO_3 assimilated; 30 °C growth negative	*T. pullulans*
b	KNO_3 not assimilated; 30°C growth positive	*T. beigelii* (syn. *T. cutaneum*)

Table 6:7 Key to the Yeasts Based Primarily on Morphology

1a	Hyaline (white, cream or tan) yeastlike colony	2
b	Pink, salmon, orange, or red yeastlike colony	8
c	Olive-brown to black yeastlike colony	13
2a	Sporangia with sporangiospores	*Prototheca* spp. (Table 6:9)
b	True hyphae with arthroconidia	3
c	Pseudohyphae or true hyphae present; arthroconidia absent; chlamydospores may be present	5
d	Pseudohyphae and true hyphae absent or rudimentary; usually blastoconidia only	6
3a	Blastoconidia present	4
b	Blastoconidia absent	*Geotrichum* spp. (Table 6:5)
4a	Annellides present; urea not hydrolyzed	*Blastoschizomyces capitatus* (Table 6:5)
b	Annellides absent; urea usually hydrolyzed	*Trichosporon* spp. (Table 6:5); *Aureobasidium* spp.
5a	Chlamydospores present	*C. albicans*; *C. stellatoidea*
b	Chlamydospores absent	*Candida* spp. (Table 6:3); *Hansenula* spp.; *Saccharomyces* spp.
6a	Urea hydrolyzed	7
b	Urea not hydrolyzed	*Candida* spp. (e.g., some strains of *C. guilliermondii*; Table 6:3; *Hansenula* spp., *Saccharomyces* spp.; *Torulopsis* spp.
7a	Creamy or mucoid colony; carbon substrates rapidly assimilated	*Cryptococcus* spp. (Table 6:4); non-pigmented strains of *Rhodotorula* spp. (Table 6:10)

b	Waxy or crumbly colony; carbon substrates not rapidly assimilated (weak, latent assimilation of glucose, glycerol and sorbitol may occur); growth may be scant without the addition of lipids or fatty acids to the medium	*Malassezia* spp.
8a	Pale pink colony turning olive-brown to black	*Aureobasidium* spp.
b	Old colonies do not turn olive-brown to black	9
9a	Filamentous colony margin	10
b	Entire colony margin	12
10a	Ballistoconidia or satellite colonies present	*Sporobolomyces* spp.
b	Ballistoconidia or satellite colonies absent	11
11a	True hyphae usually absent; phialides absent	*Rhodotorula* spp. (Table 6:10)
b	True hyphae present; indistinct phialides present	*Lecythophora hoffmannii*
12a	Ballistoconidia or satellite colonies present	*Sporobolomyces* spp.
b	Ballistoconidia or satellite colonies absent	*Cryptococcus* spp. (Table 6:4) *Rhodotorula* spp. (Table 6:10)
13a	Colony hyaline at first, dematiaceous when older	*Aureobasidium* spp.
b	Colony starts out dematiaceous (see chapter on agents of phaeohyphomycosis)	*Exophiala* spp.; *Phaeococcomyces* spp.; *Phaeoannellomyces werneckii*; *Phialophora* spp.; *Wangiella* spp.

but only in immunocompromised patients with diabetes mellitus or neoplastic diseases and those undergoing steroid or prolonged antibacterial therapy.[39] Although there has been considerable disagreement as to the proper taxonomic designation of the etiologic agent, *Torulopsis glabrata* is favored over *Candida glabrata*.[54,56,95]

Trichosporonosis

Trichosporon beigelii is the etiologic agent of white piedra, a superficial infection of the hair of the axillary and groin areas. Although relatively rare in developed countries, the disease is more prevalent in third world nations. Poor personal hygiene is suspected of playing a major role in its development. *T. beigelii* has also been associated with endocarditis, endophthalmitis, pulmonary edema and disseminated infections.[26,75] *Blastoschizomyces capitatus* (syn. *Trichosporon capitatum*)[81] is a rare etiologic agent of systemic infections (see Tables 6:2 and 6:5).

Rare Yeast Infections

Although documentation is limited, *Hansenula* spp. and *Pichia* spp. have been associated with human infections. For example, *Hansenula polymorpha* has been reported as the etiologic agent of an infection in a child with chronic granulomatous disease and *Hansenula anomala* has been mentioned sporadically in the literature as the cause of several infections of humans.[55] However, members of these two genera are rarely encountered in the clinical laboratory.

EPIDEMIOLOGY AND PUBLIC HEALTH SIGNIFICANCE

Yeasts are commonly implicated as etiologic agents of opportunistic infections associated with impairment of cellular immunity. Since *C. albicans* is both part of the normal human microflora as well as the most prevalent agent of yeast infections, the vast majority of yeast infections are endogenous in origin. Infections resulting from trauma, burns, etc. are the exceptions but constitute only a minority of cases.

C. neoformans is associated in nature with avian excreta and soil contaminated with this material. Consequently, cryptococcosis is thought to be exogenous in origin, with the respiratory tract as the portal of entry. It is interesting to note that the size of cryptococcal cells in nature are significantly smaller than those observed in vitro on nutrient media.[77] This smaller size may permit the cells to evade the natural barriers of the respiratory tract and allow them to invade and colonize the alveolar spaces. The capsules surrounding the cells may be small or absent when the organism is isolated from nature. Although A and D serotypes of *C. neoformans* have been recovered from nature, the habitat of B and C serotypes remains unknown. Recent descriptions[46,82] of two diagnostic media which provide a means of separating cryptococcal serotypes into groups A/D and B/C without the need for sophisticated serotyping tests should permit greater epidemiologic study of this pathogen and possible insight as to the natural habitat of B and C serotypes.

COLLECTION, TRANSPORT AND PROCESSING OF SPECIMENS

For information regarding the collection, transport and processing of specimens, the reader is referred to Chapter 2, "Collection, Transport and Processing of Clinical Specimens."

IDENTIFICATION OF ETIOLOGIC AGENTS

Direct microscopic examination of clinical specimens is the first critical step for the identification of yeast pathogens. Observation of yeasts in tissue or in normally sterile body fluids is highly indicative of colonization or infection.[26,55,75] There are many stains and mounting preparations used for such microscopic studies. Calcofluor white M2R (Difco Laboratories, Inc.) or Cellufluor (Polysciences, Inc.), a whitening agent which has been used in industry for a number of years, was recently described for use in direct microscopic examination of specimens.[62] This fluorescent stain is readily absorbed by polymers of N-acetyl-β-D-glucosamine in beta 1-4 glycosidic linkage such as the chitin component of fungal cell walls. Calcofluor white M2R (Cellufluor, fluorescent brightener 28) absorbs ultraviolet light in the 340-400 nm range, exhibiting maximum absorption at 345-365 nm. The use of this stain permits more rapid and accurate identification of fungal structures in clinical specimens, but other microorganisms, e.g., *Pneumocystis carinii* and *Prototheca*, as well as human connective tissue, may absorb calcofluor white M2R. Although this does diminish the specificity of the stain, laboratorians should still find that calcofluor white M2R is useful in their examination of clinical material. Other commonly used stains for the direct examination of specimens include Gram, periodic acid-Schiff, hematoxylin and eosin, mucicarmine and Gomori methenamine-silver nitrate.

Potassium hydroxide (KOH) has proved useful as a mounting medium for the examination of skin, hair and nails, as well as for viscous fluids such as sputum.[26,55,75] India ink is routinely used to detect the presence of *C. neoformans* in a variety of clinical specimens. Although the cell wall of the yeast absorbs the ink, the polysaccharide capsule surrounding the cell does not, resulting in the "halo" effect, i.e., dark cell wall surrounded by unstained capsule on the black India ink background. However, care must be exercised in interpreting India ink stained specimens. There have been reports of the erroneous diagnosis of cryptococcal meningitis due to the "halos" which may form around leukocytes or monocytes in India ink stained preparations.[69]

Not only is it possible to detect the presence of fungal pathogens in clinical specimens through direct observation, but it is also possible to make a presumptive identification of the etiologic agent for diagnosis of the infection. For example, the presence of blastoconidia and pseudohyphal elements in vaginal smears are highly suggestive of vulvovaginitis, and the morphology of these structures may permit an experienced laboratorian to identify the yeast as *C. albicans*. Similar structures in smears prepared from oral lesions would suggest thrush, stomatitis or perleche and the same etiologic agent. The "spaghetti and meatballs" structures observed in skin scrapings are created by phialides and phialoconidia of only one yeast, *M. furfur*, and are characteristic of pityriasis versicolor.

Once a yeast colony is observed on primary isolation media, a wet mount is prepared for microscopic examination. A portion of the colony is removed aseptically with a sterile loop or applicator stick and emulsified in a drop of sterile water, lactophenol cotton blue or similar mounting medium on a microscope slide. A clean coverglass is then placed over the smear and the preparation examined microscopically using low (100 X) and high (400 X) power magnification. Through such observations the laboratorian is able to assess the presence of bacterial or fungal contamination, as well as morphologic structures that may be of use in the identification of the isolate.

Upon completion of initial processing and identification procedures, the laboratorian must then determine the genus and species of the isolate. In the past, speciation of a yeast, except for *C. albicans* and *C. neoformans*, was considered to be of limited or no clinical importance. The identification of a yeast was routinely reported as "*C. albicans*" or "Yeast - not *C. albicans*." However, as noted previously, the emergence of new or saprobic species as opportunistic pathogens (Table 6:2) and their frequent resistance to commonly used antifungal drugs now necessitates the speciation of yeast isolates from clinical specimens.

PRESUMPTIVE IDENTIFICATION PROCEDURES

Germ Tube Test

A germ tube or germination hypha is the initial hyphal extension from a conidium or spore. Although similar in general appearance to a bud, a germ tube lacks a constriction at its point of origin (Fig. 6:2). Greater than 95% of all *C. albicans* isolates have been shown to produce germ tubes under appropriate conditions.[53] Since *C. albicans* is one of only three clinically

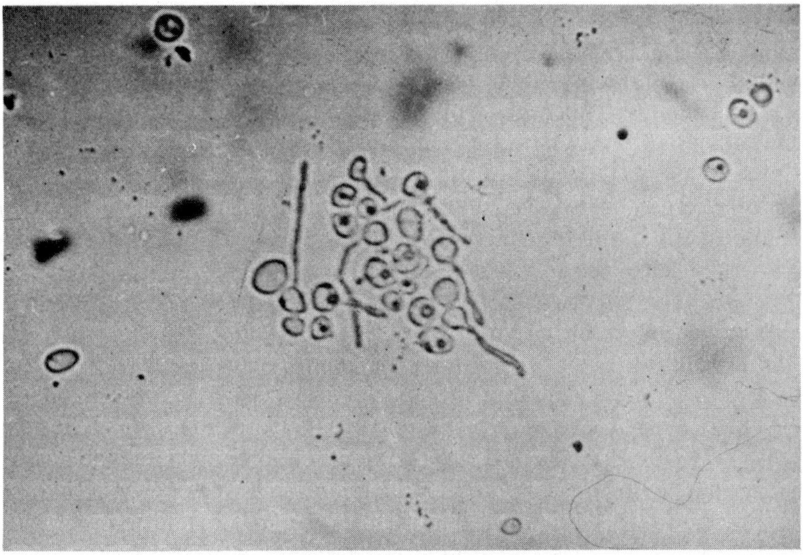

Figure 6:2. Germ tubes of *Candida albicans* formed in fetal bovine serum after 2.5 h incubation at 37°C (400X). Note that germ tubes are not constricted at their point of origin.

important yeasts to form germ tubes (*C. stellatoidea* and on rare occasions, *C. tropicalis*), and since it is the most common clinical yeast isolate, the germ tube test provides a relatively simple, inexpensive method for the presumptive identification of this ubiquitous yeast. However, there is considerable subjectivity in the interpretation of the test and its specificity and sensitivity may vary in accordance with the laboratorian's technical expertise. Therefore, identification should be confirmed by assimilation and/or fermentation tests and microscopic observation of morphologic characteristics.

To conduct the germ tube test, a portion of a yeast colony is suspended in 0.3-1.0 mL of a proteinaceous broth in a small culture tube and the tube incubated for 2-4 h at 35-37°C. A pipette may be used for inoculating the broth and then left in the tube to withdraw the yeast suspension after incubation. One drop of the suspension is removed following incubation, transferred to a microscope slide, overlaid with a coverglass and examined microscopically for the presence of germ tubes. Known isolates of *C. albicans* and *C. tropicalis* may be used as positive and negative controls, respectively. Although occasional isolates of *C. tropicalis* may form germ tubes upon initial isolation from clinical specimens,[88] this ability is quickly lost as isolates are maintained in culture.

Many broth media have been described for the germ tube test, such as various animal sera and serum derivatives, human serum, trypticase soy, peptone, glucose beef extract, etc.[53,55] Agar media such as TOC[27] and rice infusion agar[10] have also been used to demonstrate germ tube production. However, human serum may be an undesirable medium due to the presence of inhibitory factors (ferritin) or infectious agents (HIV).

Although aseptic techniques are not usually required due to the rapidity of the test, gross bacterial contamination of the tested colony may inhibit germ tube formation. Inoculum density may also affect the results of the test. Optimal germination is observed with suspensions containing from 10^5 to 10^7 cells/mL of broth. As cell density approaches 10^9 cells/mL, germ tube formation decreases drastically or may be completely absent.[53] The laboratorian should prepare a faintly turbid suspension for best results.

Phenoloxidase Test

The development of a melanin-like pigment as a result of the phenoloxidase activity of *C. neoformans* when grown on appropriate media can be used for its rapid, presumptive identification. Staib[20,55] was the first to note the production of dark brown to black pigmented colonies when *C. neoformans* was grown on a medium containing an extract of niger seed (*Guizotia abyssinica*). Subsequent investigations have described the incorporation of several other substrates, e.g., caffeic acid, diphenols, aminophenols, diaminobenzenes, indole compounds and esculin, in an agar medium to elicit phenoloxidase activity and the development of dark pigmented colonies.[16,17,25,36,47] TOC medium uses caffeic acid in oxgall agar to permit the presumptive identification of *C. albicans* through germ tube formation and *C. neoformans* by the development of dark-colored colonies.[27,42] Although pigment formation may require up to 6 days to develop on agar media, same-day results have been reported with filter papers impregnated with caffeic acid[37] or 3,4-dihydroxyphenylalanine (DOPA).[42] The sensitivity and specificity of the *Cryptococcus* diagnostic agar media may be affected by glucose concentration and nitrogen sources (ammonium sulfate, asparagine, creatinine, glutamine, glycine, tyrosine) in the nutrient base medium.[16,66] Ferric citrate has been reported to enhance phenoloxidase activity and pigment development.[36,37,42]

To perform the test, a portion of an isolated colony is streaked aseptically onto the surface of one of the agar media or an impregnated filter paper disc, which is then incubated at room temperature, 30°C or 37°C. The agar media are examined daily for the development of dark brown or black colonies, while the filter discs are observed hourly for pigment. Known isolates of *C. neoformans* and *C. laurentii* may be used as positive and negative controls, respectively.

Urea Hydrolysis

The hydrolysis of urea through the formation of urease is an important physiologic character which may be useful in the presumptive identification of several clinically important yeasts. For example, all zoopathogenic yeasts with a basidiomycetous teleomorph, e.g., *C. neoformans* or *Rhodotorula glutinis*, and a few with ascomycetous teleomorphs, e.g., *Candida lipolytica*, hydrolyze urea. Christensen urea agar slants are used in the conventional method to detect urease activity. A portion of an isolated colony is streaked aseptically over the agar surface and the medium is then incubated at 25-30°C for 1-7 days. The medium is observed daily for the development of a pink color indicative of urea hydrolysis. In contrast, urease activity may be determined within hours through the use of such rapid tests as the urea R broth (Difco Laboratories, Inc.)[76] or the rapid urea swab test.[96] In all methods, known isolates of *Cryptococcus albidus* and *C. albicans* can be used as positive and negative controls, respectively.

SPECIFIC IDENTIFICATION PROCEDURES

Morphologic Characters

Colony and microscopic features:

The identification of clinically important yeasts requires both morphologic and physiologic data (see Tables 6:6 and 6:7). The gross colony features (topography, texture, color, margination) are generally obtained with Sabouraud glucose agar. Microscopic morphology (Table 6:1) may be observed with any one of several media such as cornmeal + Tween 80, cream of rice + Tween 80, TOC and Wolin-Bevis media.[92] These media, usually in 100 mm petri plates, may be inoculated in one of two ways. In the Dalmau procedure (Fig. 6:3), a portion of the yeast colony is streaked aseptically onto the surface of the agar in a dollar sign configuration (2 or 3 parallel lines overlaid by "S"). A coverglass is then gently flame-sterilized, allowed to cool and laid over the streaked area. The streaking of the agar surface and the coverglass create an oxygen gradient from microaerophilic areas directly under the center of the coverglass to more aerobic areas at the edge of the coverglass. Alternatively, in the cut-streak method, a portion of the yeast colony is cut in two parallel lines into the agar from the surface to almost the bottom of the petri plate. Microaerophilic conditions are created over the bottom of the petri dish and aerobic conditions exist at the agar surface, so no coverglass is needed in this procedure.

Since microscopic features are observed through the coverglass on the agar surface, the Dalmau procedure permits examination of morphologic

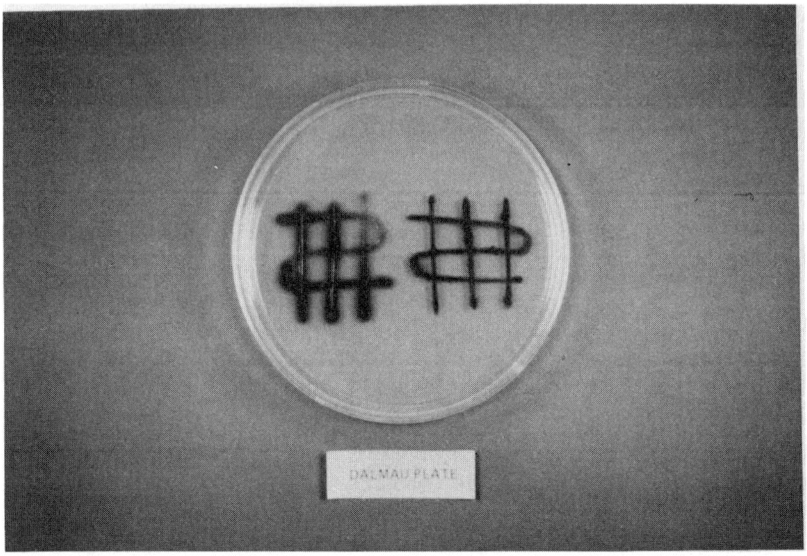

Figure 6:3. The Dalmau procedure is performed by streaking the yeast over the surface of one quadrant of the agar plate, e.g., cornmeal + Tween 80 agar, using two or three parallel lines with an "S" streak over these lines (to give a "dollar sign" configuration). A coverglass is gently flame-sterilized, allowed to cool and laid over the streak lines.

characters at any magnification. In contrast, observations must be made through the bottom of the petri plate with the cut-streak method and morphologic features may be examined only at low magnification. Although greater morphologic details may be observed with the Dalmau procedure, there is greater opportunity for contamination, since the petri plate must be opened for study. In either inoculation method, a known chlamydospore-forming isolate of *C. albicans* (see Fig. 6:4) should be used as a positive control.

Ascospore induction:

The life cycle of most, but not all yeasts, consists of three phases separated by time and space. In the vegetative phase, the fungus is actively growing, establishing its body form, absorbing and storing nutrients, etc. It is the morphology of the vegetative phase that is used to divide the fungi into molds and yeasts.

The second phase in the fungal life cycle is one in which the organism reproduces itself through asexual mechanisms, i.e., those involving mitotic nuclear divisions and no nuclear or gametic fusions. The term anamorph was formally adopted in 1980 to denote this phase of the fungus life cycle.

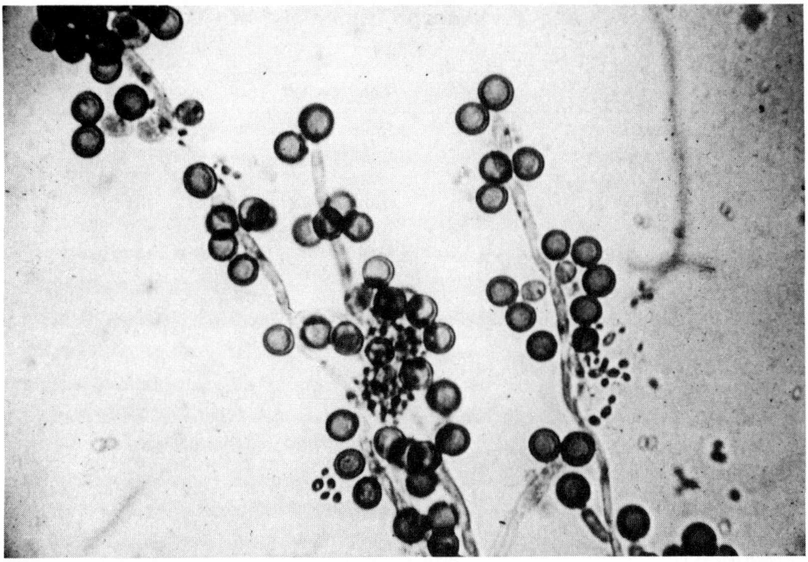

Figure 6:4. Chlamydospore production by *Candida albicans* on cornmeal + Tween 80 agar (400X). Other morphologic characters seen here include blastoconidia and pseudohyphae.

Sexual reproduction, i.e., reproduction by meiotic division of a zygotic or fused nucleus, is the third phase in the life cycle of many fungi. The first portion of this, the teleomorph phase, involves the union of two specialized or nonspecialized structures, each housing a genetically different nucleus. In some fungi, i.e., homothallic, there are no morphologic or physiologic barriers to the fusion of the sexual structures. In other fungi, i.e., heterothallic, physiologic or mating factors (probably surface antigens) or morphologic factors such as the physical separation of the structures, prohibit random union of sexual structures. In heterothallic fungi, sexual reproduction occurs only between sexual structures of opposite mating types located on morphologically distinct thalli.

Since the teleomorph and anamorph phases may be separated by time and space, and since the anamorph has historically been discovered and named first, we find that a single fungus may possess two binomials (Table 6:8). *C. guilliermondii, C. krusei* and *C. lusitaniae* all refer to the anamorphic forms of the teleomorphic phases of *Pichia guilliermondii, Issatchenkia orientalis* and *Clavispora lusitaniae*, respectively. By international convention, when sexual reproduction is found, the teleomorph binomial replaces that of the anamorph. However, the low probability that any given clinical specimen will contain both mating types virtually eliminates the observation

Table 6:8 Anamorph and Teleomorph Binomials* of Clinically Important Yeasts

Anamorph	Teleomorph
Aureobasidium pullulans	unknown
Blastoschizomyces capitatus	unknown
Candida albicans	unknown
Candida ciferrii	*Stephanoascus ciferrii*
Candida guilliermondii	*Pichia guilliermondii*
Candida krusei	*Issatchenkia orientalis*
Candida lambica	*Pichia fermentans*
Candida lipolytica	*Yarrowia lipolytica*
Candida lusitaniae	*Clavispora lusitaniae*
Candida parapsilosis	unknown†
Candida paratropicalis	unknown
Candida pelliculosa	*Hansenula anomala*
Candida pseudotropicalis	*Kluyveromyces fragilis* (*K. marxianus*?)
Candida robusta	*Saccharomyces cerevisiae*
Candida rugosa	unknown
Candida stellatoidea	unknown
Candida tropicalis	unknown
Candida zeylanoides	unknown
Cryptococcus albidus	unknown
Cryptococcus laurentii	unknown
Cryptococcus neoformans	*Filobasidiella neoformans*
Cryptococcus terreus	unknown
Cryptococcus uniguttulatus	*Filobasidium uniguttulatum*
Geotrichum spp.	*Dipodascus* spp.
Rhodotorula glutinis	*Rhodosporidium diobovatum* *Rhodosporidium sphaerocarpum* *Rhodosporidium toruloides*
Rhodotorula minuta	unknown
Rhodotorula pilimanae	unknown
Rhodotorula rubra	unknown
Sporobolomyces salmonicolor	*Aessosporon (Sporidiobolus?) salmonicolor*
Torulopsis candida	*Debaryomyces hansenii*
Torulopsis glabrata	unknown
Trichosporon beigelii	unknown
Trichosporon pullulans	unknown

*Binomials derived from references.[9,50,55]

†*Lodderomyces elongisporus* bears close physiologic and asexual morphologic similarity to *Candida parapsilosis*; however, there is considerable doubt that the two species are related as teleomorphic and anamorphic states, respectively.

of the teleomorphic form in a clinical mycology laboratory. In addition, the need for specialized media, obtaining and crossing opposite mating types and the extended incubation time required to induce the teleomorph stage, preclude its use in yeast identification in most clinical laboratories. Finally, most clinicians are only familiar with the established anamorph binomial. Consequently, laboratorians should generally use the anamorphic name when identifying yeasts isolated in clinical laboratories.

The teleomorphic phases of most clinically important yeasts are members of the Ascomycetaceae, i.e., form ascospores within specialized saclike structures termed asci. Several yeasts, e.g., *Saccharomyces cerevisiae*, commonly found in clinical laboratories will readily reproduce sexually. A number of media have been described to induce these yeasts to form ascospores. Acetate, cornmeal + Tween 80, Gorodkowa, potato glucose, V8 juice, yeast-malt extract and yeast infusion agar media, as well as vegetable wedges, gypsum and cement blocks have all been used to obtain ascospores. However, our experience has shown that acetate, cornmeal + Tween 80, V8 juice and yeast-malt extract agars are the most efficient in inducing the teleomorph phase. Although temperature can affect ascospore formation, room temperature is usually adequate for the induction of the teleomorphic phase of most clinical isolates.

A portion of growth is removed aseptically from a young actively growing culture, streaked over the surface of an ascospore induction medium, which is incubated at room temperature for up to 6 weeks. At regular intervals, a portion of growth is removed, emulsified in a drop of water on a microscope slide and either a coverglass is added for wet mount observation or the smear is air dried, heat fixed and stained for microscopic observation. Ziehl-Neelsen and the Schaeffer-Fulton modified Wirtz stains have been found to be effective ascospore stains.[50] Stained smears or wet mounts prepared from active cultures of *S. cerevisiae* (See Fig. 6:5) and *C. albicans* may be used as positive and negative controls, respectively.

Physiologic Characters (Tables 6:3 to 6:6, 6:9 and 6:10)

Although the Wickerham broth method[91] has long been accepted as the reference standard by which carbon and nitrogen utilization are measured, it is the most tedious, costly and time-consuming of all physiological tests. Consequently, though a powerful research or reference tool, it is not suitable for routine use in the clinical laboratory.

Wickerham broth fermentation test (WBF):

Many yeast species may be difficult to differentiate due to their morphologic similarity and/or overlapping carbon assimilation patterns, but may be identified through a fermentation test. The most widely used procedure for assessing an unknown yeast's fermentation pattern is the WBF test, in which the fermentation of a carbon substratum is determined by the presence or absence of gas production.

The media used in the WBF are similar to those employed in the Wickerham broth assimilation test. Here, 2.0 mL aliquots of a basal yeast extract-

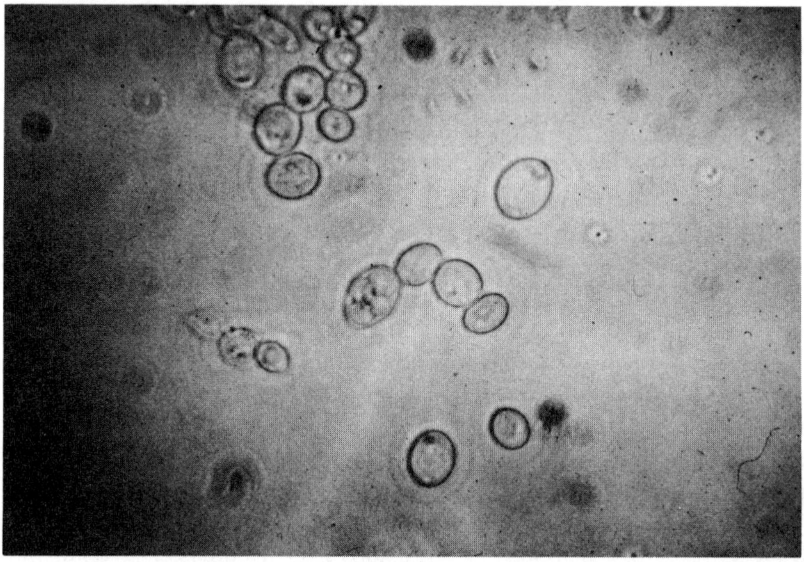

Figure 6:5. *Saccharomyces cerevisiae*. An ascus with 4 ascospores observed in a wet mount using phase-contrast microscopy (1000X). Chains of cells, i.e., rudimentary pseudohyphae, are occasionally seen.

Table 6:9 Differentiation of Clinically Common *Prototheca* Species and *Torulopsis glabrata* by Assimilation Tests

Species	Galactose	Sucrose	Trehalose
P. stagnora	+*	+/D	−
P. wickerhamii	V	−	+
P. zopfii	−/D	−	−
T. glabrata†	−	−	+

*+ = positive, − = negative, V = variable, +/D = positive or delayed positive, −/D = negative or delayed positive.

†*P. wickerhamii* and *T. glabrata* are physiologically similar when galactose is not assimilated and wet mount morphology is indicated for differentiation.

peptone broth are dispensed into screwcapped, 16 x 100 mm tubes. An inverted Durham tube (6 x 50 mm) is then inserted, the tubes are loosely capped and autoclaved for 15 min at 15 psi. Carbon sources are prepared as 6.0% aqueous solutions (except for raffinose at 12.0%) and filter-sterilized. Inulin and soluble starch should be autoclaved for 15 min at 15 psi. After the basal broths have cooled, 1.0 mL aliquots of each sterilized carbon source is added aseptically to each basal broth tube to achieve a final substratum concentration of 2.0% (or 4.0% for raffinose).

Table 6:10 Differentiation of Clinically Common *Rhodotorula* Species by Assimilation Tests

Species	Maltose	Raffinose	KNO_3
R. glutinis	+*	+/D	+
R. minuta	−	−	−
R. pilimanae	−	+	−
R. rubra	+	+	−

*+ = positive, − = negative, +/D = positive or delayed positive.

The broth tubes are inoculated with a yeast suspension in a manner similar to that outlined for the Wickerham assimilation test, then capped tightly and incubated at 25-30°C for 1-4 weeks. Incubation at temperatures above 32°C may cause the breakdown of di- and trisaccharides into their constituent monosaccharides, leading to false-positive results. The test is read on alternate days for 7 days and then at weekly intervals. Fermentation of the carbon source is indicated by the collection of gas bubbles at the top of the inverted Durham tube. Selection of appropriate quality control organisms for WBF is dependent upon the number and type of fermentation tests being conducted (see Table 6:3 or refer to the literature for more extensive test batteries).[50,55]

Wickerham broth assimilation test (WBA):

Media preparation for WBA is an involved procedure requiring filter sterilization of all components except inulin and starch. Carbon substrates are prepared in 0.67% yeast nitrogen base broth to achieve final concentrations ranging from 0.25%-1.0% (based upon carbon equivalents of glucose). Nitrogen substrates at a final concentration of 0.078% are prepared in 1.17% yeast carbon base broth. The two basal broths without substrates serve as negative controls. Vitamin requirements are assessed with a vitamin-free basal medium and sterile water serves as the negative control.

In one of several methods of preparing the WBA test, 4.5 mL of sterile basal medium is pipetted into screwcapped culture tubes along with 0.5 mL from 10X concentrated stock solutions of the carbon, nitrogen and vitamin substrates. A small portion of growth from an actively growing culture of the unknown yeast isolate is removed aseptically, emulsified in sterile water to achieve a 1+ Wickerham suspension and 0.1 mL of this suspension is then added to each tube. The inoculated tubes are incubated at 25-30°C in standing racks or on a shaker. Positive reactions may be observed in less time with agitated tubes.

Growth is assessed weekly for 3 weeks by measuring the turbidity within the tubes with a Wickerham card containing several parallel black lines, 0.75 mm thick, drawn 5 mm apart. Turbidity is assessed by holding the tube against the Wickerham card and observing the parallel lines through the yeast suspension. Turbidity is rated as 3+ if the lines are completely masked, 2+ if they are diffuse and indistinct and 1+ if the lines may be seen clearly but have rough edges. Observation of 2+ or 3+ turbidity is indicative of positive assimilation, while 1+ is recorded as a weak or negative result, depending on the turbidity of the negative control. To avoid false-positive results, WBA tests should not be incubated near fermentation tests. Ethanol vapor, a product of fermentation, may solubilize in assimilation broths, providing a carbon substrate for several yeast taxa. Selection of appropriate quality control organisms for WBA is dependent on the number and type of assimilation tests being conducted (Tables 6:3-6:5, 6:9 and 6:10).[50,55]

Auxanographic assimilation test:

The incorporation of the test yeast into an agar base medium in the auxanographic procedure permits the study of its assimilation pattern in 1-6 days as opposed to >21 days with WBA. Yeast nitrogen base agar is used in studies of carbohydrate assimilation, while yeast carbon base agar is used in tests of nitrogen assimilation. In brief, the agar media are autoclaved, cooled to approximately 50°C, inoculated aseptically through the addition of a yeast suspension in sterile water or by directly seeding the media with yeast colonies, and then poured into sterile petri plates (15 x 150 mm plates will accommodate up to 12 assimilation discs) to solidify at room temperature. Alternatively, uninoculated agar base media may be poured into petri plates, allowed to solidify and then inoculated by streaking the surface with a swab moistened with a yeast suspension. Filter paper discs impregnated with substrate solutions or the raw chemical substrates are placed onto the hardened agar surface, tamped aseptically, the plate is inverted and incubated at 25-30°C. Growth (turbidity) or color change (if a pH indicator has been included in the base media) around the substrates is assessed daily. "Carry-over", i.e., false-positive results due to assimilation of stored or carried nutrients, may be eliminated by serially maintaining the test yeast on a minimal nutrient medium prior to inoculation in the auxanographic test. Since glucose is rapidly assimilated by most yeasts, it can serve as the positive control and seeded agar plates without substrate may be used as the negative control. Although the assimilation patterns for most clinically important yeasts may be assessed at 30°C, some taxa, e.g., several *Rhodotorula* spp. and *Cryptococcus* spp., may not tolerate this incubation temperature and hence, the assimilation tests may best be incubated at 25°C for optimal growth. Incubation of auxanographic tests at temperatures greater than 30°C may cause false-positive results due to the breakdown of di- and trisaccharides into their constituent monosaccharides.

Potassium nitrate assimilation tests:

Utilization of potassium nitrate (KNO_3) is a critical physiologic character in the identification of several clinically important yeasts (see Tables 6:4 and 6:10). A number of methods have been described to evaluate KNO_3 assimilation. The WBA is tedious and time-consuming to prepare and most nitrate-positive isolates cannot be detected for up to 14 days. Although results may be obtained with auxanographic methods in a far shorter period of time, media preparation and inoculation may be time-consuming and interpretation of results may be affected by "carry-over." Rapid methods for assessing nitrate reductase activity[38,73] are occasionally limited by the presence of carotenoid pigment or complete exhaustion of the nitrate source. Unfortunately, there is no individual rapid, accurate, reliable commercial system to determine nitrate assimilation.

Commercial Yeast Identification Systems

Due to the increased incidence of opportunistic yeast infections, there are a number of commercial systems available to supplement or replace the conventional methods for identification of yeast pathogens.

Five systems—API 20C, Analytab Products; AutoMicrobic System Yeast Biochemical Card, Vitek Systems; Minitek, BBL Microbiology Systems; Quantum II Microbiology System, Abbott Laboratories; and Uni-Yeast-Tek, Flow Laboratories, Inc.,—are now available. Numerous evaluations have demonstrated that they all provide results more quickly, with less preparation and greater cost-efficiency than conventional methods.[12,15,18,19,34,45,48,49,67,84,86]

With these products, as with conventional procedures, the identification of an unknown isolate depends primarily on its carbohydrate assimilation profile. In three systems (API 20C, Minitek and Uni-Yeast-Tek), assimilation of 8-19 carbohydrates is determined directly as changes in turbidity or color of pH indicators. In two systems (Quantum II and AutoMicrobic), the carbohydrate pattern is read automatically, compared to profiles in an on-site computer data base and an identification is printed.

An alternative method of identification is the use of chromogenic substrates to measure enzyme activities. In 1985, Analytab Products introduced the Yeast-IDENT System for identification of yeastlike pathogens solely by their enzyme profiles. With this system, the laboratorian directly assesses the color changes produced by enzymatic hydrolysis of 20 substrates. Clinically important yeasts can be identified as soon as 4 h after inoculation.

REFERENCES

1. Ahearn DG. Medically important yeasts. Ann Rev Microbiol 1978;32:59-68.
2. Ahearn DG, Lawrence JB. Disseminated candidiasis caused by a sucrose-negative variant of *Candida tropicalis*. J Clin Microbiol 1984;20:187-90.

3. Ahearn DG, McGlohn MS. In vitro susceptibilities of sucrose-negative *Candida tropicalis, Candida lusitaniae* and *Candida norvegensis* to amphotericin B, 5-fluorocytosine, miconazole and ketoconazole. J Clin Microbiol 1984;19:412-16.
4. Ahearn DG, Meyer SA, Mitchell G, Nicholson MA, Ibrahim AI. Sucrose-negative variants of *Candida tropicalis*. J Clin Microbiol 1977;5:494-6.
5. Arnold AG, Gribbin B, De Leval M, Macartney F, Slack M. *Trichosporon capitatum* causing recurrent fungal endocarditis. Thorax 1981;36:478-80.
6. Bailey JW, Sada E, Brass C, Bennett JE. Diagnosis of systemic candidiasis by latex agglutination for serum antigen. J Clin Microbiol 1985;21:749-52.
7. Baker JG, Nadler HL, Forgacs P, Kurtz SR. *Candida lusitaniae*: a new opportunistic pathogen of the urinary tract. Diagn Microbiol Infect Dis 1984;2:145-9.
8. Baker JG, Salkin IF, Pincus DH, Bodensteiner DC, D'Amato RF. Pathogenicity of *Candida paratropicalis*. Arch Pathol Lab Med 1983;107:577-9.
9. Barnett JA, Payne RW, Yarrow D, eds. Yeasts: characteristics and identification. New York: Cambridge University Press, 1983.
10. Beheshti F, Smith AG, Krause GW. Germ tube and chlamydospore formation by *Candida albicans* on a new medium. J Clin Microbiol 1975;2:345-8.
11. Boom WH, Piper DJ, Ruoff KL, Ferraro MJ. New cause for false-positive results with the cryptococcal antigen test by latex agglutination. J Clin Microbiol 1985;22:856-7.
12. Bowman PI, Ahearn DG. Evaluation of the Uni-Yeast-Tek kit for the identification of medically important yeasts. J Clin Microbiol 1975;2:354-8.
13. Bowman PI, Ahearn DG. Ecology of *Cryptococcus neoformans* in Georgia. In: Proceedings of the Fourth International Conference on the Mycoses, Pan American Health Organization, Scientific Publication No. 356. 1977:258-68.
14. Bradsher RW, White FJ. Transient fungemia due to *Candida lusitaniae* (Letter). South Med J 1985;78:626-7.
15. Buesching WJ, Kurek K, Roberts GD. Evaluation of the modified API 20C system for identification of clinically important yeasts. J Clin Microbiol 1979;9:565-9.
16. Chaskes S, Tyndall RL. Pigment production by *Cryptococcus neoformans* from para- and ortho-diphenols: effect of the nitrogen source. J Clin Microbiol 1975;1:509-14.
17. Chaskes S, Tyndall RL. Pigment production by *Cryptococcus neoformans* and other *Cryptococcus* species from aminophenols and diaminobenzenes. J Clin Microbiol 1978;7: 146-52.
18. Cooper BH, Johnson JB, Thaxton ES. Clinical evaluation of the Uni-Yeast-Tek system for rapid presumptive identification of medically important yeasts. J Clin Microbiol 1978;7:349-55.
19. Cooper BH, Prowant S, Alexander B, Brunson DH. Collaborative evaluation of the Abbott yeast identification system. J Clin Microbiol 1984;19:853-6.
20. Cooper BH, Silva-Hutner M. Yeasts of medical importance. In: Lennette EH, Balows A, Hausler WJ Jr, Shadomy HJ, eds. Manual of clinical microbiology. 4th ed. Washington, DC: American Society for Microbiology, 1985:526-41.
21. Curry CR, Quie PG. Fungal septicemia in patients receiving parenteral hyperalimentation. N Engl J Med 1971;285:1221-5.
22. De Repentigny L, Marr LD, Keller JW, et al. Comparison of enzyme immunoassay and gas-liquid chromatography for the rapid diagnosis of invasive candidiasis in cancer patients. J Clin Microbiol 1985;21:972-9.
23. Dick JD, Rosengard BR, Merz WG, Stuart RK, Hutchins GM, Saral R. Fatal disseminated candidiasis due to amphotericin-B-resistant *Candida guilliermondii*. Ann Intern Med 1985;102:67-8.
24. Dyess DL, Garrison RN, Fry DE. *Candida* sepsis: implications of polymicrobial bloodborne infection. Arch Surg 1985;120:345-8.
25. Edberg SC, Chaskes SJ, Alture-Werber E, Singer JM. Esculin-based medium for isolation and identification of *Cryptococcus neoformans*. J Clin Microbiol 1980;12:332-5.

26. Emmons CW, Chapman BH, Utz JP, Kwon-Chung KJ, eds. Medical mycology. 3rd ed. Philadelphia: Lea and Febiger, 1977.
27. Fleming WH III, Hopkins JM, Land GA. New culture medium for the presumptive identification of *Candida albicans* and *Cryptococcus neoformans*. J Clin Microbiol 1977;5:236-43.
28. Fromtling RA, Abruzzo GK, Bulmer GS. *Cryptococcus neoformans*: comparisons of in vitro antifungal susceptibilities of serotypes AD and BC. Mycopathologia 1986;94:27-30.
29. Furman RM, Ahearn DG. *Candida ciferrii* and *Candida chiropterorum* isolated from clinical specimens. J Clin Microbiol 1983;18:1252-5.
30. Gold JWM. Opportunistic fungal infections in patients with neoplastic disease. Am J Med 1984;76:458-63.
31. Gordon MA. *Malassezia pityrosporum pachydermatis* (Weidman) Dodge 1935. Sabouraudia 1979;17:305-9.
32. Gordon RA, Simmons BP, Appelbaum PC, Aber RC. Intra-abdominal abscess and fungemia caused by *Candida krusei*. Arch Intern Med 1980;140:1239-40.
33. Guinet R, Chanas J, Goullier A, Bonnefoy G, Ambroise-Thomas P. Fatal septicemia due to amphotericin B-resistant *Candida lusitaniae*. J Clin Microbiol 1983;18:443-4.
34. Hasyn JJ, Buckley HR. Evaluation of the AutoMicrobic system for identification of yeasts. J Clin Microbiol 1982;16:901-4.
35. Holzschu DL, Presley HL, Miranda M, Phaff HJ. Identification of *Candida lusitaniae* as an opportunistic yeast in humans. J Clin Microbiol 1979;10:202-5.
36. Hopfer RL, Blank F. Caffeic acid-containing medium for identification of *Cryptococcus neoformans*. J Clin Microbiol 1975;2:115-20.
37. Hopfer RL, Groschel D. Six-hour pigmentation test for the identification of *Cryptococcus neoformans*. J Clin Microbiol 1975;2:96-8.
38. Hopkins JM, Land GA. Rapid method for determining nitrate utilization by yeasts. J Clin Microbiol 1977;5:497-500.
39. Horn R, Wong B, Kiehn TE, Armstrong D. Fungemia in a cancer hospital: changing frequency, earlier onset and results of therapy. Rev Infect Dis 1985;7:646-55.
40. Kaplan W. Protothecosis and infections caused by morphologically similar green algae. In: Proceedings of the 4th International Congress on Mycoses, Pan American Health Organization. Scientific Publication No. 356. 1977:218-32.
41. Kaufman L, Blumer S. Cryptococcosis: the awakening giant. In: Proceedings of the 4th International Conference on the Mycoses, Pan American Health Organization. Scientific Publication No. 356. 1977:176-82.
42. Kaufmann CS, Merz WG. Two rapid pigmentation tests for identification of *Cryptococcus neoformans*. J Clin Microbiol 1982;15:339-41.
43. Kendrick B, ed. Taxonomy of the fungi imperfecti. Toronto: University of Toronto Press, 1971.
44. Kiehn TE, Edwards FF, Armstrong D. The prevalence of yeasts in clinical specimens from cancer patients. Am J Clin Pathol 1980;73:518-21.
45. Kiehn TE, Edwards FF, Tom D, Lieberman G, Bernard EM, Armstrong D. Evaluation of the Quantum II yeast identification system. J Clin Microbiol 1985;22:216-19.
46. Kwon-Chung KJ, Polacheck I, Bennett JE. Improved diagnostic medium for separation of *Cryptococcus neoformans* var. *neoformans* (serotypes A and D) and *Cryptococcus neoformans* var. *gattii* (serotypes B and C). J Clin Microbiol 1982;15:535-7.
47. Kwon-Chung KJ, Tom WK, Costa JL. Utilization of indole compounds by *Cryptococcus neoformans* to produce a melanin-like pigment. J Clin Microbiol 1983;18:1419-21.
48. Land GA, Harrison BA, Hulme KL, Cooper BH, Byrd JC. Evaluation of the new API 20C strip for yeast identification against a conventional method. J Clin Microbiol 1979;10:357-64.
49. Land G, Stotler R, Land K, Staneck J. Update and evaluation of the AutoMicrobic yeast identification system. J Clin Microbiol 1984;20:649-52.

50. Lodder J, ed. The yeasts - a taxonomic study. 2nd ed. Amsterdam: North-Holland Publishing, 1970.
51. Long JG, Keyserling HL. Catheter-related infection in infants due to an unusual lipophilic yeast - *Malassezia furfur*. Pediatrics 1985;76:896-900.
52. Lynch JP III, Schaberg DR, Kissner DG, Kauffman CA. *Cryptococcus laurentii* lung abscess. Am Rev Respir Dis 1981;123:135-8.
53. Mackenzie DWR. Serum tube identification of *Candida albicans*. J Clin Pathol 1962;15: 563-5.
54. McGinnis MR. Recent taxonomic developments and changes in medical mycology. Ann Rev Microbiol 1980;34:109-35.
55. McGinnis MR. Laboratory handbook of medical mycology. New York: Academic Press, 1980.
56. McGinnis MR, Ajello L, Beneke ES, et al. Taxonomic and nomenclatural evaluation of the genera *Candida* and *Torulopsis*. J Clin Microbiol 1984;20:813-14.
57. McManus EJ, Jones JM. Detection of a *Trichosporon beigelii* antigen cross-reactive with *Cryptococcus neoformans* capsular polysaccharide in serum from a patient with disseminated *Trichosporon* infection. J Clin Microbiol 1985;21:681-5.
58. Melo JC, Srinivasan S, Scott ML, Raff MJ. *Cryptococcus albidus* meningitis. J Infect 1980;2:79-82.
59. Merz WG. *Candida lusitaniae*: frequency of recovery, colonization, infection and amphotericin B resistance. J Clin Microbiol 1984;20:1194-5.
60. Merz WG, Evans GL, Shadomy S, et al. Laboratory evaluation of serological tests for systemic candidiasis: a cooperative study. J Clin Microbiol 1977;5:596-603.
61. Meunier-Carpentier F, Kiehn TE, Armstrong D. Fungemia in the immunocompromised host: changing patterns, antigenemia, high mortality. Am J Med 1981;71:363-70.
62. Monheit JE, Cowan DF, Moore DG. Rapid detection of fungi in tissues using calcofluor white and fluorescence microscopy. Arch Pathol Lab Med 1984;108:616-18.
63. Morgan MA, Wilkowske CJ, Roberts GD. *Candida pseudotropicalis* fungemia and invasive disease in an immunocompromised patient. J Clin Microbiol 1984;20:1006-7.
64. Murray PR, Van Scoy RE, Roberts GD. Should yeasts in respiratory secretions be identified? Mayo Clin Proc 1977;52:42-5.
65. Myerowitz RL, Pazin GJ, Allen CM. Disseminated candidiasis - changes in incidence, underlying diseases and pathology. Amer J Clin Pathol 1977;68:29-38.
66. Nurudeen TA, Ahearn DG. Regulation of melanin production by *Cryptococcus neoformans*. J Clin Microbiol 1979;10:724-9.
67. Oblack DL, Rhodes JC, Martin WJ. Clinical evaluation of the AutoMicrobic System Yeast Biochemical Card for rapid identification of medically important yeasts. J Clin Microbiol 1981;13:351-5.
68. Polonelli L, Morace G. Specific and common antigenic determinants of *Candida albicans* isolates detected by monoclonal antibody. J Clin Microbiol 1986;23:366-8.
69. Portnoy D, Richards GK. Cryptococcal meningitis: misdiagnosis with india ink. Can Med Assoc J 1981;124:891-2.
70. Prober CG, Ein SH. Systemic tinea versicolor, or how far can furfur go? Pediatr Infect Dis 1984;3:592.
71. Randhawa HS, Paliwal DK. Occurrence and significance of *Cryptococcus neoformans* in the oropharynx and on the skin of a healthy human population. J Clin Microbiol 1977;6: 325-7.
72. Reinhardt JF, Ruane PJ, Walker LJ, George WL. Intravenous catheter-associated fungemia due to *Candida rugosa*. J Clin Microbiol 1985;22:1056-7.
73. Rhodes JC, Roberts GD. Comparison of four methods for determining nitrate utilization by cryptococci. J Clin Microbiol 1975;1:9-10.

74. Rinaldi MG, McCoy EL, Winn DF. Gluteal abscess caused by *Phialophora hoffmannii* and review of the role of this organism in human mycoses. J Clin Microbiol 1982;16: 181-5.
75. Rippon JW. Medical mycology: the pathogenic fungi and the pathogenic actinomycetes. 2nd ed. Philadelphia: W.B. Saunders, 1982.
76. Roberts GD, Horstmeier CD, Land GA, Foxworth JH. Rapid urea broth test for yeasts. J Clin Microbiol 1978;7:584-8.
77. Ruiz A, Bulmer GS. Particle size of airborne *Cryptococcus neoformans* in a tower. Appl Environ Microbiol 1981;41:1225-9.
78. Sabetta JR, Andriole VT. Cryptococcal infection of the central nervous system. Med Clin N Amer 1985;69:333-44.
79. Salkin IF. New medium for differentiation of *Candida albicans* from *Candida stellatoidea*. J Clin Microbiol 1979;9:551-3.
80. Salkin IF, Gordon MA. Polymorphism of *Malassezia furfur*. Can J Microbiol 1977;23: 471-5.
81. Salkin IF, Gordon MA, Samsonoff WA, Rieder CL. *Blastoschizomyces capitatus*, a new combination. Mycotaxon 1985;22:375-80.
82. Salkin IF, Hurd NJ. New medium for differentiation of *Cryptococcus neoformans* serotype pairs. J Clin Microbiol 1982;15:169-71.
83. Salkin IF, Martinez JA, Kemna ME. Opportunistic infection of the spleen caused by *Aureobasidium pullulans*. J Clin Microbiol 1986;23:828-31.
84. Salkin IF, Schadow KH, Bankaitis LA, McGinnis MR, Kemna ME. Evaluation of Abbott Quantum II yeast identification system. J Clin Microbiol 1985;22:442-4.
85. Schlitzer RL, Ahearn DG. Characterization of atypical *Candida tropicalis* and other uncommon clinical yeast isolates. J Clin Microbiol 1982;15:511-6.
86. Shinoda T, Kaufman L, Padhye AA. Comparative evaluation of the Iatron serological Candida Check kit and the API 20C kit for identification of medically important *Candida* species. J Clin Microbiol 1981;13:513-18.
87. Sugar AM, Stevens DA. *Candida rugosa* in immunocompromised infection, case reports, drug susceptibility and review of the literature. Cancer 1985;56:318-20.
88. Tierno PM Jr, Milstoc M. Germ tube-positive *Candida tropicalis*. Am J Clin Pathol 1977;68:294-5.
89. Walker LJ, Luecke MR. The identification of yeasts from clinical material. Proc Iowa Acad Sci 1974;81:14-22.
90. Webb CD, Papageorge C, Hall CT. Identification of yeasts. DHEW publication. Atlanta: Centers for Disease Control, 1971.
91. Wickerham LJ. Taxonomy of yeasts. U.S. Dept. Agriculture Technical Bulletin No. 1029. Washington, DC: U.S. Dept Agriculture, 1951.
92. Wolin HL, Bevis ML, Laurora N. An improved synthetic medium for the rapid production of chlamydospores by *Candida albicans*. Sabouraudia 1962;2:96-9.
93. Wong B, Kiehn TE, Edwards F, et al. Bone infection caused by *Debaryomyces hansenii* in a normal host: a case report. J Clin Microbiol 1982;16:545-8.
94. Wu TC, Koo SY. Comparison of three commercial cryptococcal latex kits for detection of cryptococcal antigen. J Clin Microbiol 1983;18:1127-30.
95. Yarrow D, Meyer SA. A proposal for the amendment of the diagnosis of the genus *Candida* Berkhout nom. cons. Int J Syst Bacteriol 1978;28:611-5.
96. Zimmer BL, Roberts GD. Rapid selective urease test for presumptive identification of *Cryptococcus neoformans*. J Clin Microbiol 1979;10:380-1.
97. Zuger A, Louie E, Holzman RS, Simberkoff MS, Rahal JJ. Cryptococcal disease in patients with the acquired immunodeficiency syndrome: diagnostic features and outcome of treatment. Ann Intern Med 1986;104:234-40.

CHAPTER 7

THE AEROBIC ACTINOMYCETES

Geoffrey A. Land, Ph.D. and Joseph L. Staneck, Ph.D.

INTRODUCTION

The aerobic actinomycetes are gram-positive to gram-variable bacteria appearing either as coccobacilli, with rudimentary branching filaments (diphtheroid-like) or as well-developed filaments (hyphae) with true branches.[17,42] The latter morphology often leads to the production of filaments which extend above the surface of the culture medium (aerial hyphae or collectively, aerial mycelium). Some members of the group are partially acidfast, which corresponds to resistance to lysis by the enzyme lysozyme.[28] These organisms have been isolated from a variety of habitats throughout the world. In the United States, the aerobic *Actinomycetales* are isolated from animals, soil and decaying vegetation and cause a variety of human and animal infections (Table 7:1).[55,70]

DEFINITION OF DISEASE AND ETIOLOGIC AGENTS

Allergic Pulmonary Disease

This is an acute to chronic pulmonary syndrome due to primary exposure followed by subsequent exposure(s) to a variety of fungal allergens, including those of the thermophilic actinomycetes.[31,64] The initial pulmonary response is a granulomatous reaction with or without accompanying interstitial fibrosis.[14,48] Restriction of airways occurs as a result of edematous changes in the lung due to histamine release and infiltration by lymphocytes. Repeated exposure causes further interstitial damage and pulmonary fibrosis, leading to severe limitation of gas transfer.

Table 7:1 Taxonomy* of the Common Pathogenic Aerobic Actinomycetes, Their Diseases and Habitats†

Aggregate group	Description	Organism	Disease	Habitat
1. Nocardioforms	Diphtheroid to filamentous, true branching mycelial elements, commonly fragmented to give coccoid to bacillary conidia: some species partially acidfast and produce aerial hyphae.	*Nocardia asteroides*	Pulmonary: transient to chronic suppurative abscesses: generalized systemic disease often with central nervous system involvement: mycetoma rare.	Ubiquitous in soil, animals and man.
		N. brasiliensis	Frequent agent of mycetoma and isolated sporotrichoid lesions: pulmonary and generalized disease rare.	Possibly ubiquitous: Predominantly North and South America, Central Plateau of Brazil and India.
		N. caviae	Generalized systemic disease: mycetoma.	Ubiquitous: Predominantly Tunisia, Japan, India, Mexico and United States.
		Rhodococcus spp.	Generalized granulomatous disease: central nervous system involvement: eczematous and granulomatous dermatitis: septicemia, pericarditis.	Ubiquitous in soil, plants and animals.
2. Maduromycetes	Filamentous with mycelial elements fragmenting to coccobacillary forms: conidia formed within a sheath on aerial or substrate mycelium.	*Actinomadura madurae*	Mycetoma	Ubiquitous in soil, plants, thorns and decaying vegetation
		A. pelletieri	Mycetoma	West Africa, Trans-Africa Belt, India: soil, thorns, plants and decaying vegetation.
		Norcardiopsis dassonvillei	Mycetoma	Ubiquitous in soil and animals.

3. Streptomycetes	Well developed, branching aerial and substrate mycelium: mycelial elements do not fragment: conidia borne in characteristic chains on extensive aerial mycelium.	*Streptomyces somaliensis*	Mycetoma	Africa, South America, United States, Arabia: soil.
		S. paraguayensis	Mycetoma	Mexico, Central America, South America: soil, thorns, plants, decaying vegetation.
		Streptomyces spp.	Nonpathogenic	Ubiquitous: common environmental isolates: usually found in soil, manure, composts, plants and air conditioning ducts.
4. Micropolysporas‡	Mycelial elements slight to no fragmentation: conidia formed singly in pairs or short chains on either aerial or substrate mycelium.	*Saccharomonospora viridis* *S. internatus*	Restrictive allergic disease.	Ubiquitous: ventilation ducts, manure and compost.
		Micropolyspora faeni *M. caesis*	Restrictive allergic disease.	Ubiquitous: soil, plant material
5. Thermoactinomycetes‡	Mycelial elements do not fragment: bacterial endospore borne singly on aerial or substrate mycelium.	*Thermoactinomyces vulgaris* *T. sacchari* *T. candidus*	Restrictive allergic pulmonary disease.	Ubiquitous: ventilation ducts, manure, composts, humidifier water, hay, moldy sugar cane.
6. Multilocular sporangia‡	Mycelial elements divide transversely and in at least two longitudinal planes to form masses of motile coccoid cells: aerial mycelium usually absent.	*Dermatophilus congolensis*	Epidemic eczema Australia, Africa, United States: man and animals.	

* Taxonomy based upon that proposed by Goodfellow and colleagues.[15,16,18]

† References 2-4,27,42,48,54,68.

‡ Groups 4-6 are beyond the scope of this chapter. The reader is directed to work by Hollick, Truehaft and M. Gordon[19-22,31,65] for details of their isolation and identification.

Pulmonary and Systemic Nocardiosis

This opportunistic infection of humans is caused primarily by members of the genus *Nocardia*.[3,56] Infection usually develops following inhalation of mycelial fragments and is characterized by the formation of abscesses which may develop into burrowing sinus tracts. Pulmonary disease may spread to other areas of the lung and may be borne hematogenously to other organs as well. Disseminated disease usually manifests itself as single to multiple brain abscesses, but infection may appear in other areas of the body.[11,39,48]

Primary Cutaneous Disease

Cutaneous nocardiosis begins as a quiescent chancriform ulcer around a broken area of the skin. The lesion may progress slightly, but at some point will become fixed and localized, resembling fixed cutaneous sporotrichosis.[53]

Actinomycotic Mycetoma

A number of the aerobic *Actinomycetales* have been reported to cause this condition, which results from progression of a chancreform ulcer beyond the fixed stage to a chronically enlarging, nodular lesion.[2,14,48,55] Mycetoma usually occurs on an extremity such as a hand or foot but may also appear on other areas of the body. Infection is characterized by abscess and sinus tract formation with occasional granulomas.[14] The purulent exudate that drains from the sinus tracts contains macroscopic grains similar in morphology and composition to the sulfur granules of anaerobic actinomycosis.[48]

HISTORICAL PERSPECTIVE

Mycetoma was first described in 1842 by a British army physician named Gill. This appears to be the first report of a clinical condition that could be attributed to one of the aerobic actinomycetes.[48] Gill called the disfigured extremity "madura foot" to denote the high endemicity of the disease in the Madurai region of India. The term "mycetoma" was proposed by Van Dyke Carter in 1860 to include all tumors of fungal origin.[14] Kanthack (1892), Vincent (1894) and Brumpt (1905-1906) recognized that there were several different combinations of grain texture, color and morphology which could appear in mycetoma and speculated that this disease involved several different etiologies.[48] Brumpt further described the genus *Indiella* (or *Streptomyces somaliensis* as it is known today) for organisms causing white-yellow grain mycetoma and *Streptothrix madurae* for the cause of mycetoma of the foot. The terms "actinomycosis" and "maduromycosis" were proposed in 1913 to refer to actinomycotic and eumycotic mycetoma, respectively. Further refinements of the basic clinical descriptions for these diseases were made by Chalmers and Archibald in 1916 and by Lavella in 1962.[48]

The first authenticated aerobic actinomycotic infection was described in 1888 by Nocard as a disseminated ulcerative disease of Guadeloupan cattle.[11] Since the syndrome was very similar to the equine farcy and glanders seen in Europe, he named the etiologic agent *Streptothrix farcinica*.[48] The genus *Nocardia* was proposed by Trevisan in 1889 to describe the delicate, branched, acidfast filaments observed by Nocard and the organism was renamed *Nocardia farcinica*.[14] Eppinger (1891) observed similar organisms in pus from brain abscesses associated with miliary pulmonary disease and peribronchial lymphadenopathy in a 52-year old glassblower. The organism, which he named *Cladothrix asteroides*, was found to fulfill all of the criteria for the genus *Nocardia* (Trevisan) and was renamed *Nocardia asteroides* by Blanchard in 1896.

Although by 1921 there were 26 documented cases of human nocardiosis, little attention was paid to the isolation and identification of the causative agents. The etiologic agent of mycetoma in a Brazilian man was first described as *Discomyces brasiliensis* by Lindenberg in 1909 and in 1913 was amended to *Nocardia brasiliensis* by Castellani and Chalmers.[56] In the ensuing years, similar organisms isolated from identical clinical situations were also placed in the genus *Nocardia* but the species name reflected the geographical derivation of the isolate,[48] viz., *N. pretoria*, *N. transvalensis* (Pijper and Pullinger, 1927) and *N. mexicanus* (Ota, 1928). All of these were reduced to synonomy with *N. brasiliensis* by Waksman and Henrici in 1943 and Ochoa in 1945.[48] However, *N. asteroides* was not definitely separated from the other actinomycetes until 1947.

Major strides were made in the definitive morphological and biochemical identification of the aerobic actinomycetes within the next two decades (1950-1970); stable taxons were established for the nocardiae, streptomycetes and the rhodochrous complex.[23-26,34,46,54,60] In 1980, Mishra, Gordon and Barnett utilized 44 biochemical and morphological tests to produce a dichotomous key for the identification of medically important nocardiae and streptomycetes.[43] This was the culmination of Dr. Ruth Gordon's 50 years of pioneering efforts to determine the identification and host-parasite relationships of the aerobic actinomycetes.[25] A recent and more technical approach to actinomycete identification has been the utilization of cell-wall chemistry to develop a chemotaxonomic separation of these organisms and the closely related genus *Mycobacterium*.[30,40,58]

The last two clinical conditions attributed to the *Actinomycetales*, farmer's lung and dermatophilosis, have only recently been recognized as human problems. Farmer's lung was first described in 1924 in Canadian farm workers with a history of acute or chronic lung disease.[48] Eight years later, Campbell reported a similar syndrome among English farmers, but it was not until 1958 that the condition was related to the inhalation of dust from decaying plant material such as moldy hay and sugar cane waste (bagasse). By 1968, the thermophilic actinomycetes were implicated as at least one of the possible etiologies of farmer's lung.[14] Serological evidence now points to *Micropolyspora faeni (Faenia rectivirgula), Thermoactinomyces vulgaris* and *T. viridis* as the most common actinomycetes involved.[31] Dermatophilosis was first reported from the Belgian Congo in 1915 as a skin disease of cattle[48] and a similar infection was described as "lumpy wool" in Australian sheep in 1929. The disease was recognized in the United States and described in 1961 by M. A. Gordon and colleagues, who, in the ensuing years, have delineated the life-cycle of the *Dermatophilaceae* and their taxonomic relationship to the *Actinomycetales*.[19-22]

CLINICAL MANIFESTATIONS

Infections caused by the aerobic actinomycetes are considered opportunistic, for the most part.[14,48,56,59] With the exception of localized primary infection, the host has a hyper- or hypo-immune status which permits the onset and spread of disease. Immune hyperactivity to cellular antigens of the thermophilic actinomycetes can induce a restrictive bronchopulmonary disease in humans, called hypersensitivity pneumonitis.[31] Organisms are found in high concentrations in closed barns, decaying silage, grain mills, old air-conditioning vents and sugar cane waste areas. This environmental association with humans leads to the pulmonary syndrome known by a variety of colloquial names, the most common being "farmer's lung." Symptoms appear when a sensitized individual is challenged through inhalation of large numbers of conidia or mycelial fragments.[64]

In acute disease, symptoms such as cough, chills, fever, malaise and dyspnea appear within 4-6 hours of exposure and are usually resolved within 18-72 hours.[48] Edematous changes occur that obstruct the airways but there is not actual invasion of lung tissue; there is occasionally some residual lung fibrosis. Interstitial lymphocytic infiltrates occur in chronic pulmonary disease, with slight to extensive fibrosis. The severe progressive form is characterized by granulomatous lesions which may develop into miliary nodules.

Pulmonary (and/or disseminated) aerobic actinomycosis is caused primarily by three species of *Nocardia (N. asteroides, N. brasiliensis* and *N. caviae*), leading to the common name of "nocardiosis."[10,11,55] However, other *Actinomycetales* have been implicated in severe human infection (Table 7:1).[6,47,48,65] The major factor contributing to invasive disease is a T-cell dysfunction which may occur as a result of lymphoproliferative malignancy, the use of drugs to combat transplanted organ rejection, intravenous drug abuse or immunodeficiency.[8,10,11,33,35,45,50,52,66] Other underlying physical problems such as endocrine disorders, severe trauma and extensive burns have acted as predisposing factors. Ironically, the therapeutic modalities developed for maintenance of critically ill patients have also contributed singly or cumulatively to invasive disease.[37,39,48,56,59] These include longterm steroid therapy, cytotoxic chemotherapy, multiple broadspectrum antibiotics, chronic peritoneal dialysis and central venous lines or catheters.[37,52,62,65,71]

Pulmonary nocardiosis may appear as a localized abscess, a pneumonia without consolidation or an acute necrotizing infection with multiple abscesses and cavitation (Fig. 7:1).[14,48,56] Symptoms include a nonproductive cough, general malaise and low fever, which may progress to productive, blood-tinged sputum, high fever, anorexia and pleurisy. Severe pneumonia is denoted by consolidation of one or more lobes with subsequent extension to the pleura and worsening of symptoms (Fig. 7:2).[2,10,11,56] With time, the pleura becomes thickened and riddled with coalescing abscesses. These large groups of abscesses may further develop into sinus tracts and fistulas which can extend throughout the abdominal cavity and penetrate the chest

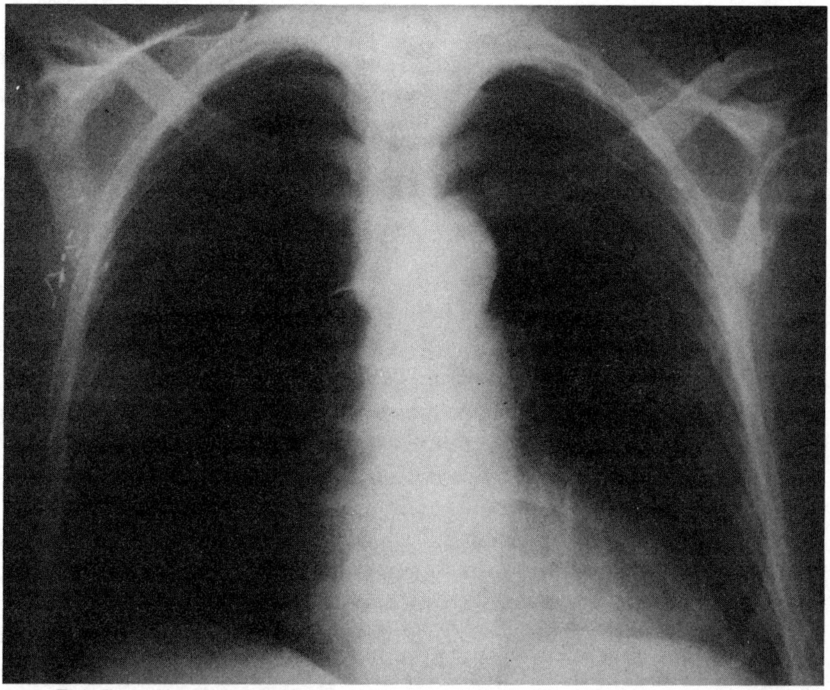

Figure 7:1 Localized pulmonary nocardiosis in a 48-year old renal transplant patient, appearing as a round mass-like density in the lower border of right upper lobe (courtesy S. Spotswood).

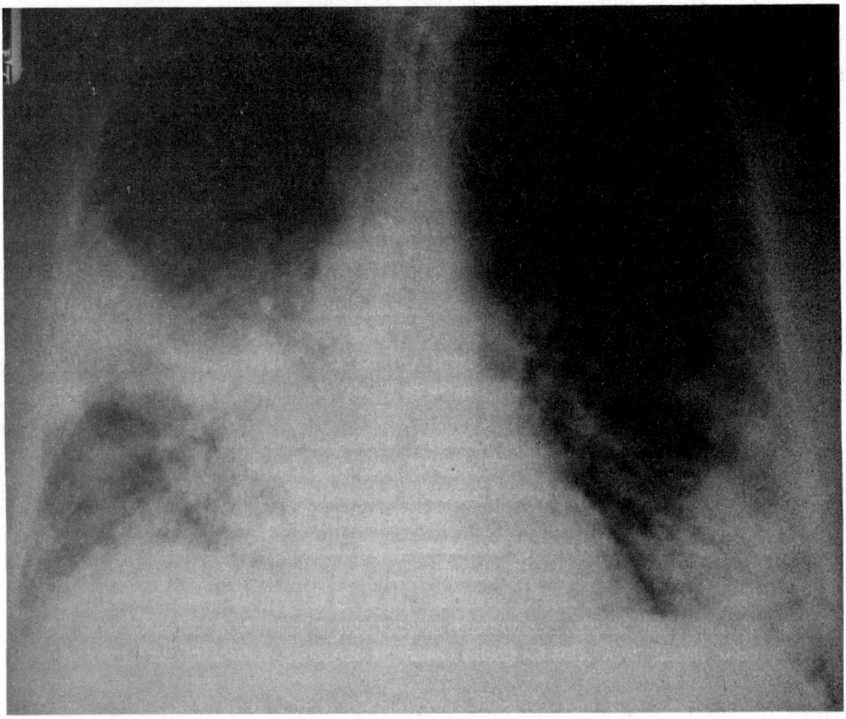

Figure 7:2 Severe disseminated pulmonary nocardiosis, characterized by right mid-lung consolidation with diffuse nodular and fluffy pattern of infiltrative density throughout the lung fields (courtesy S. Spotswood).

wall to the surface.[48] Pulmonary disease may resolve after appropriate antibiotic therapy and surgery or may reoccur one to several times over an extended period.[8,35,61]

In the immunocompromised patient, disseminated nocardiosis may occur as a result of pulmonary infection or, more rarely, from direct inoculation through the skin.[48,56] Although any organ may be attacked, the most susceptible part of the body appears to be the central nervous system and primarily the brain. The patient usually presents with classic symptoms of brain abscess; frequent to constant headaches, nuchal rigidity, convulsions and vomiting as well as varying degrees of confusion, aphasia and paresis.[10,11,48] The kidney is the second most common extrapulmonary site, with metastasis to the heart, spleen, liver and adrenal glands also reported to occur in descending order of frequency. There may be a single brain abscess or multiple lesions which subsequently coalesce into a large tunneling abscess (Fig. 7:3). Organisms may be found within the abscess and may ramify throughout the brain without an accompanying tissue response.[17,48]

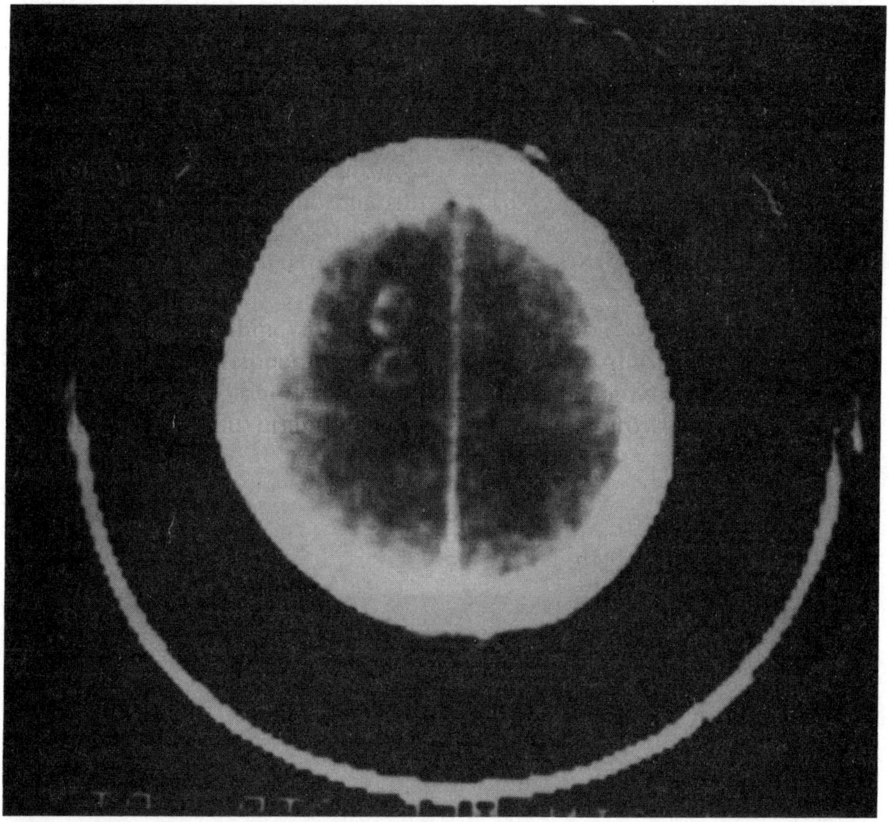

Figure 7:3 Computerized axial tomogram of brain of patient with disseminated nocardiosis, showing two large cerebral abscesses occurring just to the rear left of the mid-line.

The histological picture of pulmonary and disseminated nocardiosis is that of an acute, suppurative, pyogenic bacterial infection without caseation necrosis.[14,48] Polymorphonuclear cell infiltrates predominate, along with minor accumulations of fibrin, lymphocytes and plasma cells. Rarely is there any evidence of extensive fibrosis or progression toward a granulomatous cellular reaction. In tissue, the aerobic actinomycetes typically range in appearance from small (approximately 1-3 μm in diameter), diphtheroid coccobacillary forms to well-developed, delicately branched filaments organized into lacy aggregates called "pseudogranules" (Fig. 7:4). Organisms do not stain with hematoxylin and eosin but require a modified Gram stain or methenamine silver stain for demonstration. Bacteria may stain completely with special stains, or more commonly exhibit intermittent staining, producing a beaded, chain-like effect. If infection is caused by the

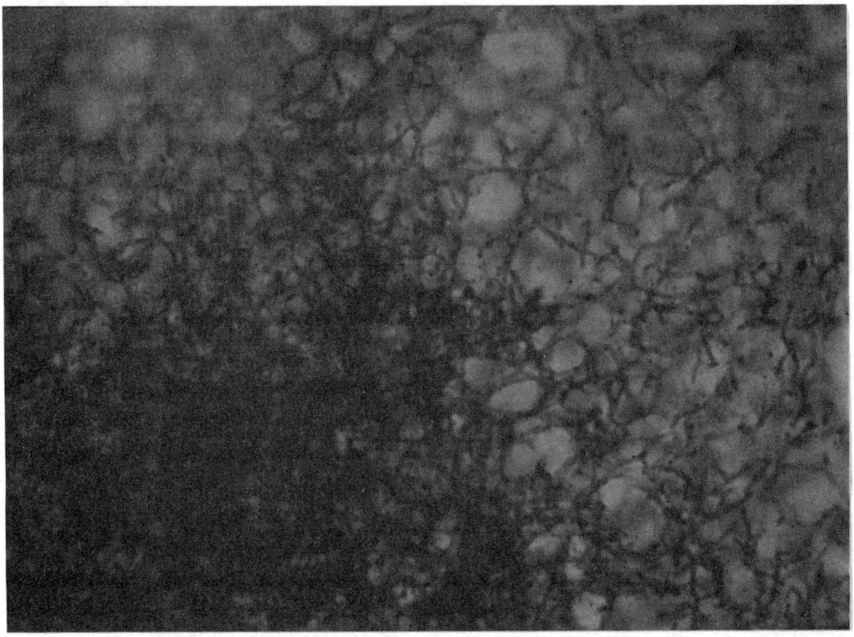

Figure 7:4 Methenamine silver stained section of lung from patient with pulmonary nocardiosis. Note the loose organization of the branching filaments into a "pseudogranule" as well as the incomplete staining of some of the filaments giving them a beaded appearance. 1000X magnification.

nocardioforms, organisms will be weakly acidfast, again with a beaded morphology.

Cutaneous nocardiosis refers to infections caused by any of the aerobic actinomycetes as a direct result of trauma to the skin. There are two forms of cutaneous nocardiosis; a primary localized form which is quite similar to fixed, localized sporotrichosis, and mycetoma. Primary cutaneous nocardiosis occurs in healthy individuals as a result of inflammation around an inoculation site, usually a thorn prick, bruise, abrasion, insect bite or splinter.[2,11,55] Lesions range from abscesses to granulomas and include cellulitis, pustules and pyoderma. Other than the primary ulceration, which may or may not heal of its own accord, there is no evidence of infection. It has been speculated that untreated primary *N. brasiliensis* infection may progress to mycetoma with a predominantly granulomatous response.[48] A similar progression was observed by Chanda and Headington,[9] who described a chronic granulomatous lesion due to *Rhodococcus* spp. in an 81-year old gardener (Fig. 7:5). Organisms are rarely found within these solitary lesions but when present, they have typical staining properties and morphology.

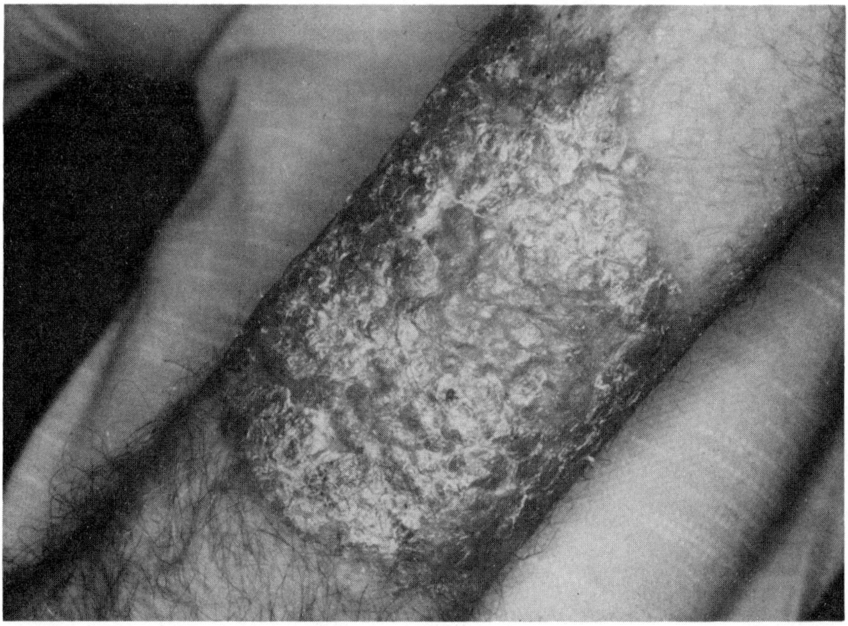

Figure 7:5 Primary cutaneous aerobic actinomycosis: erythematous, marginated and plaquelike lesion of the extensor surface of left forearm in an 81-year old man caused by a member of the *Rhodococcus* group (courtesy J. J. Chanda and J. T. Headington[9]).

Mycetoma, an infection caused by either fungi or actinomycetes, usually occurs in an extremity such as the foot, hand or arm but can appear at virtually any region of the body. Once inoculated, the organism does not remain cutaneous and localized but causes a chronic, indolent infection which eventually involves deeper tissues, including fascia and bone.[14,48] The patient with severe mycetoma has a previous history of substandard health, in contrast to the individual with primary cutaneous infection, who does not. Major contributing factors to mycetoma seem to be a decline in immunity due to advancing age and/or malnourishment and concurrent bacteriologic or parasitic diseases.

As in fixed cutaneous infection, inflammation and ulceration occur at the inoculation site. Tunneling abscesses and sinus tracts soon develop and radiate away from the primary area. As infection progresses, subcutaneous tissue, musculature, fascia and bone are destroyed, leading to swollen, convex, tumor-like areas with raised, "volcano-like" lesions (Fig. 7:6). These pustules are sites where sinus tracts have ruptured to the surface. They discharge a serosanguinous to seropurulent fluid containing grains or

Figure 7:6 Mycetoma of left foot of 78-year old hispanic woman, note the volcano-like appearance of the pustules (courtesy S. Spotswood).

granules (Fig. 7:7). Granules represent an amalgam of the bacterial (or fungal) colony and assorted host serous and cellular debris, similar to the sulfur granules of anaerobic actinomycosis.

Histopathological similarities are noted regardless of whether the mycetoma is of bacterial or fungal origin. Granules from both types of mycetoma possess distinctive colors, morphologies and textures which often lead to presumptive identification of the etiologic agent (Fig. 7:7, Fig. 7:8).[48] Actinomycotic granules occur within abscesses and are distinguishable with hematoxylin and eosin. Granules possess a very organized basophilic border and an intensely staining outer eosinophilic area, while the center remains unstained and appears poorly organized (Fig. 7:8).[48] Individual elements composing the granule may be visualized with special stains. Abscesses consist of densely packed neutrophils within an area of fibrosis and granulation tissue. The latter is characterized by proliferation of capillary vessels and infiltrates of epithelioid cells, macrophages, plasma cells and multinucleated giant cells.

EPIDEMIOLOGY AND PUBLIC HEALTH SIGNIFICANCE

Although the *Actinomycetales* are credited with causing approximately 500-1000 new cases of disease per year in the United States, they have the

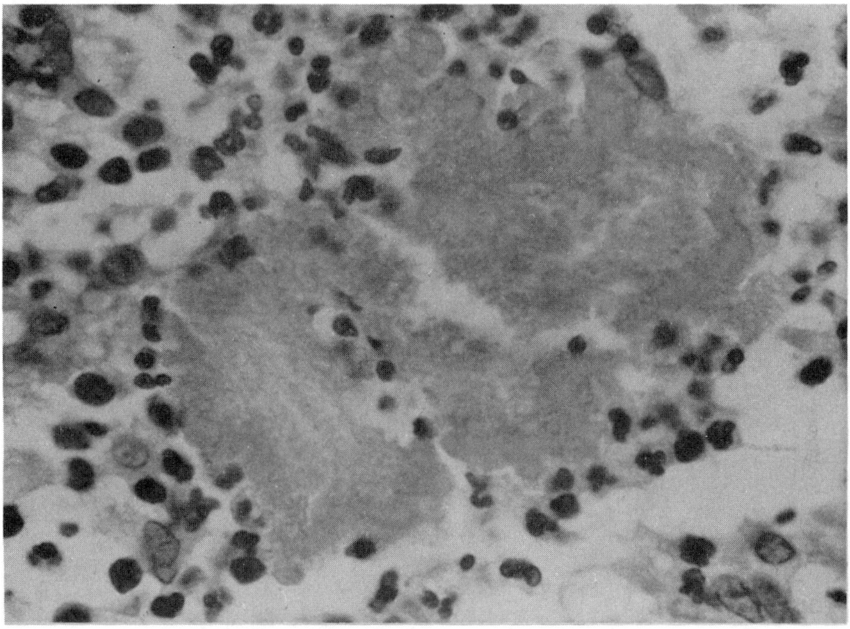

Figure 7:7 Hematoxylin and eosin staining of a biopsy of the 81-year old man with primary granulomatous dermatitis caused by *Rhodococcus* spp. The granule (the pleomorphic mass in the center) takes up the eosin stain, with more intense staining at the periphery. Immediately adjacent to the periphery is a basophilic staining infiltrate of polymorphonuclear leukocytes and plasma cell (courtesy J. J. Chanda and J. T. Headington[9]).

potential to become an even more significant public health problem in the future.[11] The reasons for this are twofold. Firstly, the aerobic actinomycetes represent some of the most common bacteria in the environment (Table 7:1).[70] They have been isolated in virtually every country of the world and are associated with a variety of habitats, including humans, animals, soil, water, living or decaying plant material, as well as ventilation systems, barns and storage facilities. Because of the close association of humans with this group of bacteria, patients undergoing invasive procedures may prove to be particularly vulnerable to pulmonary or disseminated actinomycotic infections. The rapidly growing mycobacteria and *Rhodotorula* spp., which have a similar environmental range, have already been noted as emerging pathogens for this group of patients.[12,13,38,61,70] Perhaps the most compelling reason for anticipating and preparing for an increase in actinomycotic infections in the hospitalized patient lies in the changing hospital population. Since the current fiscal climate in the healthcare field sets limits

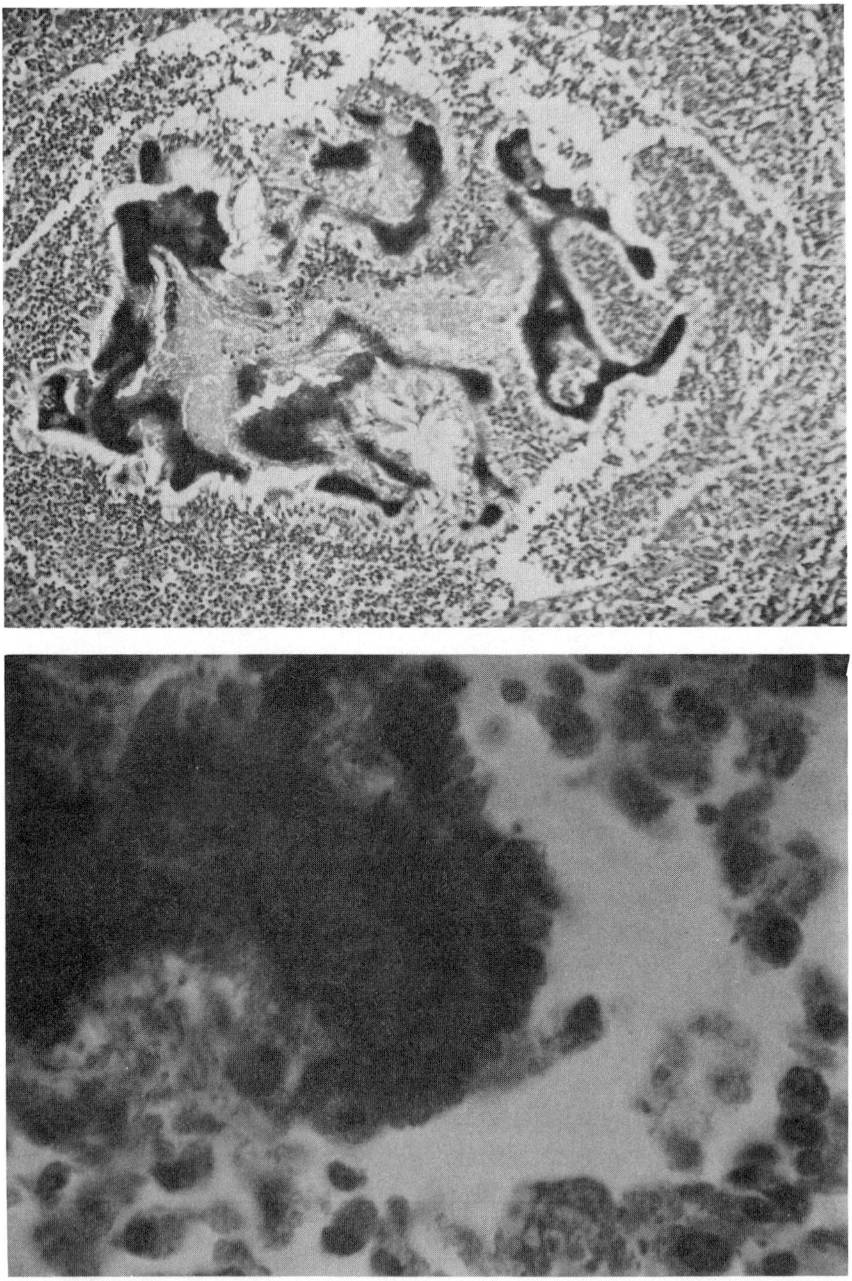

Figure 7:8 (1) Hematoxylin and eosin stained section of a granule and surrounding abscess from a patient with *N. asteroides* mycetoma. Note the inflammatory infiltrate (around) and the intense basophilic peripheral staining of the granule. 400X magnification. (2) Hematoxylin and eosin stained section of *Nocardia* granule. Note the clubbing around the peripheral filaments of the granule. 1000X magnification.

for patient stay in the hospital, hospitals have turned to programs which specialize in the longterm care of critically ill patients. Special care units for severe trauma or burns, neonatal intensive care, hematology-oncology and organ transplantation have increased the reservoir of individuals susceptible to aerobic actinomycosis. Since early diagnosis is such a critical factor in the treatment of infection in an immunocompromised patient, the laboratory must not only be aware of the clinical conditions caused by the *Actinomycetales* but must know how to rapidly isolate and accurately identify these organisms.

COLLECTION, TRANSPORT AND PROCESSING OF SPECIMENS

As with any type of specimen for microbiological examination, the specimen from actinomycosis must represent the infectious process. Respiratory secretions should be obtained from patients with signs of pulmonary infection and tissues or drainage from cases where systemic infection or mycetoma is suspected. Adequate communication between the physician and the laboratory is of paramount importance. The physician should indicate a possible diagnosis of nocardiosis or "madura foot," etc. to assist the laboratory. Without this indication, the bacteriologist may discard culture plates without allowing sufficient incubation time, may use only inhibitory media or may digest the specimen in a manner hostile toward growth of the aerobic actinomycetes. Insufficient incubation time is probably the most common factor limiting recovery of these organisms, since most appropriate specimens are normally cultured on noninhibitory media. However, cultures are incubated only 3-5 days, when aerobic actinomycetes may take up to 14 days to appear on primary isolation. Although specific details of collection and transport of specimens are covered in Chapter 2, "Collection, Transport and Processing of Clinical Specimens," there are several key points that bear repeating. With respect to pulmonary secretions, the less likely that a collection procedure contaminates the specimen with upper respiratory tract flora, the more likely the recognition and recovery of aerobic actinomycetes or any other primary pathogen. Although expectorated sputum is the most common respiratory tract specimen, logic dictates that a higher yield may be expected from specimens collected via bronchoscopy or transtracheal aspiration. Collection from a normally sterile body site should be carried out in the manner least likely to introduce contaminating flora and specimens should be centrifuged in the laboratory to concentrate organisms. Granules in drainage from mycetoma lesions are apt to provide the most diagnostic information. Rinsing the granule in distilled water or saline often removes the host cellular debris and contaminating organisms, thus improving microscopic and cultural results.

Nocardia spp. have been isolated from blood specimens inoculated into conventional or radiometric bacterial blood broth culture systems that include subculture to solid media (author's unpublished observations).[51] These species have also been recovered from vented brain heart infusion broth/agar biphasic bottles incubated at 30°C and have been found to generate enough $^{14}CO_2$ to be detected within 5 days by a radiometric system (author's unpublished observations).[49] It is also likely that chronically-vented commercial blood culture bottles with enriched media would recover bloodborne aerobic actinomycetes if the bottles were incubated at 25-30°C for up to 4 weeks with periodic subculturing.

IDENTIFICATION OF ETIOLOGIC AGENTS

Direct Examination

Respiratory secretions or pus from suspect lesions may be examined by wet mount preparations using water, saline or 10% potassium hydroxide (KOH). Smears processed by Gram and modified acidfast staining procedures are also very useful. Histological silver stains such as the Grocott may aid in the examination of granules and solid tissues. Location of macroscopically visible granules ranging in size up to 3 mm in diameter in specimens or tissue material, followed by rinsing them with sterile water permits the microscopic observation of slender, occasionally branching filaments characteristic of the aerobic actinomycetes. The relatively small diameter of the filaments, approximately 1 μm, distinguishes actinomycetes from the various eumycotic (fungal) agents of mycetoma, since the latter may measure as much as 3-4 μm in diameter. The diphtheroid to branched filamentous morphology of actinomycetes distinguishes them from other bacterial agents.

Granules should be gently crushed between glass slides and studied under the microscope. The microscope should be adjusted for maximum contrast by lowering the condenser and closing the iris diaphragm. The filaments radiate from the dense mass of bacterial cells and host cellular debris constituting the granule. Clublike projections may also be seen at the periphery of the granule in histological sections (Fig. 7:8). Gram stains of granules or specimens reveal gram-positive, intermittently stained, thin filamentous structures which often exhibit true dichotomous branching (Fig. 7:9). The delicate nature of filaments requires careful microscopic scrutiny to distinguish true microbial cells from host-cell artifacts. The *Nocardia* spp. commonly encountered in human disease—*N. asteroides, N. brasiliensis* and *N. caviae*—often exhibit a partially acidfast character when stained and decolorized with a weak mineral acid (Kinyoun modification, Fig. 7:9). The various species of *Streptomyces, Actinomadura* and *Nocardiopsis*, as well as the anaerobic *Actinomyces*, any of which may be found in material from mycetoma cases, will exhibit similar Gram stain charac-

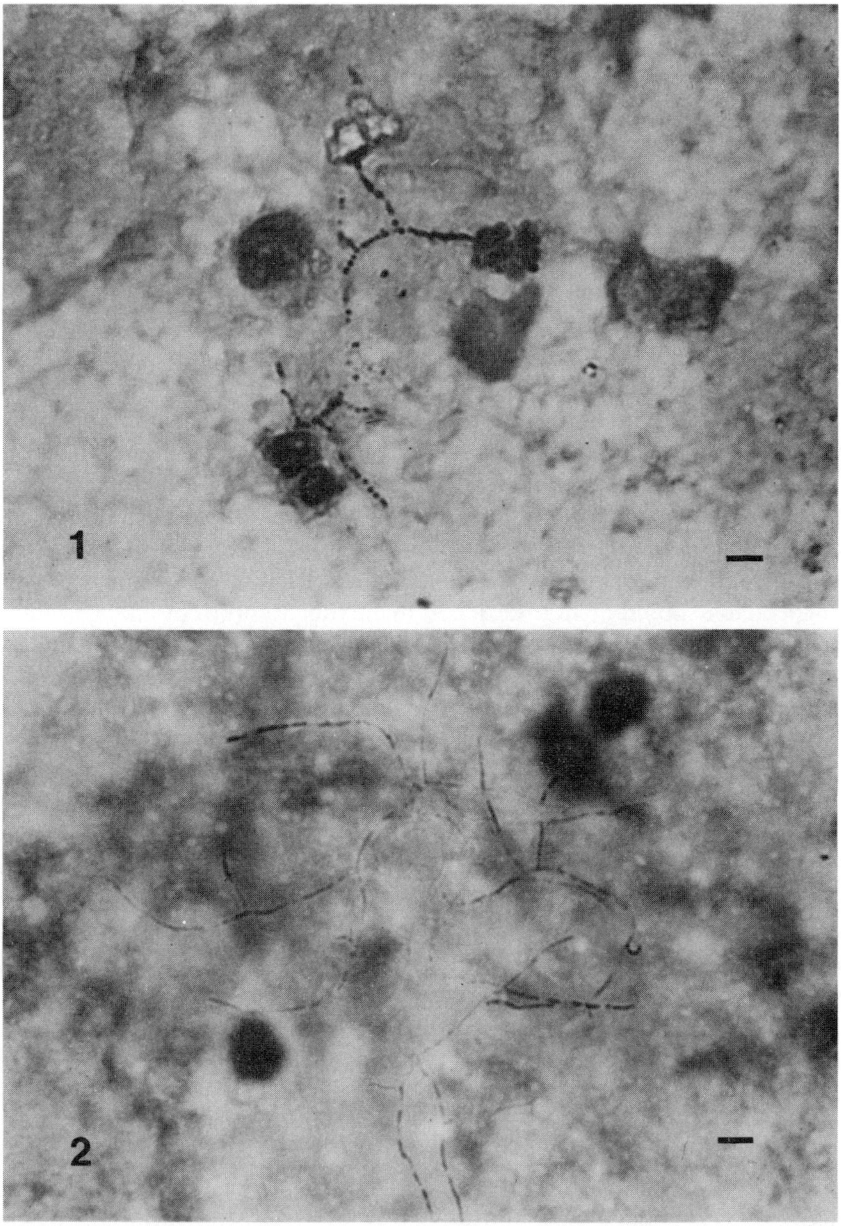

Figure 7:9 **(1) Gram stain of brain abscess material containing *Nocardia asteroides*. (2) Modified Kinyoun acidfast stain of pleural fluid containing *Nocardia asteroides*. 1000X magnification, bar equals 10 μm.**

teristics but fail to remain acidfast with the Kinyoun stain. The value of direct observation should not be underestimated as it not only provides rapid presumptive diagnosis prior to culture, but may in many cases, when culture is either nonproductive or impossible, be the only diagnostic laboratory finding possible.

Although aerobic actinomycetes do not generally have fastidious growth requirements, visible colonies may take from 2 days to 2 or more weeks to emerge on solid media. Because of the need for extended incubation at temperatures lower than 37°C, culture for aerobic actinomycetes is often done in mycology sections. This in turn leads to frequent use of Sabouraud's agar (SAB) for primary isolation, although there have been reports that SAB failed to support the growth of some *Nocardia* strains whose growth was supported by brain heart infusion (BHI) agar. Other routine noninhibitory bacteriologic media that will support growth of the aerobic actinomycetes include sheep blood and tryptic soy agars. To our knowledge, there is no routine selective medium for the primary isolation of aerobic actinomycetes, although BHI with cycloheximide and chloramphenicol may be used for heavily contaminated specimens. Since clinical signs of aerobic and anaerobic actinomycosis may be indistinguishable, the primary plating of a specimen should include enriched thioglycolate broth or chopped-meat glucose broth and noninhibitory enriched solid media which are incubated anaerobically.

Cultures are generally incubated at temperatures ranging from ambient to 37°C. Although *N. asteroides* and *N. brasiliensis* will grow at 37°C or higher, lower temperatures are felt to be more favorable for the growth of other aerobic actinomycetes.[1] Since some laboratories use 30°C for incubation of fungal cultures, they can conveniently use this temperature for incubation of actinomycete cultures. The atmosphere used may be either ambient air or 5-20% CO_2. Both sealing of plates with tape or parafilm and humidification are recommended for extended incubation. Although *Nocardia* spp. will usually appear on plated media within 2-10 days, the slower growing *Actinomadura* may require as much as 3 weeks of initial incubation for primary isolation. Plated media should be examined periodically for slowly growing colonies having either a waxy or a powdery appearance. Growth may be white, gray, buff, orange, red or purple and some species, notably *Streptomyces*, produce diffusible black or brown pigments. Colony texture may be wrinkled, glabrous or leathery. Many strains exhibit a pungent earthy or soil-like odor. Those species producing aerial hyphae develop a white, powdery, fungus-like appearance (Fig. 7:10). The variety in colors and textures of colonies (traits often influenced by the nature of the growth medium, incubation temperatures and atmospheric conditions), allow no consistent link between colonial appearance and species. Most aerobic actinomycetes will both erode and cling tenaciously to the agar surface. The texture of the colonies often requires aggressive handling of the material

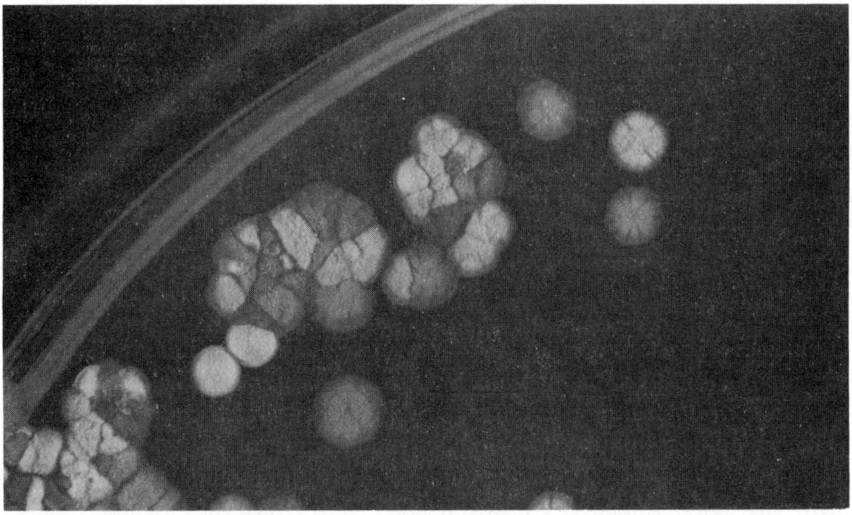

Figure 7:10 ***Nocardia brasiliensis*** **on Sabouraud agar without antibiotics after 5 days incubation. Note the powdery appearance of aerial hyphae developing on the waxy colonies.**

during smear preparation to assist in visualization of individual filaments. Unfortunately, such handling may also serve to disrupt native morphology. Because of the possibility of aerosolization of organisms, any examination or manipulation of colonies in an open plate should be carried out in a Class II biological safety cabinet. When petri dishes are used for culture, the lids should be taped shut to prevent accidental opening. Media in screwcapped tubes or bottles help to alleviate this problem.

Organism Identification

The taxonomy of the aerobic actinomycetes has historically been complicated by the phenotypic and physiologic similarity between families (as well as between genera) in the order *Actinomycetales*, i.e., *Actinomycetaceae, Mycobacteriaceae, Nocardiaceae* and *Streptomycetaceae*. Recent taxonomic strategies employing analyses of cell-wall components and computer-assisted numerical taxonomic cluster analyses have helped to clarify the distinction between taxa. Although great variety exists among the numerous species within the *Actinomycetales*, only those members of the group that are frequently encountered in human clinical specimens will be considered here. Furthermore, since thermophilic-induced hypersensitivity pneumonitis is for the most part diagnosed by serology and *Dermatophilus* has such a characteristic morphology (Table 7:1), these will not be discussed further. Hollick's recent publication on thermophiles and various papers by M. Gordon on *Dermatophilus*[19,21,31] provide complete information on the identification of these organisms.

In recognition of the limited resources and general lack of expertise for dealing with the *Actinomycetales* in many routine clinical laboratories, three progressively complex strategies will be developed for the preliminary and the definitive identification of those species commonly encountered.

Identification based on morphology and staining characteristics:

Careful microscopic examination of material from slowly growing colonies suspected of being aerobic actinomycetes will provide preliminary information regarding the nature of these organisms. Gram stain preparations examined at 1000X magnification should demonstrate gram-positive, slender, beaded, branching filaments. Growth on solid media or aging in a broth culture promotes (particularly in the *Nocardia*) fragmentation into coccobacillary forms similar in morphology to that of the *Corynebacterium*. Some isolates, notably *Streptomyces* and *Actinomadura* spp., may show short chains of conidia at the ends of filaments. In general, due to the fairly similar microscopic morphology of the aerobic actinomycetes, generic distinctions cannot be made by Gram stains. Conversely, acidfast staining may assist in making a preliminary generic assignment, i.e., a filamentous branching acidfast bacterium equates with *Nocardia* spp. There are several successful adaptations of standard acidfast techniques for use with the aerobic actinomycetes, all using less stringent decolorization procedures than those for mycobacteria. One is a modified Kinyoun method reported by Haley and Standard[29] that utilizes carbolfuchsin as a primary stain, a 95% ethanol wash prior to decolorization with 1% sulfuric acid, followed by a methylene blue counterstain. Another, reported by Haley and Callaway[28] requires steaming in the initial carbolfuchsin step, followed by a combined sulfuric acid-methylene blue solution to decolorize and counterstain concurrently. Alternatively, 5% sulfuric acid and 0.25% methylene blue solution in 30% ethanol may be used in sequence. *Nocardia*, rapidly growing *Mycobacterium* spp. and (often) *Rhodococcus* spp. (formerly called *M. rhodochrous* or the "rhodochrous complex) will show a partially acidfast character; that is, only some cells or sections of filaments will retain the primary stain and appear red against a blue background (Fig. 7:11). This acidfast character is not noted among *Actinomadura, Nocardiopsis* nor *Corynebacterium* spp. and is rarely seen among the streptomycetes.[2,4] If results from the initial smear are equivocal, passage of cultures on Middlebrook 7H10, Lowenstein-Jensen, casein or litmus milk agar enhances acidfastness and should yield more accurate results than growth from BHI, Sabouraud's or sheep blood agar.[28] It is most important, regardless of

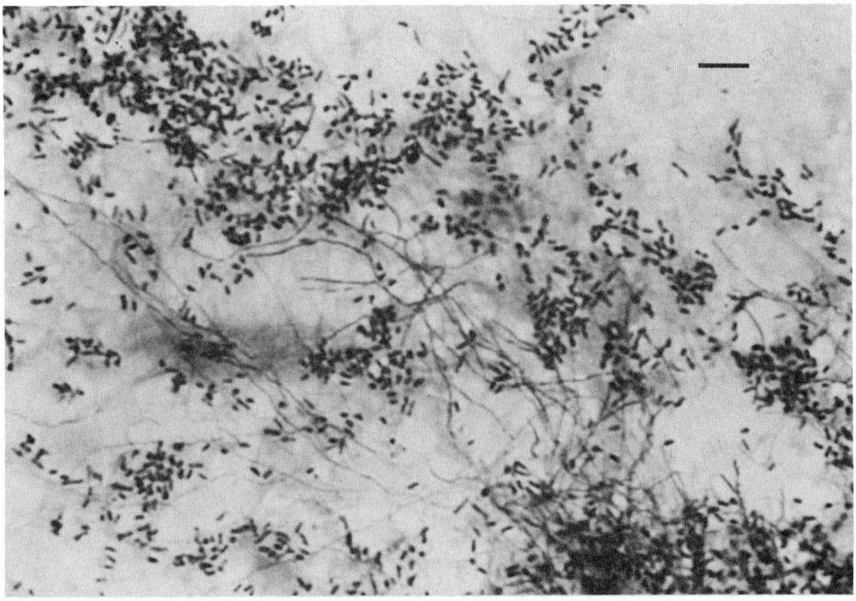

Figure 7:11 Modified Kinyoun acidfast stain of *Nocardia asteroides* culture. Note the partially acidfast staining characteristic. Only some portions of the cells retain the primary (dark) stain following decolorization and counterstain (light). 1000X magnification, bar equals 10 μm.

technique, that known acidfast positive and negative isolates (*N. asteroides* and *Streptomyces* spp.) are used as controls for each staining run.

The property of lysozyme resistance correlates with acidfastness and its detection is preferred by some to acidfast staining. Growth in a glycerol-lysozyme broth is considered a positive test and is characteristic of *N. asteroides, N. brasiliensis, N. caviae* and an occasional *Rhodococcus* spp. but not of *Actinomadura* nor *Streptomyces* spp. Known *Nocardia* and *Streptomyces* strains should serve as positive and negative controls. Lysozyme resistance is based on growth so that final results may not be available for 1-2 weeks. The acidfast stain, even with an enhancing step, gives a more immediate answer.

The microscopic morphology of aerobic actinomycetes may be studied in viable culture by lightly streaking cells from a suspect colony onto a minimal medium. Tap water agar (2% agar in plain tap water) is both useful for this purpose and simple to prepare.[4] Incubation at 25-30°C with daily examination up to 7 days is recommended. Microscopic examination can be made with a 10X objective (100X total magnification) through the unopened plate by inverting the petri dish and focusing down through the agar to the surface.

Lowering the condenser and reducing the iris diaphragm setting will increase contrast between the organism and the surrounding milieu. Typical morphology for the clinically important aerobic actinomycetes is depicted in Fig. 7:12. Note in particular the extensive tree-like branching of the surface or substrate hyphae of the *Nocardia* spp. (Fig. 7:12a). *Nocardia* often exhibits thick aerial hyphae which can be observed by shifting to a slightly

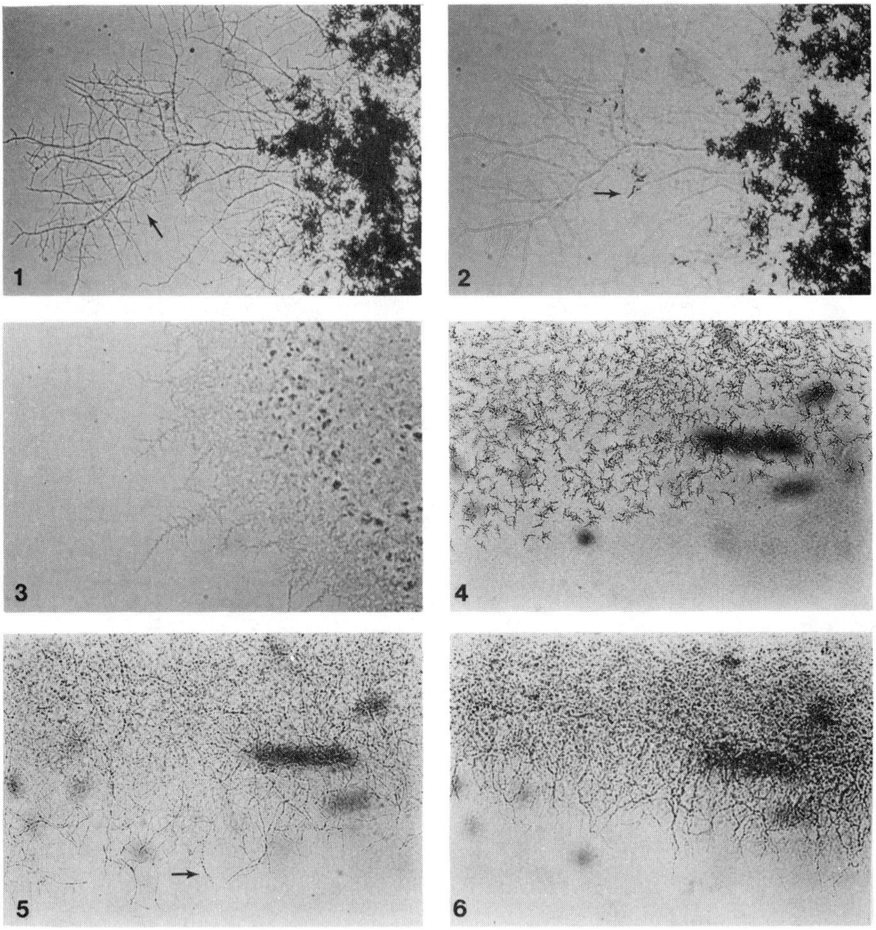

Figure 7:12 Growth of aerobic actinomycetes on tap water agar, 100X magnification. (1) *Nocardia asteroides*, 72 h incubation, arrow indicates substrate hyphae. (2) Same field as (1), altered plane of focus to illustrate aerial hyphae (arrow). (3) *Mycobacterium fortuitum*, 48 h incubation. (4) *Rhodococcus* spp., 72 h incubation. (5) *Streptomyces* spp., 72 h incubation, note chains of conidia (arrow). (6) *Actinomadura madurae*, 72 h incubation.

different plane of focus (Fig. 7:12b). Aerial hyphae are not produced by the rapidly growing mycobacterial species such as *Mycobacterium fortuitum* (Fig. 7:12c), an organism easily confused with *N. asteroides* because of physiological and morphological similarities.[57] In addition, *M. fortuitum* has a simple frost-like arrangement of substrate hyphae rather than the complex branched morphology of *Nocardia*. In contrast, extensive dichotomous branching, often with aerial hyphae, is seen in *Streptomyces* (Fig. 7:12e), *Nocardiopsis* and *Actinomadura* (Fig. 7:12f), all non-acidfast organisms. *Rhodococcus* spp. (Fig. 7:12d), organisms which phenotypically and biochemically resemble *N. asteroides*, have a predominantly coccobacillary morphology on tap water agar. They occasionally form rudimentary branches but do not form aerial hyphae. *Streptomyces* spp. tend to form chains of conidia (Fig. 7:12e) as may *Actinomadura*.

In summary, careful examination of microscopic morphology together with examination of suspect colonies by the modified Kinyoun acidfast stain will provide some basis for the preliminary grouping of *Nocardia, Mycobacterium, Streptomyces* or *Actinomadura* and *Rhodococcus*. Clearly, because of wide variation in morphology and acidfast staining even within a single species, this morphological approach should be viewed in strictly presumptive terms.

Identification based on selected biochemical characteristics:

As shown in Table 7:2, acidfastness plus the hydrolysis of casein, tyrosine and xanthine permit the separation of the three *Nocardia* species most commonly encountered in human disease. Obtaining valid results for casein and amino acid hydrolysis is often technique-dependent. The use of finely granular tyrosine and xanthine suspended evenly in basal medium cooled to a point just prior to agar solidification promotes faster recognition of positive reactions (Fig. 7:13). Placing heavy inocula into wells or troughs with a shallow layer of agar at the bottom often permits early detection of decomposition or hydrolysis activity. Incubation for a full 14 days is recommended before declaration of a negative hydrolysis reaction. For quality control and comparative purposes, a known isolate of *N. asteroides* (no hydrolysis) and an isolate of a *Streptomyces* spp. known to hydrolyze casein, tyrosine and xanthine should be run concurrently with any unknown.

The degradation of 1% ethylene glycol in solid 7H10 agar is a trait associated with the majority of *Rhodococcus* spp. but rarely noted in other taxa.[60] In a positive test, the otherwise clear agar medium is transformed to opaque (Fig. 7:13). Well characterized strains of *Rhodococcus* and *Nocardia* serve as positive and negative controls, respectively. The occasional *Rhodococcus* strain which fails to degrade ethylene glycol is easily confused with *N. asteroides* as other phenotypic and biochemical traits may be identical.[17] Likewise, the occasional ethylene glycol-positive *Nocardia*

Table 7:2 Presumptive Identification of Aerobic Actinomycetes Using Selected Biochemical Characteristics

	Acidfast Stain*	Casein Hydrolysis	Xanthine Hydrolysis	Tyrosine Hydrolysis	Urease Production	Ethylene Glycol Degradation
Nocardia asteroidesis	+	–	–	–	+	–
Nocardia brasiliensis	+	+	–	+	+	–
Nocardia caviae	+	–	+	–	+	–
Rhodococcus spp.	+/–	–	–	+/–	+/–[v]	+
Actinomadura spp.	–	+	–	+	–	–
Streptomyces or *Nocardiopsis* spp.	–	+	+/–	+	+/–	–

* Modified Kinyoun stain.
[v] Reactions variable.

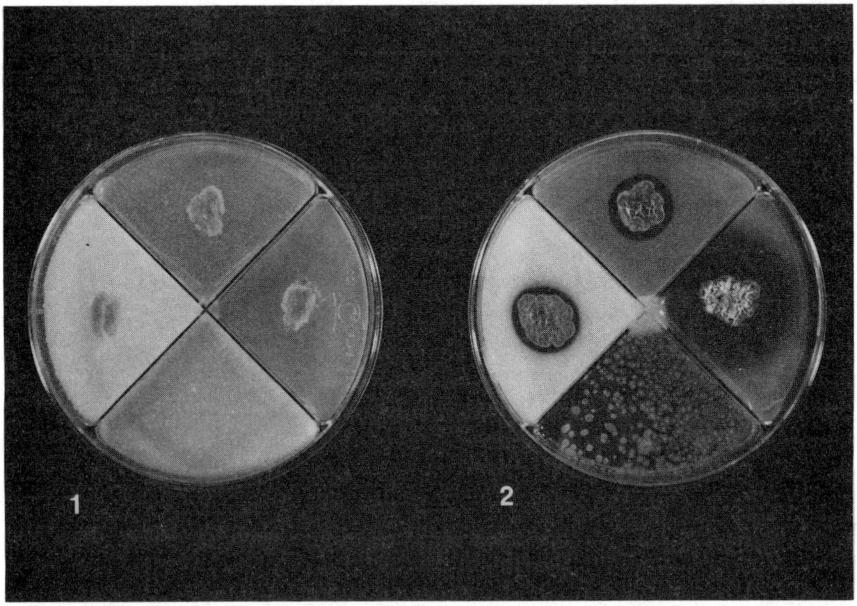

Figure 7:13 Amino acid hydrolysis and ethylene glycol degradation of aerobic actinomycetes. Clockwise beginning at the top, quadrants contain tyrosine, casein, ethylene glycol and xanthine.
(1) Inoculation with *Rhodococcus* spp. produces no hydrolysis of casein, tyrosine or xanthine (no zone) but degrades ethylene glycol, producing turbidity in an otherwise clear medium. (2) Inoculation with *Streptomyces* spp. produces distinct zones of hydrolysis surrounding growth on casein, tyrosine and xanthine but does not degrade ethylene glycol, leaving the medium clear.

further increases confusion. Reliance upon the usual creamy colony consistency and coccobacillary or minimal branching morphology of the *Rhodococcus*, as well as consideration of the specimen source and clinical presentation of the patient, help distinguish *Rhodococcus* from *Nocardia*.

The production of urease, as determined with Christensen's urea agar, will also assist in presumptive grouping of the actinomycetes. The nocardiae and rhodococci are characterized by rapid urease production, while production by *Nocardiopsis* is delayed, by *Streptomyces* variable and *Actinomadura* spp. are uniformly negative. Known strains of *N. asteroides* and *Actinomadura madurae* serve as positive and negative controls.

Use of these selective biochemical parameters allows tentative grouping of some isolates but does not reliably distinguish between the non-acidfast genera *Streptomyces, Actinomadura* and *Nocardiopsis*. If a more extensive biochemical characterization of these groups is necessary, the detailed schema developed by Mishra, Gordon and Barnett should be consulted.[43] The specific procedures for these biochemical tests can be found in an earlier article by Gordon[24] but are likely to be of greatest value to a reference laboratory.

Identification utilizing chemotaxonomic markers:

Better definition between genera of aerobic actinomycetes has been achieved by comparing the cell wall chemistry of the organisms.[40] Areas of interest include the characterization of various carbohydrates, 2,6-diaminopimelic (DAP) acid isomers, mycolic acids and lipids.[18,40,44] By using the morphological and biochemical tests already described and detecting certain carbohydrates, then using delineation between meso and LL-DAP isomers and the recognition of a "lipid characteristic of *Nocardia* type A" (LCN-A), one can discriminate with confidence between very similar groups of organisms, i.e., *Mycobacterium, Nocardia, Streptomyces* and *Actinomadura*. *Streptomyces*, for example, is the only genus among the aerobic actinomycetes that possesses the LL-DAP isomer of diaminopimelate. Likewise, the presence of LCN-A is strong evidence that the isolate belongs to one of the more commonly encountered species of *Nocardia*, although *Rhodococcus* also possesses this lipid moiety.[30,44] The *Actinomadura* are identified by the presence in their cell walls of a unique sugar, madurose. The absence of arabinose and the presence of meso-DAP in the cell wall is characteristic of *Nocardiopsis*, a non-acidfast genus virtually indistinguishable from *Streptomyces*. These as well as other chemotaxonomic characteristics of the aerobic actinomycetes commonly encountered in clinical specimens are presented in Table 7:3. Thin-layer and liquid chromatographic methods amenable to the clinical laboratory have been developed.[30,58,63] The

Table 7:3 Summary of Clinically Useful Chemotaxonomic Markers[40] for the Identification of Aerobic Actinomycetes

Species	DAP Isomer	Diagnostic Carbohydrates*	LCN-A†	Cell Wall Type
Streptomyces spp.	LL	None	−	I
Nocardia asteroides	meso	Ara + Gal	+	IV
Nocardia brasiliensis	meso	Ara + Gal	+	IV
Nocardia caviae	meso	Ara + Gal	+	IV
Rhodococcus spp.	meso	Ara + Gal	+	IV
Mycobacterium spp.	meso	Ara + Gal	−	IV
Actinomadura spp.	meso	Gal + Mad	−	IIIB
Nocardiopsis spp.	meso	Gal or none	−	IIIC

* Ara = Arabinose; Gal = Galactose; Mad = Madurose.[30] †LCN-A = Lipid characteristic of *Nocardia* type A.

information obtained from these technologies allows the classification of genera by cell wall type.[40] Review articles by Goodfellow[15] and by Goodfellow and Minnikin[18] contain detailed summaries of the chemotaxonomy of the *Actinomycetales*.

In practice, the resources and expertise necessary to pursue chemotaxonomic speciation are rarely justified for routine clinical laboratories since the frequency of need is low. A laboratory should first make smears from colonies typical of aerobic actinomycetes and stain the smears with Gram and modified acidfast procedures. This should be supplemented with observations of growth on tap water agar, casein, tyrosine and xanthine hydrolysis tests and ethylene glycol degradation. These tests are all relatively simple to perform and are quite reliable for presumptive grouping. Thus, the recovery from an appropriate specimen of a partially acidfast, gram-positive filamentous organism that fails to decompose or degrade casein, xanthine, tyrosine or ethylene glycol can be considered *N. asteroides*, the most frequently encountered human pulmonary pathogen among the aerobic actinomycetes. Isolates will occasionally fail to fit into the various patterns and should be submitted, when necessary, to reference laboratories equipped to carry out more definitive speciation. However, recent preliminary reports of success with serodiagnostic reagents and rapid enzymatic biochemical tests suggest that more definitive identification may soon be available to the routine clinical laboratory. [36,54,59]

Antimicrobial Susceptibility Testing

Evaluations of the in vitro activity of antimicrobial agents against members of the aerobic actinomycetes have been carried out in a variety of ways, none of which can be considered "standardized," due to a lack of consensus among workers. Two of the major difficulties in ascertaining the clinical merit of such testing are: (1) the inability of any study to furnish a population of infected individuals large enough to permit meaningful analysis and (2) the obvious problem in judging the predictive value of a laboratory test in a population whose "cure" is quite dependent on a complex array of factors other than the drug being used. Little can be said of the clinical utility of the variety of techniques that have been used except that, within a given system, certain drugs will appear more active than others against an individual isolate. The majority of testing has been directed at the *Nocardia*. The most frequently reported methodologies include disk diffusion and agar dilution tests against conventional antibacterial (not antifungal) agents.[5,7,41,67-69] Obtaining a homogeneous suspension of cells for standardization of the inoculum is a major technical challenge. Agitation in the presence of glass beads is a helpful strategy. Despite the slow growth of the actinomycetes, conventional zone diameter criteria have been used to interpret disk diffusion tests. Methods that determine a minimum inhibitory concentration (MIC), such as agar dilution, rely on the assumption that MIC values below achievable serum levels constitute "susceptibility" of the organism to the drug. The results of any susceptibility testing are influenced by changes in test parameters and it is likely that the aerobic actinomycetes themselves contribute more variability to each test system than would more conventional, rapidly growing bacteria. Therefore, interpretation of such testing and application to therapeutic decisions must be viewed with caution. However, as with susceptibility testing in general, the finding of high in vitro resistance is more likely to be a true predictor of negative therapeutic outcome and is more easily interpreted than a "susceptible" result.

SIGNIFICANCE OF LABORATORY FINDINGS

The aerobic actinomycetes are an innate part of the exogenous and endogenous environment of humans and all species have had, at the very least, some involvement in human infection (Table 7:1).[2,55] The acidfast organisms, *N. asteroides* and *N. brasiliensis*, are common to our biosphere and appear with fair frequency in skin and respiratory specimens.[32] Both species can cause virulent, life-threatening opportunistic infections and *N. brasiliensis* is commonly involved in infections of non-debilitated individuals. Therefore, it is imperative that identification of such organisms be reported when they are isolated in primary culture.[56] Acidfastness becomes an extremely valuable screening characteristic when dealing with aerobic pulmonary isolates, since those that are acidfast are more likely to play a pathogenic role than are the non-acidfast.

The remainder of the medically important aerobic actinomycetes, including *Actinomadura* spp., *Nocardiopsis dassonvillei, Streptomyces somaliensis* and *S. paraguayensis* are very rarely isolated from clinical specimens in the United States. However, they are predominant causes (secondary to *N. brasiliensis*) of actinomycotic mycetoma, and isolation of these organisms from specimens such as skin, wounds or tissue exudates is suggestive of infection and should be reported. As a rule, the remaining streptomycetes and *Rhodococcus* appear in skin and respiratory specimens with greater frequency than the other actinomycetes. These last two groups have previously been considered to have the least pathogenic potential of all,[55] but there have been recent reports of *Rhodococcus* spp. being involved in skin, pulmonary, bone and meningeal infections of humans.[6,9,47,65] One should, however, probably treat both groups as contaminating microflora unless an isolate represents the predominant or only flora in cultures of such specimens.

The most important caveat to remember regarding the aerobic actinomycetes is the potential of this group to infect immunocompromised patients, even to the extent of causing epidemics in intensive care facilities.[33,59] Thus, the isolation of any actinomycete, including the streptomycetes and rhodococci, from sterile body fluids, deep tissues, tissue exudates or transtracheal aspirations should be considered significant and reported. Finally, there should be open communication between the laboratory and the physician whenever a member of the *Actinomycetales* is isolated. Such physician-laboratorian interaction will increase the likelihood that the true clinical relevance of an isolate is established.

REFERENCES

1. Beadles TA, Land GA, Knezek DJ. An ultrastructural comparison of the cell envelopes of selected strains of *Nocardia asteroides* and *Nocardia brasiliensis*. Mycopathologia 1980;70:25-32.
2. Beaman BL. Actinomycete pathogenesis. In: Goodfellow M, Mordarski M, Williams ST, eds. The biology of the actinomycetes. London: Academic Press, 1984:457-79.
3. Beaman BL, Burnside J, Edwards B, Causey W. Nocardial infections in the United States - 1972-1974. J Infect Dis 1976;124:286-9.
4. Berd D. Laboratory identification of clinically important aerobic actinomycetes. Appl Microbiol 1973;25:665-81.
5. Black WA, McNellis PA. Susceptibility of *Nocardia* species to modern antimicrobials. Antimicrob Agents Chemother 1970;10:346-9.
6. Broughton RA, Wilson HD, Goodman NL, Hendrick JA. Septic arthritis and osteomyelitis caused by an organism of the genus *Rhodococcus*. J Clin Microbiol 1981;13:209-13.
7. Caroll GF, Brown JM, Haley LD. A method for determining in vitro drug susceptibilities of some *Nocardia* and *Actinomadurae*. Am J Clin Pathol 1977;68:279-83.
8. Casale TB, Macher AM, Fauci AS. Concomitant pulmonary aspergillosis and nocardiosis in a patient with chronic granulomatous disease of childhood. South Med J 1984;77:274-5.
9. Chanda JJ, Headington JT. Primary granulomatous dermatitis caused by *Rhodochrous*. Arch Dermatol 1983;119:994-7.
10. Curry WA. *Nocardia caviae*: a report of 13 new isolations with clinical correlation. Appl Microbiol 1974;28:193-8.
11. Curry WA. Human nocardiosis. Arch Intern Med 1980;140:818-26.
12. Damskey B. Nontuberculous *Mycobacteria* as an unsuspected agent of dermatological infections: Diagnosis through microbiological parameters. J Clin Microbiol 1980;11: 569-72.
13. Eisenberg ES, Alpert BE, Weiss RA, Mittman N, Soerio R. *Rhodotorula rubra* peritonitis in patients undergoing continuous ambulatory peritoneal dialysis. Am J Med 1983;75: 349-52.
14. Emmons CW, Binford CH, Utz JP, Kwon-Chung KJ, eds. Medical mycology. 3rd ed. Philadelphia: Lea and Febiger, 1977:572.
15. Goodfellow M. Characterization of *Mycobacterium, Nocardia, Corynebacterium* and related taxa. Ann Soc Belge Med Trop 1973;53:287-98.
16. Goodfellow M, Alderson G. The actinomycete-genus *Rhodococcus*: a home for the 'rhodochrous' complex. J Gen Microbiol 1977;100:99-122.
17. Goodfellow M, Cross T. Classification. In: Goodfellow M, Modarski M, Williams ST, eds. The biology of the actinomycetes. London: Academic Press, 1984:7-164.
18. Goodfellow M, Minnikin DE. Nocardioform bacteria. Ann Rev Microbiol 1977;31:159-80.
19. Gordon MA. The genus *Dermatophilus*. J Bacterial 1964;88:509-22.

20. Gordon MA. Family V *Dermatophilaceae* Austwick 1958, 42, *emend. mut. char*. Gordon 1964, 521. In: Buchanan RE, Gibbons NE, eds. Bergey's manual of determinative bacteriology. 8th ed. Baltimore: Williams and Wilkins 1974:723-6.
21. Gordon MA. Aerobic pathogenic *Actinomycetaceae*. In: Lennette EH, Balows A, Hausler WJ Jr, Shadomy HJ, eds. Manual of clinical microbiology. 4th ed. Washington, DC: American Society for Microbiology, 1985:249-62.
22. Gordon MA, Lechevalier MP, Lapa ED. Nonpathogenicity of *Frankia* sp. CpI (1) in the *Dermatophilus* pathogenicity test. Actinomycetes 1983;18:50-3.
23. Gordon RE. Some criteria for the recognition of *Nocardia madurae* (Vincent) Blanchard. J Gen Microbiol 1966;45:355-64.
24. Gordon RE, Barnett DA, Handerhan JE, Pang CH-N. *Nocardia coeliaca, Nocardia autotrophica* and the nocardia strain. Int J Syst Bacteriol 1974;24:54-63.
25. Gordon RE, Hagen WA. A study of some acidfast actinomycetes from soil with special reference to pathogenicity to animals. J Infect Dis 1936;59:200-6.
26. Gordon RE, Mihm JM. A comparative study of some strains received as nocardiae. J Bacteriol 1957;73:15-27.
27. Gordon RE, Mihm JM. A comparison of *Nocardia asteroides* and *Nocardia brasiliensis*. J Gen Microbiol 1959;20:129-55.
28. Haley LD, Callaway CS. Laboratory methods in medical mycology. Atlanta: Centers for Disease Control, Laboratory Training and Consultation Division, 1978:139-52.
29. Haley LD, Standard PG. Laboratory methods in medical mycology. Atlanta: Centers for Disease Control, Laboratory Training Section, 1973:100.
30. Hecht ST, Causey WA. Rapid method for the detection and identification of mycolic acids in aerobic actinomycetes and related bacteria. J Clin Microbiol 1976;4:284-7.
31. Hollick GE. Isolation and identification of thermophilic actinomycetes associated with hypersensitivity pneumonitis. Clin Microbiol Newsletter 1986;8:29-32.
32. Hosty TS, McDurmont C, Ajello L, George LK, Brumfield GL, Colix AA. Prevalence of *Nocardia asteroides* in sputa examined by a tuberculosis diagnostic laboratory. J Lab Clin Med 1961;58:107-14.
33. Houang ET, Lovett IS, Thompson FD, Harrison AR, Doekes AM, Goodfellow M. *Nocardia asteroides* infection - a transmissible disease. J Hosp Infect 1980;1:31-40.
34. Jensen HL. The genus *Nocardia* (or *Proactinomyces*) and its separation from other *Actinomycetales*, with some reflections on the phylogeny of the actinomycetes. Rome: Symposium on the *Actinomycetales*, 6th International Congress on Microbiology, 1953: 69-88.
35. Jonsson S, Wallace RJ Jr, Hull SI, Musher DM. Recurrent *Nocardia* pneumonia in an adult with chronic granulomatous disease. Am Rev Resp Dis 1986;134:932-4.
36. Kilian M. Rapid identification of *Actinomycetaceae* and related bacteria. J Clin Microbiol 1978;8:127-33.
37. Krick JA, Stinson EB, Remington JS. *Nocardia* infection in heart transplant patients. Ann Intern Med 1975;82:18-26.
38. Land GA, Salkin IF. Opportunistic zoopathogenic yeasts and their identification. Mycopathologia 1986;99:155-71.
39. Law BJ, Marks MI. Pediatric nocardiosis. Pediatrics 1982;70:560-5.
40. Lechevalier MP, Lechevalier HL. Chemical composition as a criterion in the classification of aerobic actinomycetes. Int J Syst Bacteriol 1970;20:435-43.
41. Lerner PI, Baum GL. Antimicrobial susceptibility of *Nocardia* species. Antimicrob Agents Chemother 1973;4:85-93.
42. McGinnis MR, D'Amato RF, Land GA. Pictorial handbook of medically important fungi and aerobic actinomycetes. New York: Praeger, 1982:160.
43. Mishra SK, Gordon RE, Barnett DA. Identification of nocardiae and streptomycetes of medical importance. J Clin Microbiol 1980;11:728-36.

44. Mordarska H, Rethy A. Preliminary studies on the chemical character of the lipid fraction in *Nocardia*. Arch Immunol Ther Exp 1970;18:455-9.
45. Palmer DL, Harvey RL, Wheeler JK. Diagnostic and therapeutic considerations in *Nocardia asteroides* infections. Medicine 1974;53:391-401.
46. Pridham TG, Hesseltine CW, Benedict RG. A guide for the classification of streptomycetes according to selected groups. Appl Microbiol 1958;6:52-79.
47. Prinz G, Ban E, Fekete S, Szabo Z. Meningitis caused by *Gordona aurantiaca (Rhodococcus aurantiacus)*. J Clin Microbiol 1985;22:472-3.
48. Rippon JW. Medical mycology: the pathogenic fungi and the pathogenic actinomycetes. 3rd ed. Philadelphia: WB Saunders, 1988:842.
49. Roberts GD, Brewer NS, Hermans PE. Diagnosis of nocardiosis by blood culture. Mayo Clin Proc 1974;49:293-6.
50. Rosett W, Hodges GR. Recent experiences with *Nocardia* infections. Am J Sci 1978; 276:279-285.
51. Ruebush TK, Goodman JS. *Nocardia asteroides* bacteremia in an immunosuppressed renal-transplant patient. Am J Clin Pathol 1975;64:537-9.
52. Sack K. Nocardial infection in a renal transplant recipient-a case report. Scand J Urol Nephrol 1985;92(suppl):59-66.
53. Satterwhite TK, Wallace RJ Jr. Primary cutaneous nocardiosis. JAMA 1979;242:333-6.
54. Schneidau JD, Shaffer MF. Studies on *Nocardia* and other *Actinomycetales*. I. Cultural studies. Amer Rev Tuberc Pulm Dis 1957;76:770-88.
55. Shaal KP, Beaman BL. Clinical significance of actinomycetes. In: Goodfellow M, Mordarski M, Williams ST, eds. The biology of the actinomycetes. London: Academic Press, 1984:389-424.
56. Smego RA, Gallis HA. The clinical spectrum of *Nocardia brasiliensis* infection in the United States. Rev Infect Dis 1984;6:164-80.
57. Staneck JL, Frame PT, Altemeier WA, Miller EH. Infection of bone by *Mycobacterium fortuitum* masquerading as *Nocardia asteroides*. Am J Clin Pathol 1981;76:216-22.
58. Staneck JL, Roberts GD. Simplified approach to identification of aerobic actinomycetes by thin-layer chromotography. Appl Microbiol 1974;28:226-31.
59. Stevens D, Pier A, Beaman B, Morozumi PA, Lovett IS, Houang ET. Laboratory evaluation of an outbreak of nocardiosis in immunocompromised hosts. Am J Med 1981;71:928-34.
60. Stottmeier KD, Molloy ME. Rapid identification of the taxon *Rhodochrous* in the clinical laboratory. Appl Microbiol 1973;26:213-14.
61. Stropes L, Bartlett M, White A. Multiple recurrences of nocardia pneumonia. Am J Med Sci 1980;280:119-22.
62. Svirbely Jr, Buesching WJ, Ayers LW, Baker PB, Britton AJ. *Mycobacterium fortuitum* infection of a Hickman catheter site. Am J Clin Pathol 1983;80:733-5.
63. Tisdall PA, Anhalt JP. Rapid differentiation of *Streptomyces* from *Nocardia* by liquid chromatography. J Clin Microbiol 1979;10:503-5.
64. Truehaft MW, Arden Jones MP. Comparison of methods for isolation and enumeration of thermophilic actinomycetes from dust. J Clin Microbiol 1982;16:995-9.
65. Tsukamura M, Kawakami K. Lung infection caused by *Gordona aurantiaca (Rhodococcus aurantiacus)*. J Clin Microbiol 1982;16:604-7.
66. Vanderstigel MV, Leclercq R, Brun-Buisson C, Schaeffer A, Duval J. Blood-borne pulmonary infections with *Nocardia asteroides* in a heroin addict. J Clin Microbiol 1986;23:175-6.
67. Wallace RJ Jr. The rapidly growing mycobacteria: characterization and susceptibility testing. Antimicrobic Newsletter 1985;2:85-92.
68. Wallace RJ Jr, Septimus EJ, Musher DM, Martin RR. Disc diffusion susceptibility testing of *Nocardia* species. J Infect Dis 1977;135:568-75.

69. Wallace RJ Jr, Weiss K, Curvey R, Vance PH, Steadham J. Differences among *Nocardia* spp. in susceptibility to aminoglycosides and beta lactam antibiotics and their potential use in taxonomy. Antimicrob Agents Chemother 1983;23:19-21.
70. Williams ST, Lanning S, Wellington EMH. Ecology of actinomycetes. In: Goodfellow M, Mordarski M, Wiliams ST, eds. The biology of the actinomycetes. London: Academic Press, 1984:480-528.
71. Woods GL, Hall GL, Schreiber MJ. *Mycobacterium fortuitum* peritonitis associated with continuous ambulatory peritoneal dialysis. J Clin Microbiol 1986;23:786-8.

CHAPTER 8

SERODIAGNOSIS OF MYCOTIC INFECTIONS

Thomas G. Mitchell, Ph.D.

INTRODUCTION

The serological measurement of host responses to mycotic infections, as well as the measurement of fungal components in sera and other fluids, is an evolving area of clinical mycology. In recent years, newer techniques and strategies and improved understanding of established procedures have resulted in significant advances in the serodiagnosis of systemic mycoses. Nevertheless, most of the tests routinely available and described in this chapter are still imperfect and considerable refinement in specificity and sensitivity are urgently needed.

Several areas of fungal immunology and serology have been reviewed recently.[1,16,36,42,47,50] In general, serodiagnosis involves two approaches—the evaluation of immunological host responses to fungal exposure and the measurement of fungal antigens and metabolites in the host. This chapter will provide an overview of the usual methodologies and the more established serological procedures.

Cell-mediated immunity is crucial for protection against a variety of fungal diseases (e.g., coccidioidomycosis, histoplasmosis, paracoccidioidomycosis, cryptococcosis and mucocutaneous candidiasis) and the assessment of various components of cell-mediated immunity may be used to define risk and predict the outcome of infection. However, with the exception of the delayed skin test reaction, this chapter will focus on the measurement of humoral responses, as these techniques are more available and standardized. Furthermore, this review does not cover immediate hypersensitivity reactions to fungal antigens.[1]

The most active area of current research is the development of tests for the detection of fungal components in the circulation. Because so many new but unproven tests have been described, only the most promising and accessible are discussed. However, breakthroughs in antigen testing may be forthcoming. Developments in the near future will feature: (1) monoclonal antibodies to diagnostic antigens, (2) new procedures with increased sensitivity and (3) possibly, rapid, over-the-counter tests for use in the home or physician's office.

PROCEDURES

In this section, techniques that are used for the detection of immunological responses to fungal antigens and methods for the detection of fungal antigens will be described. Chemical procedures to identify fungal metabolites, such as the use of gas-liquid chromatography to measure serum arabinitol produced by *Candida* and indicative of candidiasis, are available elsewhere.[25] Most serological tests are designed to measure specific antifungal antibodies in serum or spinal fluid.

Skin Tests

In general, skin testing with fungal antigens has little diagnostic value but is often useful for epidemiologic surveys of exposure or assessment of immunologic status in individual patients. Many "skin test" antigens are often used more effectively to measure antibodies in patient serum or other fluids. Antigens for the detection of delayed-type hypersensitivity to fungi were first described about fifty years ago. During the ensuing years, many of these antigen preparations have been standardized for biological testing. In some cases, specific epitopes have been defined and chemically characterized.[50]

The prototypic delayed skin test antigen was developed by Charles E. Smith and colleagues at Stanford for the detection of exposure to *Coccidioides immitis* and is termed coccidioidin.[58] Coccidioidin is a crude toluene extract of a mycelial culture filtrate of *C. immitis*. A standardized preparation (0.1 mL) is injected intradermally. A positive test is defined as induration exceeding 5 mm in diameter after 48 h. Another *C. immitis* antigen, prepared from cultured spherules and termed spherulin, is more sensitive but less specific than coccidioidin.[69] Skin testing with either antigen does not induce or boost an immune response. The skin test becomes positive within 2-3 days to 3 weeks after onset of symptoms but before the appearance of circulating antibodies. Skin test reactivity often remains positive indefinitely. A positive reaction has no diagnostic significance without a history of conversion, but a negative test can be used to exclude coccidioidomycosis, except in patients with secondary or disseminated disease who may have become anergic. Indeed, a negative skin test in confirmed cases is associated with a grave prognosis. Conversely, a positive skin test in healthy subjects implies immunity to symptomatic reinfection.

The administration, definition of a positive reaction and interpretation of other fungal skin test antigens are similar to the procedure for coccidioidin. That is, an intradermal injection of 0.1 mL eliciting a delayed, indurative (>5 mm) reaction indicates previous exposure, not necessarily active infection or disease.

Histoplasmin is a similar antigen preparation derived from a mycelial culture filtrate of *Histoplasma capsulatum.* Histoplasmin is a valuable epidemiologic tool. Most persons become skin test positive within two weeks after infection, and this reactivity usually persists for many years. As with coccidioidin, the diagnostic value of the histoplasmin skin test is minimal. A negative reaction can be used to rule out active histoplasmosis in the immunocompetent subject, but patients with anergy may be falsely negative. Without a prior history of a negative skin test, a positive reaction is meaningless except in infants, in whom a positive test can be presumed to result from recent or current infection. Because of its limited diagnostic value and the possibility that the skin test may transiently boost antibody titers, skin testing with histoplasmin should be avoided in most patients.

Epidemiologic surveys of exposure have been conducted with various antigens derived from *Paracoccidioides brasiliensis*. In the endemic areas, these paracoccidioidins exhibit crossreactivity with histoplasmin, and it is difficult to interpret double reactions of equal size in the same individual. If the reaction to one antigen is appreciably larger, that response is considered more specific.

Delayed-type hypersensitivity to *Blastomyces dermatitidis* has been detected by skin tests with blastomycins prepared from both whole yeast cell and mycelial culture filtrate antigens. However, this skin test has no diagnostic value. False-negative results are often observed in patients and false-positive crossreactions occur in many individuals sensitized to other fungi. Reactivity among patients with blastomycosis is transient and disappears with time. In individuals reactive to both blastomycin and histoplasmin, the blastomycin reaction is considered specific if it is equal to or larger than the response to histoplasmin. In at least two of the larger outbreaks of blastomycosis, blastomycin was extremely valuable because a significant number of the infections were detected by monospecific reactions, while the background level of reactivity in the population was negligible.[45]

Each of these four dimorphic, systemic pathogens is located in discreet geographic areas of endemicity, which may overlap with one or two of the other species. Within the endemic areas for coccidioidomycosis and histoplasmosis, as many as 90% or more of the resident population may be reactive to coccidioidin and histoplasmin, respectively. The percentage of reactivity to paracoccidioidin in the regions that are endemic for paracoccidioidomycosis varies from 5 to 25%. Although most patients with these three infections are males, the percentage of skin test reactivity is equal in both men and women.

Delayed skin test reactions to ubiquitous fungi, such as species of *Candida* and dermatophytes, occur in most individuals and generally indicate normal, intact cell-mediated immunity. Most of the commercially available antigen preparations are crude whole cell or culture filtrate extracts. These antigens are unpurified and tend to exhibit broad intraspecies crossreactivity.

Antibody Measurement

Methods for the serological detection of antibodies and antigens have advanced dramatically in the past decade. Nevertheless, some of the procedures described half a century ago are adequate and are routinely employed to measure antibodies to pathogenic fungi. In general, the newer techniques, such as radioimmunoassay (RIA), enzyme immunoassay (EIA), quantitative immunofluorescence, laser nephelometry and liposome immunoassay, are more sensitive than older methods, such as immunodiffusion in agarose, agglutination and complement fixation tests. Table 8:1 compares the sensitivity ranges of various immunological procedures. The amount of reactant (antigen or antibody) detectable with the newer techniques is small enough to permit the measurement of specific antigens. The sensitivity of RIA and EIA procedures can be increased by using a solid phase and indirect sandwich or capture technique.

In evaluating the ability of a serological test to detect disease, the sensitivity, specificity and predictive value of the test must be examined. These terms are defined in Table 8:2. In general, sensitivity and specificity are inversely related—increasing one of these parameters often compromises the other. The predictive value of a test is influenced by the prevalence of disease in the population tested. The positive predictive value increases as the prevalence of the disease in the population being tested increases. Of course, if the specificity of a test is 100%, its positive predictive value is also 100%.

Table 8:1 Relative Sensitivity of Immunological Tests

Test	Antibody Detectable, mL^{-1}
Immunoelectrophoresis	50–200 μg
Immunodiffusion in agar gel	3–20 μg
Counterimmunoelectrophoresis	0.5–3 μg
Crossed immunoelectrophoresis	0.1–1 μg
Agglutination	0.01–0.05 μg
Complement fixation	0.01–0.05 μg
Passive agglutination	1–10 ng
Radioimmunoassay; enzyme immunoassay	0.1–10 ng
Indirect autoradiography	100 pg

Table 8:2 Criteria for the Evaluation of Serologic Tests

Term	Definition	Calculation
TP	True Positive: number of positive reactors with disease	
FN	False Negative: number of negative reactors with disease	
TN	True Negative: number of negative reactors without disease	
FP	False Positive: number of positive reactors without disease	
Sensitivity	Percentage of true positive tests among all those with disease	$\frac{TP(100)}{TP + FN}$
Specificity	Percentage of true negative results among those without disease	$\frac{TN(100)}{TN + FP}$
Efficiency	Percentage of accurate tests in population tested	$\frac{(TP + TN)(100)}{TP + TN + FP + FN}$
Prevalence	Percentage of patients with disease among population tested	
Predictive Value Positive (PV+)	Percentage of true positive tests among all positive reactors	$\frac{TP(100)}{TP + FP}$
Predictive Value Negative (PV−)	Percentage of true negative tests among all negative reactors	$\frac{TN(100)}{TN + FN}$

Complement fixation test:

Complement fixation (CF) tests detect specific antibody that is capable of binding complement. Serum antibodies bind to antigen and immune complexes are formed. Complexes composed of appropriate immunoglobulins, primarily IgM and IgG subclasses 1, 2 and 3, will bind complement via the classical pathway. This fixed complement is unavailable to lyse sensitized sheep red cells when they are subsequently added as an indicator system to detect the unactivated complement. If the test specimen lacks antibody, the complement is not fixed and the red cells are lysed. A standard microtiter procedure for measurement of complement-fixing antibodies to fungal antigens has been developed at the Centers for Disease Control.[34] Specimens are heat-inactivated before testing to eliminate endogenous complement. Controls are required to measure the units of complement added, lysis of the sensitized sheep red cells and appropriate inhibition of lysis in the presence of positive control sera and antigens.

For the CF micromethod, wells contain dilutions of test or control serum, optimal concentrations of antigen, sensitized sheep erythrocytes and $5CH_{50}$ (complement producing 50% hemolysis of a standard suspension of erythrocytes) units of complement. Complement activity is tested by titration in wells containing erythrocytes and each antigen or diluent alone. Positive control sera are titrated in wells containing erythrocytes and complement with and without antigen; negative serum is similarly tested at various dilutions with and without antigen. Additional control wells contain erythrocytes alone and color standards (mixtures of lysed and intact erythrocytes).

Under the usual test conditions, heat-inactivated serum is diluted in veronal buffer, pH 7.3-7.4, in a volume of 25 μL per well. The initial dilution is usually 1:8. An equal volume of the optimal antigen dilution and 50 μL containing 5 CH_{50} units of complement are added to each well and the microtiter plates are held at 4°C for 15 to 18 h. To check for crossreactions, a battery of antigens are routinely employed: coccidioidin, histoplasmin, yeast cells of *H. capsulatum, B. dermatitidis* antigen and, in endemic areas, paracoccidioidin. Following overnight incubation, 25 μL of a 2.8% suspension of sensitized sheep erythrocytes is added to each well. The plates are incubated at 37°C for 30 min and centrifuged. The degree of hemolysis is interpreted by comparing each well with color standards. Positive complement fixation is defined as less than 30% hemolysis. The serum titer is the highest dilution with <30% lysis. An occasional specimen is anticomplementary as indicated by inhibition of lysis in wells without antigen.

Precipitins:

Precipitating antibodies are usually detected by immunodiffusion (ID) or counterimmunoelectrophoresis (CIE) in agarose. The ID test for precipitins to coccidioidin, histoplasmin, *B. dermatitidis, P. brasiliensis* or *A. fumigatus* employ reference antisera to define specific antigens. Reference antigens and antisera are commercially available, as are prepared agarose dishes.[37,57] The micro ID procedure developed by Kaufman et al[33] utilizes minute amounts of reactants and can be read in 24 or 48 h. The template for this method provides 17 seven-well clusters on a standard petri plate (LL Pellet Co). Alternatively, several commercial vendors offer ID test packages which include fungal antigens, control antisera and agarose preparations. Positive control antisera are essential to define specific precipitin lines.

CIE is used mainly for detection of precipitins to *Candida*. The CIE test is more sensitive but less specific than the ID test and therefore is most often used as a screen. CIE requires only 90 min and negative sera can be reported immediately. Positive sera are evaluated with the ID test, which requires 24 to 72 h.

Agglutinins:

Although agglutination tests have been described for the measurement of antibodies to many fungal pathogens, these tests are infrequently performed. Suspensions of formalin- or heat-killed yeast cells provide suitable particles for the detection of agglutinating antibodies directed against cell wall antigens. The most effective agglutinins are IgM antibodies. Tests for the detection of agglutinins to yeast cells of *C. albicans, C. neoformans* or *S. schenckii* can be performed in test tubes or in microtiter plates.[10,28,36]

Radioimmunoassay and Enzyme Immunoassay:

As indicated in Table 8:1, solid-phase radioimmunoassay (RIA) and enzyme immunoassay (EIA) are highly sensitive techniques and both methods have been applied to the measurement of a variety of mycotic antibodies and antigens. EIAs are more popular because the key immunoreactants, an enzyme-conjugated immunoglobulin and its substrate, have a reasonable shelf-life, are comparatively nontoxic and can be interpreted visually or with instrumentation.

The indirect EIA for specific antibodies is set up by coating the wells of plastic microtiter plates with the appropriate antigen. Plates can be washed, dried and stored. Dilutions of serum in a suitable buffer are subsequently added, the mixture incubated for a sufficient time and nonreactive immunoglobulins are washed away. Enzyme-conjugated antihuman immunoglobulin (which may be class-specific) is then added and allowed to combine with any antibody attached to the immobilized antigen. Excess conjugate is washed away, enzyme substrate is added and the reaction is stopped after several minutes. The amount of color generated by the reaction is determined visually or spectrophotometrically and is proportional to the antibody concentration. If relatively pure antigen is not available, antibody can be applied to the solid phase to entrap the relevant antigen; however, this antibody must not react with the conjugate.

Antigen Testing

The usual method for antigen detection is the sandwich EIA or RIA. Specific antibody is affixed to the solid phase. The test sample is added and any antigen present will bind to the immobilized antibody. After washing the wells, enzyme- or radio-labeled specific antibody is added and then nonreactive conjugate is removed. Finally, the wells are incubated in the enzyme substrate. The rate of substrate degradation is manifested by the color formed and is proportional to the amount of antigen present. The last step is unnecessary with the RIA; instead, wells are rinsed of conjugate and the amount of adherent radioactivity is measured. Reagent controls are required for all EIA and RIA procedures and standard curves must be generated to relate the results to specific amounts of antigen or antibody.

Latex agglutination (LA) tests have also been developed for the rapid detection of high molecular weight fungal polysaccharides and other antigens. Latex particles of uniform size are coated with specific immunoglobulin; normal or preimmune globulin-coated particles provide a control for nonspecific agglutination. LA tests have the advantages of being rapid, convenient and sensitive. Furthermore, since the usual antigens are heat-stable cell wall or capsular polysaccharides, specimens can be boiled, treated with reducing agents or protease to dissociate immune complexes that may block recognition of the antigen.

SYSTEMIC MYCOSES

Coccidioidomycosis

Serologic tests for the detection of antibodies to *Coccidioides immitis* were among the first mycoserologic procedures developed.[29] Standardized tests are widely available today. The interpretation of these tests is relatively straightforward and provides useful information regarding the diagnosis and prognosis of coccidioidomycosis.[20,29,48]

Two antibody tests, tube precipitin (TP) and latex agglutination (LA), measure early, IgM responses to a heat-stable component of coccidioidin (TP antigen) and denote primary or reactivation coccidioidomycosis. In most laboratories in the endemic area, the LA test has replaced the more time-consuming TP test. The TP test must be held for 5 days, while the LA test is completed in minutes. However, the LA is less specific than the TP test.

The complement fixation (CF) and immunodiffusion (ID) tests detect specific IgG and may reflect primary or progressive disease. Both the LA and ID tests are excellent screening procedures. Since all of these tests utilize the mycelial culture filtrate, coccidioidin, as antigen, they are not entirely specific.

The TP and LA tests become positive 2-4 weeks after infection and disappear in 90% of individuals after 4-6 months. The CF and ID tests become positive soon after the TP and also decline rather quickly in uncomplicated primary coccidioidomycosis.

The CF test for antibodies to coccidioidin is a powerful diagnostic and prognostic tool. Because the CF test becomes positive more slowly and persists longer, the presence of CF antibodies may reflect either active infection or the recovery stage. The CF titer correlates with the severity of disease. Most patients with disseminated coccidioidomycosis develop a titer of 1:16 or higher, whereas in nondisseminated cases, the titer is almost invariably lower. Therefore, a critical titer of 1:32 or higher reflects active, disseminated disease. However, lower titers or negative results do not exclude disseminated disease, as many patients with chronic cavitary or single extrapulmonary lesions (e.g., coccidioidal meningitis) do not develop high titers.[20,48] Low CF titers (1:2 to 1:8) are also less specific.

Serial specimens are most helpful because a change in the CF titer reflects the prognosis. The CF titer declines with recovery and eventually disappears.

A rising titer indicates active, uncontrolled infection and a poor prognosis. A stable or fluctuating titer often indicates the presence of recalcitrant or chronic disease. Only half of the patients with coccidioidal meningitis have titers of 1:32 or higher. However, most of these patients will have a positive CF test in their spinal fluid, which is equally valuable information.[48]

The CF antigen is destroyed after 30 min at 60°C but the TP antigen remains after heating. Therefore, heated coccidioidin will detect only the early antibody response to the TP antigen. The LA test employs latex particles coated with heated coccidioidin.[48] The ID method can be used to detect both TP and CF antibodies by using reference antisera, heated and unheated antigen. Antibodies to two specific heat-labile antigens, termed F (CF) and HL, may be detected.

Table 8:3 summarizes the antibody tests useful for diagnosis of coccidioidomycosis and other systemic mycoses. The values given for sensitivity and specificity in Tables 8:3, 8:4 and 8:5 represent the consensus of numerous reports. The data are subject to considerable variation for several reasons. Studies differ in many respects, including the definition of a positive result, the number and clinical forms of disease among the patients tested, the time or stage of disease when serum samples were obtained and the types of controls included (healthy subjects, patients with other mycoses, etc.). Among patients with systemic mycoses, antibody levels tend to increase with the duration and severity of infection, except in patients who are immunocompromised. Some patients, usually with sequestered disease, have consistently low titers. False-positive reactions usually result from exposure to crossreactive fungal antigens and are typically more prevalent in patients with heterologous fungal infection. Rarely, healthy individuals in an endemic area will have false-positive CF tests at lower titers. Even the ID tests, which are entirely specific, may be positive in healthy individuals in an endemic area.

Weiner[65] described an RIA that detected *C. immitis* antigen in five of nine patients with active coccidioidomycosis. These patients also produced antibody and the sera were treated to dissociate immune complexes. The antigen was a major protein fraction isolated by gel chromatography from a culture filtrate. Only 1 of 106 control sera, a patient with histoplasmosis, was falsely positive.

Histoplasmosis

The CF and ID tests for antibodies to *H. capsulatum* antigens have been widely adopted because of their convenience and availability.[55,68] Both tests may be helpful in the diagnosis and prognosis of histoplasmosis, provided the results are properly interpreted (see Table 8:3).

The CF test is routinely performed under the described conditions, and two antigens are usually employed, histoplasmin and a standardized sus-

Table 8:3 Routine Serological Tests for Antibodies—Systemic Mycoses*

Mycoses	Test†	Antigen‡	Positive Result	Sensitivity %	Specificity %	Prognostic Value	Limitations	Comments
Coccidioidomycosis	LA(TP)	C	LA titer ≥ 1:2	70–90	90–98	None		
	CF	C	titer ≥ 1:8	75–90	80–85	titer reflects severity (except in meningeal disease)	rarely crossreactive with H	titer ≥ 1:32 = secondary disease
	ID	C	precipitins F and/or HL	>90	100			
Histoplasmosis	CF	H	titer ≥ 1:8	68–83	75–99	4-fold change in titer	skin test with H may boost titer	
		Y	titer ≥ 1:8	85–94	75–95	4-fold change in titer	both H and Y crossreact with other fungal antigens	
	ID	H(10X)	precipitins m or m and h	50–85	100	Loss of h	skin test with H may boost m line	
Paracoccidioidomycosis	CF	P	titer ≥ 1:8	80–97	93	4-fold change in titer	some crossreactions with other fungal antigens	
	ID	P	precipitins 1(E), 2 and/or 3	94–98	100	loss of precipitins	precipitins 3 and m (to H) are the same	
Blastomycosis	CF	By	titer ≥ 1:8	<50	85–99	4-fold change in titer	highly crossreactive	
	ID	Bcf	precipitin A	9–79	100	loss of A		
	EIA	A	titer ≥ 1:16	77–90	92–98	titer change		

*References 16,32,36,42,48,63.

Tests†: LA = latex agglutination; CF = complement fixation; ID = immunodiffusion; TP = tube precipitin; EIA = enzyme immunoassay.

Antigens‡: C = coccidioidin; H = histoplasmin; Y = yeast phase of *Histoplasma capsulatum*; By = yeast cell homogenate of *Blastomyces dermatitidis*; Bcf = culture filtrate of *B. dermatitidis* yeasts; A = antigen A of *B. dermatitidis*; P = culture filtrate of *Paracoccidioides brasiliensis* yeasts.

pension of killed *H. capsulatum* yeast cells. Because of the possibility of crossreactivity, patient serum is also tested at the same time against other fungal antigens, such as coccidioidin, spherulin, *B. dermatitidis* or *P. brasiliensis*. CF antibodies to *H. capsulatum* antigens can be detected in serum by 2-4 weeks following exposure. Initial titers are much higher to the yeast than the mycelial antigen (histoplasmin). The CF test is performed on serial 2-fold dilutions of patient serum, beginning with a dilution of 1:8. With resolution of the infection, the antibody titer gradually declines and usually disappears (i.e., titer <1:8) by 9 months. The CF test with either *H. capsulatum* yeast or mycelial (histoplasmin) antigen is very sensitive and 90% of patients are positive (titer ≥1:8). A titer of 1:32 that persists or rises over the course of several weeks indicates active disease in patients with an established diagnosis of histoplasmosis. However, a high CF titer is not by itself diagnostic, as the results can be caused by crossreacting antibodies. In addition, low titers are not uncommon in active cases. If a patient's serum is reactive to more than one fungal antigen or if it is anticomplementary, the ID test should be conducted. Many laboratories employ the ID test as a screening procedure because it is simple to perform; however, the ID is less sensitive than the CF test.

In sensitized individuals, the skin test antigen may boost the CF antibody titer to histoplasmin and the elevated titer may remain for as long as 3 months.[38] The CF test yields results as rapidly as the skin test and is preferable for diagnostic purposes.

Precipitins are detected by routine ID or CIE procedures. The antigen is histoplasmin (in a concentration 10 times that for the CF test.) Unlike the CF test, the ID test is less sensitive (up to 80%) and requires a longer time (3 to 4 weeks) to become positive, but has a specificity of 100%. Two specific precipitins, m and h, are recognized by the formation of lines of identity with reference antiserum. The m precipitin, which is observed more frequently, appears soon after infection and may persist in the serum up to 3 years following recovery. The h precipitin, which forms closer to the serum wells, is more transient. Because it disappears soon after resolution of the disease, the presence of the h precipitin in serum denotes active infection. As with the CF titer, the m precipitin may be transiently boosted by the administration of the histoplasmin skin test and the boosting effect may last up to 3 months.[38] Both the CF and ID tests may be applied to spinal fluid for the diagnosis of meningeal histoplasmosis[36] or to pleural and other body fluids.

Paracoccidioidomycosis

The ID test is very useful in the diagnosis of paracoccidioidomycosis.[52] Nearly all patients have at least 1 of 3 specific precipitins, designated 1 (or E), 2 and 3, and detected by identity with reference antiserum.[9,51] Precipitin

1 is most frequently present. The ID test has prognostic value, because the number of precipitins is somewhat correlated with the severity of the disease and the precipitins disappear as the infection resolves.

The CF test is quantitative and useful for assessing prognosis, but crossreactions occur in sera from patients with other mycoses.[36] Most patients with active infection have CF titers ≥1:32.

Blastomycosis

Measurement of CF antibodies to various antigen preparations of *B. dermatitidis* has not proved reliable.[45] Sera from patients with blastomycosis may react with higher CF titers to heterologous antigens, especially histoplasmin, than to blastomycin, which is the analogous culture filtrate antigen. The yeast phase of *B. dermatitidis* provides a more specific antigen, but only 30-50% of patients are positive.

As indicated in Table 8:3, the ID test is more sensitive and more specific than the CF test.[35] With a yeast phase filtrate antigen and positive reference sera, antibodies to a specific antigen, designated A, can be detected. Precipitins to A are specific for *B. dermatitidis* and do not crossreact with the specific antigens of other fungal pathogens. It is not known how soon after infection the ID test becomes positive, but precipitins to A disappear within a few months after successful treatment.

An enzyme-linked immunoassay for antibodies to antigen A was recently evaluated.[39,40] The sensitivity and specificity of this test were high and titers reflected severity of disease. This EIA is a vast improvement over the conventional tests but is only available now in a few laboratories. EIA titers ≥1:32 strongly suggest active blastomycosis.[63]

OPPORTUNISTIC MYCOSES

Candidiasis

Because of the difficulty of establishing a premortem diagnosis of systemic candidiasis by microscopy or culture, the need for a diagnostic serologic test has been most urgent.[5,11,12] Different antigens and methods have been applied to the detection of diagnostic levels of antibodies. Of the numerous serologic procedures that have been proposed, the measurement of anti-*Candida* serum precipitins is probably the most acceptable and readily available test.[17,44,46,61,62] Sera are usually screened for precipitins by counterimmunoelectrophoresis (CIE) and positive specimens are then assayed by the less sensitive, but more specific immunodiffusion (ID) method. One or both of two antigens are routinely used in these tests, a crude cytoplasmic preparation of *C. albicans* or a commercially available whole culture extract. By itself, a positive test for precipitins is not diagnostic but must be interpreted along with other clinical data. Table 8:4 summarizes the most reliable antibody tests for diagnosis of candidiasis and other opportunistic mycoses.

To detect precipitins, a mixture of yeast cytoplasmic antigens can be prepared[36] or the more standardized preparation can be purchased from Hollister-Stier (HS). For the CIE test, undiluted, preheated serum is placed in a row of wells punched in 1% agarose in 50 mM barbital buffer at pH 8.6 on a 2 x 3 inch glass slide opposite wells of equal size containing dilutions of antigen, such as 1:40 to 1:5120 HS. The wells are 3 mm in diameter and the two rows are separated by a distance of 3 mm. The CIE slides are oriented with serum wells toward the anode and electrophoresed under constant current of 15 mamp/slide for 60 min and examined for precipitin lines between the wells. Controls include a positive serum and the test serum run against buffer. A specimen is considered positive if one or more precipitin lines are observed between the wells containing serum and any dilution of antigen. All CIE-positive sera are then tested by the less sensitive but more specific ID method. For the ID test, agarose slides with 3 mm well clusters are used. Test serum is placed in the central well and surrounded by wells containing barbital buffer or dilutions of antigen. The ID slides are incubated at 4°C and examined after 24 and 72 h for one or more precipitin lines between any antigen well and the serum well; one line (or more) indicates a positive test. The antigen dilutions are tested with positive control serum on a separate agarose slide. Using the optimal antigen dilution, ID titers can be determined but they rarely exceed 1:8. As alternatives to diluting the antigen, serum wells can be double-loaded or larger serum wells can be used with smaller wells for antigen.[36]

Patients with systemic candidiasis develop positive ID tests, although false-negative reactions occur in sera from patients who are severely immunocompromised or have received multiple transfusions. Healthy individuals are invariably negative, but patients with mucosal candidiasis or transient candidemia may be positive by the CIE test. The extensive data on these and other tests for *Candida* antibodies are summarized in Table 8:4. Since everyone is exposed to *Candida*, conventional tests are limited to discriminating between normal and disease levels of antibodies. More specific antigen tests are currently under development.[18] Both *Candida* surface mannan and cytoplasmic proteins can be detected in sera by enzyme immunoassay, radioimmunoassay, quantitative immunofluorescence, crossed immunoelectrophoresis or latex particle agglutination. The mannan tests usually require pretreatment of sera to dissociate antigen from antibody. Mannose (from hydrolyzed mannan) and arabinitol, a *Candida* metabolite, can also be measured in sera by gas-liquid chromatographic methods.[18,25] The detection of circulating *Candida* antigens or metabolites appears to be specific and diagnostic for systemic candidiasis. However, poor sensitivity limits most of the antigen assays.[5,18]

The EIA test for mannan was pioneered by Errol Reiss and associates at

Table 8:4 Routine Serological Tests for Antibodies—Opportunistic Mycoses*

Mycoses	Test†	Antigen‡	Positive Result	Sensitivity %	Specificity %	Prognostic Value	Limitations	Comments
Systemic Candidiasis	YA	Ca	titer ≥ 1:80; 4-fold rise in titer	33–89	82–100	4-fold change in titer	many healthy persons are positive	
	CIE	HS or S	1 or more precipitins	90–100	48–72	titer change or loss of precipitins	patients with mucosal candidiasis or transient candidemia may be positive	
	ID	HS or S	1 or more precipitins	88	90–97	titer change or loss of precipitins		
Cryptococcosis	YA	Cn	titer ≥ 1:2	38	89–100	may become positive with recovery		
	IFA	Cn	titer ≥ 1:2	39	77–100			
Invasive Aspergillosis	ID	Acf	3 or more precipitins	0–80	70–100		crossreactions with other fungi and C-reactive protein	80–100% of patients with allergic bronco-pulmonary aspergillosis are positive
Aspergilloma	ID	Acf	3 or more precipitins	80–100	70–100			
Mucormycosis	ID	Zs	1 or more precipitins	73				

*References 7,28,30,36,54,62,64,70

Tests†: YA = yeast agglutination; CIE = counterimmunoelectrophoresis; ID = immunodiffusion; IFA = indirect fluoresecent antibody.

Antigens‡: Ca = killed yeast cells of *Candida albicans*; HS = Hollister-Stier commercial *Monilia* antigen; S = cytoplasmic antigen of *C. albicans* yeast cells; Cn = killed yeast cells of *Cryptococcus neoformans*; Acf = pool of culture filtrates of 3–5 *Aspergillus* species; Zs = pool of cytoplasmic antigens of 11 zygomycetous species.

the Centers for Disease Control.[36,41] This test for *Candida* antigenemia was recently reviewed[18] and detailed procedures for setting up the EIA for *Candida* mannan have been published.[36] The test is a sandwich EIA that detects mannan from related species, including *C. albicans, C. tropicalis* and *Torulopsis glabrata*. The wells of microtiter plates are coated with rabbit anti-*C. albicans* mannan IgG to capture serum antigen; this antibody is also conjugated with enzyme for subsequent detection. Control positive serum, mannan antigen and anti-mannan IgG are available through inquiries to Dr. Reiss at CDC. Patient sera must be treated before testing by boiling with EDTA to dissociate any immune complexes. The sensitivity of this method is 1-2 ng/mL of mannan and detectable levels in serum signal active infection (Table 8:5).

Latex particles coated with rabbit anti-*Candida* mannan IgG have also been exploited for the detection of mannan antigenemia.[3,31] Optimal results are obtained after sera have been treated to dissociate immune complexes. The specificity of the LA test for *Candida* mannan, as well as its relative simplicity and speed, are excellent. However, sensitivity remains a problem. Serial specimens indicate that many patients have intermittently positive sera and tend to develop consistent or high titers only late in the course of disease. Predictably, the more specimens tested from a patient, the more likely the diagnosis will be made by positive LA. False-positives may be caused by rheumatoid factor.

A commercial LA (CLA) test (Cand-tec, Ramco) detects an undefined, heat-labile antigen and does not require pretreatment of sera. This CLA has generated findings similar to those with the LA test.[14,24,49] CLA titers of 1:2 or 1:4 may reflect mucosal candidiasis or transient candidemia and titers ≥1:8 appear to indicate systemic candidiasis. Control latex particles coated with normal rabbit immunoglobulin are required to detect nonspecific agglutination; however, this reagent is not included in the CLA package.

Many other reports of *Candida* antigen tests and potentially diagnostic antigens have presented encouraging preliminary data but require further evaluation.[2,43,59,60]

Cryptococcosis

Mycoserology for the diagnosis of cryptococcosis is very specific and sensitive. During infection, the major capsular polysaccharide of *Cryptococcus neoformans* is solubilized in the body fluids and can be titrated with anti-*C. neoformans* antiserum. The most commonly employed method is a latex agglutination (LA) test (see Table 8:5). Latex particles are coated with specific rabbit immunoglobulin and mixed with dilutions of patient serum or spinal fluid.[8] Controls include latex particles coated with normal rabbit globulin to check for nonspecific agglutination.[6] A positive agglutination at any serum or spinal fluid dilution is usually diagnostic for cryptococcosis.

Table 8:5 Serological Tests for Fungal Antigens*

Mycoses	Test†	Antigen Detected	Positive Result	Sensitivity %	Specificity %	Prognostic Value	Limitations	Comments
Systemic Candidiasis	EIA	Mannan	≥ 2 ng/mL	53–100	91–100			Mannanemia may be transient
	LA	Mannan	titer 1:2, 1:4	22–78	98			
	CLA	Unknown	titer ≥ 1:8	19–67	99	Often indicates advanced disease	Mucosal candidiasis or transient candidemia may yield positive test	
Cryptococcosis	LA	Capsular poly-saccharide	titer ≥ 1:2 (spinal fluid)	92	99	titer reflects severity	Crossreaction with *Trichosporon*	
Invasive Aspergillosis	RIA	Galacto-mannan	≥ 100 ng/mL	58–100	100			

*References 3,5,14,18,24,31,36,56,66
Tests†: EIA = enzyme immunoassay; LA = latex agglutination; CLA = commercial latex agglutination (Cand-tec); RIA = radioimmunoassay.

However, some investigators have questioned the significance of low serum titers. The serum or spinal fluid titer can be useful for prognostic evaluation.[19,22] With clinical recovery, the antigen titer declines and disappears. A rising or constant titer indicates a poor prognosis due to active or stabilized infection. Diagnosis by antigen detection is more sensitive than either direct microscopic examination with India ink or culture. Antigen in the serum of some patients (e.g., AIDS patients) may persist at high titer after prolonged therapy.[21]

Although rare, both false-positive and false-negative LA tests for cryptococcal polysaccharide may be encountered in spinal fluid as well as in serum. Most false-positive reactions are caused by rheumatoid factor, which can be eliminated by treatment of the specimen with pronase, 0.01 M dithiothreitol or boiling with EDTA.[27] Testing for the agglutination of latex particles coated with nonimmune rabbit immunoglobulin serves as a control for rheumatoid factors. Rarely, false-positive tests may occur when a crossreactive antigen is present, such as the polysaccharide of *Trichosporon beigelii*.[15] False-positive results can be caused by contamination of the specimen with a minute amount of agar or agarose, which may occur if the same pipette that is used to inoculate media for culture is reintroduced into the spinal fluid.[13] Very rarely, false-negative tests may be caused by the formation of immune complexes, a high titer of antigen (prozone) or infection with a poorly or non-encapsulated strain of *C. neoformans*.

Antibodies to *C. neoformans* are usually not detectable in the active disease state. The free polysaccharide in the serum may combine with circulating antibody, as well as inhibit antibody synthesis. However, upon clinical recovery, the serum may be positive for antibodies. These antibodies can be quantified by yeast agglutination in tubes or microtiter wells, using a killed suspension of *C. neoformans*, by agglutination of charcoal particles coated with capsular polysaccharide or by an indirect fluorescent antibody test.[7,26,28,64]

Aspergillosis

The immunologic responses detected following exposure to *Aspergillus* can sometimes clarify the pathogenesis, diagnosis and prognosis. Serologic studies involving hundreds of patients have clearly associated the presence of precipitins to various antigen preparations of *Aspergillus* species, as detected by immunodiffusion, with different forms of aspergillosis (see Table 8:4). Eighty to 100% of patients with allergic bronchopulmonary aspergillosis or aspergilloma have one or more serum precipitins to *A. fumigatus*. Unfortunately, sera from patients with invasive aspergillosis are much less frequently positive. False-positive tests are rare but may occur with serum containing crossreacting fungal antibodies or C-reactive protein. Approximately 30% or more of patients with cystic fibrosis have *Aspergillus* precipitins.[4,23] Specific IgE antibodies to *A. fumigatus* can be detected by the radioallergosorbent test in sera from patients with allergic bronchopulmonary aspergillosis. Several recent reports describing EIA's for class-specific antibodies to various antigen preparations may lead to improved methods for the detection of antibodies in invasive aspergillosis.

Detection of *Aspergillus* antigenemia in patients with invasive aspergillosis is more sensitive and specific than the test for precipitins.[18,53,66] The RIA for detecting the galactomannan of *A. fumigatus* was developed by Weiner et al.[67] Antigen was also detected in bronchoalveolar lavage fluid and in patients infected with *A. flavus*.

Mucormycosis

The literature on serological tests for mucormycosis is limited. An ID test for precipitins to a pool of zygomycetous antigens was developed by Jones and Kaufman.[30] The specificity of this test is reportedly good when reference sera are used to identify specific precipitins.

Reagents

Fungal antigens, antisera, serological test kits, reagents and equipment for the tests described in this chapter may be obtained from commercial sources such as American Scientific Products, American MicroScan, Greer Laboratories, Hollister-Stier Laboratories, Immuno-Mycologics, International Biological Laboratories, M.A. Bioproducts, Meridian Diagnostics, Nolan-Scott Biological Laboratories, Ramco Laboratories and Spectrum Diagnostics. Addresses for these companies may be found at the end of this volume.

REFERENCES

1. Al-Doory Y, Domson JF, eds. Mould allergy. Philadelphia: Lea and Febiger, 1984.
2. Araj FG, Hopfer RL, Chestnut S, et al. Diagnostic value of the enzyme-linked immunosorbent assay for detection of *Candida albicans* cytoplasmic antigen in sera of cancer patients. J Clin Microbiol 1982;16:46-52.
3. Bailey JW, Sada E, Brass C, Bennett JE. Diagnosis of systemic candidiasis by latex agglutination for serum antigen. J Clin Microbiol 1985;21:749-52.
4. Bardana EJ, Sobti KL, Cianciulli FD, Noonan MJ. Aspergillus antibody in patients with cystic fibrosis. Am J Dis Child 1975;129:1164-7.
5. Bennett JE. Rapid diagnosis of candidiasis and aspergillosis. Rev Infect Dis 1987;9:398-402.
6. Bennett JE, Bailey JW. Control for rheumatoid factor in the latex test for cryptococcosis. Am J Clin Pathol 1971;56:360-5.
7. Bindschadler DD, Bennett JE. Serology of human cryptococcosis. Ann Intern Med 1968;69:45-52.
8. Bloomfield N, Gordon MA, Elmendorf DF. Detection of *Cryptococcus neoformans* antigen in body fluids by latex particle agglutination. Proc Soc Exp Biol Med 1963;114:64-7.
9. Blumer SO, Jalbert M, Kaufman L. Rapid and reliable method for production of a specific *Paracoccidioides brasiliensis* immunodiffusion test antigen. J Clin Microbiol 1984; 19:404-7.

10. Blumer SD, Kaufman L, Kaplan W, McLaughlin DW, Kraft DE. Comparative evaluation of five serological methods for the diagnosis of sporotrichosis. Appl Microbiol 1973; 26:4-8.
11. Bodey GP, ed. Candidiasis: a growing concern. Am J Med 1984;77:1.
12. Bodey GP, Fainstein V, eds. Candidiasis. New York: Raven Press, 1985.
13. Boom WH, Piper DJ, Ruoff KL, Ferraro MJ. New cause for false-positive results with the cryptococcal antigen test by latex agglutination. J Clin Microbiol 1985;22:856-7.
14. Burnie JP, Williams JD. Evaluation of the Ramco Latex agglutination test in the early diagnosis of systemic candidiasis. Eur J Clin Microbiol 1985;4:98-101.
15. Campbell CK, Payne AL, Teall AJ, Brownell A, MacKenzie DWR. Cryptococcal latex antigen test positive in patient with *Trichosporon beigelii* infection. Lancet 1985;2:43-4.
16. Davies SF, Sarosi GA. Role of serodiagnostic tests and skin tests in the diagnosis of fungal disease. Clin Chest Med 1987;8:135-46.
17. Dee TH, Rytel MW. Clinical application of counterimmunoelectrophoresis in detection of *Candida* serum precipitins. J Lab Clin Med 1975;85:161-6.
18. deRepentigny L, Reiss E. Current trends in immunodiagnosis of candidiasis and aspergillosis. Rev Infect Dis 1984;6:301-12.
19. Diamond RD, Bennett JE. Prognostic factors in cryptococcal meningitis. Ann Intern Med 1974;80:176-81.
20. Drutz DJ, Catanzaro A. Coccidioidomycosis, Parts I and II. Am Rev Respir Dis 1978;117:559-85, 727-71.
21. Eng R, Chmel H, Corrado M, Smith SM. The course of cryptococcal capsular polysaccharide antigenemia— human cryptococcal polysaccharide elimination kinetics. Infection 1983; 11:132-6.
22. Fromtling RA, Shadomy HJ. Immunity in cryptococcosis: An overview. Mycopathologia 1982;77:183-90.
23. Gallant SP, Rucker RW, Groncy CE, Wells ID, Novey HS. Incidence of serum antibodies to several *Aspergillus* species and to *Candida albicans* in cystic fibrosis. Am Rev Respir Dis 1976;114:325-31.
24. Gentry LO, Wilkinson ID, Lea AS, Price MF. Latex agglutination test for detection of *Candida* antigen in patients with disseminated disease. Eur J Clin Microbiol 1983;2:122-8.
25. Gold JWM, Wong B, Bernard BM, et al. Serum arabinitol concentrations and arabinitol/ creatinine ratios in invasive candidiasis. J Infect Dis 1983;147:504-13.
26. Gordon MA, Lapa E. Charcoal particle agglutination test for detection of antibody to *Cryptococcus neoformans*: a preliminary report. Am J Clin Pathol 1971;56:354-9.
27. Gordon MA, Lapa EW. Elimination of rheumatoid factor in the latex test for cryptococcosis. Am J Clin Pathol 1974;60:488-94.
28. Gordon MA, Vedder DK. Serologic tests in the diagnosis and prognosis of cryptococcosis. JAMA 1966;197:961-7.
29. Huppert M. Serology of coccidioidomycosis. Mycopathologia 1970;41:107-13.
30. Jones KW, Kaufman L. Development and evaluation of an immunodiffusion test for diagnosis of systemic zygomycosis (mucormycosis): preliminary report. J Clin Microbiol 1978;7:97-101.
31. Kahn FW, Jones JM. Latex agglutination tests for detection of *Candida* antigens in sera of patients with invasive candidiasis. J Infect Dis 1986;153:579-85.
32. Kaufman L. Evaluation of serological tests for paracoccidioidomycosis: preliminary report. In: Paracoccidioidomycosis. Washington, DC: Pan American Health Organization, Science publication no. 254, 1972:221-3.
33. Kaufman L, Huppert M, Fava Netto C, Pollak L, Restrepo A. Manual of standardized serodiagnostic procedures for systemic mycoses. Part I. Agar immunodiffusion tests. Washington, DC: Pan American Health Organization, 1972.
34. Kaufman L, Huppert M, Fava Netto C, Pollak L, Restrepo A. Manual of standardized serodiagnostic procedures for systemic mycoses. Part II. Complement fixation. Washington, DC: Pan American Health Organization, 1974.

35. Kaufman L, McLaughlin DW, Clark MJ, Blumer S. Specific immunodiffusion test for blastomycosis. Appl Microbiol 1973;26:244-7.
36. Kaufman L, Reiss E. Serodiagnosis of fungal diseases. In: Rose NR, Friedman H, Fahey JL, eds. Manual of clinical laboratory immunology. Washington, DC: American Society for Microbiology, 1986:446-66.
37. Kaufman L, Standard P, Padhye AA. Exoantigen tests for the identification of fungal cultures. Mycopathologia 1983;82:3-12.
38. Kaufman L, Terry AT, Schubert JH, McLaughlin D. Effects of a single histoplasmic skin test in the serological diagnosis of histoplasmosis. J Bacteriol 1967;94:798-803.
39. Klein BS, Kuritsky JN, Chappell WA, et al. Comparison of the enzyme immunoassay, immunodiffusion and complement fixation tests in detecting antibody in human sera to the A antigen of *Blastomyces dermatitidis*. Am Rev Respir Dis 1986;133:144-8.
40. Klein BS, Vergeront JM, Kaufman L, et al. Serological tests for blastomycosis: assessments during a large point-source outbreak in Wisconsin. J Infect Dis 1987;155:262-8.
41. Lehmann PF, Reiss E. Detection of *Candida albicans* mannan by immunodiffusion, counterimmunoelectrophoresis and enzyme-linked immunoassay. Mycopathologia 1980; 70:83-8.
42. MacKenzie DWR. Serodiagnosis. In: Howard DH, ed. Fungi pathogenic for humans and animals. Part B. Pathogenicity and detection. New York: Marcel Dekker, 1983:121-218.
43. Manning-Zweerink M, Maloney CS, Mitchell TG, Weston M. Immunoblot analysis of *Candida albicans*-associated antigens and antibodies in human sera. J Clin Microbiol 1986;23:46-52.
44. Merz WG, Evans GL, Shadomy S, et al. Laboratory evaluation of serological tests for systemic candidiasis. A cooperative study. J Clin Microbiol 1977;5:596-603.
45. Mitchell TG. Blastomycosis. In: Feigin RD, Cherry JD, eds. Textbook of pediatric infectious diseases. 2nd ed. Philadelphia: WB Saunders, 1987:1927-38.
46. Odds FC, Evans EDV, Holland KT. Detection of *Candida* precipitins. A comparison of double diffusion and counterimmunoelectrophoresis. J Immunol Meth 1975;7:211-18.
47. Palmer DF, Kaufman L, Kaplan W, Cavallaro JJ. Serodiagnosis of mycotic diseases. Springfield, IL: CC Thomas, 1977.
48. Pappagianis D. Serology and serodiagnosis of coccidioidomycosis. In: Stevens DA, ed. Coccidioidomycosis. A text. New York: Plenum, 1980:97-112.
49. Price MF, Gentry LO. Incidence and significance of *Candida* antigen in low-risk and high-risk patient populations. Eur J Clin Microbiol 1986;5:416-19.
50. Reiss E. Molecular immunology of mycotic and actinomycotic infections. New York: Elsevier, 1986.
51. Restrepo A, Moncada LH. Characterization of the precipitin bands detected in the immunodiffusion test for paracoccidioidomycosis. Appl Microbiol 1974;28:138-44.
52. Restrepo A, Robledo M, Giraldo R, et al. The gamut of paracoccidioidomycosis. Am J Med 1976;61:33-42.
53. Sabetta JR, Miniter P, Andriole VT. The diagnosis of invasive aspergillosis by an enzyme-linked immunosorbent assay for circulating antigen. J Infect Dis 1985;152:946-53.
54. Schaefer JC, Yu B, Armstrong D. An *Aspergillus* immunodiffusion test in the early diagnosis of aspergillosis in adult leukemia patients. Am Rev Respir Dis 1976;113:325-9.
55. Schwarz J. Histoplasmosis. New York: Praeger, 1981.
56. Segel E, Berg RA, Pizzo PA, Bennett JE. Detection of *Candida* antigen in sera of patients with candidiasis by an enzyme-linked immunosorbent assay-inhibition technique. J Clin Microbiol 1979;10:116-18.
57. Sekhon AS, DiSalvo AF, Standard PG, Kaufman L, Terreni AA, Garg AK. Evaluation of commercial reagents to identify the exoantigens of *Blastomyces dermatitidis, Coccidioides immitis* and *Histoplasma capsulatum* species cultures. Am J Clin Pathol 1984:206-9.
58. Smith CE, Saito MT, Beard RR, et al. Serological tests in the diagnosis and prognosis of coccidioidomycosis. Am J Hyg 1950;52:1-21.

59. Stevens P, Huang S, Young LS, et al. Detection of *Candida* antigenemia in human invasive candidiasis by a new solid phase radioimmunoassay. Infection 1980;8(suppl):334-8.
60. Strockbine NA, Largen MT, Zweibel SM, Buckley HR. Identification and molecular weight characterization of antigens from *Candida albicans* that are recognized by human sera. Infect Immunol 1984;43:715-21.
61. Syverson RE, Buckley HR, Gibian JR. Increasing the predictive value positive of the precipitin test for the diagnosis of deep-seated candidiasis. Am J Clin Pathol 1978; 70:826-31.
62. Taschdjian CL, Kozinn PJ. Laboratory diagnosis of candidiasis. In: Bodey GP, Fainstein V, eds. Candidiasis. New York: Raven Press, 1985:85-110.
63. Turner S, Kaufman L, Jalbert M. Diagnostic assessment of an enzyme-linked immunosorbent assay for human and canine blastomycosis. J Clin Microbiol 1986;23:294-7.
64. Vogel RA. The indirect fluorescent antibody test for the detection of antibody in human cryptococcal disease. J Infect Dis 1966;116:573-80.
65. Weiner MH. Antigenemia detected in human coccidioidomycosis. J Clin Microbiol 1983;18:136-42.
66. Weiner MH. Immunodiagnosis of invasive aspergillosis and aspergilloma. In: Al-Doory Y, Wagner GE, eds. Aspergillosis. Springfield, IL: CC Thomas, 1985:147-55.
67. Weiner MH, Talbot GH, Gerson SL, Filice G, Gassileth PA. Antigen detection in the diagnosis of invasive aspergillosis: utility in controlled, blinded trials. Ann Intern Med 1983;99:777-82.
68. Wheat LJ, French MLV, Kohler RB, et al. The diagnostic laboratory tests for histoplasmosis. Analysis of experience in a large urban outbreak. Ann Intern Med 1982;97:680-5.
69. Woodruff WW, Buckley CE, Gallis HA, Cohn JR, Wheat RW. Reactivity to spherule-derived coccidioidin in the southeastern United States. Infect Immunol 1984;43:860-9.
70. Young RC, Bennett JE. Invasive aspergillosis. Absence of detectable antibody response. Am Rev Respir Dis 1971;104:710-16.

CHAPTER 9

ANTIFUNGAL ANTIMICROBICS: LABORATORY EVALUATION

Michael G. Rinaldi, Ph.D. and Anne W. Howell, B.S., MT(ASCP)

INTRODUCTION

In concert with an increasing incidence of human fungal infections, there has been escalated interest in and development of antifungal antimicrobics. Contempory medical practitioners have a great number of antifungal agents at their disposal and it appears that more drugs will follow. But, as with all drugs, mechanisms of action, resistance patterns, pharmacokinetics, immunomodulating properties, bioavailability and other factors must be evaluated before approval for use can be obtained. Antimycotics are currently available for therapy of all the major mycoses, from superficial through systemic infections, but the situation is far from ideal. Available agents possess distinct toxicities, vary considerably in potency and therapeutic indices and may exert static not cidal effects. Thus far, the spectacular successes achieved in development of efficacious and safe antibacterial agents have eluded antifungal antimicrobics — there are no penicillin nor cephalosporin equivalents among antifungals.

Currently available antifungal drugs include the polyenes (amphotericin B, nystatin and pimaricin), 5-fluorocytosine, several imidazoles (miconazole, ketoconazole, etc.), griseofulvin, a variety of compounds used mainly in treatment of dermatophytosis (tolnaftate, haloprogin, etc.), as well as new investigational agents and combinations. Only recently has an increase in the number of antifungals for invasive mycoses occurred. Amphotericin B was approved for clinical use in 1955-56 and remains the major agent for systemic disease: 5-fluorocytosine became available in 1971, followed by miconazole in 1978 and ketoconazole in 1981. Now itraconazole and fluconazole are undergoing clinical evaluation.

There are a variety of "over-the-counter" and prescription antifungals for superficial fungal infections and dermatomycoses. Keratinolytic agents such as Whitfield ointment, powders and sprays are widely employed. The major agents used to treat refractory dermatomycoses are griseofulvin and ketoconazole. Other imidazoles have been reported as efficacious: clotrimazole, econazole, terconazole, etc.

Laboratory evaluation of antifungals has not been widely employed due to a variety of reasons, including: a small number of available drugs to test, lack of standardization of techniques, a seeming lack of innate resistance or development of resistance by medically important fungi, the perception that invasive mycoses were not as important as bacterial infections and a lack of correlation of in vitro results with patient response. Some of these concerns are valid; however, recent developments mandate reconsideration. Some reasons for antifugal testing are: (1) availability of more broad-spectrum antifungal drugs; (2) documented resistance of fungi to amphotericin B, 5-fluorocytosine and some imidazoles; (3) differing efficacies of imidazoles against various fungal agents; and (4) efforts currently underway to achieve standardization of techniques with the anticipation of better correlation between in vitro results and patient response.

The current state of antifungal testing is perhaps equivalent to the situation with bacterial drugs some 30 years ago. Collaborative efforts by research and reference laboratories are underway and there appears to be no reason why standardization of in vitro testing cannot be achieved. This chapter will present information concerning antifugal drugs and outline some methods for their in vitro evaluation.

ANTIFUNGAL DRUGS

Polyenes

The polyene antifungal agents include amphotericin B (AMB) (Fig. 9:1), nystatin (NYS) (Fig. 9:2) and pimaricin (PIM) (Fig. 9:3). Polyenes possess a large lactone ring with a rigid lipophilic chain containing 3-7 conjugated double bonds and a flexible hydrophilic portion bearing several hydroxyl groups.[76] Most of the polyenes have a sugar unit, typically the amino-sugar, mycosamine, linked by a glycosidic bond to the alpha carbon atom of the chromophore.[100] AMB contains 7 conjugated double bonds and is referred to as a heptaene. PIM possesses 4 such bonds (a tetraene) whereas NYS is classified either as a pseudo-heptaene or as a tetraene. The length of the chromophore may contribute to the instability of some polyenes when exposed to adverse pH, light and heat.[76]

Studies of the mechanism of action of polyenes have shown that these drugs induce membrane permeability to potassium, sodium and hydrogen

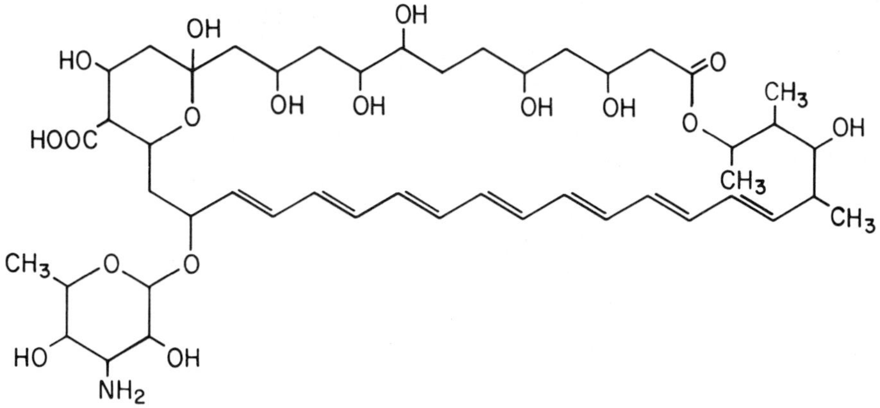

AMPHOTERICIN B (mol=924.10)

Figure 9:1. Structure of Amphotericin B.

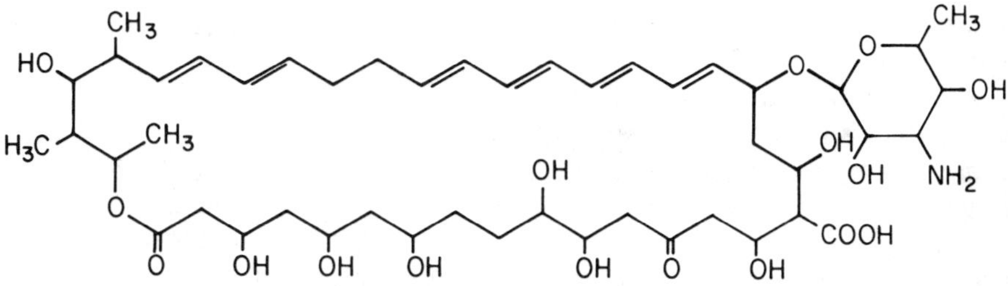

NYSTATIN (mol wt = 926.00)

Figure 9:2. Structure of Nystatin.

ions and to small non-electrolytes.[77] Most investigators indicate that polyenes form complexes with ergosterol, the main sterol in fungal membranes, which results in leakage of intracellular constituents with subsequent cell death. It seems clear, however, that polyene-membrane-cell interactions are more complex. The fungistatic effect of AMB occurs at low concentrations where drug binding is reversible and non-energy-dependent. At higher levels of AMB, a fungicidal effect occurs with irreversible and energy-dependent drug binding.[66,78] Oxidant damage may play a role in AMB action.[105] The

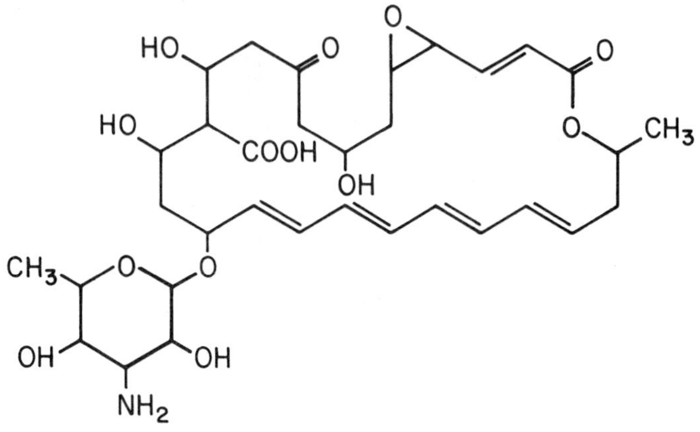

PIMARICIN (mol wt = 665.75)

Figure 9:3. Structure of Pimaricin.

size of the polyene macrolide ring and the effect of the drug appear to be correlated.[78]

AMB exhibits a broad spectrum of activity both in vitro and in vivo against the majority of invasive fungi - aspergilli, zygomycetes, blastomycetes and hyphomycetes. One major pathogenic mold, *Pseudallescheria boydii*, and one yeast species, *Candida lusitaniae*, have been found to be inherently resistant, however.[76] AMB stabilized by sodium deoxycholate is used as an intravenous preparation. The drug is notoriously toxic to the kidney, beside having other side-effects. Despite these problems, this agent remains the drug of choice for the treatment of most mycoses.

NYS is not absorbed to any degree from the gastrointestinal tract and its severe toxicity precludes intravenous use. It is used in the form of orally-administered tablets for gastrointestinal candidiasis and in various formulations of suspensions, ointments, powders and sprays for oral, vaginal and diaper dermatitis caused by yeasts, mainly *C. albicans*. NYS exhibits a similar spectrum of activity in vitro to that of AMB. The tetraene pimaricin is also too toxic for systemic administration even though its in vitro activity is broad. It is used almost exclusively to treat fungal infections of the cornea and, until the advent of the imidazoles, offered the only convincing therapy for mycotic keratitis.[96]

Resistance to polyenes by clinical isolates is low. Nevertheless, resistance has been documented and may be escalating.[76] Several *Candida* species have been reported to be resistant to AMB and NYS: *C. albicans, C. tropicalis, C. krusei, C. parapsilosis* and, perhaps most noteworthy, *C.*

lusitaniae. The incidence of AMB-resistant strains of *C. lusitaniae* is substantial enough to merit in vitro testing of isolates.[123] AMB/NYS resistance has also been reported in *Torulopsis glabrata* and *Trichosporon beigelii*.[24,79] Few molds of clinical significance have shown polyene resistance. However, the majority of clinical isolates of *Pseudallescheria boydii* are naturally resistant to AMB.[102] Occasionally, other molds show in vitro resistance which may or may not correlate with patient response. Some recently tested isolates of various fusaria from keratomycoses cases in Monterrey, Mexico, were resistant to AMB and PIM.

It has been concluded that resistance to the polyenes develops as a result of an accumulation of sterols, other than ergosterol, for which the polyenes have a lower affinity.[80] In addition to alterations in sterol and phospholipid patterns, other factors such as binding factors in the cell wall and the number of conjugated double bonds may be important in polyene resistance.[87,124] It is also of interest to ascertain whether polyene resistance is innate or develops during therapy.

5-Fluorocytosine

5-Fluorocytosine (5-FC) (Fig. 9:4) is a fluorinated pyrimidine which acts as a true antimetabolite. This cytosine analog was originally investigated for a possible role in cancer chemotherapy. It was not useful there but its antifungal properties were striking, particularly against medically important yeasts. Therefore, 5-FC has been used in the therapy of candidiasis, cryptococcosis, toruloposis, aspergillosis, chromoblastomycosis and phaeohyphomycosis.[49,64,77] Initially, 5-FC was used as monotherapy. Enthusiasm dampened as it became apparent that this antimycotic engendered some severe side-effects, that fungi quickly developed resistance or were innately resistant and that the drug was only fungistatic when used in amounts safe

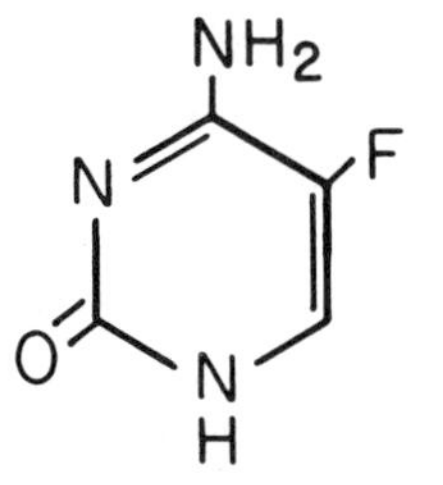

5-FLUOROCYTOSINE (mol wt = 129.10)

Figure 9:4. Structure of 5-Fluorocytosine

for systemic therapy. However, the drug was water-soluble, could be administered orally and offered an alternative to the highly toxic AMB.

In addition to its inhibition of cellular multiplication, 5-FC exerts a fungicidal effect on yeastlike fungi that is dependent upon drug concentration and the period of exposure. To a lesser extent, the same effect is observed with 5-FC and hyphomycetous dematiaceous molds, but not with aspergilli.[89] Cytosine permease actively transports 5-FC across the fungal membrane into the cell. This uptake is antagonized by cytosine, adenine and hypoxanthine, all transported across the cell membrane by the same permease.[91] Once inside the cell, 5-FC is rapidly deaminated by cytosine deaminase to 5-fluorouracil (5-FU). The specificity of the cytosine deaminase accounts for the spectrum of activity of 5-FC against fungi and for its relatively low toxicity to the mammalian host, since 5-FC is not deaminated by mammalian cells. Those fungi which lack cytosine deaminase are not susceptible to 5-FC since the drug itself is inactive.[76] The effects of 5-FC correlate directly with the amount of uracil that is replaced by 5-FU in the RNA of susceptible fungi.[88] 5-FU cannot be used as an antifungal drug because it is taken up poorly by fungi and is toxic to mammalian cells. There appear to be two independent mechanisms of action of 5-FC.[120] One mechanism involves this incorporation of 5-FU into fungal RNA, ultimately disrupting protein and carbohydrate synthesis.[62] A second mechanism involves conversion of 5-FC to 5-fluorodeoxyuridine, a potent inhibitor of thymidylate synthetase, needed to form thymidine for DNA synthesis.[91] RNA and DNA synthesis in yeasts are inhibited immediately following exposure to 5-FC. A similar, but delayed, effect is seen in some dematiaceous fungi.[89]

In vitro, 5-FC exerts a fairly narrow range of activity and clinically, 5-FC has been of value only in a restricted number of mycoses. It has become apparent that there is seldom, if ever, an indication for the use of 5-FC as a single agent. In contempory medical practice, 5-FC is used in combination with AMB. The recommended therapy of choice for the treatment of cryptococcal meningitis is a combination of AMB and 5-FC.[10] Theoretically, this combination could be synergistic. The membrane damage caused by AMB potentiates entry of 5-FC into the fungal cell; hence, lower levels of both drugs may be used. It is crucial to monitor carefully 5-FC levels in patients with renal dysfunction or receiving concurrent AMB. Since 5-FC is primarily secreted via the kidneys, dysfunction may result in serum levels toxic to the bone marrow.

Resistance to 5-FC is of genuine concern. A number of different mechanisms of resistance have been observed.[76] Resistance may occur because of a deficiency or lack of an enzyme at any step of the pathways discussed or by de novo synthesis of a surplus of normal pyrimidines that compete with the fluorinated metabolites of 5-FC.[62,91] Perhaps each mechanism occurs to some degree in resistant isolates. Numerous in vitro studies of resistance patterns have been conducted.[4,81,88,90,107] In one study, 5-FC-resistant strains were comprised of 2 different phenotypes, "totally-resistant" and "partially-resistant."[81] The former strains were unaffected by high concentrations of 5-FC, whereas the latter had a decreased growth rate when exposed to low concentrations, but following prolonged incubation, even high concentrations did not prohibit colonial development. Approximately 90% or more of isolates from susceptible species are highly susceptible to 5-FC.[88] Some spontaneous mutants with resistance will arise in susceptible populations. It appears that the degree and frequency of the development of resistance during therapy is inversely correlated with the 5-FC level in the patient and directly with the number of fungi present in the infection.[88]

If a minimum inhibitory concentration (MIC) of 25 μg/mL or greater is used as an indicator of resistance, approximately 8% of *C. albicans*, 21% of other *Candida* species, 8% of *Torulopsis glabrata* isolates and 2% of *Cryptococcus neoformans* show primary resistance.[76,30] Other reports have indicated levels of resistance in *C. albicans* ranging from 4-5% to as high as 44%, perhaps a reflection of the serotypes tested.[56,103,114] There appears to be a higher rate of resistance among isolates belonging to serotype B of *C. albicans*.[4,107] *Exophiala jeanselmei* and *Wangiella dermatitidis* exhibit frequencies of resistance comparable to that of *C. albicans*, while a relatively high frequency of resistance was found in *Fonsecaea pedrosoi*.[89] Tests of dematiaceous hyphomycetes of the genera *Bipolaris* and *Exserohilum*, as well as many aspergilli, tested in the authors' laboratory invariably reflect resistance to 5-FC.

Imidazoles

The imidazoles are a large group of synthetic compounds exhibiting antimicrobial activity against parasites, bacteria, viruses, fungi and even algae.[26,36,74,85,92,109] Compounds of this class are the most numerous of all the antifungal antimicrobics and offer beneficial alternatives to AMB and griseofulvin. Further development is in progress and release of additional imidazoles is imminent. A substantial number of imidazoles have been synthesized and evaluated in vitro and in vivo. Several have come to market initially in Europe or elsewhere before reaching the United States. The imidazoles have a broad spectrum of activity and common to all is the imidazole ring structure.

Among the first clinically applicable imidazoles was clotrimazole (CLOT) (Fig. 9:5), a tritylimidazole with chlorine substituted in one benzene ring.[17] CLOT exhibits a broad spectrum of in vitro activity against most medically important fungi. However, the drug engenders gastrointestinal disturbance, hepatic and adrenal changes and is rapidly inactivated by hepatic microsomal enzymes following oral administration.[36] Topical applications are currently used safely and effectively in oral, vaginal and superficial candidiasis and in dermatophytosis.

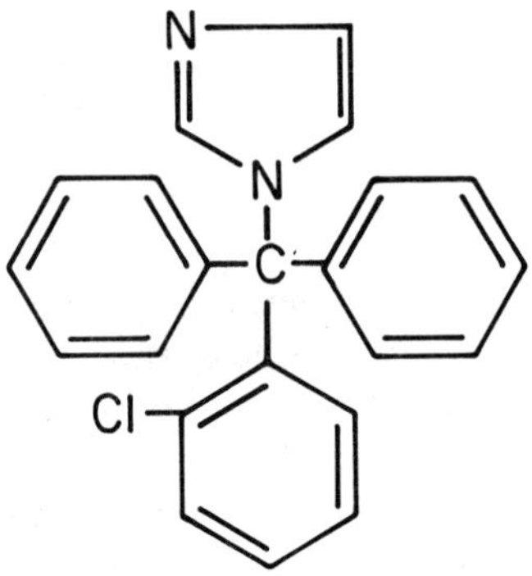

CLOTRIMAZOLE (mol wt=344.84)

Figure 9:5. Structure of Clotrimazole.

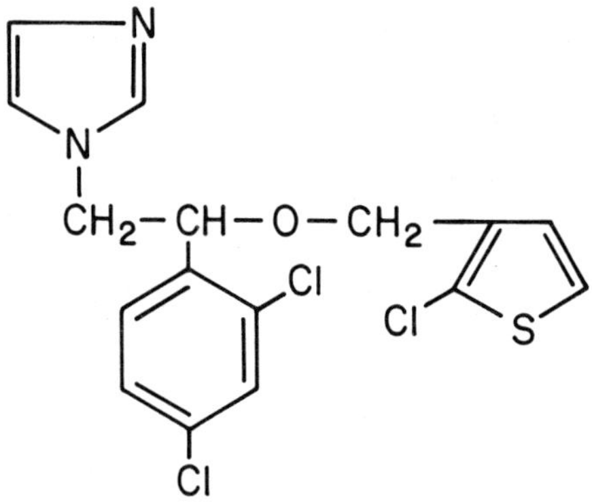

TIOCONAZOLE (mol wt=387.71)

Figure 9:6. Structure of Tioconazole.

Other N-substituted imidazole compounds like CLOT include tioconazole (Fig. 9:6), econazole (Fig. 9:7), isoconazole, sulconazole, terconazole, bifonazole, fenticonazole, oxiconazole, etc. Some of these have shown promise for therapy of invasive mycoses but the vast majority are useful only as topical preparations for candidiasis and dermatophytosis. A major interest in other imidazole antimicrobics has centered on their usefulness in the therapy of systemic mycoses. Compounds targeted for this application

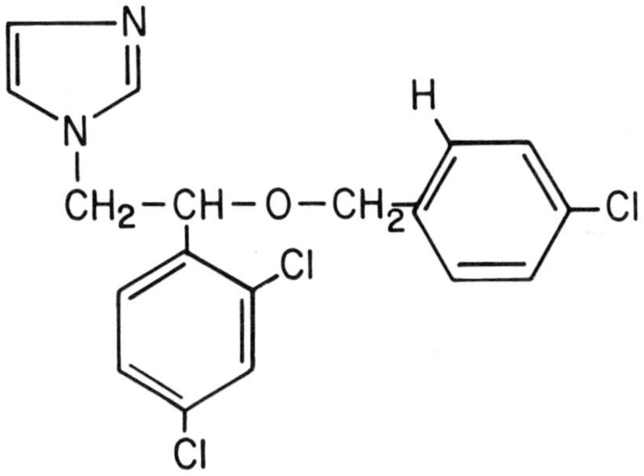

ECONAZOLE (mol wt = 381.68)

Figure 9:7. Structure of Econazole.

and for skin, nail and mucous membrane infections are miconazole (MON) (Fig. 9:8), ketoconazole (KETO) (Fig. 9:9), itraconazole (ITRA) (Fig. 9:10) and fluconazole (FLU) (Fig. 9:11). Each of these has a broad spectrum of activity against medically important fungi in vitro, although in vivo use is more restricted.

MON is a phenethyl imidazole derivative which came to clinical use in the United States in 1978.[37] The drug was initially promising but with experience it became clear that several diverse and severe side-effects (hyperlipidemia, hyponatremia, itching, cardiotoxicity) were elicited. It is believed that such toxicities are associated with Cremophor EL, the polyethoxylated castor oil used to solubilize MON for intravenous administration.[47] The drug also undergoes extensive hepatic degradation and the incidence of relapse and clinical failures was unacceptably high.[106] At present, MON is use systemically mainly for disease caused by *Pseudallescheria boydii*, a fungus with intrinsic resistance to AMB, and as an effective topical agent in dermatophytosis and yeast infections.[35]

KETO is a dioxolane-imidazole derivative with broad spectrum in vitro antifungal activity and the attribute of satisfactory oral administration. Many investigators believe the introduction of KETO for clinical use in 1981 ushered in a new era in the treatment of fungal infections[36], particularly for systemic disease. Extensive literature has been generated concerning this drug, including symposia, a book[70] and several reviews.[46] As with other imidazoles, in vitro activity is tremendously influenced by test conditions, i.e., medium, temperature and duration of incubation, pH, etc.[45,82] Clearly,

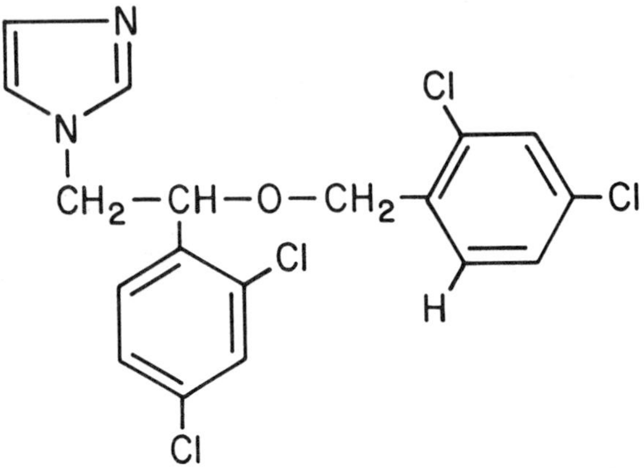

MICONAZOLE (mol wt = 479.16)

Figure 9:8. Structure of Miconazole.

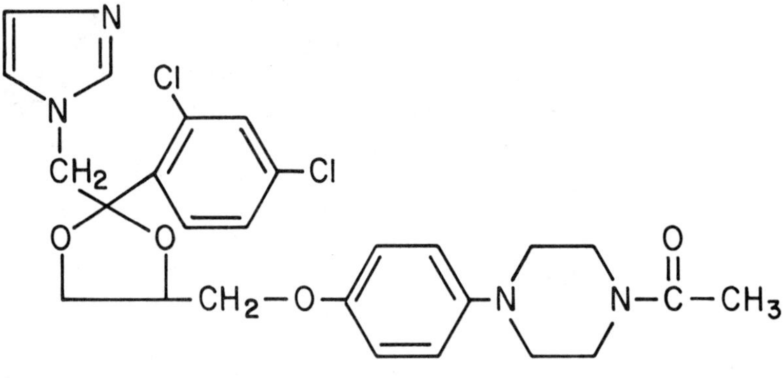

KETOCONAZOLE (mol wt = 531.43)

Figure 9:9. Structure of Ketoconazole.

acidic pH (less than pH 3) favors the bioavailability of KETO.[20,55] A wide variety of fungi are inhibited by KETO but the agent does not appear to be effective, in vitro nor in vivo, against aspergilli, the zygomycetes nor *Sporothrix schenckii*.[26,27] In vivo, KETO has been effective in therapy of pityriasis versicolor, superficial oral and vaginal candidiasis, various dermatophytoses and particularly in chronic mucocutaneous candidiasis.[36]

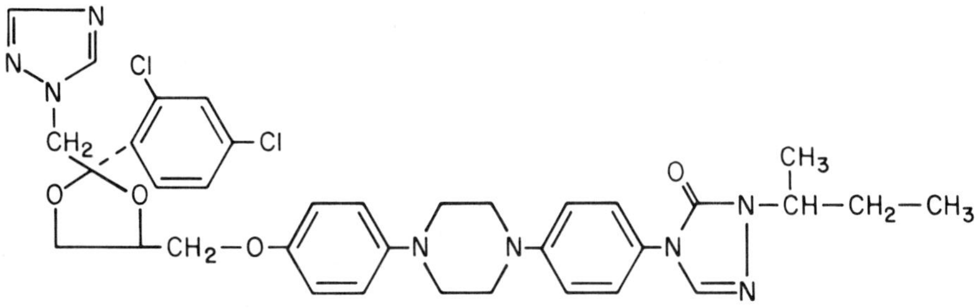

Figure 9:10. Structure of Itraconazole.

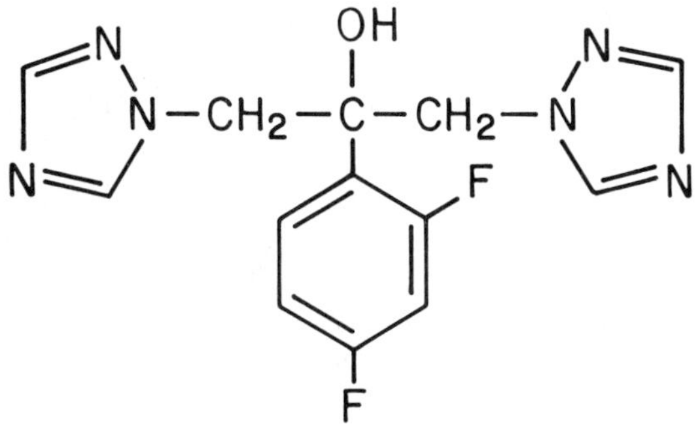

Figure 9:11. Structure of Fluconazole.

KETO therapy has shown variable effectiveness for the deep mycoses. It appears efficacious in some forms of non-life-threatening, non-meningeal coccidioidomycosis, blastomycosis, histoplasmosis and may be the therapy of choice in paracoccidioidomycosis.[36] It may also be the best alternative to AMB or AMB plus 5-FC for non-meningeal cryptococcosis. However, relapses and treatment failures have occurred. KETO is more effective in immunocompetent hosts than in those with immunological deficiencies, such as AIDS patients. The report of the collaborative study of mycoses sponsored by the National Institutes of Health indicated that KETO was more effective in patients with histoplasmosis and non-meningeal cryptococcosis than in patients with blastomycosis or non-meningeal coccidioidomycosis and was least effective in patients with sporotrichosis.[26] KETO may have a role in therapy of phaeohyphomycoses but further experience is needed to define its utility.[1]

Toxic side-effects associated with use of KETO include gastrointestinal disturbance and intolerance, gynecomastia, loss of libido and uncommon but severe liver toxicity.[22,26,60,112] All things considered, however, KETO has been a significant advance in antifungal therapy.

The two most recent imidazoles being evaluated for systemic therapy are itraconazole and fluconazole. ITRA is a triazole derivative that exhibits an in vitro spectrum similar to that of KETO but with increased activity against aspergilli (unpublished observations).[33,48,115] Like KETO, ITRA is an orally-administered drug. It has been effective in histoplasmosis, blastomycosis, coccidioidomycosis, paracoccidioidomycosis and sporotrichosis.[25,41,44] ITRA is also effective against various forms of candidiasis and dermatophytoses. It appears to be a broad-spectrum agent with greater activity than KETO.

Fluconazole is a new difluorophenyl bis-triazole derivative presently under clinical investigation. It appears to possess a broader spectrum of activity in vitro than that of KETO, but not broader than that of ITRA.[65,86,94] In vitro testing appears questionable since the data show very high inhibitory/lethal concentrations, yet FLU is very effective at much lower amounts in vivo. Apparently, this imidazole is especially influenced by in vitro testing parameters.[65] Very recent work has provided an improved method for in vitro estimation of FLU and other imidazole activity. The method involves incorporation into the antifungal test medium of antibacterial agents that bind to the 80S eukaryotic ribosome and inhibit protein synthesis. With this approach, reproducible MICs against *C. albicans* with sharp, precise endpoints were achieved.[84] FLU is a smaller molecule than either KETO or ITRA and was the only one of the triad which is partially water-soluble. Protein binding is less and cerebrospinal fluid penetration much greater with FLU than with KETO or ITRA. FLU can be administered orally or intravenously and no toxicity has been noted so far. Results of experimental studies and its pharmacokinetic properties indicate that FLU has exciting potential as an antifungal agent.

The imidazoles exert various anticellular effects on fungi and several modes of action have been proposed. These compounds appear capable of interfering with the synthesis of ergosterol via inhibition of sterol C-14 demethylation.[14,108,118] Endogenous ergosterol biosynthesis in fungi, in contrast to mammalian cells which incorporate exogenous sterol, may account for the selectivity of the imidazoles. Data generally suggest that imidazoles exert major effects on fungal cell membranes. Leakage of intracellular cations, proteins, amino acids and nucleotides, alterations of glucose metabolism and impaired uptake of amino acids have been reported.[23,58,59,67,110,111,116,119] High concentrations of MON cause direct membrane damage, whereas KETO does not exhibit this effect.[9,15] MON and KETO were also shown to inhibit hyphal elongation of germ tubes.[61] The imidazoles have been shown to affect the cytochrome oxidase systems in fungi and the cytochrome P-450 system in particular.[117,118] The major mechanisms of action of the imidazoles remain unclear. Relationships between effects on sterols, membranes, fatty acids and phospholipids, oxidative phosphorylation, cytochome oxidase or other targets remain to be revealed.

Resistance to imidazoles has been reported, albeit very rarely. An isolate of *C. albicans* was reported to develop resistance to miconazole, clotrimazole and isoconazole during therapy.[51] A series of debate-like "Letters to the Editor" concerned possible resistance to ketoconazole in *C. albicans*.[12,21,52,53,69,83,99,121] Two patients who relapsed following cessation of therapy, yielded post-treatment isolates with MIC values greater than 100 μg/mL. A similar situation was reported for an *Aspergillus* isolate.[76] Investigators feel there is genuine resistance in some of these isolates.[99]

Griseofulvin (GRIS) (Fig.9:12)

The precise mode of action of griseofulvin is not known. The target fungus must be actively growing for GRIS to have an effect. Uptake of GRIS occurs in two steps: (1) small amounts of the drug are immediately taken up by absorption and no energy is required; (2) the second step requires energy, takes a prolonged amount of time and is the step associated with fungal susceptibility.[31,32] Although there is no change in the carbohydrate, lipid, protein nor RNA content of fungal cells following exposure to GRIS, the total DNA and phosphorus levels increase.[54] GRIS inhibits synthesis of

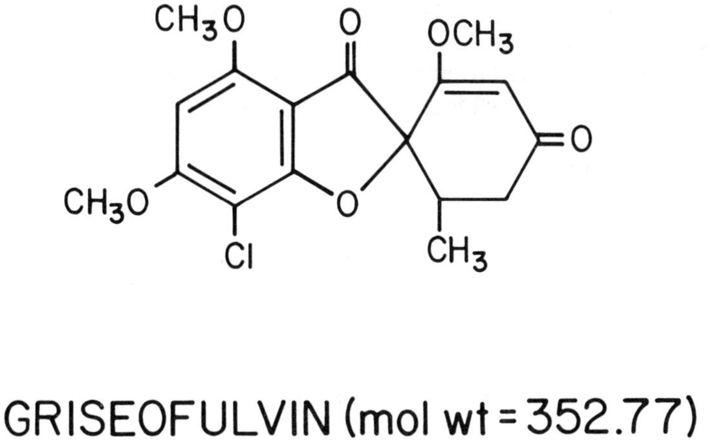

Figure 9:12. Structure of Griseofulvin.

hyphal cell wall and, through RNA binding, interferes with nucleic acid synthesis and mitosis at metaphase, as well as inhibiting microtubule function.[34,70,73,122] Electron microscopy has revealed that hyphae exposed to GRIS become swollen, contract their cytoplasmic membranes, possess disrupted endoplasmic reticulum and have no or disappearing mitochondria. Large, spherical, osmophilic, lipid bodies appear and replace much of the cytoplasm. The cell wall increases 3-4 times in thickness.[11] Substantial evidence indicates that spindle microtubule proteins are targets of GRIS.[98,112]

The use of GRIS in the therapy of mycotic diseases is limited to dermatophytosis. The drug exhibits a broad spectrum of activity, both in vivo and in vitro, against most dermatophytic fungi. GRIS remains a mainstay in the therapy of dermatophytic infections and, until the advent of the imidazoles, was the only potentially effective agent to treat chronic disease, particularly nail infections due to *Trichophyton rubrum*. GRIS is administered orally via tablet.

The incidence of GRIS-resistant dermatophytes is very rare and the significance of such isolates is unclear. Several reasons have been suggested to account for failure of dermatophytes to respond to GRIS. Therapeutic failures correlated with in vitro susceptibility data, an MIC of 3 μg/mL or greater indicating resistance.[3] It has been shown that resistance can be induced in the laboratory and that some isolates have shown spontaneously occurring resistance.[5,68]

Other Agents

Several antifungal medications, in addition to those discussed, are available for the therapy of dermatomycoses. Haloprogin (as Halotex) (Fig. 9:13) and sodium tolnaftate (as Tinactin) (Fig. 9:14) are widely-used, "over-the-counter" preparations. The literature reflects interest in evaluation of novel modifications of AMB, e.g., ornithyl amphotericin methyl ester and liposome-encapsulated amphotericin B, which offer hope for reducing toxicity of the parent congener while retaining the broad-spectrum fungicidal activity.[2,39,43,71,72,113]

The future of antifungal therapy appears bright. Increased interest by pharmaceutical firms, new knowledge of fungal biology, investigations of the modes of action of antifungals and the increased significance of mycotic diseases in contemporary medicine indicate an exciting era in medical mycology.

ANTIFUNGAL SUSCEPTIBILITY TESTING

In vitro susceptibility testing of fungi has become increasingly important. The major problems involved are the lack of standardization of techniques

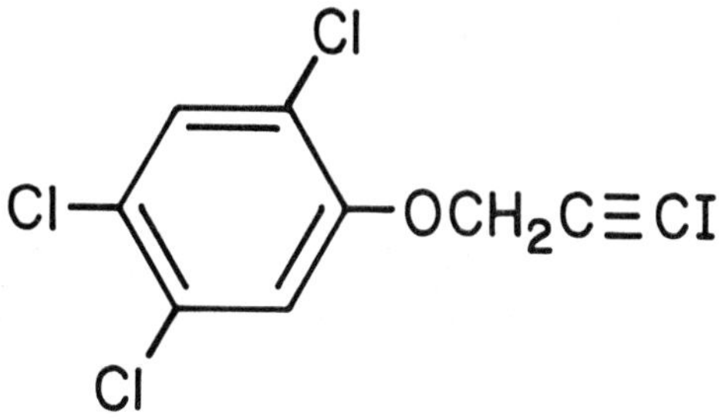

HALOPROGIN (mol wt = 361.41)

Figure 9:13. Structure of Haloprogin.

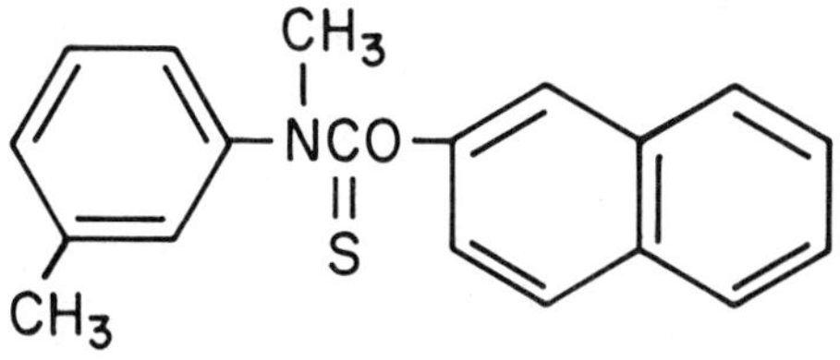

SODIUM TOLNAFTATE (mol wt = 307.43)

Figure 9:14. Structure of Sodium Tolnaftate.

and of a meaningful correlation of in vitro data with patient response to therapy. Standardized methods which provide reproducible data are necessary before clinical correlations can be made. Groups in Europe and the United States are addressing this problem.[16,19,29,38]

The various parameters of in vitro testing, methods available and interpretations thereof are discussed in detail elsewhere.[76,102] At present, there are no "right" nor "correct" methods. Therefore, the methods presented here are some that clinical and public health laboratories may perform with at least acceptable intralaboratory reproducibility, although they may not be the only, the best nor the most agreed upon ones.

The principles of susceptibility testing of fungi are essentially the same as those for testing bacteria. However, testing of fungi must deal with the fact that interpretation of results is complicated by inherent differences between fungi and bacteria in morphology, growth rate and cultural conditions. For example, the vegetative phase of medically-important fungi may be yeastlike, e.g., *Candida albicans, Cryptococcus neoformans*, or filamentous, e.g., *Aspergillus fumigatus, Rhizopus*. In addition, some organisms exist in either of two morphologic states, e.g., *Histoplasma capsulatum* variety *capsulatum*, depending upon cultural and environmental conditions. In this case, the morphologic phase of the organism may influence the test result.

Since yeasts are unicellular organisms with growth characteristics more akin to bacteria than are molds, methods for their testing have been adapted from antibacterial susceptibility test procedures. These include broth or agar dilution, as well as disk, cylinder and agar-well diffusion. The minimum inhibitory concentration (MIC) values of a drug are those that result in predetermined changes in turbidity or optical density, or in zones of inhibition of growth. The minimun lethal concentration (MLC) is determined by subculturing broth tubes showing no visible growth onto antimicrobial-free media. The MLC is that concentration of drug that results in the killing of 99.9% of the original population.

Some filamentous fungi lack ability to form distinct reproductive propagules (spores and/or conidia) and hence, do not lend themselves as readily as yeastlike fungi to laboratory testing. Reproducibility of results is a problem with filamentous organisms and perhaps the most difficult task confronting those doing studies with these fungi is the preparation of a "standardized" uniform suspension of hyphal elements.

Other factors that could adversely influence test results include stability and solubility of the antifungal agent, the character of the medium and its pH, the inoculum concentration and form and the temperature and length of incubation.

Tube Dilution Test

Dilutions of an antifungal agent in broth inoculated with a test organism provide a simple assay procedure for the quantitative assessment of the MIC for a particular organism. The dilution procedures are conventionally done in tubes but can be miniaturized for performance in microtiter plates. The MIC values obtained will vary with the type of endpoint determination used to generate the results. Visual turbidity is the most common method of determining the endpoint and requires a subjective judgment of the lowest concentration of drug that inhibits growth. For each assay, a control organism with known susceptibility must be run in parallel with unknown organisms.

Selected General Procedures for Macro-broth Dilution Testing:

Each antifungal antimicrobic agent tested requires a particular solvent. A list of the agents and their recommended solvents is shown in Table 9:1.

Table 9:1. Antifungal Antimicrobics and Their Solvents

Amphotericin B (AMB)—sterile distilled water
5-Fluorocytosine (5-FC, flucytosine)—sterile distilled water
Miconazole (MON)—sterile distilled water
Ketoconazole (KETO)—0.2N HCl
Itraconazole (ITRA)—polyethylene glycol
Fluconazole (FLU)—sterile distilled water
Pimaricin (PIM)—Dissolve in 0.5N NaOH: neutralize with 0.5N HCl
Griseofulvin (GRIS)—70% ethyl alcohol

Table 9:2. Broth Media for Antifungal Testing and Antimicrobic-Media* Combinations

Media:
Antibiotic Medium 3 (M-3)
Antibiotic Medium 12 (M-12)
Casein Yeast-Extract Glucose (CYEG)
Synthetic Amino Acid Medium—Fungal (SAAMF)
Yeast Morphology Broth, unbuffered and buffered (uYMB, bYMB)
Yeast Nitrogen Base, Shadomy's Modification of both unbuffered and buffered media (uYNB, bYNB)
Supplemented Yeast Nitrogen Base (sYNB)

Antimicrobic-Media Combinations:
AMB and PIM = SAAMF, M-3, M-12, bYMB, CYEG
5-FC = SAAMF, uYMB, uYNB, sYNB, CYEG
MON, KETO, ITRA and FLU = SAAMF, bYMB, bYNB, CYEG
GRIS = SAAMF, bYMB, bYNB, CYEG

*See Table 9:1 for definitions of antimicrobials.

The selection of appropriate broth media is of critical importance in the testing of all antifungal agents. A list of acceptable broth media for each of eight antifungals is shown in Table 9:2. Of particular note is Synthetic Amino Acid Medium - Fungal (SAAMF) (Gibco Laboratories; Otisville BioPharm). SAAMF is a synthetic, totally-defined, iso-osmolal, well-buffered, nutritionally-adequate, nonantagonistic, tissue culture-like medium, designed to reflect the biochemical milieu of human serum.[50] It has been shown, however, that the MOPS-Tris buffer components of SAAMF may interfere with the activity of some antifungal agents.[18] SAAMF is an appropriate choice for susceptibility testing with the caution that the MOPS Tris buffering system be replaced with alternative compounds, e.g., HEPES, at times. SAAMF is prepared as per the manufacturer's instructions and then must be filtered through a 0.22 μm filter. The filtrate may be stored at 4-8°C for up to 3 months.

Some antifungal antimicrobials may be obtained from hospital pharmacies, although others (i.e., pure, assayed, reagent powders) must be obtained from the pharmaceutical producer. It is important to consider the stated drug potency listed by the company when preparing stock drug solutions, since potency may vary with each lot. However, if this information is not given, one must assume 100% potency for testing purposes. Use the following formula to determine the amount of drug to weigh, substituting the correct potency:

$$\frac{\text{Desired Volume (mL) X Desired Concentration } (\mu g/mL)}{\text{Amount to weigh(mg)} = \text{Drug Potency } (\mu g/mL)}$$

For convenience in the clinical laboratory, it has been determined that antifungal drug dilutions may be prepared in advance and aliquoted into prelabeled sets of tubes, which are stored frozen at -70°C for up to 3 months. Dilution sets are thawed at room temperature when needed. Table 9:3 shows the serial dilution ranges and the final concentrations after inoculum dilution in μg/mL for the following antifungals.

1. Amphotericin B (Fungizone I.V., E. R. Squibb): AMB comes as a sterile, lyophilized cake. Reconstitute 1 vial with 10 mL of sterile distilled water to give a colloidal suspension of 5,000 μg/mL. Further dilute in sterile distilled water to a concentration of 184.7 μg/mL. Serially dilute in 2-fold steps in sterile distilled water through 7 dilutions to achieve a dilution range of 184.7 to 1.4 μg/mL. Aliquot 0.1 mL of each dilution into small freezer-safe tubes and freeze at -70°C.
2. 5-Fluorocytosine (Flucytosine, Hoffman-LaRoche): 5-FC is a pure assayed powder. Weigh the appropriate amount and dissolve in 5 mL sterile distilled water for a drug concentration of 3,227.5 μg/mL; filter sterilize through an 0.45 μm filter.If necessary, quickly warm the solution in a 50°C water bath to completely dissolve the drug. Prepare 2-fold serial dilutions to achieve a 6-tube dilution scheme ranging from 3,227.5 to 100.8 μg/mL. Aliquot in sets and freeze.
3. Miconazole (Janssen Pharmaceutica; Ortho Pharmaceutical): MON is available as the pure, assayed powder or as Monistsat I.V. solution. The solution is recommended for antifungal testing. Each ampoule

Table 9:3. Serial Drug Dilution Ranges and Final Concentrations*

Drug	Range (μg/mL)	Final Concentration (μg/mL)
AMB	184.7 – 1.4	18.47 – 0.14
PIM	192 – 1.5	19.2 – 0.15
5-FC	3,227.5 – 100.8	322.75 – 10.08
MON	200 – 3.12	20 – 0.3
KETO	128 – 0.125	12.8 – 0.012
ITRA	100 – 0.18	10 – 0.018
FLU	800 – 12.5	80 – 1.25
GRIS	400 – 0.7	40 – 0.07

*Following 0.1 mL drug + 0.9 mL inoculum.

contains 10 mg/mL MON in 20 mL polyethoxylated castor oil. Store ampoules at room temperature. To prepare dilution tubes, dilute the Monistat I.V. to a concentration of 200 μg/mL in sterile distilled water; then make serial 2-fold dilutions through 5 steps to give a range of 200 to 3.125 μg/mL. Aliquot and freeze in sets.

4. Ketoconazole (Janssen Pharmaceutica): KETO is supplied as a pure, assayed powder. Prepare a 50,000 μg/mL solution in 1 mL 0.2N HCl. Dilute in sterile distilled water to a concentration of 128 μg/mL. Prepare 11 serial 2-fold dilutions in sterile distilled water to achieve concentrations ranging from 128 to 0.125 μg/mL. Aliquot and freeze in sets.
5. Itraconazole (Janssen Pharmaceutica): ITRA is suppled as a pure, assayed powder. Assume 100% potency unless otherwise stated. To prepare a stock solution of ITRA, place 10 mL polyethylene glycol (mol. wt. 200) in a small beaker with a stirring rod, heat while continually stirring to 75°C, then add 50 mg ITRA. Continue to heat until clear; this usually requires 1-4 h. Do not filter-sterilize. The concentration of this stock is 5,000 μg/mL. Store at room temperature in a sterile, sealed glass tube for up to 4 weeks. For testing, further dilute the stock solution to 100 μg/mL in polyethylene glycol and prepare nine 2-fold dilutions to give a dilution set ranging from 100 to 0.18 μg/mL. Aliquot and freeze in sets.
6. Fluconazole (Pfizer Central Research): FLU is supplied as a pure, assayed powder with 100% potency unless otherwise stated. Weigh 10 mg FLU and dissolve in 5 mL sterile distilled water to give a concentration of 2,000 μ/mL. Warm the solution in a 50°C water bath to facilitate complete dissolution. Prepare an 800 μg/mL solution, then 2-fold serial dilutions in sterile distilled water through 6 steps, giving a range from 800 to 12.5 μg/mL. Aliquot and freeze in sets.
7. Griseofulvin (Ortho Pharmaceutical): GRIS is available as an assayed powder. Prepare a 2,000 μg/mL working solution in 70% ethyl alcohol. Dilute 1:5 in sterile distilled water to a concentration of 400 μg/mL. Prepare 9 dilution tubes with 1 mL sterile distilled water each and make 2-fold serial dilutions to give a concentration range of 200 to 0.7 μg/mL. Aliquot in sets and freeze.
8. Pimaricin (Natamycin; Alcon Laboratories): PIM as a pure, assayed powder varies in potency with each lot number. Consider the stated potency and weigh out enough power to achieve a concentration of 6,400 μg/mL when dissolved in 3 mL of 0.5N NaOH. Filter-sterilize through an 0.22 μm filter. Within 30 min, neutralize with approximately 3 mL of 0.5N HCl to achieve a pH of 6.0-7.0. Bring the volume to 10 mL with sterile distilled water. The resulting solution has a concentration of 1,922 μg/mL. Dilute 1:10 in sterile distilled water to prepare a stock solution of 192 μg/mL. Store protected from light at 4°C. To prepare the dilution set, serially dilute the stock solution in 2-fold steps through

8 tubes to give a range of concentrations, 192 to 1.5 μg/mL. Aliquot in sets and freeze.

Considerations and Procedure for Inoculum Preparation

The yeast inoculum should be obtained from an actively growing culture to ensure the use of viable, reproducing cells. Choose several well-isolated colonies from a 24-48 h pure culture on an antimicrobic-free agar plate to inoculate 5 mL broth medium (SAAMF, Yeast Nitrogen Base broth, Yeast Morphology broth, etc.). Incubate the inoculated broth overnight at 35°C. Resuspend the broth culture by vigorous vortexing. Use a hemacytometer to count the yeast cells, counting all cell clumps as one unit. It may be necessary to dilute the broth culture 1:100 for counting. Calculate cells per mL and prepare a 1 x 10^4 cells/mL stock in broth. Add 0.9 mL of stock cells to each drug tube containing 0.1 mL of diluted drug. Place 1 mL of stock cells in a growth control tube. Thus, each drug concentration has been further diluted 1:10 by the addition of the inoculum. Incubate the inoculated set in a 35°C non-CO_2 incubator 24 h or until the growth control tube shows visible growth.

Molds may present a number of problems in preparing a homogenous inoculum. Wetting agents, such as a 0.05% solution of Tween 80 (Remel Laboratories), may be needed to reduce the surface tension surrounding spores/conidia and hyphae. A conidial or spore suspension is prepared by shaking a few mL sterile distilled water over the surface of a potato flakes agar slant. Use this medium to promote growth with reproductive propagules.[95] Transfer the suspension with a sterile pasteur pipette to a sterile tube. Determine the concentration of colony-forming units per mL (CFU/mL) with a hemacytometer. Count clumped propagules as one unit. Also set up plate counts to ascertain the validity of the hemacytometer count. Prepare the dilutions necessary to achieve a 1 x 10^4 CFU/mL broth suspension. Add 0.9 mL of this inoculum suspension to each drug tube and place 1 mL in the growth control tube.

Some moldlike fungi do not readily (or ever) produce spores, conidia or other easily countable inocula. Therefore, to prepare a homogenous standardized inoculum, it is necessary to harvest a small portion of mycelial growth and gently homogenize it in a tissue grinder or with glass beads on a rotary shaker. Allow the larger hyphal fragments to settle for about 5-10 min. Prepare a suspension having 95% transmission as determined by a spectrophotometer set at 530 nm and dilute this suspension 1:10 in broth as the inoculum.

To prepare a broth inoculum of *Histoplasma capsulatum* variety *capsulatum, Blastomyces dermatitidis, Coccidioides immitis, Xylohypha bantiana* or other agents felt to present risk to laboratory personnel, perform hemacytometer counts on merthiolate-killed fungi. Prepare a 1:2 dilution of the conidial/spore/hyphal suspension in 10% merthiolate. Let stand at least 15 min for killing before counting. Multiply counts by 2 to take into account this dilution. Incubate mold tests at room temperature and read results when the growth control tube is positive, which may be several days for many filamentous fungi. The MIC for either yeastlike or moldlike fungi is read as the first tube concentration showing no growth or a marked reduction in growth from the previous tube.

Take subcultures for determination of MLC when the first MIC is read. Remove 0.1 mL from each tube with no evidence of growth, or at the MIC value and all greater concentrations, and from the growth control tube. Spread each 0.1 mL over the surface of a drug-free agar plate. Incubate plates at 35°C for yeasts and at room temperature for molds. Any agar medium which readily supports growth of the fungi may be used, e.g., Sabouraud dextrose agar (Difco Laboratories). The plates are read when the growth control plate is positive. Any plate with 5 or fewer colonies is considered negative. The first concentration with a negative subculture is the MLC.

It is important to include a polyethylene glycol growth control when testing ITRA; most yeasts grow satisfactorily but some molds do not. It may also be prudent to include an ethyl alcohol control when testing GRIS. As always, it is important to consider special growth requirements of the fungi being tested. For example, it is possible to test the susceptibility of isolates of *Malassezia furfur*, but it is necessary to recognize that this organism requires a source of fatty acids, e.g., olive oil, for growth.

Quality Control

Controls for each drug tested must be included in every run to ensure that the test is providing reproducible data. Thus, each susceptibility test must have a quality control organism which gives known and reproducible MIC/MLC values. The laboratorian must look for trends in MIC/MLC values with control organisms as a means of monitoring drug potency. The growth control tubes should ensure that the medium is functioning properly, that the inoculum is viable and actively growing and that any solvents used to prepare drug solutions are not influencing test results. Several reports have listed appropriate control organisms for antifungal testing;[76,102] the following fungi are suggested as suitable: *Saccharomyces cerevisiae* ATCC 9763, *Candida albicans* ATCC 10231, *Candida tropicalis* ATCC 13803 and *Candida pseudotropicalis* ATCC 28838.

Synergism Studies

In vitro susceptibility testing may involve testing of combinations of antifungal agents, usually AMB + 5-FC, although other drug combinbations are also possible. Such studies are usually conducted in a "checkerboard" titration, in which the two antifungal agents are tested together in all concentrations of serial dilutions of each and compared to the results of testing each drug separately. Synergism may be defined as a 4-fold or greater reduction in the MIC of one drug when used in combination with a subinhibitory concentration of a second antifungal agent.[76] In vitro synergism studies may be valuable tools in assessing the potential use of combined antifungal therapy.

Prepare drug dilutions as for single agent testing. Arrange and label 12 x 75 mm double snap-cap plastic tubes (Evergreen Scientific) in a checkerboard pattern (Table 9:4). Dispense 0.05 mL of each drug dilution into the appropriate tubes, so the final volume will be 0.1 mL/tube. This effectively dilutes each drug 1:2, therefore the concentrations used should be twice the desired drug concentration. Inoculate each tube with 0.9 mL of a 1 x 10^4 CFU/mL fungal suspension. Incubate and read as for single drug MIC/MLC testing. Results are reported as the MIC or MLC of one antimicrobic plus the MIC or MLC of the other drug, i.e., for AMB + 5-FC, a final result might read 0.58 μg/mL AMB + 20.17 μg/mL 5-FC.

ASSAYS OF ANTIMYCOTIC ACTIVITY IN BIOLOGICAL FLUIDS

The determination of levels of antifungal antimicrobics in serum, cerebrospinal fluid or other body fluids may provide the clinician valuable information that will aid in adjusting chemotherapy. A variety of methods have been developed, including microbiological assay, gas-liquid chromatography, high performance liquid chromatography, thin-layer chromatography, fluorometry and spectrophotometry; these methods are described in detail elsewhere.[6,42,76] Although chemical methods are more specific, sensitive and rapid than bioassays, chemical techniques are, at present, not

Table 9:4 Synergy Checkerboard Broth Dilution Pattern for Amphotericin B (AMB) and 5-Fluorocytosine (5-FC)

5-FC (μg/mL)	322.75	*	X	X	X	X	X	X	X	X
	161.38	*	X	X	X	X	X	X	X	X
	80.69	*	X	X	X	X	X	X	X	X
	40.34	*	X	X	X	X	X	X	X	X
	20.17	*	X	X	X	X	X	X	X	X
	10.08	*	X	X	X	X	X	X	X	X
			**	**	**	**	**	**	**	**
			0.14	0.29	0.58	1.16	2.31	4.62	9.24	18.47
			AMB (μg/mL)							

* = Tubes with only 5-FC.
** = Tubes with only AMB.
X = Tubes with both 5-FC and AMB.

practical in most clinical laboratories, since these methods often require extraction methods and specialized instrumentation not readily available there. It has also been noted that the physiochemical methods do not evaluate the biological activity of the antifungal agent in the biological fluid, as do bioassays.[14] Bioassay methods include agar diffusion, radiometric and turbidometric methods and potassium or rubidium ion efflux.[7,8,13,63,101,104] The following agar diffusion bioassay methodology is for use in the clinical or public health laboratory. The antifungals discussed are AMB, 5-FC and KETO, but other drugs can be assayed in an analogous manner.

Preparation of Antimycotic Standards

1. AMB: Reconstitute 1 vial of Fungizone I.V. by adding 10 mL sterile distilled water. Shake the vial until clear. This results in a colloidal suspension of 5,000 μg/mL. Dilute 1:5 with sterile distilled water to obtain a stock solution of 1,000 μg/mL. This solution may be frozen at -70°C for up to 6 months or held at 4°C for a week.
2. 5-FC: Prepare a stock solution containing 5,000 μg/mL in distilled water and sterilize by filtering through a 0.22 μm filter. This solution may be stored indefinitely at -70°C.
3. KETO: Prepare a stock solution of 5,000 μg/mL in 0.2N HCl. Store frozen at -70°C indefinitely.

Serum-Drug Standards for Bioassays

Each bioassay described will incorporate three or four different serum-drug concentrations for preparing standard curves. All of the bioassays require pooled, normal human sera (TC Serum; Difco Laboratories) as final diluents. Individual lots of such sera must be pretested against the various assay organisms to determine that the sera are free of any inhibitory factors. It is important that all test concentrations of drug produce zones of inhibition in the linear regions of the dose response curves.[101] In preparing dilutions, use a fresh pipette for each step. Standards may be prepared in advance, aliquoted and frozen at -70°C for up to 6 months.

A. AMB Serum Standards: Standard concentrations = 1.0, 0.5, 0.25, 0.125 μg/mL. Thaw AMB stock solution (1000 μg/mL) and dilute in sterile distilled water to 50 μg/mL. Continue to dilute in pooled, normal human serum to 5 μg/mL. Dilute 1:4 in serum to prepare the first standard, 1.0 μg/mL. Then make serial 2-fold dilutions through 3 more steps, using 1 mL volumes of pooled serum as diluent.

B. 5-FC Standards: Standard concentrations = 160, 80, 40 μg/mL. Dilute the 5,000 μg/mL stock 1:5 in sterile distilled water to give a 1,000 μg/mL solution. Prepare a 320 μg/mL solution in pooled, normal human serum by diluting 1:3.2. Make 2-fold serial dilutions through 3 more dilutions, using 1 mL serum volumes.

C. KETO Standards: Standard concentrations = 20, 5, 2, 0.5 μg/mL. Dilute the 5,000 μg/mL stock 1:10 in sterile distilled water to prepare a 500 μg/mL solution. Further dilute 1:5 in pooled, normal human serum to give a 100 μg/mL solution. Prepare a dilution series in serum to achieve the 4 standards.

Preparation of Inoculum of Test Organisms

A. *Paecilomyces variotii* ATCC 36257 is maintained on antimicrobic-free Sabouraud dextrose agar slants by weekly subculture. This organism is used in the bioassay of AMB. Use only fully mature, tan-colored colonies to prepare the inoculum; four days is usually a sufficient growth period. Collect the mature conidia by gently washing the surface of the mycelial growth with a few mL sterile distilled water. Remove the conidial-water suspension and allow to stand until the larger mycelial fragments settle to the bottom of the tube. Prepare a suspension equal in turbidity to that of a No.2 McFarland standard (65-70% T at 590 nm) in sterile distilled water.

When it is necessary to assay AMB in the presence of 5-FC, substitute *Chrysosporium pruinosum* ATCC 36374 for *Paecilomyces variotii*. This particular isolate of *C. pruinosum* exhibits high resistance to 5-FC. Add the antagonist cytosine at a final concentration of 10 μg/mL to the medium to prevent 5-FC from interfering with the determination of AMB content.[75]

B. *Candida pseudotropicalis* ATCC 46764 or 28838 is maintained by weekly passage on antimicrobic-free Sabouraud dextrose agar slants. This yeast is used in the bioassay of 5-FC and KETO. Suspend growth from a 2-7 day old slant in Yeast Nitrogen Base-Glucose broth (Difco Laboratories) to match the turbidity of a 0.5 McFarland standard (88-91% T at 590 nm). Vortex 1 min, then incubate at 35°C 4-6 h. Adjust the turbidity with sterile distilled water to that of a No.2 McFarland standard (4×10^6 cells/mL; 65-70% T at 590 nm).

Medium Preparation

A. AMB Assay Medium: Antibiotic Medium M-12 (Difco Laboratories; BBL Microbiology Systems): Prepare 1 L of Antibiotic Medium M-12 in distilled water, heat to boiling and sterilize by autoclaving at 121°C, 15 psi for 15 min. Final pH should be approximately 6.1. Aseptically dispense 30 mL each into sterile 25 x 200 mm round-bottom, screwcapped tubes. Store at 4°C for up to 3 months.

B. 5-FC and KETO Assay Medium: Yeast Nitrogen Base-Glucose Agar (or Bodet's Modified Agar[13]): The test medium is prepared by adding 7 g of Trypticase Peptone (BBL Microbiology Systems), 15 g glucose and 15 g granulated agar (BBL Microbiology Systems) to 1 L distilled water. Final pH is 6.0. Heat to boiling to dissolve ingredients and ensure complete mixing, then autoclave and aseptically dispense in 25 mL amounts into 25 x 200 mm screwcapped tubes. Store at 4°C for up to 3 months.

Preparation and Inoculation of Assay Plates

1. AMB: Melt the appropriate number of tubes of M-12 agar in a boiling water bath. Allow the tubes to cool to 48°C and add 0.5 mL of the No.2 McFarland conidia/spore suspension to each tube. Mix gently by inversion and pour each tube into a 150 x 15 mm round plastic dish on a level surface. Allow the agar to solidify at room temperature. The agar surface should be free of moisture. Use two seeded agar plates for each assay. Use a 5 mm cork borer to prepare 11 wells around the periphery of a plate. Fill duplicate wells with 50 μL of each of the 4 standards, the patient's serum and serum control. Incubate plates at room temperature for 24-48 h until clear zones appear around the wells.
2. 5-FC and KETO: Melt the appropriate number of YNB-glucose agar (or Bodet's Modified Agar) tubes in a boiling water bath. Cool to 48°C and add 0.5 mL of the No.2 McFarland standard fungal suspension to each tube. Pour plates as with AMB medium. Use two seeded plates for each assay. Use a 5 mm cork borer to prepare 9 wells around the periphery for 5-FC and 11 wells for KETO. Fill duplicate wells with 10 μL of each of the standards, the patient's serum and a serum control. Incubate the plates overnight (15-20 h) at 35°C.

Reading the Plates

Visually inspect the bioassay plates for zones of inhibition around the standards and unknowns. There should be a clearly defined circular area of inhibition around each standard well. Carefully measure the diameter of each zone to the nearest 0.1 mm with a metric caliper-micrometer. Determine the mean value for each standard and unknown. Plot the standard curve as mean zone diameters (in mm) on the abscissa versus drug concentrations (in μg/mL) on the ordinate of 2 or 3 cycle semilogarithmic paper. Chose and plot the "best line of fit." To determine the level of antimycotic agent in the unknown specimen, plot the zone size value from the abscissa on the dose response curve. Read directly across the ordinate to find the drug level in μg/mL.

Quality Control

Include a control fluid of known drug concentration with each test. The control value must be within 15% of the known value. The control fluid concentrations and the limits of their 15% values are:

AMB = 0.25 μg/mL; 0.21-0.29 μg/mL
5-FC = 80 μg/mL; 68-92 μg/mL
KETO = 2 μg/mL; 1.7-2.3 μg/mL

Other Body Fluids

It is possible to determine drug levels in urine, cerebrospinal, joint and other body fluids by using the same protocol. Control fluid of the type being tested must be used as a control. It is desirable to collect and save frozen samples of these fluids for use as the need arises.

CONCLUSIONS AND COMMENTS

Since it appears that the number and importance of fungal diseases will continue to escalate, antifungal chemotherapy and the laboratory evaluation of antifungal antimicrobics represent significant areas of increasing interest in modern medicine. With an increase in the number of antifungal agents, continuing investigations on effective use of antifungals and new insights into their mechanisms of action, the laboratorian will undoubtedly be asked more frequently to provide in vitro assessment of antimycotic activity. Standardization of methodology remains the chief difficulty in obtaining meaningful correlation between in vitro and in vivo data, but this issue is being vigorously addressed.[28,40,84,93,97]

REFERENCES

1. Adam RD, Pacquin ML, Petersen EA, et al. Phaeohyphomycosis caused by the fungal genera *Bipolaris* and *Exserohilum*. A report of 9 cases and review of the literature. Medicine 1986;65:203-17.
2. Ahrens J, Graybill JR, Craven PC, Taylor RL. Treatment of experimental murine candidiasis with liposome-associated amphotericin B. Sabouraudia 1984;22:163-6.
3. Artis WM, Odle BM, Jones HE. Griseofulvin-resistant dermatophytosis correlates with in vitro resistance. Arch Dermatol 1981;117:16-19.
4. Auger P, Dumas C, Joly J. A study of 666 strains of *Candida albicans*: correlation between serotype and susceptibility to 5-fluorocytosine. J Infect Dis 1979;139:590-4.
5. Aytoun RSC, Campbell AH, Napier EJ, Seiler DAL. Mycological aspects of action of griseofulvin against dermatophytes. Arch Dermatol 1960;81:650-6.
6. Bach PR. Quantitative extraction of amphotericin B from serum and its determination by high-pressure liquid chromatography. Antimicrob Agents Chemother 1984;26:314-17.
7. Bannatyne RM, Cheung R. Discrepant results of amphotericin B assays on fresh versus frozen serum samples. Antimicrob Agents Chemother 1977;12:550.
8. Bannatyne RM, Cheung R, Devlin HR. Microassay for amphotericin B. Antimicrob Agents Chemother 1977;11:44-6.
9. Beggs WH. Comparison of miconazole- and ketoconazole-induced release of K^+ from *Candida* species. J Antimicrob Chemother 1983;11:381-3.

10. Bennett JE, Dismukes WE, Duma RJ, et al. A comparison of amphotericin B alone and in combination with flucytosine in the treatment of cryptococcal meningitis. N Engl J Med 1979;301:126-31.
11. Blank H, Taplin D, Roth FJ. Electron microscopic observations of the effects of griseofulvin on dermatophytes. Arch Dermatol 1960;81:667-80.
12. Blatchford NR, Emanuel MB, Cauwenbergh G. Ketoconazole resistance. Lancet 1982;2:770-1.
13. Bodet CA III, Jorgensen JH, Drutz DJ. Simplified bioassay method for measurement of flucytosine or ketoconazole. J Clin Microbiol 1985;22:157-60.
14. Borgers M. Mechanisms of action of antifungal drugs, with special reference to the imidazole derivatives. Rev Infect Dis 1980;2:520-34.
15. Borgers M, Waldron HA. The action of ketoconazole on fungi. Clin Res Rev 1981;1:165-71.
16. British Society for Mycopathology. Report of a working group. Laboratory methods for flucytosine (5-fluorocytosine). J Antimicrob Chemother 1984;14:1-8.
17. Buchel KH, Draber W, Regel E, Plempel M. Synthesis and properties of clotrimazole and other antimycotic 1-triphenylmethyl imidazoles. Drugs Made in Germany 1972;15: 79-94.
18. Calhoun DL, Galgiani JN. Analysis of pH and buffer effects on flucytosine activity in broth dilution susceptibility testing of *Candida albicans* in two synthetic media. Antimicrob Agents Chemother 1984;26:364-7.
19. Calhoun DDL, Roberts GD, Galgiani JN, et al. Results of a survey of antifungal susceptibility tests in the United States and interlaboratory comparison of broth dilution testing of flucytosine and amphotericin B. J Clin Microbiol 1986;23:298-301.
20. Carlson JA, Mann HJ, Canafax DM. Effect of pH on disintegration and dissolution of ketoconazole tablets. Am Hosp Pharm 1983;40:1334-6.
21. Church JA, Neff DN, Marbut C. Resistance to ketoconazole. Lancet 1982;2:211.
22. DeFelice R, Johnson DG, Galgiani JN. Gynecomastia with ketoconazole. Antimicrob Agents Chemother 1981;19:1073-4.
23. DeNollin S, Borgers M. The ultrastructure of *Candida albicans* after in vitro treatment with miconazole. Sabouraudia 1974;12:341-51.
24. Dick J, Merz WG, Saral R. Incidence of polyene-resistant yeasts recovered from clinical specimens. Antimicrob Agents Chemother 1980;18:158-63.
25. Dismukes WE, Bradsher R, Girard W, et al. Itraconazole therapy for blastomycosis and histoplasmosis. [Abstract 798.] In: Abstracts of the 26th Interscience Conference on Antimicrobial Agents and Chemotherapy. Washington, DC: American Society for Microbiology, 1986:242.
26. Dismukes WE, Stamm AM, Graybill JR, et al. Treatment of systemic mycoses with ketoconazole: emphasis on toxicity and clinical response in 52 patients. Ann Intern Med 1983;98:13-20.
27. Dixon DM, Shadomy S, Shadomy HJ, Espinel-Ingroff A, Kerking TM. Comparison of the in vitro antifungal activities of miconazole and a new imidazole, R 41,400. J Infect Dis 1978;138:245-8.
28. Doern GV, Tubert TA, Chapin K, Rinaldi MG. Effect of medium composition on results of macrobroth dilution antifungal susceptibility testing of yeasts. J Clin Microbiol 1986;24:507-11.
29. Drouhet E, Barcale T, Bastide J, et al. Standardisation de l'antibiogramme antifongique rapport du groupe d'etudes de la Societe Francaise de Mycologie Medicale. Bull Soc Fr Med Mycol 1984;10:131-4.
30. Drouhet E, Mercier-Soucy L, Montplaisir S. Sensibilité et resistance des levures pathogènes aux 5-fluoropyrimidines. I. Relation entre les phenoypes de resistance à la 5-fluorocytosine, le sèrotype de *Candida albicans* et l'écologie de différentes espèces de *Candida* d'origine humaine. Ann Microbiol (Instit Pasteur) 1975;126B:25-39.
31. El-Nakeeb MA, Lampen JO. Uptake of ^{3}H-griseofulvin by microorganisms and its correlation with sensitivity to griseofulvin. J Gen Microbiol 1965;39:285-93.

32. El-Nakeeb MA, McLellan WL, Lampen JO. Antibiotic action of griseofulvin on dermatophytes. J Bacteriol 1965;89:557-63.
33. Espinel-Ingroff A, Shadomy S, Gebhart RJ. In vitro studies with R 51,211 (Itraconazole). Antimicrob Agents Chemother 1984;26:5-9.
34. Evans G, White NH. Effect of the antibiotics radicicolin and griseofulvin on the fine structure of fungi. J Exp Biol 1967;18:465-70.
35. Fredriksson T. Treatment of dermatomycoses with topical tioconazole and miconazole. Dermatol 1983;166(Suppl 1):14-19.
36. Fromtling RA. Imidazoles as medically important antifungal agents: an overview. Drugs of Today 1984;20:325-49.
37. Galgiani JN. Antifungal susceptibility testing. Recent findings and experience. Antimicrobic Newsletter 1986;3:17-22.
38. Galgiani JN, Drutz DJ, Feingold DS, et al. Committee report. Antifungal susceptibility testing. Villanova, PA: National Committee for Clinical Laboratory Standards, 1985;5:433-81.
39. Galgiani JN, VanWyck DB. Ornithyl amphotericin methyl ester treatment of experimental candidiasis in rats. Antimicrob Agents Chemother 1984;26:108-9.
40. Galgiani JN, Yturralde CA, Dugger KO. Susceptibility of *Candida albicans* to flucytosine when tested in different formulations of yeast nitrogen base broth. Diagn Microbiol Infect Dis 1986;5:273-6.
41. Ganer A, Arathoon E, Stevens D. Initial clinical experience with itraconazole, an oral triazole antifungal. [Abstract 799.] In: Abstracts of the 26th Interscience Conference on Antimicrobial Agents and Chemotherapy. Washington, DC: American Society for Microbiology, 1986:242.
42. Granich GG, Kobayashi GS, Krogstad DJ. Sensitive high-pressure liquid chromatographic assay for amphoericin B which incorporates an internal standard. Antimicrob Agents Chemother 1986;29:584-8.
43. Graybill JR, Craven PC, Taylor RL, Williams DM, Magee WE. Treatment of murine cryptococcosis with liposome-associated amphotericin B. J Infect Dis 1982;145:748-52.
44. Graybill JR, Stevens DA, Galgiani JN, et al. Itraconazole treatment of coccidioidomycosis. [Abstract 788.] In: Abstracts of the 26th Interscience Conference on Antimicrobial Agents and Chemotherapy. Washington, DC: American Society for Microbiology, 1986:240.
45. Haller I. Modern aspects of testing azole antifungals. Postgrad Med J 1979;55:681-2.
46. Heel RC, Brogden RN, Carmine A, Morley PA, Speight TM, Avery GS. Ketoconazole: a review of its therapeutic efficacy in superficial and systemic fungal infections. Drugs 1982;23:1-36.
47. Heel RC, Brogden RN, Parkes GE, Speight TM, Avery GS. Miconazole: a preliminary review of its therapeutic efficacy in systemic fungal infections. Drugs 1980;19:7-30.
48. Heeres J, Backx LJJ, Van Cutsem J. Antimycotic azoles. 7. Synthesis and antifungal properties of a series of novel triazole-3-ones. J Med Chem 1984;27:894-900.
49. Hoeprich PD. Chemotheapy of systemic fungal diseases. Ann Rev Pharmacol Toxicol 1978;18:205-31.
50. Hoeprich PD, Finn PD. Obfuscation of activity of antifungal antimicrobics by culture media. J Infect Dis 1972;126:353-61.
51. Holt RJ, Azmi A. Miconazole-resistant *Candida*. Lancet 1978;1:50-1.
52. Horsburgh CR, Kirkpatrick CH, Teutsch CB. Ketoconazole and the liver. Lancet 1982;1:860.
53. Horsburgh CR, Kirkpatrick CH. Long-term therapy of chronic mucocutaneous candidiasis with ketoconazole: experience with twenty-one patients. Am J Med 1983;74:23-9.
54. Huber FM, Gottlieb D. The mechanisms of action of griseofulvin. Can J Microbiol 1968;14:111-18.
55. Hume AL, Kerkering TM. Ketoconazole (Nizoral, Janssen Pharmaceutica, Inc.). Drug Intell Clin Pharm 1983;17:169-73.

56. Hutlet J. Evaluation of two methods of in vitro susceptibility testing of *Candida albicans* against 5-fluorocytosine. Can J Med Technol 1976;38:169-72.
57. Ichise K, Tanio T, Saji I, Okua T. Activity of SM-4470, a new imidazole derivative, against experimental fungal infections. Antimicrob Agents Chemother 1986;30:366-9.
58. Iwata K, Kanda Y, Yamaguchi H, Osumi M. Electron microscope studies on the mechanism of action of clotrimazole on *Candida albicans*. Sabouraudia 1973;11: 2205-9.
59. Iwata K, Yamaguchi H, Hiratani T. Mode of action of clotrimazole. Sabouraudia 1973;11:158-66.
60. Janssen PAJ, Symoens JE. Hepatic reactions during ketoconazole treatment. Am J Med 1983;74(1B):80-5.
61. Johnson EM, Richardson MD, Warnock DW. Effect of imidazole antifungals on the development of germ tubes by strains of *Candida albicans*. J Antimicrob Chemother 1983;12:303-16.
62. Jund R, Lacroute F. Genetic and physiological aspects of resistance to 5-fluoropyrimidines in *Saccharomyces cerevisiae*. J Bacteriol 1970;102:607-15.
63. Kaspar RL, Drutz DJ. Rapid, simple bioassay for 5-fluorocytosine in the presence of amphotericin B. Antimicrob Agents Chemother 1975;7:462-5.
64. Kobayashi GS, Medoff G. Antifungal agents: recent developments. Ann Rev Microbiol 1977;31:291-308.
65. Kobayashi GS, Travis S, Medoff G. Comparison of the in vitro and in vivo activity of the bis-triazole derivative UK 49,858 with that of amphotericin B against *Histoplasma capsulatum*. Antimicrob Agents Chemother 1986;29:660-2.
66. Kotler-Brajtburg J, Medoff G, Schlessinger D, Kobayashi GS. Amphotericin B and filipin effects on L and HeLa cells: dose response. Antimicrob Agents Chemother 1974;11:803-8.
67. Kuroda S, Uno J, Arai T. Target substances of some antifungal agents in the cell membrane. Antimicrob Agents Chemother 1978;13:454-9.
68. Lenhart K. Griseofulvin-resistant mutants in dermatophytes. 2. Physiological and genetic studies. Mykosen 1970;13:139-44.
69. Levine HB. Resistance to ketoconazole. Lancet 1982;2:211.
70. Levine HB, ed. Ketoconazole in the management of fungal disease. Sydney: ADIS Press, 1982:154.
71. Lopez-Berestein G, Hopfer RL, Mehta R, et al. Liposome-encapsulated amphotericin B for treatment of disseminated candidiasis in neutropenic mice. J Infect Dis 1984;150: 278-83.
72. Lopez-Berestein G, Mehta R, Hopfer RL, et al. Treatment and prophylaxis of disseminated infection due to *Candida albicans* in mice with liposome-encapsulated amphotericin B. J Infect Dis 1983;147:939-45.
73. Malawista SE, Sato H, Bensch KG. Vinblastine and griseofulvin reversibly disrupt the living mitotic spindle. Science 1968;160:770-2.
74. McGabe RE, Araujo EG, Remington JS. Ketoconazole protects against infection with *Trypanosoma cruzei* in a murine model. Am J Trop Med Hyg 1983;32:960-2.
75. McGinnis MR. Laboratory handbook of medical mycology. New York: Academic Press, 1980.
76. McGinnis MR, Rinaldi MG. Antifungal drugs: mechanisms of action, drug resistance, susceptibility testing, and assays of activity in biological fluids. In: Lorian V, ed. Antibiotics in laboratory medicine. 2nd ed. New York: Academic Press, 1985:223-81.
77. Medoff G, Brajtburg J, Kobayashi GS. Antifungal agents useful in therapy of systemic fungal infections. Ann Rev Pharmacol Toxicol 1983;23:303-30.
78. Medoff G, Kobayashi GS. Mode of action of antifungal drugs. In: Howard D, ed. Fungi pathogenic for humans and animals. Part B. Pathogenicity and detection I. New York: Marcel Dekker, 1983:325-55.

79. Melcher GP, Rinaldi MG, Drutz DJ, Frey CI. Disseminated mycoses caused by *Trichosporon beigelii*: clinical and mycological aspects. [Abstract 841.] In: Abstracts of the 26th Interscience Conference on Antimicrobial Agents and Chemotherapy. Washington, DC: American Society for Microbiology, 1986;252.
80. Merz WG, Sanford GR. Isolation and characterization of a polyene resistant variant of *Candida tropicalis*. J Clin Microbiol 1979;9:677-80.
81. Normark S, Schonebeck J. In vitro studies of 5-fluorocytosine resistance in *Candida albicans* and *Torulopsis glabrata*. Antimicrob Agents Chemother 1972;2:114-21.
82. Odds FC. Laboratory evaluation of antifungal agents: a comparative study of five imidazole derivatives of clinical importance. J Antimicrob Chemother 1980;6:749-61.
83. Odds FC. Ketoconazole resistance. Lancet 1982;2:771.
84. Odds FC, Abbot AB, Pye G, Troke PF. Improved method for estimation of azole antifungal inhibitory concentrations against *Candida* species, based on azole/antibiotic interactions. J Med Vet Mycol 1986;24:305-11.
85. Pegrem PS Jr, Kerns FT, Wasilauskas BL, Hampton KD, Scharyji M, Burke JG. Successful ketoconazole treatment of protothecosis with ketoconazole-associated hepatotoxicity. Arch Intern Med 1983;143:1802-5.
86. Perfect JR, Savani DV, Durak DT. Comparison of itraconazole and fluconazole in treatment of cryptococcal meningitis and *Candida* pyelonephritis in rabbits. Antimicrob Agents Chemother 1986;29:579-83.
87. Pierce AM, Pierce HD Jr, Unrau AM, Oehlschlager AC. Lipid composition and polyene antibiotic resistance of *Candida albicans* mutants. Can J Biochem 1978;56:135-42.
88. Polak A. 5-Fluorocytosine - current status with special reference to mode of action and drug resistance. Contrib Microbiol Immunol 1977;4:158-67.
89. Polak A. Mode of action of 5-fluorocytosine and 5-fluorouracil in dematiaceous fungi. Sabouraudia 1983;21:15-25.
90. Polak A, Scholer HJ. Mode of action of 5-fluorocytosine and mechanisms of resistance. Chemother 1975;21:113-30.
91. Polak A, Scholer HJ. Mode of action of 5-fluorocytosine. Rev Inst Pasteur Lyon 1980;13:233-44.
92. Pottage JC, Kessler HA, Goodrich JM, et al. In vitro activity of ketoconazole against herpes simplex virus. Antimicrob Agents Chemother 1986;30:215-19.
93. Radetsky M, Wheeler RC, Row MH, Todd JK. Microtiter broth dilution method for yeast susceptibility testing with validation by clinical outcome. J Clin Microbiol 1986;24:600-6.
94. Richardson K, Brammer KW, Marriott MS, Troke PF. Activity of UK 49,858, a bis-triazole derivative, against experimental infections with *Candida albicans* and *Trichophyton mentagrophytes*. Antimicrob Agents Chemother 1985;27:832-5.
95. Rinaldi MG. Use of potato flakes agar in clinical mycology. J Clin Microbiol 1982;15:1159-60.
96. Rippon JW. Medical mycology. The pathogenic fungi and the pathogenic actinomycetes. 2nd ed. Philadelphia: WB Saunders, 1982:730-1.
97. Rogers TE, Galgiani JN. Activity of fluconazole (UK 49,858) and ketoconazole against *Candida albicans* in vitro and in vivo. Antimicrob Agents Chemother 1986;30:418-22.
98. Roobol A, Gull K, Pogson CI. Evidence that griseofulvin binds to microtubule associated protein. FEBS Letter 1977;75:149-53.
99. Ryley JF, Wilson RG, Barrett-Bee KJ. Azole resistance in *Candida albicans*. Sabouraudia 1984;22:53-63.
100. Ryley JF, Wilson RG, Gravestock MB, Poyser JP. Experimental approaches to antifungal chemotherapy. Adv Pharmacol Chemother 1981;18:49-176.
101. Shadomy S, Espinel-Ingroff A. Methods for bioassay of fungal agents in biologic fluids. In: Laskin AI, Lechevalier HA, eds. CRC handbook of microbiology. 2nd ed. Boca Raton, FL: CRC Press, 1984:327-37.

102. Shadomy S, Espinel-Ingroff A, Cartwright RY. Laboratory studies with antifungal agents: susceptibility tests and bioassays. In: Lennette EH, Balows A, Hausler WJ Jr, Shadomy HJ, eds. Manual of clinical microbiology. 4th ed. Washington, DC: American Society for Microbiology, 1985:991-9.
103. Shadomy S, Kirchoff CB, Espinel-Ingroff A. In vitro activity of 5-fluorocytosine against *Candida* and *Torulopsis* species. Antimicrob Agents Chemother 1973;3:9-14.
104. Shadomy S, McCay JA, Schwartz SI. Bioassay for hamycin and amphotericin B in serum and other biological fluids. Appl Microbiol 1969;17:497-503.
105. Sokol-Anderson ML, Brajtburg J, Medoff G. Sensitivity of *Candida albicans* to amphotericin B administered as a single or fractionated doses. Antimicrob Agents Chemother 1986;29:701-2.
106. Stevens DA. Miconazole in the treatment of coccidioidomycosis. Drugs 1983;26:347-54.
107. Stiller RL, Bennett JE, Scholer HJ, Wall M, Polak A, Stevens DA. Susceptibility to 5-fluorocytosine and prevalence of serotype in 402 *Candida albicans* isolates from the United States. Antimicrob Agents Chemother 1982;22:482-7.
108. Sud IJ, Feingold DS. Heterogeneity of action mechanisms among antimycotic imidazoles. Antimicrob Agents Chemother 1981;20:71-4.
109. Sud IJ, Feingold DS. Action of antifungal imidazoles on *Staphylococcus aureus*. Antimicrob Agents Chemother 1982;22:470-4.
110. Swamy KHS, Sirsi M, Rao GR. Studies on the mechanism of action of miconazole: effect of miconazole on respiration and cell permeability of *Candida albicans*. Antimicrob Agents Chemother 1974;5:420-5.
111. Swamy KHS, Sirsi M, Rao GR. Studies on the mechanism of action of miconazole: II. Interaction of miconazole with mammalian erythrocytes. Biochem Pharmacol 1976;25:1145-50.
112. Tkach JR, Rinaldi MG. Severe hepatitis associated with ketoconazole therapy for chronic mucocutaneous candidiasis. Cutis 1982;29:482-4.
113. Tremblay C, Barza M, Fiore C, Szoka F. Efficacy of liposome-intercalated amphotericin B in the treatment of systemic candidiasis in mice. Antimicrob Agents Chemother 1984;26:170-3.
114. Utz C, Shadomy S. A diffusion disc susceptibility test for 5-fluorocytosine. Chemother 1973;3:9-14.
115. Van Cutsem J, Van Gerven F, Van de Ven MA, Borgers M, Janssen PAJ. Itraconazole, a new triazole that is orally active in aspergillosis. Antimicrob Agents Chemother 1984;26:527-34.
116. Vanden Bossche H. Biochemical effects of miconazole on fungi: 1. Effects on the uptake and/or utilization of purines, pyrimidines, nucleosides, amino acids and glucose by *Candida albicans*. Biochem Pharmacol 1974;23:887-99.
117. Vanden Bossche H, Willemsens G. Effects of the antimycotics, miconazole and ketoconazole, on cytochrome P-450 in yeast microsomes and rat liver microsomes. Arch Int Physiol Biochem 1982;90:218-19.
118. Vanden Bossche H, Willemsens G, Cools W, Cornelisen F, Lauwers WF, Van Cutsem JM. In vitro and in vivo effects of the antimycotic drug ketoconazole on sterol synthesis. Antimicrob Agents Chemother 1980;17:922-8.
119. Vanden Bossche H, Willemsens G, Van Cutsem JM. The action of miconazole on the growth of *Candida albicans*. Sabouraudia 1975;13:63-73.
120. Waldorf AR, Polak A. Mechanisms of action of 5-fluorocytosine. Antimicrob Agents Chemother 1983;32:79-85.
121. Warnock DW, Johnson EM, Richardson MD, Vickers CFH. Modified response to ketoconazole of *Candida albicans* from a treatment failure. Lancet 1983;1:642-3.
122. Weber K, Wehland J, Herzog W. Griseofulvin interacts with microtubules both in vivo and in vitro. J Mol Biol 1976;102:817-29.

123. Winn RE, Hadfield TL, Smith MB, Rinaldi MG, Guerra C. *Candida lusitaniae* mycoses. [Abstract F35.] In: Abstracts of the 86th Annual Meeting of the American Society for Microbiology. Washington, DC: American Society for Microbiology, 1986:403.
124. Woods RA, Bard AM, Jackson IE, Drutz DJ. Resistance to polyene antibiotics and corrected sterol changes in two isolates of *Candida tropicalis* from a patient with an amphotericin B-resistant funguria. J Infect Dis 1974;129:53-8.

CHAPTER 10

GLOSSARY FOR MEDICAL MYCOLOGY

Barbara E. Robinson, Ph.D.

ACHLOROPHYLLOUS: lacking chlorophyll.

ACROGENOUS: developing at the apex of a conidiophore.

ACROPETAL: produced successively toward the apex (in a chain of conidia, the youngest is at the apex).

ACTINOMYCETES: a common name for the group of diphtheroid-like to filamentous bacteria classified within the order *Actinomycetales*.

ACTINOMYCOTIC: describing disease caused by members of the actinomycetales.

AERIAL HYPHAE: hyphae or filaments growing above the medium or substrate.

AGGLUTINATION: aggregation or clumping of particulate matter, as of particulate antigens by antibodies specific for one or more surface sites.

AGGLUTININ: antibody to an antigen on the surface of particles (such as microbes, erythrocytes or latex beads); capable of causing agglutination.

ALLANTOID: slightly curved or kidney-shaped.

ALOPECIA: loss of hair.

ANAMORPH: asexual portion of the fungal life cycle characterized by mitotic division of somatic nuclei; a polymorphic fungus may have more than one anamorph.

ANERGY: inability to respond to an antigen in the normal manner.

ANNELLATION: a ring of cell wall material remaining at the apex of an annellide after the release of an annelloconidium.

ANNELLIDE: a conidiogenous cell in which blastic (enteroblastic) conidia are formed in a basipetal manner, typically resulting in the elongation of the cell with the production of each conidium; with ring-like scars (annellation) on the outer surface near the conidiogenous locus; with the formation of each new conidium, the annellide must grow through the scar left by the previous conidium.

ANNELLOCONIDIUM (pl. annelloconidia): a conidium produced from an annellide.

ANTHROPOPHILIC: infecting humans, either exclusively or principally.

ANTIBODY: serum immunoglobulins induced by and reacting with specific antigen.

ANTIFUNGAL AGENTS: antimicrobials; compounds that inhibit the in vitro and/or in vivo development of fungi.

ANTIFUNGAL SUSCEPTIBILITY TESTING: evaluation of the activity of antifungal agents in vitro or in vivo.

ANTIGEN: a determinant group capable of inducing a detectable immune response by the formation of specific antibodies or by lymphocyte activation.

ANTISERUM: serum containing antibodies.

APICULATE: having a short projection at one or both apices.

ARTHROCONIDIUM (pl. arthroconidia): thallic conidium formed by fission through a double septum (fragmentation) or by lysis of a disjunctor cell.

ARTHUS REACTION: type III hypersensitivity reaction that follows skin testing of a sensitized subject with an antigen; type of necrotic reaction mediated by immune complexes, activation of complement and local inflammation.

ASCOCARP: complex, protective structure within which asci and ascospores develop.

ASCOMYCETOUS: having characteristics of the class Ascomycotina.

ASCOMYCOTINA (ASCOMYCETES): subdivision of higher fungi in which sexual reproduction results in formation of ascospores within an ascus.

ASCOSPORE: one of usually eight sexual, haploid spores formed within a round or elongated sac (ascus) as a result of nuclear fusion and meiosis; may be uni- or multicellular.

ASCUS (pl. asci): sac-like structure in which ascospores are formed.

ASEPTATE: having no cross-walls (septa).

ASSIMILATION: ability to use a carbon or nitrogen source for growth with oxygen as the final electron acceptor; incorporation or conversion of nutrients in culture as evidenced by growth.

ASTEROID BODY: eosinophilic antigen-antibody complex formed around a cell of *Sporothrix schenckii*; stellate in shape and $\geqslant 10$ μm in diameter.

ATOPY: predisposition to IgE-mediated allergies.

AUTOTROPHIC: capable of producing primary source of energy from light or inorganic compounds.

AUXANOGRAPHIC TECHNIQUE: method for determining utilization of carbon or nitrogen sources by placing the substrate onto the surface of a basal agar medium seeded with a test organism.

BALLISTOCONIDIUM (pl. ballistoconidia): a forcibly discharged conidium.

BASIDIOMYCETOUS: having characteristics of members of the Basidiomycotina.

BASIDIOMYCOTINA (BASIDIOMYCETES): subdivision of higher fungi in which sexual reproduction results in formation of basidiospores on a basidium.

BASIDIOSPORE: sexual spore formed on the outside of the basidium following karyogamy and meiosis.

BASIDIUM (pl. basidia): a specialized cell, characteristic of the Basidiomycotina, upon which basidiospores are formed.

BASIPETAL: produced successively from the base (in a chain of conidia, the youngest is at the base).

BINOMIAL: name of a taxon composed of a generic designation and a species epiphet.

BIOASSAY: test designed to assess the biological activity of an agent in biological fluids such as serum or cerebrospinal fluid, as for assay of antifungal agents.

BISERIATE: with reference to *Aspergillus* spp., two layers of cells borne upon the vesicle; uppermost are the phialides, below are the supportive metulae.

BLASTIC: one of two basic modes of conidium development in which the conidium initial enlarges as de novo growth and then differentiates from the parent cell by a septum.

BLASTOCONIDIUM (pl. blastoconidia): an asexual, mitotically-derived conidium produced solitarily, synchronously or in acrogenous chains; typically released by fission of a double septum.

BULBOUS: bulb-like; enlarged at the base.

BULLA (pl. bullae): a blister.

CANDIDIASIS: disease in man or animals caused by *Candida* spp.

CANDIDOSIS: see candidiasis.

CARRY-OVER: ability of an organism in culture to carry nutrients intra- or extracellularly while continuing to divide for a few generations; causes appearance of background growth when continued growth is absent.

CAROTENOID: having orange, red or yellow carotene pigments(s).

CELL-MEDIATED IMMUNITY: immune responses mediated by T-lymphocytes and phagocytes; cellular immunity.

CHANDELIER: structure resembling antlers or candelabra; formed by repeated terminal branching of filaments in cultures of *Trichophyton schoenleinii*; often referred to as favic chandelier.

CHEMICAL ASSAY: relatively sensitive, specific and rapid chemical method, such as gas-liquid chromatography or fluorometry, used as a quantitative assay, as for assay of antifungal agents in biological fluids.

CHEMOTAXONOMIC: describes an approach to identification using the biochemical characteristics of a fungus as a means of taxonomic classification.

CHITIN: major polysaccharide component of the fungal cell wall, consisting of β-1,4-linked N-acetyl-D-glucosamine residues.

CHLAMYDOCONIDIUM: thick-walled cell as in *Fusarium* spp. that functions as a reproductive body or as a survival spore under adverse conditions.

CHLAMYDOSPORE: thick-walled, asexual, reproductive propagule of the Zygomycetes released by lysis of the cell wall; a thick-walled, nonreproductive storage vesicle of *Candida albicans*.

CLASSICAL COMPLEMENT PATHWAY: series of antigen-antibody interactions leading to activation of complement enzymes.

CLAVATE: club-shaped.

CLEISTOTHECIUM (pl. cleistothecia): round, closed (cleisto-) fruiting body (ascocarp) of an ascomycete, within which asci and ascospores develop.

COLLARETTE: small collar of cell-wall material at the tip of a phialide remaining from the outer cell wall of the first formed conidium.

COLUMELLA (pl. columellae): sterile, dome-like structure at the tip of a sporangiophore within a sporocarp or a sporangium.

COMMENSALISM: form of symbiosis in which one organism benefits by association with another organism which is neither benefited nor harmed by the association.

COMPLEMENT: system of serum proteins which when activated mediate antigen-antibody reactions to cause cell lysis.

COMPLEMENT FIXATION: serologic assay which detects activation of the classical complement pathway, i.e., the binding of complement by an antigen-antibody complex.

COMPLEMENT PATHWAY, CLASSICAL: see classical complement pathway.

CONIDIOGENOUS: giving rise to conidia.

CONIDIOPHORE: specialized hypha that supports the conidial apparatus and/or bears conidia.

CONIDIUM (pl. conidia): asexual, exogenous propagule that forms in any manner other than cytoplasmic cleavage, free-cell formation or conjugation.

CONSPECIFICITY: state (degree) of the same species.

COREMIUM (pl. coremia): a loosely bound synnema.

COUNTERIMMUNOELECTROPHORESIS: movement of an antigen and an antibody toward each other in an electrophoretic field to promote their precipitation.

C-REACTIVE PROTEIN: serum protein often elevated during inflammation.

CROSSREACTION: reaction of an antibody with an antigen other than the antibody-inducing antigen.

CRYPTOCOCCOSIS: disease in animals or humans caused by *Cryptococcus* spp.

CYCLOHEXIMIDE: antifungal agent produce by species of *Streptomyces*; actidione.

CYLINDROFUSIFORM: essentially cylindrical but with a slight tapering at both ends.

DACRYOCYSTITIS: inflammation of the lacrimal sac.

DELAYED HYPERSENSITIVITY: type IV cell-mediated hypersensitivity reaction mediated by T-lymphocytes that develops slowly with a cellular infiltrate and edema.

DEMATIACEOUS: describes a member of the family Dematiaceae having pigmented conidia and/or conidiophores that are generally olivaceous, dark-brown or black.

DENTICLE: small tooth-like projection that may bear a spore.

DERMATOMYCOSIS: infection of the skin, hair or nails by a fungus.

DERMATOPHYTE: plant (-phyte) capable of infecting the skin (dermato-); generally applied to fungi which exclusively infect the skin, hair and nails.

DERMATOPHYTID: allergic pustular eruption induced by and usually distant from a primary infection by a dermatophyte.

DERMATOPHYTOSIS: term applied to infection by a generally recognized group of fungi that infect the skin, hair or nails; ringworm.

DEUTEROMYCOTINA (DEUTEROMYCYETES): Fungi Imperfecti; subdivision of fungi which reproduce by asexual means only.

DICHOTOMOUS: hyphae which branch into two branches somewhat equal in size.

DIMORPHIC: having two different forms.

DISARTICULATE: to separate at the septa.

DISCOID: disc-like, i.e., solid, round and flat (or only slightly raised).

DISJUNCTOR CELL: empty cell that fragments or, more commonly, lyses, releasing the conidium.

ECHINULATE: covered with small spines; prickly.

ECTOTHRIX: outside (ecto-) the hair (-thrix); refers to the location of arthroconidia on infected hairs.

ELLIPTICAL: as opposed to globose, noticeably longer along one axis.

ENDOCARDITIS: inflammation of the lining of the heart and its valves.

ENDOGENOUS: originating from within.

ENDOPHTHALMITIS: inflammation of the internal structures of the tissues of the eyeball.

ENDOSPORE: spore produced within a protective structure.

ENDOTHRIX: inside (endo-) the hair (-thrix); refers to the location of arthroconidia in infected hairs.

EN GRAPPE: in a grape-like cluster; refers to arrangement of microconidia.

ENTEROBLASTIC: blastic conidium having walls formed from the inner cell-wall layers of the conidiogenous cell.

EN THYRSE: arranged singly; refers to microconidia along a hypha.

ENTIRE: not torn; having no teeth.

ENZYME IMMUNOASSAY: immunologic test for detection of antigen or antibody using an enzyme-substrate system to detect antigen-antibody binding.

EPONYCHIAL: on top of (epi-) the nail (-onych-).

EUMYCOTIC: refers to a disease caused by true fungi.

EVANESCENT: having a short existance; fugacious.

EXOGENOUS: originating from without or due to external causes.

EXUDATE: oozed matter, as at the edge of a lesion or colony.

FARMER'S LUNG: restrictive, allergic bronchopulmonary disease caused by several true fungi and some thermophilic actinomycetes.

FASICULATE: existing in fasicles or bundles.

FAVIC CHANDELIER: see chandelier.

FAVUS: disease caused by *Trichophyton schoenleinii* with lesions that may be covered with scutula.

FERMENTATION: enzymatically-controlled anaerobic breakdown of a carbohydrate that produces gas (carbon dioxide) and alcohol.

FILAMENTOUS: thread-like.

FLOCCOSE: wooly or cotton-like in texture.

FLUORESCENCE: property of emitting light when exposed to light, the wavelength of the emitted light being longer than the wavelength of the absorbed light.

FLUORESCENT ANTIBODY: immunoglobulin conjugated to a fluorescent dye that can be detected by ultraviolet microscopy or fluorometry.

5-FLUOROCYTOSINE: fluorinated analog of cytosine with antifungal activity, particularly against yeastlike fungi.

FOLLICULITIS: inflammatory reaction in hair follicles.

FOMITE: inanimate object able to harbor pathogenic microorganisms that may be transmitted to susceptible hosts.

FOOT CELL: cell that has a bent, foot-like extension at one end.

FRUITING BODY: large, complex, fungal structure that contains conidia or sexual spores.

FUNGEMIA: presence of a fungus in the bloodstream.

FUSIFORM: spindle-shaped; narrowing toward both ends.

GENICULATE: having a zig-zag appearance resulting from sympodial proliferation.

GEOPHILIC: term applied to fungi whose natural habitat is principally the soil.

GEOTRICHOSIS: disease in humans or animals caused by *Geotrichum* spp.

GERM TUBE: initial hypha that originates from a germinating cell, e.g., blastoconidium, and lacks constrictions at its point of origin; is used to presumptively identify *Candida albicans* (after 2-4 h incubation in serum at 37°C).

GLABROUS: smooth.

GLOBOSE: round.

GRAINS: see granules.

GRANULES: macroscopically visible particles found in purulent exudates from mycetoma lesions; composed of host cell debris and either actinomycotic or eumycotic organisms tightly intertwined in a subglobose mass; may be bound together by a cement-like material.

GRISEOFULVIN: antimicrobial agent having specific activity against dermatophytes and related fungi; obtained from *Penicillium griseofulvum* and other fungi.

GYMNOTHECIUM: fruiting body (ascocarp) composed of a loose network of interwoven hyphae, through which mature ascospores sift out.

HEMOLYSIN: antibody that specifically triggers lysis of erythrocytes in the presence of complement.

HEMOLYSIS: lysis of erythrocytes.

HETEROTHALLIC: requiring the union of either morphologically and/or physiologically dissimilar sexual structures.

HILUM (pl. hila): disjunctor, discernable as a scar, between a conidium and a conidiophore.

HOLOBLASTIC: blastic conidium whose walls are formed from all cell-wall layers of the conidiogenous cell.

HOLOMORPH: complete form of a fungus; term used to describe the combination of the anamorphic and teleomorphic states of the life cycle of one fungus.

HOMOTHALLIC: describes the union of sexual strucures not morphologically or physiologically dissimilar.

HÜLLE CELL: very thick-walled, elliptical cell with a prominent lumen, used primarily in identification of *Aspergillus* spp.

HUMORAL IMMUNITY: immune response mediated by immunoglobulins.

HYALINE: colorless or transparent.

HYALOHYPHOMYCOSIS: heterogenous group of infections caused by opportunistic molds; denoted by the presence of hyaline hyphal elements in tissue.

HYPHA (pl. hyphae): filamentous, tubular, branching vegetative element of a fungus.

IATROGENIC: induced inadvertently by treatment.

IMIDAZOLES: large group of synthetic antifungal agents with the imidazole ring structure, e.g., miconazole, ketoconazole, fluconazole.

IMMEDIATE HYPERSENSITIVITY: type I, IgE-mediated hypersensitivity manifested by rapid erythematous skin test reaction.

IMMUNODIFFUSION: diffusion of soluble antigens and antibodies within a gel; a precipitate is formed when specific reactants meet.

IMMUNOSUPPRESSION: depression of immune responses by chemical, physical or biological means.

INOCULUM: any biological material (e.g., spores, conidia or hyphal fragments) used to seed a culture medium for growth of microorganisms.

INTERCALARY: being formed as an integral part of a vegetative hypha, other than at the tip of the hypha.

INTERTRIGINOUS: occurring on adjacent skin surfaces.

KARYOGAMY: fusion of two nuclei.

KERATIN: scleroprotein containing large amounts of cystine.

KERION: mass of severe inflammatory pustules involving hair follicles, especially on the scalp.

LAGENIFORM: flask-shaped.

LANOSE: wooly.

LENTICULAR: similar in form to a double convex lens.

LOCULE: cavity, especially one in a stroma.

MACROCONIDIUM (pl. macroconidia): larger of two distinctly different-sized conidia produced by the same fungus in a similar way; may be uni- or multicellular.

MACULE: area of perceptible change in skin color, not raised above or depressed below the surrounding skin.

MEGASPORE: large-spore (5-10 μm) ectothrix morphology on hair, as with *Trichophyton verrucosum* infection.

MEIOSIS: nuclear reduction division resulting in haploid nuclei.

MELANIN: colored complex of protein, mostly tan, brown or black, resulting from polymerization of tyrosine and hydroxyphenylalanine by tyrosinase.

MELANOID: dark or black.

MENINGITIS: inflammation of the meninges.

METULA (pl. metulae): conidiophore branch bearing phialides, as in *Penicillium* spp.

MICROAEROPHILIC: growing best under reduced oxygen tension.

MICROCONIDIUM (pl. microconidia): the smaller of two distinctly different-sized conidia produced by the same fungus in a similar way; usually unicellular and spherical, ovoid, pyriform or clavate in shape.

MICROIDES: small (2–3 μm) ectothrix conidia in chains, as with *Trichophyton mentagrophytes* scalp infections.

MINIMUM INHIBITORY CONCENTRATION: lowest concentration of a drug at which no growth of a microorganism is noted.

MINIMUM LETHAL CONCENTRATION: lowest concentration of a drug that kills 99.9% of an original microbial inoculum.

MITOSIS: nuclear division resulting in formation of two new nuclei having the same number of chromosomes as the parent cell.

MURIFORM: refers to a spore divided in more than one plane by intersecting septa.

MYCELIUM: mass of hyphae which constitutes the thallus.

MYCETOMA: chronic infection characterized by nodules, draining sinus tracts and granules in tissues and tract fluids.

NICHE: ecological place or position particularly suitable for an organism.

OBOVATE: inversely ovate, narrower at the base; obovoid.

ONTOGENY: development, as of a conidium.

ONYCHOMYCOSIS: fungal infection of the nail.

OPPORTUNISTIC INFECTION: infection caused by a relatively avirulent organism in an immunocompromised or otherwise debilitated host.

OPSONIN: antibody or other molecule that promotes phagocytosis.

OVATE: narrower at the top like a hen's egg; inversely obovate; ovoid.

PAPULE: small swelling or lesion raised above the surface of the skin.

PARENCHYMA: soft tissue of higher plants.

PARONYCHIA: inflammation of tissue folds surrounding the nails.

PECTINATE HYPHAE: hyphae with unilateral tooth-like projections like those of a comb.

PENULTIMATE: next to the last.

PERCURRENT: growth of a conidiogenous cell through a scar left by the release of a previous terminal conidium.

PERFECT STATE: see teleomorph.

PHAEOHYPHOMYCOSIS: heterogenous group of fungal infections characterized by the presence of dematiaceous septate fungal elements in tissues.

PHENOLOXIDASE: enzyme that oxidizes hydroxyl derivatives or aromatic hydrocarbons; determination of phenoloxidase activity is used to presumptively identify *Cryptococcus neoformans*.

PHENOTYPIC: pertaining to the observable characteristics of an organism.

PHIALIDE: conidiogenous cell in which a blastic (enteroblastic) conidium is produced at the apex in a basipetal manner that typically does not result in an increase in the length of the cell; a collarette may be present.

PHIALOCONIDIUM (pl. phialoconidia): conidium produced within a phialide.

PIEDRA: stony concretion or nodules on infected hair.

PILAR: related to hair.

PITYRIASIS VERSICOLOR: superficial skin disease of humans caused by *Malassezia furfur*.

POLYENES: antifungal agents containing alternating carbon-carbon double bonds, e.g., amphotericin B, nystatin and pimaricin.

POLYMORPHIC: having more than one form (should be used in place of "pleomorphic"); among the Fungi Imperfecti, having more than one anamorph.

POLYPHIALIDE: phialide having more than one conidiogenous locus.

PRECIPITATION: formation of an insoluble complex by appropriate concentrations of soluble reagents, such as antigens and antibodies.

PROPAGULE: individual unit that can give rise to another organism.

PROTEAN: displaying great diversity.

PROTEINACEOUS: relating to or being a protein.

PROTOTROPH: naturally occurring morphologic form of a species.

PROZONE: non-reactive combination of antigen and antibody due to excess antibody; false-negative serologic reaction.

PSEUDOHYPHA (pl. pseudohyphae): false hypha; series of cells that remain attached to each other forming a hypha-like structure.

PSEUDOPARENCHYMA: false parenchyma; morphologically similar to true parenchyma but formed from interwoven hyphae.

PSEUDOSEPTUM: septum-like structure; see pseudohypha.

PYCNIDIUM: multicellular subglobose enclosure in which conidia are formed; distinguished from ascocarp by a lack of sexual structures (asci) and by the small size of its conidia.

PYRIFORM: pear-shaped.

RADIOIMMUNOASSAY: immunologic test using radiolabelled antigen, antibody or other component for assay.

REFLEX BRANCHING: branching backwards, that is, in a direction opposite to that of the elongating filament, forming a reflex angle; characteristic of *Trichophyton soudanense*.

REFRACTORY: resistant to treatment.

RHEUMATOID FACTOR: serum IgM with specificity for IgG; can cause false results in serologic tests.

RHIZOID: hyphal root-like structure that may act as a feeding organ.

RINGWORM: superficial fungus infection; derived from the ancient belief that these infections were caused by worm-like organisms which induced the elevated borders of lesions that were often circular or circinate.

RUDIMENTARY: primitive or elementary.

SAPROBIC: ability to obtain nutrients from dead organic matter.

SATELLITE COLONIES: colonies which appear outside of streak lines in culture, as those resulting from ballistoconidia being discharged into the surrounding medium.

SCLEROTIC CELLS: subglobose, thick-walled fungal cells normally characterized by septa in more than one plane, characteristic of chromoblastomycosis.

SCUTULUM (pl. scutula): concave crust, seen especially in favus, composed essentially of fungal filaments and arthroconidia.

SEPTATE: having cross-walls or septa.

SEPTUM (pl. septa): cross-wall, as in a hypha or conidium.

SPHEROID: round.

SPHERULE: sporangium-like structure with a thick wall; produced by *Coccidioides immitis*.

SPHERULIN: skin test reagent made from spherules of *Coccidioides immitis*.

SPICULE: small projection.

SPORANGIOLUM (pl. sporangiola): small sporangium producing a small number of sporangiospores.

SPORANGIOPHORE: special hypha on which a sporangium develops.

SPORANGIOSPORE: asexual spore borne within a sporangium.

SPORANGIUM (pl. sporangia): sac-like structure in which the contents are cleaved into asexual spores (sporangiospores).

SPORE: propagule formed by sexual means or by an asexual cleavage process.

SPORODOCHIUM: highly compact clusters of conidiogenous cells and conidia scattered upon the vegetative hyphae.

STELLATE: star-like.

STERIGMA (pl. sterigmata): any spore-bearing projection.

STOLON: a runner; horizontal hypha giving rise to other specialized structures, e.g., rhizoids in Zygomycetes.

STROMA: mass of fungal tissue within which fruiting bodies are formed.

SUBSTRATE HYPHAE: hyphae or filaments growing along the surface of agar media.

SUBUNGUAL: under the nail.

SUPPURATION: process of forming or discharging pus.

SUPRAFOLLICULAR: situated above the hair follicle.

SYCOSIS BARBAE: figgy condition of the beard.

SYMPODIAL: describes growth in which, after formation of the initial conidium, the conidiophore continues to elongate by a series of new growth points, each of which arises beneath and lateral to the previous conidium.

SYNANAMORPH: each anamorph in the life cycle of a fungus having two or more anamorphs.

SYNERGISM TESTING: in vitro susceptibility testing involving combinations of reagents, such as antifungal agents, to determine whether the combination is more effective than each individual agent: synergy is defined as a 4-fold or greater reduction of the MIC of one drug when used in combination with a subinhibitory concentration of a second antifungal agent.

SYNNEMA (pl. synnemata): compact group of erect and sometimes fused conidiophores bearing conidia at the apices and/or along the sides of the upper portion.

TAXON: any taxonomic group.

TELEOMORPH: sexual (perfect) form of the life cycle of a fungus, characterized by meiotic division of a zygotic or fused nucleus.

THALLIC: one of two basic modes of conidium development in which the conidium is derived from the conversion of pre-existing hyphal elements and enlargement of the conidium initial occurs only after the initial has been cut off by a septum.

THALLUS: vegetative growth of a fungus; a colony.

THRUSH: oral candidiasis; found primarily in infants and immunocompromised individuals.

TINEA: used generically to refer to a superficial fungal infection.

TINEA VERSICOLOR: see pityriasis versicolor.

TITER: measure of relative concentration, as of antibody and antigen; usually the highest dilution of serum capable of reacting in a serologic test.

TORULOPSOSIS: disease in humans or animals caused by *Torulopsis* spp.

TRICHOSPORONOSIS: disease in humans or animals caused by *Trichosporon* spp.

TURBINATE: in the form of a top.

UNISERIATE: phialides of *Aspergillus* spp. arising directly from the vesicle in one series or row; metulae are lacking.

VERRUCOSE: bearing wart-like bulges.

VERRUCULOSE: delicately verrucose.

VESICLE: swollen apex of a conidiophore; blister-like protuberance on a cell.

VESTIGIAL: rudimentary or degenerate structure.

VULVOVAGINITIS: coincident inflammation of the vulva and vagina.

WOOD'S LIGHT: ultraviolet lamp with a nickel oxide-containing filter that permits the passage of light with a maximum wavelength of about 365 nm; used in dermatology and medical mycology to detect, by their fluorescence, hairs infected with some fungi and lesions of pityriasis versicolor.

YEAST: fungus, member of the Hemiascomycetes, which proliferates asexually by unicellular forms and which, under appropriate conditions, produces sexual spores (ascospores), e.g., *Saccharomyces cerevisiae.*

YEAST CARBON BASE: basal preparation that provides all nutrients, except a nitrogen source, required for growth of yeasts; is used to determine differential utilization of various nitrogen sources in vitro.

YEASTLIKE: morphologically similar to yeasts; vegetative state consists of a unicellular form that reproduces by asexual means.

YEAST NITROGEN BASE: basal preparation which provides all nutrients, except a carbon source, required for growth of yeasts; used to determine differential utilization of various carbon sources in vitro.

ZOOGLEA: bacteria or fungi embedded in a gelatinous matrix.

ZOOPATHOGEN: organism capable of causing disease in humans and/or animals.

ZOOPHILIC: fungi that have a predilection to infect lower animals rather than humans.

ZYGOMYCOTINA (ZYGOMYCETES): subdivision of fungi in which asexual reproduction results in formation of sporangiospores within a sporangium and sexual reproduction results in the formation of zygospores.

ZYGOSPORE: resting sexual spore produced by members of the Zygomycetes by the fusion of two haploid cells.

CHAPTER 11

MEDIA AND STAINS FOR MYCOLOGY

Ira F. Salkin, Ph.D.

MEDIA

Aspergillus Differential Medium[14]

Tryptone	15.0 g
Yeast extract	10.0 g
Ferric citrate	0.5 g
Agar	15.0 g
Distilled water	1000.0 mL

Mix reagents. Add water to bring volume to 1000.0 mL and dissolve reagents. Dispense 7.0 mL portions into sterile screwcapped 20 x 150 mm tubes. Autoclave for 15 min at 15 psi. Allow to solidify in slants.

Assimilation Media, Auxanographic Carbon Test[19]

Carbon assimilation basal agar for auxanographic test:

Yeast nitrogen base	0.67 g
Noble agar	20.0 g
Distilled water	1000.0 mL

While mixing, bring agar and distilled water to boil. Add yeast nitrogen base and mix to dissolve. Dispense 25.0-30.0 mL portions into sterile screwcapped test tubes (25 x 200 mm). Autoclave for 15 min at 15 psi. After solidification, tubes of basal agar are melted and cooled to 48-52°C prior to use. Selection of quality control organisms is dependent upon the number and type of assimilation tests being performed.

Final pH at 25°C = 5.4 ± 0.2.

Carbon disks for auxanographic assimilation:

Carbon source	3.0 g
Distilled water	100.0 mL

Dissolve carbon source in distilled water and sterilize by filtration. Saturate 6 mm filter paper disks in carbon source solution and dry in a sterile petri dish. Quality control of carbon disks is performed using carbon assimilation basal agar for auxanographic test. Selection of quality control organisms is dependent upon the number and type of assimilation tests being performed.

Assimilation Media, Auxanographic Nitrate Test[5,14]

Nitrate disks for auxanographic assimilation:

Potassium nitrate	3.0 g
Distilled water	100.0 mL

Dissolve potassium nitrate in distilled water and autoclave for 15 min at 15 psi. Saturate 6 mm filter paper disks and dry in a sterile petri dish. Quality control is performed in conjunction with peptone disks and nitrogen assimilation basal agar for auxanographic test. *Cryptococcus albidus* and *C. neoformans* can be used as positive and negative quality control organisms, respectively.

Nitrogen assimilation basal agar for auxanographic test:

Yeast carbon base	11.7 g
Noble agar	20.0 g
Distilled water	1000.0 mL

While mixing, bring agar and distilled water to a boil. Add yeast carbon base and mix to dissolve. Dispense 15.0 mL portions into screwcapped test tubes (20 x 150 mm). Autoclave for 15 min at 15 psi. After solidification, tubes of basal agar are melted and cooled to 48-52°C prior to use. *C. albidus* and *C. neoformans* can be used as positive and negative quality control organisms respectively.

Final pH at 25°C $= 5.5 \pm 0.2$.

Peptone disks for auxanographic nitrate assimilation:[14]

Peptone	3.0 g
Distilled water	100.0 mL

Dissolve peptone in distilled water and autoclave for 15 min at 15 psi. Saturate 6 mm filter paper disks and dry in a sterile petri dish. Quality control is performed in conjunction with nitrate disks and nitrogen assimilation basal agar for auxanographic test. *C. albidus* and *C. neoformans* can be used as positive and negative quality control organisms, respectively.

Assimilation Media, Nitrate Broth[14]

10X basal medium for Wickerham broth nitrate assimilation (WBNA):[14]

Yeast carbon base	11.70 g
Potassium nitrate	0.78 g
Distilled water	100.00 mL

10X basal medium for WBNA negative growth control:[14]

Yeast carbon base	11.7 g
Distilled water	100.0 mL

Both basal solutions are separately prepared by dissolving ingredients in distilled water and sterilizing by filtration. Final test and control tubes contain 0.5 mL of the respective basal medium and 4.5 mL of sterile distilled water. *C. albidus* and *C. neoformans* can be used as positive and negative quality control organisms, respectively.

Assimilation Medium, Potassium Nitrate Agar[16]

Potassium nitrate	1.40 g
Yeast carbon base	1.60 g
Bromthymol blue	0.12 g
Noble agar	16.00 g
Distilled water	1000.00 mL

While mixing, bring agar and distilled water to a boil. Add remaining ingredients and mix to dissolve. Adjust to pH 5.9-6.0. Dispense 5.0-7.0 mL portions into screwcapped test tubes (16 x 125 mm). Autoclave for 15 min at 15 psi. Cool at a slant until agar has solidified. *C. albidus* and *C. neoformans* can be used as positive and negative quality control organisms respectively.

Assimilation Media, Wickerham Broth Carbon Test[22]

10X basal medium for Wickerham broth carbon assimilation (WBCA):

Yeast nitrogen base	6.7 g
Distilled water	100.0 mL

Dissolve yeast nitrogen base in distilled water and sterilize by filtration. For testing, add 0.5 mL of 10X WBCA basal medium and 4.5 mL of carbon source solution to a sterile test tube. Add 0.5 mL of 10X WBCA basal medium and 4.5 mL of sterile water to a sterile test tube to serve as a negative control tube for each isolate tested. Note: an alternate method for WBCA: Prepare 10X concentrated carbon source solutions and a 0.74% yeast nitrogen base (YNB) and sterilize by filtration. Dispense 0.5 mL of

10X carbon source solution (e.g., 5.0% glucose) and 4.5 mL of 0.74% YNB. Add 0.5 mL sterile distilled water and 4.5 mL of 0.74% YNB to a sterile test tube to serve as a negative control tube for each isolate tested. Selection of quality control organisms is dependent upon the number and type of assimilation tests being performed.

Carbon sources for Wickerham broth assimilation:

Carbon source,	
monosaccharide hexoses (e.g., glucose)	0.50 g
monosaccharide pentoses (e.g., xylose)	0.60 g
disaccharides (e.g., sucrose)	0.25 g
trisaccharides (e.g., raffinose)	0.34 g
Distilled water	90.00 mL

Dissolve carbon source in distilled water and sterilize by filtration. Carbon sources (e.g., raffinose) which are not readily soluble at room temperature can be dissolved by mixing over low heat. Quality control of carbon source solutions is performed with 10X WBCA basal for liquid assimilation tests. Selection of quality control organisms is dependent upon the number and type of assimilation tests being performed.

Auxanographic Assimilation, Carbon Disks - see Assimilation Media, Auxanographic Carbon Test.

Auxanographic Assimilation, Nitrate Disks - see Assimilation Media, Auxanographic Nitrate Test.

Auxanographic Assimilation, Peptone Disks - see Assimilation Media, Auxanographic Nitrate Test.

Auxanographic Test, Carbon Basal Agar - see Assimilation Media, Auxanographic Carbon Test.

Auxanographic Test, Nitrogen Basal Agar - see Assimilation Media, Auxanographic Nitrate Test.

Blood Agar[5]

Heart infusion base:

Beef heart muscle infusion	500.0 g
Tryptose or Thiotone peptic digest of animal tissue	10.0 g

Sodium chloride	5.0 g
Agar	15.0 g
Distilled/demineralized water	1000.0 mL

Suspend 40.0 g of heart infusion base in 1000.0 mL of water; heat to boiling until agar is dissolved. Autoclave at 121°C for 15 min. Cool to 45-50°C and add 5% sterile, defibrinated sheep blood (vol/vol). Dispense aseptically.

Final pH at 25°C = 7.4 ± 0.2.

Brain Heart Infusion Agar[5,15]

Calf brain, infusion from	200.0 g
Beef heart, infusion from	250.0 g
Proteose or Gelysate (BBL) pancreatic digest of gelatin	10.0 g
Glucose	2.0 g
Sodium chloride	5.0 g
Disodium phosphate	2.5 g
Agar	15.0 g
Distilled/demineralized water	1000.0 mL

Mix ingredients and boil to dissolve. Dispense and autoclave at 121°C for 15 min. Slant and cool.

Final pH at 25°C = 7.4 ± 0.2.

Caffeic Acid Agar[14]

Glucose	5.000 g
Ammonium sulfate	5.000 g
Yeast extract	2.000 g
Potassium phosphate	0.800 g
Magnesium sulfate	0.700 g
Caffeic acid	0.180 g
Ferric citrate	0.002 g
Purified (Noble) agar	20.000 g
Distilled water	1000.000 mL

While mixing, bring agar and distilled water to a boil. Add remaining ingredients and mix until dissolved. Autoclave for 12 min at 15 psi. Dispense 15.0-30.0 mL portions into sterile petri dishes (15 x 100 mm). Cool until agar has solidified. To prepare tubed agar slants, dispense 5.0-7.0 mL portions into screwcapped test tubes (16 x 125 mm) and sterilize by autoclaving as above. Cool tubed media at a slant until agar has solidified. *C. neoformans* and *C. laurentii* can be used as positive and negative quality control organisms, respectively.

Final pH = 5.5. ± 0.2

Carbon Assimilation Basal Agar - see Assimilation Media, Auxanographic Carbon Test.

Carbon Assimilation, Wickerham Broth - see Assimilation Media, Wickerham Broth Carbon Test.

Carbon Disks, Auxanographic Assimilation - see Assimilation Media, Auxanographic Carbon Test.

Carbon Sources, Wickerham Broth Assimilation - see Assimilation Media, Wickerham Broth Carbon Test.

Carbon Sources, Wickerham Broth Fermentation - see Wickerham Broth Fermentation (WBF).

Casein Agar[15]

Skim milk (dried)	75.0 g
Agar	20.0 g
Demineralized water	1000.0 mL

Add the milk to 500.0 mL of water, a little at a time, stirring constantly; do not leave lumps. Autoclave at 115°C for 15 min. Dissolve the agar in 500.0 mL of water and autoclave. Cool the solutions to 60-65°C and pour the agar solution into the skim milk suspension. Mix gently and pour 20.0 mL per 100 mm petri plate or approximately 5.0 mL per quadrant of quadrant plate while hot.

Converse Liquid Medium (Levine Modification)[14]

Ammonium acetate	1.23 g
Glucose	4.00 g
Dipotassium phosphate	0.52 g
Monopotassium phosphate	0.40 g
Magnesium sulfate	0.40 g
Zinc sulfate	0.002 g
Sodium chloride	0.014 g
Sodium carbonate	0.012 g
Tamol	0.50 g
Calcium chloride	0.002 g
Agarose	10.00 g
Distilled/demineralized water	1000.0 mL

Mix reagents and bring to a boil. Autoclave for 15 min at 121°C and dispense 15 mL aliquots into sterile 100 mm petri dishes.

Cornmeal Agar[12,14]

Cornmeal	50.0 g
Agar	15.0 g
Distilled water	1000.0 mL

Mix cornmeal in 500.0 mL distilled water and heat for 1 h (or autoclave for 10 min at 15 psi). Filter through cheesecloth, bring volume up to 1000.0 mL and add agar. Bring to a boil. Dispense into culture tubes or flasks and autoclave for 15 min at 15 psi. Final pH at 25°C = 6.0 ± 0.2.

Cornmeal - 1% Tween 80 Agar[12,14]

Cornmeal	50.0 g
Noble agar	15.0 g
Tween (Polysorbate) 80	10.0 mL
Distilled water	1000.0 mL

Note: dehydrated cornmeal agar can be substituted for first two ingredients.

Add cornmeal to 500.0 mL of distilled water and heat for 1 h. Filter suspension through cheesecloth and make up to 1000.0 mL with distilled water. While mixing, add agar and bring to a boil. Add tween 80 and mix to dissolve. Autoclave for 15 min at 15 psi. Dispense 15.0-30.0 mL portions into sterile petri dishes (15 x 100 mm). Cool until agar has solidified. Alternatively, medium can first be dispensed into screwcapped test tubes and then autoclaved as above. After solidification, tubes of agar are melted, cooled to 48-52°C, poured into sterile petri dishes and allowed to resolidify prior to use. *Candida albicans* can be used as a quality control organism for chlamydospore production. Final pH at 25°C = 6.0 ± 0.2.

Cystine Heart Hemoglobin Agar[5,23]

Cystine heart agar:

Beef heart infusion	500.0 g
Proteose peptone	10.0 g
Glucose	10.0 g
Sodium chloride	5.0 g
L-cystine	1.0 g
Agar	15.0 g
Distilled/demineralized water	250.0 mL

Mix dehydrated ingredients thoroughly. Suspend 25.5 g of the cystine heart agar base in 250.0 mL of distilled water. Mix well, heat with frequent agitation and boil for 1 min. Autoclave at 15 psi for 15 min (121°C). Cool to 50°C. Aseptically add 250.0 mL of sterile 2% hemoglobin solution which has been cooled to 50°C. Dispense into screwcapped tubes and slant. Seal caps after cooling. Final pH at 25°C = 6.8 ± 0.2.

Hemoglobin solution:

Hemoglobin	5.0 g
Distilled/demineralized water, cold	250.0 mL

Place hemoglobin in a dry flask and vigorously swirl the flask as the cold water is added. Shake together for 15 min until the hemoglobin is completely in solution. Autoclave at 15 psi (121°C) for 15 min. Cool to 50 °C.

Czapek-Dox-Solution Agar[18]

$NaNO_3$	3.0 g
K_2HPO_4	1.0 g
$MgSO_4 \cdot 7H_2O$	0.5 g
KCl	0.5 g
$FeSO_4 \cdot 7H_2O$	0.01 g
Sucrose	30.0 g
Distilled water	1000.0 mL

Add reagents to water and bring to a boil after all solids have dissolved. Dispense 8 mL per 16 x 125 mm tubes, autoclave for 15 min at 121°C. Allow tubes to cool in a slanted position. Final pH at 25°C = 7.3 ± 0.2.

Dermatophyte Test Medium[2,15]

Phytone	10.0 g
Glucose	10.0 g
Agar	20.0 g
Phenol red solution	40.0 mL
Hydrochloric acid, 0.8 N	6.0 mL
Cycloheximide	0.5 g
Gentamicin	0.1 g
Chlortetracycline hydrochloride	0.1 g
Distilled/demineralized water, to	1000.0 mL

Add phytone, glucose and agar to 900.0 mL of water and boil to dissolve. Add phenol red solution (0.5 g in 15.0 mL of 0.1 N NaOH made up to 100.0 mL with water) while stirring. Dissolve cycloheximide in 2.0 mL acetone and add while stirring the hot medium. Dissolve gentamicin in 2.0 mL of distilled water and add to medium while stirring. Add water to make 1000.0 mL. Autoclave at 12 psi for 10 min and allow to cool to 47-50°C. Dissolve chlortetracycline in 25.0 mL of sterile distilled water and add while stirring.

Dispense aseptically into tubes or bottles. Slant (if necessary) and cool. Final pH at 25°C = 5.5 ± 0.2.

DOPA - see Phenoloxidase Medium.

Emmons Modification of Sabouraud's Dextrose Agar - see Sabouraud's Dextrose (Glucose) Agar.

Ethylene Glycol Medium (for aerobic actinomycetes)[20]

Control agar:

Middlebrook-Cohn 7H10 mycobacterial agar in petri plates or plate sections.

Test agar:

Prepare 7H10 agar, autoclave at 15 psi for 15 min, cool, add ethylene glycol (1% vol/vol.) and dispense at 20.0 mL per 100 mm plates or at 5.0 mL per quadrant into quadrant plates. Suspend test organism in broth or saline to match a 0.5 MacFarland turbidity standard. Dilute 1:10 in broth or saline. Flood one control and one test agar plate (or section) with diluted suspension. Let dry and incubate at 35-37°C for 5-10 days. Use known *Rhodococcus* spp. and *Nocardia asteroides* as positive and negative control strains respectively.

Interpretation:

1. *N. asteroides* will grow on both control and test agar; however, most isolates will fail to degrade ethylene glycol leaving each agar relatively clear.
2. Most isolates of *Rhodococcus* spp. will grow on both control and test agar but will degrade the ethylene glycol in the test agar causing a milky turbidity throughout the plate while the control agar remains clear.
3. Compare reactions of unknown isolates with those of control strains.

Fermentation, Wickerham Broth - see Wickerham Broth Fermentation (WBF).

Gorodkowa Ascospore Agar[14]

Glucose	0.63 g
Sodium chloride	1.30 g

Beef extract	2.50 g
Agar	2.50 g
Distilled water	250.00 mL

While mixing, bring agar and distilled water to a boil. Add remaining ingredients and mix to dissolve. Dispense 5.0-7.0 mL portions into screw-capped test tubes (16 x 125 mm). Autoclave for 15 min at 15 psi. Cool at a slant until agar has solidified. *Saccharomyces cerevisiae* and *C. albicans* can be used as positive and negative quality control organisms, respectively.

7H10 Agar - see Middlebrook and Cohn 7H10 Agar.

Inhibitory Mold Agar[23]

Tryptone	3.0 g
Beef extract	2.0 g
Yeast extract	5.0 g
Glucose	5.0 g
Soluble starch	2.0 g
Dextrin	1.0 g
Chloramphenicol	0.125 g
Gentamicin	5.0 mg
Salt A	10.0 mL
Salt C	20.0 mL
Agar	17.0 g
Distilled/demineralized water	970.0 mL

Salt A:

NaH_2PO_4	25.0 g
Na_2HPO_4	25.0 g
Distilled/demineralized water	250.0 mL

Salt C:

$MgSO_4 \cdot H_2O$	10.0 g
$FeSO_4 \cdot H_2O$	0.5 g
NaCl	0.5 g
$MnSO_4 \cdot H_2O$	2.0 mL
Distilled/demineralized water	250.0 mL

Hydrochloric acid, concentrated (1-2 drops can be used to solubilize Salt C)

Suspend dry ingredients except chloramphenicol in water. Boil to dissolve. Dissolve chloramphenicol in 2.0 mL of 95% ethyl alcohol and add to the medium. Adjust pH to 6.7. Autoclave for 15 min at 15 psi (121°C) and dispense after autoclaving.

Final pH = 6.7 ± 0.2 at 25°C.

Kelley's Agar[14]

Hemoglobin solution	20.0 mL
Glucose	10.0 g
Peptone	10.0 g
Sodium chloride	5.0 g
Beef extract	3.0 g
Agar	15.0 g
Distilled/demineralized water	980.0 mL

Prepare the hemoglobin solution by adding 5.0 mL of citrated sheep blood to 15.0 mL of distilled water. Add all ingredients except the hemoglobin solution to the distilled water and bring medium to a boil. Add the hemoglobin solution and mix. Dispense 8.0 mL per 16 x 125 mm screwcapped tube, autoclave for 10 min at 121°C, allow to cool in a slanted position.

Final pH at 25°C = 6.6 ± 0.2.

Kleyn's Acetate Ascospore Agar[14]

Tryptose	2.50 g
Glucose	0.62 g
Sodium chloride	0.62 g
Sodium acetate trihydrate	5.00 g
Agar	15.00 g
Distilled water	1000.00 mL

While mixing, bring agar and distilled water to a boil. Add remaining ingredients and mix to dissolve. Dispense 5.0-7.0 mL portions into screw-capped test tubes (20 x 125 mm). Autoclave for 15 min at 15 psi. Cool at a slant until agar has solidified. *S. cerevisiae* and *C. albicans* can be used as positive and negative quality control organisms, respectively.

Lactrimel Medium[10]

Whole wheat flour	20.0 g
Skim milk	200.0 mL
Honey	10.0 g
Agar	18.0 g

Mix reagents to dissolve. Autoclave for 15 min at 15 psi.

Levine Modification of Converse Liquid Medium - see Converse Liquid Medium.

Lowenstein-Jensen Medium[1,15]

Monopotassium phosphate, anhydrous	2.40 g
Magnesium sulfate · $7H_2O$	0.24 g
Magnesium citrate	0.60 g
Asparagine	3.60 g
Potato flour	30.00 g
Glycerol	12.00 mL
Distilled water	600.00 mL
Homogenized whole eggs	1000.00 mL
Malachite green, 2% aqueous	20.00 mL

Dissolve the salts and asparagine in the water. Admix the glycerol and potato flour, autoclave at 121°C for 30 min and cool to room temperature. Scrub eggs (not more than 1 week old) in 5% soap solution and then rinse thoroughly in cold running water. Immerse in 70% ethyl alcohol for 15 min. Break eggs into a sterile flask. Homogenize by hand-shaking and filter through 4 layers of sterile gauze. Add 1 liter of homogenized eggs to the potato-salt mixture. Prepare the malachite green and admix thoroughly. Dispense 6-8 mL into 20 x 150 mm screwcapped tubes. Slant and inspissate at 85°C for 50 min. Incubate for 48 h at 37°C to check sterility and store at 4-6°C with caps tightly closed. Final pH at 25°C = 7.0 ± 0.2.

Lysozyme-Glycerol Broth (for actinomycetes)[9,12]

Glycerol (control) broth:

Peptone	5.0 g
Beef extract	3.0 g
Glycerol	70.0 mL
Distilled water	1000.0 mL

Autoclave at 115°C for 15 min.

Final pH at 25°C = 7.0 ± 0.2.

Lysozyme solution

Lysozyme	100.0 mg
0.0l N HCl	100.0 mL

Sterilize by filtration. Final pH at 25°C = 6.8 ± 0.2.

Suspend several fragments of growth in 5.0 mL of glycerol broth as a control. Prepare lysozyme test broth by adding 5.0 mL of lysozyme solution to 95.0 mL of glycerol broth and aliquot in 5.0 mL quantities. Inoculate lysozyme test broth and incubate both test and control broth at room temperature until good growth is noted in control (up to 14 days). Determine resistance to lysozyme by comparison of test with control broth.

Middlebrook and Cohn 7H10 agar[15]

Stock solutions:

Solution 1. Store at room temperature.	
Monopotassium phosphate	15.0 g
Disodium phosphate	15.0 g
Distilled water	250.0 mL
Solution 2. Store at 4-10°C.	
Ammonium sulfate	5.0 g
Monosodium glutamate	5.0 g
Sodium citrate · $2H_2O$ USP	4.0 g
Ferric ammonium citrate	0.4 g
Magnesium sulfate · $7H_2O$ ACS	0.5 g
Biotin, in 2 mL of 10% NH_4OH	0.005 g
Distilled water, to	250.0 mL
Solution 3. Store at 4-10°C.	
Calcium chloride · $2H_2O$ ACS	0.05 g
Zinc sulfate · $7H_2O$ ACS	0.10 g
Copper sulfate · $5H_2O$ ACS	0.10 g
Pyridoxine hydrochloride	0.10 g
Distilled water, to	100.0 mL
Solution 4. Glycerol	
Solution 5. Malachite green, 0.01% aqueous.	

Preparation:

To 975.0 mL of distilled water, add:

Solution 1	25.0 mL
Solution 2	25.0 mL
Solution 3	1.0 mL
Solution 4	5.0 mL

Adjust the pH to 6.6 by adding approximately 0.5 mL of 6 N HCl. Add 2.5 mL of solution 5 and 15.0 g of agar. Autoclave at 121°C for 15 min. Cool to 56°C and add 100.0 mL of OADC Enrichment (Difco). Dispense the completed medium into petri plates or tubes. Store at 4-8°C; unsealed and unprotected containers should be used for no longer than 1 week. During preparation and storage, protect from light.

Final pH at 25 °C = 6.6 ± 0.2.

Nitrate Assimilation Broth - see Assimilation Media, Nitrate Broth.

Nitrate Assimilation, Potassium - see Assimilation Medium, Potassium Nitrate.

Nitrate Assimilation, Wickerham Broth - see Assimilation Media, Nitrate Broth.

Nitrate Disks, Auxanographic Assimilation - see Assimilation Media, Auxanographic Nitrate Test.

Nitrogen Basal Agar - see Assimilation Media, Auxanographic Nitrate Test.

Peptone Disks, Auxanographic Assimilation - see Assimilation Media, Auxanographic Nitrate Test.

Phenoloxidase Medium[4]

Solution A	
Noble agar	15.0 g
Distilled water	500.0 mL
Solution B	
Monobasic potassium phosphate	4.0 g
Magnesium sulfate · 7 H_2O	2.5 g
Thiamine hydrochloride	0.01 g
Biotin	0.00002 g
DL-3,4-dihydroxyphenylalanine (DOPA)	0.2 g
Glucose	5.0 g
Asparagine	1.0 g
Glutamine	1.0 g
Glycine	1.0 g
Distilled water	500.0 mL

Solution A - While mixing, bring agar and distilled water to a boil. Autoclave for 15 min at 15 psi. Cool Solution A to 55°C. Solution B - Dissolve ingredients in distilled water and sterilize by filtration. Adjust to pH 5.5 using 1M mono- or dibasic potassium phosphate. Bring Solution B to 55°C and mix together with Solution A. Dispense 15.0-30.0 mL portions into sterile petri dishes (15 x 100 mm). Cool until agar has solidified. *C. neoformans* and *C. laurentii* can be used as positive and negative quality control organisms, respectively.

Pine-Drouhet Agar[17]

Use distilled/demineralized water in all preparations.

Solution A:

Primary mineral solution:

KH_2PO_4	8.00 g
$(NH_4)_2SO_4$	8.00 g
$MgSO_4{\cdot}7H_2O$	0.80 g
$CaCl_2{\cdot}2H_2O$	0.08 g
$ZnSO_4{\cdot}7H_2O$	0.09 g

Dissolve reagents in the listed order in 750.0 mL of water. Do not add the next reagent until the one before is completely dissolved. Make up to 1000.0 mL with water.

Add casein acid hydrolysate 4.0 g

Secondary mineral solution: *10.0 mL*

1. $FeSO_4{\cdot}7H_2O$	5.70 g
$MnCl_2{\cdot}4H_2O$	0.80 g

Dissolve in 500.0 mL of water containing 1% HCl.

2. $NaMoO_4{\cdot}2H_2O$	0.15 g

Dissolve in 500.0 mL of water.

Mix solutions #1 and #2.

D-glucose	10.0 g
Alpha-ketoglutaric acid	1.0 g
L-cysteine, HCl	1.0 g
Glutathione, reduced	0.5 g
L-asparagine	0.1 g
L-tryptophan	0.02 g

Vitamin solution (5x): *10.0 mL*

Thiamine, HCl	0.1 g
Myo-inositol	0.1 g
Pantothenate, Ca	0.1 g
Niacin	0.1 g
Riboflavin	0.1 g
(+) Biotin	0.005 g

Dissolve in 400.0 mL of water and add water to 500.0 mL.

Filter sterilize. Aliquot in 100.0 mL in dark bottles and store at 4°C.

Hemin chloride solution: 1.0 mL

To a solution of 100.0 mL of water containing 2-3 drops of concentrated NH_4OH, add 0.2 g of hemin. Store at 4°C.

Coenzyme A, trilithium salt 1.0 mL

Prepare solution at a concentration of 1.0 mg/mL in water. Store at −70°C.

Thioctic acid 1.0 mL

Dissolve 0.02 g DL-thioctic acid in 50.0 mL of 95% ethanol. Add 50.0 mL of water containing 1% Na_2S by volume; filter sterilize and store at 4°C.

Adjust the pH of Solution A to 6.5 with 20% (w/v) KOH and adjust volume to 500.0 mL with water.

Solution B:

Starch suspension 500.0 mL

Suspend 2.0 g of purified starch in 50.0 mL of cold water. Add slowly to 450.0 mL of boiling water.

Noble agar 12.5 g

Add agar and stir until dissolved.

Oleic acid 1.0 mL

Dissolve 1.0 g in 10.0 mL 95% ethanol and store at -20°C.

Autoclave Solution B at 121°C for 20 min and allow to cool in a 50°C water bath.

Allow solution A to warm in a 50°C waterbath. Mix solution A with B and dispense 30.0 mL per petri plate or 8.0 mL per sterile disposable 16 x 100 mm screwcapped tube.

Potassium Nitrate Assimilation - see Assimilation Medium, Potassium Nitrate.

Potato Dextrose Agar[5,15]

Potatoes - white	200.0 g
Glucose	10.0 g
Agar	18.0 g
Water	1000.0 mL

Peel and dice potatoes. Simmer them in water for 1 h. Filter through coarse paper. Add the glucose and agar, dissolve by heat and filter through cotton and gauze. Restore volume to 1 L. Dispense into culture tubes and autoclave at 121°C (15 psi) for 10 min. Final pH at 25°C = 5.6 ± 0.2.

Rice Grain Medium[15]

White rice (not enriched)	8.0 g
Distilled water	25.0 mL

Place rice and water into a 125 mL Erlenmeyer flask and autoclave at 15 psi for 15 min. Although *Microsporum canis, M. gypseum* and many other dermatophytes grow and form conidia on this medium, *M. audouinii* grows poorly and does not form conidia.

Sabhi agar[5,14]

Glucose	21.0 g
Neopeptone	5.0 g
Proteose peptone	5.0 g
Calf brain infusion	100.0 g
Beef heart infusion	125.0 g
Sodium chloride	2.5 g
Disodium phosphate	1.25 g
Agar	15.00 g
Distilled/demineralized water	1000.0 mL

Suspend all components in 1000 mL of water, bring to a boil to dissolve completely, dispense into tubes or bottles. Autoclave for 15 min at 15 psi (121°C). Slant (if necessary) and cool.

Final pH at 25°C = 7.0 ± 0.2.

Sabouraud Dextrose (Glucose) Agar, Emmons Modified[15]

Dextrose	20.0 g
Neopeptone or Polypeptone (BBL)	10.0 g
(Pancreatic digest of casein USP	5.0 g)
(Peptic digest of animal tissue USP	5.0 g)
Agar	20.0 g
Distilled/demineralized water	1000.0 mL

Suspend all components in 1000 mL water. Boil to dissolve. Dispense into tubes and autoclave at 118-121°C for 15 min. Avoid overheating which will result in a softer agar. Slant tubes if necessary and cool. Final pH at 25°C = 7.0 ± 0.2.

Sabouraud Dextrose Agar with Cycloheximide and Chloramphenicol[7]

Sabouraud dextrose agar	1000.0 mL
Chloramphenicol (in 10.0 mL of 95% ethanol)	0.050 g
Cycloheximide (in 10.0 mL acetone)	0.050 g

Prepare Sabouraud dextrose agar as described. Add chloramphenicol solution, remove from boiling and mix. Add cycloheximide solution, mix and return complete medium to boiling. Autoclave and dispense as for Sabouraud Dextrose (Glucose) Agar, Emmons Modified. Do not allow medium to cool before pouring, since reheating medium may cause inactivation of the antibiotics.

Sheep Blood Agar - see Soybean-Casein Digest Agar.

Soil Extract Agar[14]

Soil	500.0 g
Glucose	2.0 g
Yeast extract	1.0 g
Monopotassium phosphate	0.5 g
Agar	15.0 g
Tap water	1000.0 mL

Mix 500.0 g of soil and 1000 mL of tap water. Autoclave for 3 h at 121°C. Filter through Whatman No. 2 filter paper. Add reagents to the filtrate and add water to 1000.0 mL. Adjust pH to 7.0. Dispense 8.0 mL in 16 x 125 mm tubes. Autoclave for 15 min at 121°C. Allow to cool in a slanted position.

Soybean-Casein Digest Agar[5,15] (Trypticase soy agar base, BBL; or Tryptic soy agar base, Difco)

Pancreatic digest of casein	15.0 g
Papaic digest of soybean meal	5.0 g
Sodium chloride	5.0 g
Agar	15.0 g
Distilled water	1000.0 mL

Suspend and heat to boiling to dissolve completely. Sterilize in the autoclave for 15 min at 15 psi (121°C). Cool to 45-50°C and dispense. Final pH at 25°C = 7.3 ± 0.2. For sheep blood agar: Aseptically add 5% sterile defibrinated sheep blood to agar at 45-50°C, mix and dispense.

Tap Water Agar[3]

Agar	2.0 g
Tap water	100.0 mL

Suspend agar in tap water. Autoclave at 121°C for 15 min. Cool and dispense.

Trichophyton Agars[14]

These media are commercially available in dehydrated form from Difco Laboratories (Detroit, MI) and as agar slants from Remel Laboratories

(Lenexa, KA). They should be prepared according to manufacturers' directions.

Tyrosine or Xanthine Agar[12,15]

Nutrient agar	23.0 g
Tyrosine	5.0 g
(or Xanthine	4.0 g)
Demineralized water	1000.0 mL

Dissolve the nutrient agar in the water. Add tyrosine or xanthine and mix to distribute the crystals evenly. Adjust the pH to 7.0 and autoclave at 121°C for 15 min. Dispense into plates, 20.0 mL per plate, or 5.0 mL per quadrant of a quadrant plate with the crystals evenly distributed. Final pH at 25°C = 7.0 ± 0.2.

V-8 Juice Agar[14]

V-8 juice	180.0 mL
Calcium carbonate	2.0 g
Agar	15.0 g
Distilled water	805.0 mL

Dissolve reagents in water. Autoclave for 15 min at 15 psi.

V-8 Juice Ascospore Agar[14]

V-8 juice	350.0 mL
Compressed yeast	5.0 g
Agar	14.0 g
Distilled water	340.0 mL

While mixing, bring agar and distilled water to a boil. Mix V-8 juice and compressed yeast and adjust to pH 6.8. Heat V-8/yeast solution for 10 min in flowing steam, cool and readjust to pH 6.8. Mix the two solutions together and dispense 5.0-7.0 mL portions into screwcapped test tubes (16 x 125 mm). Autoclave for 15 min at 15 psi. Cool at a slant until agar has solidified. *S. cerevisiae* and *C. albicans* can be used as positive and negative quality control organisms, respectively.

Wickerham Broth Assimilation, Carbon Sources - see Assimilation Media, Wickerham Broth Carbon Test.

Wickerham Broth Carbon Assimilation (WBCA) - see Assimilation Media, Wickerham Broth Carbon Test.

Wickerham Broth Fermentation (WBF)[21]

Carbon sources for Wickerham broth fermentation (WBF):[21]

Carbon source (except raffinose)	6.0 g
for raffinose	12.0 g

Distilled water	100.0 mL

Dissolve carbon source in distilled water and sterilize by filtration. Quality control of carbon solutions is performed with WBF basal medium. Selection of quality control organisms is dependent upon the number and type of fermentation tests being performed.

Basal medium for Wickerham broth fermentation:[14]

Yeast extract	4.5 g
Peptone	7.5 g
Bromthymol blue*	0.04 g
Distilled water	1000.0 mL

*Optimal; pH indicator not required since positive fermentation is based on accumulation of gas bubbles in the Durham tube insert.

Dissolve ingredients in distilled water and dispense 2.0 mL aliquots into screwcapped test tubes (16 x 100 mm). Place an inverted Durham tube (6 x 50 mm) in each test tube and autoclave for 15 min at 15 psi. After tubes of basal medium have cooled, add 1.0 mL of 6% (12% for raffinose) carbon source solution (presterilized by filtration). Selection of quality control organisms is dependent upon the number and type of fermentation tests being performed.

Wickerham Broth Nitrate Assimilation (WBNA) - see Assimilation Media, Nitrate Broth.

Wickerham Broth Nitrate Assimilation, (WBNA) Negative Growth Control - see Assimilation Media, Nitrate Broth.

Wolin-Bevis Medium[12,24]

Tween 80	3.00 mL
Glucose	0.25 g
L-histidine	0.25 g
Ammonium sulfate	1.00 g
Monobasic potassium phosphate	1.00 g
Agar	15.00 g
Distilled water	1000.00 mL

While mixing, bring agar and distilled water to a boil. Add remaining ingredients and mix to dissolve. Autoclave for 15 min at 15 psi. Dispense 15.0-30.0 mL portions into sterile petri dishes (15 x 100 mm). Cool until agar has solidified. *C. albicans* can be used as a quality control organism for chlamydospore production. Final pH at 25°C = 5.4 ± 0.2.

Xanthine Agar - see Tyrosine or Xanthine Agar.

Yeast Extract - Malt Extract Agar[14]

Yeast extract	3.0 g
Malt extract	3.0 g
Peptone	5.0 g
Glucose	10.0 g
Agar	15.0 g
Distilled water	1000.0 mL

While mixing, bring agar and distilled water to a boil. Add remaining ingredients and mix to dissolve. Dispense 5.0-7.0 mL portions into screw-capped test tubes (16 x 125 mm). Autoclave for 15 min at 15 psi. Cool at a slant until agar has solidified. *S. cerevisiae* and *C. albicans* can be used as positive and negative quality control organisms for ascospore production, respectively.

Yeast Extract Phosphate Agar [14]

Solution A:

Yeast extract	1.0 g
Agar	15.0 g
Distilled/demineralized water	1000.0 mL

Solution B:

Disodium phosphate, anhydrous	4.0 g
Monopotassium phosphate	6.0 g
Distilled/demineralized water	30.0 mL

Solution C:

Ammonium hydroxide, concentrated	1.0 mL

Mix reagents for Solution A and bring to a boil. Mix reagents for Solution B. Adjust pH to 6.0 with either 1 N hydrochloric acid or 1 N sodium hydroxide. Add 2.0 mL of Solution B to Solution A. Dispense into tubes or bottles. Autoclave for 15 min at 15 psi (121°C). Slant (if necessary) and cool. One drop of Solution C is added to the edge of the inoculated medium.

Chloramphenicol (0.05 mg/mL) can be added to the medium to improve selectivity; however, cycloheximide should not be added as it is inactivated by ammonium hydroxide.

Final pH (before addition of ammonium hydroxide) = 6.0 ± 0.2.

STAINS AND DYES

Acidfast Stain, Kinyoun's Method - cold (modified for actinomycetes)[9]

Reagents:

1. Kinyoun carbolfuchsin	
Basic fuchsin	4.0 g
Phenol	8.0 mL
Alcohol, 95%	20.0 mL
Distilled water	100.0 mL
Dissolve dye in alcohol and then add water and phenol.	
2. 1.0% Aqueous sulfuric acid	
Sulfuric acid, concentrated	1.0 mL
Distilled water	99.0 mL
3. Methylene blue counterstain:	
Methylene blue	2.5 g
Ethanol, 95%	100.0 mL

Procedure:

1. Prepare a smear and fix over flame or by slide warmer.
2. Flood slide with Kinyoun carbolfuchsin for 5 min.
3. Pour off excess stain.
4. Flood with 50% alcohol and immediately wash with water.
5. Decolorize with 0.5% aqueous sulfuric acid solution for 2 min.
6. Immediately wash with water.
7. Counterstain with methylene blue for 1 min.
8. Rinse with water, dry and examine under oil immersion objective.

The filaments of *Nocardia* usually appear partially acidfast with this staining procedure.

Acidfast Stain, Kinyoun's Method - warm (modified for actinomycetes)[8]

Carbolfuchsin reagent:

Solution 1	
Basic fuchsin	3.0 g
95% ethyl alcohol	100.0 mL
Solution 2	
Phenol (conc.)	5.0 mL
Distilled water	95.0 mL

Add 90.0 mL of solution 2 (phenol solution) to 10.0 mL of solution 1 (carbolfuchsin solution). Store at 37°C in a tightly sealed bottle.

Hank's sulfuric acid-methylene blue reagent:

Solution 3	
Methylene blue	1.0 g
Distilled water	100.0 mL
Solution 4	
Sulfuric acid (conc.)	5.0 mL
Distilled water	95.0 mL

Add 20.0 mL of solution 3 (methylene blue) to 80.0 mL of solution 4 (sulfuric acid). Store at room temperature in a tightly sealed bottle.

Procedure:

1. Prepare smear and fix over flame or by slide warmer.
2. Flood slide with carbolfuchsin solution and steam gently for 60-80 s. DO NOT BOIL STAIN.
3. Cool and rinse gently under running tap water.
4. Flood slide with Hank's sulfuric acid-methylene blue solution for 60-90 s.
5. Wash slide with water, dry and examine under oil immersion objective.

Calcofluor White[11]

Calcofluor white M2R	1.0 g
Distilled water	100.0 mL

Prepare stock solution of calcofluor white by mixing it with water. Dissolve by heating gently. Prepare 0.1% working solution of calcofluor white in water. Evans blue (0.05%) may be added to the working solution for a counterstain.

Giemsa Stain[15]

Giemsa stock stain solution:

Geimsa powder	0.5 g
Methyl alcohol, absolute, acetone-free	33.0 mL

Mix thoroughly, allow to sediment and store at room temperature.

Stock buffer solutions:

1. Alkaline buffer:

Na_2HPO_4	9.5 g
Distilled water	1000.0 mL

2. Acid buffer:

$Na_2HPO_4 \cdot H_2O$	9.2 g
Distilled water	1000.0 mL

Buffered water (pH 7.0-7.2):

Acid buffer	39.0 mL
Alkaline buffer	61.0 mL
Distilled water	900.0 mL

Mix in well cleaned glassware. Stable for several weeks if sealed.

Triton X-100 stock solution (10%):

Triton X-100	10.0 mL
Distilled water	90.0 mL

Mix and store in a tightly stoppered bottle. Keeps indefinitely at room temperature. For working Triton X buffered water solution, add 1.0 mL of Triton X stock solution to 1000.0 mL of buffered water.

Prepare working solution of Giemsa stain by diluting stock Giemsa solution 1:20 with Triton X buffered water. The working solution must be prepared fresh daily. The final pH must be 7.0-7.2 for correct differentiation of cellular components.

Procedure:

1. Fix air-dried smear in methanol for 30-60 s and air dry.
2. Flood slide with stain and allow to remain for 30 min.
3. Rinse with buffered water without Triton X. Air dry.

Gomori Methenamine Silver Stain, Grocott Modification[6]

Chromic acid, 5%:

Chromic acid	5.0 g
Distilled water	100.0 mL

Silver nitrate solution, 5%:

Silver nitrate	5.0 g
Distilled water	100.0 mL

Methenamine solution, 3%:

Hexamethylenetetramine (methenamine)	3.0 g
Distilled water	100.0 mL

Borax solution, 5%:

Borax (photographic grade)	5.0 g
Distilled water	100.0 mL

Stock Methenamine-silver nitrate solution:

Silver nitrate, 5% solution	5.0 mL
Methenamine, 3% solution	100.0 mL

Note: a white precipitate may form but will dissolve upon shaking the solution. This solution should be stored at 4°C and is stable for months at this temperature.

Working Methenamine-silver nitrate solution:

Borax, 5% solution	2.0 mL
Distilled water	25.0 mL
Mix well and add:	
Stock Methenamine-silver nitrate solution	25.0 mL

Sodium bisulfite solution, 1%:

Sodium bisulfite	1.0 g
Distilled water	100.0 mL

Gold chloride solution, 0.1%:

Gold chloride, 1% solution	10.0 mL
Distilled water	90.0 mL

Note: this solution may be used repeatedly.

Sodium thiosulfate solution, 2%:

Sodium thiosulfate	2.0 g
Distilled water	100.0 mL

Stock Light green solution:

Light green, S.F. (yellow)	0.2 g
Distilled water	100.0 mL
Glacial acetic acid	0.2 mL

Working Light green solution:

Light green, stock solution	10.0 mL
Distilled water	50.0 mL

Procedure:

1. Prepare slides, including a control slide.
2. Deparaffinize slide through 2 changes of xylene, absolute alcohol, 95% ethyl alcohol and distilled water as usual. Slides stained previously by the hematoxylin and eosin procedure can be restained with the GMS stain by removal of the coverslip (with xylol) and running through the alcohols to water.
3. Oxidize in 5% chromic acid solution for 60 min.
4. Wash in tap water for 10-15 s.
5. Rinse in 1% sodium bisulfite for 60 s to remove residual chromic acid solution.
6. Wash in tap water for 5-10 min.
7. Wash with 3-4 changes distilled water.
8. Place in working methenamine-silver nitrate solution in oven at 58-60°C for 30-60 min until section turns yellowish-brown. Use paraffin-coated forceps to remove control slide from this solution. Dip slide in distilled water. Examine microscopically for adequate silver impregnation. The fungi should be dark brown. Repeat with other slides if impregnation is sufficient.
9. Rinse in 6 changes of distilled water.
10. Tone in 0.1% gold chloride solution for 2-5 min.
11. Rinse in distilled water.
12. Remove unreduced silver with 2% sodium thiosulfate solution for 2-5 min.
13. Wash thoroughly in tap water.
14. Counterstain with working light green solution for 30-45 s.
15. Dehydrate with 2 changes of 95% alcohol, absolute alcohol, clear with 2-3 changes of xylene and mount coverslip with Permount.

Results:

1. Fungi should be sharply outlined in black.
2. Mucin will be taupe to dark grey.
3. Inner parts of the mycelia and hyphae will be old rose.
4. Background should be light green.

Note: variable staining of *Nocardia* spp. occurs. This is dependent upon the type or strain of *Nocardia* present. Therefore, if *Nocardia* is suspected, two slides should be prepared, one heated at 58-60°C for 60 min and the other for 90 min.

Gram Stain, Hucker Modification[15]

Crystal violet:

1. Solution A:

Crystal violet (certified)	2.0 g
Ethyl alcohol, 95%	20.0 mL

2. Solution B:

Ammonium oxalate	0.8 g
Distilled water	80.0 mL

Mix solutions A and B. Store for 24 h before use and filterd when transferring into working stain bottle.

Gram iodine:

Iodine	1.0 g
Potassium iodide	2.0 g
Distilled water	300.0 mL

Grind the dry iodine and potassium iodide in a mortar. Add water a few mL at a time and grind thoroughly after each addition until solution is achieved. Rinse the solution into an amber glass bottle with the remainder of the distilled water.

Decolorizer:

Ethyl alcohol, 95%	100.0 mL
Acetone	100.0 mL

Mix and store in a tightly capped bottle.

Counterstain:

1. Stock solution:

Safranin O (certified)	2.5 g
Ethyl alcool, 95%	100.0 mL

2. Working solution:

Stock solution	10.0 mL
Distilled water	90.0 mL

Procedure:

1. Flood fixed and cooled smear with crystal violet; let stand 1 min.
2. Rinse with tap water and drain off excess.
3. Flood smear with iodine solution; let stand 1 min.
4. Wash with tap water and decolorize until no more stain runs off the slide.
5. Wash with tap water and add counterstain for 10-30 s.
6. Wash with tap water. Blot or air dry and examine. Gram-positive organisms are blue; gram-negative organisms are red.

Grocott Modification of Gomori Methenamine Silver Stain - see Gomori Methenamine Silver Stain.

Hucker Modification of Gram Stain - see Gram Stain.

Kinyoun's Method - see Acidfast Stain, Kinyoun's Method, Cold or Warm Stain.

Lactophenol Cotton Blue[14]

Phenol, concentrated	20.0 mL
Lactic acid	20.0 mL
Glycerol	40.0 mL
Cotton blue (aniline blue)	0.05 g
Distilled water	20.0 mL

Dissolve cotton blue in distilled water. Add phenol, lactic acid and glycerol. Mix well. Store at room temperature.

Mayer's Egg Albumin - see Periodic Acid-Schiff Stain.

Mayer's Mucicarmine Stain[19]

Weigert's iron hematoxylin:

1. Solution A:	
Hematoxylin	1.0 g
Alcohol, 95%	100.0 mL
2. Solution B:	
Ferric chloride, 29% aqueous solution	4.0 mL
Distilled water	95.0 mL
Hydrochloric acid, concentrated	1.0 mL
3. Working solution:	
Equal parts of Solutions A and B	Prepared fresh

Metanil yellow solution:

Metanil yellow	0.25 g
Distilled water	100.0 mL
Glacial acetic acid	0.25 mL

Mucicarmine stain:

Carmine	1.0 g
Aluminum chloride, anhydrous	0.5 g
Distilled water	2.0 mL

Mix stain in a small test tube. Heat over small flame for 2 min. Liquid will become almost black and syrupy. Dilute with 100.0 mL of 50% alcohol and allow to stand for 24 h. Filter and dilute 1:4 with tap water for use.

Procedure:

1. Deparaffinize sections and rehydrate by passing through xylene, absolute alcohol, 95% alcohol and distilled water. If necessary, mercury precipitates can be removed through iodine and photographic hypo solution washes.
2. Stain for 7 min in working solution of Weigert's hematoxylin.
3. Wash in tap water for 5-10 min.
4. Rinse in diluted mucicarmine solution for 30-60 min or longer. Check control slide after 30 min.
5. Rinse in distilled water.
6. Stain in metanil yellow solution for only 60 s.
7. Rinse quickly in distilled water.
8. Rinse quickly in 95% alcohol.
9. Dehydrate in 2 changes of absolute alcohol; clear with 2-3 changes of xylene and mount coverslip with permanent mounting fluid.

Results:

1. Mucin will be deep red to rose.
2. Nuclei will be black.
3. Background will be yellow.

Methenamine Silver Stain - see Gomori Methenamine Silver Stain, Grocott Modification.

Mucicarmine Stain - see Mayer's Mucicarmine Stain.

Periodic Acid-Schiff Stain (PAS)[14,19]

Periodic acid solution:

Periodic acid	5.0 g
Distilled water	100.0 mL

Mix reagents in a brown bottle and tighten cap. Store at 4°C.

Basic fuchsin solution:

Basic fuchsin	0.1 g
Ethyl alcohol, absolute	5.0 mL
Distilled water	95.0 mL

Mix ethyl alcohol and distilled water in a brown bottle. Add basic fuchsin and mix by rotating container. Tighten the cap. Store at 4°C.

Sodium metabisulfite solution:

Sodium metabisulfite	1.0 g
Hydrochloric acid, 1 N	10.0 mL
Distilled water	190.0 mL

Add the hydrochloric acid to the distilled water in a brown bottle. Add sodium metabisulfite and mix well. Tighten the cap of the bottle and store at 4°C.

Light green stain:

Light green crystals	0.2 g
Distilled water	100.0 mL
Glacial acetic acid	0.2 mL

Mayer's egg albumin:

Egg white	50.0 mL
Glycerin	50.0 mL

Beat the egg white and glycerin together and filter through several thicknesses of gauze. A crystal of thymol can be added to preserve the albumin and inhibit the growth of mold.

Direct mount procedure:

1. Fix specimen on slide.
2. Place slide in absolute ethyl alcohol for 60 s. Liquid or homogenized specimens can be applied to the slide in the usual manner, air-dried and heat-fixed. Albumin can be used to fix skin, hair and nail specimens to a slide. Commercially available egg albumin is appropriate for this purpose. Alternately, use Mayer's Egg Albumin.
3. Drain alcohol and immediately place in 5% periodic acid solution for 5 min.
4. Wash in running water for 2 min.
5. Place in basic fuchsin solution for 2 min.
6. Wash in running water for 2 min.
7. Immerse slide in sodium metabisulfite solution for 3-5 min.
8. Wash in running water for 5 min.

9. Place in light green solution for 5 s.
10. Wash in running water for 5-10 s.
11. Place slide at 5 s intervals into 85%, 95% and absolute alcohols.
12. Dip into xylene and mount coverslip with mounting fluid.

Results:

1. Fungi will stain magenta (purplish-red).
2. Background will be light green.

Potassium hydroxide[14]

Potassium hydroxide (crystals)	10.0 g
Distilled/demineralized water	100.0 mL

Mix well. Store at room temperature.

Schaeffer-Fulton Modification of Wirtz Stain - see Wirtz Stain.

Weigert Iron Hematoxylin - see Mayer's Mucicarmine Stain.

Wirtz Stain, Schaeffer-Fulton Modified[13]

Aqueous malachite green, 5%:

Malachite green	5.0 g
Distilled water	100.0 mL

Heat while mixing to dissolve malachite green in distilled water. Filter to remove any remaining particulate matter.

Basic fuchsin - stock solution, 2.5%:

Basic fuchsin	2.5 g
Ethanol	100.0 mL

Basic fuchsin - working solution, 0.25%:

Basic fuchsin - stock solution	10.0 mL
Distilled water	90.0 mL

Procedure:

1. Prepare a smear of the yeast by emulsifying growth in a drop of water. Allow the smear to air-dry and heat fix.
2. Flood heat-fixed smear with 5.0% aqueous malachite green for 60-90

s. Heat to steaming 3-4 times. Additional malachite green can be added if necessary.

3. Wash with running tap water for 60 s.
4. Counterstain with 0.25% basic fuchsin (working solution) for 30 s.
5. Wash with running tap water and dry.
6. Observe under oil immersion for ascospores which stain blue-green while vegetative cells stain red.

REFERENCES

1. BBL quality control and production information manual for tubed media. Cockeysville, MD: BBL Microbiology Systems, 1987:T-21257.
2. Beneke ES, Rogers AL. Medical mycology manual. 3rd ed. Minneapolis: Burgess Publishing, 1970:43-4.
3. Berd D. Laboratory identification of clinically important aerobic actinomycetes. Appl Microbiol 1973;25:665-81.
4. Chaskes S, Tyndall RL. Pigment production by *Cryptococcus neoformans* from para- and ortho-diphenols: effect of the nitrogen source. J Clin Microbiol 1975;1:509-14.
5. Difco manual. 10th ed. Detroit: Difco Laboratories, 1984:1155.
6. Emmons EW, Binford CH, Utz JP, Kwon-Chung KJ. Medical mycology. 3rd ed. Philadelphia: Lea and Febiger, 1977:559-61.
7. Georg LK, Ajello L, Papegeorge C. Use of cycloheximide in the selective isolation of fungi pathogenic to man. J Lab Clin Med 1954;44:422-8.
8. Haley LD, Callaway CS. Laboratory methods in medical mycology. Atlanta: Centers for Disease Control, 1978:139-52.
9. Haley LD, Standard PG. Laboratory methods in medical mycology. Atlanta: Centers for Disease Control, 1973:100.
10. Honbo S, Padhye AA, Ajello L. The relationship of *Cladosporium carrionii* to *Cladiophialophora ajelloi*. Sabouraudia. J Med Vet Mycology 1984;22:209-18.
11. Koneman EW, Roberts GD. Practical laboratory mycology. 3rd ed. Baltimore: Williams and Wilkins, 1985:188.
12. MacFaddin JF. Media for isolation - cultivation - identification - maintenance of medical bacteria; vol.1. Baltimore: Williams and Wilkins, 1985:928.
13. McGinnis MR. Laboratory handbook of medical mycology. 1st ed. New York: Academic Press, 1980:351-2.
14. McGinnis MR. Laboratory handbook of medical mycology. 1st ed. New York: Academic Press, 1980:526-84.
15. Phillips E, Nash P. Culture media. In: Lennette EH, Balows A, Hausler WJ Jr, Shadomy HJ, eds. Manual of clinical microbiology. 4th ed. Washington, DC: American Society for Microbiology, 1985:1056-1102.
16. Pincus DH, Salkin IF, Hurd NJ, Levy IL, Kemna MA. Modification of potassium nitrate assimilation test for identification of clinically important yeasts. J Clin Microbiol 1988;26: 366-8.
17. Pine L, Drouhet E. Sur l'obtention et la conservation de la phase levure d'*Histoplasma capsulatum* et d'*H. dubosi* en milieu chimiquement defini. Ann Inst Pasteur 1963;105: 798-804.
18. Raper KB, Fennell DI. The genus *Aspergillus*. Malabar, FL: Robert E. Krieger, 1965.
19. Rippon J. Medical mycology: the pathogenic fungi and the pathogenic actinomycetes. 2nd ed. Philadelphia: WB Saunders, 1982:790-5.

20. Stottmeir KD, Molley ME. Rapid identification of the taxon *Rhodochrous* in the clinical laboratory. Appl Microbiol 1973;26:213-14.
21. van der Walt JP. Criteria and methods used in classification. In: Lodder J, ed. The yeasts - a taxonomic study. 2nd ed. Amsterdam: North-Holland Publishing, 1970:72.
22. van der Walt JP. Criteria and methods used in classification. In: Lodder J, ed. The yeasts - a taxonomic study. 2nd ed. Amsterdam: North-Holland Publishing, 1970:80-1.
23. Washington JA III. Laboratory procedures in clinical microbiology. 1st ed. New York: Springer-Verlag, 1981:789-90.
24. Wolin HL, Bevis ML, Laurora N. An improved synthetic medium for the rapid production of chlamydospores by *Candida albicans*. Sabouraudia 1962;2:96-9.

CHAPTER 12

METHODS FOR PARASITOLOGY

Earl G. Long, Ph.D., Kenneth W. Walls, Ph.D.
and Dorothy Mae Melvin, Ph.D.

INTRODUCTION

Parasitic diseases constitute major health problems in underdeveloped and emerging countries in tropical and subtropical areas. Malaria, a leading cause of morbidity and mortality that had been fairly well controlled in many countries, is once again increasing in prevalence because of economic constraints, resistance of *Anopheles* mosquito vectors to insecticides and resistance of the parasites themselves to chemotherapeutic agents. As worldwide travel by United States citizens increases and more foreign nationals visit this country each year, opportunities for exposure to parasitic infections increase. In the developed countries, parasitic infections are being found with increasing frequency in patients with acquired or induced immunodeficiencies. Parasites such as *Pneumocystis carinii* and *Cryptosporidium* spp., once unusual, are now common and deadly in patients with the acquired immune deficiency syndrome (AIDS). In the United States, there have been several epidemics of gastroenteritis caused by *Giardia lamblia* or *Cryptosporidium* and transmitted through municipal water systems.

There are no good figures on the prevalence of parasites in the United States because many parasitic diseases are not reportable. A survey of the types of parasites seen in fecal specimens submitted to state public health laboratories[14] gives some index of the relative frequency of different parasites (Table 12:1). It should be emphasized that these figures are based on specimens submitted for parasitic examination because of some clinical suspicion of parasitic diseases; they are not prevalence figures. Data are not available on whether these infections were acquired in the United States or in foreign countries. In addition, laboratories differ in their proficiency

Table 12:1. Frequency of Intestinal Parasites in 388,745 Fecal Specimens Examined by State Health Department Laboratories, 1976*

	Number	% of Specimens†	% of All Identifications
Protozoa	48,353	3.8	61.1
Giardia lamblia	14,773	0.6	18.7
Entamoeba histolytica	2,486	0.4	3.1
Dientamoeba fragilis	1,588		2.0
Balantidium coli	21		
Isospora belli	3		
Nonpathogenic	29,482	7.6	37.3
Nematodes	29,107		36.8
Ascaris lumbricoides	9,207	2.4	11.6
Trichuris trichiura	8,796	2.3	11.1
Enterobius vermicularis	7,088	1.8	9.0
Hookworm	3,216	0.8	4.1
Strongyloides stercoralis	757	0.2	1.0
Trichostrogylus spp.	22		
Heterodera spp.	21		
Trematodes	396		
Clonorchis/Opisthorchis	210	0.05	0.3
Schistosoma mansoni	143	0.04	0.2
Fasciolopsis buski	5		
Heterophyes heterophyes	3		
Fasciola hepatica	1		
Metagonimus yokogawai	1		
Paragonimus westermani	1		
Cestodes	1,205		1.5
Hymenolepis nana	946	0.2	1.2
Taenia spp.	209	0.5	0.3
T. solium	6		
T. saginata	45		
Taenia spp. unknown	158	0.04	0.2
Hymenolepis diminuta	23		
Diphyllobothrium latum	25		
Diphylidium caninum	2		

*Adapted from Center for Disease Control, Intestinal Parasite Surveillance Annual Summary 1976, issued August 1977. Does not include laboratories in Guam, Puerto Rico or Virgin Islands.

†Percentages are not calculated for parasites identified less than 100 times.

in performing parasitology tests, so figures vary from state to state. A recent summary of malaria cases in the United States[15] shows that there has been a gradual and continuing increase in the number of malaria cases seen each year.

The diagnosis of most parasitic infections must be confirmed in the laboratory, usually by demonstrating a stage of the parasite in a patient's specimen and identifying the parasite on the basis of morphologic characteristics. In some instances, serologic or culture procedures may be helpful. Serodiagnosis is especially useful for parasitic diseases which involve tissue invasion and in which there is no diagnostic stage in blood or stool.

This chapter outlines satisfactory methods of specimen collection, methods for examining specimens and criteria for identifying parasites. Histories of each parasite and the disease it causes are available in many excellent textbooks of parasitology and tropical medicine.[1,5,6,8,24,26,28,45,50,54] Pathology is also thoroughly covered in other publications.[8,27,41,46,53]

GENERAL CONSIDERATIONS

The specimens sent most frequently for parasitologic examination are fecal specimens for helminths and protozoa; perianal preparations for *Enterobius* eggs; and blood films for malaria parasites, trypanosomes and microfilaria. Table 12:2 lists many of the types of specimens which can be expected in the parasitology laboratory, techniques for proper collection and criteria for "unacceptable" specimens (those which are not suitable for examination unless others are unavailable). Additional information on specimen collection and processing is provided in several manuals.[20,23,31,49] Details are provided on methods which have proved to be highly satisfactory for fecal and blood specimens and are widely used in clinical laboratories in the United States.

SPECIMEN PROCESSING AND EXAMINATION

Collection of Fecal Specimens for Intestinal Parasites

Feces should be collected in clean, dry, waterproof containers with close-fitting lids (for example, waxed cardboard ice cream cartons). Contamination with soil, urine or water must be avoided. Specimens may contain mineral oil, antidiarrheal or radiopaque compounds for at least one week after their administration and therefore may be difficult if not impossible to examine. Antibiotic therapy may reduce the number of excreted protozoa. Because some organisms are shed intermittently, multiple specimens should be collected, e.g., 3 specimens taken at 2 to 3 day intervals are recommended.[47]

Processing of Fecal Specimens

Specimens should be delivered to the laboratory and examined as soon as possible—preferably within 2 hours. If delay is unavoidable, specimens should be preserved in a fixative. A two-vial preservation technique (one vial of 10% formalin and one of Polyvinyl alcohol (PVA) - fixative) is valuable for specimens which must be mailed or be delayed in submission. The two-vial preservation system allows optimal material for both concen-

Table 12:2. Checklist for Specimen Collection

Specimen	Proper Collection	Unsuitable Specimens
Abscess fluid	Aspirate aseptically process immediately.	Cotton swabs, frozen specimens.
Blood	Thick and thin smears from fingersticks or venipuncture.	Clotted blood, blood with anticoagulants.
Bronchial brush, Pulmonary aspirates Cerebrospinal fluid	Transport in sterile saline sealed tubes; see abscess fluid.	Dried specimen.
Duodenal fluid	Use "Entero-test" string as alternative to tube aspiration.	Cotton swabs.
Feces	Clean waterproof container for fresh specimens; formalin and polyvinyl alcohol (PVA) for preserved specimens; three stools taken two days apart. Specimens with barium, bismuth, mineral oil may be difficult to examine for a week.	Stools frozen, contaminated with urine or water; or from patients taking broad-spectrum antibiotics.
Hair	Pluck near root with forceps; carry in clean envelope or petri dish.	
Perianal specimens	For pinworm, use clear "scotch" tape.	Frosted tape.
Proctoscopic/Sigmoidoscopic specimens	Use sterile serological pipette to aspirate or curette.	Cotton swab, dry swab.
Serum (for serodiagnosis)	Allow blood to clot at room temperature before refrigeration or centrifugation.	Hemolyzed blood or blood with anticoagulant.
Skin	Clean dry envelope or petri dish. Alternatively, place one drop of mineral oil on lesion, scrape gently with scalpel, apply scrapings to glass slide and add coverslip.	
Sputum, saliva	Sputum cup or covered sterile container.	Saliva for *Pneumocystis*.
Tissue	Portion should be fixed for sectioning and staining; remainder for touch or squash preparations.	Bone, cotton swabs.
Urine	First morning collection, 20–50 mL.	Swabs.
Urogenital	Moist swab in culturette or aspirate into sterile container.	Dry swab, frozen specimen.

tration and staining;[14] with each preservative, 1 part of feces must be thoroughly mixed with 3 parts of fixative.

Specimens should always be examined macroscopically for adult worms or tapeworm proglottids and portions of the specimen that contain mucus or blood should be selected for preservation and examination. The consistency of the stool (liquid, soft, formed) and the presence of blood or mucus should be noted.

Preparation of Wet Mounts

Prepare direct wet mounts to look for motile trophozoites from fresh specimens, especially soft or liquid ones. Wet mounts prepared from formalin-fixed specimens and from concentrates obtained by various concentration techniques can be used to detect protozoan cysts, helminth eggs and larvae, but are not satisfactory for detecting protozoan trophozoites.

To prepare a wet mount, a bit of specimen should be removed with an applicator stick and mixed in a drop of physiological saline on a 2 x 3 inch glass slide. A second portion can be mixed with a drop of iodine (Dobell and O'Connor or Lugol's, diluted 1:5). Each portion should be covered with a #1 thickness, 22 mm^2 coverslip. Mounts can be sealed with a 50:50 petroleum jelly-paraffin mixtures (Vaspar) to retard drying and allow oil immersion to be used if needed. The density should be such that newspaper print can just be read through the mount.

Saline and iodine mounts can be prepared also from concentrates. Direct mounts of formalin-fixed feces as an additional evaluation probably add little.[17]

Examination of Wet Mounts

Entire wet mount preparations should be systematically scanned with a 10X or 20X objective for detection of parasites. Higher magnifications (40X or 100X objectives) should be used to identify protozoan cysts.

Saline wet mounts of fresh feces are particularly useful for detecting motile organisms such as larvae or protozoan trophozoites. The type of motility of flagellate trophozoites is often sufficient to determine species. Motility of amebic trophozoites can also be evaluated in this type of mount, but permanent stains are usually necessary for detailed morphologic study and species identification. Protozoan cysts can often be more readily detected in unstained saline mounts because of their refractility; however, stained preparations (either temporary iodine stains or permanent stains) are needed to see morphologic details. In wet mounts, helminth eggs can best be identified from the unstained preparates as iodine tends to obscure characteristics of eggs. Although permanent stains are prepared primarily for detection of protozoa, helminth eggs may be detected in them.[51]

Unstained wet mounts of formalin-fixed feces are useful also in detecting and identifying helminth eggs and larvae and in detecting protozoan cysts. Chromatoid bodies, fibrils and sometimes nuclei can be seen in formalin-fixed cysts, but iodine stains are usually needed for observing nuclear structure. Iodine brings out nuclei and fibrils of cysts but does not stain chromatoid bodies. Glycogen stains brown or red-brown. Cysts are less refractile in iodine mounts than in saline or unstained mounts and may be more easily overlooked. Oocysts of *Cryptosporidium* can be readily detected in iodine-stained wet mounts, especially if they are numerous. The oocysts differ from yeasts in that they are more nearly spherical, have a thinner wall, may contain four crescentic sporocysts and have a single, conspicuous, refractile granule. Yeast cells stain immediately with iodine; oocysts of *Cryptosporidium* remain unstained for about 15 min.

Protozoan trophozoites sometimes can be recognized in formalin-fixed material and, in the case of flagellates, can be identified on the basis of visible morphology. Amebic trophozoites, however, cannot be reliably identified in this type of mount, so permanent stains are necessary.

Fecal Concentration Procedures

Various fecal concentration procedures have been described. They allow easier detection of parasitic forms by decreasing the amount of background material and by concentrating the parasites on the basis of differences in specific gravity of parasitic forms and fecal material. Concentration procedures allow enriched recovery of helminth eggs and larvae as well as protozoan cysts.[30] The two general methods are sedimentation (in which parasites are concentrated in the sediment by gravity or centrifugation) and flotation (in which parasites float on a solution of high specific gravity).

A disadvantage of the flotation methods is that the high specific gravity of the solution causes opercula to open and may cause distortion of protozoan cysts and larvae; however, this can be prevented by first fixing the specimen in formalin, as described for the zinc sulfate procedure.

The two methods described here are the formalin-ethyl acetate method[37,52] and a modification of the zinc sulfate flotation method,[12] which uses formalin-fixed feces.[2] The formalin-ethyl acetate method generally allows good recovery of parasite forms, including *Cryptosporidium* oocysts, although *Giardi lamblia* cysts and *Hymenolepis nana* eggs may not concentrate well. The formalin-zinc sulfate method allows most parasites to be recovered but is not satisfactory for recovering schistosome eggs, because they do not float. This method is not recommended, therefore, for laboratories receiving large numbers of specimens from patients in trematode endemic areas. However, this method has the advantage of providing a clearer background without the grit often found in sedimented specimens. Either method should be satisfactory for routine diagnostic parasitology.

Formalin-ethyl Acetate Method for Fresh Specimens

1. Mix a sample of stool about the size of a marble in saline so that the resulting suspension will yield approximately 1 mL of sediment from 10 mL of suspension.
2. Strain 10 mL of suspension through wet gauze into a 15 mL conical centrifuge tube. (Conical paper cups with the tips cut off are convenient funnels.)
3. Centrifuge at 2500 rpm (650 x g) for 1-2 min. Decant supernate. Sediment should be approximately 1 mL. If there is not enough sediment, add more suspension from original specimen. If there is too much sediment, resuspend and pour off a portion, add additional saline and centrifuge again.
4. Resuspend the sediment in fresh saline, centrifuge and decant.
5. Add about 9 mL of 10% formalin to the sediment. Mix thoroughly and allow to stand at least 5 min.
6. Add 3 mL of ethyl acetate, stopper tube and shake tube in an inverted position for at least 30 s. Remove stopper carefully.
7. Centrifuge at 2000 rpm (450-500 x g) for 1-2 min. After centrifugation, there will be 4 layers (from top to bottom: ethyl acetate, debris, formalin and sediment).
8. Ring the debris plug with an applicator stick and pour off the top 3 layers. Debris that clings to the sides of the tube can be removed with a cotton-tipped applicator stick.
9. Mix the sediment with the few drops of fluid remaining in the tube and transfer drops to a slide to make unstained and iodine wet mounts for microscopic examination. Mounts can be sealed with Vaspar if desired.

Formalin-ethyl Acetate Method for Preserved Specimens

1. Mix formalinized specimen.
2. Depending upon the amount of specimen, strain a quantity through wet gauze into a conical 15 mL tube to give about 0.5-0.75 mL of sediment.
3. Add water to make 10 mL of suspension, mix and centrifuge at 2000-2500 rpm (500-650 x g) for 1-2 min.
4. Decant supernate. If desired, wash again. If sediment amount is incorrect, adjust by adding more specimen or pouring off specimen and adding water.
5. Add 9 mL of 10% formalin and mix thoroughly.
6. Add 4 mL of ethyl acetate, stopper tube and shake in an inverted position for at least 30 s. Remove stopper with care.
7. Centrifuge at 2000 rpm (450-500 x g) for 1-2 min.
8. After centrifugation, there will be 4 layers (ether, debris, formalin and sediment). Ring the debris plug with an applicator and pour off the top 3 layers. Remove debris from sides of tube with a cotton swab.

9. Mix sediment in the small amount of fluid remaining in tube. Prepare unstained and iodine wet mounts from sediment for microscopic examination. Mounts can be sealed with Vaspar, if desired.

Formalin-zinc Sulfate Flotation Method

The original zinc sulfate flotation procedure was developed by Faust et al[12] in 1938 for recovering protozoan cysts, helminth eggs and larvae. Since that time, various modifications have been described. Use feces-formalin suspensions which have stood at room temperature at least 30 min for the modified procedure described here.

The specific gravity of the zinc sulfate solution is critical to the efficiency of this technique. It should be 1.200 but must not be lower than 1.195 nor higher than 1.200.

Place feces-formalin suspensions in round-bottomed tubes.

1. Centrifuge feces-formalin suspensions at approximately 1800 rpm (400-450 x g) for 3.5 min. Allow the centrifuge to come to a full stop.
2. Decant the supernate from each tube and drain the last drop against a clean section of paper towel.
3. Place the tubes of sediment in a rack that holds them upright and steady.
4. Add zinc sulfate to each tube to within 1 cm of the rim.
5. Using two applicator sticks, mix the packed sediment thoroughly so that no coarse particles remain.
6. Immediately centrifuge at 1500 rpm (400 x g) for 1 min.
7. As soon as the centrifuge stops, carefully transfer the tubes to the rack. Avoid disturbing the surface films, which now contain the floating parasites.
8. Allow the tubes to stand for 1 min.
9. Place a drop of saline and a drop of iodine on a 2 x 3 inch glass slide.
10. Transfer a loopful of the surface film to the slide with a wire loop that has a horizontal bend. Deposit next to the drop of saline. Deposit a second loopful of the surface film next to the drop of iodine. In making the transfer, carefully touch the surface film but do not penetrate the surface.
11. Using the heel of the loop, mix first the fecal drop and the saline, then the fecal drop and the iodine.
12. Flame the loop before proceeding to the next tube.
13. Superimpose a clean 22 mm square, No. 1 coverslip on each drop. Avoid trapping air bubbles.
14. Place each prepared slide in a petri dish containing moist paper toweling to retard evaporation, or seal with Vaspar.
15. Examine the mounts within an hour. A longer holding period may cause distortion and make identifying some stages more difficult.

Permanent Stains

Either fresh or PVA-fixed feces can be used to prepare fecal smears for staining. If smears are prepared from fresh feces, they must not be allowed to dry before placing them in Schaudinn's fixative. Prepare smears by spreading a portion of feces in an even film on a 1 x 3 inch clean glass slide. If the material is too watery to adhere to the slide, a drop of Mayer's albumin or normal (parasite-free) feces can be mixed with the specimen. Smears should be left in Schaudinn's fixative at room temperature for at least 1 h or at 50°C for 5 min. Slides may be left in fixative for several days.

Films of PVA-fixed feces are prepared by thoroughly mixing the material before removing the samples. The smear should be spread in an even square so that it extends to the top and bottom edges of the slide. Slides of PVA-fixed material should dry at least 4 h at 35°C (preferably overnight) before staining.

Staining

The Wheatley modification of the trichrome stain is most widely used in the United States, although hematoxylin stains are used in many laboratories. The trichrome stain method which has the advantage of using stable reagents and requiring a relatively short time is described here.

Trichrome Stain Schedule for PVA-preserved feces:

Reagent	*Time*	*Purpose*
1. 70% Alcohol plus iodine	10–20 min	Remove mercuric chloride (hydration)
2. 70% Alcohol	3–5 min	Remove iodine (hydration)
3. 70% Alcohol	3–5 min	Wash (hydration)
4. Trichrome stain	6–8 min	Stain
5. 90% Alcohol acidified	2 s	Destain
6. 95% Alcohol	Rinse	Stop destaining
7. 95% Alcohol	5 min	Dehydration
8. Carbol-xylene	10 min	Clearing and dehydration
9. Xylene	10 min	Clearing

Trichrome stain schedule for unpreserved fecal smears:

Stain as described for PVA-fixed fecal smears, except use times noted:

1. Schaudinn's fixative	5 min at 50°C or 1 h at room temp.

2. 70% Alcohol plus iodine	1 min
3. 70% Alcohol	1 min
4. 70% Alcohol	1 min
5. Trichrome stain	2–8 min
6. 90% Alcohol acidified	Brief dip
7. 95% Alcohol	Rinse
8. 95% Alcohol	Rinse
9. Carbol-xylene	1–3 min

A modification of the Trichrome technique which employs a substitute for xylene has been described.[32] Trichrome stains should be examined with oil immersion objectives. Specimens can be screened with a 50X immersion objective, but identification should be made with 100X oil immersion objective.

Trichrome stain reaction:

The appearance of trichrome-stained fecal films varies with the specimen, thickness of the smear and age of the stain. Nuclear structures and cytoplasm should be readily differentiated. Usually chromatin stains red to red-purple and cytoplasm stains green to blue. Inclusions in cytoplasm may stain green to red. Background material usually stains green to blue-green.

There are several modified acidfast and fluorescent stains now available for detection of oocysts of *Cryptosporidium*[12,43] and *Isospora*.[33] The modified acidfast stains are not as sensitive as the fluorescent stains but are quicker and do not require a fluorescence microscope. The procedure described here is a modification of the Kinyoun carbolfuchsin stain.

Smears to be stained for oocysts of *Cryptosporidium* and *Isospora belli* can be made from fresh or formalin-preserved feces. These smears should be spread quite thinly.

Kinyoun carbolfuchsin stain for Cryptosporidium and *Isospora* oocysts in fresh and formalin-preserved feces:

Reagents

1. Absolute methanol
2. 50% Ethanol
3. Kinyoun carbolfuchsin stain

a. Basic fuschin	4 g
b. 95% Ethanol	20 mL
c. Phenol, liquid	8 mL
d. Distilled water	100 mL

4. Acid-alcohol

a. Sulfuric acid, concentrated	10 mL
b. 95% Ethanol	90 mL

5. Counterstain

a. Malachite green dye	3 g
b. Distilled water	100 mL

Loeffler's methylene blue may be used for counterstain in place of malachite green. Most laboratories doing modified acidfast stains for *Nocardia* have this stain available.

Stain Schedule

1. Make a thin smear from fresh or preserved feces. Smears from formalized material should be made from the upper layers of sediment after centrifuging the specimen for 2 min at 300-400 x g (1000-1500 rpm on a table-top centrifuge).
2. Dry the smear with gentle heat; do not exceed 60°C.
3. Fix with absolute methanol for 30 s; allow to drain and dry.
4. Flood the smear with carbolfuchsin for 1 min at room temperature.
5. Rinse thoroughly with tap water.
6. Decolorize with acid-alcohol for 2 min.
7. Rinse with tap water.
8. Counterstain with malachite green for 2 min.
9. Rinse with tap water, dry thoroughly and mount with coverslip.
10. Examine at 40X and 100X magnifications. Oocysts of *Cryptosporidium* cysts stain red, contain refractile granules and are 4-6 μm in size. Yeasts and fecal debris stain green. Occasionally, small spherical, encapsulated yeasts may stain red. These yeasts usually occur in tetrads, are smaller (2-3 μm) an stain uniformly. Oocysts of *Isospora* are ellipsoidal and 28-30 μm long. Their contents stain deep red and appear granular, but many oocysts will remain only partially stained.

Egg Counts

Egg counts have been used to estimate the intensity of infection with helminths which have a relatively constant output of eggs. Although developed primarily for estimating hookworm burden, egg counts have been used for other nematodes including *Ascaris* and *Trichuris*. Counts are an estimate at best, because the daily output of eggs can vary and the number of eggs recovered will be affected by the consistency of the specimen as well as the diet and immune system of the host. Egg counts can be helpful in determining the role of helminth infection in the disease (e.g., anemia) and the desirability of treatment or in assessing the success of treatment. Stoll's egg-counting technique[44] is a dilution technique. Both the standard smear of Beaver[4] and the Kato smear[29] depend upon a smear containing a measured amount of feces. If standard smears[4] contain about 2 mg of feces per coverslip, the number of eggs per gram can be estimated by multiplying the number of eggs per coverslip by 500. If no eggs are detected in a 2 mg preparation, the estimate should be reported at <500 eggs/g. Detailed methods for performing counts are outlined in the Centers for Disease Control manual.[31]

Perianal Preparations

Collection of perianal preparations:

Perianal preparations are most useful in detecting *Enterobius* (pinworm) infection,[10] although they can also be helpful in diagnosing *Taenia* infections. Either cellulose tape (Fig. 12:1) or vaseline-paraffin swabs can be used. Specimens should be collected between 9 p.m. and midnight or in the early morning before the patient has bathed or defecated. Several specimens should be collected over a number of days to rule out infection, because the female worms may not migrate each day. Only 55% of pinworm infections will be detected[38] with 1 cellulose-tape preparation, although 90% can be diagnosed with 3 specimens. Persons with more severe infections will have a higher proportion of positive specimens.

Examination of cellulose-tape preparations:

Cellulose-tape preparations should be examined with the 10X objective and reduced light. Any finding should be confirmed with the high dry objective. The entire tape (including the edges) should be screened. For easier examination, tape can be cleared with toluene (i.e., pull the tape back, add a drop of toluene and smooth the tape into position with applicator sticks, a piece of gauze or cotton). *Enterobius* eggs remain viable and infective for weeks, so all specimens must be handled carefully.

Examination of Duodenal Material

Material obtained by duodenal aspiration should be centrifuged and the sediment examined. Both direct wet mounts and permanently-stained smears may be prepared. Although in most instances specimens are collected to detect *G. lamblia* trophozoites, strongyloidiasis or other parasite infections may be diagnosed. An alternative to aspiration is the string test (Enterotest, Hedico Corp.).[3] Material from the bile-stained portion of the string should be scraped away and used for wet mount and stain preparation. At least 3 stool specimens should be examined before collection of duodenal material as in most instances it is possible to detect organisms in stool.[21]

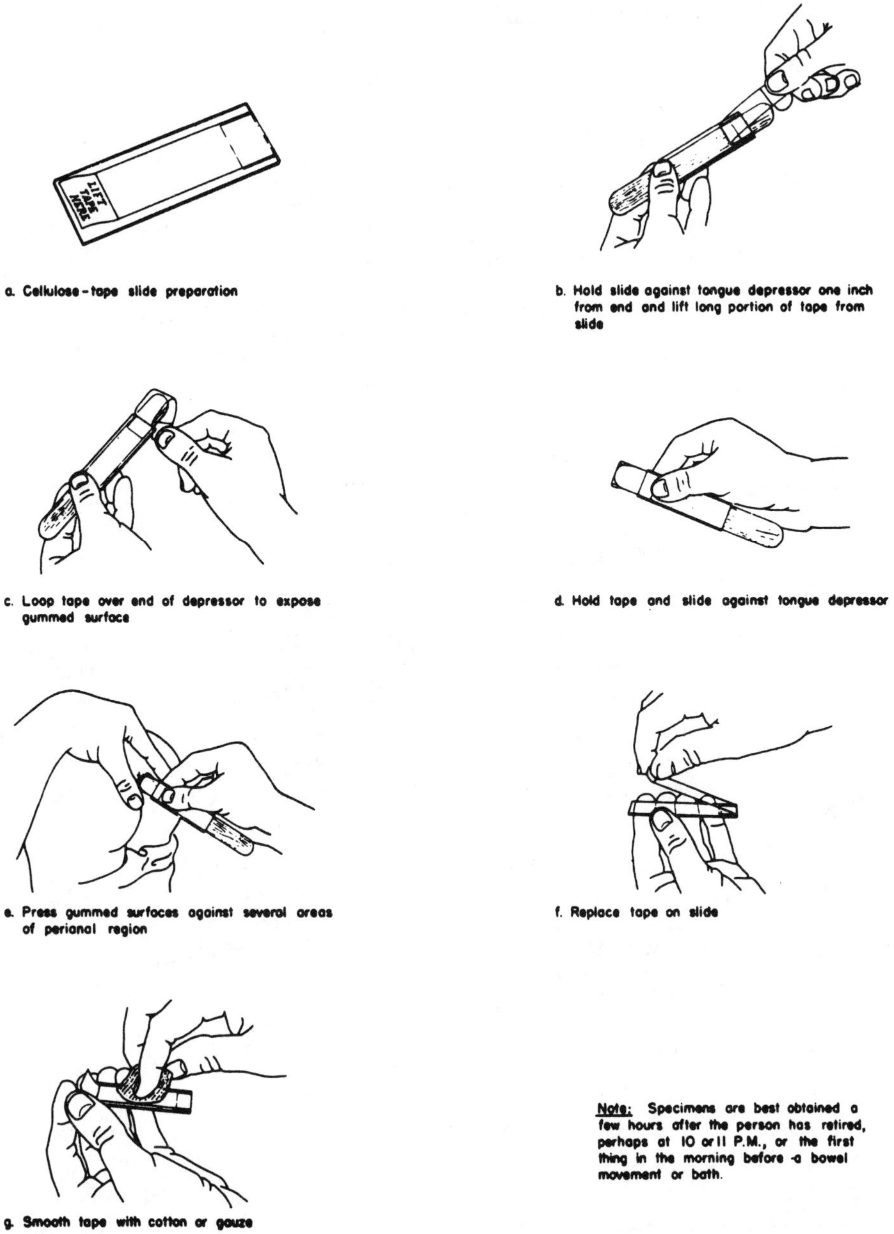

a. Cellulose-tape slide preparation
LIFT TAPE HERE
b. Hold slide against tongue depressor one inch from end and lift long portion of tape from slide
c. Loop tape over end of depressor to expose gummed surface
d. Hold tape and slide against tongue depressor
e. Press gummed surfaces against several areas of perianal region
f. Replace tape on slide
g. Smooth tape with cotton or gauze
Note: Specimens are best obtained a few hours after the person has retired, perhaps at 10 or 11 P.M., or the first thing in the morning before a bowel movement or bath.

Figure 12:1. Use of cellulose-tape slide preparation for diagnosis of pinworm infections (adapted from Brooke, Donaldson, and Mitchell, 1949).

Culture Methods for Protozoa

Culture methods for various intestinal, blood and tissue protozoa have been described,[48] although they are not widely used by diagnostic laboratories in the United States. Many amebae, including *Entamoeba histolytica*, can be grown in various diphasic media that have a slant overlaid with a broth. When inoculated with feces, bacteria are supplied by the inoculum, but when culturing bacteria-free material such as amebic liver abscess drainage, the medium must also be inoculated with bacteria such as *Clostridium perfringens*. Culture may enhance detection of *Trichomonas vaginalis*.[19] Cultural procedures for intestinal protozoa are reviewed in the Centers for Disease Control manual.[31] Cultural methods for malaria are research techniques rather than diagnostic procedures, but cultures can aid in diagnosing leishmanial and trypanosomal infections.

Blood Specimens

Collection of blood for thick and thin smears:

The use of films made from blood without added anticoagulant, such as those prepared with blood from a finger stick, are preferred because staining will be better. It is important to note that, although blood containing an anticoagulant can be used, films should be prepared within 1 hour. Malarial parasite stages in peripheral blood vary in number and type according to the part of the life cycle and some specimens contain so few parasites that they may be overlooked; at other times, ring forms may predominate and make species identification difficult. The best time to obtain blood for smears is about midway between paroxysms of chills and fever. With suspected *Plasmodium falciparum* malaria (especially in the early acute stage), the parasites may be more numerous between peaks of fever. Some microfilariae (e.g., *Wuchereria bancrofti*) may be found only at certain times of the day and films should be prepared appropriately.

Preparation of blood films:

Prepare both thick and thin blood films. Use only clean, unscratched slides. A thick film should cover an area about the size of a dime (15-20 mm) and should not be so thick that it will flake off during the staining process. Newspaper print can be read through a properly made thick film before drying. To prepare a thick film, place 2-3 small drops of blood on a clean glass slide and swirl or "puddle" the drops into a thick film with the corner of another slide. Allow the film to dry at room temperature in a horizontal position. Thick films must not be heated because heat will fix the erythrocytes. Thick films can be stained after they have been dried overnight or at least for several hours.

Thin films should have a large portion that is one-cell-layer thick, with erythrocytes slightly separated. Prepare a thin film by placing 1 drop of blood toward the end of a clean glass slide. Use a second slide, held at an acute angle, to spread the film. Back the spreading slide into the drop of blood and allow the blood to spread along the width of the slide. With a smooth, even motion, push the spreader slide toward the other end, leaving a thin film.

Thin films must be fixed before they are stained. They should be immersed in methyl alcohol for a few seconds and allowed to dry. Thick films should not be fixed so that the aqueous stain will lyse the unfixed red cells. It is easiest to make thick and thin films on separate slides, although the thick film may be on one end and the thin film on the other of the same slide. In such combined films, care must be taken not to expose the thick films to alcohol vapors, which might fix them.

Staining of blood films:

Giemsa is preferred over Wright's stain for blood films for two reasons: 1) Giemsa stains malarial parasites better than does Wright's stain, and 2) Wright's stain, which contains alcohol, requires separate dehemoglobinization of thick films, whereas the aqueous Giemsa stain allows dehemoglobinization during the staining process. For best staining of malarial parasites, blood films should be stained within 3 days after preparation. Erythrocytes in thick films become fixed and will not lyse if staining is delayed too long.

Fresh, buffered Giemsa stain should be prepared from stock Giemsa each day. Stock solution is troublesome to make and is best purchased already prepared. Giemsa stain is prepared by mixing 1 part stock Giemsa and 50 parts phosphate-buffered water, pH 7.0-7.2 (add 100 mL of M/15 phosphate buffer, pH 7.0-7.2, to 900 mL of distilled water). Stain 45 min. Rinse thin films briefly in buffered water and air dry. Thick films should be rinsed for a longer time.

The magnification selected for examining films is determined by the type of parasites expected. Films should be screened with low power (10X) to detect microfilariae and with 50X and 100X oil immersion to detect blood protozoa. Thick films should be examined for no less than 5 min and thin films for no less than 30 min before they are reported as negative.

Giemsa stain reaction:

Giemsa stains leukocyte nuclei purplish-blue, cytoplasm and cytoplasmic granules various colors, depending on the type of leukocyte. Malarial parasites have blue cytoplasm and red to red-violet chromatin. Pigment varies from golden-brown to brown-to-black. Schuffner's stippling appears as fine red or pink granules if the pH of the stain is correct.

Miscellaneous Procedures

Specimens other than those discussed must be treated individually depending on the type of specimen and organisms that may be present (Table 12:3). In most cases, direct preparations should be examined. Fluids should be centrifuged and the sediment examined. Bronchoalveolar lavage specimens should be centrifuged and sediment stained with Giemsa and a methenamine silver nitrate[9,25] or toluidine blue O stain[16,22] for *Pneumocystis carinii*. Bronchial brushings may be examined, but generally have proved less useful than lavage.[34] Induced sputum has been used successfully for diagnosis of *P. carinii* infection in AIDS patients.[7,36] Sputum can be examined directly for the presence of *Paragonimus* eggs or larvae of *Strongyloides* and can be mixed with PVA-fixative on a slide for later staining to detect the presence of *E. histolytica*. Urine sediment can be examined directly for the presence of *T. vaginalis* or *Schistosoma haematobium* eggs. Direct wet mounts of cerebrospinal fluid should be examined for the presence of free-living amebae and trypanosomes. Giemsa-stained films can be prepared from various fluids with or without previous centrifugation. Direct mounts of vaginal and urethral swabs, duodenal drainage and aspirates and sigmoidoscopic material should be examined. Permanent stains can be made from smears of these materials fixed in Schaudinn's fixative or portions can be preserved in PVA-fixative for later staining. Tissue should be examined by preparing both direct mounts of crushed fragments of the tissue in saline and stains of impression smears. Histologic sectioning and staining methods can be used for biopsied tissue obtained for diagnosis of trichinosis or other tissue parasite infections.

QUALITY CONTROL

Quality control for procedures in the parasitology laboratory is similar to that in other laboratory areas. New reagents should be checked for proper reactivity, preferably in parallel with known reagents, before being used. Specific gravity of zinc sulfate solutions should be checked and appropriately adjusted at least once a week. Permanent stains of fecal smears should be controlled with a control smear of known staining qualities each time a batch of slides is stained. Positive material is recommended for a positive control and feces that do not contain parasites as negative control. Host cells such as epithelial cells and inflammatory cells in the negative specimen can be used to check the stain quality.

The laboratory should have a set of reference materials that are reviewed periodically. Positive slides and formalin-fixed materials are particularly useful, but color atlases[1,35,42,53,54] and collections of color photomicrographs[39,40,41] are also helpful. Participation in a proficiency testing program provides external control of the quality of work performed.

Table 12:3. Parasites Which Can be Found in Nonfecal Specimens for Diagnosis

Specimen	Infectious Agent	Stage
Sputum	*Paragonimus westermani*	Eggs
	Ascaris lumbricoides, hookworm or *Strongyloides stercoralis* (rare).	Larvae
Bronchoalveolar lavage, bronchial brushings or induced sputum	*Pneumocystis carinii*	Trophozoites and cysts
Anal swab or cellulose tape preparation	*Enterobius vermicularis*	Eggs
	Taenia spp.	Eggs
Duodenal aspirate	*Strongyloides stercoralis*	Larvae
	Isospora belli	Oocysts
	Giardia lamblia	Trophozoites
Muscle biopsy	*Trichinella spiralis*	Larvae
Cyst fluid	*Echinococcus granulosis*	Hydatid sand
	Entamoeba histolytica	Trophozoites
Rectal biopsy	*Cryptosporidium*	Cysts
	Schistosoma mansoni	Eggs
Urine or urinary bladder biopsy	*Trichomonas vaginalis*	Trophozoites
	Schistosoma haematobium, *S. mansoni*	Eggs
Liver biopsy	*Schistosoma* spp.	Eggs
	Nonhuman ascarids (visceral larval migrans)	Larvae
	Entamoeba histolytica	Trophozoites
	Echinococcus granulosus	Hydatid cyst
	Capillaria hepatica	Eggs
	Trypanosoma cruzi	Amastigotes
	Leishmania donovani	Amastigotes
Other tissue biopsy	*Taenia solium*	Cysticercus
	Multiceps multiceps	Coenurus
	Echinococcus granulosus	Hydatid cyst
	Onchocerca volvulus	Adult or microfilaraia
	Other filariae	Adult
	Trypanosoma cruzi	Amastigotes
	Entamoeba histolytica	Trophozoites
	Pneumocystis carinii	Cysts and trophozoites
	Toxoplasma gondii	Cysts and tachyzoites
	Leishmania	Amastigotes
	Microsporidea	Cysts
Cerebrospinal fluid	*Trypanosoma* spp.	Trypomastigotes
	Free-living amebae	Trophozoites
	Toxoplasma gondii	Tachyzoites
Blood	*Plasmodium* spp.	Trophozoites, schizonts, gametocytes
	Babesia spp.	Trophozoites
	Toxoplasma gondii (rare)	Tachyzoites
	Leishmania donovani (rare)	Amastigotes
	Filariae	Microfilariae
	Trypanosoma spp.	Trypomastigotes

SAFETY

Many specimens submitted for parasitic examination are potentially infectious. Blood or tissue specimens containing malarial parasites, trypanosomes, *Leishmania* species or *Toxoplasma* organisms may cause infections if there is a break in the skin. Fresh fecal specimens containing protozoan cysts, trophozoites of *Dientamoeba fragilis*, eggs of *Enterobius vermicularis*, *Hymenolepis nana* or *Taenia solium*, and specimens containing filariform larvae of *Strongyloides stercoralis* can be infectious. Older fecal specimens may contain filariform larvae of hookworm *S. stercoralis*, or *Trichostrongylus* spp. and embryonated eggs of *Ascaris lumbricoides* or *Trichuris trichiura*. *Ascaris* eggs sometimes survive and embryonate even in formalin. In addition, fecal specimens can contain bacteria such as *Salmonella* and *Shigella* or viruses. All blood and fecal specimens should be considered infectious, as should all materials used in processing specimens, and all such specimens and used reagents should be disposed of in a manner similar to that recommended for hazardous bacterial specimens.

Storage of ether presents a special problem. The Centers for Disease Control currently recommend[13] that ether be stored in small containers (0.25 lb or 1 lb depending on workload) in a storage cabinet for flammable solvents or, after opened, on open shelves in a well-ventilated area. Storage of ether in an explosion-proof refrigerator is not recommended because the flash point of ethyl ether (-49°F) is significantly lower than refrigerator temperature. Vapors can accumulate in the refrigerator and, when the door is opened, can be ignited by a spark or flame in the room to cause an explosion.

REFERENCES

1. Ash LR, Orihel TC. Atlas of human parasitology. 2nd ed. Chicago: American Society of Clinical Pathologists Press, 1984.
2. Bartlett MS, Harper K, Smith N, Verbanac P, Smith JW. Comparative evaluation of a modified zinc sulfate flotation technique. J Clin Microbiol 1978;7:524-58.
3. Beal CB, Viens P, Grant RGI, Hughs JM. A new technique for sampling duodenal contents. Am J Trop Med Hyg 1970;19:349-52.
4. Beaver PC. The standardization of fecal smears for estimating egg production and worm burden. J Parasitol 1950;36:451-5.
5. Beaver PC, Jung RC, eds. Animal agents and vectors of human diseases. 5th ed. Philadelphia: Lea and Febiger, 1985.
6. Beaver PC, Jung RC, Cupp EW. Clinical parasitology. 9th ed. Philadelphia: Lea and Febiger, 1984.
7. Bigby TD, Margolskee D, Curtis JL, et al. The usefulness of induced sputum in the diagnosis of *Pneumocystis carinii* pneumonia in patients with the acquired immunodeficiency syndrome. Am Rev Respir Dis 1986;133:515-18.
8. Binford CH, Connor DH, eds. Pathology of tropical and extraordinary diseases; vol 1. Washington, DC: Armed Forces Institute of Pathology, 1976.
9. Brinn NT. Rapid metallic histological staining using the microwave oven. J Histotech 1983;6:125-9.

10. Brooke MM, Donaldson AW, Mitchell RB. A method supplying cellulose tape to physicians for diagnosis of enterobiasis. Pub Health Rep 1949;64:897-901.
11. Carroll MJ, Cook J, Turner JA. Comparison of polyvinyl alcohol and formalin-preserved fecal specimens in the formalin-ether sedimentation technique for parasitological examination. J Clin Microbiol 1983;18:1070-2.
12. Casemore DP, Armstrong M, Sands RL. Laboratory diagnosis of cryptosporidiosis. J Clin Pathol 1985;38:1337-41.
13. Centers for Disease Control. Laboratory safety at the Centers for Disease Control. Atlanta: Centers for Disease Control, 1974; HEW publication no. (PHS)76-8118.
14. Centers for Disease Control. Intestinal parasite surveillance annual summary, 1985. Atlanta: Centers for Disease Control, 1977.
15. Centers for Disease Control. Malaria Surveillance Annual Summary, 1985. Atlanta: Centers for Disease Control, 1986.
16. Chalvardjian AM, Grawe LA. A new procedure for the identification of *Pneumocystis carinii* cysts in tissue sections and smears. J Clin Pathol 1963;16:383-4.
17. Estevez EG, Levine JA. Examination of preserved stool specimens: lack of value of the direct wet mount. J Clin Microbiol 1985;22:666-7.
18. Faust EC, D'Antoni JS, Odom V, et al. A critical study of clinical laboratory techniques for the diagnosis of protozoan cysts and helminth eggs in feces. Am J Trop Med Hyg 1938;18:169-83.
19. Fouts AC, Kraus SJ. *Trichomonas vaginalis*: Reevaluation of its clinical presentation and laboratory diagnosis. J Infect Dis 1980;141:137-43.
20. Garcia LS, Ash LR. Diagnostic parasitology: clinical laboratory manual. 2nd ed. St. Louis: CV Mosby, 1979.
21. Gordts B, Hemelhof W, Van Tilborgh K, Retore P, Coudranel S, Butzler JP. Evaluation of the new method for routine in vitro cultivation of *Giardia lamblia* from human duodenal fluid. J Clin Microbiol 1985;22:702-4.
22. Gosey L, Howard RM, Witebaky FG, et al. Advantages of a modified toluidine blue-O stain and bronchoalveolar lavage for the diagnosis of *Pneumocystis carinii* pneumonia. J Clin Microbiol 1985;22:803-7.
23. Lennette EH, Balows A, Hausler WJ Jr, Shadomy HJ, eds. Manual of clinical microbiology. 4th ed. Washington, DC: American Society for Microbiology, 1985.
24. Maegraith B. Adams and Maegraith's clinical tropical diseases. 7th ed. London: Blackwell, 1980.
25. Mahan CT, Tale GE. Rapid methenamine silver stain for *Pneumocystis* and fungi. Arch Pathol Lab Med 1978;102:351-2.
26. Manson-Bahr PH, Apted FIC, eds. Manson's tropical diseases. 16th ed. Baltimore: Williams and Wilkins, 1982.
27. Marcial-Rojas RA. Pathology of protozoal and helminthic diseases. Baltimore: Williams and Wilkins, 1971.
28. Markell EK, Voge M. Medical parasitology. 5th ed. Philadelphia: WB Saunders, 1981.
29. Martin LK, Beaver PC. Evaluation of Kato thick-smear technique for quantitative diagnosis of helminth infections. Am J Trop Med Hyg 1968;17:382-91.
30. McMillan A, McNeillage GJC. Comparison of the sensitivity of microscopy and culture in the laboratory diagnosis of intestinal protozoal infection. J Clin Pathol 1984;37:809-11.
31. Melvin DM, Brooke MM. Laboratory procedures for the diagnosis of intestinal parasites. Atlanta: Centers for Disease Control, 1985; HHS publication no. (PHS)PB 85-8282.
32. Neimeister R, Logan AJ, Egleton JH. Modified trichrome staining techniques with a xylene substitute. J Clin Microbiol 1985;22:306-7.
33. NG E, Markell EK, Fleming RL, Fried M. Demonstration of *Isospora belli* by acidfast stain in a patient with acquired immune deficiency syndrome. J Clin Microbiol 1984;20:384-6.

34. Ognibene FP, Shelhamer J, Gill V, et al. The diagnosis of *Pneumocystis carinii* pneumonia in patients with the acquired immunodeficiency syndrome using subsegmental bronchoalveolar lavage. Am Rev Respir Dis 1984;129:929-32.
35. Peters W, Gilles HM. A color atlas of tropical medicine and parasitology. London: Wolfe Medical Publications, 1977.
36. Pitchenik AE, Ganjei P, Torres A, Evans DA, Rubin E, Baier H. Sputum examination for the diagnosis of *Pneumocystis carinii* pneumonia in the acquired immunodeficiency syndrome. Am Rev Respir Dis 1986;133:226-9.
37. Ritchie LS. An ether sedimentation technique for routine stool examinations. Bull US Army Med Dept 1948;8:326.
38. Sadun EH, Melvin DM. The probability of detecting infections with *Enterobius vermicularis* by successive examination. J Pediatr 1956;48:438-41.
39. Smith JW, Ash LR, Thompson JH, McQuay RM, Melvin DM, Orihel TC. Diagnostic parasitology - intestinal helminths. Chicago: American Society of Clinical Pathologists, 1976.
40. Smith JW, Melvin DM, Orihel TC, Ash LR, McQuay RM, Thompson JH. Diagnostic parasitology - blood and tissue parasites. Chicago: American Society of Clinical Pathologists, 1976.
41. Smith JW, Melvin DM, Orihel TC, Thompson JH. Diagnostic parasitology - intestinal protozoa. Chicago: American Society of Clinical Pathologists, 1976.
42. Spencer FM, Monroe LS. The color atlas of intestinal parasites. Springfield, IL: CC Thomas, 1982.
43. Sterling CR, Arrowood MJ. Detection of *Cryptosporidium* spp. infections using a direct immunofluorescent assay. Pediatr Infect Dis 1986;5(suppl):139-42.
44. Stoll NR. Investigations on the control of hookworm disease. XV: an effective method of counting hookworm eggs in feces. Am J Hyg 1923;3:59-70.
45. Strickland GT. Hunter's tropical medicine. 6th ed. Philadelphia: WB Saunders, 1984.
46. Sun T. Pathology and clinical features of parasitic diseases. New York: Masson Publishing USA, 1982.
47. Thomson RB Jr, Hass RA, Thompson JH. Intestinal parasites: the necessity of examining multiple stool specimens. Mayo Clin Proc 1984;59:641-2.
48. Trager W. Cultivation of parasites in vitro. Am J Trop Med Hyg 1978;27:216-22.
49. US Naval Medical School. Medical protozoology and helminthology. Bethesda, MD: National Naval Medical Center, 1965.
50. Warren KS, Mahmoud AAF, eds. Geographic medicine for the practitioner: algorithms in the diagnosis and management of exotic diseases. Chicago: University of Chicago Press, 1978. 150 p.
51. Wood JC, Friendly G, De LaMaza LM. Detection of helminth and larvae in trichrome-stained stool smears. J Clin Microbiol 1982;16:1137.
52. Young KH, Bullock SL, Melvin DM, Spruill CL. Ethyl acetate as a substitute for diethyl ether in the formalin-ether sedimentation technique. J Clin Microbiol 1979;10:852-3.
53. Yamaguchi T. Color atlas of clinical parasitology. Philadelphia: Lea and Febiger, 1981.
54. Zaman V. Atlas of medical parasitology. 2nd ed. Sydney, Australia: Adis Health Science Press, 1984.

CHAPTER 13

SEROLOGY OF PARASITIC INFECTIONS

Kenneth W. Walls, Ph.D.

INTRODUCTION

Many parasitic infections are characterized by having rare or no diagnostic forms of the parasite in blood or fecal samples. Some of the intestinal parasites have prepatent periods in which the diagnostic stages have not yet appeared, or in some cases, therapy has resulted in amorphous forms or interference by nonrelated material which makes visualization of the parasite difficult. Blood and tissue parasites may be sequestered within tissue and few parasites are found free in the circulation. Serologic methods have been developed for most of the diseases in which these problems occur, making diagnosis possible in otherwise questionable cases. A wide variety of serologic tests are available and tests are being improved constantly. These tests vary in both sensitivity and specificity, depending upon the test methodology, the antigen used and the purpose for which the test was designed. Screening tests are highly sensitive but often lack specificity; diagnostic tests are highly specific but sacrifice some sensitivity. Tests for speciation use highly purified antigens and more difficult procedures but clearly identify the infecting agent. Reagents are commercially available for some tests but other reagents must be produced in specialized laboratories. Laboratories undertaking serologic tests should become thoroughly familiar with them before testing clinical specimens. Appropriate controls must be included in every test run. Reference laboratories are available and should be used rather than initiating testing if the anticipated number of requests for tests is low.

Criteria for interpretation of each test should be considered carefully. The Centers for Disease Control (CDC) is the national reference laboratory, but, because of special reagents developed and new procedures initiated by workers there, their criteria will not always agree with those for commercial reagents. CDC evaluates new reagents as they appear on the market to determine their acceptability.[11] Interpretation of tests using commercial reagents should be based on manufacturers' information and on literature specific for the tests. The current CDC methods and minimum significant reactivity levels are shown in Table 13:1. As with other serologic procedures, rising titers are more significant than a single result, but in many cases parasitic infections are chronic or have long prepatent periods such that diagnosis is not attempted until antibody titers have been well established and are no longer rising. In these cases it is frequently possible to obtain more definitive diagnostic data by performing additional different tests on a single sample. In the United States, commercial reagents are available for toxoplasmosis, amebiasis and trichinosis. Sera for tests available only at CDC must be submitted through state health department laboratories.

As can be seen by referring to Table 13:1, the introduction of new tests continues to change the reference diagnostic program. Immunoblot offers specificity not seen before in conventional serology and FAST-ELISA (Becton-Dickinson and Co.) is a rapid, simple screening method. In spite of the availability of these new methods, the more classical serological tests are still the tests of choice for clinical laboratories. Complement fixation

Table 13:1 Test of Choice and Significant Titers for Selected Parasitic Infections as Performed at the Centers for Disease Control

Disease	IHA*	IIF	IB†	CF	ELISA	OTHERS
Amebiasis	1:64					
Babesiosis	1:8					
Cysticercosis			Pos			
Echinococcosis	1:128‡		Pos			
Leishmaniasis				1:8		
visceral cutaneous		1:16				
Paragonimiasis			Pos			
Schistosomiasis			Pos			FAST-ELISA‡ 1:10
Strongyloidiasis					1:8	
Toxocariasis (DLM, VLM)					1:32	
Toxoplasmosis — IgG		1:16				Reversed-EIA
— IgM		1:16				1:16
Trichinosis						BFT 1:5
Trypanosomiasis (Chagas' Disease)		1:16		1:8		

IHA* — Indirect Hemagglutination
IIF — Indirect Immunofluorescence Test
IB — Immunoblot (Western Blot)
CF — Complement Fixation
ELISA — Enzyme Linked Immunosorbent Assay
BFT — Bentonite Flocculation Test
† — Immunoblot used either for confirmation or to speciate.
‡ — These tests used for screening purposes and confirmed by the second test.

(CF), although rapidly being superceded, is still used; indirect hemagglutination (IHA), long the "workhorse" of parasitic serology, is a major test; indirect immunofluorescence (IIF) continues to find use; and the newer enzyme-linked immunosorbent assay (ELISA) procedures are now standard for some diseases.

COMPLEMENT FIXATION

In spite of its many disadvantages, complement fixation continues to be the test of choice for leishmaniasis and Chagas' disease. This procedure requires more training and experience than most other tests and inherent errors are more easily committed. In addition, the reagents are labile and numerous substances, including drugs, lipidic material or contamination, may cause anticomplementary activity and false positive results. Excessive hemolysis in the patient's serum makes reading of results impossible. All of these problems can be solved by the experienced technician but other procedures are obviously preferable.

INDIRECT HEMAGGLUTINATION

Perhaps the simplest and most adaptable procedure in parasitic serology is the IHA test in which erythrocytes coated with antigen agglutinate in the presence of antibody. Few reagents are required and the variables are limited and easily defined. The condition of the patient's serum is not critical; hemolysis and minor bacterial contamination have little effect on the reaction. Occasionally, the presence of complement or therapeutic agents may cause the carrier erythrocytes to hemolyze or agglutinate. Heat inactivation of the serum should solve the problem of hemolysis but problems resulting from therapy can usually only be overcome by testing a subsequent serum.

Various erythrocytes have been used. Sheep cells have been used most commonly but have the disadvantage that many infectious diseases and other conditions cause the development of "heterophile" antibodies that produce nonspecific agglutination of sheep erythrocytes. Human O^-Rh^+ cells are frequently used to overcome this nonspecific agglutination and turkey cells, because of their large size and unique settling patterns, have been used.

One of the advantages of the IHA is the ability to use a variety of antigens including excretions, secretions, extracts, metabolic products and fractions of organisms. After erythrocytes are treated with tannic acid to alter surface characteristics, they will adsorb proteins, carbohydrates and some lipids, making possible the selection of specific antigens for each test. Since soluble components are adsorbed to erythrocytes, antigens can be more highly purified than for some other procedures and can be smaller molecular units than utilized in CF. Unfortunately, not all antigens can be adsorbed to erythrocytes and some produce autoagglutination of the cells in the absence of antibody.

Other agglutination procedures such as bentonite flocculation, cholesterol-lecithin agglutination and direct agglutination have been used for specific diseases but have not been as widely accepted as have CF, IHA and IIF.

INDIRECT IMMUNOFLUORESCENCE

Although IIF has been shown to be superior in sensitivity and specificity to IHA in many cases, technical difficulties and the need for specialized equipment often not available in smaller laboratories has limited universal acceptance of IIF. Procedural problems with IIF are not as great as those in CF but more prevalent than with IHA. The condition of the patient's serum is not critical and neither minor hemolysis nor bacterial contamination interferes with the IIF reaction. As a result, specimens collected under adverse conditions in the field or on filter paper are adequate. Since the IIF reaction is read visually, the antigens used can be less refined than those needed for CF or IHA and may consist of whole organisms or microtome sections of adult worms. The type of fluorescence or the portion of the antigen that fluoresces may indicate the stage of infection or may signal a nonspecific result or crossreaction which could not be identified with any other procedure. A disadvantage of IIF is that antigens must be visualized, so that extracts or fractionated antigens cannot be used unless attached to a suitable carrier. Two such methods have been described— the defined antigen substrate spheres (DASS)[1] in which latex particles are coated with antigen for use in a conventional IIF procedure, and the solid-phase fluorescent immunoassay (FIAX) (Whittaker M. A. Bioproducts)[10] in which cellulose acetate surfaces on a special dip-stick are coated with antigen and processed in a modified IIF test read in a special fluorometer. Both methods showed promise but because of technical difficulties and the need for specialized equipment were never widely accepted. Conventional IIF continues to be the method of choice for several parasitic infections.

ENZYME-LINKED IMMUNOSORBENT ASSAY (ELISA)

ELISA and its modifications, including the FAST-ELISA, and immunoblot techniques are rapidly reshaping serodiagnosis of parasitic infections. Although earlier thought to be too complex and difficult for field use, modifications of reagents and development of more practical procedures have shown ELISA to be quite adaptable. At present, ELISA is considered to be a reference test which is generally performed only in larger laboratories.

Procedurally, ELISA differs little from IIF except that the ELISA indicator system consists of an enzyme and its substrate rather than fluorescein activated by ultraviolet light. The ELISA enzymatic reaction is multiplicative and thus results in higher sensitivity; it permits the use of small molecular antigens not useful for IIF. The test has been used for detection of both antigens and antibodies in viral, bacterial, fungal and parasitic infections.[9,12] Some advantages are that the test is simple, suitable for processing large numbers of specimens and the reagents have long shelf lives. Disadvantages include long incubation times and the tendency to magnify errors because of the enzyme amplification.[4] ELISA has been shown to be uniquely useful in the detection of IgM antibody, a measure of early infection. In addition, by reversing the system and adsorbing specific antibody to the solid-phase, ELISA can be used for the identification of circulating antigens in amebiasis,[8] toxoplasmosis,[3] schistosomiasis,[6] pneumocytosis[5] and other infections.

FAST-ELISA

The recently introduced FAST-ELISA is a simplified rapid method for testing.[2] It uses a newly developed polystyrene Falcon Assay Screening Test (FAST) system in which a specially designed lid for a microtitration plate has 96 styrene beads on sticks such that the beads are suspended just above the bottom of each well when the lid is placed on a plate. The test is performed by adsorbing antigen to the beads and successively placing the cover on plates containing patient's serum, enzyme conjugate and finally substrate. Washing between each step is accomplished by spraying the tips of the sticks with buffer and shaking off the excess before introducing the beads into the next plate. The color reaction in the substrate is terminated by simply taking off the plate cover, which removes the beads containing enzyme. All reaction times are shortened to 5 min and the entire test is completed in less than 20 min. Since the FAST-ELISA TEST is primarily designed as a screening test, a single dilution of patient's serum is added to each well so that one plate can be used to test up to 94 sera plus positive and negative controls. If properly performed using a colorimeter to read the color reactions, the test is a quantitative kinetic assay. The potential for the FAST-ELISA as both a diagnostic laboratory and as a field test is virtually unlimited.

IMMUNOBLOT TECHNIQUE

Until recently, the immunoblot (or Western Blot) technique was used only for research purposes to characterize antigens and antibodies. Now techniques and equipment have progressed to a level which makes possible diagnostic uses. As described by Tsang et al[7], antigen is electrophoresed on polyacrylimide gel and electrophoretically transferred to cellulose acetate; then an ELISA procedure is run using patient's serum as the source of antibody. By using highly purified antigens and detecting specific bands, exquisite specificity can be obtained. In many cases this specificity has permitted speciation of the infecting agent. Since the introduction of this procedure for the diagnosis of the acquired immunodeficiency syndrome (AIDS), more laboratories are prepared to offer such a test, but the major disadvantage remains the necessity for specialized equipment and training. In its present form, immunoblot can only be performed in laboratories with access to the proper antigens needed for it.

REFERENCES

1. Deelder AM, Ploem JS. An immunofluorescence reaction for *Schistosoma mansoni* using the defined antigen substrate spheres (DASS) system. J Immunol Meth 1974;4:239-51.
2. Hancock K, Tsang VCW. Development and optimization of the FAST-ELISA for detecting antibodies to *Schistosoma mansoni*. J Immunol Meth 1986;92:167-76.
3. Knappen van F, Panggabean SO, Leusden van J. Demonstration of *Toxoplasma* antigen containing complexes in active toxoplasmosis. J Clin Microbiol 1985;22:645-50.
4. Landon J. Enzyme immunoassay: techniques and uses. Nature 1977;268:483-4.
5. Leggiadro RJ, Yolken RH, Simkins JH, Hughes WT. Measurement of *Pneumocystis carinii* antigen by enzyme assay. J Infect Dis 1981;144:484.
6. Stek M Jr. Erythroadsorption and enzyme-linked immunoassays (EAIA and ELISA) for specific circulating antibodies and antigens in schistosomiasis. Ann Immunol (Inst Pasteur) 1984;135:13-23.
7. Tsang VCW, Peralta JM, Simon R. The enzyme-linked immuno-electro transfer blot technique (EITB) for studying the specificities of antigens and antibodies separated by gel electrophoresis. Meth Enzymol 1983;92:377-91.
8. Ungar BL, Yolken RH, Quinn TC. Use of a monoclonal antibody in an enzyme immunoassay for the detection of *Entamoeba histolytica* in fecal specimens. Am J Trop Med Hyg 1985;34:465-72.
9. Voller A, Bartlett A, Bidwell DE. Enzyme immunoassays for parasitic diseases. Trans R Soc Trop Med Hyg 1976;70:98-106.
10. Walls KW, Barnhart ER. Titration of human serum antibodies to *Toxoplasma gondii* with a simple fluorometric assay. J Clin Microbiol 1978;7:234-5.
11. Wilson M, Ware DA, Walls KW. Evaluation of commercial serodiagnostic kits for toxoplasmosis. J Clin Microbiol 1987;25:2262-5.
12. Yolken RH. Enzyme immunoassays for the detection of infectious antigens in body fluids: current limitations and future prospects. Rev Infect Dis 1982;4:35-68.

CHAPTER 14

BLOOD AND TISSUE PARASITES

Kenneth W. Walls, Ph.D.

INTRODUCTION

The blood and tissue parasites are a diverse group, which includes *Plasmodium, Toxoplasma, Pneumocystis, Babesia, Trichinella*, hemoflagellates, filarias and both parasitic and free-living amebae. In addition, the tissues of humans may become parasitized accidentally by the larval stages of helminths such as *Echinococcus* and *Toxocara*, which normally infect lower animals.

Laboratory procedures for the diagnosis of these parasites vary to some degree with the specific body location, the phase or stage of infection and the growth pattern of the organism. Excellent photomicrographs are presented in the "Atlas of Human Parasitology"[5] as an aid to the morphological identification of both protozoa and helminths.

PROTOZOA

Malaria

Four species of malaria parasites infect humans: *Plasmodium vivax, P. falciparum, P. malariae* and *P. ovale*. In the United States, autochthonous cases of malaria are unusual. Of the 1091 reported cases of malaria among foreign visitors, immigrants and Americans who had lived or visited in endemic areas in 1986,[19,22] only 36 were acquired in the United States. In addition to the large number of foreign cases reported in the United States between 1978 and 1982, the total number of cases reported also appears to rise each year (Fig. 14:1). Transfusions with infected blood and needle-sharing among drug addicts continue to contribute occasional cases acquired

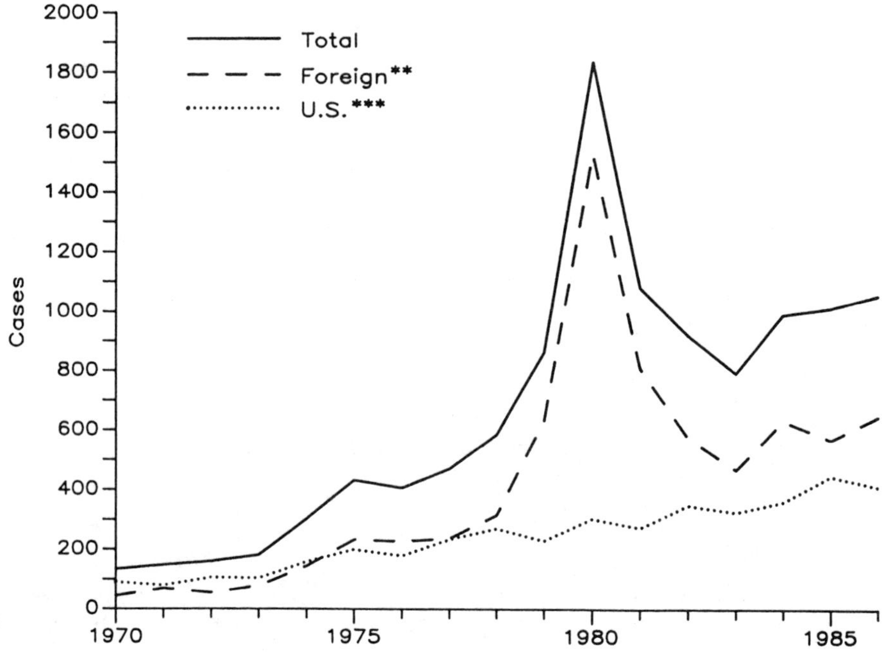

* Includes Puerto Rico, the Virgin Islands, and Guam
** Includes 26 cases acquired in the United States
*** Includes 10 cases acquired in the United States

Figure 14:1. Cases of malaria in U.S. and foreign civilians, United States, 1970-1986*.[22]

within the U.S. The most severe infections are caused by *P. falciparum* and may be fatal if untreated. *P. vivax* infection is most common, followed by *P. falciparum, P. malariae* and *P. ovale*, in that order. The appearance of drug-resistant strains of *P. falciparum* has complicated both prophylaxis and therapy.[20]

Life cycles of the four *Plasmodium* species are similar and involve transmission by a vector with sexual development (sporogony) in the vector and asexual development (schizogony) in humans. The vector for malaria is the female *Anopheles* mosquito; over 60 species have been incriminated in different parts of the world.

Fig. 14:2 shows the basic stages in the life cycle of malaria parasites. Sporozoites, which are the end product of sporogony in the mosquito, are introduced into the human host as the mosquito bites. These forms enter

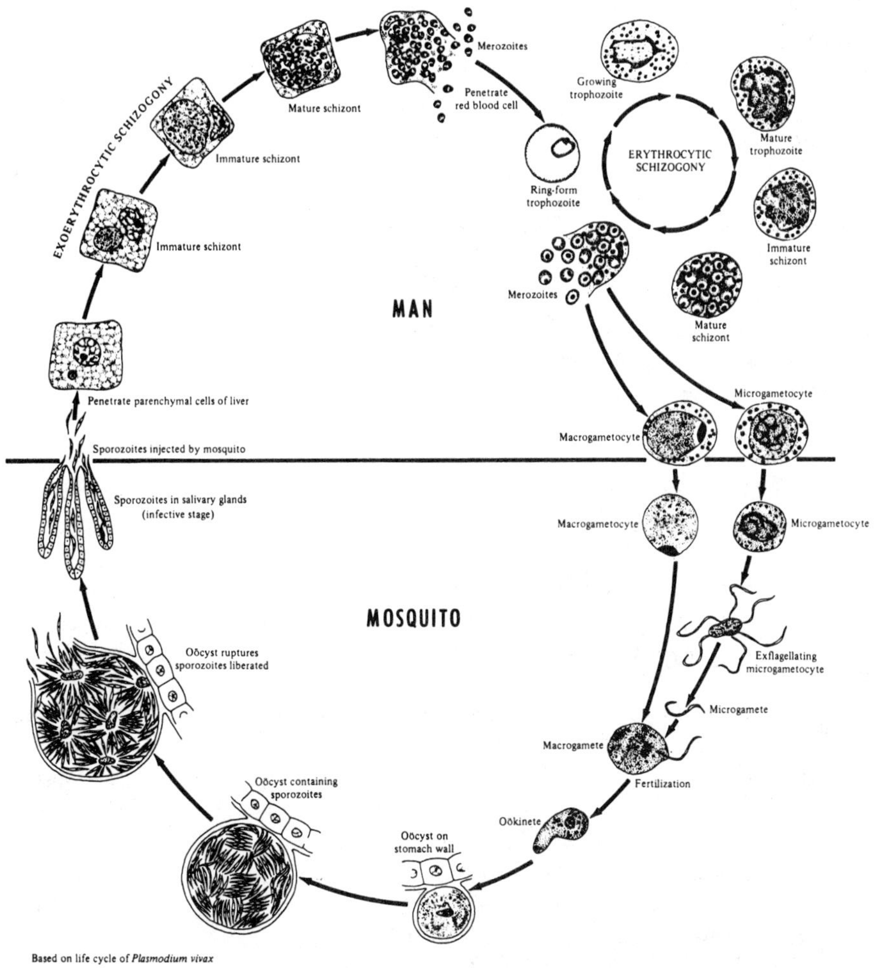

Figure 14:2. Life cycle of malaria. Courtesy of Parasitology Training Materials, Training and Laboratory Program Office, Centers for Disease Control, Atlanta, GA.

the blood stream and are transported to the liver where they penetrate the parenchymal cells to undergo extensive development and multiplication. This is the exoerythrocytic phase of the cycle and causes no damage or symptoms.

After one complete growth phase, which requires from one to several weeks depending upon the species, the merozoites (end products of asexual growth) are released from the liver cells into the circulating blood and initiate the erythrocytic phase of the malaria life cycle. Although the exoerythrocytic phase in the liver is not cyclic (that is, the merozoites produced do not enter other liver cells),[26] residual organisms (hypnozoites) are present in the livers of persons infected with *P. vivax* and *P. ovale*. These residual exoerythrocytic stages are responsible for the relapses which may occur upon infection with these two species.[24] Residual liver stages are not present in *P. malariae* or *P. falciparum* infections, nor do these species cause relapsing illness. Symptoms may recur but they stem from subpatent blood infections and are called recrudescences. Recrudescences are commonly associated with *P. malariae* infections and may occur years after initial clinical episodes. In blood-induced infections, such as those acquired through transfusions or contaminated hypodermic needles, there is no liver phase and thus relapses do not occur, but recrudescences are possible.

The symptoms of malaria are produced by the cyclic growth of the asexual stages in circulating blood. *P. malariae* completes its growth in 72 h; thus, chills and fever occur on days 1 and 4, the so-called quartan pattern. The other three species complete their growth in approximately 48 h and have a tertian pattern of chills and fever on days 1 and 3.

Laboratory diagnosis of clinical infection is based on morphology of the parasitic stages found in stained films of peripheral blood. Because clinical disease, therapy and epidemiology may vary with species, identification to the species level is important. Since the appearance of chloroquin-resistant *P. falciparum* strains, the need for consideration of alternative therapy has made rapid identification of this species very important. However, if the infecting species cannot be immediately determined, the laboratory should notify the physician at once that the patient has malaria and that species identification will be made as soon as possible. Serologic tests, which are useful for population surveys and epidemiological studies (e.g., to find the donor responsible for transfusion-induced malaria), may be of value when parasitemia is low and slide identification difficult.

In properly stained blood films, the three components of a malaria parasite are demonstrated: cytoplasm, which stains blue; chromatin (nucleus), which stains red or purple-red; and pigment, which does not stain and varies from golden-brown to black depending upon the species. All three components should be seen (except with young ring forms in which the pigment usually is not visible) in order to identify a structure as a parasite.

Parasitology training materials from the Centers for Disease Control (CDC) list the following as stages found in circulating blood.

Asexual stages

1. Trophozoites—growing, undivided parasite;
   a. Ring—youngest trophozoite stage consisting of a vacuole surrounded

by cytoplasm with a chromatin mass: occasionally has 2 chromatin dots (called fragmentation of the chromatin, does not represent nuclear division).

b. Growing Trophozoite—older form with distinctive cytoplasm and one chromatin mass: may be ameboid or compact: pigment granules present.

c. Mature Trophozoite—largest and oldest trophozoite stage: usually fills the red blood cell (rbc): has a single chromatin mass: pigment is present.

2. Schizonts—dividing form;
   a. Immature Schizont—stage with 2 or more chromatin masses: cytoplasm is undivided: pigment beginning to clump.
   b. Mature Schizont—division of nucleus and cytoplasm is complete and stage is composed of a cluster of individual parasites called merozoites: pigment is clumped.
   c. Merozoites—individual parasites produced as a result of asexual division (schizogony): consists of a mass of cytoplasm and a single chromatin mass: stage released when rbc ruptures and which penetrates other rbc: rarely seen outside rbc in blood films.

Sexual stage or gametocyte

Develops in rbc and circulates until it dies. This is the infective stage for the mosquito and in the mosquito it develops into a gamete:

1. Macrogametocyte—female sex cell;
2. Microgametocyte—male sex cell.

Speciation is based on the appearance of asexual and sexual forms in stained blood films and recognition of the growth stages is important for reliable diagnoses (Table 14:1).

In stained thin films, the features used to identify species are:

1. Appearance of the red cell;
   a. Size—enlarged or normal
   b. Schüffner's stippling—present or absent
2. Appearance of the parasite;
   a. Cytoplasm of growing trophozoite—ameboid or compact, outline and staining intensity (light or dark blue)
   b. Number of merozoites in mature schizont
   c. Amount and color of pigment
   d. Shape of gametocyte.

Detailed morphologic criteria for identifying malaria species in thin films are shown in Table 14:1. Features such as double chromatin dots and multiple infected red cells can be seen with any species, even though they are more common in *P. falciparum* infections, the examiner should not place undue weight on these characteristics in identification. In addition to

Table 14:1. Comparison of *Plasmodium* Species Affecting Humans[81]

	Appearance of Erythrocyte		Appearance of Parasite			
Species	Size	Schüffner's Stippling	Cytoplasm	Pigment	Number of Merozoites	Stages Found in Circulating Blood
Plasmodium vivax	Enlarged. Maximum size (attained with mature trophozoites and schizonts) may be 1½–2 times normal erythrocyte diameter.	+ With all stages except early ring forms.	Irregular, ameboid in trophozoites. Has "spread-out" appearance.	Golden-brown, inconspicuous.	12–24 Average is 16.	All stages. Wide range of stages may be seen on given film.
Plasmodium malariae	Normal.	– (Ziemann's dots rarely seen.)	Rounded, compact trophozoites with dense cytoplasm. Band-form trophozoites occasionally seen.	Dark-brown, coarse, conspicuous.	6–12 Average is 8. "Rosette" schizonts occasionally seen.	All stages. Wide variety of stages usually not seen. Relatively few rings or gametocytes generally present.
Plasmodium ovale	Enlarged. Maximum size may be 1¼–1½ times normal red blood cell diameter. Approximately 20% or more of infected red blood cells are oval and/or fimbriated (border has irregular projections).	+ With all stages except early ring forms.	Rounded, compact trophozoites. Occasionally slightly ameboid. Growing trophozoites have large chromatin mass.	Dark-brown, conspicuous	6–14 Average is 8.	All stages.

Plasmodium falciparum	Normal. Multiple infected red blood cells are common.	– (Maurer's dots occasionally seen.)	Young rings are small, delicate, often with double chromatin dots. Gametocytes are crescent or elongate.	Black, coarse and conspicuous in gametocytes.	6–32 Average is 20–24.	Rings and/or gametocytes. Other stages develop in blood vessels of internal organs but are not seen in peripheral blood except in severe infections.

the appearance of the parasite and the red cell, the stages present may also aid in species identification. For example, in *P. falciparum* infections, only ring-form trophozoites and gametocytes are found in peripheral blood, except in cases of overwhelming infection when growing trophozoites and schizonts may appear. All of the asexual stages as well as gametocytes are found with infection by the other three species and a wide range of growth stages is common in *P. vivax* infections.

The most difficult species to identify is *P. ovale*, since it has characteristics that resemble both *P. vivax* and *P. malariae*. Both the parasitized red cell and the parasite must be considered; therefore, *P. ovale* cannot be reliably identified from thick films. Examination of thin films, often of several, is essential to establish an accurate identification. The trophozoites, like those of *P. malariae*, tend to be dense and compact and to contain dark brown pigment. The number of merozoites produced is also about the same as that by *P. malariae*. On the other hand, the parasitized cells are enlarged and contain Schüffner's stippling, resembling those found in *P. vivax* infection. However, in *P. ovale* infections, 20% or more of those parasitized may be oval and/or fimbriated (having irregular projections of the cell border). In *P. vivax* infections, 6% or less of the cells may be oval; thus, an occasional oval parasitized cell is of no diagnostic significance. Although both species cause enlargement of the red cell, enlargement is less marked with *P. ovale*, where it is usually 1.25–1.5 times normal size, than with *P. vivax*, where enlargement may be 1.5–2.0 times normal. This variation can sometimes be helpful in identification. Because maximal erythrocyte enlargement usually accompanies the mature trophozoite and schizont stages, measurements should be made on erythrocytes containing these stages. Measure 25–50 parasitized cells for a reliable determination. Erythrocytes with *P. vivax* will probably measure 12–15 μm and those with *P. ovale*, from 10–12 μm. When identification of *P. ovale* is in doubt, additional blood films should be made at 6- or 12-hour intervals in an effort to obtain diagnostic forms. *P. ovale* infections are often confused with those caused by *P. vivax* unless sufficient distinctive characteristics are present. However, since these two infections are handled and treated in much the same way, this is not a serious clinical error.

Mixed infections with two (or rarely three) species of *Plasmodium* occur occasionally. However, they must be diagnosed cautiously since artifacts and atypical appearances may be confusing. Characteristic forms of both suspected species must be seen for diagnosis of mixed infection.

Morphology in thick films differs from that in thin films. In thick films, red cells are lysed and identification depends primarily on morphology of the parasites. Rings are often incomplete and appear like punctuation marks (",!,?). Trophozoites are more compact, which makes it difficult to differentiate those of *P. vivax*. Gametocytes of *P. falciparum* are stubbier. Although red cells are generally lysed, there may be some that are partially fixed near the periphery of the thick film. Enlargement of parasitized red cells and Schüffner's stippling may be evident in the partially fixed red cells. In areas where red cells are lysed, Schüffner's stippling may sometimes be apparent as a pink halo around trophozoites, without showing the distinct granularity seen in thin films.

Occasionally, only a few rings without any older forms may be seen in a blood smear. Since the appearance of the ring forms is essentially the same for all four species, "Malaria, species undetermined" should be reported. Species identification should not be made unless species-specific characteristics are seen. Additional films should be made at 12 and 24 h, since older stages will probably be found in later films from *P. vivax, P. malariae* and *P. ovale* infections. Infections with numerous rings and no other forms can be diagnosed as *P. falciparum*. Young rings of *P. falciparum* are smaller than those of other species—as small as ⅙ the diameter of an erythrocyte—whereas those of other species are rarely less than ⅓ the erythrocyte diameter.

Artifacts

Patients are sometimes erroneously diagnosed as having malaria when artifacts or host cells are confused with parasites. Platelets are probably the most troublesome host component. Single platelets adhering to red cells in thin films can be confused with rings or young growing trophozoites. Clumps of platelets may be mistaken for mature schizonts, or if elongated, for gametocytes of *P. falciparum*. In thick films, fragments of white cells and platelets may be mistakenly identified as parasites.

Artifacts which present problems include precipitated stain, dust, dirt, bacteria, yeasts, molds and occasionally fibers from cotton or gauze used to clean the patient's finger. These problems can be reduced or eliminated if the films are correctly prepared and stained and the examiner identifies as parasites only structures which have blue cytoplasms and red or purple-red chromatin. Brown or black pigment should also be present in all stages except rings.

Malaria Serology

Although new serologic procedures and antigens are being introduced regularly, most are so specialized and so limited in availability that their use in diagnosis is questionable at present. Indirect immunofluorescence (IIF) using either infected human or simian blood cells as antigen remains the test of choice for most laboratories doing diagnostic serologic testing for malaria. The test can be used for speciation in early primary infections, but in endemic areas where exposure has been extensive and prolonged the test becomes less specific and can be relied on only for generic diagnosis.

Speciation must still be determined by blood film examination. IIF is excellent for diagnosis of individual cases as it is highly specific and much faster than blood film examination. Since the test is unaffected by hemolysis, specimens can be collected in the field, dried onto filter paper and mailed or transported to a central laboratory where testing is done. A titer of 1:128 or greater is considered indicative of having had malaria at some time. As with speciation, the test cannot identify current active disease in persons from endemic areas.

Once considered a potentially useful procedure, indirect hemagglutination (IHA) is now used only for epidemiological purposes.[34,72] IHA is much simpler and more rapid to perform than IIF, and large numbers of specimens can be tested rapidly to indicate the amount of malaria in a population or to monitor changes in prevalence due to control measures or reintroduction. IHA is less sensitive and specific than the IIF procedure and as such has never gained popularity as a diagnostic test.

Two innovations which will have future impact on the diagnosis of malaria are worthy of note. As in other areas of serological diagnosis of infectious diseases, the introduction of enzyme immunoassays (EIA) has made possible the use of new antigens and the development of new procedures. Recently, highly specific synthetic antigens consisting of 12–18 amino acids arranged to mimic the active areas of the sporozoite antigens have been developed.[16] As this technology progresses it is likely that specific antigens for the detection and identification of malaria, including speciation will be possible.

FAST-ELISA[45] has the potential for more immediate application and offers a simple rapid procedure for the performance of EIA tests for malaria. Although special microtitration plates are required, the test is simple and the procedure can be performed with little or no specialized training. The FAST-ELISA, then, appears to overcome two of the major deterrents to the serodiagnosis of malaria: because of its simplicity it does not need to be restricted to large laboratories and large numbers of samples can be processed in very little time. The test can be used for both diagnosis and epidemiological purposes.

BABESIOSIS

Babesia species are intraerythrocytic parasites of various animals, including horses, dogs and rodents. Only recently have cases of babesiosis in humans been reported but the incidence is apparently rising.[74,75]

The morphology of *Babesia* organisms varies slightly with the species, but in general they resemble malaria parasites. It is not unusual for babesiosis to be misdiagnosed as malaria from stained smears. The species most commonly reported as infecting humans is *B. microti*, a rodent parasite. The ring-form parasites are sometimes slightly ameboid or oval, but are morphologically similar to those of *P. falciparum* (see color plate Fig.14:3). It is difficult to differentiate the rings of these two parasites, but those of *Babesia* tend to be smaller and more delicate. There are no gametocytes nor any pigment seen in *Babesia* infections. Forms with 4 tiny trophozoites—tetrads—may be found, which are helpful in identification of *Babesia*.

To date, the serodiagnosis of babesiosis with the IIF is in a developmental stage, but the IIF test appears to be an acceptable procedure for diagnosis. Antigen prepared from *B. microti* from the blood of infected hamsters has been used successfully to measure antibody in proven cases.[75] The limited number of cases restricts the evaluation of sensitivity but tests on sera from a large number of patients with other diseases indicate that the specificity is excellent. Serologic procedures are available in reference laboratories such as those at the CDC.

TOXOPLASMOSIS

A sporozoan of the coccidian group, *Toxoplasma gondii* is a ubiquitous intracellular parasite found in all warm-blooded animals throughout the world. Toxoplasmosis has been recognized in a variety of clinical forms in humans, including: (1) adult acquired disease, usually similar to infectious mononucleosis; (2) congenitally acquired disease causing severe fetal damage and death or resulting in delayed physical, mental or ocular damage; (3) ocular disease of posterior chorioretinitis and blindness; (4) severe disseminated and central nervous system (CNS) involvement of immunosuppressed patients, including those with acquired immune deficiency syndrome (AIDS); and (5) perhaps the most insidious, chronic asymptomatic infection causing no problems for the patient but providing a source for transmission of infection from mother to fetus. Adult acquired infections rarely result in more than a transient mild disease which is self-limiting except in the immunocompromised patient, where the results can be catastrophic. The severity of congenitally acquired infection depends upon the time during pregnancy when the infection is acquired by the mother. Infections early in pregnancy are infrequently transmitted to the fetus but transmission usually results in major damage or death of the fetus. Later infections are readily transmitted but usually result in mild or asymptomatic infections without apparent fetal damage. Retrospective studies have shown, however, that these asymptomatic newborns are likely to develop adverse sequelae later in life.[91] Toxoplasmosis is one of the most common causes of chorioretinitis and it has been suggested that ocular toxoplasmosis is primarily the expression of a much earlier infection, perhaps of congenital origin.[64] Toxoplasmic encephalitis is now recognized as one of the major causes of death among AIDS patients[58] and it has also been suggested that these cases result from reactivation of earlier acquired infections.

Toxoplasma have both sexual and asexual stages. The sexual stages are limited to cats and other felinidae. Gametogeny, oocyst production and sporogony occur in the feline intestinal mucosal epithelium and the oocysts, the infective stage, are passed in the feces; thus, this stage of the disease resembles isosporiasis. In all other animals, including humans, where the infection is extraintestinal, the *Toxoplasma* organisms (tachyzoites) divide by a process of endodyogeny in which two daughter cells are formed in a mother cell; then the mother cell disintegrates. The organisms are able to invade and proliferate in almost any cell. With the development of immunity, cysts containing numerous organisms (bradyzoites) develop in many tissues but especially in skeletal and cardiac muscle and in the CNS (see color plate Fig. 14:4). These cysts may persist in a latent form for many years, perhaps for life.

In addition to congenital transmission, infection can occur from the ingestion of undercooked meat containing tissue cysts, ingestion of oocysts from cat feces, by blood transfusion or from tissue transplants. Transmission has been demonstrated clearly in the laboratory to occur through ingestion of oocysts.[40] Natural outbreaks have been reported to result from ingestion of contaminated meat[50], drinking contaminated water[7] and from inhaling dust in a riding horse ring where the soil was contaminated by cat feces containing oocysts.[84] Occasional cases have been reported that resulted from blood transfusions or tissue transplants.[71,81]

Accurate and early diagnosis of toxoplasmosis is important, especially for pregnant women and AIDS patients. Early treatment can reduce fetal damage and ameliorate the encephalitic symptoms of the immunocompromised patient.[93] The parasite can be isolated by inoculating blood or biopsy specimens into animals or cell culture, but positive results may occur in the absence of active disease because of the chronic nature of infection. Whole blood, or preferably the buffy coat of white cells obtained by centrifugation, is inoculated intraperitoneally into clean mice or directly into cell cultures. Biopsy specimens, particularly of enlarged lymph nodes, frequently yield isolates. Such tissues are simply triturated in buffered saline and inoculated, or in some cases they may need to be predigested with trypsin to release organisms and the digested material washed before injection. Cell cultures are observed daily for intracellular multiplication of the parasite and cell destruction. Mice are monitored for illness or death, but these seldom occur. Mice that die and any remaining after 2 weeks are necropsied and histopathology is done on the brain, liver and spleen to look for characteristic cysts. Blind passage of apparently negative tissues to additional mice will occasionally result in detection of less virulent strains.

Histopathology, especially utilizing the newer enzyme-labeled techniques, is very useful for identifying active disease. With conventional staining procedures, it is difficult to recognize the small tachyzoites indicative of active multiplication but these and even deposits of antigen become evident with the enzyme stain.[27] This procedure is especially useful in the determination of active cerebral involvement in AIDS patients. However, the presence of latent organisms in chronic disease makes careful interpretation essential.

Most cases are diagnosed by serology. Good serodiagnostic methods have been available since the introduction of the Sabin-Feldman Dye Test.[76] Detection of both IgG and IgM are essential to proper diagnosis. The detection of IgG is now used primarily to determine susceptibles, particularly among pregnant women. For smaller laboratories, IIF is perhaps the test of choice; although an enzyme-linked immunosorbent assay (ELISA) is more practical for larger laboratories testing many specimens. Other tests, such as indirect hemagglutination and latex agglutination, are simple and inexpensive but have been shown to have problems.[89] The newer methods of direct agglutination and latex-bound ELISA appear to be highly effective for screening.[62,82] In the U.S., up to 1/3 of the adult population has persistent antibody but titers of 1:256 or higher are suspicious and should be confirmed by an IgM test.

IIF using an IgM-specific conjugate has been used for some time but with the knowledge of three problems: low sensitivity, false-positive reactions in the presence of rheumatoid factor and false-negative reactions caused by interference from competing IgG antibody. A capture ELISA procedure which separates the IgM from the IgG is preferred,[39] since this decreases the possibility of detecting rheumatoid factor and prevents interference of IgG with IgM reactivity. Similar results can be obtained with IIF tests that utilize a preseparation on minicolumns. Titers of 1:64 or above in adults and any level in newborns are suggestive of current disease.

Commercial reagents are available for all these procedures but should be used with caution. The user should be aware of the many problems associated with some of these kits.[92] Reference sera and controls should be run to monitor the performance of every test. Equally important, the person performing the test should be thoroughly familiar with all aspects of the test and the person interpreting the results should be aware of all the innuendos associated with toxoplasmosis serology.

PNEUMOCYSTOSIS

Pneumocystis carinii is an organism of uncertain taxonomy, but is considered by many to be a sporozoan. It may cause pulmonary disease among malnourished and premature infants. However, in the U.S., most disease is among immunocompromised persons, including those with AIDS, transplant recipients and those being treated for leukemia, lymphoma and other malignancies. Because more than 50% of patients with AIDS have *P. carinii* infection at some time during the course of their disease, the number of cases of *P. carinii* reported each year has increased dramatically. In addition, more aggressive chemotherapy for malignancies and prior to transplantation has contributed to increased numbers of *P. carinii* infections in these patient groups.

Because the organism is difficult to grow in culture[9,56,66] and serologic tests are not well developed, the epidemiology of these infections has not been clearly defined. Most overt infections probably arise from latent infections that are activated when the patient becomes immunocompromised. Spread of infection by the respiratory route has been demonstrated in experimental animals[46,90] and there are probably asymptomatic carriers who serve as sources of infection.

The organisms develop in two forms—cysts and trophozoites. The growing trophozoites are extracellular but are attached to cells and divide by binary fission (Fig. 14:5). The cysts are formed when trophozoites become rounded and produce a cyst wall. Division continues within the cyst so that the mature cyst contains 8 organisms. Sexual stages with synaptonemal complexes have been demonstrated by electron microscopy.[61]

Infection usually produces an interstitial pneumonia, with the alveoli filled with foamy exudate containing organisms. The amount and character of the cellular response vary with the underlying condition of the patient and the severity and length of infection. *Pneumocystis* pneumonia must be differentiated from other opportunistic pulmonary infections such as cytomegalovirus pneumonia, tuberculosis, histoplasmosis, nocardiosis, aspergillosis, cryptococcosis and legionnaires' disease, as well as from noninfectious conditions such as tumor infiltration and reaction to chemotherapeutic agents.

Diagnosis is usually established by demonstrating organisms in material from the lungs. The optimal specimen depends upon the type of patient. AIDS patients typically have large numbers of organisms, so that bronchoalveolar lavage specimens may be used to detect them.[14,51,65] Induced sputum specimens have been used for diagnosis by some groups[9,68] but have been reported as unreliable by others.[13] Transbronchial biopsy performed on immunosuppressed patients other than AIDS patients may be positive when organisms are not found in washes and lavages.

Biopsy specimens can be examined as impression smears and/or sections depending upon the size of the specimen and the capabilities of the laboratory. Impression smears stained with Giemsa allow the internal structure of cysts and trophozoites to be seen. Prepare impression smears by blotting excess fluid from the tissue on sterile paper or gauze and touching the tissue repeatedly and firmly to a slide. If the tissue is too wet, the organisms may remain in the alveoli and not adhere to the slide. If the amount of tissue is limited, a small circle of impressions should be made in the center of multiple slides and each circled with a diamond-tipped marker. Bronchoalveolar lavage specimens or washes should be centrifuged and smears made

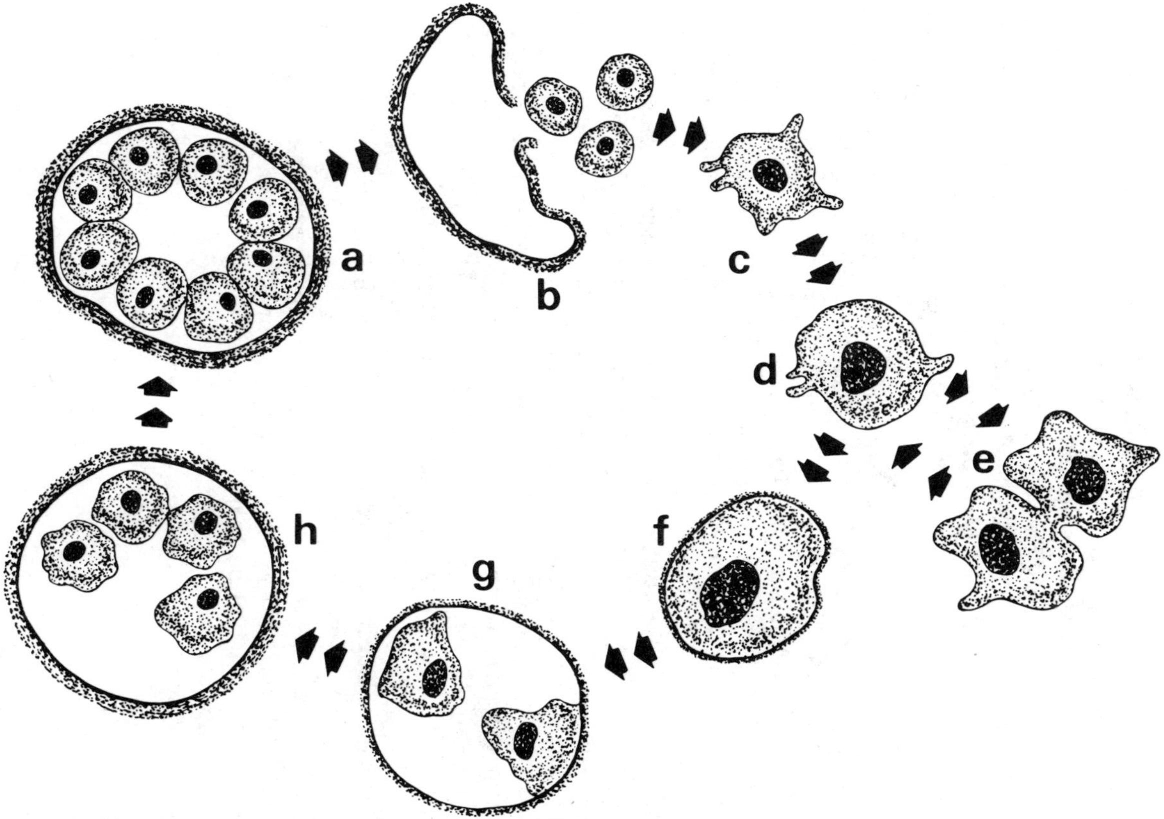

Figure 14:5. **Life cycle of *Pneumocystis carinii*: a. Mature cyst. b. Ruptured cyst. c. and d. Growing trophozoites. e. Dividing trophozoite. f.,g. and h. Immature cysts.**

from the sediment. Likewise, induced sputum may be sedimented; however, prior treatment with a mucolytic agent may be required. Bloody specimens may be more easily examined if a small amount of sterile distilled water is added to lyse the red blood cells.

Various stains have been used. Most examiners stain the cyst walls to allow ready detection of organisms under low-power microscopy; however, the morphology of cysts does not differ greatly from that of yeasts, and if occasional organisms are found, a definite identification may not be possible. *Pneumocystis* cysts do not show budding and are often cup-shaped with an indentation on one side (see color plate Fig. 14:6a,b). Stains for cyst walls have the advantage of staining empty cysts which have discharged their contents. Stains used for staining cyst walls include Gram-Weigert,[73] methenamine silver,[48] methenamine silver modified for rapid staining,[12,59] cresyl echt violet,[10] toluidine blue O[48] and modified toluidine blue O.[44]

Giemsa stains the contents of cysts and free trophozoites but does not stain cyst walls. Organisms have red nuclei and blue cytoplasm (see color plate Fig. 14:6c). Since differentiating a few trophozoites from fragments of cells may be difficult, diagnosis is most reliably based on finding a typical cyst containing 6-8 organisms. Specimens from AIDS patients typically contain large clumps of trophozoites which are easily demonstrated by the usual Giemsa stain or by a rapid type Giemsa, Diff Quick (Baxter-Travenol Co.), which allows detection in minutes. The characteristic clumping of trophozoites is shown in Figure 14:6d (see color plate).

Neither tests for detecting antigen nor tests for antibody have proved helpful in the diagnosis of *Pneumocystis* pneumonia to date. Improved procedures measuring changes in antibody levels and utilizing panels of specific monoclonal antibodies may enhance the diagnostic techniques currently available. At present, diagnosis still requires morphologic demonstration of organisms best accomplished with histologic stains or with immunospecific staining.[42,57]

Although organisms from animals, primarily rats, are being cultured for development of diagnostic reagents, the uncertainty of relationships between organisms from different hosts may complicate the use of these materials for the detection and study of organisms in human infections and the results of these tests should be interpreted with caution.

AMEBIASIS

Free-living Amebae

At least two groups of free-living amebae, *Naegleria* and *Acanthamoeba* are capable of causing disease in humans. Of the two, organisms of the genus *Naegleria* appear to be more virulent. They are ameboflagellates—

that is, they exist either as amebae or as flagellates but can divide only in the ameboid form. They divide by promitosis (during division the nuclear membrane persists). *Naegleria* species cause an acute meningoencephalitis which usually has a rapidly fatal course. Butt[15] used the term "primary amebic meningoencephalitis" (PAM) to describe this disease, thus differentiating it from the CNS disease caused by *Entamoeba histolytica* which is secondary to infection of the colon, liver or lung. About 100 cases of PAM have been reported throughout the world. Most of the patients acquired the infection from swimming in warm water, but face washing with contaminated water has also caused infection. The amebae penetrate the olfactory bulbs and progress to the subarachnoid space. To date, few individuals with PAM have survived. A 14-year old Australian boy survived after treatment with amphotericin B,[18] a 6-year old boy from England recovered after being treated with amphotericin B and sulfadiazine[4] and a 9-year old girl from California recovered after treatment with amphotericin B and miconazole. Amebae were detected in the cerebrospinal fluid (CSF) of two of these patients early in the disease and treatment was begun immediately. Treatment was initiated for the third before symptoms began. This suggests that early treatment is essential for recovery. The pathogenic *Naegleria* are 10-15 μm in size and have prominent nucleoli, contractile vacuoles and lobose pseudopodia.

Amebae of the genus *Acanthamoeba* have been implicated in at least 5 cases of encephalitis or meningoencephalitis and in chronic skin ulcers.[60] The route of infection has not been definitely established, but in one instance a cutaneous lesion may have been the primary focus of infection, with subsequent hematogenous spread to the CNS. The disease is more slowly progressive than that caused by *Naegleria*, but the outcome is fatal. The *Acanthamoeba* are 15-35 μm in diameter, have large central nucleoli and exhibit classic mitosis. Often the pseudopods are sharp and spiny—hence the name, *Acanthamoeba*. There is no flagellate form.

At least two species of *Acanthamoeba* have been implicated as the cause of keratitis in soft contact lens wearers.[21,83] Although the exact means by which the amebae entered the cornea could not be determined, it is clear that improper hygienic procedures were used by patients as compared to those used by healthy soft lens wearer controls. Whether the lens provides a microenvironment for the growth and development of the amebae or causes microtraumatic areas where the amebae can penetrate into the eye was not established. Approximately 100 cases have been reported to the CDC and at least one progressed to encephalitis.

If a patient has signs of meningeal irritation, headache, vomiting and a stiff neck, has a history of swimming in warm water and a CSF with elevated proteins and low glucose levels but without detectable bacteria, free-living amebae should be considered.

These two genera have different drug susceptibilities. *Naegleria* organisms are susceptible to amphotericin B but resistant to sulfadiazine, whereas sulfadiazine is quite active against some strains of *Acanthamoeba* in vitro. In order to establish appropriate therapy, it is important to document infection by detecting amebae microscopically, to confirm this in subsequent culture and to differentiate the amebae on the basis of nutritional requirements, type of mitosis and presence or absence of a flagellate stage. Amebae are best demonstrated by direct microscopy of CSF with a phase contrast microscope; care must be taken to differentiate amebae from macrophages.

Culture methods are useful for isolating and identifying organisms. *Naegleria* parasites require a medium containing a low concentration of NaCl (0.4% rather than 0.85%) and living bacterial cells. Cultured mammalian cells and other enrichment materials such as serum have been used but are not as practical. *Acanthamoeba* organisms can be cultured in tryptic digest soy broth.[29] All cultures must be aerobic and incubated at 37°C.

Entamoeba histolytica

Although *E. histolytica* is described in detail in Chapter 15, "Intestinal and Atrial Protozoa," and the organism is most commonly seen as an intestinal pathogen, frequent cases occur in which the parasite penetrates the intestinal wall and is disseminated to cause severe extraintestinal disease. This most commonly results in liver abscess, brain abscess or invasion of the lungs or eyes. In most cases the intestinal phase of the disease was not recognized nor diagnosed. This disease is frequently confused with other diseases and conditions of these tissues, so that proper therapy depends on an accurate differential diagnosis.

Diagnosis rests upon identification of the organism in material from the lesion, or when this is not possible, on serologic results. In cases of liver abscess, fluid withdrawn from the abscess can be described as an "anchovy paste" fluid due to its consistency and color. Amebae can frequently be seen microscopically in this material or cultured from it. Giemsa or other appropriate stains can be used to demonstrate the invasive amebae in biopsied material from lung or brain.

The IHA test has been shown to be a highly sensitive and specific serologic test that is simple to use. Commercial enzyme immunoassays (EIA) are available and have been shown to be equally effective though somewhat more difficult to perform. Double diffusion and counterimmunoelectrophoresis tests are available and are excellent, but are primarily screening tests since titer levels cannot be determined. Titers from IHA and EIA procedures usually can be correlated with recency of acquisition and activity of infection, but titers may persist for long periods of time. It is important that titers be interpreted carefully in relation to the clinical evaluation of the patient.

Microsporidia

Microsporidia are small unicellular organisms which had been reported as tissue parasites in rare instances prior to AIDS. Several genera and species had been proposed as the etiologic agents. As with other previously little known organisms which have gained importance in the immunocompromised, microsporidia are being found increasingly in AIDS patients. Intracellular forms varying in size from 1.0 to 8.0 μm have been detected with electron micrographs or Gram and Giemsa stains of biopsy tissue sections. The most frequent site of infection has been the small intestine and the variable-sized forms were found in the cytoplasm of enterocytes.

HEMOFLAGELLATES

Hemoflagellates include the trypanosomes and *Leishmania* species. Hemoflagellates may be present in any of four different morphologic stages, with the stages varying among species (Fig. 14:7 and Table 14:2). Several years ago terminology was revised and the forms were named according to their flagellation. The amastigote (leishmania) stage does not have a flagellum and is an obligate intracellular parasite. The promastigote (leptomonas) has a free flagellum but no undulating membrane. The epimastigote (crithidia) has a free flagellum and an undulating membrane which arises anterior to the nucleus. The trypomastigote (trypanosome) has a free flagellum and an undulating membrane that extends the length of the organism. Four species of trypanosomes infect humans. Two are found in Africa and two in the Western Hemisphere.

AFRICAN TRYPANOSOMIASIS

Trypanosoma gambiense (West Africa) and *T. rhodesiense* (East Africa) are etiologic agents of African sleeping sickness. According to Hoare,[47] the correct species names should be *Trypanosoma brucei gambiense* and *Trypanosoma brucei rhodesiense*, but because this is not common usage and for simplicity, the "*brucei*" will be omitted from these discussions.

T. rhodesiense appears to be more virulent than *T. gambiense*, often causing death before the classical sleeping stage of encephalitis is reached. *T. rhodesiense* parasitizes large game animals and cattle in Africa and is probably a zoonosis of humans.

The two African trypanosomes have similar life cycles. They are transmitted by blood-sucking flies of the genus *Glossina* (tsetse flies) and humans acquire infection by receiving the trypomastigote stage during a fly bite. Only the trypomastigote has been found in the human host. It may be

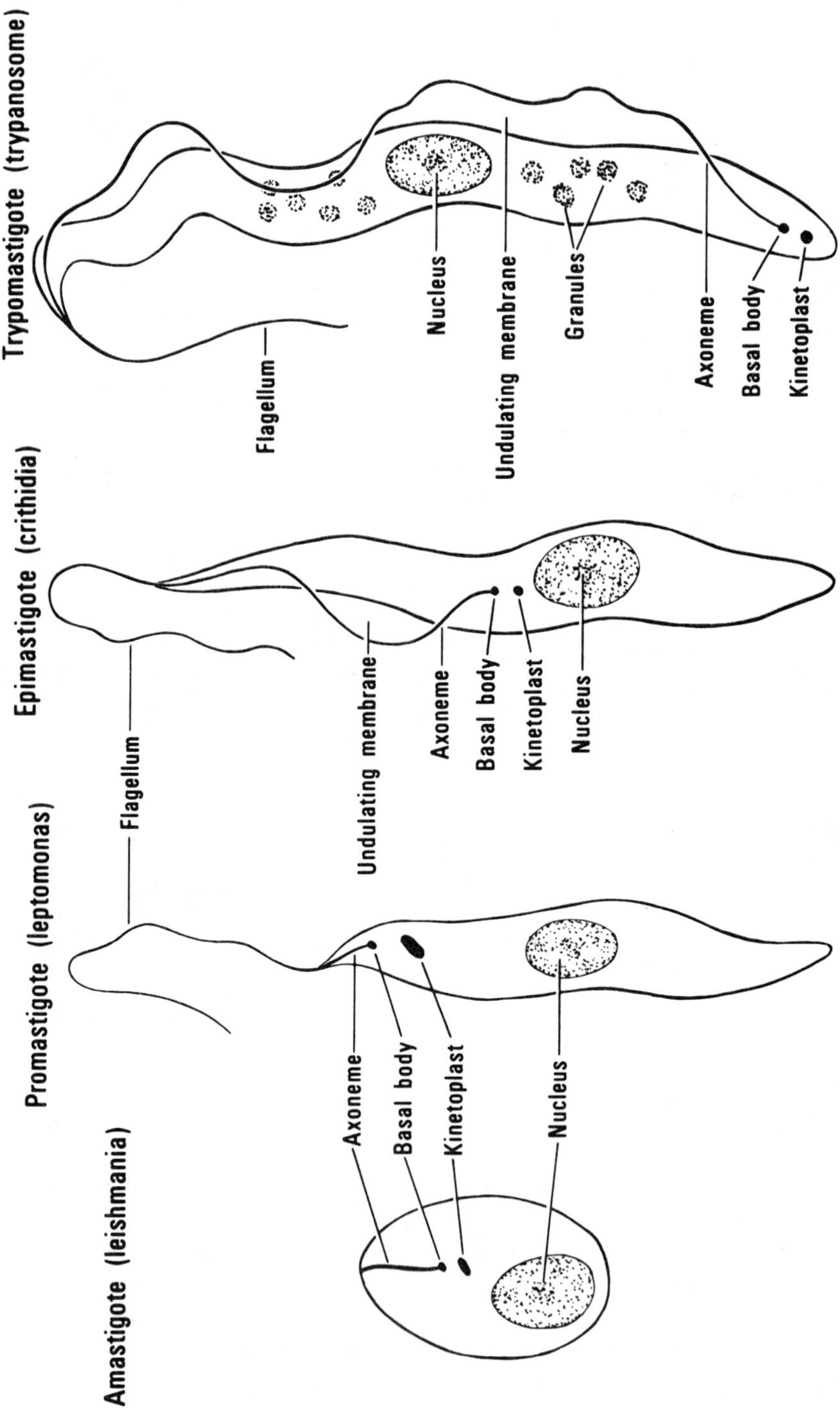

Figure 14:7. Hemoflagellate forms.[81]

Table 14:2. Blood and Tissue Flagellates Found in Humans[81]

Species	Developmental Stages				Transmission	Vectors
	Amastigote	Promastigote	Epimastigote	Trypomastigote		
Leishmania *L. donovani*	Intracellular, in reticuloendothelial system, lymph nodes, liver, spleen, bone marrow, etc. Culture.	Midgut and pharynx of vector. Culture.	—	—	Bite	Sand flies *Phlebotomus*, *Lutzomyia*
L. tropica and *L. braziliensis*	Intra- and extracellular in skin and mucous membranes of humans.	Midgut and pharynx of vector. Culture.	—	—	Bite	Sand flies *Phlebotomus*, *Lutzomyia*
Trypanosoma *T. gambiense*	—	—	Salivary glands of vector. Culture.	Blood, lymph nodes, cerebrospinal fluid of final host. Intestine and salivary glands of vector.	Bite	Tsetse fly *Glossina*
T. rhodesiense	—	—	Salivary glands of vector. Culture.	Blood, lymph nodes, cerebrospinal fluid of final host. Intestine and salivary glands of vector.	Bite	Tsetse fly *Glossina*
T. cruzi	Intracellular in viscera, myocardium, brain of humans. Tissue culture.	Intracellular in man but transitional.	Midgut of vector. Culture.	Blood (temporary) of humans. Intestine and rectum, feces of vector. Culture.	Feces of vector into wound.	Reduviid bugs (*Triatominae*)
T. rangeli	—	Transitional only.	Midgut of vector. Culture.	Blood of humans. In hemolymph, salivary glands and proboscis of vector. Culture.	Bite	Reduviid bugs (*Triatominae*)

present in blood, lymph nodes, spleen and spinal fluid. The parasites do not actually invade cells but cause damage to many tissues of the body.

Diagnosis of African trypanosomiasis is based on finding the trypomastigotes in wet mounts or stained films of peripheral blood or of lymphatic material during the early phases of infection. In the later, chronic phases of *T. gambiense* infections, organisms may be present in spinal fluid. Patients with *T. rhodesiense* infections rarely survive long enough for the organisms to reach the central nervous system. Although organisms can be detected in wet mounts, their specific characteristics are visible only in stained preparations. Giemsa stain is preferred. Thick blood films are usually recommended, but the organisms may be somewhat distorted. More distinct morphology can be seen in thin films (see color plate Fig. 14:8). Since the organisms are intercellular, the thicker portions of the thin film should be examined. Repeated daily blood films may be needed to establish a diagnosis.

In stained films, stumpy forms (averaging 20 μm in length) and slender forms (approximately 30 μm long) may be seen. In general, the parasites are elongated with the nucleus located in the center or slightly posterior. A small discrete kinetoplast is located subterminally. Although variations occur, the posterior end is often blunt or bluntly tapered. An organism without a free flagellum is found occasionally. *T. gambiense* and *T. rhodesiense* are morphologically identical. When few organisms are present, the blood may be concentrated by fractional centrifugation and wet mounts or stained slides prepared from the sediment.

Cultivation on special blood agar and animal inoculation have been used in some cases but are generally less successful than direct examination of blood or lymph.

Serologic diagnosis of African trypanosomiasis continues to be unsuccessful, except when used on an experimental basis. These organisms are notoriously variable in antigenic makeup and as a consequence specific tests have been elusive. Indirect immunofluorescent procedures have been most widely used but with poor results. ELISA using purified antigens shows promise and is used in some areas but cannot be recommended for routine use. At least one report[87] has shown that ELISA can be used successfully to diagnose African sleeping sickness by selecting the proper variable antigen type. When this antigen was used, the test was 100% specific and 98.6% sensitive.

AMERICAN TRYPANOSOMIASIS

Two species of trypanosomes affect humans in the Western Hemisphere—*Trypanosoma cruzi* (Central and South America, from Mexico to Argentina) and *T. rangeli* (Central America and northern South America). *T. rangeli* is apparently a nonpathogenic parasite in humans while *T. cruzi* is the causative agent of Chagas' disease, a very serious human infection.

In South America, *T. cruzi* is often found in humans and wild animals, but this is not the case in the U.S. Although a number of wild animals, including opossums, raccoons and armadillos in the U.S. have been found to be infected, only a few indigenous human cases have been reported. It is suspected that other cases may have gone undetected.

T. cruzi is transmitted by species of triatomid bugs ("kissing" bugs). The infective forms (young trypomastigote stage) are located in the hind gut of the bug and are passed in the feces. As the bug feeds, it defecates and the organisms are rubbed into the wound when the area is scratched. In the human host, *T. cruzi* occurs in the blood in the trypomastigote stage and in tissue (reticuloendothelial cells, CNS or myocardium) as the amastigote stage. The trypomastigotes can be detected in wet mounts of blood and occasionally in lymphatic material during the initial, acute phase of the disease (which may be 2-4 weeks) and during later febrile periods. However, the probability of detecting parasites during later febrile periods is rather low since organisms are usually rare at this time.

The trypomastigotes have a large kinetoplast and typically have "C" and "S" shapes rather than the more delicate curves of typical African trypanosomes (see color plate Fig. 14:8b). Organisms are rather badly distorted in thick films, so thin films made somewhat thicker than usual are more useful. Since the parasites are intercellular, overlapping and piling up of red cells will not interfere with diagnosis and the thicker portions of the films should be examined.

Direct examination of blood often fails to reveal organisms, and other more reliable methods of diagnosis include cultivation of blood (usually centrifuged sediment), lymphatic material or biopsied tissue; animal inoculation with the same materials; and serologic procedures. Neither cultivation, animal inoculation nor serology are done in most clinical laboratories, so specimens from patients suspected of Chagas' disease are usually sent to reference laboratories such as the CDC.

The serodiagnosis of Chagas' disease has been extensively studied. Both IHA and CF tests have been used successfully. Although the IHA test is more easily performed and requires fewer reagents, it is less specific than the CF test[28] and as a consequence, IHA has been most extensively used as an epidemiological tool.

The Pan American Health Organization has recommended complement fixation with a delipidized antigen from cultured organisms as the test of choice. CF is highly sensitive and specific and of particular value for analysis of specimens from patients with acute Chagas' disease. Rising CF titers are highly indicative of current infection. However, due to the chronicity of this disease, stable to moderately high titers in either CF or IHA are difficult to interpret. Other tests such as IIF[38] and direct agglutination[1,88] have been reported to be of value; neither, however, offers sufficient advantage over CF and IHA to be recommended as a substitute.

Xenodiagnosis, wherein uninfected triatomid bugs are fed on patients, is considered an effective and reliable procedure but is not practical nor feasible for the average laboratory. Amastigote forms may be recovered from bone marrow or other biopsied tissue but must be distinguished from similar stages of *Leishmania* which may inhabit the same tissues. Tissue examinations are not usually reliable for diagnosis of Chagas' disease.

T. rangeli is also transmitted by triatomid bugs, but the organisms are in the foregut of the bug and are injected as the bug bites. The young trypomastigote is the infective stage for vertebrate hosts, including monkeys, dogs, cats and other animals, as well as humans. The process of development of the infection in humans is not completely known, but so far only the trypomastigote stage in the blood has been described and presumably it is the only stage present in the mammalian host. Parasitemia tends to be scanty and the organisms are difficult to detect in blood films. In Giemsa stained smears, the parasite resembles the African trypanosomes. *T. rangeli* is undulated, about 30 μm long, with a long free flagellum. The nucleus is usually anterior to the center of the body; the subterminal kinetoplast is smaller than that of *T. cruzi* but is a distinct, readily seen structure. The posterior of the *T. rangeli* trypomastigote tapers to a more or less pointed end. *T. rangeli* can usually be differentiated from *T. cruzi* without much difficulty. Cultivation techniques and animal inoculation can also be used for diagnosis.

LEISHMANIASIS

The classification and nomenclature of *Leishmania* species affecting humans are not clearly defined.[55] Regardless of the differences in nomenclature and species, the organisms fall into two broad categories: the visceral forms and the cutaneous forms. Diagnostic procedures generally relate to the area of the body affected rather than to the species suspected. For simplicity, only 4 species are listed here:

Visceral Form and Its Location

1. *Leishmania donovani*—India, Mediterranean area, other parts of the Eastern Hemisphere, South and Central America (the species in the Western Hemisphere may not be *L. donovani*).

Cutaneous Forms and Their Locations

1. *Leishmania tropica*—Eastern Hemisphere.
2. *Leishmania braziliensis*—Western Hemisphere (Central and South America).
3. *Leishmania mexicana*—Mexico and Central America.

The clinical manifestations of the cutaneous forms vary somewhat and a number of specific names have been given to these variations. It is possible that all the organisms are variants or subspecies of the same primary species; for more details refer to parasitology textbooks and specific literature. Diagnostic procedures are similar despite different species designations.

There are only two morphologic stages of *Leishmania*: the amastigote, which is the stage found in humans, and the promastigote stage found in the vector. The life cycles of the *Leishmania* are basically the same regardless of species. In visceral infection, the organisms are present in the reticuloendothelial cells of the viscera and (less commonly) in monocytes and leucocytes in the blood or in monocytes or macrophages of the skin. In cutaneous infection, the parasites are found primarily in monocytes and macrophages of the skin. The amastigotes are acquired by the vector, i.e., species of *Phlebotomus* or *Lutzomyia* (sandflies),[11] when biting an infected host. In the vector, the parasite transforms to promastigotes and multiplies. In approximately one week, the infective promastigotes are present in the proboscis of the fly and are injected when it bites. After gaining access to the human host, the *Leishmania* develop in either cutaneous or visceral tissue.

Although serologic methods are available for the diagnosis of visceral leishmaniasis, the most conclusive evidence of infection is demonstration of organisms in biopsied tissue of spleen, liver, bone marrow or lymph nodes. Bone marrow is probably the most common tissue examined. Impression smears can be prepared and stained with Giemsa. Amastigotes may be visible intracellularly or extracellularly if the cells have ruptured. The presence of the rod-shaped kinetoplast allows differentiation from *Toxoplasma* and *Histoplasma* (see color plate Fig. 14:9). The biopsied tissue can be sectioned in the usual histological fashion. Cutaneous forms may be found in stained preparations of material scraped or aspirated from the edges of cutaneous lesions. Both the visceral and the cutaneous forms may be cultured on blood agar. Cultures are grown at room temperature and should be examined for promastigote stages after 10 days. Cultures should be examined at weekly intervals up to 60 days before being discarded as negative. The importance of culture cannot be over emphasized. In one autochthonous case in the U.S., the correct diagnosis could easily have been overlooked without culture. Serologic data were inconclusive and, since leishmaniasis is rare in the U.S., could have been considered noncontributory; however, a diagnosis of dermal leishmaniasis was reached[79] by culture and identification of the *Leishmania*. Hamster inoculation has been used for diagnosis but the procedure requires 4-6 weeks for demonstration of organisms. Both cutaneous and visceral forms cause visceral infection in the hamster. The animal is sacrificed for impression smears or sections of the spleen and liver. Amastigotes of all species are identical and species cannot be determined on the basis of morphology.

Speciation is useful for a variety of reasons including therapeutic and epidemiological considerations. A simplified procedure to identify isoenzymes has been perfected and is now available in some reference laboratories.[52,53] Cultured organisms can be rapidly and precisely identified by these methods. The procedure requires only a minimal number of organisms so that original cultures are sufficient in many cases. Additional cultures may be necessary for definite determination of slow-growing species. Although the special equipment and training needed for the test precludes its use in smaller laboratories, the test is suitable for larger laboratories in endemic areas. An intradermal test, the Montenegro test, may be useful in diagnosing cutaneous and mucocutaneous leishmaniasis but it is unsatisfactory for visceral infections.

The sharing of common antigens by the three major *Leishmania* species makes species differentiation by serologic means difficult. The homologous reaction is usually stronger than heterologous ones, but this criterion cannot be used as a basis for speciation. A variety of serological procedures have been used but the most successful are IHA, IIF, CF and direct agglutination (DA). All use cultured promastigotes as antigen but treat them in slightly different ways. IHA is the simplest procedure but is less specific and sensitive than either IIF or DA. Since IHA is rapid and inexpensive, it can be used satisfactorily when speciation is not necessary, as in endemic areas where a large number of tests are done. DA using trypsinized promastigotes has proved to be more sensitive than any of the other tests but it requires more antigen. The CDC currently uses IIF to diagnose cutaneous leishmaniasis, considering titers ≥1:16 as significant; and uses CF to diagnose the visceral disease, where titers ≥1:8 are considered diagnostic. Crossreactivity with sera from Chagas' disease cases occurs, but usually only at low levels, causing only minor confusion. Chagas' disease can usually be eliminated from consideration by clinical and/or geographical evidence.

HELMINTHS

The blood and tissue helminths include the filariae, *Trichinella* and the larval stages of *Taenia solium*, as well as *Echinococcus granulosus* and *Toxocara canis*, which parasitize lower animals.

FILARIASIS

Filariae are helminths belonging to the class Nematoda. Adult males and females inhabit the tissues or serous cavities of humans. Seven species are considered to be human parasites: *Wuchereria bancrofti, Brugia malayi, Loa loa, Mansonella ozzardi, M. perstans, M. streptocerca* and *Onchocerca volvulus*. The latter two species have microfilariae in tissue only and are diagnosed by examining skin snips or biopsies. The other five release the diagnostic stage into peripheral blood and so are considered with the blood parasites (Table 14:3).

The filariae have similar life cycles and are transmitted by blood-sucking insects—mosquitos for *W. bancrofti* and *B. malayi* and flies for the other species. The stage acquired by the vector is a prelarval form called a microfilaria, which develops into the larval stage in the thoracic muscles of the vector. In 1-2 weeks this stage is present in the sheath of the proboscis. When the fly or mosquito bites, the infective larvae actively migrate from the arthropod into the bite wound to infect the human host. In about a year, the parasites are mature in the human host and the female begins to discharge microfilariae. Unlike infections caused by other blood parasites, filarial infections are not transferred by blood transfusions from infected donors or by contaminated needles. The microfilariae are not infective for humans and filariasis can be acquired only from vectors. Geographical distribution of each species is well known (Table 14:3), so areas of residence and travel can be used to determine the filarial infection to which the patient may have been exposed.[77]

The diagnostic stage is a microfilaria, a worm-like organism whose body consists of a column of nuclei interspersed with anatomical features such as a nerve ring, excretory pore, excretory cell, innerbody, anal pore, anal cell and a series of 4 R-cells that are precursors of the lower digestive tract system (Fig. 14:10). In addition, 3 of the species are enveloped in a sheath.

Laboratory diagnosis of species other than *O. volvulus* or *M. streptocerca* depends on finding microfilariae in either wet mounts or stained films of blood. With 3 of the 5 species, the presence of microfilariae in peripheral blood is periodic and blood for diagnosis must be taken at the time of peak parasitemia. Microfilariae of *L. loa* tend to be more numerous during the day (diurnal periodicity) and blood should be taken between 10 A.M. and 2 P.M. Both *W. bancrofti* and *B. malayi* have a nocturnal periodicity and blood should be collected at night. Specimens for *W. bancrofti* should be obtained around midnight (between 10 P.M. and 2 A.M.). Blood can be collected by finger puncture or venipuncture. Small amounts of anticoagulants such as EDTA do not seem to adversely affect the microfilariae so blood with EDTA can be used for wet mounts and stained films. Occasionally, however, the sheath may be shed if the blood stands for several hours.

Wet mounts are used in initial screening to detect organisms. Blood can be examined directly or when diluted with saline. Since organisms are intercellular, they may be readily seen moving among the cells. If no microfilariae are found in direct mounts, concentrations such as those by Knott's technique, the saponin technique or the membrane filter technique[23,31] can be used. Knott's technique with 2% formalin is probably the most widely used. It is an effective procedure but the microfilariae are killed and could be confused with artifacts such as fibers. The microfilariae remain viable in the saponin method. Membrane filter concentration enhances the

Table 14:3. Filariae that Parasitize Humans[81]

	Wuchereria bancrofti	*Brugia malayi*	*Loa loa*	*Mansonella ozzardi*	*Mansonella perstans*	*Mansonella streptocerca*	*Onchocerca volvulus*
Geographic distribution	Cosmopolitan; tropics and subtropics	Asia	West and Central Africa	South and Central America	Africa, South and Central America	West Africa	Africa, Central and South America
Adult habitat	Lymphatic system	Lymphatic system	Subcutaneous tissues	Mesenteries, body cavities	Mesenteries, perirenal, retroperitoneal tissues	Subcutaneous tissues	Subcutaneous tissues
Vector	Mosquitoes	Mosquitoes	*Chrysops* (deer fly)	*Culicoides Simulium*	Culicoides (midge)	Culicoides (midge)	*Simulium* (black fly)
Location of microfilariae	Blood	Blood	Blood	Blood	Blood	Skin	Skin
Periodicity	Nocturnal*	Nocturnal†	Diurnal	None	None	None	None
Morphology of microfilariae							
Sheath	Present	Present	Present	Absent	Absent	Absent	Absent
Length (μm)	230–300	175–260	250–300	175–240	190–200	180–240	Two sizes 285–370 150–290
Width (μm)	7.5–10	5–6	6–8.5	4–5	4–5	5–6	5–9
Tail and tail nuclei	Tapered to point; no nuclei in end of tail.	Tapered; terminal and subterminal nuclei.	Tapered; nuclei irregularly spaced to end of tail.	Long, slender tail; no nuclei in end of tail.	Tapered, bluntly rounded; nuclei to end of tail.	Tapered, bluntly rounded; nuclei to end of tail. Tail bent in hook shape.	Tapered to point; no nuclei in end of tail.

*Subperiodic in Pacific Islands.
†Subperiodic form as well.

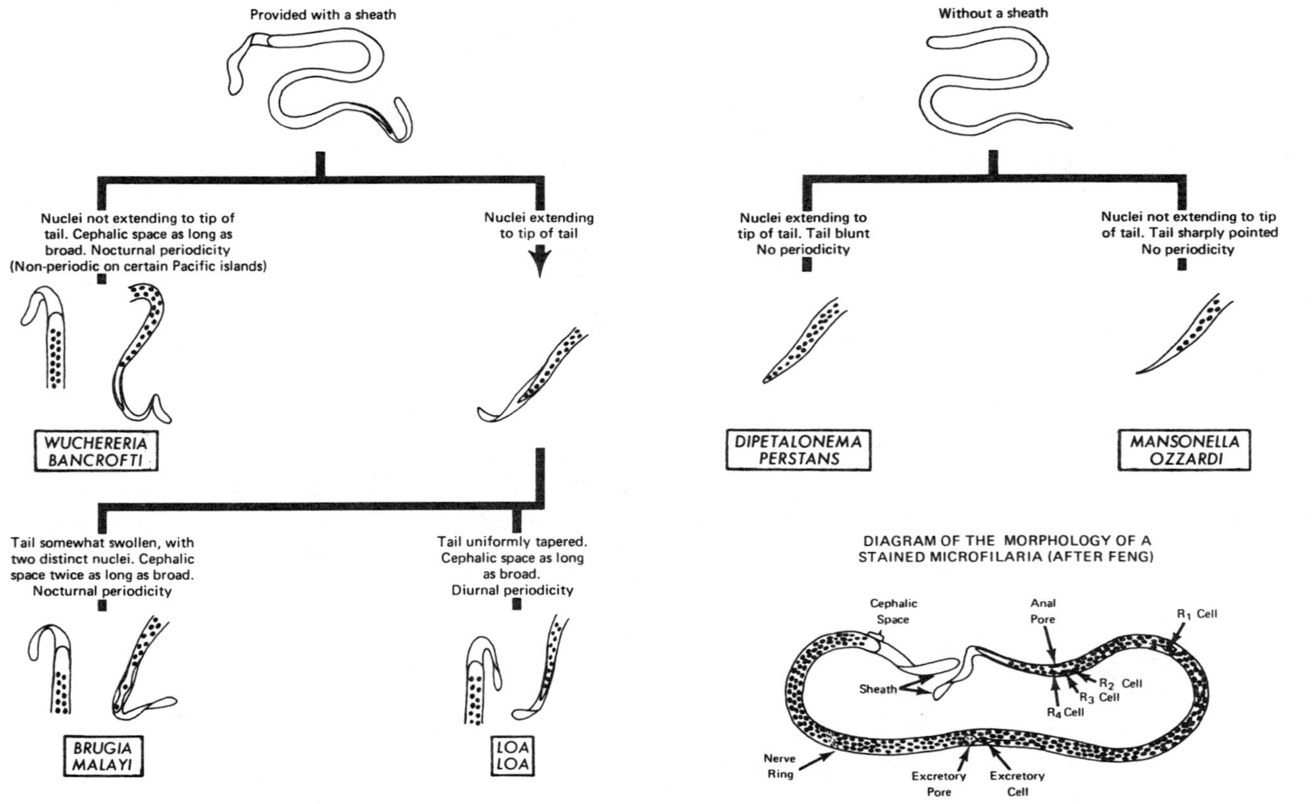

Figure 14:10. Key to microfilariae found in peripheral blood of humans. Courtesy of Training and Laboratory Program Office, Centers for Disease Control, Atlanta, GA.

detection of microfilariae but identification of the species requires other preparations that permit morphological differentiation. Species identification cannot be made from wet mounts so that stains of thick blood films or concentrates to demonstrate anatomical features are essential to establish a specific diagnosis. Thin films are of little or no value. Not all the anatomical characteristics can be seen in a single type of stain but rarely would all these features be needed to determine species. Differential diagnosis is generally based on the presence or absence of a sheath, the presence or absence of nuclei in, and the configuration of, the tip of the tail. Occasionally, the length of the cephalic space and/or the staining reaction can be useful. Preparations should be scanned under low power; then high-dry and oil immersion magnification should be used for more detailed study. The characteristics for each species are presented in the key in Figure 14:10.

The visibility of these characteristics varies with the stain used. A variety of staining techniques have been developed, including Giemsa, hematoxylin, azure, methyl green and pyronine. Giemsa is the most commonly used but in doubtful cases other stains may be needed for detailed staining of particular structures. The Giemsa procedure is the same as that used for thick films for diagnosis of other blood parasites. Cells are often indistinct in Giemsa stained films. Giemsa stains the sheath of *B. malayi* but not the sheath of *L. loa*. The sheath of *W. bancrofti* may or may not stain with Giemsa but does not stain as intensely as that of *Brugia*. Thus, if a sheath is not visible in a Giemsa stained film, the presence of the structure cannot be ruled out. Sheaths are better demonstrated in hematoxylin stains, which should be used if the presence of a sheath is in question.

Although the tip of the tail is very flexible and sometimes bent or twisted so it cannot be seen, its appearance can be used for differentiating microfilariae. When the tail of *W. bancrofti* is bent, the nuclear column may appear to extend to the tip of the tail. Several organisms should be examined for reliable identification.

Lymph obtained from swollen nodes can occasionally be used to diagnose *W. bancrofti* and *B. malayi* infections. Microfilariae have been reported in fluid from Calabar swellings, the nonspecific swellings which can occur on any part of the body in loaiasis. Patients with chronic filariasis may not have detectable microfilariae in blood or lymph smears. Serologic tests can be used for diagnosis but may not be conclusive.

O. volvulus adults are coiled in the subcutaneous tissues (Fig. 14:11) and microfilariae migrate through the skin. They may migrate to the eye and are a significant cause of blindness in some parts of the world. Diagnosis is established by detecting microfilariae in a piece of shaved skin teased in saline. Infection caused by *M. streptocerca* is diagnosed in a similar fashion. Microfilariae of these two species do not occur in the peripheral blood.

Serologic diagnosis of filariasis has never been successful. Several serologic tests including double diffusion and IIF have been used, but the IHA

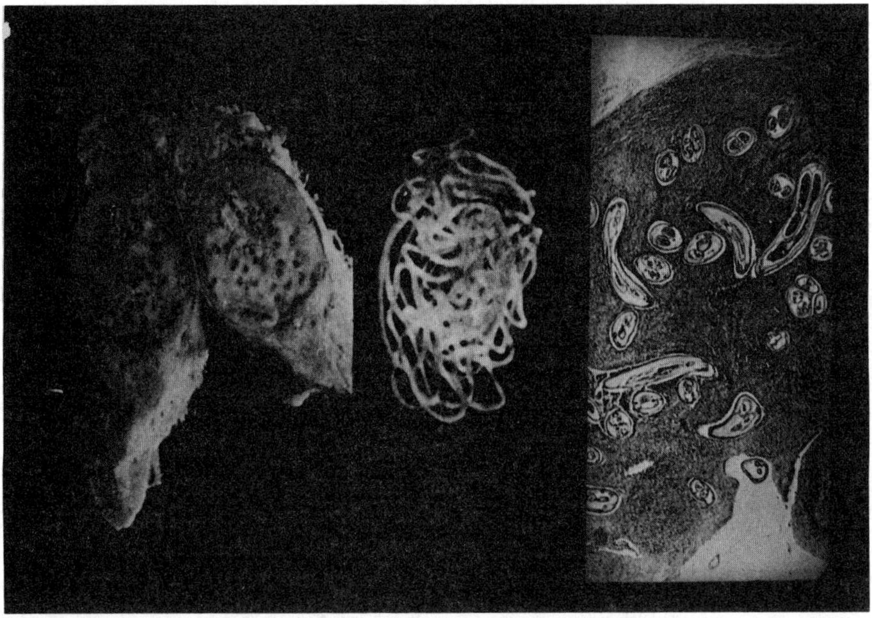

Figure 14:11. Onchocerciasis. Left, a gross photograph of a bisected subcutaneous nodule of onchocerciasis. Center, an adult worm which has been digested from a similar nodule. Right, similar nodule showing sections of the worm encased in fibrous tissue.[81]

is the most popular. Serology is plagued by lack of sensitivity and specificity. The chief problem is an inability to obtain pure homologous antigens. An extract of adult *Dirofilaria immitis*, the dog filarial worm, is used. Kagan and Norman[49] reported only 73% sensitivity for an IHA with this antigen, which reacts with all genera of filarial worms. Ambroise-Thomas and Kien-Truong[3] reported 99% sensitivity for the use of an IIF with frozen sections of *M. vitae*. It is not unusual to find a low or negative serologic result in the presence of measurable parasitemia. This could be explained by an apparent adsorption or blocking of antibody by an excess of antigen in the blood. To overcome some of these problems, the bentonite flocculation test (BFT), which is much less sensitive but more specific, is usually included. When both the IHA and BFT are positive, there is an increased assurance of their significance. In spite of problems, the IHA and BFT can assist the clinician faced with a patient with compatible travel history and symptomatology but no microfilariae. Because of the rare occurrence of filariasis in the U.S. and the minimal quality of serologic procedures, the CDC no longer offers these tests routinely.

TRICHINOSIS

Trichinosis is not uncommon in the U.S. and is usually acquired by eating raw or improperly cooked meat containing infective larvae of *Trichinella spiralis*. Outbreaks are most often associated with eating undercooked pork or hamburger containing pork, but outbreaks regularly occur due to ingestion of raw or undercooked wild bear meat.

T. spiralis has both an intestinal phase (which is very short) and a tissue phase. The same host serves as both the definitive and intermediate host. Humans become infected by ingesting viable larvae in contaminated meat. The parasites mature in the mucosa of the small intestine in about 2-4 days and the females begin to deposit larvae. These gain access to the blood and are carried to all parts of the body. Those reaching skeletal or striated muscle continue to develop and in about 3 weeks become encysted in the muscle fibers. In humans, this is a blind alley for the parasite, but in other animals such as hogs, rats and bears, these encysted larvae are a source of infection when the tissue is eaten by a susceptible host.

Trichinosis can be diagnosed by direct examination of biopsied tissue or, more commonly, by serologic methods. Biopsied tissue can be examined in 3 ways: (1) by teasing it apart with needles, compressing the teased tissue between 2 slides and examining them microscopically for larvae; (2) digesting the tissue, centrifuging and examining the sediment microscopically for larvae; and (3) sectioning and staining the tissue in the usual fashion for histology. Larvae are large and can be readily recognized with low-power microscopy.

Virtually every serologic technique has been used for diagnosis of trichinosis. The most popular have been bentonite flocculation, indirect immunofluorescence and direct hemagglutination.[54,63,69] Although BFT is a very specialized technique and not available in many laboratories, it has persisted for many years as the test of choice because it is highly specific and sensitive. The major drawback (as with most serologic tests for trichinosis) is that it does not show positive results until at least 3 weeks after infection and then shows positive results for years. With the development of the ELISA[36] and its improvement by the use of purified antigens,[41] it is now apparent that BFT has both false-positive and false-negative reactions. In addition, CIE has been improved by the development of purified antigens such that Despommier[35] now recommends that CIE be used to screen sera and positive results be confirmed with ELISA.

LARVA MIGRANS

Although larva migrans is caused by the larval stages of a number of species of parasites infecting lower animals, the most frequent infection in humans is that caused by *Toxocara canis*, the common roundworm of dogs, resulting in visceral larva migrans (VLM) and ocular larva migrans (OLM). The infection is usually seen in young children and is acquired when embryonated eggs of *T. canis* are ingested. The eggs hatch in the intestine, penetrate the wall and blindly migrate through the tissues—frequently causing major mechanical damage. The larvae may persist for prolonged periods of time but finally die in the tissues of the patient, unable to complete their normal life cycle. Diagnosis is often difficult. The patient may have a variety of responses, varying from an asymptomatic state to disease with skin rash, pneumonitis and CNS involvement. Eosinophilia is common.[6]

OLM, resulting from invasion of the eye, produces lesions frequently indistinguishable from retinoblastoma and other retinal lesions. Incorrect diagnosis can lead to enucleation when therapy would have sufficed. Diagnosis by direct tissue examination is not practical but serologic methods are reliable and available.

The almost 100% crossreactivity between *T. canis* and *A. lumbricoides* prevented development of good serologic tests for many years. IHA and BFT were both insensitive and nonspecific. A highly successful ELISA procedure using an embryonated egg antigen was described by Cypess.[30] Crossreactivity with *Ascaris* remained but could be lowered or removed by preadsorbtion of sera with *Ascaris* antigen, leaving a specific *Toxocara* reaction. Following the introduction by de Savigne[32] of an antigen prepared from the excretions-secretions of cultured *T. canis* larvae, Glickman et al[43] showed almost 100% specificity without need for adsorption. This is now the test of choice used by the CDC.

Other larvae also cause human infection. Animal hookworm species cause "creeping eruption," where skin inflammation with serpiginous tracks develop in the area of larval penetration and migration. Since there are no good laboratory methods, neither for direct examination nor serology, infections must be diagnosed clinically.

CYSTICERCOSIS

Cysticercosis is caused by the larval stage of *T. solium*, which in the adult form parasitizes the human intestinal tract. The usual intermediate host for *T. solium* is the pig and larval stages or cysticerci are found in various tissues. Man, however, can also serve as an intermediate host with cysticerci developing in skeletal muscle, heart, subcutaneous tissue, eye and the CNS. Cysticerci are fluid-filled cysts approximately 1 cm in diameter, containing an invaginated scolex (Fig. 14:12). Humans may acquire the infection by ingesting infective eggs passed in human feces. Relatively few cysticercosis cases are seen in the U.S. and the majority of cases are imported from Mexico and Central America, where the disease is most common. Cysticercosis is most often diagnosed by serologic means, although

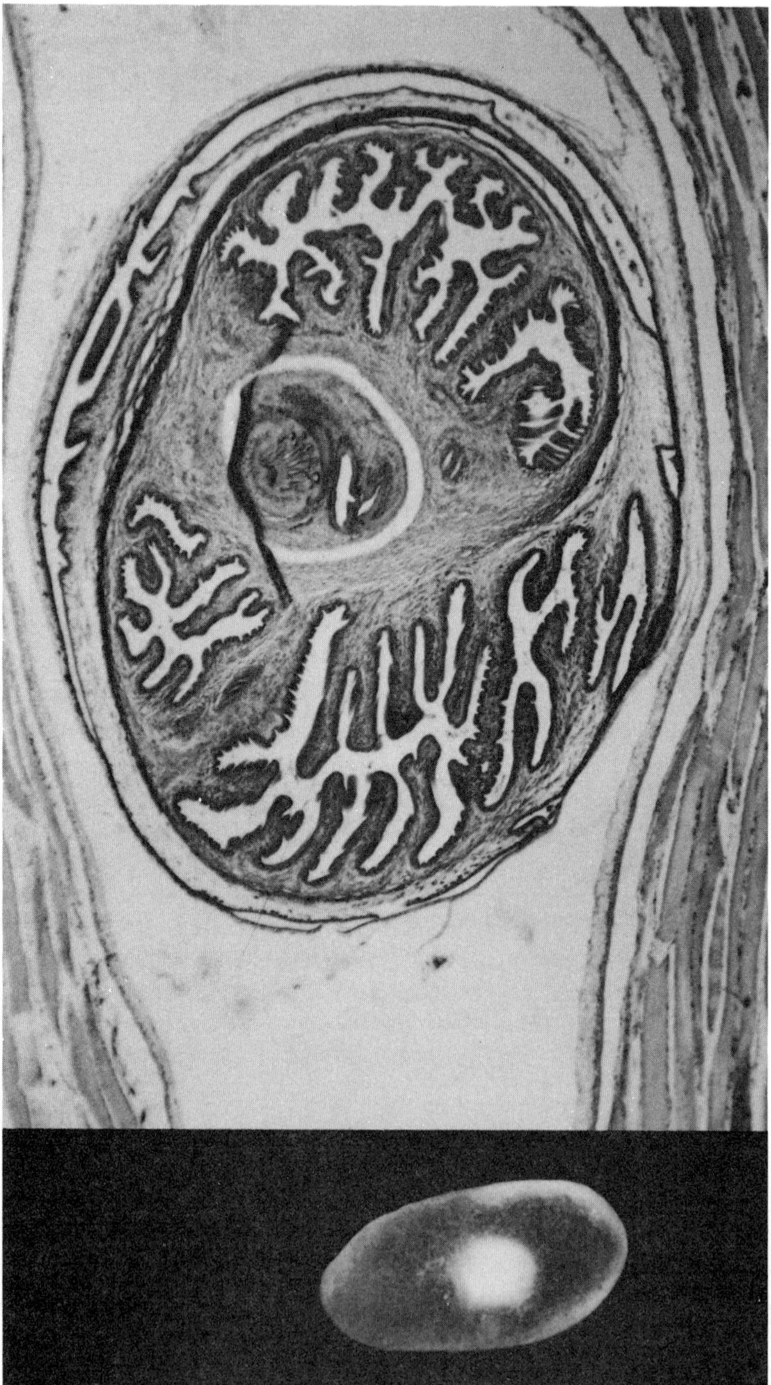

Figure 14:12. Cysticercosis. Left, photomacrograph of an intact cysticercus. The white area in the center is the inverted scolex. Right, photomicrograph of a section of a cysticercus in tissue. Hooklets may be seen in the inverted scolex. The clear areas to the sides of the inverted scolex are fluid filling the cyst.

it may occasionally be diagnosed by biopsy. Serologic techniques continue to be developed. Proctor et al.[70] in South Africa and Biagi et al.[8] in Mexico have reported excellent results when testing sera from man and animals with IHA. Significant crossreactions occur between cysticercosis and echinococcosis sera and antigens. Recently, Tsang et al,[85,86] described a transfer blot (Western Blot) procedure using column-fractionated cysticerci. They reported 100% specificity and 96% sensitivity. At present, however, the IHA is the most commonly used procedure. If the clinician is alerted to the crossreactivity with echinococcosis, IHA titers of 1:128 can be considered clinically important.

ECHINOCOCCOSIS

The definitive hosts for *Echinococcus* species are dogs, foxes, coyotes and related animals. The usual intermediate hosts are sheep, cattle or rodents, depending upon the species. Man becomes infected (hydatid disease) by ingesting eggs from feces of the definitive host and thus functions as an intermediate host. It generally takes years for cysts to develop to a size where they produce symptoms. Larval stages (hydatid cysts) develop in liver, lungs and other areas of the body.

Diagnosis can be made either by finding scoleces in fluid from cysts or by serologic testing. Immunodiagnostic procedures are the common means of diagnosis and the immunodiagnosis of echinococcosis has been thoroughly reviewed by Schantz and Gottstein.[78] The IHA test is a sensitive and reactive procedure but shows marked crossreactivity between echinococcosis sera and cysticercosis sera. Other tests, such as the IIF[2,17] and CIE,[67] have been used to increase specificity but each has its own drawbacks. IHA reactivity may remain at low levels for long periods (years), although IIF tends to become negative within one year.[2] Scoleces from viable cysts are used as antigen for IIF and are somewhat more difficult to obtain than the hydatid fluid used for antigen in the IHA. The IHA is also a more rapid and simpler procedure. IHA titers of 1:128 should be considered significant and titers of 1:16,000 are not unusual (see Table 13:1 in Chapter 13, "Serology of Parasitic Infections").

Counterelectrophoresis (CEP) and double-diffusion tests have been used to demonstrate an antigen-antibody precipitin line known as "band 5," which is specific for echinococcosis.[17,25] Although relatively easy to perform, these tests require large amounts of serum and antigen and frequently need concentration of the serum to attain sufficient reactivity. The test is positive in only about 40% of proven cases. ELISA tests using a variety of purified antigens have been reported and show promising results. A purified "band 5" antigen has been prepared for ELISA and several evaluations indicate it gives excellent results.[37] CEP or ELISA with "band 5" antigen can be used effectively to differentiate hydatidosis from cysticercosis.

Patients in whom the hydatid cyst is limited to the lung or in whom the cyst is old and calcified may have negative or low serologic titers. Complete surgical removal of a cyst usually leads to a rapid decline in antibody after 2-3 months and failure to observe a decline usually indicates an incomplete removal. Rupture of a cyst, either naturally or at surgery, may cause an allergic response with very high titers of antibody.

A skin test (Casoni) for echinococcosis is available but is of little diagnostic value because it is undersensitive and reactivity is lifelong. Negative reactions are questionable due to the low sensitivity of the test and positive reactions give no temporal information; thus, skin test data are difficult to interpret.

SPARGANOSIS

Sparganosis is caused by the spargana (or plerocercoid) stages of diphyllobothrid species of lower animals, probably *Spirometra* species. The spargana of these species develop in various animals, including frogs and snakes. The habit of using raw frog or snake tissue to treat ''black eyes'' may lead to ocular infections if the parasite migrates from the animal tissue into the eye. Humans may also become infected by swallowing infected water crustacea (*Cyclops* spp.). Infections are usually located in the cutaneous tissue or the eye and are generally diagnosed by examining surgically removed tissues.

COENUROSIS

Coenurosis is the disease produced by the larvae of the genus *Multiceps*—cestode parasites of dogs, wolves and foxes. The larval stage, a coenurus, is a cystic form similar to the hydatid cyst of *Echinococcus* species. Multiple scoleces are produced, as in the hydatid cyst, but there is only one cyst cavity and no daughter cysts. The usual intermediate hosts are herbivores (such as sheep) or rodents, depending on the species. The location of the coenurus also varies with the species; it may be located in the brain, eye, subcutaneous tissue or somatic musculature. Humans are infected by ingesting eggs from dog feces. Diagnosis is based on examination of surgically removed tissue.

REFERENCES

1. Allain DS, Kagan IG. An evaluation of the direct agglutination test for Chagas' disease. J Parasitol 1974;60:179-84.
2. Ambroise-Thomas P, Kien-Truong T. L'immuno-fluorescence dans le diagnostic sérologique et le contrôle post-opératoire de l'hydatidose humaine - bilan de 300 cas. I. Materiel et methods. Cah Med Lyon 1970;46:2955-62.

3. Ambroise-Thomas P, Kien-Truong T. Application of the indirect fluorescent antibody test on sections of adult filariae to the serodiagnosis, epidemiology and post-therapeutic surveillance of human filariasis. Geneva: World Health Organization, WHO/FIL/72-101, 1972.
4. Apley J, Clarke SKR, Roome APCH, et al. Primary amoebic meningoencephalitis in Britain. Br Med J 1970;1:596-9.
5. Ash LR, Orihel TC. Atlas of human parasitology. 2nd ed. Chicago: American Society of Clinical Pathologists, 1984.
6. Beaver PC, Snyder CH, Carrera GM, Dent JH, Lafferty J. Chronic eosinophilia due to visceral larva migrans. Report of three cases. Pediatrics 1952;9:7-19.
7. Benenson MW, Takafuji ET, Lemon SM, Greenup RL, Sulzer AJ. Oocyst-transmitted toxoplasmosis associated with ingestion of contaminated water. N Engl J Med 1982;307:666-9.
8. Biagi F, Navarrete F, Pina A, Santigo AM, Tapia L. Estudio de tres reacciones serologicas en el diagnostico de la cisticercosis. Rev Med Hos Gen (Mex) 1961;24:501-8.
9. Bigby TD, Margoskee D, Curtis JL, Sheppard MD, Hadley WK, Hopewell PC. The usefulness of induced sputum in the diagnosis of *Pneumocystis carinii* pneumonia in patients with the acquired immunodeficiency syndrome. Am Rev Respir Dis 1986;133:515-18.
10. Bowling MC, Smith I, Wescott SL. A rapid staining procedure for *Pneumocystis carinii*. Am J Med Technol 1973;39:267-8.
11. Bray RS. Leishmaniasis. Ann Rev Microbiol 1974;28:189-217.
12. Brinn NT. Rapid metallic histological staining using the microwave oven. J Histotechnol 1983;6:126-9.
13. Brown SM, Sithale L, Aranda C. Induced sputum in the diagnosis of *Pneumocystis carinii* pneumonia. Am Rev Respir Dis 1986;133:180.
14. Broaddus C, Dake MD, Stulbarg S, et al. Bronchoalveolar lavage and transbronchial biopsy for the diagnosis of infection in the acquired immunodeficiency syndrome. Ann Intern Med 1985;102:747-52.
15. Butt C. Primary amebic meningoencephalitis. N Engl J Med 1966;274:1473-6.
16. Campbell GH, Aley SB, Hall T, et al. Use of synthetic and recombinant peptides in the study of host-parasite interactions in malaria. Am J Trop Med Hyg 1987;37:428-44.
17. Capron A, Yarzabal LA, Vernes A, Fruit J. Le diagnostic immunologique de l'echinococcose humaine. Pathol-Biol 1970;18:357-65.
18. Carter RF. Primary amebic meningoencephalitis, an appraisal of present knowledge. Trans R Soc Trop Med Hyg 1972;66:193-213.
19. Centers for Disease Control. Chemoprophylaxis of malaria. Morbid Mortal Weekly Rep 1978;27(suppl):81.
20. Centers for Disease Control. Prevention of malaria in travelers - 1984. Morbid Mortal Weekly Rep 1984;33(suppl 2):75-103.
21. Centers for Disease Control. *Acanthamoeba* keratitis associated with contact lenses—United States. Morbid Mortal Weekly Rep 1986;35:405-7.
22. Centers for Disease Control. Malaria surveillance annual summary 1986. Atlanta: U.S. Public Health Service, 1987.
23. Chulaerk P, Desowitz RS. A simplified membrane filtration technique for diagnosis of microfilaremia. J Parasitol 1970;56:623.
24. Coatney GR, Collins WE, Warren McW, Contacos PG. The primate malarias. Washington, DC: U.S. Government Printing Office, 1971.
25. Coltorti EA, Varela-Diaz VM. Detection of antibodies against *Echinococcus granulosus* arc 5 antigens by double diffusion test. Trans R Soc Trop Med Hyg 1978;72:226-9.
26. Contacos PG, Collins WE. Malaria relapse mechanism. Trans R Soc Trop Med Hyg 1973;67:617-18.
27. Conley FK, Jenkins KA, Remington JS. *Toxoplasma gondii* infection of the central nervous system: use of the peroxidase-antiperoxidase method to demonstrate *Toxoplasma* in formalin fixed paraffin embedded tissue sections. Hum Pathol 1981;12:690-8.

28. Cuadrado RR, Kagan IG. The prevalence of antibodies to parasitic diseases with sera of young army recruits from the United States and Brazil. Am J Epidemiol 1967;86: 330-40.
29. Culbertson CG. Soil amoeba infection. In: Lennette EH, Spaulding EH, Truant JP, eds. Manual of clinical microbiology. 2nd ed. Washington, DC: American Society for Microbiology, 1974:602-4.
30. Cypess RH, Karol MH, Zidian JL, Glickman LT, Gitlin D. Larva-specific antibodies in patients with visceral larva migrans. J Infect Dis 1977;135:633-40.
31. Dennis D, Kean BH. Isolation of microfilariae: report of a new method. J Parasitol 1971;57:1146-7.
32. de Savigny DH, Tizard IR. Toxocaral larva migrans: the use of larval secretory antigens in haemagglutination and soluble antigen fluorescent antibody tests. Trans R Soc Trop Med Hyg 1977;71:501-7.
33. Desmonts G, Couvreur J. Congenital toxoplasmosis: a prospective study of 378 pregnancies. N Engl J Med 1974;290:1110-16.
34. Desowitz RG, Saave JJ, Stein B. Application of the indirect hemagglutination test in recent studies on the immunoepidemiology of human malaria and the response in experimental malaria. Mil Med 1966;131(suppl):1157-66.
35. Despommier DD. Trichinellosis. In: Walls KW, Schantz PM, eds. Immunodiagnosis of parasitic diseases; vol I. Helminthic diseases. Orlando, FL: Academic Press, 1986:163-81.
36. Engvall D. Ljungstrom I. Detection of human antibodies to *Trichinella spiralis* by enzyme-linked immunosorbent assay, ELISA. Acta Pathol Microbiol Scand C. 1975;83:231-7.
37. Farag H, Bout D, Capron A. Specific immunodiagnosis of human hydatidosis by the enzyme linked immunosorbent assay (ELISA). Biomedicine 1975;23:276-8.
38. Fife EH Jr, Muschel LH. Fluorescent antibody technique for serodiagnosis of *Trypanosoma cruzi* infection. Proc Soc Exp Biol Med 1959;101:540-3.
39. Franko EL, Walls KW, Sulzer AJ. Reverse enzyme immunoassay for the detection of immunoglobulin M antibodies against *Toxoplasma gondii.* J Clin Microbiol 1981;13:859-64.
40. Frenkel JK, Dubey JP, Miller NL. *Toxoplasma gondii* in cats: fecal stage identified as coccidian oocysts. Science 1970;167:893-6.
41. Gamble HR, Graham CE. A monoclonal antibody-purified antigen for the immunodiagnosis of trichinosis. Am J Vet Res 1984;45:67-73.
42. Gill VJ, Evans G, Stock F, Parillo JE, Masur H, Kovacs JA. Detection of *Pneumocystis carinii* by fluorescent-antibody stain using a combination of three monoclonal antibodies. J Clin Microbiol 1987;25:1837-40.
43. Glickman LT, Grieve RB, Lauria SS, Jones DL. Serodiagnosis of ocular toxocariasis: a comparison of two antigens. J Clin Pathol 1985;38:103-7.
44. Gosey LL, Howard RM, Whitebsky FG, Gill VJ, MacLowry JD. Advantage of modified toluidine blue O and bronchoalveolar lavage for diagnosis of *Pneumocystis carinii* pneumonia. J Clin Microbiol 1985;22:803-7.
45. Hancock K, Tsang VCW. Development and optimization of the FAST-ELISA for detecting antibodies to *Schistosoma mansoni.* J Immunol Meth 1986;92:167-76.
46. Hendley JO, Weller TH. Activation and transmission in rats of infection with *Pneumocystis.* Proc Soc Exp Biol Med 1971;137:1401-4.
47. Hoare CA. The trypanosomes of mammals; a zoological monograph. Oxford, England: Blackwell Scientific Publications, 1972.
48. Hughes WT. Current status of laboratory diagnosis of *Pneumocystis carinii* pneumonitis. Boca Raton, FL: CRC Critical Reviews in Clinical Laboratory Sciences, 1975:145-70.
49. Kagan IG, Norman LG. Serodiagnosis of parasitic diseases. In: Rose NR, Friedman HF, eds. Manual of clinical immunology. Washington, DC: American Society for Microbiology, 1976:382-409.
50. Kean BH, Kimball AC, Christenson WN. An epidemic of acute toxoplasmosis. JAMA 1969;208:1002-4.

51. Kelly J, Landis JW, Davis GS, Trainer TD, Jakab GJ, Green GM. Diagnosis of pneumonia due to *Pneumocystis* by subsegmental pulmonary lavage via the therapeutic bronchoscope. Chest 1978;74:24-8.
52. Kreutzer R, Semko ME, Hendricks LD, Wright N. Identification of *Leishmania* spp. by multiple isozyme analysis. Am J Trop Med Hyg 1983;32:703-15.
53. Kreutzer R, Sourath N, Semko ME. Biochemical identities and differences among *Leishmania* species and subspecies. Am J Trop Med Hyg 1987;36:22-32.
54. Labzoffsky NA, Baratawidjaja RK, Kuitunen E, Lewis FN, Kavelman DA, Morrissey LP. Immunofluorescence as an aid in the early diagnosis of trichinosis. Can Med Assoc J 1964;90:920-1.
55. Lainson R, Shaw JJ. Leishmaniasis of the New World: taxonomic problems. Br Med Bull 1972;28:44-8.
56. LaTorre CR, Sulzer AJ, Norman LG. Serial propagation of *Pneumocystis carinii* in cell line culture. Appl Environ Microbiol 1977;33:1204-6.
57. Linder E, Lundin L, Vorma H. Detection of *Pneumocystis carinii* in lung-derived samples using monoclonal antibodies to an 82KDa parasite component. J Immunol Meth 1987;98:57-62.
58. Luft BJ, Brooks RG, Conley FK, McCabe RE, Remington JS. Toxoplasmic encephalitis in patients with acquired immune deficiency syndrome. JAMA 1984;256:913-17.
59. Mahan CT, Sale GE. Rapid methenamine silver stain for *Pneumocystis* and fungi. Arch Pathol Lab Med 1978;102:351-2.
60. Martinez JA, Sotelo-Avila C, Garcia-Tamayo J, Moron JT, Willaert E, Stamm WP. Meningoencephalitis due to *Acanthamoeba* spp.: pathogenesis and clinico-pathological study. Acta Neuropathol 1977;37:183-91.
61. Matsumoto Y, Yoshida Y. Sporogony in *Pneumocystis carinii*: synaptonemal complexes and meiotic nuclear divisions observed in precysts. J Protozool 1984;31:420-8.
62. Moyer NP, Hudson JD, Hausler WJ Jr. Evaluation of MUREX SUDS Toxo Test. J Clin Microbiol 1987;25:2049-53.
63. Norman LG, Kagan IG. Bentonite, latex and cholesterol flocculation tests for the diagnosis of trichinosis. Public Health Rep 1963;78:227-32.
64. O'Connor R. Manifestations and management of ocular toxoplasmosis. Bull NY Acad Sci 1974;50:192-210.
65. Ognibene FP, Shelhamer J, Gill V, et al. The diagnosis of *Pneumocystis carinii* pneumonia in patients with the acquired immunodeficiency syndrome using subsegmental bronchoalveolar lavage. Am Rev Respir Dis 1984;129:929-32.
66. Pifer LL, Hughes WT, Murphy MJ. Propagation of *Pneumocystis carinii* in vitro. Pediatr Res 1977;11:305-16.
67. Pinon JM, Sulahian A, Remy G, Dropsy G. Immunological study of hydatidosis. Am J Trop Med Hyg 1979;28:318-24.
68. Pitchenik AE, Ganjei P, Torres A, Evans DA, Rubin E, Baier H. Sputum examination for the diagnosis of *Pneumocystis carinii* in the acquired immunodeficiency syndrome. Am Rev Respir Dis 1986;133:226-9.
69. Plonka WS, Gancarz Z, Zawadzka-Jedrzejewska B. A rapid screening hemagglutination test in the diagnosis of human trichinosis. J Immunol Meth 1972;1:309-12.
70. Proctor EM, Powell SJ, Elsdon-Dew R. The serological diagnosis of cysticercosis. Ann Trop Med Parasitol 1966;60:146-51.
71. Remington JS, Cavanaugh EN. Isolation of the encysted form of *Toxoplasma gondii* from human skeletal muscle and brain. N Engl J Med 1965;273:1308-10.
72. Rogers WA Jr, Fried JA, Kagan IG. A modified indirect microhemagglutination test for malaria. Am J Trop Med Hyg 1968;17:804-9.
73. Rosen PP, Martini N, Armstrong D. *Pneumocystis carinii* pneumonia: diagnosis by lung biopsy. Am J Med 1975;58:794-801.
74. Ruebush TK II, Cassaday PB, Marsh HJ, et al. Human babesiosis on Nantucket Island. Ann Intern Med 1977;86:6-9.

75. Ruebush TK II, Juranek DD, Chisholm ES, Snow PC, Healy GR, Sulzer AJ. Human babesiosis on Nantucket Island. Evidence for self-limited and subclinical infections. N Engl J Med 1977;297:825-7.
76. Sabin AB, Feldman HA. Dyes as microchemical indicators of a new immunity phenomenon affecting a protozoon parasite (*Toxoplasma*). Science 1948;108:660-3.
77. Sasa M. Human filariasis: a global survey of epidemiology and control. Baltimore: University Park Press, 1976.
78. Schantz PM, Gottstein B. Echinococcosis (hydatidosis). In: Walls KW, Schantz PM, eds. Immunodiagnosis of parasitic diseases; vol I. Helminthic diseases. Orlando, FL: Academic Press, 1986:69-107.
79. Shaw PK, Quigg LT, Allain DS, Juranek DD, Healy GR. Autochthonous dermal leishmaniasis in Texas. Am J Trop Med Hyg 1976;25:788-96.
80. Siegal SE, Lunde MN, Gelderman AH, et al. Transmission of toxoplasmosis by leucocyte transfusion. Blood 1971;37:388-94.
81. Smith JW, Melvin DM, Orihel TC, Ash LR, McQuay RM, Thompson JH. Diagnostic parasitology—blood and tissue parasites. Chicago: American Society of Clinical Pathologists, 1976.
82. Smith SB, Repetti CF. Evaluation of a rapid screening immunoassay for antibodies to *Toxoplasma gondii*. J Clin Microbiol 1987;25:2207-8.
83. Stehr-Green JK, Bailey TM, Brandt FH, Carr JH, Bond WW, Visvesvara GS. *Acanthamoeba* keratitis in soft contact lens wearers. JAMA 1987;258:57-60.
84. Teutsch SM, Juranek DD, Sulzer A, Dubey JP, Sikes RK. Epidemic of toxoplasmosis associated with infected cats. N Engl J Med 1979;300:695-9.
85. Tsang VCW, Brand JA, Boyer AE, Wilson M, Schantz PM, Maddison SE. Human cysticercosis (*Taenia solium*). I: Systematic fractionation, characterization of relevant antigens and their behaviors on polystyrene and nitrocellulose matrices. 1988; (in press).
86. Tsang VCW, Brand JA, Boyer AE. Human cysticercosis (*Taenia solium*). II: EITB (immunoblot) assay and diagnostic glycoprotein antigens. 1988; (in press).
87. van Meirvenne N, Magnus E, Verroort T. Comparison of variable antigen types produced by trypanosoma strains of the subgenus *Trypanozoon*. Ann Soc Belge Med Trop 1977;57:409-23.
88. Vattuone NH, Yanovsky JF. *Trypanosoma cruzi*: agglutination activity of enzyme treated epimastigotes. Exp Parasitol 1971;30:349-55.
89. Walls KW, Remington JS. Evaluation of a commercial latex agglutination method for toxoplasmosis. Diagn Microbiol Infect Dis 1983;1:265-71.
90. Walzer PD, Schnelle V, Armstrong D, Rosen PP. The nude mouse: a new experimental model for *Pneumocystis carinii* infection. Science 1977;197:177-9.
91. Wilson CB, Remington JS, Stagno S, Reynolds DW. Development of adverse sequelae in children born with subclinical congenital toxoplasma infection. Pediatrics 1980;66:767-74.
92. Wilson M, Ware DA, Walls KW. Evaluation of commercial serodiagnostic kits for toxoplasmosis. J Clin Microbiol 1987;25:2262-5.
93. Wong B, Gold JWM, Brown AE, et al. Central nervous system toxoplasmosis in homosexual men and parenteral drug abusers. Ann Intern Med 1984;100:36-42.

CHAPTER 15

INTESTINAL AND ATRIAL PROTOZOA

James W. Smith, M.D. and Marilyn S. Bartlett, M.S.

INTRODUCTION

Several protozoa may inhabit the intestine and the atrial areas of humans. Disease can be caused by *Entamoeba histolytica, Dientamoeba fragilis, Giardia lamblia, Trichomonas vaginalis, Balantidium coli, Isospora belli, Cryptosporidium* spp. or possibly *Blastocystis hominis*. Diagnosis is usually based on morphologic demonstration of protozoan cysts or trophozoites in appropriate specimens. In addition to these disease-causing protozoa, there are a number of commensal nonpathogenic protozoa which must be recognized and differentiated from the potential pathogens.

In the United States, certain groups of immunologically intact people are at high risk for acquiring protozoan infections because of environmental factors. They include children in day-care centers, male homosexuals and persons who drink unfiltered surface water. Persons who live in places with poor sanitation and hygiene are at increased risk. Patients who are immunocompromised may develop more severe and protracted infections with intestinal protozoa. This is particularly noted in patients with acquired immune deficiency syndrome (AIDS), who frequently develop severe cryptosporidiosis or isosporiasis.

Reviews of results of three parasitology proficiency testing programs in the United States[55,67] show that many laboratories have difficulty detecting and correctly identifying protozoa in fecal specimens.

The only atrial or intestinal protozoan for which serologic tests have been widely applied is *E. histolytica*. However, tests for giardiasis, cryptosporidiosis and trichomoniasis have been described. In addition, antigen detection tests for diagnosis of amebiasis and giardiasis are being evaluated and immunofluorescent reagents have been described for detection of *Cryptosporidium* and *Giardia*.

AMEBAE

Disease

The amebae belong to the class Rhizopoda and move by means of pseudopodia. *E. histolytica* is the most significant pathogen in this group and amebiasis is still the leading fatal intestinal parasitic infection in the United States.[13] *D. fragilis* can cause a less severe disease. It is taxonomically a flagellate but because it appears to be an ameba by light microscopy and must be differentiated from the amebae, it is discussed here. In most instances, *E. histolytica* is a commensal in the intestinal lumen and does not cause disease. It can cause amebic colitis, which is characterized by alternating periods of constipation and diarrhea and asymptomatic intervals; occasionally it causes amebic dysentery, in which there is extensive ulceration of the colon with severe bloody diarrhea and toxicity. Amebic dysentery can be fatal if accompanied by such complications as perforation and peritonitis.[1,40] In instances of invasive amebiasis, such as colitis and dysentery, the organisms may reach the blood stream and may cause metastatic abscesses in other organs, usually the liver. Liver abscesses develop in approximately 5% of people with untreated symptomatic intestinal disease.[2] Amebiasis is acquired by ingesting cysts in contaminated food or water.

Dientamebiasis is a milder disease[80] not accompanied by liver abscesses. It is increasingly recognized as a cause of diarrhea, especially in California.[13] There is no cyst stage and it has been suggested that the infection may sometimes be acquired by ingesting *Enterobius vermicularis* eggs which contain *D. fragilis* trophozoites.[10] Diagnosis is based on demonstrating the organism in stained smears of fecal material.

It has been suggested that *B. hominis*, long considered to be a yeast, is an ameba capable of causing disease.[28,46,66,81] Diagnosis depends upon finding large numbers of organisms in a specimen and ruling out other possible causes of gastrointestinal distress and diarrhea. It is important to differentiate *B. hominis* from disease-causing amebae.

The other intestinal amebae (*E. coli, E. hartmanni, Endolimax nana* and *Iodamoeba butschlii*) are important in that they must be differentiated from the potentially pathogenic amebae (*E. histolytica* and *D. fragilis*). Size and morphologic characteristics of cysts and trophozoites allow identification of species. Size ranges of intestinal protozoa are shown in Figure 15:1.

Morphology

Morphologic characteristics of intestinal amebae are outlined in Tables 15:1 and 15:2 and illustrated in Figures 15:2 and 15:3, as well as in color plate Figures 15:4, 15:5 and 15:6. A self-study course is available.[13]

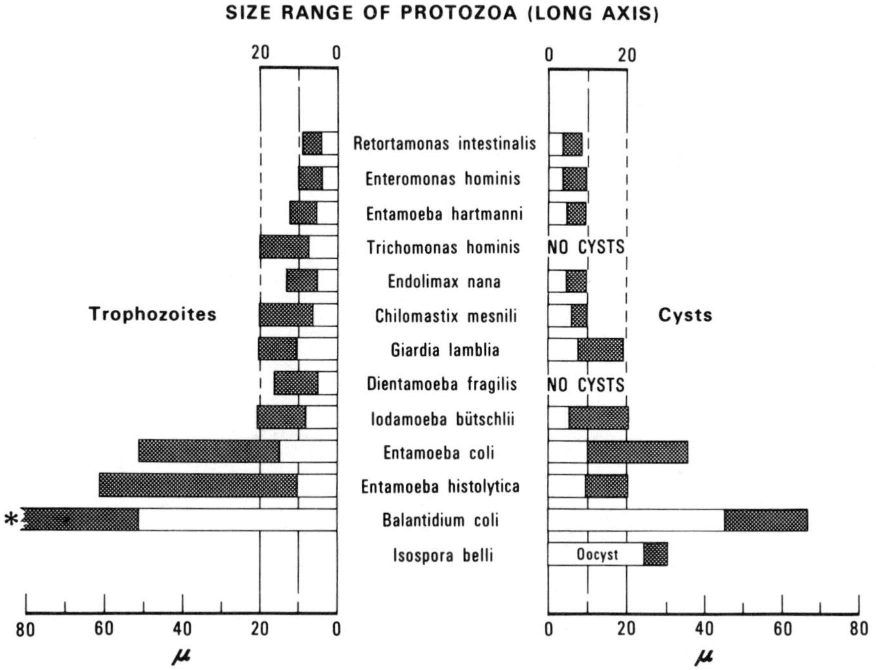

Figure 15:1 Size range of protozoa.

Members of the genus *Entamoeba* have chromatin granules deposited on the nuclear membrane (peripheral nuclear chromatin) and are the only amebae with this characteristic that infect the intestinal tract of humans. *E. hartmanni* and *E. histolytica* are very similar in appearance and must be differentiated primarily on the basis of size. Trophozoites of *E. hartmanni* are smaller than 12 μm at the widest point and the cysts are smaller than 10 μm. Conversely, trophozoites of *E. histolytica* generally are larger than 12 μm and the cysts are larger than 10 μm. Because of random variability in size, it is possible for specimens that contain numerous organisms of one of these species to have occasional organisms which slightly overlap the size range of the other species. A second species identification should not be made unless there is a distinct population of organisms of the second size. Trophozoites of *E. histolytica* vary from 12-60 μm at the widest point with the largest organisms generally associated with invasive disease. Patients with amebic dysentery often harbor *E. histolytica* trophozoites containing phagocytized erythrocytes in their cytoplasm. In general, the cytoplasm of invasive *E. histolytica* does not contain ingested bacteria or yeasts; these inclusions are more common in commensal forms.

In saline wet mounts of fresh material, trophozoites of *E. histolytica* show directional movement, with elongated pseudopods of ectoplasm sharply

Table 15:1. Morphology of Trophozoites of Intestinal Amebae[9]

Species	Size (Length)	Motility	Nucleus: Number	Nucleus: Peripheral Chromatin	Nucleus: Karyosomal Chromatin	Cytoplasm: Appearance	Cytoplasm: Inclusions
Entamoeba histolytica	10–60 μm. Usual range, 15–20 μm—commensal form.* Over 20 μm—invasive form.†	Progressive with hyaline, fingerlike pseudopods.	1 Not visible in unstained preparations.	Fine granules. Usually evenly distributed and uniform in size.	Small, discrete. Usually centrally located, but occasionally is eccentric.	Finely granular.	Red blood cells occasionally. Noninvasive organisms may contain bacteria.
Entamoeba hartmanni	5–12 μm. Usual range, 8–10 μm.	Usually nonprogressive but may be progressive occasionally.	1 Not visible in unstained preparations.	Similar to *E. histolytica*.	Small, discrete, often eccentric.	Finely granular.	Bacteria.
Entamoeba coli	15–50 μm. Usual range, 20–25 μm.	Sluggish, nonprogressive, with blunt pseudopods.	1 Often visible in unstained preparations.	Coarse granules, irregular in size and distribution.	Large, discrete, usually eccentric.	Coarse, often vacuolated.	Bacteria, yeasts, other materials.
Endolimax nana	6–12 μm. Usual range, 8–10 μm.	Sluggish, usually nonprogressive with blunt pseudopods	1 Visible occasionally in unstained preparations.	None	Large, irregularly shaped, blot-like.	Granular, vacuolated.	Bacteria.
Iodamoeba butschlii	8–20 μm. Usual range, 12–15 μm.	Sluggish, usually nonprogressive.	1 Not usually visible in unstained preparations.	None.	Large, usually central. Surrounded by refractile, achromatic granules. These granules are often not distinct even in stained slides.	Coarsely granular, vacuolated.	Bacteria, yeasts or other material.
Dientamoeba fragilis‡	5–15 μm. Usual range, 9–12 μm.	Pseudopods are angular, serrated or broad lobed, and hyaline, almost transparent.	2 (In approximately 20% of organisms only 1 nucleus is present.) Nuclei invisible in unstained preparations.	None.	Large cluster of 4–8 granules.	Finely granular.	Bacteria; occasionally red blood cells.

* Commensal form—usually found in asymptomatic or chronic cases; may contain bacteria.
† Invasive form—usually found in acute cases; often contain red blood cells.
‡ Flagellate—included with amebae for diagnostic purposes.

Table 15:2. Morphology of Cysts of Intestinal Amebae[9]

			Nucleus			Cytoplasm	
Species	Size (Diameter or length)	Shape	Number	Peripheral Chromatin	Karyosomal Chromatin	Chromatoid Bodies	Glycogen
Entamoeba histolytica	10–20 μm. Usual range, 12–15 μm.	Usually spherical.	4 in mature cyst. Immature cysts with 1 or 2 occasionally seen.	Peripheral chromatin present. Fine, uniform granules, evenly distributed.	Small, discrete, usually centrally located.	Present. Elongated bars with bluntly rounded ends.	Usually diffuse. Concentrated mass often present in young cysts. Stains reddish brown with iodine.
Entamoeba hartmanni	5–10 μm. Usual range, 6–8 μm.	Usually spherical.	4 in mature cysts. Immature cysts with 1 or 2 often seen.	Similar to *E. histolytica.*	Similar to *E. histolytica.*	Present. Elongated bars with bluntly rounded ends.	Similar to *E. histolytica*
Entamoeba coli	10–35 μm. Usual range, 15–25 μm.	Usually spherical. Occasionally oval, triangular or other shapes.	8 in mature cyst. Occasionally supernucleated cysts with 16 or more are seen. Immature cysts with 2 or more occasionally seen.	Peripheral chromatin present. Coarse granules irregular in size and distribution, but often appear more uniform than in trophozoites.	Large, discrete, usually eccentric but occasionally centrally located.	Present, but less frequently seen than in *E. histolytica*. Usually splinter-like with pointed ends.	Usually diffuse, but occasionally well defined mass in immature cysts. Stains reddish brown with iodine.
Endolimax nana	5–10 μm. Usual range, 6–8 μm.	Spherical, ovoidal or ellipsoidal.	4 in mature cysts. Immature cysts with less than 4 rarely seen.	None.	Large (blot-like), usually central.	Occasionally granules or small oval masses seen, but bodies as seen in *Entamoeba* spp. are not present.	Usually diffuse. Concentrated mass seen occasionally in young cysts. Stains reddish brown with iodine.
Iodamoeba butschlii	5–20 μm. Usual range, 10–12 μm.	Ovoidal, ellipsoidal, triangular or other shapes.	1 in mature cysts.	None.	Large, usually eccentric. Refractile, achromatic granules on one side of karyosome. Indistinct in iodine preparations.	Occasionally granules present, but chromatoid bodies as seen in *Entamoeba* spp. are not present.	Compact, well-defined mass. Stains dark brown with iodine.

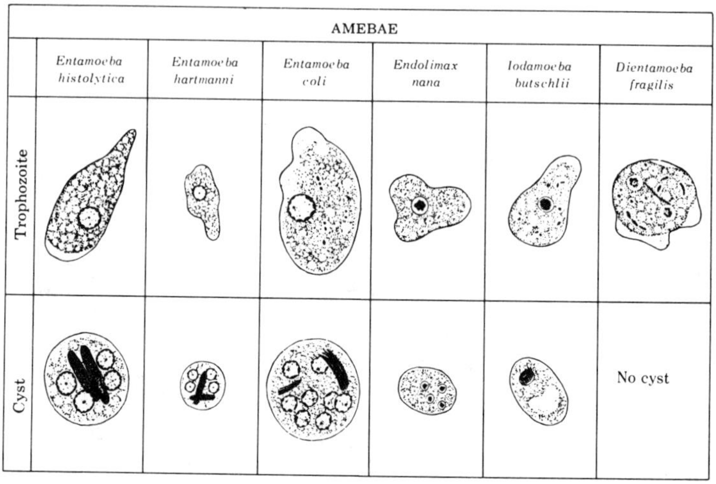

Figure 15:2 Amebae found in stools of humans.[9]
***Flagellate - included with amebae for diagnostic purposes.**

demarcated from the more granular endoplasm; the organisms are often elongated. The nuclei are not visible. Trophozoites of *E. hartmanni*, although smaller, move similarly but more sluggishly. In contrast, trophozoites of *E. coli* are more round and do not show directional movement. In addition, there may be multiple blunt pseudopodia and the ectoplasm is less sharply demarcated from the endoplasm. Nuclei of *E. coli* may be visible in unstained saline wet mounts.

In stained preparations, the nuclei of *E. histolytica* typically have evenly distributed peripheral chromatin with a rounded central karyosome; those of *E. hartmanni* are similar in appearance (Figs. 15:2, 15:3, 15:4 color plate). In contrast, the nuclei of *E. coli* trophozoites have more irregularly distributed chromatin on the nuclear membrane and have larger, more irregular karyosomes, sometimes with fragments of chromatin material in the karyolymph space (Figs. 15:2, 15:3, 15:4 color plate). However, nuclei do vary in appearance. *E. histolytica* or *E. hartmanni* may have karyosomes which are not central and peripheral chromatin which is unevenly distributed; conversely, *E. coli* trophozoites may have evenly distributed peripheral chromatin or central karyosomes. Thus, no single characteristic is pathognomonic of the nuclei of any of the *Entamoeba* species. The cytoplasm of *E. histolytica* generally has few phagocytized inclusions and is delicately stained, as is that of *E. hartmanni*; however, degenerating organisms may have vacuolated cytoplasm. The cytoplasm of *E. coli* trophozoites typically contains numerous ingested bacteria and yeasts and may be quite vacuolated. The cytoplasm of *E. coli* also tends to stain more intensely than that of *E.

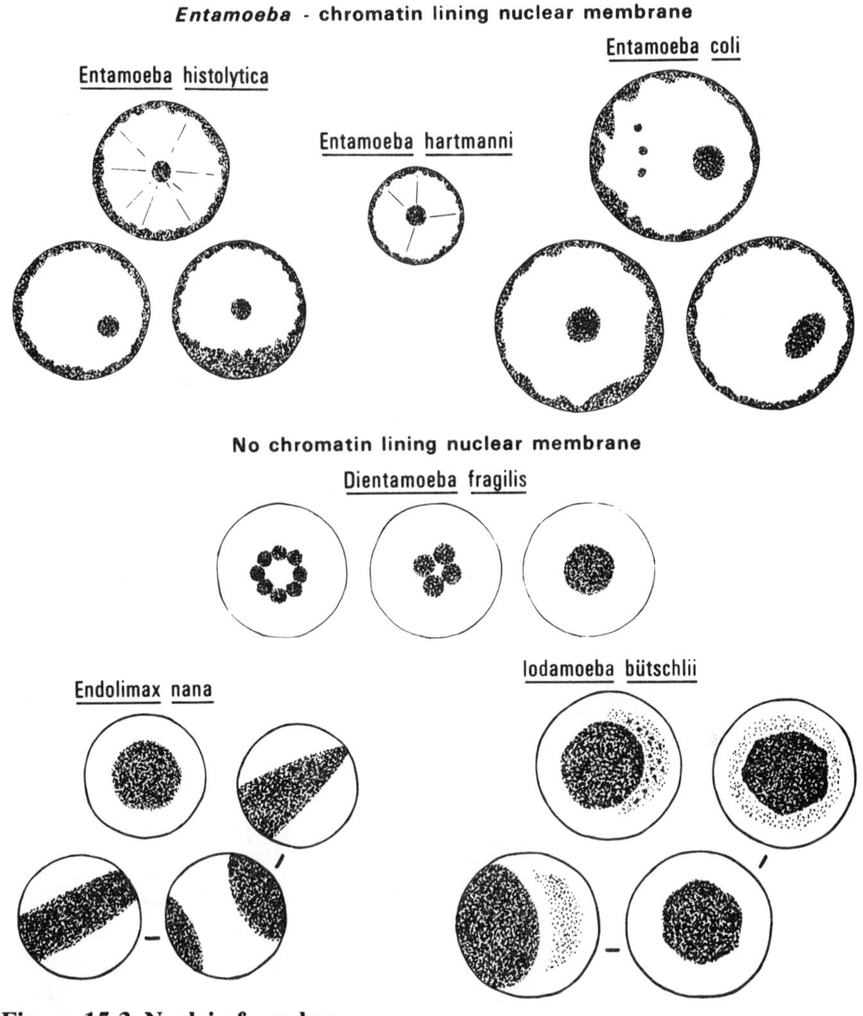

Figure 15:3 Nuclei of amebae.
***Flagellate - included with amebae for diagnostic purposes.**

histolytica. Nuclear and cytoplasmic characteristics of a number of trophozoites must be examined before making an identification.

Cysts of *E. histolytica* and *E. hartmanni* are similar in appearance (Fig. 15:4 color plate, Table 15:2). Cysts may contain glycogen, which stains reddish-brown in fresh iodine wet mounts. In preparations made from preserved material, the glycogen may have dissolved and only the vacuoles remain. Cysts may contain chromatoid bodies of various sizes and shapes (they generally are rods with rounded ends). Mature cysts contain 4 nuclei, each approximately 1/6 the diameter of the cyst.[15] As division takes place, the cyst nuclei become smaller. Immature cysts containing 1 to 3 nuclei may also be found. *E. coli* cysts (Figs. 15:4, 15:6 color plates) may also contain glycogen vacuoles, particularly when immature, and may have chromatoid bodies, but these typically are fibers or have angular or splintered ends. Mature *E. coli* cysts contain 8 nuclei, but careful focusing up and down may be required to see them all. Hypernucleate forms containing 16 or even 32 nuclei are occasionally seen. Quadrinucleate *E. coli* cysts are uncommon and the size of the nuclei in quadrinucleate cysts is generally 1/4 rather than 1/6 the diameter of the cyst.[15] Nuclei of *E. histolytica* cysts tend to be distributed within the cytoplasm whereas those of *E. coli* are often clumped together toward the center. The distribution of chromatin on the nuclear membrane and the location of the karyosome are similar to those of trophozoites but are less reliable characteristics for identifying cysts than for identifying trophozoites. Cysts of the *Entamoeba* species are generally round, although there may be some irregularity and indentation, particularly in fixed specimens, possibly as a result of shrinkage. This is seen especially with *E. coli*. Cytoplasm of *E. coli* cysts often stains intensely with trichrome stain and is granular; thus, these cysts can easily be missed when screening a slide. It may be necessary to make the light very bright to see the nuclei.

In saline wet mounts, cysts appear as refractile structures and chromatoid bodies may be evident. Nuclei of unfixed *E. histolytica* cysts are not visible, whereas those of unfixed *E. coli* may be. In formalin-fixed specimens, nuclei of both species may be visible. In iodine-stained wet mounts, nuclei are visible, chromatoid bodies do not stain, glycogen masses stain reddish-brown (unless glycogen has been washed out by prolonged fixation) and cysts are less refractile.

Identifying non-*Entamoeba* species may be difficult (Tables 15:1, 15:2). The trophozoites of three species, *D. fragilis, E. nana* and *I. butschlii*, overlap in size and the nuclei share the characteristic of not having peripheral chromatin. The karyosomes of *D. fragilis* nuclei consist of 4-8 granules (Figs. 15:2, 15:3, 15:5 color plate) which may be seen in particularly well-stained organisms; however, in many trophozoites the granules appear to be one large karyosome. If the organism is binucleate, as 80% of *D. fragilis* trophozoites are, identification may still be possible; however, if it is uninucleate, differentiation from other amebae may not be possible unless multiple trophozoites are examined. The nucleus of *E. nana* typically has a large rounded central karyosome and a clear karyolymph space, without peripheral chromatin on the nuclear membrane (Figs. 15:2, 15:3, 15:5 color plate), but *E. nana* also may have nuclei with triangular-shaped, band-shaped and split karyosomes (Fig. 15:5 color plate) with portions on the nuclear membrane on opposite sides of the nucleus. The presence of these different forms of nuclei are an identifying characteristic of *E. nana*. *I.*

butschlii also has a large karyosome with no chromatin on the nuclear membrane. In some instances, the nuclei of *I. butschlii* and *E. nana* cannot be differentiated. However, *I. butschlii* nuclei typically contain achromatic granules in a group as a crescent on one side of the karyosome or as a circle surrounding the karyosome so that the karyolymph space is hazy (Figs. 15:2, 15:3, 15:5 color plate). Unfortunately, these may not be demonstrable in many organisms and differentiation from *E. nana* may be difficult. Finding various forms of karyosomes in nuclei, as described for *E. nana*, favors that identification as does finding cysts of the species.

The cytoplasms of the trophozoites of the three species also differ. *I. butschlii* is a particularly voracious scavenger, with the cytoplasm generally being filled with numerous ingested bacteria, yeasts and vacuoles. The cytoplasm of *E. nana* often has a similar appearance, although the amount of phagocytized material may be slightly less. The cytoplasm of *D. fragilis* may stain less intensely than that of the other two species, but it also may contain ingested bacteria.

D. fragilis does not have a cyst stage. Differentiation of the cysts of *I. butschlii* and *E. nana* (Fig. 15:5, 15:6 color plates) is usually not difficult. The nucleus of an *I. butschlii* cyst often has an eccentric karyosome with a crescent of achromatic granules adjacent to it, although various forms may also be seen (some with central karyosomes and some without demonstrable achromatic granules). The cysts themselves vary in shape with a "kidney-bean" type being common. The cyst of *E. nana* when mature contains four nuclei which are smaller than the single nucleus of the *I. butschlii* cyst. In wet mounts with iodine, the karyosomes of *I. butschlii* cysts do not stain, while those of the *E. nana* cysts stain dark brown. *I. butschlii* cysts typically contain large glycogen masses stained with iodine. Glycogen is washed out in materials stored in fixative for an extended period and thus may not be evident in wet mounts of such material and glycogen is removed during permanent staining procedures. It should be emphasized that glycogen masses are also common in cysts of *Entamoeba* species and there may be large amounts of glycogen in young *E. nana* cysts; thus, the presence of glycogen alone is not enough to identify *I. butschlii*. *E. nana* cysts may contain some chromatoid fibrils but generally do not contain the prominent chromatoid bodies associated with the *Entamoeba* species.

B. hominis is described as having three stages of trophozoites[81] and no cyst stage. In identifying the organism in wet mounts or stained preparations, the most significant characteristics are size variation of individual organisms and organisms with a central vacuole surrounded by granules (Fig. 15:7).[46] Typical *B. hominis* in a trichrome-stained fecal film are shown in Figure 15:8 color plates.

Diagnosis of Amebiasis

Intestinal amebiasis is usually diagnosed by demonstrating organisms in fecal specimens or in aspirates of colonic lesions obtained during colonoscopy

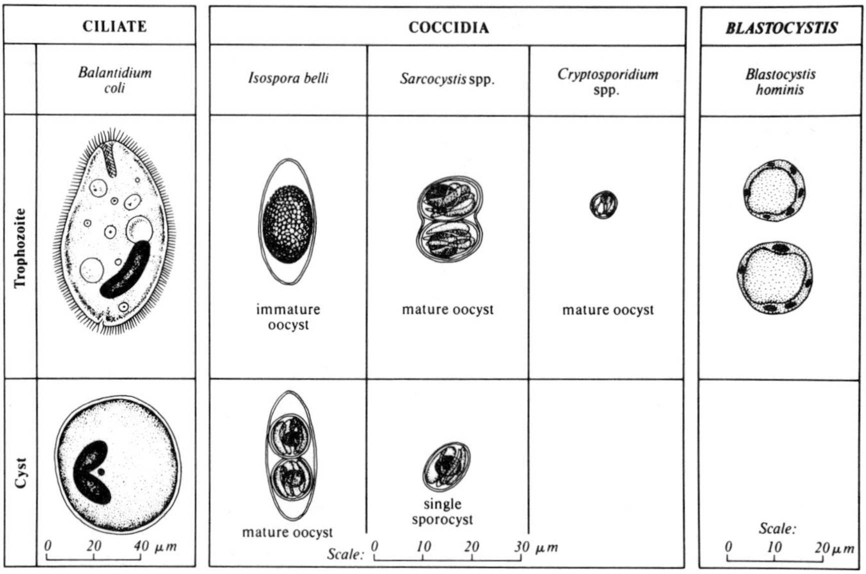

Figure 15:7 Ciliate, coccidia and *Blastocystis* found in stool specimens of humans.[9]

or sigmoidoscopy. There can be variability of shedding of organisms, so multiple specimens may need to be examined. Purgation may assist in diagnosis in some patients.[4] Broad spectrum antimicrobial therapy may suppress amebae for days or weeks and examination may need to be delayed until the patient has been off antimicrobials for 1-2 weeks.

Amebic liver abscesses[2] may develop in up to 5% of patients with intestinal disease and in some patients with no history of intestinal illness. Diagnosis is usually established by serologic tests, which are positive in over 90% of patients. Only half of patients with liver abscesses will have *E. histolytica* detected in stool examinations at the time they present with liver disease. Sometimes amebae can be demonstrated either by wet mounts and stains or by culture in material aspirated from the abscess. The last material aspirated is most likely to contain amebae.

Culture of stool or abscess material may increase sensitivity in detecting amebae but is not widely used.[49] When culturing abscess material, the culture must be inoculated with bacteria such as *Clostridium perfringens* or the amebae will not grow. There are numerous bacteria in fecal specimens so bacteria are not added to such cultures.

A system to subclassify *E. histolytica* strains based on their zymodemes (isoenzyme patterns) has shown 22 different zymodemes which can be classified into three groups.[45,64] Only *E. histolytica* of group II are pathogenic for humans and those of the other two groups appear to be nonpathogenic. This useful epidemiologic tool shows that most *E. histolytica* found in stool are nonpathogenic strains. Homosexuals generally have nonpathogenic strains.[3] There is debate about whether the zymodeme is a fixed genotypic characteristic or a phenotypic one which can vary depending on associated microbial flora.[52] To determine the zymodeme, the *E. histolytica* must be cultured.

Although new serologic tests are continually being introduced and evaluated, the indirect hemagglutination test (IHA) introduced in 1961[38] is the procedure most frequently used for diagnosing amebiasis. Details of this procedure are described by Kagan.[37] The IHA is a highly sensitive and specific procedure involving a sonic lysate of axenically grown amebae. Several reports[30,47,50] have shown that IHA is positive for more than 96% of patients with amebic liver abscess and for more than 85% of those with acute amebic dysentery. Specificity was verified by a reactivity rate of less than 6% in nonamebic diseases and healthy individuals. Antibody can persist for more than two years in some patients.[31] Negative amebic serologic results rule out an amebic liver abscess. It has been suggested that serologic tests for amebae be performed on all patients suspected of having ulcerative colitis to rule out amebiasis before beginning therapy.[40] The Centers for Disease Control (CDC) consider an IHA titer of 1:128 to be clinically significant. Because the test is simple and the reagents are commercially available, serodiagnosis by IHA at the local level is possible.

Counterelectrophoresis (CEP) is an alternative procedure which can be performed in the small laboratory. Available from at least two manufacturers, CEP is a sensitive, simple and rapid test. Although not quantitative, CEP results correlate well with those of IHA for clinically apparent cases.[42] Some sera produce more than one reactivity band, the significance of which is not yet known.

At this time, diagnosis of intestinal amebiasis depends upon the skills of the person examining the stool specimen. Pseudo-outbreaks have occurred because inflammatory cells were misidentified as *E. histolytica*[41] or outbreaks have been missed because *E. histolytica* were not recognized. A more objective and more sensitive method would be desirable. Recently, immunologic methods such as ELISA [enzyme-linked immunosorbent assay] which can detect *E. histolytica* antigens in fecal specimens have been described and show promise, but none are yet commercially available.[58,61,73]

ATRIAL AMEBA

The atrial ameba, *Entamoeba gingivalis*, occurs in the oral cavity only as a trophozoite and closely resembles *E. histolytica*. The two species might be confused when attempting to diagnose pulmonary amebiasis with expectorated sputum. *E. gingivalis* typically contains numerous cytoplasmic inclusions, including bacteria, erythrocytes and leukocytes.[18,57]

FLAGELLATES

Giardia lamblia

Intestinal flagellates belong to the class Mastigophora. The intestinal pathogen *Giardia lamblia* is recognized more and more often as a cause of disease in the United States and elsewhere in the world.[78] It is the most common pathogen found in fecal specimens in the United States, being present in 4% of all specimens submitted to state public health laboratories.[13]

Giardiasis is an intestinal parasitic infection of the duodenum and jejunum with alternating diarrhea and constipation accompanied by abdominal discomfort. The abdominal discomfort is likely to be described as being above rather than below the umbilicus, as is commonly seen with colonic disease. *G. lamblia* has caused several large outbreaks in the United States, including ones in New York[65] and Colorado,[53] and is a continuing problem as a cause of sporadic diarrhea.[32] It has also caused outbreaks in various foreign countries and has been a problem among travelers to Moscow and Leningrad.[34] It has caused outbreaks in nursery schools.[7,56] Campers in the western United States have acquired giardiasis from contaminated water in lakes and streams.[6] It appears that various animal species may be reservoirs for *G. lamblia.*

Large outbreaks of giardiasis have generally been traced to water supplies. *G. lamblia* cysts apparently are not killed by the level of chlorine in many municipal water supplies and major outbreaks can occur unless filtering systems are adequate to remove organisms.[33]

The trophozoites of *Giardia* inhabit the duodenum and jejunum, where they cause inflammation of the mucosa with a malabsorption-type syndrome. The symptoms may vary somewhat from day to day, but giardiasis should be suspected in anyone who has had diarrhea for 10 days or longer.

Diagnosing giardiasis depends on demonstrating the parasites in fecal specimens or in the contents of the small intestine.[69] Studies have shown that the number of organisms in feces may vary significantly from day to day[17], and it is suggested that obtaining a specimen every second or third day for a total of three specimens might be advisable if the diagnosis is suspected but the fecal examination is negative. Occasionally, diagnosis may still not be established and duodenal aspiration may be required to demonstrate the presence of *G. lamblia.* Alternately, a string test may be helpful.[5] A string taped to the cheek is swallowed. The patient is instructed to take small sips of water and after several hours the string is retrieved.

Material expressed from the bile-stained portion is examined for *Giardia* in wet mount or permanent stain preparations.

Antigen detection has been pursued as a method for diagnosing giardiasis and various tests have been evaluated.[54,62,63,74] In addition, immunologic stains have been described.[23] Serologic tests to detect antibody have been proposed but are not widely available.[51,75,76] At this time, no commercially available immunodiagnostic tests have been sufficiently evaluated to be recommended for diagnostic use.

Morphologic characteristics of intestinal flagellates are illustrated in Figure 15:9, in Figures 15:6 and 15:10 color plates and described in Tables 15:3 and 15:4. Morphologically, *G. lamblia* organisms have a distinctive appearance. The trophozoite is shaped like half a pear, with a sucking disk occupying the anterior half of the flattened surface. The organism has two nuclei with large karyosomes, usually central, and no peripheral chromatin. There are four pairs of flagella. The two axonemes extend through the organism in the long axis and there is a median body in the center of the organism. In saline mounts of fresh material, the *Giardia* trophozoites have a movement described as a "falling leaf in a stream," which differs markedly from the ameboid movement of *E. histolytica*. Cysts of *G. lamblia* are oval and have 2-4 nuclei, the quadrinucleate form being mature. Fibrils (axonemes and median body) are visible in wet mounts. Portions of the cytoplasm may have pulled away from the cyst wall. In fecal specimens there may be some cysts which show degenerative changes. Diagnosis should not be based on

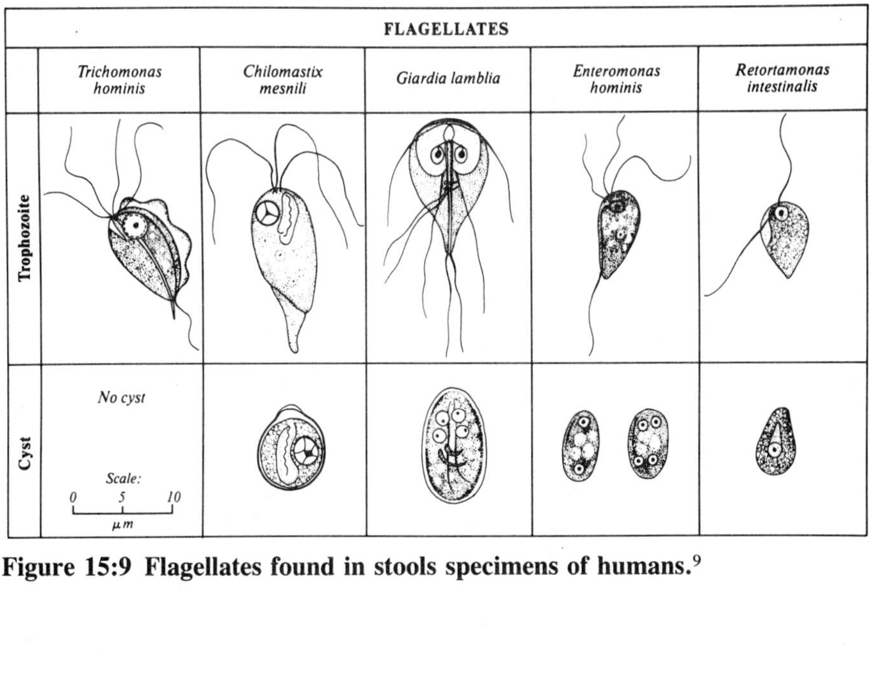

Figure 15:9 Flagellates found in stools specimens of humans.[9]

Table 15:3. Morphology of Trophozoites of Intestinal Flagellates[9]

Species	Size (Length)	Shape	Motility	Number of Nuclei	Number of Flagella*	Other Features
Trichomonas hominis	8–20 μm. Usual range, 11–12 μm.	Pear shaped.	Nervous, jerky.	1 Not visible in unstained mounts	3–5 anterior. 1 posterior.	Undulating membranes extending length of body.
Chilomastix mesnili	6–24 μm. Usual range, 10–15 μm.	Pear shaped.	Stiff, rotary.	1 Not visible in unstained mounts.	3 anterior. 1 in cytosome.	Prominent cytostome extending ⅓–½ length of body. Spiral groove across ventral surface.
Giardia lamblia	10–20 μm. Usual range, 12–15 μm.	Pear shaped.	"Falling leaf."	2 Not visible in unstained mounts.	4 lateral. 2 ventral. 2 caudal.	Sucking disk occupying ½–¾ of ventral surface. Median bodies lying horizontally or obliquely in lower part of body.
Enteromonas hominis	4–10 μm. Usual range, 8–9 μm.	Oval.	Jerky.	1 Not visible in unstained mounts.	3 anterior. 1 posterior.	One side of body flattened. Posterior flagellum extends free posteriorly or laterally.
Retortamonas intestinalis	4–9 μm. Usual range, 6–7 μm.	Pear shaped or oval.	Jerky.	1 Not visible in unstained mounts.	1 anterior 1 posterior.	Prominent cytosome extending approximately ½ length of body.

* Not a practical feature for identification of species in routine fecal examinations.

Table 15:4. Morphology of Cysts of Intestinal flagellates[9]

Species	Size (Length)	Shape	Number of Nuclei	Other Features
Trichomonas hominis	No cyst.			
Chilomastix mesnili	6–10 μm. Usual range, 8–9 μm.	Lemon shaped with anterior hyaline knob.	1. Not visible in unstained preparations	Cytostome with supporting fibrils. Usually visible in stained preparations.
Giardia lamblia	8–19 μm. Usual range, 11–12 μm.	Oval or ellipsoidal.	Usually 4. Not distinct in unstained preparations. Usually 1 located at one end.	Fibrils or flagella longitudinally in unstained cysts. Deep staining fibers or fibrils may be seen lying laterally or obliquely across fibrils in lower part of cyst. Cytoplasm often retracts from a portion of cell wall.
Enteromonas hominis	4–10 μm. Usual range, 6–8 μm.	Elongated or oval.	1–4, usually 2 lying at opposite ends of cyst. Not visible in unstained mounts.	Resembles *E. nana* cyst. Fibrils or flagella are usually not seen.
Retortamonas intestinalis	4–9 μm. Usual range, 4–7 μm.	Pear shaped or slightly lemon shaped.	1. Not visible in unstained mounts.	Resembles *Chilomastix* cyst. Shadow outline of cytostome with supporting fibrils extends above nucleus.

cysts which do not contain demonstrable nuclei and fibrils, but finding such should suggest that a careful search be made for the typical cysts of *G. lamblia*.

Chilomastix mesnili

This flagellate is nonpathogenic and may be found in stool specimens (Figs. 15:9, 15:6 and 15:10 color plates; Tables 15:3, 15:4). In saline mounts of fresh material, it exhibits stiff and wobbling movement. It has three anterior flagella and there is a flagellum in the cytostome. The nucleus usually has a large, granular karyosome and there may be chromatin on the nuclear membrane usually arranged in a "lop-sided" manner, with a concentration of chromatin along one side. The karyosome may be central. The cytostome—which extends ⅓ to ½ the length of the body—is adjacent to the nucleus and can be a helpful diagnostic characteristic. Careful focusing may aid in demonstrating the cytostome. A spiral groove extends the length of the body and is usually visible only as an irregularity of the outline of the organism. The organisms are elongated and have a tapered posterior end, with the nucleus located near the anterior end. In stained smears, these organisms may be confused with trophozoites of various species of amebae, but *Chilomastix* should be suspected if nuclei are always present at one end of the organism and the opposite end tapers. Flagella are not generally visible in fresh or stained materials. The type of motion in wet mounts allows these organisms to be differentiated from the amebae and other intestinal flagellates. Cysts of *C. mesnili* are elongated and lemon-shaped, with a nipple-like hyaline structure at one end. The nucleus is generally at the side and resembles the nucleus of the trophozoites. The primitive cytostome and cytostomal fibrils are evident in wet mounts or in permanently stained smears. The configuration of the cytostomal fibrils has been likened to that of a safety pin.

Trichomonas hominis

This flagellate is a commensal which can inhabit the intestinal tract of humans (Figs. 15:9, Table 15:3). It does not form cysts and is small (12 μm). In saline mounts, it has quick, jerky motions and when trapped in fecal material, an undulating membrane and flagella may be visible. The undulating membrane arises at the anterior end and runs the entire length of the organism, ending in a free posterior flagellum. Extending through the center of the organism are an axostyle and a curved rod, the costa, which marks the attachment of the undulating membrane. *T. hominis* trophozoites generally do not stain well in permanently stained preparations and may be difficult to differentiate from amebae. Organisms are usually delicately stained and the nucleus varies in appearance—sometimes appearing to have peripheral chromatin. The presence of multiple organisms with nuclei toward one end, the presence of the axostyle and the lack of a prominent cytostome may lead one to suspect *T. hominis*. The undulating membrane is usually not evident in stained preparations or is seen as a wavy line extending down the body.

Enteromonas hominis and *Retortamonas intestinalis* are two very small, infrequently seen, nonpathogenic flagellates. Their characteristics are included in Tables 15:3 and 15:4 (Fig.15:9).

Trichomonas vaginalis

This flagellate is a common cause of vaginitis and is similar to *T. hominis*. *T. vaginalis* may also occasionally cause prostatitis in males, although males are more often asymptomatic carriers who may spread the infection.[24] It has been isolated from the respiratory tracts of infants with respiratory diseases.[432]

T. vaginalis is larger (30 μm) than *T. hominis* and the undulating membrane extends only half the length of the body. Because they have different habitats, it is generally not necessary to differentiate *T. hominis* from *T. vaginalis*. Diagnosis is usually based on detecting organisms with the typical jerky movement in wet mounts of vaginal materials, urine or urethral materials. Culture methods which are somewhat more sensitive than direct wet mounts have been developed and may prove helpful,[26,49] but cost effectiveness should be evaluated.[70] An antigen detection ELISA test has been described.[77]

CILIATES

The only ciliate which parasitizes humans is *Balantidium coli*. This organism is quite large (Figs.15:7, 15:8 color plate; Table 15:5) and is covered with cilia which are slightly longer near the cytostome. Both cysts and trophozoites occur; each contains a large macronucleus and a smaller micronucleus. Cilia may still be evident in immature cysts. Trophozoites have prominent contractile vacuoles. Infected persons often have been in contact with swine, a reservoir host for this organism. The disease is similar to amebiasis of the colon and there may be invasive disease with ulcerations of the colon. Metastatic disease does not occur in other organs. It is important to differentiate *B. coli* from some of the free-living protozoa found in pond water, etc., with which they can be confused.

INTESTINAL SPOROZOA

There are three human intestinal sporozoan infections which in many ways are similar to the feline stage of toxoplasmosis.[25] They are cryptos-

Table 15:5. Morphology of Intestinal Ciliate, Coccidia and *Blastocystis*[9]

Species	Size (Length)	Shape	Motility	Number of Nuclei	Other Features
CILIATE					
Balantidium coli Trophozoite	40–70 μm or more. Usual range, 40–50 μm.	Ovoid with tapering anterior end.	Rotary, boring.	1 large, kidney shaped macronucleus. 1 small subspherical micronucleus immediately adjacent to macronucleus. Macronucleus occasionally visible in unstained preparations as hyaline mass.	Body surface covered by spiral, longitudinal rows of cilia. Contractile vacuoles are present.
Cyst	45–65 μm. Usual range, 50–55 μm.	Spherical or oval.	—	1 large macronucleus visible in unstained preparations as hyaline mass.	Macronucleus and contractile vacuole are visible in young cysts. In older cysts, internal structure appears granular.
COCCIDIA					
Isospora belli	Oocyst: 25–30 μm. Usual range, 28–30 μm.	Ellipsoidal.	Nonmotile.		Usual diagnostic stage is immature oocyst with single granular mass (zygote) within. Mature oocyst contains 2 sporocysts with 4 sporozoites each.
Sarcocystis hominis	Sporocyst: 13–17 μm. Usual range, 14–16 μm.	Oval.	Nonmotile.		Mature oocysts with thin wall collapsed around 2 sporocysts or free fully mature sporocysts with 4 sporozoites inside are usually seen in feces.
S. suihominis	11–15 μm. Usual range, 12–13 μm.				
Cryptosporidium	Oocyst: 3–6 μm. Usual range, 4–5 μm.	Spherical or oval.	Nonmotile.		Mature oocyst contains 4 "naked" sporozoites. No sporocysts are present.
BLASTOCYSTIS					
Blastocystis hominis					
Vacuolated form	5–30 μm. Usual range, 8–10 μm.	Spherical, oval or ellipsoidal.	Nonmotile.	1 usually, but 2–4 may be present. Located in "rim" of cytoplasm. In binucleated organisms, the 2 nuclei may be at opposite poles. In quadrinucleated forms, the 4 nuclei are evenly spaced around periphery of cell.	Cell contains large central body or "vacuole" with a thin band or "rim" of cytoplasm around the periphery. Occasionally a ring of granules may be seen in cytoplasm and the cell appears to have a "beaded rim."

poridiosis, isosporiasis and sarcocystis. Organisms parasitize the mucosal cells of the small intestinal tract.

Cryptosporidiosis is caused by *Cryptosporidium* spp. and it appears that organisms which infect calves, dogs or other mammals may infect humans.[21] Although the disease was first described in immunocompromised individuals,[48] reports of infection in the immunocompetent rapidly followed,[16,79] including reports of waterborne and day-care center outbreaks.[12] *Cryptosporidium* appears to cause a significant portion of cases of endemic diarrhea.[20,36,59] Infection is acquired by ingestion of the thick-walled oocysts with subsequent invasion of epithelial cells by the liberated sporozoites. Diarrhea caused by the infecting organism is usually severe, although self-limited in the non-immunosuppressed.[35] In the immunosuppressed, such as patients with AIDS, it may cause severe protracted diarrhea of up to 7 liters of stool per day. There is no effective therapy. Some immunosuppressed patients can have decreasing severity of symptoms or even resolution with time, but many have continued diarrhea for the rest of their lives. Liquid, frothy specimens and those from known immunosuppressed individuals should be evaluated for *Cryptosporidium* oocysts. Reports of *Crytosporidium* spp. in bronchial epithelial cells and associated with abnormalities of the biliary ducts and with pancreatitis[21,29] suggest sputum and bile from immunocompromised patients as other specimens to be evaluated if clinically indicated.

A plethora of stains have been described for demonstrating oocysts (Fig.15:7); most stains can be used with practice and with good control material. It is best to select one method and gain experience with it. The modified acidfast stain described in the method section stains oocysts red (Fig.15:8 color plate). Organisms may be detected in wet mounts of concentrates, particularly by flotation, but in most instances confirmation should be by stain.[27,44] Immunofluorescent stains have been described[71,72] and reagents are becoming available commercially. Serologic tests have also been described, but have not been widely used for diagnosis because a large number of individuals (50%) with no known infection have antibodies, probably reflecting previous infection.[11]

Isosporiasis[8] is caused by the coccidian *Isospora belli* and the disease is acquired by fecal-oral contamination. Infection in the intestinal mucosa is both asexual, with schizogony, and sexual with formation of oocysts. In freshly passed feces, there are immature oocysts which contain zygotes. Each zygote then divides to form two sporoblasts; each of these sporoblasts develops a cyst wall and becomes a sporocyst in which there are four curved, sausage-shaped sporozoites. Diagnosis is established by demonstrating the oocysts (Figs.15:7, 15:8 color plate, Table 15:5) by direct examination, by concentration procedures, or occasionally, in peroral small intestinal biopsies. Infection can cause a malabsorption syndrome and is generally self-limited. Isosporiasis has been reported in patients with AIDS,

Table 15:6. Nonparasitic Objects[68]

Artifact*	Resemblance	Saline Mount	Differential Characteristics of Artifact in Permanent Stain: Cytoplasm	Differential Characteristics of Artifact in Permanent Stain: Nucleus
Polymorphonuclear leukocytes (Seen in dysentery and other inflammatory bowel diseases.)	*E. histolytica* cyst	Usually not a problem. Granules in cytoplasm. Cell border irregular.	Less dense, often frothy. Border less clearly demarcated than that of ameba.	More coarse. Large, relative to size of organism. Irregular shape and size. Chromatin unevenly distributed. Chromatin strands may link nuclei.
Macrophages (See in dysentery and other inflammatory bowel diseases. May be present in purged specimens.)	Amebic trophozoite, especially *E. histolytica*	Nuclei larger and of irregular shape, with irregular chromatin distribution. Cytoplasm granular; may contain ingested debris. Cell border irregular and indistinct. Movement irregular and pseudopodia indistinct.	Coarse, may contain inclusions.	Large and often irregular in shape. Chromatin irregularly distributed. Many granules. Nucleus may be disintegrated.
Squamous epithelial cells (from anal mucosa)	Amebic trophozoite	Nucleus refractile and large. Cytoplasm smooth. Cell border distinct.	Stains evenly, no inclusions.	Large and single. Large chromatin mass may resemble karyosome.
Columnar epithelial cells (from intestinal mucosa)	Amebic trophozoite	Nucleus refractile and large. Cytoplasm smooth. Cell border distinct.	Variable appearance, may be vacuolated.	Large with heavy chromatin on nuclear membrane. Often large central chromatin mass resembling karyosome.
Yeasts (Normal constituent of feces)	Protozoan cyst	Oval. Thick wall. No internal structure. Budding forms may be seen.	Oval. Little internal structure. Refractile cell wall. Budding forms may be seen.	Not confused with amebic nucleus.
Starch granules	Protozoan cyst	Rounded or angular. Very refractile. No internal structure. Stain pink to purple in iodine mounts.	Not a problem in permanently stained slides.	

* Other artifacts such as contaminating plant cells and pollen grains are occasionally seen. These should not be difficult to differentiate.

where disease can be severe[19,39] and occasionally is extraintestinal.[60] Cysts of *Isospora* may morphologically resemble hookworm eggs but are much smaller. They can be differentiated by measurement with an ocular micrometer. In addition, they stain acidfast in a manner similar to the oocysts of *Cryptosporidium*.

Various *Sarcocystis* species which infect the muscles of domestic animals may infect humans who eat undercooked, infected meat.[22] Humans develop the sexual stage in the epithelium of the intestinal tract. In sarcocystosis, the form found in fresh feces is the mature sporocyst (either singly or in pairs) (Fig.15:7) rather than the immature oocyst that is typically associated with *I. belli* infections. Infections caused by *Sarcocystis* spp. were formerly called *Isospora hominis* infections.

NONPARASITIC OBJECTS

Numerous objects found in feces may be confused with parasites; many are described in Table 15:6. Various blood and tissue cells are likely to be confused with amebae, but generally do not have the detailed structure of amebae and have nuclei which are relatively larger than those of amebae. Several pseudoepidemics of amebiasis have resulted from laboratories misidentifying white blood cells as amebae.[41] Yeasts are frequently present in stools, but are not usually difficult to identify; however, they may be confused with *Cryptosporidium*.

REFERENCES

1. Adams EB, MacLeod IN. Invasive amebiasis. I. Amebic dysentery and its complications. Medicine 1977;56:315-23.
2. Adams EB, MacLeod IN. Invasive amebiasis. II. Amebic liver abscess and its complications. Medicine 1977;56:325-34.
3. Allason-Jones E, Mindel A, Sargeaunt P, Williams P. *Entamoeba histolytica* as a commensal intestinal parasite in homosexual men. N Engl J Med 1986;315:353-6.
4. Andrews J. The diagnosis of intestinal protozoa from purges and normally passed stools. J Parasitol 1934;20:253-4.
5. Beal CD, Viens P, Grant RGL, Hughes JM. A new technique for sampling duodenal contents. Am J Trop Med Hyg 1970;19:349-52.
6. Barbour AG, Nichols CR, Fukushima T. An outbreak of giardiasis in a group of campers. Am J Trop Med Hyg 1976;25:384-9.
7. Black RE, Dykes AC, Sinclair S, Wells JG. Giardiasis in day-care centers: evidence of person-to-person transmission. Pediatrics 1977;60:486-91.
8. Brandborg LL, Goldberg SB, Breidenbach WD. Human coccidiosis —a possible cause of malabsorption. N Engl J Med 1970;283:1306-13.
9. Brooke MM, Melvin DM. Morphology of diagnostic stages of intestinal parasites of humans. 2nd ed. Atlanta: Centers for Disease Control; HHS publication no. (CDC)84-8116.

10. Burrows RB, Swerdlow MA. *Enterobius vermicularis* as a probable vector of *Dientamoeba fragilis*. Am J Trop Med Hyg 1956;5:258-65.
11. Campbell PN, Current WL. Demonstration of serum antibodies to *Cryptosporidium* spp. in normal and immunodeficient humans with confirmed infections. J Clin Microbiol 1983;18:165-9.
12. Centers for Disease Control. Cryptosporidiosis among children attending day-care centers, Georgia, Pennsylvania, Michigan, California, New Mexico. Mortal Morbid Weekly Rep 1984;33:599-601.
13. Centers for Disease Control. Intestinal parasite surveillance annual summary 1978. Atlanta: Centers for Disease Control, 1979.
14. Centers for Disease Control. Laboratory Training and Consultation Branch. Amebiasis: laboratory diagnosis—a self-instructional lesson, parts I, II and III. Washington, DC: U.S. Department of Health, Education and Welfare publication no. (PHS)77-8327, 1976.
15. Copeland BE, Kimber J. Nuclear size in diagnosis of *Entamoeba histolytica* on stained smears. Am J Clin Pathol 1968;50:664-8.
16. Current WL, Reese NC, Ernst JV, Bailey WS, Hegman MB, Weinstein WM. Human cryptosporidiosis in immunocompetent and immunodeficient persons. N Engl J Med 1983;308:1252-7.
17. Danciger M, Lopez M. Numbers of *Giardia* in the feces of infected children. Am J Trop Med Hyg 1975;24:237-42.
18. Dao AH. *Entamoeba gingivalis* in sputum smear. Acta Cytol 1985;29:632-3.
19. DeHovitz JA, Pape JW, Boncy M, Johnson WD Jr. Clinical manifestations and therapy of *Isospora belli* infection in patients with the acquired immunodeficiency syndrome. N Engl J Med 1986;315:87-90.
20. DuPont HL. Cryptosporidiosis and the healthy host. (Editorial). N Engl J Med 1985; 312:1319-20.
21. Fayer R, Ungar BL. *Cryptosporidium* spp. and cryptosporidiosis. Microbiol Rev 1986; 50:458-83.
22. Fayer R. Other protozoa: *Eimeria, Isospora, Cystoisospora, Besnoitia, Hammondia, Frenkelia, Sarcosystis, Encephalitozoa* and *Nosema*. In: Jacobs L, ed. Handbook series in zoonoses, section C: parasitic zoonoses. Boca Raton, FL: CRC Press, 1982;1:187-196.
23. Fleck SL, Hames SE, Warhurst CD. Detection of *Giardia* in human jejunum by the immunoperoxidase method. Specific and non-specific results. Trans R Soc Trop Med Hyg 1985;79:110-13.
24. Fouts AC, Kraus SJ. *Trichomonas vaginalis*: reevaluation of its clinical presentation and laboratory diagnosis. J Infect Dis 1980;141:137-43.
25. Frenkel JK. Advances in the biology of sporozoa. Z Parasitenkd 1974;45:125-62.
26. Garber GE, Sibau L, Ma R, Proctor EM, Shaw CE, Bowie WR. Cell culture compared with broth for detection of *Trichomonas vaginalis*. J Clin Microbiol 1987;25:1275-9.
27. Garcia LS, Bruckner DA, Brewer TC, Shimizu RY. Techniques for the recovery and identification of *Cryptosporidium* oocysts from stool specimens. J Clin Microbiol 1983; 18:185-90.
28. Garcia LS, Bruckner DA, Clancy MN. Clinical relevance of *Blastocystis hominis*. Lancet 1984;1:1233-4.
29. Gross TL, Wheat J, Bartlett MS, O'Connor KW. AIDS and multiple system involvement with *Cryptosporidium*. Am J Gastroenterol 1986;81:456-8.
30. Healy GR. The use and limitations to the indirect hemagglutination test in the diagnosis of intestinal amebiasis. Health Lab Sci 1968;5:174-9.
31. Healy GR, Visvesvara GS, Kagan IG. Observations on the persistence of antibodies to *E. histolytica*. Arch Invest Med 1974;5:495-500.
32. Hopkins RS, Shillam P, Gaspard B, Eisnach L, Karlin RJ. Waterborne disease in Colorado: three years' surveillance and 18 outbreaks. Am J Public Health 1985;75:254-7.

33. Jacobowski W, Hoff JC, eds. Proceedings of the national symposium on waterborne transmission of giardiasis. Springfield, VA: National Technical Information Service, U.S. Environmental Protection Agency, 1979:306.
34. Jokipii AMM, Hemila M, Jokipii L. Prospective study of acquisition of *Cryptosporidium, Giardia lamblia*, and gastrointestinal illness. Lancet 1985;2:487-9.
35. Jokipii L, Jokipii AMM. Timing of symptoms and oocyst excretion in human cryptosporidiosis. N Engl J Med 1986;315:1643-8.
36. Jokipii L, Pohjola S, Jokipii AMM. *Cryptosporidium*: a frequent finding in patients with gastrointestinal symptoms. Lancet 1983;3:358-61.
37. Kagan IG. Serodiagnosis of parasitic diseases. In: Rose NR, Friedman RF, Fahey JL, eds. Manual of clinical immunology. Washington, DC: American Society for Microbiology, 1986:467-87.
38. Kessel JF, Lewis WP, Pasquel CH, Turner JA. Indirect hemagglutination and complement fixation tests in amebiasis. Am J Trop Med Hyg 1965;14:540-50.
39. Koyashi LM, Kort MP, Berlin GW, Bruckner DA. *Isospora* infection in a homosexual man. Diagn Microbiol Infect Dis 1985;3:363-6.
40. Krogstad DJ, Spencer HC, Healy GR. Current concepts in parasitology—amebiasis. N Engl J Med 1978;298:262-5.
41. Krogstad DJ, Spencer HC, Healy GR, Gleason NN, Sexton DJ, Herron CA. Amebiasis: epidemiologic studies in the United States, 1971-1974. Ann Intern Med 1978;88:89-97.
42. Krupp IM. Comparison of counter-immunoelectrophoresis with other serologic tests in the diagnosis of amebiasis. Am J Trop Med Hyg 1974;23:27-30.
43. McLaren L, Davis L, Healy G, James G. Isolation of *Trichomonas vaginalis* from the respiratory tract of infants with respiratory disease. Pediatrics 1983;71:888-90.
44. McNabb SJN, Hensel DM, Welch DF, Heijbel H, McKee GL, Istre GR. Comparison of sedimentation and flotation techniques for identification of *Cryptosporidium* spp. oocysts in a large outbreak of human diarrhea. J Clin Microbiol 1985;22:587-99.
45. Mathews HM, Moss DM, Healy GR, Visvesvara GS. Polyacrylamide gel electrophoresis of isoenzymes from *Entamoeba* species. J Clin Microbiol 1983;17:1009-12.
46. Matsumoto Y, Yamada M, Yoshida J. Light-microscopical appearance and ultrastructure of *Blastocystis hominis*, an intestinal parasite of man. Zentral Bakteriol Hyg A 1987; 264:379-85.
47. Meerovitch E, Ali KZ. Preliminary report on the serological response of amebiasis patients from an endemic area in N.W. Saskatchewan. Can J Public Health 1967;58:270-4.
48. Meisel JL, Perera DR, Meligro C, Rubin CE. Overwhelming watery diarrhea associated with a *Cryptosporidium* in an immunosuppressed patient. Gastroenterology 1976; 70:1156-60.
49. Melvin DM, Brooke MM. Laboratory procedures for the diagnosis of intestinal parasites. Atlanta: Laboratory Training and Consultation Division, Centers for Disease Control, 1974.
50. Milgram EA, Healy GR, Kagan IG. Studies on the use of the indirect hemagglutination test in the diagnosis of amebiasis. Gastroenterology 1966;50:645-9.
51. Miotti PG, Gilman RH, Santosham M, Ryder RW, Yoken RH. Age-related rate of seropositivity of antibody to *Giardia lamblia* in four diverse populations. J Clin Microbiol 1986;24:972-5.
52. Mirelman D. Ameba-bacterium relationship in amebiasis. Microbiol Rev 1987;51:272-84.
53. Moore GT, Cross WM, McGuire D, et al. Epidemic giardiasis at a ski resort. N Engl J Med 1969;281:402-7.
54. Nash TE, Herrington DA, Levine MM. Usefulness of an enzyme-linked immunosorbent assay for detection of *Giardia* antigen in feces. J Clin Microbiol 1987;25:1169-71.
55. National Research Council and the Institute of Medicine. The U.S. capacity to address tropical infectious disease problems. Washington, DC: National Academy Press, 1987:95.

56. Polis MA, Tuazon CU, Alling DW, Talmanis E. Transmission of *Giardia lamblia* from a day care center to the community. Am J Public Health 1986;76:1142-4.
57. Rachman R, Rosenberg M. Distinction between *Entamoeba gingivalis* and *Entamoeba histolytica*, revisited. Acta Cytol 1986;30:82.
58. Randall GR, Goldsmith RS, Shek J, Mehalko S, Heyneman D. Use of the enzyme-linked immunosorbent assay (ELISA) for detection of *Entamoeba histolytica* antigen in fecal samples. Trans R Soc Trop Med Hyg 1984;78:593-5.
59. Ratnam S, Paddock J, McDonald E, Whitty D, Jong M, Cooper R. Occurrence of *Cryptosporidium* oocysts in fecal samples submitted for routine microbiological examination. J Clin Microbiol 1985;22:402-4.
60. Restrepo C, Macher AM, Radany EH. Disseminated extraintestinal isosporiasis in a patient with acquired immune deficiency syndrome. Am J Clin Pathol 1987;87:536-42.
61. Root DM, Cole FX, Williamson JA. The development and standardization of an ELISA method for the detection of *Entamoeba histolytica* antigens in fecal samples. Arch Invest Med 1978;9:203-10.
62. Rosoff JD, Stibbs HH. Isolation and identification of a *Giardia lamblia*-specific antigen useful in the coprodiagnosis of giardiasis. J Clin Microbiol 1986;23:905-10.
63. Rosoff JD, Stibbs HH. Physical and chemical characterization of a *Giardia lamblia*-specific antigen useful in the coprodiagnosis of giardiasis. J Clin Microbiol 1986;24:1079-83.
64. Sargeaunt PC, Williams JE. A comparative study of *Entamoeba histolytica* (NIH:200,/HK9, etc.), "*E. histolytica*-like" and other morphologically identical amebae using isoenzyme electrophoresis. Trans R Soc Trop Med Hyg 1980;74:469-74.
65. Shaw PK, Brodsky RE, Lyman DO, et al. A community-wide outbreak of giardiasis with evidence of transmission by a municipal water supply. Ann Intern Med 1977;87:426-32.
66. Sheehan DJ, Raucher BG, McKitrick JC. Association of *Blastocystis hominis* with signs and symptoms of human disease. J Clin Microbiol 1986;24:548-50.
67. Smith JW. Identification of fecal parasites in the special parasitology survey of the College of American Pathologists. Am J Pathol 1979;72:371-3.
68. Smith JW, McQuay RM, Ash LR, Melvin DM, Orithel TC, Thompson JH. Diagnostic parasitology. Intestinal protozoa. Chicago: American Society of Clinical Parasitology, 1976.
69. Smith JW, Wolfe MS. Giardiasis. Ann Rev Med 1980;31:373-83.
70. Smith RF. Incubation time, second blind passage and cost considerations in the isolation of *Trichomonas vaginalis*. J Clin Microbiol 1986;24:139-40.
71. Sterling CR, Arrowood MJ. Detection of *Cryptosporidium* spp. infections using a direct immunofluorescent assay. Pediatr Infect Dis 1986;5:S139-42.
72. Stibbs HH, Ongerth JE. Immunofluorescence detection of *Cryptosporidium* oocysts in fecal smears. J Clin Microbiol 1986;24:517-21.
73. Ungar BLP, Yoken RH, Quinn TC. Use of a monoclonal antibody in an enzyme immunoassay for the detection of *Entamoeba histolytica* in fecal specimens. Am J Trop Med Hyg 1985;34:465-72.
74. Vinayak VK, Chandna R, Kenkateswarlu K, Miita S. Detection of *Giardia lamblia* antigen in the feces by counterimmunoelectrophoresis. Pediatr Infect Dis 1985;4:383-6.
75. Visvesvara GS, Healy GR. Antigenicity of *Giardia lamblia* and the current status of serologic diagnosis of giardiasis. In: Erlandsen SL, Meyer EA, eds. *Giardia* and giardiasis. New York: Plenum Publishing, 1984:219-21.
76. Visvesvara GS, Healy GR. The possible use of an indirect immunofluorescence test using axenically grown *Giardia lamblia* antigens in diagnosis of giardiasis. In: Jacobowski W, Hoff JC, eds. Proceedings of the national symposium on waterborne transmission of giardiasis. Springfield, VA: National Technical Information Service, Environmental Protection Agency, 1979:53-63.
77. Watt RM, Philip A, Wos SM, Sam GJ. Rapid assay for immunological detection of *Trichomonas vaginalis*. J Clin Microbiol 1986;24:551-5.

78. Wolfe MS. Current concepts in parasitology. Giardiasis. N Engl J Med 1978;293:319-21.
79. Wolfson JS, Richter JM, Waldron MA, Weber DJ, McCarthy DM, Hopkins CC. Cryptosporidiosis in immunocompetent patients. N Engl J Med 1985;312:1278-82.
80. Yang J, Scholten TH. *Dientamoeba fragilis*. A review with notes on its epidemiology, pathogenicity, mode of transmission and diagnosis. Am J Trop Med Hyg 1977;26:16-22.
81. Zierdt CH. Studies of *Blastocystis hominis*. J Protozool 1973;20:114-21.

CHAPTER 16

INTESTINAL HELMINTHS

James W. Smith, M.D., Marilyn S. Bartlett, M.S., and Kenneth W. Walls, Ph.D.

INTRODUCTION

There are three major groups of helminths which infect the intestinal tract of humans. They are nematodes (roundworms), trematodes (flukes) and cestodes (tapeworms). Two genera of trematodes, *Paragonimus*, which infects the lung, and *Schistosoma*, which lives in the blood vessels, are also discussed in this chapter because the eggs are often found in feces. Nematode infections are still a significant problem in the United States[3,9,52] and the world[12] (Table 16:1), although the incidence and severity of infections have decreased as a result of improved sanitation. Trematode infections usually are acquired outside the United States, although there are endemic foci of schistosomiasis in Puerto Rico. Cestode infections continue to be a problem and the incidence of *Taenia* infections may have risen during the past 50 years.[9] The life cycles vary, ranging from simple ones in which eggs are infective when passed or shortly thereafter, to complex life cycles in which one or sometimes two or more intermediate hosts are required. Knowledge of the life cycles is important for understanding the ways in which infections might be acquired and prevented. Many parasitic infections are acquired by ingesting embryonated eggs or infective larvae from appropriate intermediate hosts. Other infections are acquired when larvae penetrate the skin. Some parasites remain localized in the gastrointestinal tract, whereas others migrate through various tissues of the body before lodging in the final sites of infection. The time from entry of the parasite into the host until infection can be detected by routine diagnostic techniques is known as the prepatent period. For intestinal helminths, it is the time between infection by the parasite and presence of eggs or larvae in feces. Most of the helminths discussed here are exclusively human parasites, although some may have animal reservoirs.

Table 16:1. Incidence of Intestinal Parasites in 388,745 Fecal Specimens Examined by State Health Department Laboratories.* 1976

	Number	% of Specimens†	% of All Identifications
Protozoa	48,353		61.1
Giardia lamblia	14,773	3.8	18.7
Entamoeba histolytica	2,486	0.6	3.1
Dientamoeba fragilis	1,588	0.4	2.0
Balantidium coli	21		
Isospora belli	3		
Nonpathogenic	29,482	7.6	37.3
Nematodes	29,107		36.8
Ascaris lumbricoides	9,207	2.4	11.6
Trichuris trichiura	8,796	2.3	11.1
Enterobius vermicularis	7,088	1.8	9.0
Hookworm	3,216	0.8	4.1
Strongyloides stercoralis	757	0.2	1.0
Trichostrongylus spp.	22		
Heterodera spp.	21		
Trematodes	396		0.5
Clonorchis/Opisthorchis	210	0.05	0.3
Schistosoma mansoni	143	0.04	0.2
Fasciolopsis buski	5		
Heterophyes heterophyes	3		
Fasciola hepatica	1		
Metagonimus yokogawai	1		
Paragonimus westermani	1		
Cestodes	1,205		1.5
Hymenolepis nana	946	0.2	1.2
Taenia spp.	209	0.5	0.3
T. solium	6		
T. saginata	45		
Taenia spp. unknown	158	0.04	0.2
Hymenolepis diminuta	23		
Diphylloboththrium latum	25		
Dipylidium caninum	2		

*Adapted from Center for Disease Control: Intestinal Parasite Surveillance Annual Summary 1976, Issued August 1977. Does not include laboratories in Guam, Puerto Rico, or Virgin Islands.

†Percentages are not calculated for parasites identified less than 100 times.

Signs and symptoms of helminth infections vary and may be caused by adults, larvae or eggs. Parasitic infections should be included in the differential diagnosis of patients with elevated eosinophil counts. Eosinophilia is particularly common when parasites are in tissue, and at this time diagnostic stages are often not present in feces. Knowledge of geographic distributions, epidemiology and life cycles is important for ascertaining what infections a patient might develop, who might acquire the infection from an infected patient and what measures might be effective in controlling spread of the parasite.

Diagnosis is usually based on demonstrating a stage of the parasite in fecal specimens, although demonstrating parasites in tissues, detecting serologic response to the parasites[49] or detecting parasite antigens[29] may be effective in some instances. There is now investigational work on use of genetic probes. Adult worms, or portions of adult worms, may be passed in fecal specimens and can be identified by careful examination and comparison to characteristics described and illustrated in atlases and standard textbooks on parasitology. It is also important to differentiate human parasites from free-living helminths, insect larvae, fragments of tissue and foreign debris.

The principal means of diagnosing intestinal helminth infection is identifying eggs in fecal specimens. Identification should be approached systematically using the following characteristics. Size of eggs is particularly important (Figs. 16:1, 16:2). The eggs of different parasites vary in size, and careful measurement of eggs with a calibrated ocular micrometer allows differentiation of those which are similar in overall configuration but different in size. The shape of the egg is important, as are the thickness and structure of the shell. The presence of specialized structures such as hooks, spines, opercula and shoulders on the shell may aid in identification. A mammillated albuminoid outer covering is present on *Ascaris lumbricoides* eggs, and polar plugs are present in the eggs of *Trichuris trichiura* and some related organisms. The contents of the eggs should be examined to determine the level of development. Some eggs contain undeveloped ova when passed, whereas others are dividing and still others contain larvae. If these features are considered and if the examiner is familiar with the appearance of fecal background material, there should be little difficulty in identifying eggs in fecal specimens.

NEMATODES

Nematodes are the most common helminths which parasitize the intestinal tract of humans and infections are usually diagnosed by finding adults, eggs or larvae in feces.

Enterobius vermicularis

This parasite causes enterobiasis (oxyuriasis, pinworm infection), which is the most common helminth infection in the United States and affects people of all social strata and age but particularly children. The infection rate is not known, although there are estimates that 30% of all children and 16% of all adults are infected, with rates exceeding 50% in some institutions.[52] When embryonated eggs are ingested, the larvae hatch and in approximately one month mature to adults in the caecum and adjacent areas. They are not attached to mucosa. The female measures 13 mm long, has a sharply pointed

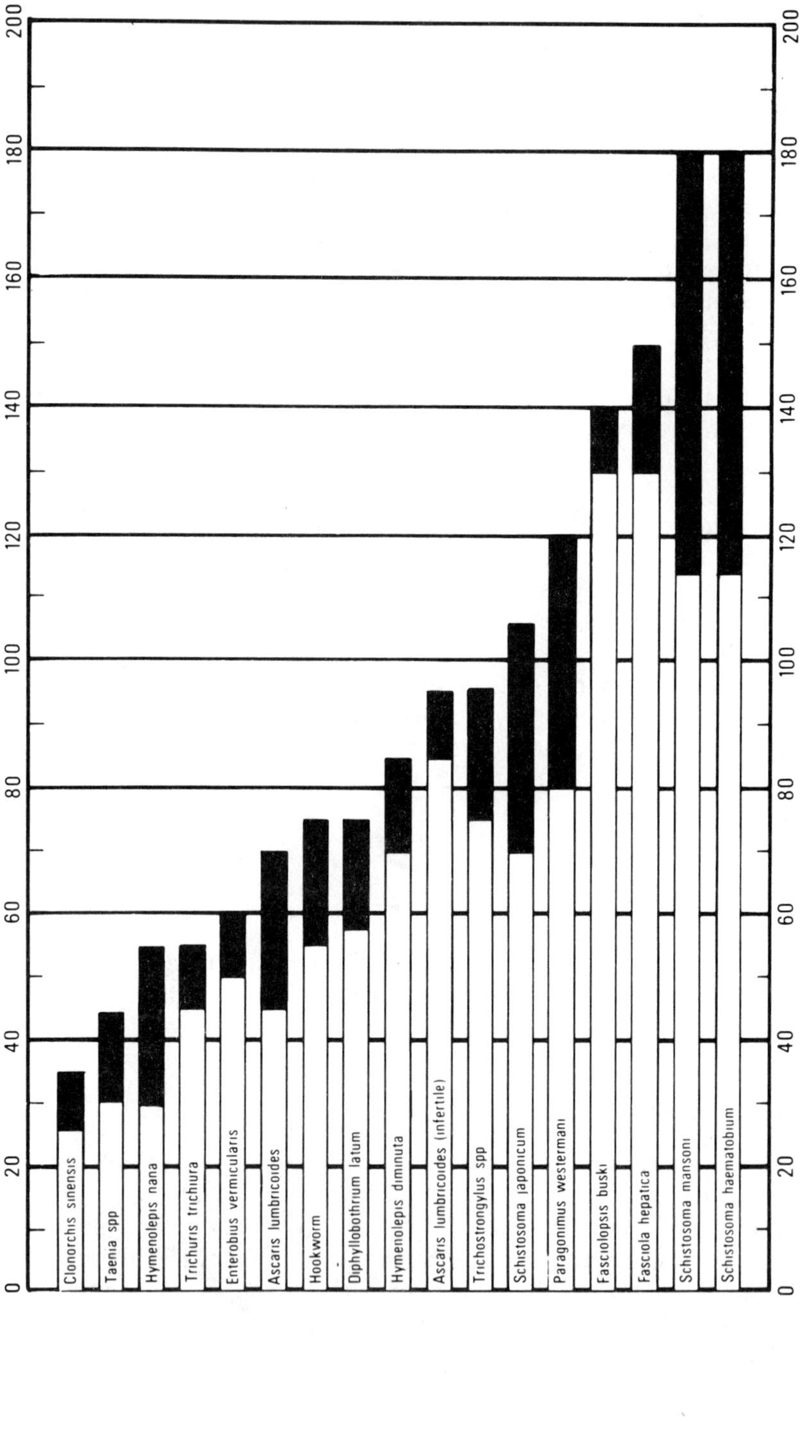

Figure 16:1. Sizes of helminth eggs. Size ranges of lengths are shown by dark bars.[39]

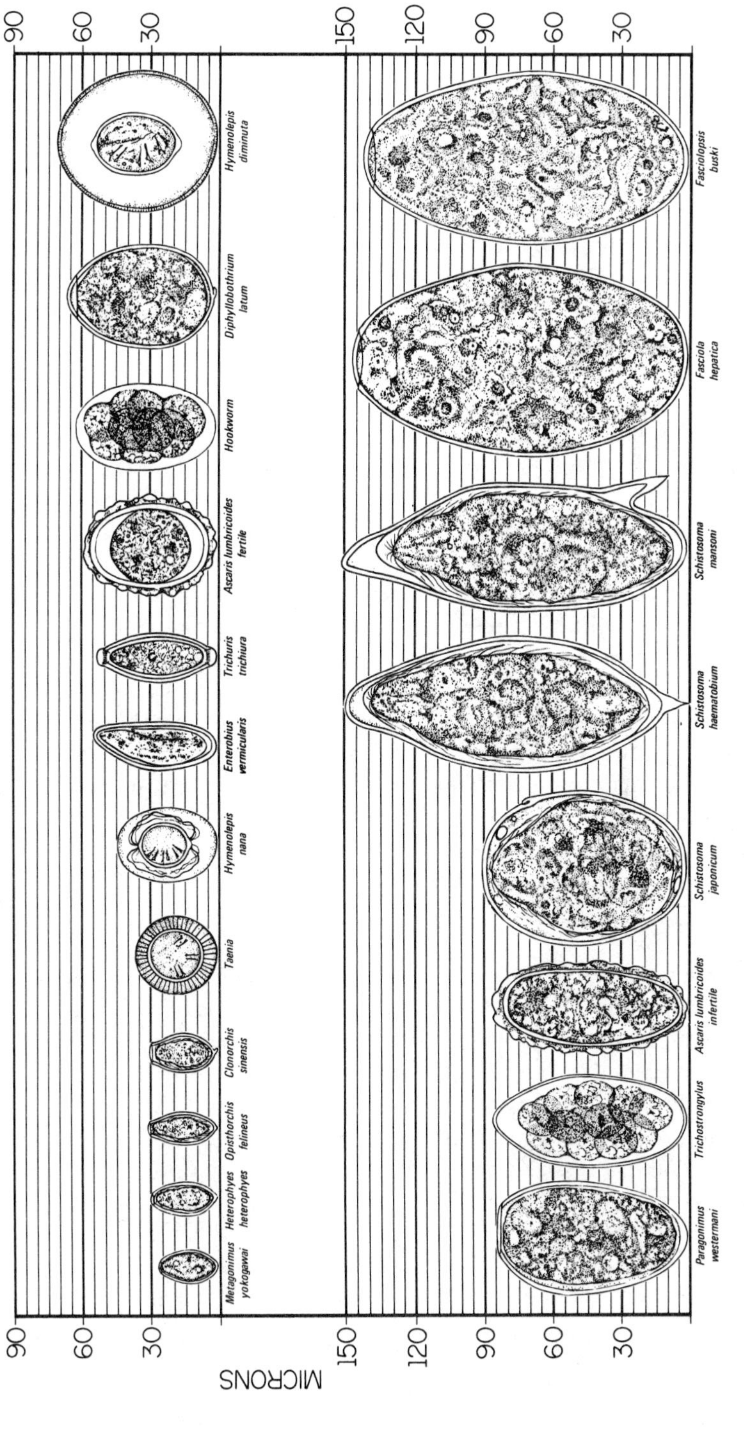

Figure 16:2. Relative sizes of helminth eggs.[10]

posterior end and has cephalic alae; the male is much smaller (2-5 mm). The gravid female migrates to the anus, usually when the patient is sleeping, and deposits eggs in the perianal folds or ruptures, releasing numerous embryonated eggs. The eggs are infective when passed or shortly thereafter, and infection is acquired by ingesting them. Eggs are distributed in clothing, bed linen, etc., and other persons may become infected by accidentally ingesting infective eggs from these objects. Symptoms ascribed to pinworm infections include pruritus ani and irritability, but in most cases, no symptoms are evident. Occasionally, *Enterobius* is found in ectopic sites such as the vagina or peritoneal cavity and may even lead to the formation of granulomas.[11,43]

Infection is diagnosed by demonstrating the typical eggs by the cellophane-tape or Vaseline-swab techniques. Because the eggs are deposited outside the anus, fecal specimens are generally not satisfactory for diagnosis (it is estimated that only 5-10% of cases will be diagnosed on fecal examination).

The frequency of positive cellophane-tape examinations is an index of the severity of infections.[38] If examinations are done daily for 6 days, persons with the lightest infections will be positive only 1 or 2 days, whereas those with severe infections will be positive every day. Approximately 90% of infections will be detected if 3 specimens are examined.

The egg of *E. vermicularis* (Figs. 16:2, 16:3) is elongate, measures 50-60 μm long by 20-30 μm wide and has a moderately thick shell. The shell is flattened on one side and contains a folded larva. Sometimes gravid female worms or occasionally immature and male worms may be found in the perianal area or in feces; they are identified on the basis of their gross morphology and if gravid, on the morphology of the eggs contained in the uteri.

Trichuris trichiura

This parasite causes trichiuriasis (whipworm infection), which is a common helminth infection, worldwide in distribution and especially prevalent in tropical and subtropical areas. The adult worm measures up to 50 mm long, has a long slender anterior portion and a shorter, thick posterior portion, thus leading to the common name "whipworm." The posterior portion of the male is curled. Infection is acquired by ingesting embryonated eggs. Larvae hatch and in approximately 3 months mature in the lower small intestine and colon. The adults infect the colon, with their anterior portions entwined in the intestinal mucosa where they apparently remain attached for their entire lifetime (4-5 years). Most infections are asymptomatic, but heavy infections (150 or more worm pairs) may be accompanied by diarrhea and very heavy infections by dysentery.[23] The unembryonated eggs are passed in the stools and require a maturation period of several weeks (3 weeks under optimal conditions) before infective larvae develop.

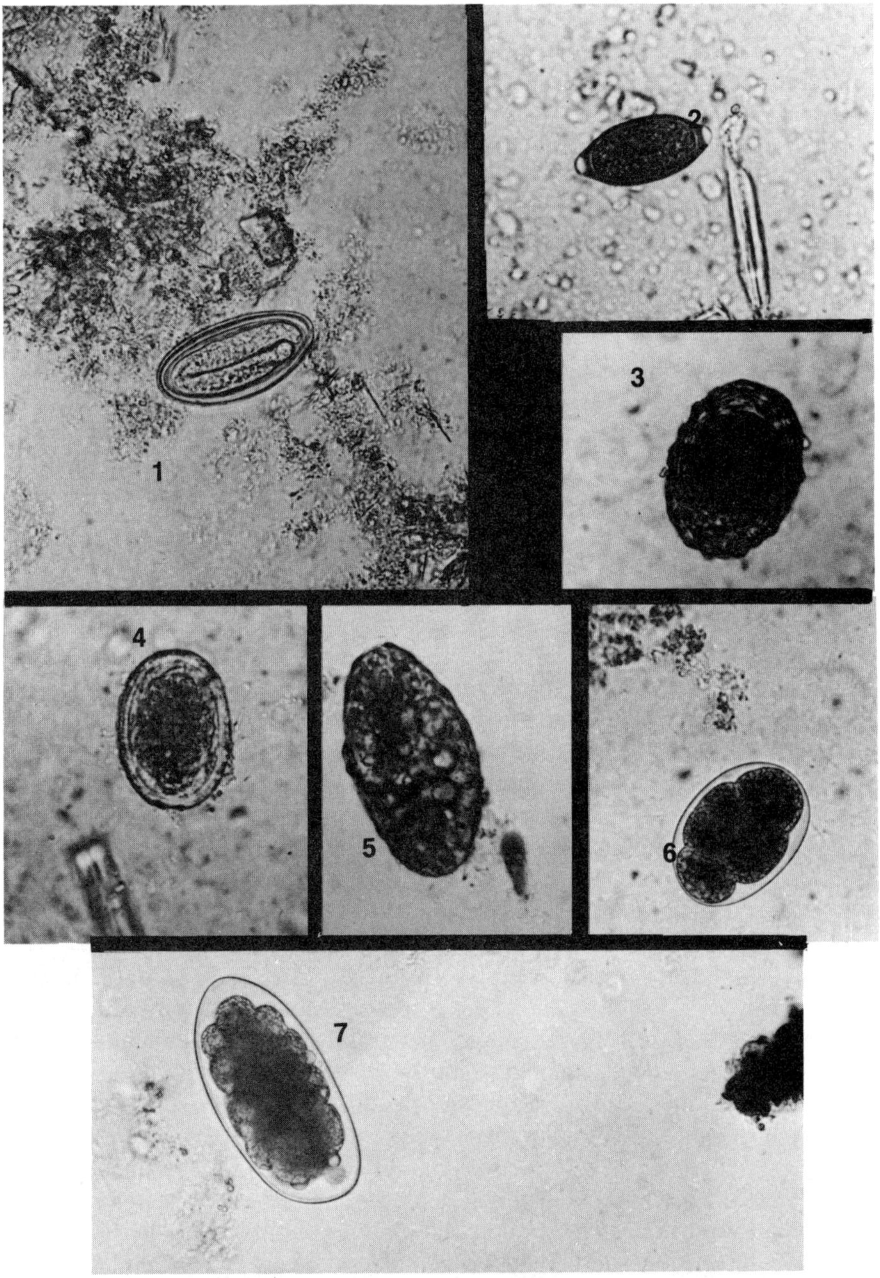

Figure 16:3. Nematode eggs. These eggs in wet mount were photographed at the same magnification. (1) *Enterobius vermicularis*. (2) *Trichuris trichiura*. (3) *Ascaris lumbricoides*, fertile. (4) *A. lumbricoides*, fertile, decorticate. (5) *A. lumbricoides*, unfertilized. (6) Hookworm. (7) *Trichostrongylus* spp.

Diagnosis is established by finding the typical elongated, undeveloped eggs (Figs. 16:2, 16:3), measuring 52-57 by 22-23 μm, with distinctive refractile polar plugs. The shell is moderately thick. A simple method of estimating worm burden is to prepare a standard smear (equivalent to 2 mg of stool). Infections are designated as light when there are fewer than 5 eggs per coverslip and heavy when there are more than 25 eggs. Atypical eggs may be found in specimens from patients receiving antihelminthic therapy.

Ascaris lumbricoides

This parasite is the largest nematode which parasitizes the intestinal tract of humans and causes ascariasis (roundworm infection), which is found worldwide. The female measures up to 35 cm long by 6 mm in diameter; the male is slightly smaller. The posterior end of the male has a copulatory organ and tends to be curled. The adults live in the upper small intestine and are not attached to the mucosa. The female lays approximately 200,000 eggs per day. The eggs are undeveloped when passed but, when deposited in a satisfactory environment, they develop and contain infective larvae after 2-4 weeks. When infective eggs are ingested, the larvae are released in the small intestine, penetrate the intestinal mucosa and enter the blood stream, which carries them to the lungs. They undergo some maturation in the lungs and migrate upward in the respiratory tree, are swallowed and finally reach the small intestine where they grow to maturity. Time from ingestion of the embryonated egg to presence of an egg-laying adult (prepatent period) is 2-3 months. The migration phase may cause Loeffler syndrome (diffuse pulmonary infiltrates with peripheral eosinophilia) or eosinophilia alone.

Adult worms can cause various symptoms.[35,53] They are not attached to mucosa and can migrate up and down the intestine. Migration is often stimulated by fever or drug therapy—especially anesthetics. With severe infections, worms in the intestine may cause various symptoms, including intestinal obstruction. Worms may migrate into the bile duct and cause obstruction or cholangitis or may migrate into the lumen of the appendix and cause appendicitis. Adult worms occasionally migrate into the stomach and may be vomited out or they may be passed in feces.

Diagnosis is established by identifying an adult worm or by finding eggs in stool specimens. One adult female can produce enough eggs to provide approximately 5 eggs per coverslip containing 2 mg of feces. Infections with more than 100 eggs per coverslip are considered heavy. *Ascaris* worms can cause serious disease because of their propensity to migrate and infection with even one worm probably should be treated. Adult *Ascaris* are nonsegmented and have 3 pseudolips at the anterior end.

Fertile *Ascaris* eggs are round to slightly oval and have an irregular albuminoid mammillated covering which is stained yellow-brown by bile. They measure 45-70 by 35-50 μm and have a very thick shell. In recently passed specimens, the undivided ovum is rounded and there is usually a clear area between the ovum and the shell at each end of the egg (Figs. 16:2, 16:3). Occasionally, eggs lose the mammillated covering (decorticated) (Fig. 16:3) and in such instances, the thick shell is quite prominent. Decorticated eggs must be differentiated from the thin-shelled eggs of hookworm—especially in older fecal specimens in which *Ascaris* may be segmented. Infertile eggs may also be found; they are generally larger and more elongated than fertile eggs (Fig. 16:3), measuring up to 94 μm long by 44 μm wide. The external mammillated layer is more irregular and the egg contains irregular globules of yolk material which fill the egg so there are no clear areas at the ends. *Ascaris* eggs can survive in 5-10% formalin for extended periods and may become embryonated. Warming the specimens to 60°C for 15-20 min during fixation will prevent this.

In general, serologic tests for ascariasis in humans are unimportant except in the differential diagnosis of *A. lumbricoides* larval migration from *Toxocara* spp. larval migration (visceral larva migrans, VLM). The newer serologic tests using larval excretory-secretory antigens aid in this differentiation.[13,18] This is discussed more thoroughly under visceral larval migrans. Humans are occasionally infected by the pig *Ascaris, A. suum.*

Hookworms

Hookworm infections are cause by two species, *Ancylostoma duodenale* (the so-called Old World hookworm) and *Necator americanus* (the so-called New World hookworm). Infections are most prevalent in tropical and subtropical areas, but they also occur in temperate areas and are particularly common in persons who do not wear shoes. This parasite infects the small intestine, where the adults attach to the intestinal mucosa and cause blood loss estimated at 0.15-0.25 mL of blood per day per adult *A. duodenale* and 0.03 mL per day per adult *N. americanus*. The female worm measures up to 12 mm long; the male is slightly shorter. The male has a fan-shaped copulatory bursa at the posterior end. The female lays eggs which are passed in the feces. Under appropriate conditions, the eggs embryonate, hatch and release rhabditiform larvae (Fig. 16:4) which molt and develop to the filariform infective stage in about 7 days. If filariform larvae contact the skin of an appropriate host, as when someone steps on fecally contaminated soil with bare feet, the larvae penetrate the skin, gain access to the host's circulation, travel to the lungs and penetrate into the alveoli. They then migrate through the tracheobronchial tree to the epiglottis, are swallowed and finally reach the small intestine. During the initial invasion and migration stage, there may be marked inflammation around the area of larval penetration (a condition known as ground itch), particularly in sensitized individuals.

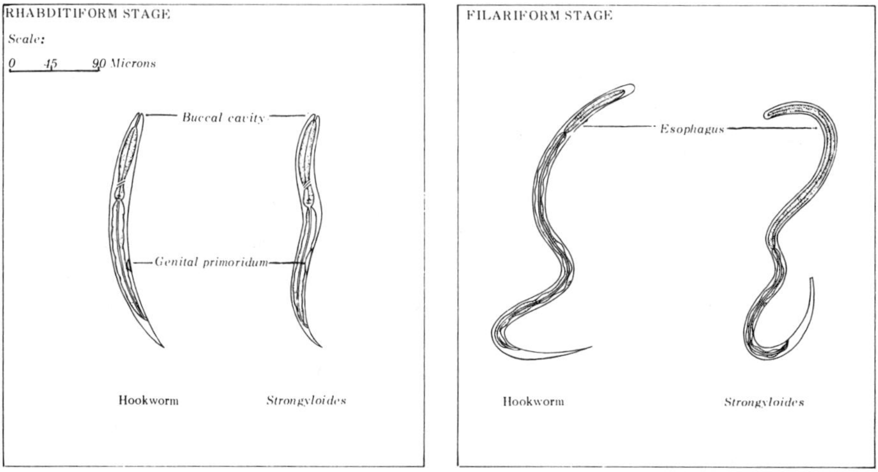

Figure 16:4. Nematode larvae. Comparison of rhabditiform and filariform larvae of hookworm and *Strongyloides stercoralis*.[4]

Larvae migrating through the lungs, if particularly numerous, may cause Loeffler syndrome. The prepatent period is estimated to be 5 weeks. Gastrointestinal symptoms are usually minimal, although severe infections may be accompanied by diarrhea, abdominal pain and nausea. The major manifestation of infection is iron-deficiency anemia caused by blood loss. An estimate of the number of worms present may give some indication as to whether anemia is caused by hookworms and whether the infection needs to be treated.[25]

The diagnosis is established by finding the oval eggs (Figs. 16:2, 16:3, 16:5), which measure 58-76 by 38-40 μm and have a thin shell. The egg is usually in the 4-8 cell segmented stage when passed but may vary from unsegmented in diarrheic specimens to embryonated in specimens from severely constipated patients. In direct examinations of 2 mg of feces, more than 25 eggs per coverslip indicates heavy infection. Unfixed older specimens may contain embryonated eggs and larvae. In such instances, the larvae must be differentiated from those of *Strongyloides stercoralis*. Hookworm rhabditiform larvae (Fig. 16:4) have a long buccal cavity and an inconspicuous genital primordium. Hookworm filariform larvae (Fig. 16:4) have an esophagus which is approximately one-fourth the length of the larva and a pointed tail. In addition, the larvae must be differentiated from those of free-living nematodes and the eggs must be differentiated from those of *Trichostrongylus* spp. (Figs. 16:3, 16:5) and *Meloidogyne (Heterodera)* spp. (Fig 16:5). The latter are parasitic nematodes of plants whose eggs are ingested when eating incompletely washed root vegetables such as carrots.

The eggs of various hookworm species cannot be differentiated and

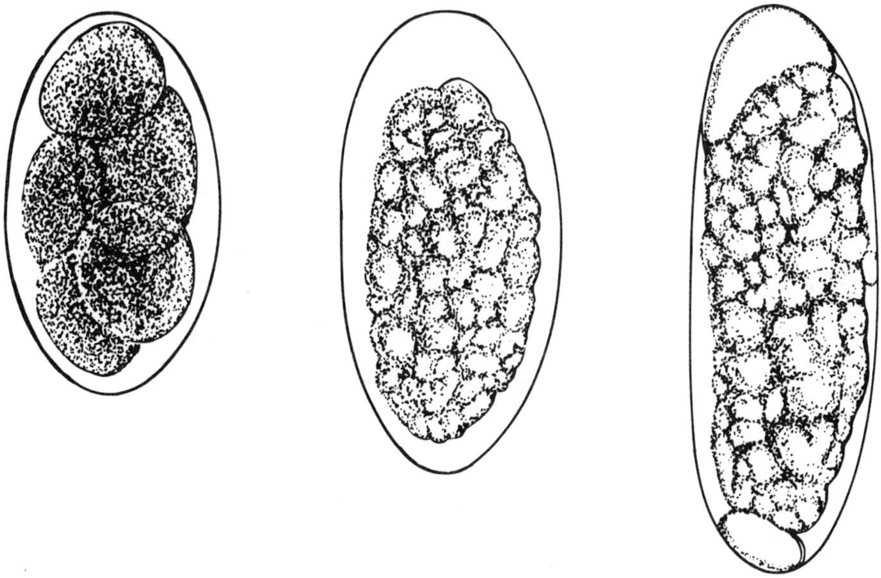

Figure 16:5. Comparison of eggs of hookworm (left), *Trichostrongylus* spp. (center), and *Meloidogyne (Heterodera)* spp. (right).

species identification must be based on the examination of adult worms. Adults are occasionally found in feces and can be identified on the basis of mouth parts and on configuration of the copulatory bursa of male worms. Mouths of adult *N. americanus* have cutting plates whereas those of *A. duodenale* have two pairs of teeth.

Trichostrongylus spp.

These parasites are small nematodes of various species which parasitize the small intestine of humans and various animals. Infections are particularly common in the Far East, India and Russia, but there have been sporadic cases in Latin America and the United States.[31] Females lay eggs which are passed in the segmented stage in stools, become embryonated and then hatch. The infective larvae crawl on vegetation and infect a new host when ingested with contaminated food or water. Adults develop in the intestine without migrating through the lung. The disease is usually asymptomatic, although unusually heavy infections may produce abdominal pain and diarrhea.

The diagnosis is established by finding the typical eggs (Figs. 16:2, 16:3), which measure 79-98 by 40-50 μm and have somewhat pointed ends. Eggs of the different species cannot be differentiated. Their larger size and tapered ends help to differentiate them from hookworm eggs (Fig. 16:5).

Strongyloides stercoralis

This parasite is a small nematode which lives buried in the mucosa of the upper small intestine. Infection is most common in warm climates but also occurs in temperate areas. The parasitic female is only 2 mm long and reproduces parthenogenetically (there is no parasitic male). The embryonated eggs are laid in the mucosa and hatch just before or as they reach the lumen. The rhabditiform larvae find their way to the intestinal lumen and are passed in the feces. The larvae mature to infective filariform larvae in a short time and upon contact with the skin of a susceptible host penetrate and migrate through the circulatory system to the lungs and on to the small intestine. This process is similar to the migration of hookworm. Under certain conditions, larvae may develop a free-living cycle in the soil with male and female adults, eggs and larvae.

The infection may be accompanied by skin lesions at the site of penetration or Loeffler syndrome while the parasites migrate through the lungs. The intestinal phase is usually asymptomatic or mild; however, patients with severe infections may have diarrhea and the infection may show remissions and exacerbations.[2]

Autoinfection commonly occurs and thus low grade infection can persist for over 35 years.[34] Autoinfection may result from development of rhabditiform larvae to infective filariform larvae while still in the intestine with subsequent penetration to the circulation and completion of the life cycle. Autoinfection may also occur from perianal fecal material in which larvae mature to the infective filariform stage and penetrate the perianal skin. In persons with a degree of immunity to the infection, only a small number of larvae survive to complete the life cycle but the infection persists. If the individual becomes immunosuppressed as with malnutrition, antineoplastic therapy, immunosuppression for transplantation or acquired immunodeficiency syndrome, then increasing numbers of parasites complete the cycle and hyperinfection develops.[21,36] If adults live in tissues other than intestine, it is called disseminated infection. Hyperinfection or disseminated infection often manifest as diffuse interstitial pneumonia which may be diagnosed by demonstrating larvae in pulmonary material.

Diagnosis of intestinal infection is generally established by detecting larvae in feces (eggs are very rarely seen in feces). *Strongyloides* rhabditiform larvae (Fig. 16:4) have a short buccal cavity and a prominent genital primordium. *Strongyloides* filariform larvae have a notched tail and an esophagus approximately one-half the length of the body. Occasionally, larvae may not be found even with repeated fecal examinations. In such instances, organisms may be demonstrated in duodenal aspirates or with the Baermann concentration technique.[33] This procedure is particularly useful for detecting infection in persons with a history of strongyloidiasis or from an endemic area who are to be immunosuppressed for transplantation or by antineoplastic therapy. It can also be useful to evaluate success of therapy. Superinfection should be suspected when filariform larvae are present in freshly passed specimens[15] or when larvae are numerous.

Strongyloidiasis may be difficult to differentiate from ascariasis and sometimes serologic tests can be useful. The indirect hemagglutination test (IHA) uses an extract of adult worms. Crossreactions with ascariasis occur in about 15% of sera tested, but the homologous titer is usually considerably higher than the heterologous. A titer of 1:64 is considered clinically significant (Table 16:2).

Ascaridoid Nematodes

Anisakiasis is the name commonly applied to infections of the digestive tract caused by ascaridoid nematodes of the genera *Anisakis, Phocanema, Terranova* and *Contracaecum*. Usual hosts are marine mammals such as whales, porpoises and dolphins. Eggs laid by the mature worms are passed in feces, hatch in water and are eaten by small crustaceans where they develop into third-stage larvae infective for fish or squid. When infected

Table 16:2. The Test of Choice and Significant Titers for Selected Parasitic Infections as Performed at the Center for Disease Control

	IHA	BFT	IIF	CF	DAT	Others
Amebiasis	1:128					
Ascariasis						1:32—ELISA
Chagas disease				1:32		
Cysticerosis	1:128					
Echinococcosis	1:128					
Filariasis	1:128*	1:5*				
Leishmaniasis					1:64	
Malaria			1:64			
Paragonomiasis				1:16		
Pneumocystosis			1:16			
Schistosomiasis			Pos			
Strongyloidiasis	1:64					
Toxocariasis						1:32—ELISA
Toxoplasmosis			1:256			
Trichinosis		1:5				

*Both must be positive.
IHA—indirect hemagglutination
BFT—bentonite flocculation
IIF—indirect immunofluorescence
CF—complement fixation
DAT—direct agglutination
ELISA—enzyme linked immunosorbent assay

fish or squid are eaten by mammalian hosts, the larvae develop to mature worms in clusters in the mucosa of the digestive tract. Humans are accidental hosts, acquiring the infection by eating raw or inadequately preserved marine fish or squid. The clinical presentations of patients with anisakiasis vary somewhat with the species causing the infections. Although most early reports of human infection were from Japan,[42] anisakiasis has been reported in the United States.[24] Diagnosis is by identification of larvae which are coughed, vomited or discovered at surgery or endoscopy. Identification depends on differentiating structures of the esophagus and intestine of the larval worms which may be 50 mm long by 1 mm wide. Patent infection does not develop in humans so eggs are not passed in feces.

TREMATODES

Trematodes, or flukes, are flattened dorsoventrally and all are hermaphroditic except for the schistosomes which have separate sexes. The mature worms vary greatly in length from 1 mm (*Metagonimus*) to 70 mm (*Fasciolopsis*) and have two suckers—one oral, through which the digestive tract opens, and one ventral for attachment. The cuticle may be smooth or rough depending on the species.

Hermaphroditic Flukes

The adult hermaphroditic flukes which infect humans inhabit the biliary tree, intestine and lungs and lay eggs which have opercula. The larger eggs are unembryonated and the smaller eggs contain larvae when passed. Eggs are found in feces and/or sputum, depending on species. They complete their life cycles in freshwater and require two intermediate hosts—the first of which is usually certain specific snails, and the second of which is a plant, crab, fish, ant, etc. Feces must reach the water to allow the larval stages to enter the appropriate first intermediate host. Dietary custom must allow the second intermediate host to be ingested in order for the life cycle to be completed. The incidence of these infections in the United States has increased as consumption of raw fish has become fashionable.[1]

Most trematode infections occur in tropical and subtropical areas of the world with specific trematodes being present in specific areas. Dams and irrigation systems are particularly important in the epidemiology of the diseases. Diagnosis is usually established by demonstrating eggs in fecal specimens by direct mount or formalin-ethyl acetate concentration. Zinc sulfate methods are unsatisfactory unless the specimens are first fixed in formalin to prevent the opercula from popping.

Although serologic tests for most hermaphroditic trematodes are being evaluated—particularly those for *Fasciola, Clonorchis* and *Paragonimus*—

none are used routinely.[49] In the United States, the prevalence, even among travelers, does not justify providing serodiagnostic procedures for these infections even in large centers, although such new "universal" tests as ELISA may make the tests practical.

Fasciolopsis buski

This large intestinal fluke measures up to 70 mm long by 20 mm wide and 3 mm thick. This worm attaches to the mucosa of the upper small intestine. Eggs mature in freshwater and release larvae, which invade the tissues of a particular species of snail. In the snail, large numbers of larval forms are produced, which leave the snail and encyst on water plants. When these water plants are ingested by appropriate hosts, the parasites excyst and grow to adulthood in the small intestine. The infections are usually asymptomatic but there may be diarrhea, epigastric pain and eosinophilia. Diagnosis is established by finding the large operculate eggs (Figs. 16:2, 16:6), which measure 130-140 by 80-85 μm in feces. The eggs are unembryonated and cannot be reliably differentiated from those of *Fasciola hepatica.*

Heterophyes spp. and *Metagonimus* spp.

These parasites are minute worms (1-3 mm long) of numerous species which can infect the intestine of humans. The first intermediate hosts are various molluskan species and the second intermediate hosts are freshwater fish. Infection is acquired from ingesting uncooked or inadequately cooked fish and is generally asymptomatic, but occasional patients may have symptoms similar to those associated with *Fasciolopsis* infection. The diagnosis is established by finding the embryonated operculate eggs in feces. These small eggs (Fig. 16:2), which measure less than 30 by 17 μm, may be difficult or impossible to differentiate from those of *Clonorchis* and *Opisthorchis* species, although the former generally do not have an abopercular knob or hook and are wider at the opercular end.

Fasciola hepatica and *Fasciola gigantica*

These parasites may infect the biliary tract of various domestic and wild animals such as cattle, sheep and goats, and may occasionally parasitize humans. *F. hepatica* is distributed worldwide, whereas *F. gigantica* is restricted to Africa and the Orient. Both parasites live in the biliary tree and the eggs are passed in feces. Life cycles are similar to that of *Fasciolopsis* and the second intermediate host is an aquatic plant such as watercress. Ingested infective larvae are released from their cysts to penetrate through the intestinal wall and reach the liver, where they bore through the capsule into the liver parenchyma and finally reach the bile ducts. There they mature

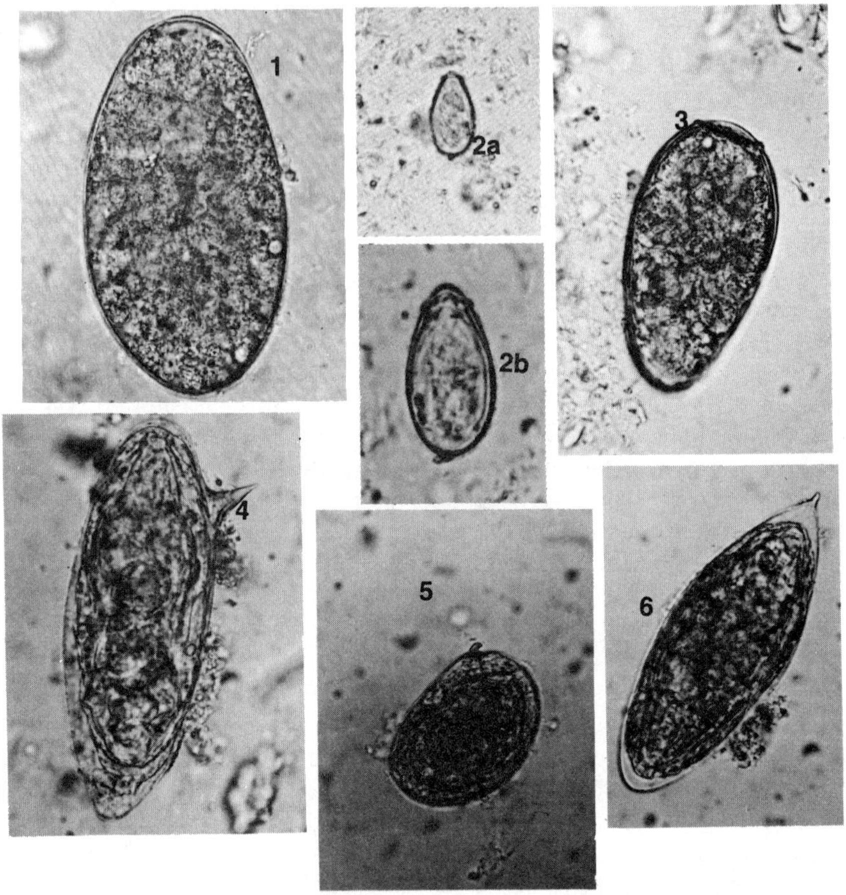

Figure 16:6. Trematode eggs. Except for (2b), these eggs in wet mounts were photographed at the same magnification. (1) *Fasciola hepatica* or *Fasciolopsis buski*. (2a, 2b) *Clonorchis sinensis*; (2b) is an oil immersion photomicrograph to show the structure of this small egg. (3) *Paragonimus westermanni*. (4) *Schistosoma mansoni*. (5) *S. japonicum*. (6) *S. haematobium*.

and live. The young larvae produce little or no symptomatology, but there may be eosinophilia during migration. The adult worms may cause fibrosis of the bile ducts with hyperplasia of the epithelium and may contribute to stone formation and development of cirrhosis. Diagnosis is established by finding the operculate unembryonated eggs (Figs. 16:2, 16:5) measuring 130-150 by 60-90 μm in the stool. Eggs cannot be reliably distinguished from those of *Fasciolopsis*. Occasionally, eggs may be present after infected liver of other hosts is ingested (spurious infection). If such a situation is suspected, the patient should not eat liver for 3-5 days, after which additional stools can be examined. If eggs are still found, the person is actually infected.

Clonorchis sinensis and *Opisthorchis*

These species are closely related. It has been proposed that *C. sinensis* should be in the genus *Opisthorchis*. *C. sinensis* is sometimes known as the Chinese liver fluke. These parasites inhabit the biliary tree of humans and other animals such as cats and dogs. *Clonorchis* is found in the Orient, whereas *Opisthorchis* is usually found in Central and Eastern Europe but has been found sporadically in the Orient. The adult parasites measure as long as 25 mm. The life cycle is similar to that of other trematodes, with freshwater fish being the second intermediate host. Infection is acquired by eating uncooked or undercooked fish. The infective larvae migrate through the ampulla of Vater into the bile ducts of the liver where they mature and live. Infection is generally asymptomatic although heavy infections are sometimes accompanied by various gastrointestinal disturbances.[41] Infection is diagnosed by demonstrating the embryonated eggs (23-35 by 12-20 μm) (Figs. 16:2, 16:6) in stools. Eggs of these two genera cannot be reliably differentiated and it is acceptable to report a generic identification of *Clonorchis*/*Opisthorchis* species. The eggs of *Clonorchis*/*Opisthorchis* often have a hook on the abopercular end and are narrower at the end with the flattened operculum than are the eggs of *Heterophyes*/*Metagonimus* species.

Paragonimus spp.

Several species may parasitize the lungs of humans and other carnivores. *Paragonimus westermani* is the most common and is found in Asia. Occasionally, human infections with *Paragonimus* have been reported from Africa, South and Central America. Some have been caused by species other than *P. westermani*. The adult parasites measure up to 20 mm long by 6 mm wide and live in the lungs, where they are encapsulated by the host's fibrous reaction. Capsules usually communicate with the bronchi and eggs are passed into the respiratory tree, where they may be expectorated in sputum or swallowed and passed in the stools. The life cycle is similar to that of other trematodes, with infection following ingestion of encysted larvae in muscles of the second intermediate hosts, i.e., crayfish or crabs. Larvae migrate through the intestinal wall, across the peritoneal cavity, and bore through the diaphragm to reach the lungs. Symptoms may accompany migration. The presence of adult worms in the lungs leads to the production of large amounts of mucus and to episodes of hemoptysis. Parasites occasionally do not reach the lungs and have been found in various ectopic locations. Diagnosis is established by finding the operculate unembryonated eggs (Figs. 16:2, 16:6), which measure 80-120 by 48-60 μm and have a moderately thick shell. The operculum is flattened and is usually set off from the rest of the shell by prominent shoulders. The abopercular end may be thickened but does not have a knob. Measuring eggs with a properly calibrated ocular micrometer aids in distinguishing them from the eggs of *Diphyllobothrium latum* and *Fasciola* or *Fasciolopsis*.

Schistosoma spp.

Schistosomes have male and female worms and inhabit the blood vessels of humans and many animals. They are the most important trematodes which infect humans in terms of incidence and severity of infection. The species that infect humans are *Schistosoma mansoni, S. haematobium* and *S. japonicum* and its close relative *S. mekongi*.[48] The incidence of infection is rising in many areas of the world,[51] although control efforts in mainland China seem to be successful.[6] The adult parasites are slender—females measure up to 26 mm long and 0.5 mm in diameter. Males are slightly shorter and have a folding of the body which forms a gynecophoral canal in which the female resides. Adults live in small venules and elicit little inflammatory reaction. Eggs are deposited in venules near the mucosa and eventually find their way into the lumen of the viscus. There is significant inflammatory reaction around the eggs. The eggs are fully developed when they are passed in feces or urine. When exposed to freshwater, the larvae escape from the eggs and find appropriate freshwater snail intermediate hosts. In the snail, the organisms proliferate and many thousands of fork-tailed larvae known as cercariae are released. These cercariae swim in the water for a short period of time, and if they contact the skin of a susceptible host, they actively penetrate, enter the circulation, establish themselves in the liver and develop. During this stage there may be enlargement and tenderness of the liver, which may be accompanied by urticaria and eosinophilia.[22] The worms then migrate to the system of vessels to which they are specialized, reach sexual maturity and begin laying eggs. The disease in humans is caused by inflammatory reaction to the eggs. In addition, some eggs are carried to the liver by the portal vein or to the lungs by the systemic circulation and cause granulomatous inflammation and fibrosis in these organs. In the liver, they can cause the eventual development of cirrhosis and portal hypertension. Diagnosis is established by demonstrating eggs in feces (*S. mansoni, S. japonicum*) or urine (*S. haematobium*). The eggs can be demonstrated in direct mounts or with formalin-ether, formalin-ethyl acetate or acid-ether concentrations.[33] Zinc sulfate flotation is not satisfactory for the heavy schistosome eggs. In some instances, especially patients with light or chronic infections, hatching techniques may be used,[33] permitting large amounts of feces to be examined. Feces are mixed with distilled water in a flask (usually a side-arm flask) which is painted or covered to keep out light. The eggs hatch and the swimming larvae (miracidia) seek light in the neck or the side arm, where they can be detected with a hand lens, removed with a capillary pipette and examined microscopically. Because miracidia of the species are indistinguishable, species differentiation cannot be accomplished with this technique. Eggs sometimes can be demonstrated in crush preparations or sections of material from biopsies of rectal or bladder tissue. Movement of larvae in unfixed eggs shows that they are alive and that the infection is active.

Schistosoma mansoni

Infection with this parasite occurs in tropical Africa, the Nile Valley, Brazil, Venezuela, West Indies and Puerto Rico. In the United States, the infection is generally seen in Puerto Ricans who have moved to the United States. *S. mansoni* adults live in the mesenteric veins of the large intestine and in the early stages of infection, can cause abdominal pain and diarrhea with blood and mucus in the stools.[8] Occasionally, worms may reach atypical sites and cause manifestations such as transverse myelitis.[7] These symptoms usually subside and the chronic phase begins. There may be fistulas in the perianal area and hepatic cirrhosis may develop.

Diagnosis is usually made on the basis of identifying the large eggs (116-180 by 45-58 μm), which are oval and have a large distinctive lateral spine protruding from the side near one end (Figs. 16:2, 16:6). If the spine is not readily evident, the egg can be rotated by tapping the coverslip.

Schistosoma japonicum and *Schistosoma mekongi*

S. japonium is found primarily in China, Korea and the Philippines and *S. mekongi* in the Mekong delta in Vietnam.[48] They cause a disease similar to that associated with *S. mansoni* but it is generally more severe because more eggs are laid per worm. The eggs readily reach the liver and liver problems are common in people infected with these parasites. The eggs (Figs. 16:2, 16:6) of *S. japonicum* are slightly oval, measuring 70-105 by 50-65 μm, and characteristically have a short rudimentary lateral spine which may be difficult to demonstrate even with tapping of the coverslip to turn the egg. The eggs of *S. mekongi* have a similar appearance but are smaller (51-78 by 39-66 μm).

Schistosoma haematobium

This parasite is found primarily in Africa, particularly along the Nile River delta, although there are foci in the Middle East and Madagascar. Parasites migrate from the liver through the hemorrhoidal veins to the venous plexes of the urinary bladder, prostate, uterus and vagina, and the initial symptom is often hematuria, most commonly at the end of micturition. The host's reaction to eggs in the bladder wall causes fibrosis and hyperplasia of the epithelium. Association between *S. haematobium* infections and squamous cell carcinoma of the urinary bladder has been noted but is not conclusively proven. Diagnosis is usually established by finding eggs in urinary sediment. The eggs (Figs. 16:2, 16:6) measure 112-180 by 40-70 μm and have a prominent terminal spine. Mid-day urine specimens are usually best for detection of eggs.

Serologic Diagnosis

Even though the immunology of schistosomiasis has been extensively studied, available serologic tests are inadequate. New diagnostic serologic procedures are introduced frequently and many different approaches have been suggested. Buck and Anderson[5] evaluated complement fixation (CF) and cholesterol-lecithin tests and found each lacking. Tanaka et al[44] reported that CF is highly sensitive and specific for detecting *S. japonicum* infections. Fiorillo et al[17] reported that indirect hemagglutination (IHA) is sensitive in acute infections. Wilson et al[54] demonstrated that the indirect immunofluorescence (IIF) test with cryostat sections of adult worms was highly sensitive and specific. Earlier IIF tests showed marked crossreactivity with sera from patients with trichinosis, but the test using adult sections reacted less frequently with these sera and visual evaluation of the location and type of fluorescence frequently allowed false-positive reactions to be detected.

More recently, the enzyme-linked immunosorbent assay (ELISA) and the Western blot or immunoblot test (IBT) have solved some problems. The development and evaluation of ELISA and its modifications has been reviewed by Maddison[28] and it is apparent that improvement of both sensitivity and specificity has been accomplished. Unquestionably, this is due to the use of purified antigens in the ELISA test. McLaren et al[32] showed the advantages of a purified egg antigen. Tsang et al[45] developed the purified *S. mansoni* microsomal antigen (MAMA) which has proven to be highly specific and can differentiate *S. mansoni* infections from *S. japonicum* infections. A similar preparation from *S. japonicum* (JAMA) appears to offer equivalent homologous specificity.

Immunoblot is more difficult to perform than ELISA but offers exquisite specificity.[46] By electrophoresing the antigen on polyacrylamide gel, transferring it to cellulose acetate strips and reacting it with patient's sera, specific bands can be identified. IBT, then, is an excellent test to confirm ELISA-positive sera.

Still unsolved is the problem that in endemic areas, up to 80% of the population has chronic disease lasting for years and resulting in persistent antibody, often at significant levels. Lunde et al[27] suggested that the comparison of reactivity to cercerial (C) and adult (A) antigens in the ELISA test could indicate acute or chronic disease: a C/A ratio of greater than 1 indicates acute disease. This has been confirmed by some[16] but not by others.[14] Some investigators[26,40] have evaluated the significance of IgG and IgM antibodies and report a correlation between the levels of IgM and acute disease. Further evaluations are necessary.

Recently, Hancock and Tsang[19] described the FAST-ELISA test using the MAMA antigen. FAST-ELISA is a rapid, simple procedure using a special microtitration plate. With a processing time of less than 20 min, no special equipment other than the plates is required and a single dilution of serum is sufficient for screening. With the addition of a simple ELISA reader, quantitative data can be obtained. Usable under both field and laboratory conditions, FAST-ELISA promises to be a major serologic test for diagnostic and epidemiologic evaluations in schistosomiasis. For laboratories not prepared to use ELISA tests, IIF remains the test of choice. However, due to the lack of correlation with the type of clinical disease,[29] titers are of no value and results should be reported simply as "positive" or "negative" (Table 16:2). Serology is most useful for ruling out schistosomiasis as a diagnosis.

Cercarial Dermatitis

Swimmer's itch, also known as schistosomal cercarial dermatitis, is a skin infection caused by the cercariae of non-human schistosomes such as those which infect birds.[8,20] The disease occurs after a person's skin is exposed to fresh or saltwater containing cercariae. In hosts other than the natural ones, the cercariae may cause erythema and urticaria and sometimes papules in the skin. Some degree of prior sensitization is required. Reaction is most intense 2 or 3 days after the initial contact and subsides spontaneously. Cercarial dermatitis has been described in many parts of the world, including the United States, and diagnosis is established on clinical grounds.

CESTODES

Cestodes[50] are flattened parasites with an anterior portion known as the scolex (which has structures for attachment to the intestinal mucosa) and a body or strobila composed of a chain of segments. The individual segments, proglottids, develop from the neck; each proglottid has both male and female organs. The point at which the male and female sex organs meet is known as the genital pore. The proglottids develop from immature to mature egg-producing proglottids, with those farthest from the scolex being most fully developed. In some species, eggs are laid from the individual proglottids, but in most the eggs are stored in the uterus. Such proglottids filled with eggs are known as gravid proglottids. Adult and larval stages of cestodes contain basophilic laminated bodies known as calcareous corpuscles which aid in recognizing tissue as cestode tissue in histologic sections. Eggs or gravid proglottids are passed in feces. Eggs of most species contain larvae, though some species are undeveloped. An intermediate host(s) is (are) required for the asexual larval stages to develop. The infective stage develops in the tissues of the intermediate host, and the life cycle is completed when the infective larva is ingested.

Humans sometimes serve as intermediate hosts for cestode larval stages which cause such diseases as hydatid disease, cysticercosis, coenurosis and sparganosis. (These infections are discussed in Chapter 14, "Blood and Tissue Parasites.")

Taenia spp.

Two *Taenia* spp. cause human infections. Humans are the sole definitive hosts for *Taenia saginata*, the beef tapeworm, and *Taenia solium*, the pork tapeworm. *T. solium* is most common in Eastern Europe, Latin America, China, Pakistan and India; most cases in the United States are imported. *T. saginata* is more widely distributed and prevalent in the Middle East, Africa, Europe and Latin America, but is occasionally seen in the United States. The frequency with which *Taenia* infections were found with stool examinations in 1976 was approximately twice that in previous surveys in 1931-35 and 1963-67,[9] but the reason for this increase is not known.

Both *Taenia* spp. live in the small intestine and produce worms with a strobila up to 7 meters long. The eggs are stored in the uterus and reach the outside when gravid proglottids drop from the strobila and either rupture in the intestine, freeing the eggs, or pass intact in the stools. In the latter instance, the proglottids may move in the fashion of an inchworm and may sometimes be present in specimens submitted to the laboratory. The appropriate intermediate hosts (cattle for *T. saginata* and pigs for *T. solium*) swallow the eggs, and larval cysts (cysticerci) develop in the tissues. Infection is acquired by ingesting cysticerci in poorly cooked meat. Usually there is only one adult worm in each infection.

Symptoms in humans are minimal and the disease is usually discovered by observing individual proglottids or a portion of strobila in a fecal specimen. Individual gravid *Taenia* spp. proglottids may exit from the anus and crawl on perianal skin. Untrained persons may suspect that these proglottids are pinworms (*Enterobius*). Humans may be an intermediate host for *T. solium* and develop a disease known as cysticercosis; they cannot be an intermediate host of *T. saginata*.

Diagnosis is based on finding eggs or proglottids. Eggs can be detected in stools with direct wet mount examination or concentration techniques, or found in perianal folds using the cellophane-tape technique. The eggs (Figs. 16:2, 16:7) are spherical (31-34 μm in diameter), have a thick radially-striated shell and contain a 6-hooked embryo. Eggs of the two species

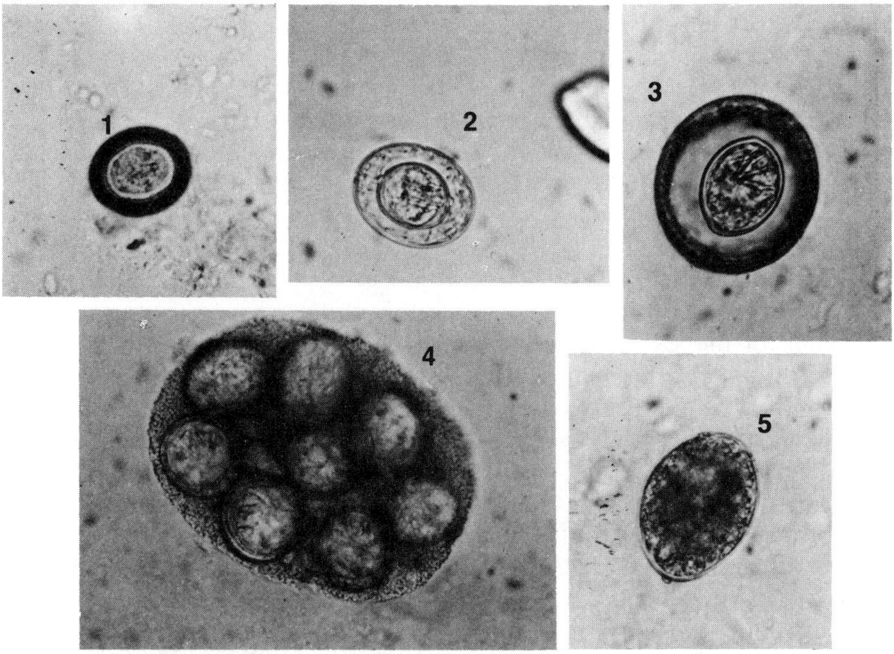

Figure 16:7. Cestode eggs. These eggs in wet mounts were photographed at the same magnification. (1) *Taenia* spp. (2) *Hymenolepsis nana*. (3) *H. diminuta*. (4) *Dipylidium caninum* egg packet. (5) *Diphyllobothrium latum*.

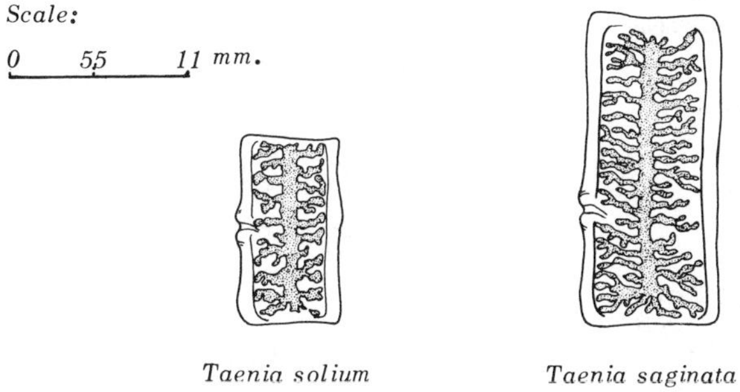

Figure 16:8. Proglottids of *Taenia solium* and *T. saginata*.[4]

cannot be differentiated. Passed proglottids can be cleared overnight in cooled glycerol to allow the uterine branches to be counted and a specific identification to be made. Alternatively, proglottids can be cleared in carbol-xylol (1:3), or the uterus can be injected with India Ink through the genital pore. The gravid proglottids of *T. saginata* have 15-20 lateral branches and those of *T. solium* have 7-13 (Fig. 16:8). In the past, after therapy, it was important to demonstrate that the scolex had been passed. The current therapy with niclosamide leads to degeneration of the worm and the scolex is not found. Species can also be differentiated by the morphology of the scolex. The scolex of each species has 4 suckers. That of *T. solium* has a rostellum with 2 rows of hooks, whereas that of *T. saginata* has no rostellum and no hooks (Fig. 16:9).

Hymenolepsis nana

This dwarf tapeworm of humans causes the most prevalent tapeworm infection in the United States.[9] The adult has a delicate strobila which measures up to 40 mm long, and the scolex has a prominent rostellum with hooks. Humans are usually infected by ingesting freshly passed eggs. The larvae are released from the eggs, penetrate the mucosal villi and develop into infective larvae. The larvae then emerge into the lumen and develop into adults in direct proportion to the number of eggs ingested. The tissue larval stage, which only lasts for a few days, is apparently sufficient to illicit strong immunity in the host. *H. nana* is a common parasite of the house mouse, and its eggs may develop into infective larvae in various intermediate arthropod hosts. When infection develops by accidentally ingesting an infected arthropod, immunity is not developed to the tissue phase, and eggs from adult worms may hatch in the small intestine of the same host to produce a hyperinfection. This is one hypothesis of the pathogenesis in the occasional individuals in whom large numbers of worms are found. Patients with large numbers of worms may have gastrointestinal symptoms. Diagnosis is established by finding the oval colorless eggs (Figs. 16:2, 16:7), which measure 30-55 μm in the greatest dimension. These eggs contain a 6-hooked embryo and typically have prominent polar thickenings from which 4-8 long, thin filaments project. The size of the eggs and the presence of polar filaments are particularly helpful in differentiating *H. nana* from *H. diminuta*.

Hymenolepsis diminuta

This parasite usually infects rats, mice and other rodents, but sporadic human infections may occur. The embryonated eggs are ingested by fleas or other arthropods and infective larvae develop in their tissues. Humans are infected by accidentally ingesting infected arthropods. Adult tapeworms measure 10-70 cm long, and the scolex has no hooks. Infected humans are usually asymptomatic. Eggs are released when the gravid proglottids detach from the adult worm and disintegrate in the intestine. Diagnosis is based on finding the oval or round eggs (Figs. 16:2, 16:7), which measure 60-82 by 72-86 μm. These eggs have an inner membrane with 2 rudimentary polar thickenings but do not have polar filaments. They contain a 6-hooked embryo.

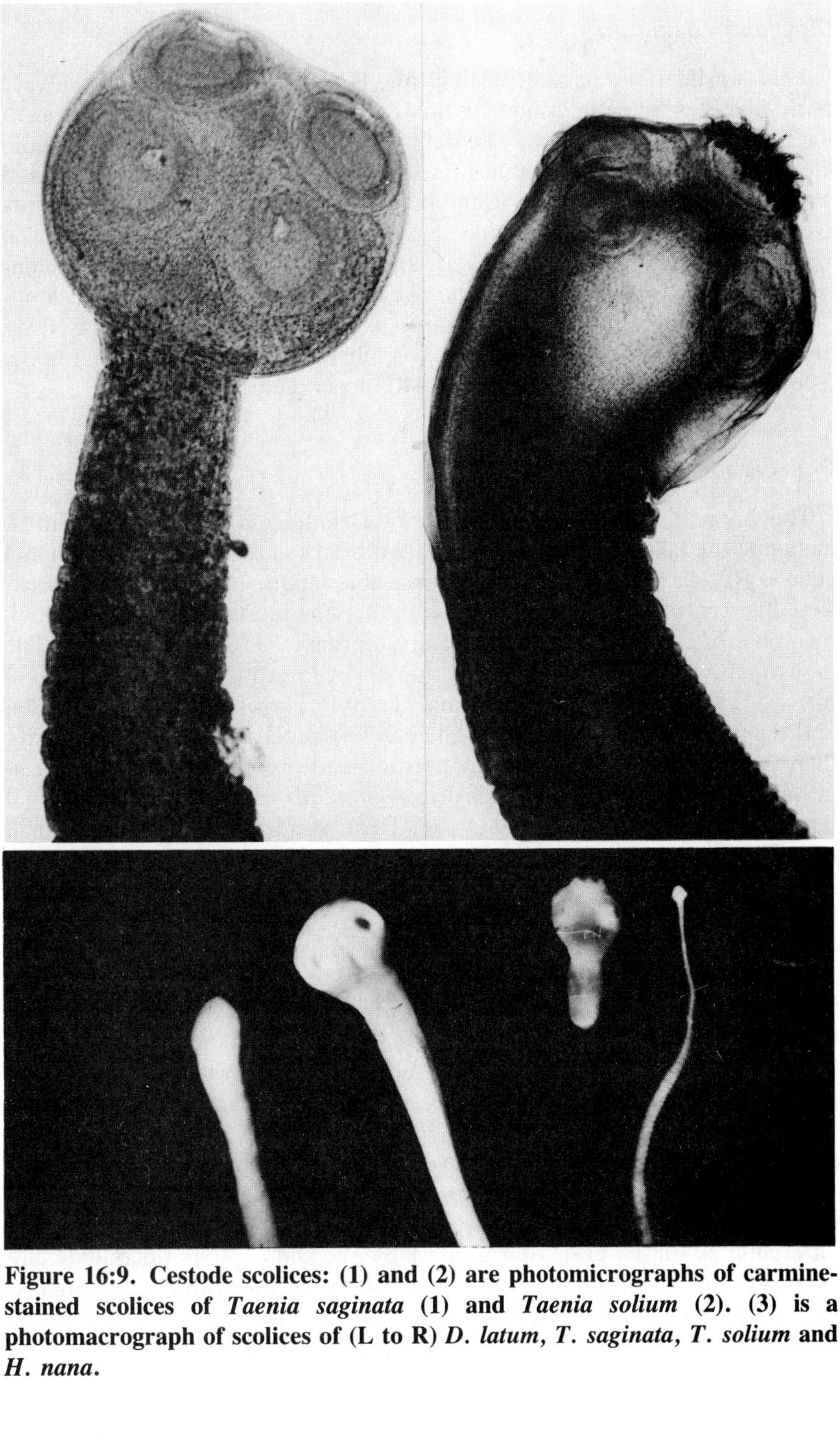

Figure 16:9. Cestode scolices: (1) and (2) are photomicrographs of carmine-stained scolices of *Taenia saginata* (1) and *Taenia solium* (2). (3) is a photomacrograph of scolices of (L to R) *D. latum*, *T. saginata*, *T. solium* and *H. nana*.

Dipylidium caninum

This is primarily a parasite of dogs and is distributed worldwide. The life cycle involves a flea intermediate host which ingests the egg and in which the infective larva develops. The definitive host is infected by ingesting an infected flea. The larva attaches itself to the small intestinal mucosa and matures to the adult parasite which measures up to 70 cm long. Proglottids are barrel-shaped, have a genital pore and sex organs on each side. The gravid proglottid contains numerous small capsules or packets, each containing 8-20 eggs. Humans are usually accidental hosts and remain asymptomatic, although occasionally symptoms result.[30] Diagnosis is based on finding gravid proglottids or packets of embryonated eggs (Fig. 16:7) in the feces. The spherical eggs measure 20-40 μm in diameter.

Diphyllobothrium latum

The broad fish tapeworm is widely distributed in temperate climates, including the lakes of the northern United States and Canada, northern and central Russia and Europe, Manchuria and Japan. Infections have occasionally been reported in other areas. In the United States, there have been a number of cases associated with consumption of uncooked salmon.[37] The parasite attaches to the small intestine and can grow to be as long as 10 meters. The scolex is elongated and has two long lateral grooves called bothria which are used for attachment. The proglottids which are wider than they are long have active oviposition and the eggs are passed unembryonated. The eggs must reach freshwater in order to embryonate and release the swimming 6-hooked larvae. The first intermediate host is a small aquatic arthropod, and the second intermediate host is a small fish which ingests the arthropod. The third intermediate host is a larger fish which ingests the smaller fish.[47] In the muscles of the larger fish, a larva known as a sparganum or plerocercoid, develops to the infective stage. The life cycle is completed when a larva in raw or inadequately cooked fish is ingested by any of the definitive hosts. Human infections are usually caused by *D. latum*, but other species of *Diphyllobothrium* occasionally infect humans. Infected humans are usually asymptomatic, with the passage of a portion of the strobila causing a patient to seek medical attention. Occasionally, patients develop Vitamin B_{12} deficiency with resulting megaloblastic anemia because the parasite selectively competes for the vitamin.

Diagnosis is based on finding the characteristic eggs (Figs. 16:2, 16:7), which measure 58-76 μm long by 40-51 μm wide. Species cannot be differentiated on the basis of egg morphology. The eggs are operculate and unembryonated, in contrast to those of other cestodes infecting humans. Eggs may have a small nipple-like protrusion (abopercular knob) on the posterior pole. There are no shoulders adjacent to the operculum.

REFERENCES

1. Adams KO, Jungkind DL, Bergquist EJ, Wirts CW. Intestinal fluke infection as a result of eating sushi. Am J Clin Pathol 1986;86:688-9.
2. Berry AJ, Long EG, Smith JH, Gourley WK, Fine DP. Chronic relapsing colitis due to *Strongyloides stercoralis*. Am J Trop Med Hyg 1983;32:1289-93.
3. Blumenthal DS. Intestinal nematodes in the United States. N Engl J Med 1977;297: 1437-9.
4. Brooke MM, Melvin DM. Morphology of diagnostic stages of intestinal parasites of man. U.S. Department of Health, Education and Welfare publication no. (HSM)72-8116, 1969.
5. Buck AA, Anderson RI. Validation of the complement fixation and slide flocculation tests for schistosomiasis. Geographic variations of test capacities. Am J Epidemiol 1972;96: 205-14.
6. Bueding E. Report of the American schistosomiasis delegation to the Peoples Republic of China. Am J Trop Med Hyg 1977;26:427-57.
7. Centers for Disease Control. Acute schistosomiasis with transverse myelitis in American students returning from Kenya. Morbid Mortal Weeky Rep 1984;33:445-7.
8. Centers for Disease Control. Cercarial dermatitis among bathers in California; Katayama syndrome among travelers in Ethiopia. Morbid Mortal Weekly Rep 1982;31:435-8.
9. Centers for Disease Control. Intestinal parasite surveillance annual summary. 1976. Atlanta: Centers for Disease Control, 1977.
10. Centers for Disease Control. Parasitology training materials. Parasitology Training Branch, 1981.
11. Chandrasoma PT, Mendis KN. *Enterobius vermicularis* in ectopic sites. Am J Trop Med Hyg 1977;26:644-9.
12. Cook GC. Gastrointestinal helminth infections: the clinical significance of gastrointestinal helminths—a review. Trans Royal Soc Trop Med Hyg 1986;80:675-85.
13. Cypess RH, Karol MH, Zidian JL, Glickman LT, Gitlin D. Larva specific antibodies in patients with visceral larva migrans. J Infect Dis 1977;135:633-40.
14. Deelder AM, Kornelis D. Immunodiagnosis of recently acquired *Schistosoma mansoni* infections. A comparison of various immunological techniques. Trop Geogr Med 1981; 33:36-41.
15. Eveland LK, Kenney M, Valentin Y. Laboratory diagnosis of autoinfection in strongyloidiasis. Am J Clin Pathol 1975;63:421-5.
16. Feldmeier H, Büttner DW. Immunodiagnosis of schistosomiasis haematobium and schistosomiasis mansoni in man. Application of crude extracts from adult worms and cercariae in the IHA and ELISA. Bakteriol Hyg I Abt Orig A 1983;225:413-21.
17. Fiorillo AM, Costa JC, Passos J. Identification of hemagglutinating antibodies in chronic schistosomiasis. Rev Inst Med Trop 1973;15:371-6.
18. Glickman LT, Schantz PM, Dembroske R, Cypess RH. Evaluation of serodiagnostic tests for visceral larva migrans. Am J Trop Med Hyg 1978;27:492-8.
19. Hancock K, Tsang VCW. Development and optimization of the FAST-ELISA for detecting antibodies of *Schistosoma mansoni*. J Immunol Meth 1986;92:167-76.
20. Hoeffler DF. Cercarial dermatitis: its etiology, epidemiology and clinical aspects. Arch Environ Health 1974;29:225-9.
21. Igra-Siegman Y, Kapila R, Sen P, Kaminski ZC, Louria DB. Syndrome of hyperinfection with *Strongyloides stercoralis*. Rev Infect Dis 1981;3:397-407.
22. Istre GR, Fontaine RE, Tarr J, Hopkins RS. Acute schistosomiasis among Americans rafting the Omo River, Ethiopia. JAMA 1984;251:508-10.
23. Jung RC, Beaver PC. Clinical observations on *Trichocephalus trichirus* (whipworm) infestation in children. Pediatrics 1951;8:548-77.
24. Klicks MM. Anisakiasis in the western United States; four new case reports from California. Am J Trop Med Hyg 1983;32:526-32.

25. Layrissi M, Roche M. The relationship between anemia and hookworm infection. Results of surveys of rural Venezuelan population. Am J Hyg 1964;79:279-301.
26. Lunde MN, Ottesen EA. Enzyme-linked immunosorbent assay (ELISA) for detecting IgM and IgE antibodies in human schistosomiasis. Am J Trop Med Hyg 1980;29:82-5.
27. Lunde MN, Ottesen EA, Cheever AW. Serological differences between acute and chronic schistosomiasis mansoni detected by enzyme-linked immunosorbent assay (ELISA). Am J Trop Med Hyg 1979;28:87-91.
28. Maddison SE. Schistosomiasis. In: Walls KW, Schantz PM, eds. Immunodiagnosis of parasitic disease; vol 1. Helminthic disease. Orlando, FL: Academic Press, 1986:1-37.
29. Maddison SE. Parasitic infections. In: Wicher K, ed. Microbial antigenodiagnosis. Boca Raton, FL: CRC Press, 1987;2:110-123.
30. Margolis B. Dog tapeworm infestation in an infant. Am J Dis Child 1983;137:702.
31. Markell EK. Pseudohookworm infection—trichostrongyliasis. N Engl J Med 1968;278: 831-2.
32. McLaren ML, Lillywhite JE, Dunne DW, Doenhoff MJ. Serodiagnosis of human *S. mansoni* infections: enhanced sensitivity and specificity in ELISA using a fraction containing *S. mansoni* egg antigens, $omega_1$, and $alpha_1$. Trans Royal Soc Trop Med and Hyg 1981;75:72-9.
33. Melvin DM, Brooke MM. Laboratory procedures for the diagnosis of intestinal parasites. 3rd ed. Atlanta: Centers for Disease Control, 1982.
34. Pelletier LL Jr. Chronic strongyloidiasis in World War II far east ex-prisoners of war. Am J Trop Med Hyg 1984;33:55-61.
35. Piggott J, Hansbarger EA, Neafie RC. Human ascariasis. Am J Clin Pathol 1970;53: 223-34.
36. Purtilo DT, Meyers WM, Connor DH. Fatal strongyloidiasis in immunosuppressed patients. Am J Med 1974;56:488-93.
37. Ruttenber AJ, Weniger BG, Sorvillo F, Murray RA, Ford SL. Diphyllobothriasis associated with salmon consumption in pacific coast states. Am J Trop Med Hyg 1984;33:455-9.
38. Sadun EH, Melvin DM. The probability of detecting infections with *Enterobius vermicularis* by successive examination. J Pediatr 1956;48:438-41.
39. Smith JW, Ash LR, Thompson JH, McQuay RM, Melvin DM, Orihel TC. Diagnostic parasitology—intestinal helminths. Chicago: American Society for Microbiology, 1976.
40. Stek M Jr. Erythroadsorption and enzyme-linked immunoassays (EAIA and ELISA) for specific circulating antibodies and antigens in schistosomiasis. Ann Immunol (Inst Pasteur) 1984;135:13-23.
41. Strauss WG. Clinical manifestations of clonorchiasis. A controlled study of 105 cases. Am J Trop Med Hyg 1962;11:625-30.
42. Sugimachi K, Inokuchi K, Ooiwa T, Fujino T, Ishii Y. Acute gastric anisakiasis: analysis of 178 cases. JAMA 1985;253:1012-13.
43. Symmers WStC. Pathology of oxyuriasis. Arch Pathol 1950;50:475-516.
44. Tanaka H, Dennis DT, Kean BH, Matsuda H, Sasa M. Evaluation of a modified complement fixation test for schistosomiasis. J Exp Med 1972;42:537-42.
45. Tsang VCW, Hancock K, Kelley MA, Wilson BC, Maddison SE. *Schistosoma mansoni* adult microsomal antigens: a serologic reagent. II. Specificity of antibody responses to the *S. mansoni* microsomal antigen (MAMA). J Immunol 1983;130:1366-70.
46. Tsang VCW, Peralta JM, Simon R. The enzyme-linked immuno-electro transfer blot technique (EITB) for studying the specificities of antigen and antibodies separated by gel electrophoresis. Meth Enzymol 1983;92:377-91.
47. Vik R. The genus *Diphyllobothrium*. An example of systematics and experimental biology. Exp Parasitol 1964;15:361-80.
48. Voge M, Bruckner D, Bruce JI. *Schistosoma mekongi* spp. nova from man and animals, compared with four strains of *Schistosoma japonicum*. J Parasitol 1978;64:577-84.
49. Walls KW, Schantz PM. Immunodiagnosis of parasitic diseases; vol 1. Helminthic disease. New York: Academic Press, 1986.

50. Wardle RA, McLeod JA. The zoology of tapeworms. Minneapolis: University of Minnesota Press, 1952.
51. Warren KS. Regulation of the prevalence and intensity of schistosomiasis in man. Immunology or ecology. J Infect Dis 1973;127:595-609.
52. Warren KS. Helminthic disease endemic in the United States. Am J Trop Med Hyg 1974;23:723-30.
53. Warren KS, Mahmoud AAF. Algorithms in the diagnosis and management of exotic diseases. XXII. Ascariasis and toxocariasis. J Infect Dis 1977;135:868-72.
54. Wilson M, Sulzer AJ, Walls KW. Modified antigens in the indirect immunofluorescence test for schistosomiasis. Am J Trop Med Hyg 1974;23:1072-6.

CHAPTER 17

ARTHROPODS OF PUBLIC HEALTH IMPORTANCE

Harry D. Pratt, Ph.D. and James W. Smith, M.D.

INTRODUCTION

Arthropods affect human health in many ways. Some are involved in mechanical transmission of pathogens. Others play important roles in biological transmission, development and survival of viruses, bacteria, protozoa and metazoa that cause human disease. Arthropods also injure humans directly by envenomization, vesication, blood sucking, irritation and invasion of tissues and stimulation of allergic responses.

Mosquitoes, certain flies, fleas, bed bugs, spiders, ticks and mites bite with their mouthparts. Bees, wasps and scorpions sting with an apparatus at the posterior end. For most people, such bites or stings cause only temporary swelling or itching that can be treated with calamine or other soothing lotions, although some become secondarily infected from scratching. A few persons become sensitized to protein from the insect saliva or sting.[13] If they are bitten or stung at a later date when the body has had time to develop antibodies to these foreign proteins, these people develop larger, more severe welts around each bite or sting. A person repeatedly infested with scabies mites may develop a generalized rash and intense itching over much of the body and not simply in the area of the scabies mite lesions. The most severe reactions affect a few people who are sensitized by the stings of bees, wasps or hornets. If these people are stung again, they may have a severe reaction to this same foreign protein and may die from anaphylactic shock, respiratory edema, vascular occlusion or damage to the nervous system. Such deaths usually occur rapidly, often within an hour after the sting.[12,13,14,20]

Arthropods occur in great abundance in the environment, so fragments and excrement of these animals are present in dust and soil. Persons may become hypersensitive to these materials (such as house dust mites) and develop allergic manifestations, i.e., asthma and hay fever.[13,14,20,24,27]

Parasitism by arthropod parasites which are attached to the skin or temporarily invade tissues is generally called infestation. In contrast, parasitism of body tissues, intestine or atria by protozoa or helminth parasites is generally called infection.

It is beyond the scope of this chapter to deal with all of the arthropods of public health importance or the therapy and control of arthropod infestations. Some of the most common arthropods submitted to laboratories for identification in the United States are discussed and pictorial keys are presented which show the principles of classification. More detailed information on arthropods of public health importance and their identification, treatment and control is available in many references.[2,4,6,7,12,14,18,20,24,27,29,33]

Phylum Arthropoda

The phylum Arthropoda comprises invertebrate animals with a segmented body, several pairs of jointed appendages and a rigid chitinous exoskeleton which is molted periodically and renewed as the animal grows. The classes Insecta and Arachnida are of importance to public health workers.

Arthropods develop from egg to adult through a process known as metamorphosis (literally a change in form). Insects with gradual metamorphosis pass through three stages during their life: egg, nymph and adult. Examples are sucking lice, bed bugs and cockroaches. They change (metamorphose) gradually through a succession of molts until they become adults. In some with gradual metamorphosis, such as cockroaches and kissing bugs, the nymphs have wing pads while adults have wings. Nymphs are always sexually immature, while adults are sexually mature, ready to mate and reproduce. Nymphs have the same type of mouthparts and live in the same environment as adults, for example, head lice nymphs and adults on the scalp.

Insects with complete metamorphosis pass through four stages during their life: egg, larva, pupa and adult. Examples are flies, mosquitoes, fleas and bees. Insects with complete metamorphosis differ greatly in the immature and adult stages. Typical larvae are wigglers of mosquitoes, maggots of flies and caterpillars of moths and butterflies. The pupa is a nonfeeding stage during which the larva undergoes profound external and internal changes to become an adult. Typically larvae have different mouthparts and live in different environments than the adults. For example, the mosquito wiggler lives in water, while the adult is aerial.

Ticks and mites have a still different type of metamorphosis, with four stages: egg, larva, nymph (or several nymphal stages) and adult. Tick and mite larvae have three pairs of legs, while nymphs and adults have four pairs.[6,12,14,20,23,29,33]

Phylum Arthropoda

Class Insecta:

Adults have three body regions (head, thorax and abdomen), one pair of antennae, three pairs of legs and wings in many species.

Order Orthoptera, Family Blatellidae (cockroaches)
The German cockroach (*Blattella germanica*)
The brown-banded cockroach (*Supella longipalpa*)

Order Orthoptera, Family Blattidae (cockroaches)
The American cockroach (*Periplaneta americana*)
The Oriental cockroach (*Blatta orientalis*)

Order Hemiptera, Family Cimicidae (bed bugs)
The bed bug (*Cimex lectularius*)

Order Anoplura, Family Pediculidae (human lice)
The head louse (*Pediculus humanus capitis*)
The body louse (*Pediculus humanus humanus*)

Order Anoplura, Family Pthiridae (pubic lice)
The pubic or crab louse (*Pthirus pubis*)

Order Siphonaptera, Family Pulicidae (fleas)
The Oriental rat flea (*Xenopsylla cheopis*)
The cat flea (*Ctenocephalides felis*)
The dog flea (*Ctenocephalides canis*)

Order Diptera (flies)
Fly maggots—many species of the families Calliphoridae, Gasterophilidae, Hypodermatidae, Muscidae, Piophilidae, Sarcophagidae, Stratiomyiidae, and Syrphidae.

Class Arachnida:

Adults with one or two body regions (cephalothorax and abomen, or one body region in ticks and mites), no antennae, four pairs of legs and no wings. Larvae of ticks and mites have only three pairs of legs.

Order Acarina, Family Ixodidae (hard ticks)
The American dog tick (*Dermacentor variabilis*)
The Rocky Mountain wood tick (*Dermacentor andersoni*)
The lone star tick (*Amblyomma americanum*)
The brown dog tick (*Rhipicephalus sanguineus*)
The northern deer tick (*Ixodes dammini*)

Order Acarina, Family Argasidae (soft ticks)
Relapsing fever ticks (*Ornithodoros* spp.)

Order Acarina, Family Sarcoptidae (Scabies mites)
Scabies mite (*Sarcoptes scabiei*)

Order Acarina, Family Demodicidae (Follicle mites)
Hair follicle mite (*Demodex folliculorum*)
Sebaceous gland mite (*Demodex brevis*)

Order Acarina, Family Trombiculidae (Chiggers)
Chigger (*Eutrombicula alfreddugesi*)

Order Acarina, Family Pyemotidae (Straw itch mites)
Hay itch mite (*Pyemotes tritici*)

Order Acarina, Family Dermanyssidae (Rodent and bird mites)
Rodent and bird mites (*Ornithonyssus, Dermanyssus* spp.)
House mouse mite (*Liponyssoides sanguineus*)

Order Acarina, Family Acaridae (= Tyroglyphidae) (Cheese mites)
Cheese mite (*Acarus siro*)

Order Acarina, Family Pyroglyphidae (House dust mites)
House dust mites (*Dermatophagoides pteronyssinus* and *D. farinae*)

Order Acarina, Family Theridiidae (Combfooted spiders)
Black widow spider (*Latrodectus mactans*)

Order Araneida, Family Loxoscelidae
Brown recluse spider (*Loxosceles reclusa*)

These arthropods can be identified by using the characters in the pictorial keys (Figs. 17:1-17:11) or specialized literature.[6,7,14,20,29]

Specimens of ectoparasites should be preserved and shipped for identification in 70% ethyl alcohol rather than formalin which is irritating to the eyes of the identifier. Fly larvae may have to be washed in water to remove debris, particularly those collected from fecal samples or wounds. After they are washed in clean water, fly larvae should be killed in hot (not boiling) water, about 80°C, for a few minutes so that they do not become black when placed in 70% alcohol.

Most arthropods can be identified with a dissecting microscope with magnifications of 25-75 power. Some fleas, fly larvae and mites may have to be treated overnight with cold 10% potassium or sodium hydroxide to remove the internal flesh and permit better visibility of key characters. Such specimens can then be washed in water, dehydrated in ethyl alcohol, cellosolve, clove oil and mounted in Canada balsam as permanent slides. Quick nonpermanent mounts can be made with commercially available clearing and mounting medium, such as Hoyer mounting medium, frequently used in identifying mites.

Assistance in identifying arthropods may be obtained from entomologists in public health laboratories, educational institutions, natural history museums or agriculture extension agencies.

COCKROACHES

Although some 3500 species of cockroaches have been described for the entire world, less than 1% are domiciliary, and only about 8 are commonly found in buildings in the United States (see Fig 17:1). Cockroaches are often more important pests in buildings such as hospitals and laboratories, than are flies, fleas or lice.[11,12,23,33]

Cockroaches are usually classified in the order Orthoptera, although some authorities place them in the orders Dictyoptera or Blattaria. Cockroaches are flattened dorsoventrally like a pancake, with long, filiform antennae, and legs adapted for running. They vary in size from immatures 1-2 mm long to giant tropical species 75 mm or more in length. Cockroaches develop by gradual metamorphosis with 3 stages in their life cycle: eggs in a capsule or ootheca, sexually immature nymphs without wings and sexually mature adults usually with wings. They have chewing mouthparts and feed on many types of organic matter including fecal material loaded with pathogens.

Causative Organisms

The four species of cockroaches most commonly found in buildings throughout the United States and much of the world may be divided into two groups: the small species with adults less than 15 mm long include the German and the brown-banded cockroaches; and the large species with adults more that 15 mm long include the American and Oriental cockroaches (see Fig. 17:1).

The German cockroach (*B. germanica*) is easily recognized by its small size, adults 10-13 mm long, and the two black longitudinal bars on the pronotum of both adults and nymphs. This is the most common cockroach in kitchens and bathrooms.

The brown-banded cockroach (*S. longipalpa*, formerly *Supella supellectilium*) receives its name from the brown bands across the wings of the adult or the thorax and abdomen of the nymph. Adults, averaging 10-13 mm long, have a single, broad, dark, longitudinal stripe in the middle of the pronotum. It is found in many parts of buildings besides the kitchen and bathroom.

The American cockroach (*P. americana*) is the largest cockroach commonly found in buildings, adults 27-35 mm long. This is a chestnut brown species with indistinct yellowish markings around the edge of the pronotum, wings of the adult extending beyond the tip of the abdomen and cerci tapering, the last segment twice as long as wide. It may be found throughout buildings but prefers warm, damp basements, steam tunnels and sewers.

The Oriental cockroach (*B. orientalis*) is a large, entirely dark insect, adults 18-24 mm long. The male has short truncate wings extending 2/3 to 3/4 the length of the abdomen and the female has short triangular wing pads.

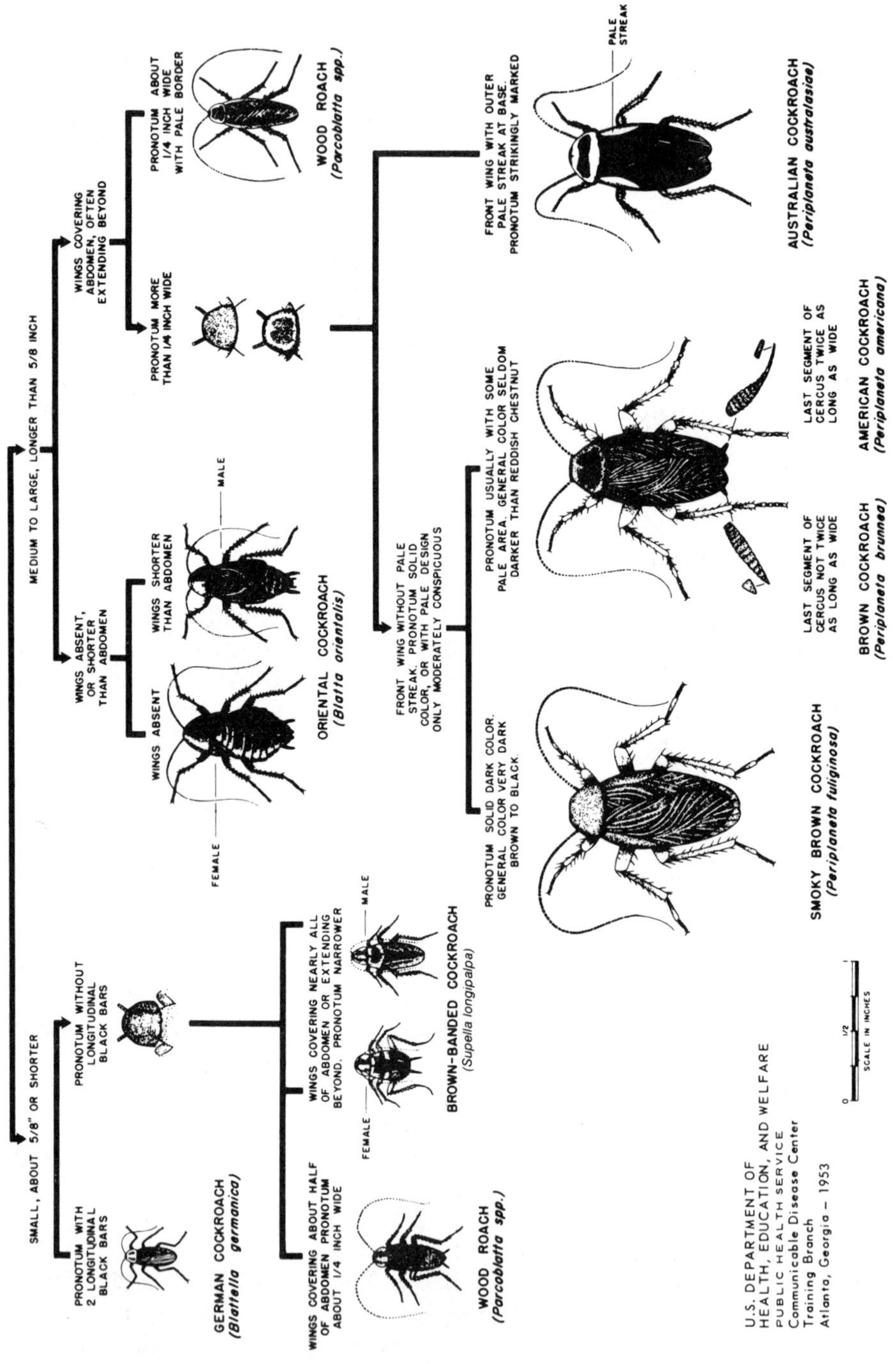

Figure 17:1. Pictorial key to some common adult cockroaches.[7]

The Oriental cockroach is more abundant in the cooler parts of the United States, where it is found particularly in kitchens, basements and sewers.

Clinical Manifestations

Cockroaches may cause asthmatic attacks, nervous disorders and sleeplessness in sensitive children and adults. They are obnoxious pests, which taint an area with a characteristic repugnant odor and foul with excrement all surfaces and food with which they come in contact. Cockroaches have been implicated as carriers of *Salmonella*, a cause of food poisoning. There may be a relationship between cockroaches and hepatitis A infection. Some people sensitive to "house dust" may have asthmatic attacks and be allergic to cockroaches, fragments of cockroach bodies or cockroach feces.[11,23]

Epidemiology

At the present time the German cockroach is considered by the American pest control industry to be the number one pest in buildings, more important than house flies, fleas or lice. The many studies cited by Cornwell[11] and Mallis[23] indicate that cockroaches (a) increase their populations very rapidly if left undisturbed, (b) may transmit a wide variety of pathogens, particularly *Salmonella*, (c) are often abundant in food storage areas and food-handling establishments such as kitchens, restaurants and hospitals and (d) may occur in large numbers in toilets, privies and sewers.

Private householders usually attempt to control cockroaches with a variety of commercially available aerosols, dusts, sprays and traps. However, in institutions such as hospitals and in food-handling establishments, cockroach control should be carried out by licensed professional pest control technicians using approved pesticides. In recent years the development of small, plastic bait boxes with amidino-hydrazone has given effective control of small cockroaches, such as the German cockroach, in computers, TV sets and delicate electronic equipment where no aerosols, dusts or sprays should be used.

BED BUG

Causative Organism

The bed bug (*C. lectularius*) is a reddish-brown insect about 5 mm long with a pair of 4-segmented antennae and a 3-segmented, blood sucking probascis which lies in a groove on the underside of the head and thorax (Fig. 17:2). It has short wing pads but cannot fly. The pronotum is broad

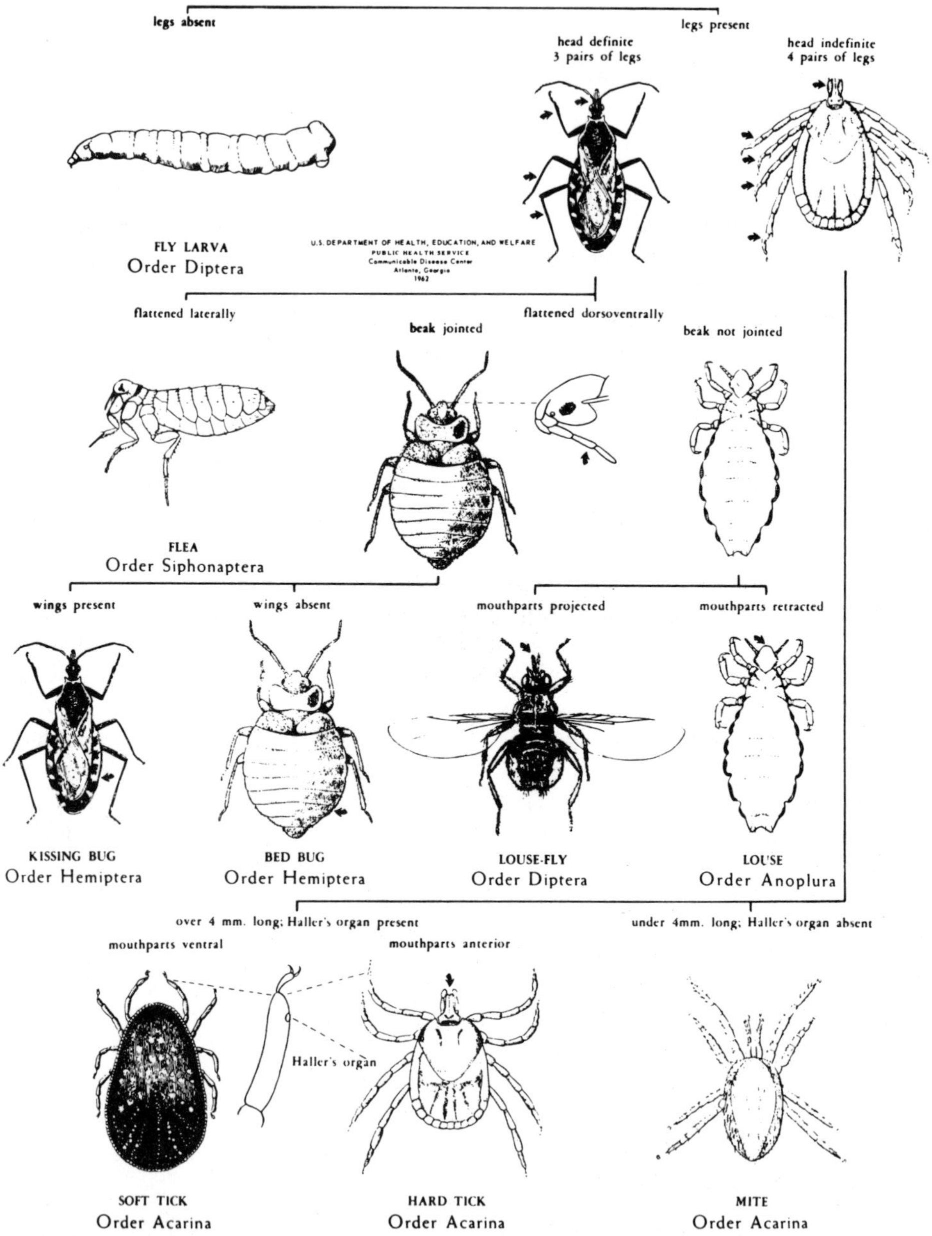

Figure 17:2. Pictorial key to groups of human ectoparasites.[7]

and has a concave anterior margin. The abdomen is flattened and somewhat heart-shaped.

Clinical Manifestations

The public health importance of bed bugs is associated with their bites. Some people are very sensitive to bed bug bites and will have reddish wheals as big as a half dollar which itch intensely, whereas other people are hardly aware of them. Theoretically, the bed bug should be an ideal vector of human diseases since it feeds repeatedly on humans. However, these insects have not been incriminated in the transmission of any communicable disease despite many careful experiments. Bed bugs cause nervous disorders and sleeplessness among sensitive children and adults.

Epidemiology

Before DDT, bed bugs were very common pests in homes, hotels and institutions. Since World War II, residual spraying with DDT and other synthetic insecticides has almost eradicated bed bugs from the United States. However, since 1972 when DDT was banned in the United States, sporadic reports of bed bugs have been more numerous. Health authorities and pest control operators indicate that these recent bed bug infestations are often associated with shipping of infested furniture or movement of people from infested to uninfested buildings. Large numbers of these nocturnal pests may be present in a bedroom. During the day, they hide in mattresses, bedsteads, cracks in the wall or behind loose wallpaper. They are intermittent feeders on sleeping victims and, when surfeited, retire to their hiding places. Frequently the first signs of bed bug infestations are tiny spots of red blood or the flattened dead insects which were killed as the victim rolled while asleep.

Bed bugs develop by gradual metamorphosis with three stages in their life cycle: egg, nymph and adult. The eggs, which are glued to surfaces in daytime resting places such as cracks in a wall or buttons on a mattress, hatch in 1-7 weeks, depending on temperature. There are five nymphal stages, each of which must follow a blood meal. The adults, which can live as long as 9-18 months, can survive for several months without feeding.

HUMAN LICE

Three types of sucking lice in the order Anoplura are specific parasites of humans. They are usually, but not always, found on a certain part of the body. They are named according to the region of the body that they infest and/or their general appearance: head louse, body louse and pubic or crab louse. In other parts of the world, particularly before synthetic insecticides such as DDT were used, the body louse was involved in outbreaks of epidemic or louse-borne typhus, trench fever and louse-borne or epidemic relapsing fever. Fortunately these diseases do not occur in the United States today. However, increasing numbers of louse infestations have been reported in the United States and many other countries in recent years, particularly reports of head lice in children.[21,24,27]

Causative Organisms (See Fig. 17:3)

Although some authorities consider the head and body lice as separate species, they are considered subspecies in this chapter. The head louse (*P. humanus capitis*) is 1-2 mm with 3 pairs of legs of approximately equal size and an elongate abdomen with slightly darkened margins but without lateral hairy tufts. The adults and immatures called nymphs are found on the head and neck, particularly behind the ears and on the back of the neck. The eggs, called "nits," are glued to the hairs. (See Figs. 17:4 and 17:5)

The body louse (*P. humanus humanus*) is usually 2-4 mm long, with 3 pairs of legs of approximately equal size and elongate abdomen with pale margins and no lateral hairy tufts. The adults and nymphs are found on the hairy parts of the body below the neck and frequently rest on clothing when they are not feeding. Typically the eggs are laid on clothing, particularly along the seams.

The pubic or crab louse (*P. pubis*) (See Fig. 17:3) is 0.8-1.2 mm long, with the first pair of legs much smaller and more slender than the second and third pairs and a short, crab-like abdomen with lateral hairy tufts. The adults and immatures are typically found on the pubic and anal areas of the body or on other parts of the body with widely spaced hairs (e.g., chest, armpits, moustache, beard, eyebrows or eye lashes). The eggs are glued to hairs.

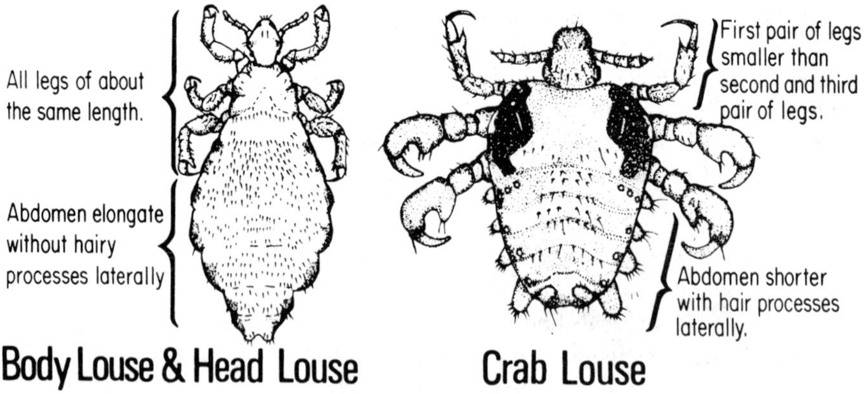

Figure 17:3. Lice commonly found on man.[7]

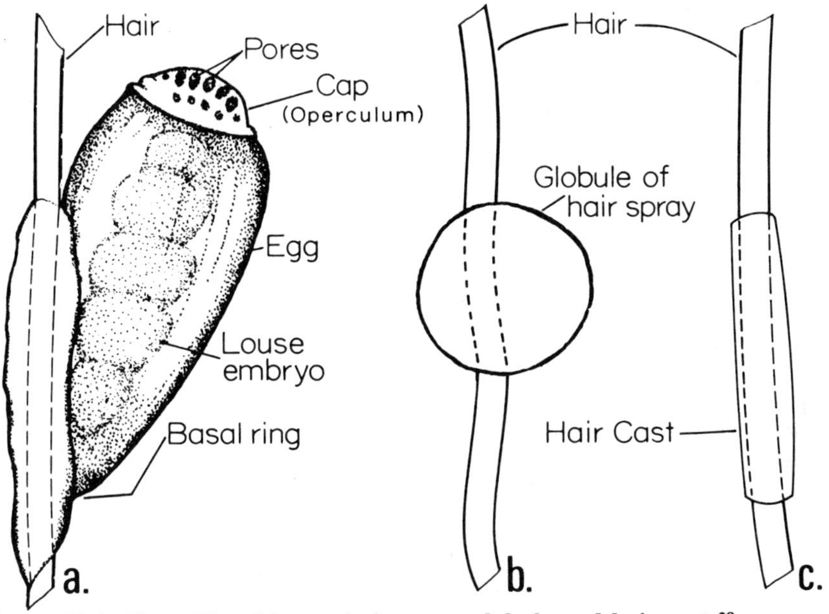

Figure 17:4. Egg of head louse, hair spray globule and hair cast.[28]

Clinical Manifestations

Infestations of human lice are called pediculosis from the generic name of the head and body lice, *Pediculus*. Frequently such infestations lead to scratching, secondary infections and scarred, hardened, or pigmented skin—the classic signs of pediculosis.

Epidemiology

Pediculosis usually occurs among people who live in crowded locations and have limited facilities for regular bathing and laundering.[21] The three types of human lice grow by a process known as gradual metamorphosis, with three stages in their life histories: eggs, nymphs and adults (See Fig. 14:5). The eggs, frequently called "nits," are attached to hairs (head and crab lice) or clothing (body lice). The immature nymphs suck blood and molt three times before becoming adults. The adults may live for a month or more. They can live 24-48 h away from their human host. Eggs of body lice laid on clothing can survive for a longer time. However, infested clothing not worn for a month would probably be free of both living lice and viable eggs.[24] Lice are transmitted from an infested to an uninfested person by direct contact and indirectly by infested materials such as clothing or bedding. Head lice (which are a major problem in the United States) can

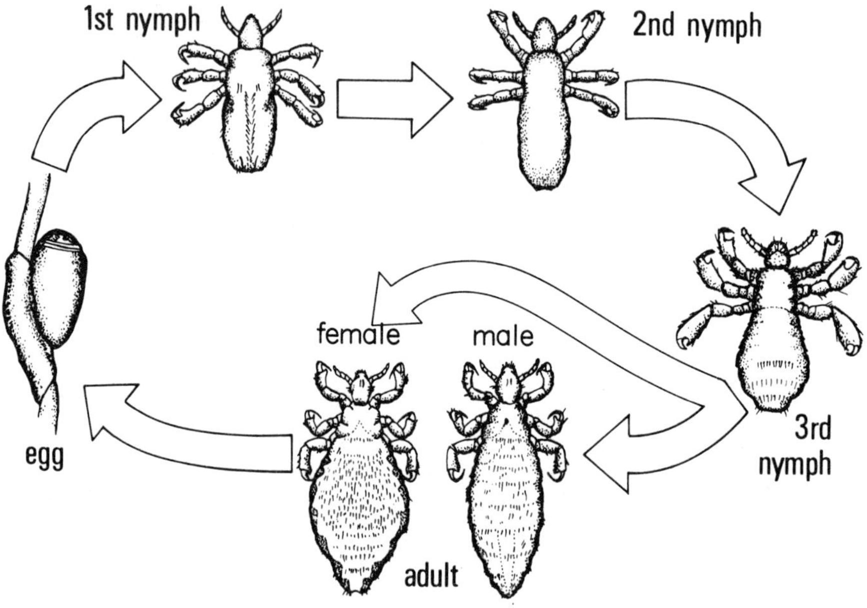

Figure 17:5. Life cycle of the head louse.[7]

be acquired from infested garments such as coats, caps and scarves, when such items are piled together (particularly in school cloak rooms), from infested combs and brushes, from lying on infested carpets or beds or from having the head come in contact with upholstered furniture that has been used by an infested person. Hairs with nits attached may be blown considerable distances and serve as a means of transmission. Usually one person will have only 10-20 head lice, but infestations with hundreds or thousands of body lice have been reported.[21,24,26] The distribution of head lice is influenced by such factors as age, sex, crowding at home, family size, method of closeting clothes, socioeconomic status and race. In the United States, some recent studies[24] suggest that blacks are less frequently infested than whites. Crab lice are usually spread by sexual contact. Rarely, they may be acquired from toilet seats.

Diagnosis

Head louse infestations are diagnosed by directly inspecting the head and neck, particularly behind the ears and the back of the neck, for crawling forms and nits. Body and crab louse infestations are diagnosed by examining for lice and their eggs in those hairy parts of the body below the neck, particularly the chest, armpits, pubic and anal areas. Body lice and their eggs are often found on clothing, especially the seams of collars, waists and armpits. Care should be taken to differentiate between an active infestation—live adults or nymphs—and an inactive infestation—the presence of empty egg cases. Pseudoepidemics, particularly of head lice in school children, have been reported[25] in which globules of hair spray or hair casts have been misidentified as nits (See Fig. 17:4). It is best to remove the hairs and examine them under a microscope to confirm the presence of eggs glued to hairs. Live eggs contain an embryo and have a cap with pores; empty eggs usually do not have the cap.

FLEAS

Fleas are among the most common ectoparasites which affect humans in the United States. Fleas cause irritation, loss of blood and extreme discomfort. Public health workers are concerned with fleas as vectors of bubonic plague and fleaborne or murine typhus (from rats to humans) and as vectors of sylvatic plague among wild rodents and occasionally among humans. In addition, fleas are intermediate hosts for the double-pored dog tapeworm (*Dipylidium caninum*), and two rodent tapeworms (*Hymenolepis diminuta* and *H. nana*) that occasionally infect humans.

Causative Organisms

Fleas are small, wingless insects with bodies compressed from side to side and long legs adapted for jumping. They vary from 1-8.5 mm in length, averaging about 2-4 mm (see Fig. 17:6). The name of the flea order, "Siphonaptera," refers to their blood-sucking or "siphoning" mouthparts and to their lack of wings. The presence or absence of the genal and pronotal combs, the shape of the head and of the spermatheca in the female, the length of the labial palpi, and the number and position of bristles on the head, abdomen and tarsi offer important characters for the identification of 10 common fleas found in the United States, as shown in Figure 17:7.

The Oriental rat flea (*X. cheopis*) is the chief vector of transmission of fleaborne or murine typhus from rats to humans and probably was the most important species transmitting plague bacteria during the period 1900 to 1925 when urban plague occurred in the United States.[12,14] This flea is easily identified by the characters shown in the pictorial key (see Fig. 17:7). It has no genal or pronotal combs, the front margin of the head is rounded and the thorax is of normal length, the mesopleuron is divided by a vertical thickening, the ocular bristle is inserted in front of the eye and the female has a large, dark, C-shaped spermatheca.

The cat flea (*C. felis*) and the dog flea (*C. canis*) have both a genal comb and a pronotal comb, each with about 7 or 8 pointed black teeth on each

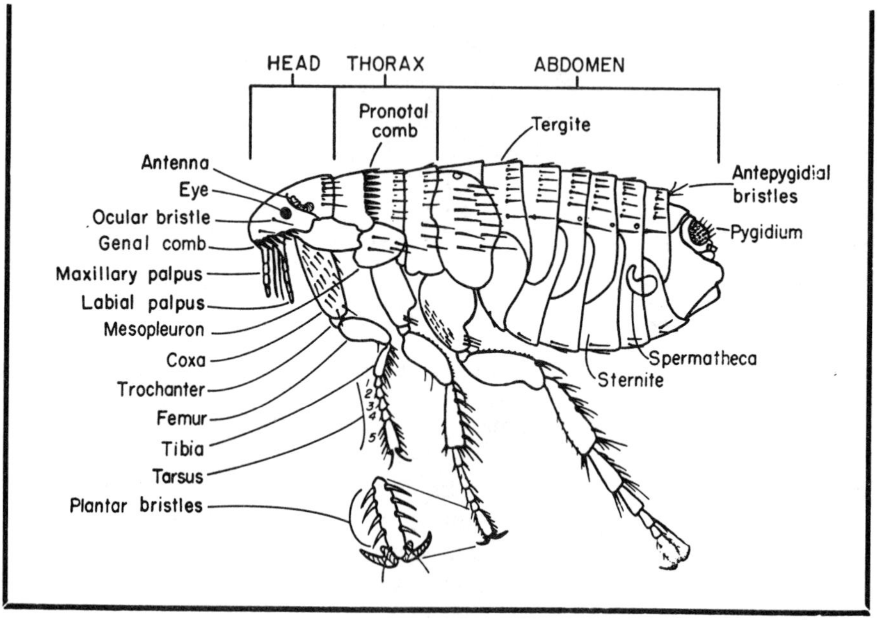

Figure 17:6. Flea.[7]

side of the body. As shown in the pictorial key (see Fig. 17:7) the head of the cat flea is about twice as long as high and the first two teeth of the genal comb are about the same length, whereas the head of the dog flea is less than twice as long as high and the first spine of the genal comb is definitely shorter than the second. It is usually not necessary to distinguish between these two closely related species because they have similar habits. Both species attack cats, dogs, rats and humans. In many parts of the United States, the cat flea is more abundant than the dog flea.

A number of other species of fleas do attack humans, particularly the human flea (*Pulex irritans*) in the San Francisco Bay area and the sticktight flea (*Echidnophaga gallinacea*) in the southern United States. Other species of fleas not included in the pictorial key can be identified with specialized literature.[14,15,20,29]

Clinical Manifestations

Flea bites are almost unbearable for some persons, whereas others are relatively undisturbed by them. People who are bitten by fleas have reactions ranging from small red spots (where the mouthparts of the fleas have penetrated the skin) surrounded by a slight swelling and reddish discoloration to a very severe generalized rash. The development of sensitivity to flea

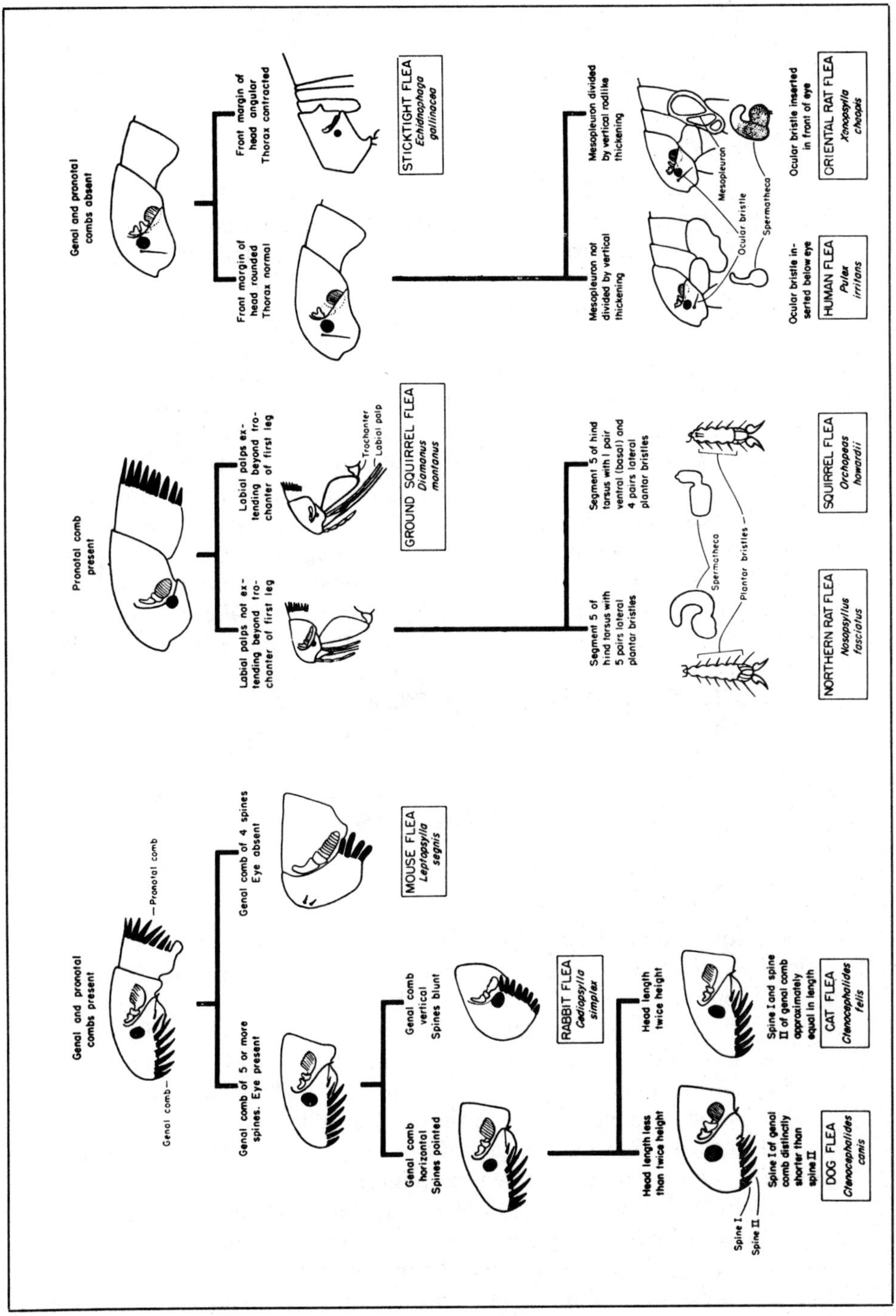

Figure 17:7. Pictorial key to some common fleas in the United States.[7]

bites requires an initial sensitization. Then a latent period occurs between the time of the first exposure and the time when subsequent flea bites elicit skin reactions.

Epidemiology

Fleas are bloodsucking insects seldom found far from their host animals—often cats, dogs or rats. Fleas develop by a process known as complete metamorphosis through four life stages: eggs, larvae, pupae and adults. The adult fleas suck blood from warm-blooded animals and usually mate on the host. The eggs of many fleas are not sticky or attached to the host, so they drop onto the ground, carpet or bedding material of the host. This explains the large number of fleas in dog or cat boxes, in kennels or in carpets on which a pet sleeps. Depending on the flea species, temperature and moisture, the eggs hatch in a few days or weeks, and the larvae develop—sometimes as rapidly as 2 weeks—and molt to become pupae. Then, the adults emerge from the pupal cocoons, ready for a blood meal. Many human infestations occur in homes with heavy infestations of cat or dog fleas. Often, as long as the cat or dog is present, the newly emerged adult fleas simply hop onto and feed on these pets and complaints of people being bitten by fleas are infrequent. However, if people leave their homes and take their pets with them, or board their cats or dogs for 2-4 weeks or longer, numerous fleas may mature in the vacant houses or apartments. Such fleas have had no opportunity for a blood meal. When people enter such a dwelling, they may be attacked by hundreds of hungry fleas and suffer excruciating pain.[14,20,23,33]

FLY MAGGOTS

Causative Organisms

Flies belong to the insect order Diptera, or two-winged insects. Flies develop by complete metamorphosis and have four stages in their life cycle: egg, larva, pupa and adult. Fly larvae are frequently called maggots. A brief discussion of fly maggots is included in this manual because some species are found in wounds and sinuses or in the umbilical area of newborn babies, and because fly larvae are sometimes found in stool samples submitted to laboratories for examination. More thorough discussions of fly larvae, and keys and illustrations to aid in identification are included in a publication by James[17] and in a number of general references.[14,20,29]

A typical muscoid fly larva is legless and somewhat cone-shaped, with a narrow anterior end bearing the mouthparts and anterior spiracles and a broader posterior end with two prominent posterior spiracles. The structures of the mouthparts and the anterior and posterior spiracles are characters used in identification. Most muscoid fly larvae are rather similar and are classified using characters shown in Fig. 17:8. Two types easily recognized with the naked eye are the lesser house fly and latrine fly (*Fannia* spp.), which have prominent lateral processes, and the rat-tailed maggot (*Eristalis tenax*), which has a long, telescopic respiratory tube.

Clinical Manifestations and Epidemiology

Infestation with fly maggots causes a condition known as "myiasis," in which the fly larvae feed on living, necrotic or dead tissues of humans or on the food in the human alimentary canal. Depending on the location of the fly larvae, a number of terms have been used to describe the various types of myiasis as gastrointestinal or enteric (digestive tract); dermal, subdermal or cutaneous (skin); auricular (ear); ocular (eye); nasopharyngeal (nose); and urinary or urogenital (urogenital tract).

Enteric Myiasis

Maggots in the digestive tract may cause queasiness, nausea, pain in the abdomen, diarrhea, dysentery (with actual discharge of blood resulting from injury to the intestinal mucosa) and nervousness. Fifty species of fly larvae have been reported, either positively or questionably, from cases of "enteric myiasis" among humans.[17] The flies involved are usually species which lay their eggs or larvae on cold meats, fish, cheese, ripe fruits and other foods. Most of the eggs and larvae are undoubtedly destroyed by normal digestive juices in the human alimentary tract. However, there is documentation of living larvae which were expelled either in the stool or vomit, or both. Some of these cases involved children who drank "dirty water" from a ditch containing rat-tailed maggots (*E. tenax*) and children and adults who ate meat or fish containing larvae of the flesh flies (*Sarcophaga* spp.) or ripe fruits with soldier fly larvae (*Hermetia illucens*).

Laboratory workers should be very careful in reporting enteric myiasis. Stool samples can easily be contaminated in the laboratory, particularly by species of flesh flies (*Sarcophaga* spp.) which are strongly attracted by the smell of feces. These insects lay larvae rather than eggs and the first two larval instars are often completed in a day. It is possible to find first instar larvae in a cardboard carton containing a fresh stool sample on a laboratory bench one day and third instar larvae in the material when it is examined the following day. In such cases of questionable "enteric myiasis," a second stool sample should be passed in a fly-free room and the material examined at once.

Dermal and Subdermal Myiasis

The infestation of cuts and open sores by living calliphorid and sarcophagid larvae that feed on bleeding, festered or malodorous tissues has been known

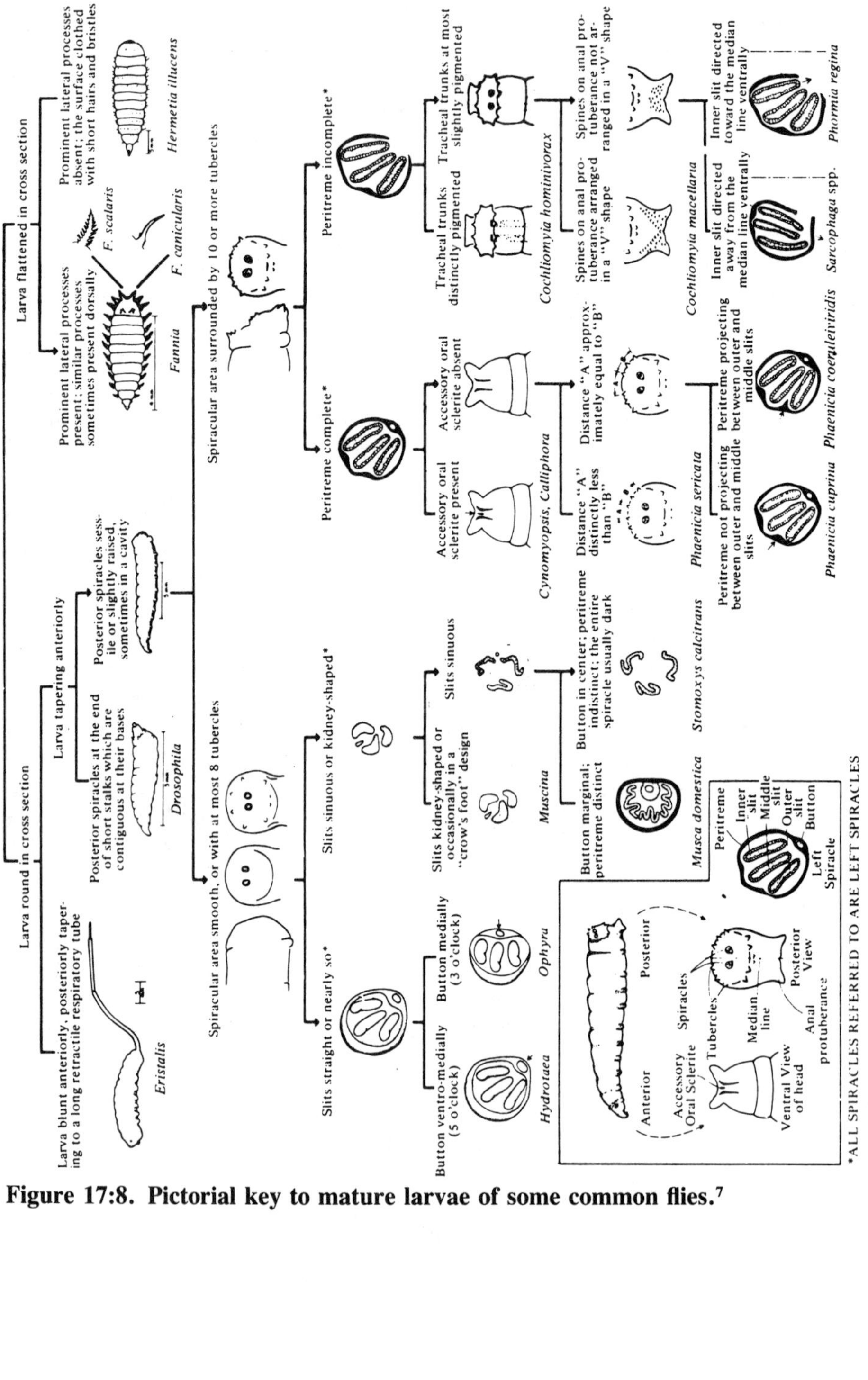

Figure 17:8. Pictorial key to mature larvae of some common flies.[7]

for many years—most often in wounded persons on battlefields. Civilian cases following snagging on barbed wire, gun-shot wounds and open sores on the scalp and other parts of the body have been reported.[14,17] Many of these cases should be considered as facultative myiasis caused by species of green-bottle flies (*Phaenicia*), black blow flies (*Phormia regina*) or flesh flies (*Sarcophaga*) that are attracted by the smell of blood or diseased tissues and that normally lay their eggs or larvae on dead animals. Sometimes such infestations are benign or even beneficial in "cleaning" suppurating wounds. Several infestations by larvae of the green-bottle fly (*Phaenicia sericata*) have been reported in the umbilical region of newborn babies, the flies having been attracted by the smell of blood of the tied cord.

Other more serious cases of dermal myiasis in humans are caused by larvae of the primary screw-worm (*Cochliomyia hominivorax*) or *Wohlfahrtia,* whose larvae are normally obligatory parasites of mammals. Their larvae are imbedded in living tissue where they develop and can cause very painful, serious wounds—particularly in the eye, nose, mouth and vaginal regions.[14,17] In recent years travelers returning from Central and South America have had painful wounds caused by the human bot fly (*Dermatobia hominis*). Other tourists who lay on beaches in Africa returned to the United States infested with larvae of the tumbu fly (*Cordylobia anthropophaga*) which caused furuncular swelling and infected sores.

TICKS

Ticks are bloodsucking arachnids that are ectoparasites of many vertebrates including humans. Ticks have a four-stage life cycle: egg, 6-legged larva, 8-legged sexually immature nymph and 8-legged sexually mature adult. Usually a blood meal is necessary for the larva to molt to become a nymph and another blood meal for the nymph to molt to become an adult. Hard ticks have only one nymphal stage, whereas soft ticks may have as many as four or five.

There are two families of ticks (see Fig. 17:9)—hard ticks in the family Ixodidae and soft ticks in the family Argasidae. Hard ticks have a hard dorsal plate or scutum and the mouth parts are located at the anterior end, clearly visible from above. Soft ticks have a leathery body but no hard plate on the dorsal part of the body, and the mouthparts are located ventrally and are not visible from above. Both hard and soft ticks can bite humans and cause painful itching lesions.

Causative Organisms

The five most important species of hard ticks in the United States are the American dog tick (*D. variabilis*), the Rocky Mountain wood tick (*D.*

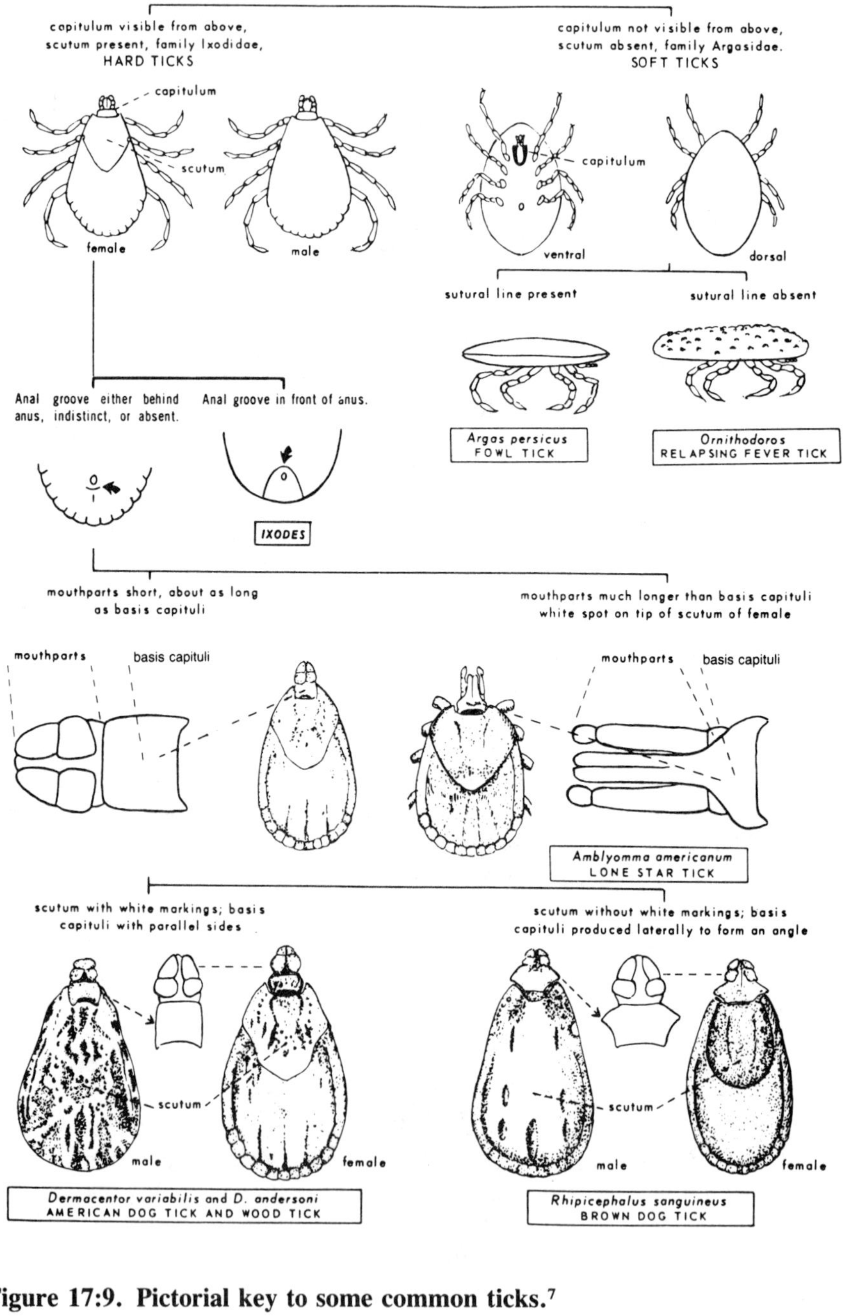

Figure 17:9. Pictorial key to some common ticks.[7]

andersoni), the lone star tick (*A. americanum*), the brown dog tick (*R. sanguineus*) and the northern deer tick (*I. dammini*). Ticks in the genera *Dermacentor* and *Amblyomma* are often called "ornate" ticks because they have whitish markings on the scutum easily seen with the naked eye, a hand lens or a stereoscopic microscope. The ticks in the other genera of North American ticks do not have these whitish markings on the scutum and are called "inornate" ticks.

The American dog tick (*D. variabilis*) is found in most of the eastern United States, in a limited area on the Pacific Coast, in northern Idaho and in eastern Washington. Small males of these brownish ticks may be only 3 mm long, whereas engorged females may grow 5-13 mm or more. As shown in the pictorial key (see Fig. 17:9), the mouthparts (palpi and hypostome) are about as long as the basis capituli and the sides of the basis capituli are parallel. The scutum has diffuse whitish markings that may be faint or well-defined. The spiracular plates on the underside of the abdomen, behind and lateral to the fourth pair of coxae, have finer punctuation, called "goblets," than in *D. andersoni*. The American dog tick is the important vector of Rocky Mountain spotted fever in the eastern United States. It may also be involved in the transmission of tularemia and Q fever and may cause tick paralysis.[12,14,20,29,31,33,34]

The Rocky Mountain wood tick (*D. andersoni*) is similar to the American dog tick. However, it generally has more whitish markings on the scutum and the "goblets" on the spiracular plate are larger and less numerous than in the American dog tick. *D. andersoni* is the major vector of Rocky Mountain spotted fever and Colorado tick fever in the Rocky Mountain region. It may also be involved in the transmission of tularemia and Q fever and may cause tick paralysis.[14,16,31,34]

The lone star tick (*A. americanum*) has mouthparts much longer than the basis capituli. The female has a conspicuous whitish marking at the tip of the scutum, from which is derived the name "lone star tick" for the Lone Star State of Texas. This tick may be involved in the transmission of Rocky Mountain spotted fever and tularemia and may cause tick paralysis. Unlike the American dog tick and the Rocky Mountain wood tick, whose larvae and nymphs do not normally feed on humans, the larvae, nymphs and adults of the lone star tick all feed on humans. Many cases in which people were bitten by lone star tick larvae and nymphs and had severe itching and redness comparable to attacks of chiggers have been reported.[14,16,20]

The brown dog tick (*R. sanguineus*) rarely bites humans in North America. However, it is so commonly found on dogs and in buildings where both dogs and humans live that it should be mentioned. The brown dog tick is an entirely brownish tick which varies in size from 3-13 mm or more in engorged females. It has no whitish markings as do the three ticks previously mentioned. The sides of the basis capituli are angled and the palpi are about as long as the basis capituli. It is not uncommon to find larvae, nymphs and adults of the brown dog tick in a home, kennel or laboratory because this tick obtains its blood meal from dogs.

The northern deer tick (*I. dammini*) in the nymphal stage occasionally feeds on humans in the warm portion of the year. It may transmit the protozoan parasite, *Babesia microti*, that causes human babesiosis in northeastern United States and Canada and the spirochete (*Borrelia burgdorferi*) that causes Lyme disease. Several other related species of *Ixodes*, (*I. pacificus, I. ricinus* in Europe, and *I. scapularis*) may also be involved in the transmission of Lyme disease. Other species of *Ixodes* are of interest because of their role in the transmission of Powassan encephalitis. *Ixodes* ticks are easily recognizable because they have the anal groove curved in a U-shape in front of the anus, whereas other genera of hard ticks have the anal groove behind the anus or absent. Species identification of ticks in the genus *Ixodes* is difficult and the services of a specialist may be required.[10,19,30]

Soft ticks in the genus *Ornithodoros*, family Argasidae, transmit relapsing fever spirochetes (*Borrelia*) in limited areas of 13 western states. These ticks are dull-colored, leathery species with mouthparts on the ventral side of the body which are not visible from above. In the United States at least four species of *Ornithodoros* (*O. hermsi, O. parkeri, O. talaje* and *O. turicata*) are proven vectors of the *Borrelia* that cause tickborne relapsing fever. Tick identification requires specialized literature.[7,9] Many cases of relapsing fever have been contracted in rural cabins inhabited by small rodents such as chipmunks. As they slept, people were bitten by infected relapsing fever ticks which came from the rodent nests in the cabins.[16,32]

Clinical Manifestations

Tick bites:

The bite of a tick may be serious or annoying or both. Although soft ticks feed for only a few minutes or hours, some hard ticks remain fastened for several days. The bite is seldom painful, at least at the outset, but by the time a person notices it, the tick may have done considerable damage. Tick bites often heal slowly and itch intensely—leading to scratching, redness and sometimes secondary infections. Some patients have allergic manifestations with blister formation at the site of the bite.

Tick paralysis:

In some cases, if a tick remains fastened and engorged for several days, patients develop a peculiar flaccid paralysis beginning at the extremities and gradually spreading to other parts of the body. If the tick is not removed, the muscles of respiration are affected and the patient may die of respiratory insufficiency. If the tick is found and removed, the patient usually recovers spontaneously within a few hours— suggesting that the paralysis is caused by a toxic substance in the tick's saliva.

Epidemiology

People usually encounter ticks in grassy or brushy areas where ticks find their normal hosts among small animals such as field mice and dogs (*D. variabilis*), large mammals (*D. andersoni*) or deer (*I. dammini* and *A. americanum*). The rising incidence of Rocky Mountain spotted fever and Lyme disease in the eastern United States may be associated with the fact that more people live in the suburbs with brushy areas infested with ticks or may indicate that more dogs in homes are infested with ticks.[14,16,30] Most of the cases of relapsing fever in the West are associated with people sleeping in rustic cabins infested with small rodents and soft ticks (*Ornithodoros*).[14,16,32]

Control

If a tick becomes attached, the simplest method of removal is by slowly and steadily pulling with forceps so as not to break off the mouthparts and leave them in the wound. Sometimes touching the tick with the lighted end of a cigarette, or with a drop of chloroform, ether, benzene or vaseline will cause it to release its hold. An antiseptic should be applied to the tick wound. If the hands have touched the tick during removal, they should be washed thoroughly with soap and water, because the secretions may be infective. Tick control in the home usually requires treating both the animals and building with approved insecticides.[12,23,33]

MITES

Mites are tiny arthropods with eight legs in the adult stage, a saclike body and no antennae. Public health workers may be asked to identify mites which parasitize humans. Because of their tiny size, mites are usually difficult for the nonspecialist to identify. However, well-mounted specimens of the more common species may be identified by reference to the pictorial key (see Fig 17:10) or to illustrations in specialized literature.[1,7,14,20,22,24,27]

Scabies

The scabies, mange or itch mite (*S. scabiei*) causes a contagious skin disease called scabies, mange or "the itch." Diagnosis is confirmed by locating a mite in the epidermis in papules or vesicles or in a tiny burrow in the skin containing the female and her eggs. Some authorities recommend

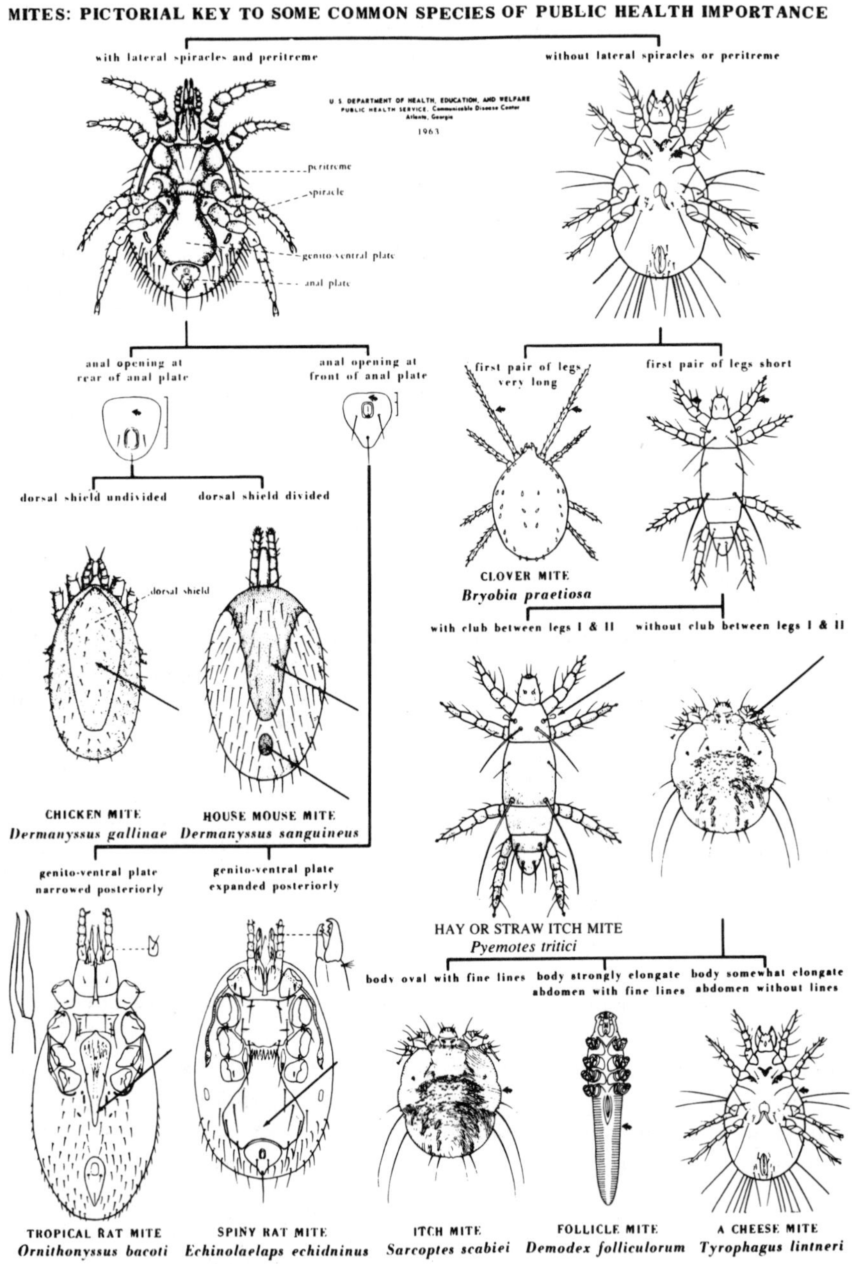

Figure 17:10. Pictorial key to some common species of mites of public health importance.[7]

cutting a tiny bit of skin from a papule with a safety razor blade or removing the mite from the burrow with a teasing needle, placing the skin or mite in a drop of sodium or potassium hydroxide on a microscope slide, applying a coverglass and examining with a microscope. Other workers prefer scraping the skin with a sharp scalpel with a drop of mineral oil or immersion oil, and then examining this liquid on a microscope slide with a microscope having a magnification of at least 40 diameters. The liquid may contain the mite, an egg or the dark mite feces from the burrow. Scabies mites are oval, saclike and less than 1 mm long. Females average 400 μm x 300 μm and males 250 μm x 150 μm. The mouthparts at the anterior end contain chelicerae and paired palpi. The body surface is finely wrinkled and has a number of rather conspicuous spines and many backward-projecting triangular scales on the dorsal surface. The legs are short and stocky, and the two anterior pairs are widely separated from the two posterior pairs. The anus is located at the posterior end (see Fig. 17:10).[1,7,14,20,22,24,27]

Clinical Manifestations

Scabies mites are found in papules around the finger webs, anterior surfaces of the wrists and elbows, anterior axillary folds, belt line, thighs and external genitalia of men, and nipples, abdomen and lower buttocks of women. Itching is intense, especially at night, but complications are rare except as lesions become secondarily infected by scratching. The mites are often present on a person for several days to weeks before itching begins.[24,27] Many workers have shown that the "scabies rash" has a characteristic distribution particularly in the armpits, waist, buttocks, inner thigh and ankle areas that does not correspond with the location of the mites, which are primarily in the webbing of the fingers and folds of the wrists. Most authorities believe that a true sensitization occurs with scabies. If the first infestation with scabies mites is eradicated with treatment, the rash will appear with reinfestation sometimes within a matter of hours and even with the skin penetration of only a single mite.[24,27]

Epidemiology

Since 60% of the mites have been reported in lesions on the hands or wrists, touching or shaking the hands of infested people appears to be one of the primary methods of transmitting scabies mites. Female mites are transferred from one person to another by simple contact and, less commonly, from undergarments or soiled bedclothes freshly contaminated by infested persons. Scabies is often considered to be a "family disease"—with mites transferred from husband to wife or from one child to another, particularly if more than one child sleeps in the same bed. The disease is frequently acquired from sexual contact. Outbreaks can also occur in hospitals or schools. Scabies epidemics occur particularly during times of war, poverty or social upheaval when facilities for bathing and laundering are limited. The disease occurs among certain groups who bathe or launder their clothing infrequently, but scabies is uncommon among persons who bathe daily and regularly launder clothing and bedding. If one member of a family or group of persons has scabies, it is advisable to treat all close contacts.[3,24,27]

Follicle Mites

Follicle mites (*Demodex*) are tiny, cigar-shaped creatures, 0.3-0.4 mm long with short stubby legs and annulate abdomens (see Fig 17:10). Nutting and others[24,27] have demonstrated that two species are normal inhabitants of the adult human face, the hair follicle mite (*Demodex folliculorum*) and the sebaceous gland mite (*Demodex brevis*). Probably in half of middle-aged adults, a few specimens of these mites can be pressed out of the facial skin, particularly the nose, with an ordinary microscope slide and demonstrated with a microscope. However, women who avoid daily washing to the face with soap and water and use cosmetics such as cleansing cream excessively, may have large populations of follicle mites, leading to blackheads and two conditions which dermatologists call "pityriasis folliculorum," a dry follicular scaling with slight burning sensation, and "rosacealike demodicosis," a mild to moderate erythema with superficial papules and dry scaling skin. Follicle mites are often an incidental finding in histologic sections of the facial skin.

Chiggers

Trombiculid mites infest grasses and bushes and their six-legged larvae i.e., chiggers (red bugs, harvest mites) may attack humans. *Eutrombicula alfreddugesi* is one of the most common chiggers attacking humans in the United States. The chiggers attach to the skin, usually in areas where clothing is tight, such as the part of the ankle at the top of the sock or skin touched by belts and elastic bands. Sensitive individuals react to the secretions of the chiggers with swollen, reddish, itching areas at the sites of attachment which persist for days. Excoriations may become secondarily infected. The eight-legged nymphs and adults are nonparasitic vegetarians. In Asia, *Trombicula akamushi* and *T. deliensis* are vectors of rickettsiae causing scrub typhus.

Hay or Straw Itch Mite

The hay or straw itch mite (*Pyemotes tritici*) is normally a parasite of grain-infesting Lepidoptera (moth larvae) and Coleoptera (beetles). Humans are bitten by female mites while sleeping on straw or straw mats, working in infested fields or handling stored grain products infested with insect larvae that the mites attack. These persons may have a severe reddish rash or dermatitis with secondary infection due to scratching which is called hay, straw or grain itch. Other species of *Pyemotes* which attack wood borers may cause a similar dermatitis in person crawling under houses. Recent studies of *Pyemotes* mites suggest that many cases of hay, straw or grain itch attributed to *P. ventricosus* may actually have been caused by *P. tritici*.[27] Adult mites have a characteristic club-shaped hair between the first and second pairs of legs (see Fig. 17:10). In gravid females, the abdomen becomes enormously enlarged, resembling a small pearl.

Rodent and Bird Mites

Mites in the family Dermanyssidae are normally parasites of rodents and birds but may attack people, causing dermatitis. People raising domestic fowl may suffer from attacks by the chicken mite (*Dermanyssus gallinae*), northern fowl mite (*Ornithonyssus sylviarum*) or tropical bird mite (*Ornithonyssus bursa*). These same species are often parasites of birds nesting in buildings, such as sparrows, starlings or pigeons, and may leave a bird nest and attack people.

The tropical rat mite (*Ornithonyssus bacoti*) is a common parasite of rats, particularly in the southern United States. There are many reports of this species attacking humans when rats are killed in buildings. The mites leave the dead animal and seek a blood meal from the nearest warmblooded animal, people.

The house mouse mite (*Liponyssoides sanguineus*, sometimes reported as *Dermanyssus* or *Allodermanyssus*) is a parasite of house mice. It is sometimes abundant in buildings in the northeastern United States and may transmit the rickettsiae causing rickettsialpox from mice to humans.

Grain and Cheese Mites

Grain and cheese mites are often found in tremendous numbers in flour, grain, cheese and dried fruits, particularly when humidity and temperature are high. Some of these mites, such as the cheese mite (*Acarus siro*), may cause dermatitis among persons who handle these infested foods.

House Dust Mites

House dust mites (*Dermatophagoides pteronyssinus*, *D. farinae* and *Euroglyphus maynei*) have been demonstrated as allergenic agents of house dust. House dust mite populations are highest during the warm, humid summer months, particularly in floor dust of bedrooms and family rooms and in dust in couches and overstuffed furniture. Recent research suggests that the allergens are located in mite exoskeletons and excreta as well as in the actual mites. This observation may explain why killing the mites with pesticides fails to offer immediate relief to house dust allergy symptoms. Suggested measures to control house dust allergy in humans include desensitizing patients with mite extracts and keeping general house dust levels low with frequent cleaning.[27]

SPIDERS

Spiders in the order Araneida are easily identified by a number of characters: 8 legs, no antennae, body divided into 2 regions (cephalothorax and abdomen) joined by a narrow pedicel and unsegmented abdomen with spinnerets at the tip. Spiders kill their prey by biting and injecting venom. Therefore, it is not surprising that there are reports of some 50 species of spiders biting humans, usually in self-defense.[12] Usually the bite is no worse that the sting of a bee or wasp, with some redness, swelling and pain for a few hours, but some cases can be more severe. The bite of 2 spiders, the black widow spider and the brown recluse spider (and their relatives), can cause serious illness or death.

The black widow spider (*Latrodectus mactans* and related species) is easily identified in all stages by the orange or reddish hourglass marking on the ventral side of the abdomen (see Fig. 17:11). Females vary from 5-13.5 mm in body length, with a globose, shiny black abdomen with one or more reddish dots on the dorsal side and all-black legs. Males and immatures are smaller than females and have additional reddish or pale spots on the abdomen and banded legs.

The brown recluse, violin or fiddleback spider (*Loxosceles reclusa*) is a brownish species with 2 characteristics in combination: a dark fiddle or violin marking on the tan or brownish cephalothorax and 6 eyes arranged in 3 pairs forming a semicircle (Fig. 17:11). The body is 8-9 mm long.

Clinical Manifestations

The venom of *L. mactans* injected by the female is a neurotoxin and can cause a burning or stinging sensation at the site of the bite, severe cramplike pain within an hour, a rigid "boardlike abdomen" and severe illness for a day or more. Most cases recover spontaneously in 2 or 3 days, but occasionally the bite can be fatal to children or the elderly. Many cases are treated with antivenom and calcium gluconate. In the United States, the venom of 4 other species of *Latrodectus: L. bishopi, L. geometricus, L. hesperus* and *L. variolus*, can also cause illness.

The venom injected by the female or male *L. reclusa* is a necrotoxin

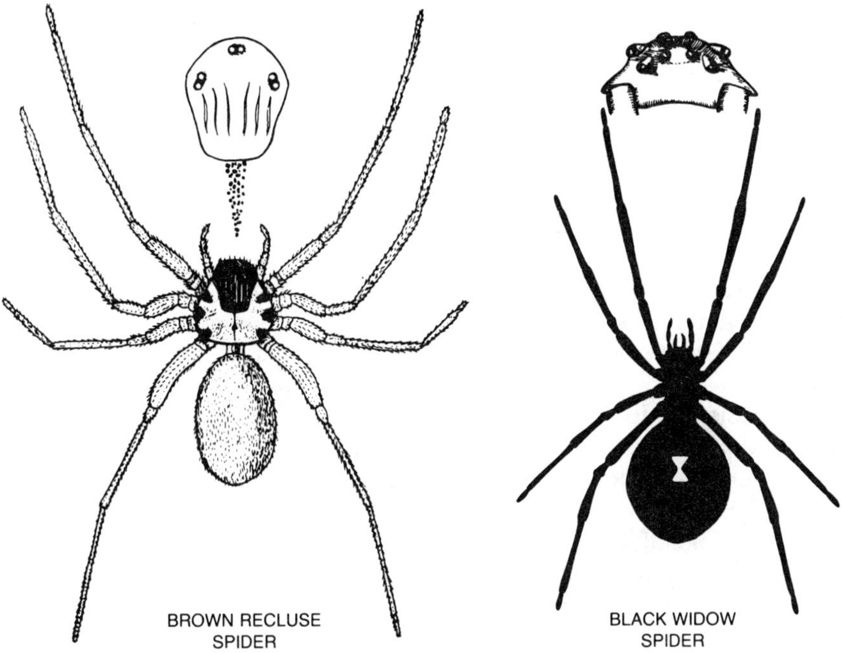

Figure 17:11. Brown recluse and black widow spiders.[7]

which causes necrotic lesions, with deep tissue damage, that extend for several days after the initial bite, heal very slowly (often 6-8 weeks or longer) and leave disfiguring scars. This is the cutaneous form of loxoscelism. The cutaneovisceral form of the disease may involve the kidneys and liver, with passage of haemoglobin and albumin in the urine and renal failure that may occasionally end in death. The venom of three other species of *Loxosceles: L. arizonica, L. deserta* and *L. laeta*, can also cause serious illness and leave ugly scars. Treatment with corticosteroids or surgical excision of the bite site may be recommended for some victims.

Epidemiology

The black widow spider normally builds a coarse web in areas such as dark corners of basements, storerooms, garages or privies, under stones, in woodpiles or in water meter boxes along the edges of streets. The female usually does not bite without provocation but will defend herself or her eggs when cornered. Most bites occur in outdoor privies.

The brown recluse spider spins an irregular, whitish web in cupboards, bathrooms, outhouses and tornado cellars, in stumps, under rocks or in piles of debris. These spiders are shy and retiring (as the species name *reclusa* implies) but do bite humans when cornered. Most bites occur on the hands, arms or legs when people are putting on clothing or rolling over in bed.[4,7,14,18,20,23,29,33]

DELUSORY PARASITOSIS

Laboratory personnel are frequently the "court of last appeal" for the person with illusions or delusions of parasitosis, sometimes called "delusory parasitosis"—an emotional disorder in which the patient has an unwarranted belief that live organisms are present on or in his body; or entomophobia, the irrational, unfounded fear of insects or other arthropods. Examples include: the secretary or entire secretarial staff who start itching or scratching at the sight of a pest control operator and believe the cause is "paper fleas," insects or mites never found in office files; the elderly, often senile, patients who bring in a collection of dirt, lint or debris and believe it contains insects which are biting or annoying them; or the people who believe every spider is a black widow or brown recluse spider after seeing or hearing reports of these spiders attacking someone, often hundreds of miles away.

Laboratory personnel can help allay these fears by allowing such patients to look through a microscope at actual insects, mites or spiders, showing them illustrations of these arthropods, or by letting the patients examine with a microscope their sample of debris and asking them if they see an arthropod, or a segmented leg or antenna, rather than dirt or lint. Laboratory personnel can also help by allowing such patients, frequently sensitive, well-educated people, to read articles about "delusory parasitosis" in well-illustrated books.[12,23]

REFERENCES

1. Baker EW, Evans TM, Gould DJ, Hull WB, Keegan HL. A manual of parasitic mites of medical or economic importance. New York: National Pest Control Association, 1956:170.
2. Beaver PC, Jung RC, Cupp EW. Clinical parasitology. 9th ed. Philadelphia: Lea and Febiger, 1984:825.
3. Benenson AS. Control of communicable diseases in man. 14th ed. Washington, DC: American Public Health Association, 1985:485.
4. Borror DJ, DeLong DM, Triplehorn CA. An introduction to the study of insects. 5th ed. Philadelphia: Saunders College Publishing, 1981:827.
5. Brown HW, Neva FA. Basic clinical parasitology. 5th ed. Norwalk, CT: Appleton-Century-Crofts, 1983:339.
6. Busvine, JR. Insects and hygiene. 3rd ed. London: Chapman and Hall, 1980:568.
7. Communicable Disease Center. Pictorial keys: arthropods, reptiles, birds and mammals of public health significance. Atlanta: US Public Health Service, CDC, 1967:192.
8. Cooley RA. The genera *Dermacentor* and *Otocentor* (Ixodidae) in the United States, with studies in variation. Nat Inst Health Bull 1938;171:89.

9. Cooley RA, Kohls GM. The Argasidae of North America, Central America and Cuba. Notre Dame, IN: University Press, American Midland Naturalist Monograph No. 1, 1944:152.
10. Cooley RA, Kohls GM. The genus *Ixodes* in North America. Nat Inst Health Bull 1945;184:246.
11. Cornwell PB. The cockroach, a laboratory insect and an industrial pest. London: Rentokil Library, Hutchinson, 1968:391.
12. Ebeling, W. Urban entomology. Richmond, CA: University California, Division Agriculture Sciences, 1975:695.
13. Frazier CA, Brown PK. Insects and allergy and what to do about them. Norman, OK: University of Oklahoma Press, 1980:350.
14. Harwood RF, James MT. Entomology in human and animal health. 7th ed. New York: MacMillan, 1979:548.
15. Holland GP. The Siphonaptera of Canada. Kamloops, Br Col: Can Dept Agri Tech Bull 1949;70:306.
16. Hoogstraal H. Changing patterns of tickborne diseases in modern society. Ann Rev Entomol 1981;26:75-99.
17. James MT. The flies that cause myiasis in man. USDA Misc Pub 1948;631:175.
18. Kaston BJ. How to know the spiders. 3rd ed. Dubuque, IA: WC Brown, 1978:272.
19. Keirans JE, Clifford CM. The genus *Ixodes* in the United States: a scanning electron microscope study and key to the adults. J Med Entomol 1978;(Suppl)2:149.
20. Kettle DS. Medical and veterinary entomology. New York: Wiley, 1982:658.
21. Kim KC, Pratt HD, Stojanovich CJ. The sucking lice of North America. University Park, PA: Pennsylvania State Press, 1986:241.
22. Kratz GW. A manual of acarology. 2nd ed. Corvallis, OR: Oregon State University Book Store, 1978:509.
23. Mallis A. Handbook of pest control. 6th ed. Cleveland, OH: Franzak and Foster, 1982:1101.
24. Orkin W, Maibach HI, eds. Cutaneous infestations and insect bites. New York: Dekker, 1985:321.
25. Osgood SB, Jellison WB, Kohls GM. An episode of pseudopediculosis. J Parasitol 1961;47:985-6.
26. Pan-American Health Organization. Proceedings of the international symposium on the control of lice and louse-borne diseases. Washington, DC: Pan-American Health Organization, 1973: Scientific Publication 263:311.
27. Parish LC, Nutting WB, Schwartzman RM, eds. Cutaneous infestations of man and animal. New York: Prager, 1983:392.
28. Pratt HD, Littig KS. Lice of public health importance and their control. Atlanta: US Public Health Service, CDC, 1973:27.
29. Smith KGV, ed. Insects and other arthropods of medical importance. London: British Museum (Natural History), 1973:561.
30. Spielman A, Wilson ML, Levine JF, Piesman J. Ecology of *Ixodes dammini*-borne human babesiosis and Lyme disease. Ann Rev Entomol 1985;30:439-60.
31. Strickland GT, ed. Hunter's tropical medicine. 6th ed. Philadelphia, PA: Saunders, 1984:1057.
32. Thompson RS, Burgdorfer W, Russell R, Francis BJ. Outbreak of tickborne relapsing fever in Spokane, WA. JAMA 1969;210:1045-50.
33. Truman LC, Bennett GW, Butts WL. Scientific guide to pest control operations. 3rd ed. Cleveland, OH: Harvest Publishing, 1976:276.
34. Yunker CE, Keirans JE, Clifford CM, Easton ER. *Dermacentor* ticks (Acari: Ixodoidea: Ixodidae) of the new world: a scanning electron microscope atlas. Proc Entomol Soc Wash 1986;88:609-27.

CHAPTER 18

CARE AND USE OF THE MICROSCOPE

Anthony D. Oldham, B.Sc.

REQUIREMENTS FOR GOOD MICROSCOPY

1. A microscope equipped with a light source adequate at the highest magnification, yet capable of having its light level reduced so it will not blind at low power.
2. Most users require at least 3 objectives, commonly 10X, 40X and 100X; or 4 objectives, 4X, 10X, 40X and 100X. Other magnifications are often seen, such as 3.5X instead of 4X, 43X or 45X instead of 40X, and 97X instead of 100X. These variations are only significant if calibrating an eyepiece reticle. It is preferable that the objectives be those designed for use on that microscope. Objectives of the wrong type or of a different make may reduce the efficiency of the microscope.
3. A pair of eyepieces, sometimes called oculars, available in a variety of magnifications, are also required. Generally, they will be 10X which, with the objectives noted in (2), will give total magnifications of about 40X, 100X, 400X and 1000X. These eyepieces should also be of the right make and type for the particular microscope. The user of a microscope with a miscellany of objectives and eyepieces of various types will have problems.
4. Easily adjusted coarse and fine focuses and a smoothly moving stage are essential.
5. A brightfield condenser that can take full advantage of the best objective's light-gathering power and yet allow the use of the 3.5X or 4X objective with a full field of view is a necessity. In some cases this calls for an auxiliary lens.
6. Some microscopes have a device that prevents the objective from breaking a slide, but many do not. Some objectives are spring-loaded for the same reason. If neither safety feature exists, one must exercise great care when first setting up the microscope.

7. In virtually all brightfield condensers there is a diaphragm whose purpose is to produce "contrast" in the specimen being viewed. Skillful use of this diaphragm enables the user to reach a compromise between resolution and contrast. A field-stop diaphragm in the base illuminator which allows Koehler illumination to be set can improve the performance of the microscope greatly, but if that diaphragm is not present, the Koehler effect can still be approximated and reasonable performance maintained.

RESOLUTION

One of the most important formulae concerns resolution. Figure 18:1 illustrates resolution as the ability to distinguish space between objects. The dots on the left are single ones arranged in groups, but as you look to the right it is more difficult to be certain the dots are separated. At a distance, it is even less certain. When the space between the dots cannot be resolved and they blur together, the limit of resolution has been exceeded.

The ability of a microscope to resolve the space between features is determined by this formula:

$$\text{RESOLUTION} = \frac{\text{Wavelength of light from 400-700 nm (visible)}}{\text{N.A. of objective} + \text{N.A. of condenser}}$$

N.A., or numerical aperture, will be a very small number and for ease of solving this equation, N.A. can be considered 1. Several possible resolution values can be calculated, according to the wavelength of the light.

$$\text{RESOLUTION} = \frac{700\text{ nm}}{1 + 1} = 350\text{ nm } (0.35\ \mu\text{m})$$

That number, 0.35 μm, is the smallest space between two features that can be resolved when the light is red (700 nm) and the N.A. of the objective and the condenser is only 1. If the wavelength of the light is lowered to 400, then the result is:

$$\text{RESOLUTION} = \frac{400\text{ nm}}{1 + 1} = 200\text{ nm } (0.20\ \mu\text{m})$$

There is now almost twice the resolution because the shorter wavelength of 400 nm (blue light), was used. The smallest space which can now be distinguished between two features is down from 0.35 μm to 0.20 μm.

Figure 18:1. Resolution as ability to resolve space between features.

Wavelength

The theory is that the shorter the wavelength of light, the better the resolution. Wavelength can be shortened by placing a blue filter in the lightpath between the illuminator and the condenser. The light coming from the illuminator contains all the visible and, sometimes, the ultraviolet (UV), wavelengths. The blue filter cuts out much of the red light, so it is a key to higher resolution.

Something else affects the wavelength of light from an illuminator. When the power is turned down, the lamp filament cools and tends more to the red end of the spectrum. If the power is turned up, the filament becomes hotter and gives more blue light. The blue filter takes care of much of the shift, but to achieve better resolution, the light source should be kept at high brightness and neutral density filters used to lower the light to an acceptable level. However, the bulb will burn out more quickly if the power is kept high. In addition, the next user may be temporarily blinded by the high intensity if the neutral density filter is not in the lightpath. For these reasons, the power should always be turned down when leaving the microscope.

Numerical Aperture

The formula for resolution shows that raising the value of the numerical apertures will increase resolution even further. The simplest description of numerical aperture is "light-gathering power." Visualize an objective into which a bundle of light rays is entering and assume that rays could be added one at a time until eventually not one more would enter. Then the limit of that objective's light-gathering power would have been reached. One less ray and theoretically the full light-gathering power of the objective has not quite been reached. Numerical aperture is a measure of the light-gathering power of a lens system and a means of comparing one system with another. The higher the numerical aperture, the greater the light-gathering power. Figure 18:2 illustrates a cone of light entering an objective. The formula for N.A. is:

$$\text{N.A.} = n \times \text{Sine } A$$

where n is the refractive index of the medium through which the light rays pass. For example, $n = 1.0$ for air and $n = 1.515$ for immersion oil. A is half the aperture angle as shown in Figure 18:2. It is obvious that the N.A. and, therefore, the light-gathering power increases as the angle increases. In Figure 18:2 all the objectives are the dry types so the calculations are:

$$\text{N.A.} = 1 \times \text{Sine } 6 = 0.10$$
$$\text{N.A.} = 1 \times \text{Sine } 13 = 0.22$$
$$\text{N.A.} = 1 \times \text{Sine } 40 = 0.65$$

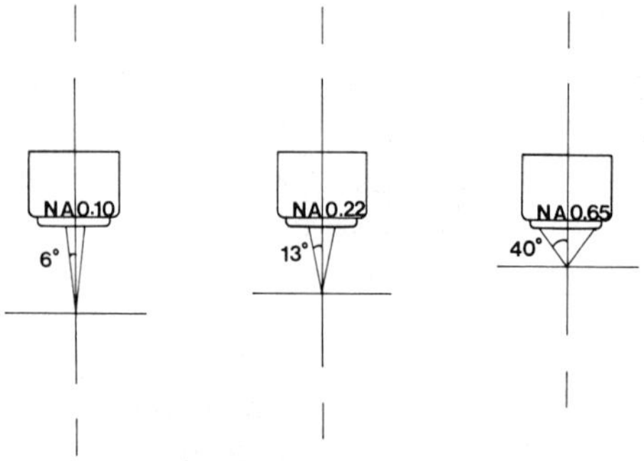

Figure 18:2. Cones of light entering three objectives showing effect of aperture angles on numerical aperture values.

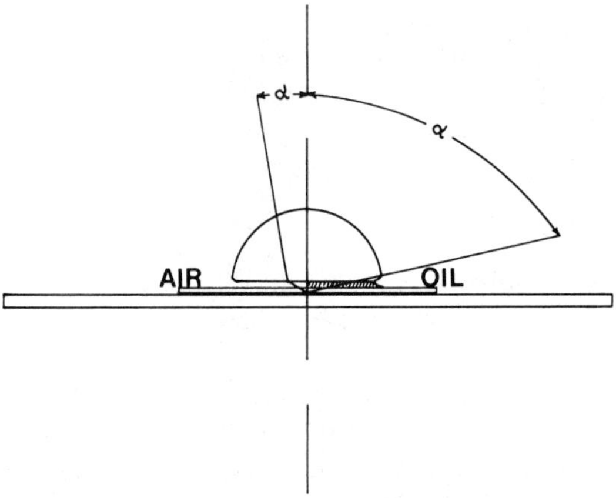

Figure 18:3. The effect of immersion oil on numerical aperture.

It can be seen from these calculations that the N.A. can only be as large as the sine of the angle A so long as the light is passing through air. This angle can never be greater than 90 degrees, as shown in Figure 18:3. The Sine of 90 degrees is 1.

Thus the numerical aperture cannot be greater than 1 in air. But when immersion oil is placed between the specimen and the front lens of an objective designed for that purpose, the refractive index of air will no longer be a restriction. Instead, as the formula and Figure 18:3 show, the angle will increase greatly due to the presence of the oil, which has roughly the same refractive index as the glass of the objective's front lens. Therefore, very high resolution can be obtained with blue light, a high N.A. objective and a high N.A. condenser. In the very best high-power (63-100X) oil immersion objectives and condensers the N.A. can be as high as 1.4. It can be as low as about 0.08 with a 2.5X dry objective. There are lower power objectives with even smaller N.A. values but they are not common.

To take full advantage of the high N.A. of an objective, the condenser should have an N.A. equal to that of the objective. Furthermore, if the condenser has an N.A. greater than 1, then the rules discussed here will apply; the condenser will only give its highest light output if there is immersion oil between its top lens and the underside of the slide. But the real determinant of any numerical aperture is the sample itself. Very high N.A. may only make a particular specimen look washed out, thus forcing a reduction in N.A. to achieve a proper compromise between resolution and contrast.

There are other substances which have different refractive indices from air and immersion oil. Water, for instance, has a refractive index of 1.333, that of glycerin is 1.455 and that of methylene iodide is 1.740. These are not used as commonly as immersion oil but they do serve for special applications. Although oil is not usually placed on the condenser's top lens for routine work, it is an available resource when striving for ultimate resolution.

MECHANICAL TUBE LENGTH

An extrememly important part of microscopy theory is the concept of "Mechanical Tube Length." This refers, in the mechanical sense, to the distance from the shoulder of the objective to the shoulder of the eyepieces and in virtually all modern microscopes is 160 mm. The parfocality of the microscope's various objectives depends on the tube length remaining constant. But the tube length will vary if users have different interpupillary distances. Each objective puts its aerial image in exactly the same plane. If the tube length is 160 mm, the unadjusted focus of the eyepieces will be in exactly the right place. However, if the interpupillary distance is changed, causing the tube length to increase or decrease as shown by the dotted lines of Figure 18:4, the eyepiece focus will have to be changed to compensate for this change in tube length. Some microscopes make this adjustment automatically and others, called Siedentopf binoculars after the Zeiss scientist who invented them, are designed so there is no change in tube length as interpupillary distance is changed. Where adjustment is needed, the microscope will have a scale engraved on the binocular, usually running from 55 to 75 mm. Each user sets the correct distance between the eyepieces for his or her eyes, reads the number from the scale and sets each of the eyepieces to the same number. Now, providing the objectives in the microscope are the parfocalized set supplied by the manufacturer, there will be parfocality as the user goes from one power to another, with only small shifts of the fine focus required to bring the image to maximum sharpness.

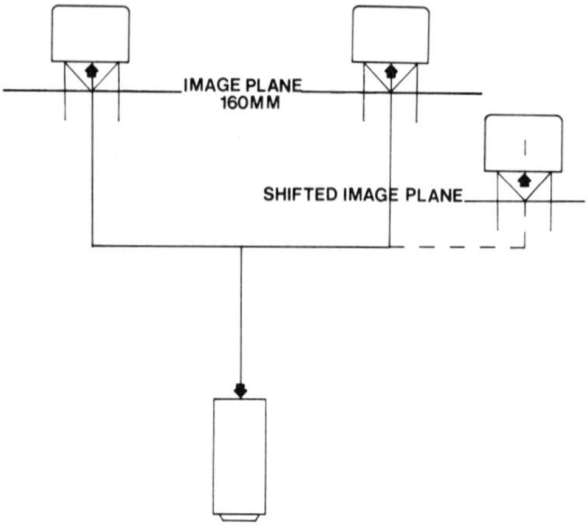

Figure 18:4. The effect of interpupillary distance on tube length and image plane.

Infinity Corrected Optics

All good microscopes have "Infinity Corrected Optics." Infinity-corrected refers to the fact that the image created by the objective is focused at infinity and not at the tube length. This means that the imaging rays leaving the objective are parallel, thus allowing different optical elements to be inserted while maintaining consistent image quality. These elements may be polarizers, wave plates, vertical illuminators, magnification changers, etc. A tube lens is necessary to bring the parallel beams into focus again, so the eyepieces can magnify this image as they do in a microscope equipped with objectives corrected for a "finite" tube length. The presence of this tube lens means that one cannot casually swap eyepieces or objectives between microscopes without causing difficulties.

COVERGLASS EFFECT

The coverglass plays a substantial role in the theory of microscopy. The optical calculations for many objectives are based on the presence of a coverglass of close to 0.17-0.18 mm thickness. When there is no coverglass, or the coverglass differs from the prescribed value in thickness, there will be some slight loss of image quality due to under or over correction of spherical aberration. In the absence of a coverglass, the loss of image quality will be particularly noticeable with the high-dry objective. If the loss of image quality is significant, objectives designed to work without a coverglass can be used, as well as objectives with a special compensating collar which can be adjusted for maximum image quality with coverglasses varying in thickness from 0.12-0.22 mm.

EMPTY MAGNIFICATION

The whole raison d'etre of a microscope is to magnify an image to the optimum extent. Normally, it is considered wise to limit magnification to <1000 times the numerical aperture. With a microscope that has a high-power objective of 100X and a numerical aperture of 1.4, the total magnification should be limited to 1400X. If 15X eyepieces are used, the total magnification would be 1500X, but one would be using so-called "empty magnification." This refers to the fact that beyond 1000X N.A. the differentiation effects break up the image and may lead to seeing image structures which are not real. When merely measuring or counting things without concern for structure, it is possible to work in the empty magnification range.

OPTICAL ABERRATIONS AND TYPES OF OBJECTIVES

Chromatic Aberration

There are many different optical aberrations that the manufacturer must overcome to produce an objective which will perform almost perfectly. Chromatic aberration refers to the effect of a lens on light of different wavelengths, causing each to focus at different points after passing through the lens (Fig. 18:5). A lens which corrects the focus for two of the three basic colors is called an achromat. To correct for all three colors requires

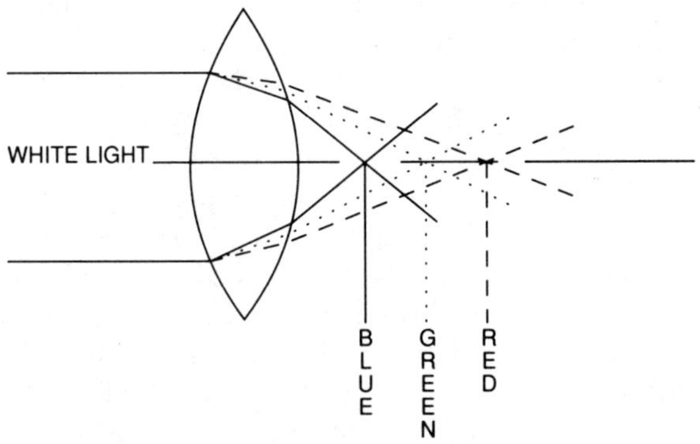

Figure 18:5. The effect of a lens on light of differing wavelengths.

a much more complex and therefore more expensive objective called an apochromat.

Field Curvature

Curvature of field is easy to see. The effect is that the center of the field is in sharp focus while the periphery is not. Adjusting the focus can reverse the effect, making the periphery sharp while the center is fuzzy. Objectives which correct this are called plan objectives or simply flat-field objectives. The plan correction can be applied to achromats, creating planachromats, or to apochromats, creating planapochromats, which are among the finest objectives available, since they are corrected for essentially all other aberrations as well.

Compensating Eyepieces

Some aberrations are extremely difficult to correct in the objective itself and it has been a practice for many years to make so-called compensating eyepieces which correct, for example, lateral chromatic aberrations. The compensating eyepieces must be used with a microscope designed in this way; the use of just any eyepiece on such a unit will result in loss of image quality.

No lens system is absolutely perfect nor is it likely that any ever will be, but the best objectives made, corrected or partially corrected for most aberrations, work so well that only a true expert can see the remaining error. The residual error is, in a moderately priced system of recent vintage, less than most can detect.

KOEHLER ILLUMINATION

"Koehlering" is the most widely accepted way of setting up the illumination path to take full advantage of the numerical aperture of the optics, while reducing the inherent disadvantages of the light source to a minimum. This is not a new discovery, but was invented by Dr. August Koehler of Zeiss in 1893. Although the system is moderately complex to describe, its effects are simple and can be explained in simple terms. If the condenser is set to the appropriate height, it puts a greatly enlarged image of the lamp filament into the specimen plane. This is called critical illumination and is used in fluorescence work. For brightfield microscopy, however, it has the disadvantage that, although the light intensity is maximized, the filament itself can be seen as alternating light and dark bars, often with quite a lot of color. This can be eliminated by using a ground glass diffuser in the light source, but this reduces the available light as well. The Koehler system places the condenser at such a height that the field-stop diaphragm and not the filament is imaged into the specimen plane. This gives virtually all the available light as even illumination without an image of the filament. Furthermore, the field-stop diaphragm is used to limit the field of light to the size of the eyepiece field-stop. This eliminates scattered light which would otherwise reduce contrast. In addition, the light in the specimen plane is parallel, thus reducing other optical distortions. Lastly, the aperture or illumination angle of the microscope as well as the contrast are controlled simply by the condenser diaphragm.

HOW TO SET UP AND ADJUST A MICROSCOPE FOR OPTIMUM PERFORMANCE

These instructions apply to almost any microscope. Numbered instructions are steps common to all instruments. Instructions identified by both number and letter deal with variations on different makes and models of microscopes.

1. Find a specimen slide that has a coverslip and clean it with lens tissue dampened with lens cleaning fluid (Sparkle, Kodak lens fluid or any other similar commercial lens cleaner).
2. Place the slide on the stage, being sure it is right side up.
3. If you know your interpupillary distance and settings for difference in visual acuity, set them.
4. Turn on the light source and set it to a low light level.

4a. If the light source has only a simple on/off switch, there is no way to increase the light level, but the level can be decreased with neutral density filters.

5. If there is a base diaphragm, open it all the way.

5a. If there is no such diaphragm in the base illuminator, you will not be able to set up true Koehler illumination but you can approximate it (see items 8 and 12a).

6. Open the condenser diaphragm all the way.
7. If the instrument has an auxiliary lens, usually located under the condenser, swing it out of the light path.

Note: The purpose of the auxiliary lens is to create a full field of view with low-power objectives. It should generally be out of the light path when using objectives of 10X or higher, because its presence will slightly decrease the quality of the higher power images. Some auxiliary lenses are fixed and cannot be removed; some can be unplugged, some are slide-out types, and some are swing-out but do not have to be removed from the light path except for the most demanding high-power work. In the latter case, try looking at a detail in the slide under highest power with and without the auxiliary lens; if there is no appreciable difference, the lens may be left in place.

8. Rack the condenser all the way up.
9. Rotate the nosepiece to bring the 10X objective into position. If there is no 10X, use the most similar one, such as 8X, 16X or 20X. Much higher or lower power objectives will make "Koehlering" more difficult.
10. If the condenser has a swing-out top lens, it generally must be in for setting up. Its normal purpose is to increase the numerical aperture of the condenser when using the high-power objectives and it is usually out for low-power work.
11. Focus the 10X objective on the specimen, using the coarse and fine focus knobs.
12. Partially close the field-stop diaphragm (in the base) while looking into the binocular. The diaphragm should appear as a dark, and perhaps fuzzy, ring around the field of the specimen.

12a. If there is no diaphragm, the condenser should normally be left all the way up or moved down about a millimeter, if doing so gives a clearer illumination.

13. Depending on make and model, the swing-out top lens may improve or deteriorate a 10X image. Try both ways to determine which is best.
14. Now set interpupillary distance and tube length to compensate for the difference in visual acuity between left and right eyes if this was not done at step 3. If done then, fine tune the settings if necessary.

If the microscope has a scale on the binocular, read and memorize your interpupillary distance. The scale is usually graduated from 55 to 75 mm. To compensate for the tube length error, set each eyepiece to the same number if they have a similar scale engraved on the barrels. If not, one of the following will apply to this microscope:

14a. If the microscope has a fixed right or left eyepiece, cover the other eyepiece with a piece of paper and focus on the specimen using one eye only. Then cover the fixed eyepiece with the paper and adjust the adjustable eyepiece for sharp focus using the other eye. Using a piece of paper enables you to keep both eyes open. Try not to close one eye when doing these adjustments because tightly closing an eye will temporarily deform it and change its focus. It will take some time for the eye to return to normal.

14b. If the microscope has two adjustable eyepieces, they will probably be graduated from −5 through zero to +5. If your eyes have significant performance difference, set the eyepiece on the better eye to zero, cover the other eyepiece with a piece of paper and focus your slide sharply with the coarse and fine focus knobs using this eye. Then change the paper to the other side and adjust the second eyepiece for the other eye.

15. Continue the Koehlering at this point by raising and lowering the condenser slightly until the diaphragm edge is as sharp as possible. The slide must also be in sharp focus; if not, adjust the fine focus until it is.
16. Look at Figure 18:6. If your diaphragm image is not centered, as shown in Figure 18:6B, center it with the pair of centering screws found in one of these locations: left and right front of base illuminator, left and right of microscope base, left and right front of condenser or left and right rear of condenser. Different microscopes have their own systems and even different models from the same manufacturer can be radically different. Now open the centered diaphragm until it just disappears—do not open it farther.

Some auxiliary lenses are centerable and if so, will have knurled centering screws on the carrier. If the microscope has this, center the lens after centering the diaphragm image in step 16. Then the diaphragm image will remain centered whether the auxiliary lens is in or out.

For most purposes the settings which have just been made should remain undisturbed for all work at 10X and above. The base diaphragm may have to be opened when a 4X objective is used if inserting the auxiliary lens of the condenser does not provide a full field of view. Do not disturb the condenser height, however, and if opening the diaphragm, reset the Koehler illumination when moving to a higher power again. A person taking photomicrographs may prefer to Koehler for each individual objective.

17. Having Koehlered the microscope, close the contrast diaphragm slowly while watching the effect of varying contrast. Set the diaphragm to give optimum contrast. If you cannot decide what is best, take out one of the eyepieces and look into the tube. You will be able to see the diaphragm; set it as shown in Figure 18:7.

Remember the blue filter and that it is better to use a higher light level, since blue means higher resolution, and reduce the brightness with neutral

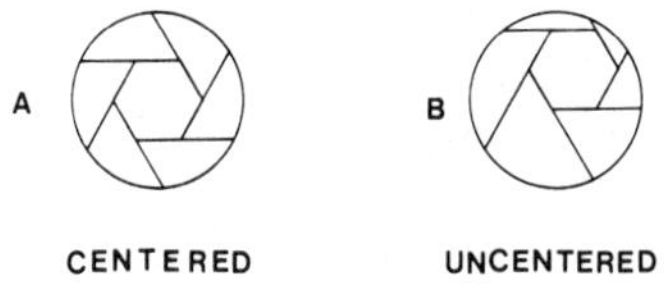

Figure 18:6. Centering of the condenser diaphragm. A = centered; B = uncentered.

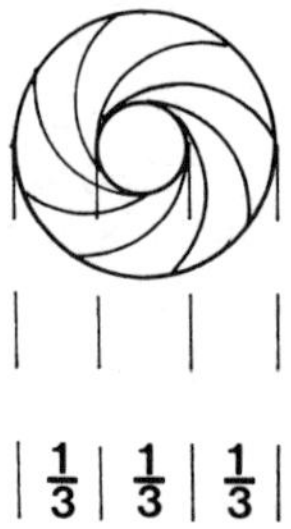

Figure 18:7. Proper setting of the diaphragm in "Koehlering."

density filters. Above all, remember that much of the advantage of Koehler illumination is wasted unless microscope optics are scrupulously clean.

TROUBLESHOOTING

This section deals with some of the common problems encountered by microscopists, how they may recognize the causes and eliminate the problems.

Problem:

Light source flickers on and off.

There are several possible causes for this problem. They include loose plug at wall sòcket; loose plug at transformer or power supply; loose or improperly installed light bulb; bulb about to burn out; defective switch; and frayed or broken wiring. The last two can be dangerous and should be repaired as soon as possible.

Problem:

Specimen is unevenly illuminated.

Provision is made on many microscopes for centering the light source. The ways in which it is done for different makes and models are so numerous that it is not feasible to describe them all. There are adjustment screws or adjustment levers which smoothly move the lamp or a mirror for centering. There are Allen screw adjustments not always labelled with their intended use. Sometimes the whole lamp socket is to be rotated, moved in and out, and then locked, often with a single screw. Hold a piece of paper over the base illuminator to align the lamp for the most even illumination that can be achieved. Some microscopes allow removal of the illuminator so it can be pointed at a wall a few feet away for adjustment of the centering and focus of the filament. Only fine tuning of the centering is needed when the illuminator is reinserted. Some units are prealigned and require no adjustment.

Problem:

The field stop (illuminator) diaphragm cannot be imaged when trying to Koehler the microscope.

The most common cause for this problem is the presence of a diffusing filter (ground glass) between the base illuminator and the condenser. If this filter is temporarily removed, the microscope can be Koehlered. If the illumination is still somewhat uneven the filter can be reinserted after Koehlering. The auxiliary lens is also a common cause of this problem. Swing it out when Koehlering.

Problem:

The fine focus control does not work.

The actual cause of this problem depends on the make of the microscope. Basically, most microscopes require that the fine focus be in the center of its range when just at the focal point of one of the objectives, preferably the 40X. To solve the problem, you need to know whether the fine focus has a limited travel (from half a turn to as many as 20 full turns) or has unlimited travel. Many microscopes have fine focus markers as shown in Figure 18:8A or 18:8C, depending on whether the focus is at one extreme or the other, or is at the center of its range as shown in Figure 18:8B. To set this type of focus, operate the fine control to bring the marker to the center as in Figure 18:8B. Focus the specimen with the coarse focus control only. Now there will be enough range on the fine focus for all objectives.

Some microscopes have a system where the coarse focus is brought to a preset stop. The fine focus is then forced to be in the center of its range at the focal point. If it is not, bring the coarse focus to its stop, set the fine focus to mid-range by counting the number of turns to go from one extreme

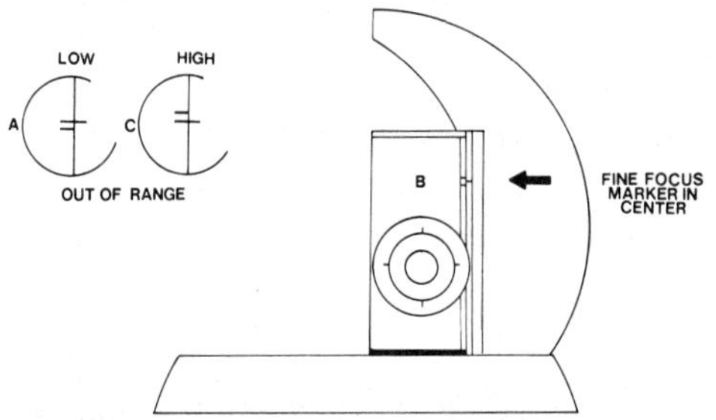

Figure 18:8. Adjustment of the fine focus control. A = out of range, low; B = centered; C = out of range, high.

to the other and then go back to the mid-point. If the specimen is not now in focus with the 40X objective, loosen the stage and adjust its position up or down until focus is achieved. Lock the stage at that point.

Lastly, a great many microscopes have a simple lock on the coarse focus knob which can be released, the microscope brought to focus with the coarse focus at 40X, then the lock reset at that point. The fine focus on these units is often free-running with no limit stops.

Problem:

When you are scanning a slide under high power, the specimen goes out of focus more than usual.

It is normal to make small focus corrections when scanning under high power and every microscopist develops a feel for what is normal and what is not. The common cause for this problem is the accumulation of dirt, tiny bits of glass and other debris under the slide holder. These make the slide holder move up and down slightly, which in turn, makes the slide do the same. Remove the slide holder and thoroughly clean it as well as the stage surface. If thorough cleaning does not solve the problem, it is possible that the slide holder is deformed. Replace the holder or have a service person refit it to the stage.

It is also not uncommon for a heavily used stage to develop a somewhat wavy surface, which can be easily seen upon examination. Replace the stage or have it reground if this happens.

Problem:

Sudden deterioration of the image quality of the 40X objective.

In many microscopes the 40X objective is the same length as the 100X oil immersion objective, so the 40X is often inadvertently dragged through the oil. Unless the oil is efficiently cleaned off the 40X objective, it will not function properly. The oil may harden on the front of the lens and cleaning must be repeated until all the oil is removed. There are instructions on proper cleaning of an oily objective in the section on care and cleaning of the microscope.

If the objective performs no better after thorough cleaning, the objective may have had excess oil on it so long that it has broken through the seal and found its way into the objective. This requires refurbishment or replacement of the objective altogether.

Problem:

The 10X objective can be focused beautifully but as soon as a higher power objective is used, it cannot be focused.

Either the slide is upside down or the coverslip/mounting medium thickness is so great that focus at high power is prevented. Turn the slide right side up if this is the problem. If not, try another slide and/or coverslip.

Problem:

The 50X, 63X or 100X oil immersion objective does not give a clear image.

These oil immersion objectives tend to become dirty unless they are occasionally removed and cleaned. Clean them until they are immaculate again. See the instructions in the cleaning section.

Another common cause of this problem is the formation of an air bubble under the front lens of the objective. When the oil immersion objective is brought straight down into the oil, there is the risk of trapping a tiny air bubble under the lens. This can be avoided by sweeping the objective into the oil from the side.

Problem:

The power supply is turned all the way up but the specimen can scarcely be seen.

There are several possibilities. The nose piece may not be quite in its click-stop; the condenser may be too low; one or both of the diaphragms may be closed or nearly so. Check the nose piece position; rack the condenser up; check the diaphragm; or, better, reset the Koehlering.

Problem:

Objects are seen moving across the field of view.

These mobile artifacts are caused by separations in the vitreous humor of the eyes of the microscopist. The key word is "moving." Streaks and artifacts that are not moving can have myriad causes. The easiest way to avoid this problem is to be very careful when setting the interpupillary distance, tube length, visual acuity compensation and, above all, the contrast. Too much contrast will lead to this problem. The correct setting of these parameters is described in the section on setting up the microscope.

Problem:

The microscope "pulls my eyes."

This can have a simple cause or can be a very serious problem. First, determine whether the tube length adjustment is properly set. If there is any doubt, check it very carefully according to the instructions in the set-up section. It is very common to find a microscope where one of the adjustable eyepieces is screwed all the way in and the other all the way out.

Second, if you have a microscope where it is easy to open the eyepieces and remove the lenses, someone may have done so, often from only one eyepiece. The eyepiece was then not reassembled properly, so there is one correctly assembled eyepiece and one incorrect one. This will result in "pulling of the eyes." Professional help is probably needed to correct this situation, but you may be able to find an identical eyepiece that has definitely not been disturbed. Disassemble it very, very carefully, noting how each lens and spacer is placed; make a drawing for future use. Then clean and reassemble the original eyepiece identically.

Lastly, the worst of all problems, the binocular head is out of collimation. Collimation means the ability of the binocular system to cause the line of vision of each eye to travel via separate optical pathways to exactly the same spot on the slide. If the separate pathways do not converge at exactly the same spot, the binocular is out of collimation. If the error is strictly from left to right, most eyes, especially young ones, can accommodate quite readily. If, however, the error requires one eye to look down while the other looks up, very severe eyestrain can ensue. A single crosshair on the stage will appear as shown in Figure 18:9. If you are confronted with this problem, do not consider trying to work around it. Acquire another microscope and have the defective one repaired. If you are not sure you have this problem, ask for a professional opinion. This problem can only be solved by a person trained in all phases of microscope repair. In many instances, a good microscope service engineer can correct the problem in a short time.

CARE AND CLEANING

There are five areas on the microscope that need special care and cleaning. From top to bottom, they are:

1. The eyepiece lenses;
2. The front lens of each objective;
3. The stage and slide holder, particularly the underside of the slide holder;
4. The top lens (both movable and fixed) of the condenser.

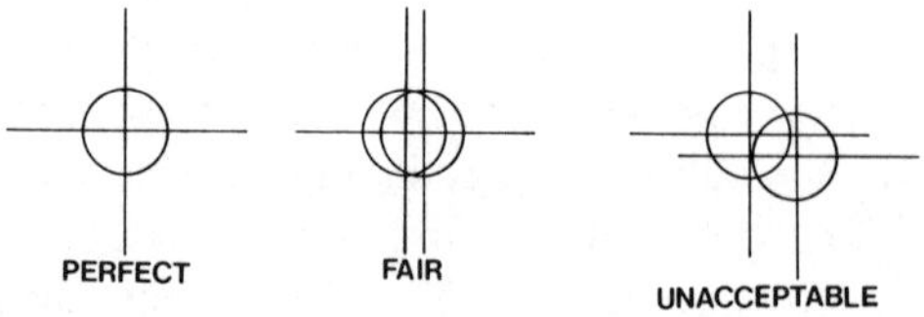

Figure 18:9. Problem of "pulling of eyes." Crosshair in perfect, in only fair and in unacceptable alignment.

5. The blue filter and the window of the illuminator in the base.

If the microscope is kept covered when not in use, less dust will gather on its various optical surfaces. Likewise, if immersion oil is wiped off the objectives after each use, the need for frequent special cleaning will be reduced. Use a good lens cleaning fluid for normal cleaning of optical surfaces. Every laboratory supply company makes one. All are water-based and not likely to harm the anti-reflex coatings of lenses. Use lens paper, not facial tissues, for cleaning. Lens tissue is manufactured with special care to keep abrasive particles out of the paper. When an optical surface is particularly dusty, blow or brush the dust away before applying lens cleaner. If this is not done, the dust may scratch the surface, since it may contain abrasive material. Do not "scrub" optical surfaces when cleaning them with lens tissue or a Q-tip dampened with cleaning fluid. A lens will tolerate many cleanings if cleaned gently. An objective or any lens coated with immersion or mounting oil will require the use of a suitable solvent for the oil, since normal lens cleaner will not remove oils. Formerly, ethyl ether was considered an ideal solvent for this purpose. It does work well and will not harm the cements and sealers in the optics, but it has some drawbacks. Not least among these is the fact that it is extremely flammable and can be explosive. A better choice is inhibited 1,1,1-trichloroethane. This compound is neither flammable nor explosive; it evaporates quickly like ether; it is a very good solvent for immersion oil and mounting media, but it will not soften the sealers and cements. It is available from laboratory and specialty chemical supply companies. Several sources recommend using xylene. It works but it is not easy to use and is toxic.

Do not use any of the commonly available alcohols; methyl, ethyl or isopropyl: do not use acetone nor any other ketone. They may damage the objectives.

To determine whether the front lens of an objective is clean, stand under a good light or by a window. Hold the objective in one hand so that its front lens is directly under the strong light. In the other hand hold an eyepiece upside down and let your hands touch to bring the eyepiece into focal range. Aim the eyepiece at the front lens of the objective (Fig. 18:10), look through the upside down eyepiece and, with minor adjustment of the position of your hands, you should see a greatly enlarged view of the front lens. Now it can readily be determined whether it is clean. A truly clean front lens will have no streaks, no oil droplets, no stains, nothing but the clear surface of the glass. The cleaning may need to be repeated more than once to attain a perfectly clean surface.

Professionals commonly use so-called canned air when cleaning microscopes. If you do the same, you should be aware of some hazards. Tiny chips of glass are often present in the nooks and crannies of a microscope stage. Take care not to blow any into anyone's eyes. These chips are not

Figure 18:10. Examination of an objective for cleanliness.

only sharp but could be infectious. Do not use canned air to blow dust from optics in the vicinity of the diaphragm, especially a camera diaphragm. The air pressure is powerful enough to bend the delicate leaves out of shape. Instead, use a hand squeeze bulb or brush dust away with a soft camel's hair brush.

The slide holder on most microscopes can be removed easily. Remove the holder occasionally to give holder and stage a thorough cleaning. If using canned air, be careful not to blow glass chips in anyone's face and take care not to be pricked by such a chip. This glass comes mostly from coverslips so the pieces are very small. If they are left under the slide holder, they cause the holder to ride up and down a few microns, moving the slide with it, and making focusing difficult when scanning a field under high power.

Lower the condenser or remove it and clean the top lens with lens cleaner. If the lens is oily, use a solvent first and then lens cleaner. If the condenser has a swing-out top lens, clean this lens as well as the lens or window underneath.

Clean the blue filter and the window of the illuminator in the base, since it is easy to image dust in that area into the field of view.

Occasionally it may be necessary to go "dirt-chasing," as photomicrographers do, whenever an annoying speck or specks appear in the field of view or on films. It is usually wise to start at the eyepiece and clean everything in the light path all the way down to the source lamp. Start with an obvious and easy thing. Rotate each eyepiece. If the dirt rotates with an eyepiece, it is on one of the eyepiece lenses. If not, move the slide; if the dirt moves with the slide, clean the slide. Make sure that the condenser diaphragm is not too far closed. High contrast emphasizes dirt that would not be seen otherwise. Too much contrast can even make striations in the vitreous humor of the eyes appear as moving artifacts in the field of view.

You can also use a low power objective (10X, etc.) to go "through" the focus, i.e., focus on a slide, preferably one with clear areas, then pass through the focal point and, as it were, come out on the other side. This brings the surfaces below the slide into focus: the base illuminator window, blue filter, mirror in the base and the lens in the illuminator itself. You should be able to see the dirt on one of these, and once you know where the dirt is, you can remove it.

Microscopes that have a camera mounted on a trinocular head can have special "dirt-chasing" problems. The inside of the camera should be blown out with a rubber squeeze bulb. Do not use a canned air blower; this may damage delicate shutter diaphragm leaves. Below the camera there is often a third eyepiece whose front lens collects a lot of debris. If the camera is removed to reach this lens, the camera's film plane focus may be inadvertently disturbed. It is better to open the camera shutter and keep it open with a time exposure. Then the lens will be exposed and the dirt can be gently removed. Similarly, some microscope camera systems have a built-in lens (not an eyepiece) below the camera. The same approach applies, open the shutter on a time exposure and gently remove the collected dirt.

It is not uncommon for sufficient bits of celluloid from films and other detritus to block almost all light. When this accumulation is removed, the automatic exposure settings may change dramatically because the system has compensated for the presence of the interfering dirt. Recalibrate the exposure times after cleaning.

If the five principal points are kept clean and, in addition, a thorough cleaning and preventive maintenance is done once or twice a year by a competent professional, a microscope should give very little trouble.

POLARIZED LIGHT MICROSCOPY

To be precise, this discussion concerns linearly polarized light that can be imagined as follows: assume that light is strictly a wave motion and there are only two waves, one vertical as shown in Figure 18:11, the other the same waveform but lying on its side (the straight line in Fig. 18:11). If a polarizing element (known as a Nicol prism) is set in the correct position,

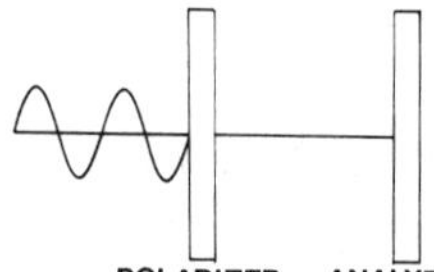

Figure 18:11. Polarization of light with polarizer and analyzer.

it will totally stop the vertical wave and only light of the horizontal wave will pass through. Another polarizing element can now be inserted into the light path and when rotated 90° relative to the first element, it will totally stop the horizontal wave. No light now passes through and only darkness is seen. This effect is sometimes called "crossed Nicols." Although the two elements are identical, for position identification they are known as "polarizer" and "analyzer." If something is placed between them, such as a crystal that also rotates the plane of polarized light, the crystal is clearly seen against a black background.

Anything that rotates a plane of polarized light can be singled out from the myriad things that do not. Identification of crystals in urine is an example. Some organisms in tissue can also be seen in polarized light. Many other clinically significant substances rotate the plane of polarized light and there are numerous applications of this technique.

The simplicity of polarized light lies in the fact that the polarizing elements are simple devices, easy to insert into most microscopes. The elements look like neutral density filters and can be purchased inexpensively in film form for cutting your own polarizer and analyzer. One goes where the blue glass is normally placed; the other goes somewhere above the specimen, most conveniently under the binocular head. It is important not to allow this analyzer to warp or tilt when installing it under the binocular or image quality will suffer. Since you cannot reach the analyzer once it is installed under the binocular, do not worry about its position. To achieve the "crossed Nicols" effect, look through the binocular and rotate the polarizer in the illuminator window until the darkest possible field is obtained. A slide should be in place and in focus; if there is anything on it that rotates the plane of polarized light, it will be clearly seen.

Strains in glass are among the things that do rotate the plane of polarized light. These strains result from inadequate annealing during manufacture or from pressure on an element built into a mount. For this reason, microscopes used for critical polarized light work such as petrography are equipped with strain-free optics. In clinical work, however, such expensive optics are not usually necessary.

When examining crystalline materials (the most common use of polarized light), it helps to be able to change the color of the background. Some microscopes have wave plates that permit such a change and have the precise function of identifying crystals by their color.

PHASE CONTRAST MICROSCOPY

Phase contrast allows examination of specimens which cannot be stained in the conventional way. It is a way of temporarily "staining" a specimen by causing light waves to interfere with one another to produce differentiating densities and, therefore, contrast in the specimen that otherwise would have little or none. The interference effect is due to the presence of two rings; one in the objective, called a phase ring, and the other, known as a phase annulus, in the special phase condenser. When these two rings are perfectly aligned—superimposed—the phase effect will be at a maximum and the unstained specimen will show the greatest detail.

A special optical tool called a phase telescope is required to achieve this alignment. This tool exists in several forms. Some manufacturers build one into the microscope. When it is swung into place, focused with its built-in focusing device, the phase ring and the phase annulus will be seen simultaneously. There will be two adjusting knobs or keys on the microscope or supplied with it. These are used for alignment. As a rule, each phase annulus must be separately aligned. Some microscopes have a discrete telescope that looks like a focusing eyepiece. It should temporarily replace one of the eyepieces, be focused on the rings and the alignment done. Remove the telescope when alignment is completed and replace the normal eyepiece.

Some microscopes are used exclusively for phase and are seldom far out of alignment. When microscopes are used primarily for brightfield work and only occasionally in the phase mode, there is a common problem of trying to align the phase system without first aligning the microscope in the brightfield mode. If the latter alignment is poor (see set-up instructions), phase alignment will be difficult. Go back to brightfield if there is difficulty with phase alignment. Go through the Koehlering process. When a good brightfield image has been achieved, replace the brightfield specimen with a phase specimen. You should be close to focus with this specimen. Swing in the phase annulus, align the rings and remove the telescope. Only minor adjustments should now be needed to obtain a perfect phase image.

Some slight variations are common in phase contrast set-up. First, it is usual to completely open the condenser diaphragm if there is one. In the normal phase condenser, the diaphragm is automatically removed when going to any phase annulus. In some microscopes, however, there is a single "slide-in" phase annulus. Such a microscope will still have its condenser diaphragm in the operating position. It is also usual to exchange the blue filter for a green one. The phase contrast effect is maximized at the wavelength of green light. However, the blue filter may be retained or no filter at all used, if preferable, without significant image deterioration.

Quite often simple problems occur when phase is only rarely used. For example, the objectives will be labelled in some way, such as Phase 1, Phase 2, etc., and the phase annuli in the condenser must be matching ones. If a phase 1 objective is used with a phase 2 annulus, or vice versa, they will not produce a good phase image. Similarly, if you return to brightfield and forget to rotate the phase annulus carrier in the condenser to its brightfield position, you will not have a good brightfield image. If you return to brightfield for more than just alignment you should repeat the Koehlering.

FLUORESCENCE MICROSCOPY

Perhaps it may be said that in the various modes of microscopy such as brightfield, phase contrast and polarized light, strict rules of operation based on scientific facts can be followed, and if followed conscientiously, the best possible results can be achieved. In fluorescence microscopy this is only partially true. In addition to the strict rules, there is also a "touch of art." Experienced fluorescence users who have acquired this art can obtain much better performance from their instruments than those who only follow the strict rules.

Fluorescence systems generally use mercury lamps as ultraviolet (UV) and blue light sources. These lamps, which usually range in power from 50-250 watts, can be dangerous. First, they obviously contain mercury, which is a hazardous substance. If a cold lamp breaks, the mercury spillage should be dealt with in accordance with standard procedures for spills. Secondly, when a mercury source lamp is lit, it puts out very high levels of UV light. Never look at it with the naked eye. Its brilliance alone is enough to temporarily blind. Its UV content can cause severe eye irritation and even permanent blindness. A third hazard is the possibility of the lamp exploding when hot. Therefore, observe these precautions. Wear protective goggles when working with an exposed cold mercury source and keep them on when ready to ignite the source. Never ignite a mercury source with its housing open. Before changing a source lamp, make a note of how the old one was installed. Look at the wiring. Determine whether the contacts need to be cleaned or the wiring or insulation replaced. Call for service if they do. The lamp will have two or perhaps three terminals. It is important to handle them safely. Each terminal is a metal screw and nut assembly attached directly to the glass. When tightening the terminals, hold the metal portion gently in a pair of pliers and tighten the nuts firmly but gently by hand. Do not tighten the nuts while relying on the glass to take the strain; the glass may crack and the first time you ignite the lamp it will be destroyed. Lastly, try not to get fingerprints on the bulb; if you do, wipe them off with an alcohol-dampened cloth before igniting the bulb. When ready to install the carefully mounted lamp in its housing, guide it in gently and do not force it if it hangs up.

A transmitted-light fluorescence microscope nearly always has a darkfield condenser; occasionally one uses a brightfield. To adjust the system, go to a 10X or other reasonably low power objective. If there is a brightfield condenser, Koehler the microscope with a normal, nonfluorescent slide and the usual light source. Then substitute a fluorescent slide, turn on the mercury source and expect to be very close to focus for the fluorescent material.

If there is a darkfield condenser, a darkfield specimen is needed. If none is available, make a buccal smear and cover with a coverslip. Put a drop of immersion oil on the top lens of the condenser and raise the condenser until it touches the underside of the slide. Focus the low power objective until the cells of the buccal smear are sharply imaged. You may now substitute a fluorescent slide, turn on the mercury source and expect to be close to focus, needing only minor adjustment of the fine focus. If you begin with a fluorescent slide and not a buccal smear, be sure that the fluorescent slide is one that does fluoresce brightly. Problems may arise if it does not fluoresce or only does so weakly.

Once you have come this far, it is time for "art." Open all diaphragms fully. If the mercury lamp housing has an adjustable collector lens, adjust it for maximum intensity of fluorescence. Now, move the condenser up or down very gradually and slowly until the fluorescent image is as bright as possible. You will probably see a greenish halo in the field. If you do, adjust the condenser up or down slightly more till the green haze is as round as possible. If the now round halo is not centered, center it with the condenser controls.

A mercury source puts out a great deal of UV and blue light. The choice of an appropriate exciter filter allows the selection of wavelengths which excite a particular specimen. The specimen has been treated with a fluorochrome that seeks out certain characteristics. The fluorochrome will attach itself to those parts of the specimen where these characteristics are present and will fluoresce when UV or blue light is directed to it. The fluorescence will almost always be at longer, visible wavelengths. Theoretically, the specimen should now be clearly seen. But there will be much blue light overlying the fluorescence as well as nonspecific fluorescence of various hues which tends to obscure the true color of the fluorescing substance. A barrier filter is needed to cut out this background but leave the specific fluorescence. Many believe that the barrier filter is there solely to protect eyes, but this is not so. The main purpose of a barrier filter is to remove nonspecific fluorescence and blue background. Since glass blocks UV light completely, it is mainly the glass of the optical system that prevents short wave UV from harming the eyes. The excitation of most fluorochromes is done by low wavelength blue light. Both mercury and xenon sources have high levels of blue light of appropriate wavelengths.

Immersion oil causes the most problems with fluorescence work. Oil tends to accumulate all over the microscope and must constantly be removed.

Epifluorescence or reflected light fluorescence is now considered a better system, since light passes down from a separate illuminator through the excitation filter and through the objective to the specimen rather than coming through a base illumination system and the darkfield condenser, then up through the specimen. The specimen is observed in the normal way and only the longer wavelength fluorescence is seen. A barrier filter is still used to ensure the purest fluorescent image. Many modern epifluorescence systems incorporate groups of three carefully matched elements, sometimes referred to as clusters, containing the excitation filter, the barrier filter and a dichroic reflector which has the unique function of reflecting shortwave light (the blue to be eliminated) but transmitting longer wavelength light (the fluorescence desired). In epifluorescence it is also helpful that the blue light is directed downward and some of it disappears after passing through the specimen. The cluster elements only need to deal with that part of the blue light that is reflected. Since no darkfield condenser is used, no oil is needed except when observing a specimen through an oil immersion objective, when only a small quantity of oil need be placed on the top of the slide. This eliminates most of the problems with oil in fluorescence work.

Source alignment is the most difficult problem in fluorescence work. Difficulties arise because every microscope has a different source alignment procedure. Many microscopes use a high intensity halogen bulb instead of a mercury source. This gives adequate light for many procedures but not for all. The alignment rules are the same for these light sources as for mercury. Critical illumination is used, where the light source is imaged into the specimen plane. It is easy to align an epi system. Remove the objective in the viewing position, place a piece of white paper on the stage and adjust the collector lens control and diaphragms, if any, until a sharp image of the halogen source filament and electrode tips can be seen on the paper. Adjust the lamp position with the available controls until the image is centered. Controls to move the lamp sideways and controls to move it up and down vary from quite sophisticated devices to some as simple as a single screw which, when loosened, allows the entire lamp to be moved. Some microscopes have a ground-glass screen on which the lamp image can be seen at any time upon removal of a cover. Many fluorescence systems have a mirror behind the light source which provides for a significant increase in the energy level. In the alignment procedure the mirror image is made to appear side by side with the "real" image. The controls will adjust the two images to equal size (see Fig. 18:12).

Alignment of the transmitted light system is more difficult because of the

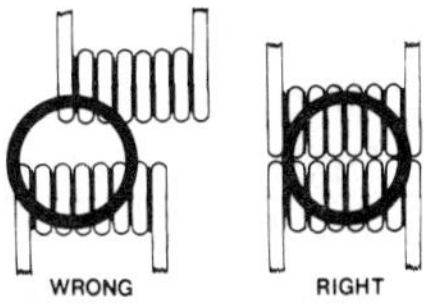

Figure 18:12. Alignment of the mirror image in fluorescence microscopy.

presence of the darkfield condenser. It is necessary to align the light source independently of the microscope and then to put the two together for fine tuning. These are the basic rules:

First, look in the instruction manual to determine whether there is a mirror in the UV lamp housing. If you do not have a manual, put on goggles, open the housing when the lamp is turned off and cold and look for a mirror. The alignment procedure is roughly the same, with or without a mirror. Find the screws or controls which allow adjustment of the lamp itself, usually in two directions, up/down and left/right. Now find the controls for the mirror, usually an in/out control so the size of the mirror image can be matched to the size of the actual lamp image. A tilt control is also common and allows placement of the mirror image side by side with the lamp image. Before adjusting any of these controls, use the collector lens focus control to obtain a reasonably sharp image of the source lamp electrodes or filament. So far this procedure differs little from the one for epi illumination, but here you cannot display the image on white paper on the stage, the paper must be below the darkfield condenser. Be cautious. Even with filters in place, the light can be so intense that you cannot look at the paper. Put extra filters in or add several thicknesses of paper until the light level is comfortable.

Once you have achieved the alignment of Fig. 18:12 on the paper, touch up the condenser alignment for optimum centering and adjust the condenser height very slightly for maximum intensity. Much of the "art" lies in this last adjustment. You may be astonished at the brilliance gained by a small delicate adjustment. You are "stretching" the oil layer between the slide and the condenser—almost like a flexible lens. A touch too far down with the condenser and the oil film will detach, so that nothing is seen. A touch too high and the image intensity will be greatly reduced. It helps to have an alignment specimen that has a permanent high level of fluorescence. Biological specimens tend to fade in a matter of days or even hours, so they cannot be relied upon for demonstrating the quality of alignment. Permanent fluorescence slides in a variety of wavelengths are available.

MEASURING RETICLE

Although the fluorescence microscope is not the usual place to find a measuring reticle in the eyepiece, they are found commonly in other microscopes in the bacteriology laboratory. A good stage micrometer, available from many manufacturers, is needed for calibration of the reticle, which will generally have an arbitrary scale with 50 or 100 divisions. You need to know what each division really measures when used with different objectives. Lay the micrometer right side up on the stage. Focus on its scale with the lowest power to be used, e.g., 10X. Rotate the eyepiece if necessary to bring the reticle scale parallel to the micrometer scale. Align the zero end of the two scales so they coincide exactly. Find the next line of the reticle that coincides exactly with a line on the micrometer, perhaps the 20th line (do not count the zero line). Read how much of the micrometer lies between zero and line 20. If it were 0.2 mm, 20 lines represent 20 divisions; 0.2 mm divided by 20 is 0.01 mm or 10 μm per division at 10X magnification. Follow the same procedure carefully for each of the other objectives and the reticle will be calibrated for all magnifications.

GLOSSARY

Aberrations:

The various errors and distortions found in microscope optics that need compensation or partial compensation to produce true images of material studied.

Achromats:

Microscope objectives whose back focal planes are compensated for blue and red only. Other aberrations such as spherical aberration and coma are also corrected to the degree that the lower price will allow.

Aerial image:

The image formed at the tube length or by the tube lens in infinity-corrected systems, magnified by its power and then viewed and further magnified by the optics of the eyepiece.

Apochromats:

Microscope objectives whose back focal planes are compensated for more than two colors. The correction for other aberrations has also been taken much farther than for achromats. Correction for curvature of field is often built into these expensive objectives, thus making them "planapochromats." Compensating eyepieces are used with apochromats to compensate for chromatic difference of magnification.

Auxiliary lens:

A lens found below the condenser in many microscopes whose purpose is to expand the field of view to full or greater size with low power objectives.

Brightfield:

Microscopy in which the specimen is fully illuminated by direct light.

Compensating eyepiece:

An eyepiece designed to correct some of the aberrations in the objective that are difficult to correct in the objective itself. Objectives and eyepieces designed in this way must be used together to correct such errors.

Contrast:

A means of making the general structure of a specimen more visible by slightly reducing the aperture with the condenser diaphragm. There is an accompanying loss of resolution but an acceptable compromise can usually be made.

Critical illumination:

Projecting the image of the light source into the specimen plane. It has the advantage of great brightness but the disadvantage that the light source (filament or plasma) can be seen superimposed on the structure of the specimen. It is commonly used in fluorescence microscopy where that effect is of little importance.

Empty magnification:

Images that are not really there are created when the image of an object is magnified too far. New details of structure are not revealed but rather the excess magnification begins to break up the detail already revealed.

Epi illumination:

Sometimes called reflected light microscopy because the light is led to the specimen from above, passing through the objective and reflected (as the focused image) back up through the objective to the viewing optics. This is the only way a solid specimen such as a polished block of steel can be illuminated and is thus widely used in industry. It is also the preferred means of illuminating (exciting) a fluorescence specimen because it requires no oil between the condenser top lens and the bottom of the slide, and uses no substage condenser at all.

Field-stop diaphragm:

Sometimes called the illumination diaphragm, it is found in the base illuminator and in the epi illumination system. By imaging the light source into its plane, it becomes the light source, thus eliminating several disadvantages of critical illumination. (See also Koehler illumination.)

Flat field:

A term describing an objective that has been corrected for curvature of field thus allowing the entire field of view to be in sharp focus. Although almost always incorporated into expensive objectives, it can be found on simpler ones. It is a virtual necessity for hematology applications.

Fluorescence:

In microscopy, a term used to define a method of revealing certain organisms or conditions where a fluorochrome rather than a stain bonds to the object. Presence or absence of the organism or condition is indicated by the presence or absence of specific fluorescence. The fluorochrome is excited to fluoresce by some form of UV or blue light and reveals itself by emitting some longer wavelength of visible light.

Halogen lamp:

Sometimes called a Quartz-Tungsten-Halogen or QTH lamp. A modern improvement of the tungsten lamp, it has a quartz envelope instead of glass, whereby it can emit a considerable amount of UV light. It has a tungsten filament as does a tungsten lamp but, whereas the tungsten normally boils off the filament and deposits on the cooler lamp envelope, the presence of a trace of iodine vapor inhibits this action in the QTH lamp. The tungsten returns instead to the filament and the lamp does not blacken but remains bright to the end of its life.

Infinity-corrected optics:

An optical system which allows the image from the objectives to be projected into infinity instead of at the tube length of the microscope. The rays leaving the objective are parallel and can therefore have other optical elements interposed, such as polarizers, waveplates, magnification changers, etc. Special lenses must be used to bring the images to focus at the normal (tube length) position. Optics from an infinity-corrected system cannot be used on a noncorrected system and vice versa.

Interpupillary distance:

Sometimes called interocular distance and quite literally the distance between the centers of the two eyes. Most microscopes allow an adjustment of this distance between 55-75 mm.

Koehler illumination:

A method of imaging the light source into the plane of the field-stop diaphragm, thus allowing it to become, essentially, the light source itself. This has several advantages, the most obvious being that the filament of the light source is not seen in the specimen plane and so alternating light and dark bars that the filament exhibits cannot deteriorate the image.

Linearly-polarized light:

In microscopy, a means of using a pair of so-called Nicol prisms (usually referred to as the polarizer and analyzer) to successively block the path of light waves vibrating at 90° to each other. These waves, sometimes called the ordinary and the extraordinary waves, when blocked leave the field in nearly complete darkness. A specimen which also rotates the plane of polarized light when placed between the polarizer and analyzer will be revealed against this otherwise dark background. This is a commonly used method for studying crystalline material.

Mechanical tube length:

The distance, usually 160 mm, from the shoulder of the objective to the shoulder of the eyepiece. Optically, it is essentially the point to which the image from the objective is projected. The eyepieces must also focus on that point to see and magnify the aerial image from the objectives.

Mercury lamp:

A powerful source of UV light used in fluorescence microscopy. It consists of two electrodes sealed in a quartz envelope containing a small amount of mercury. When the lamp is ignited, the mercury vaporizes and intensely bright plasma flows between the electrodes. There is sometimes a third electrode to which high voltage is applied to ignite the lamp. The power ratings of such lamps commonly range between 50-250 watts.

Neutral density filter:

In microscopy, a filter used to reduce the level of light without affecting the color of that light. Neutral density filters affect all wavelengths of light to roughly the same degree and therefore do not change the color significantly.

Numerical aperture:

The "light gathering power" of an optical system expressed in a simple mathematical formula that allows comparison of one system with another.

Parcentricity:

A feature of a microscope that maintains centering when a user goes from a higher to a lower power objective. When a feature within a specimen has been centered under a high-power objective, it should remain centered or at least still within the field of view as the user goes to each successively lower power. This feature is generally built into the microscope at the time of manufacture. In some microscopes it depends to a large degree upon which position of the nosepiece each objective occupies. When the objectives are moved in the nosepiece, the parcentricity can be lost. In the best microscopes, however, the parcentricity is independent of objective position.

Parfocality:

A feature of a microscope that maintains focus when the user goes from low-power objectives to successively higher power ones or vice versa. The microscope should remain very close to focus when a user changes objectives. If the microscope is still equipped with its original optics, parfocality is under the user's control. There should be no parfocality problems if adjustments are made to compensate for any change in tube length with change of interpupillary distance.

Plan achromat:

A method for achieving excellent contrast in a specimen which is not stained. The optics of the microscope are designed to cause the various wavelengths of light to interfere with one another especially in the green region. This interference causes the various regions and boundaries in the specimen to exhibit differentiating shades they would not otherwise show. This enhances the contrast and makes the almost invisible plainly visible.

Plan achromat:

An objective corrected for chromatic aberration in the red and blue range, for spherical aberration in the green and for curvature of field.

Plan apochromat:

An objective chromatically corrected for more than two colors as well as for virtually all the other aberrations that can be compensated, as well as being corrected for curvature of field.

Resolution:

The ability of an optical system to recognize space between any group of features.

Transmitted light microscopy:

The most common system of microscopy in which the light is brought from a base illuminator through a condenser and up through the specimen. Virtually all types of microscopy can be performed in transmitted light, the only exception being the study of solid objects through which light cannot pass.

Tungsten lamp:

Any incandescent lamp in a microscope is probably a tungsten lamp, so-called because the filament is made of tungsten, which can be heated to brilliant white heat without melting. The envelope is usually made of glass that effectively stops the lamp from emitting much UV light.

COLOR PLATES

Chapters 14 and 15

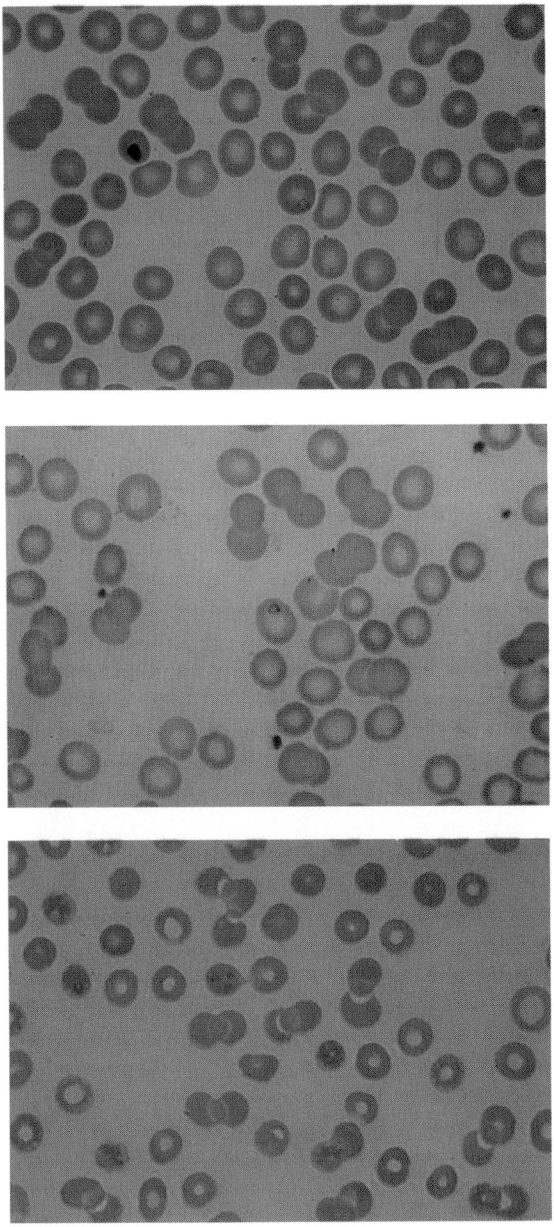

a.

b.

c.

Figure 14:3. ***Babesia*** **spp. Giemsa stain, oil immersion. Individual ring forms are seen in a and b while 4 nuclei are visible in one cell and individual rings in others in c. Courtesy Dr. G.R. Healy.**

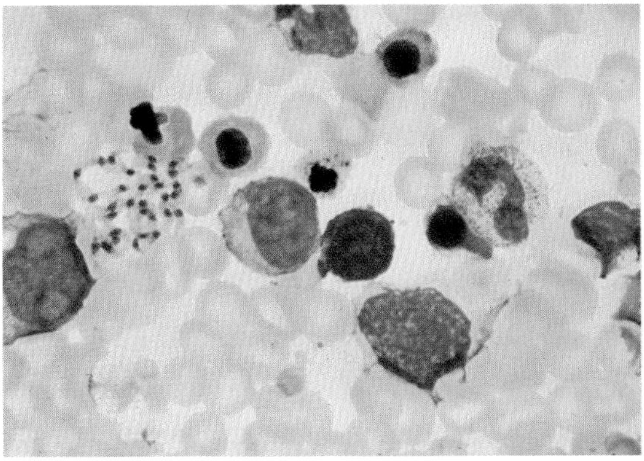

a.

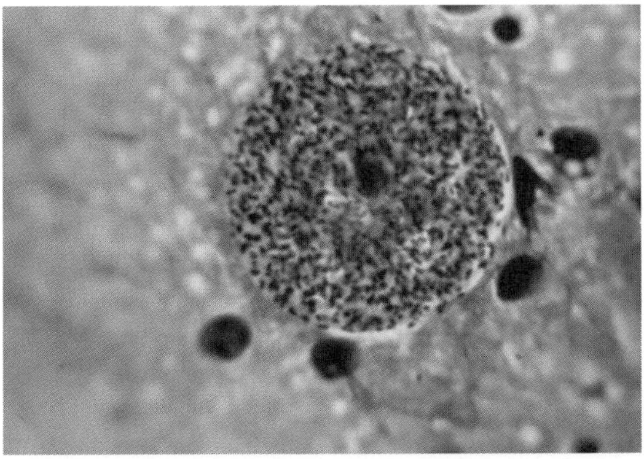

b.

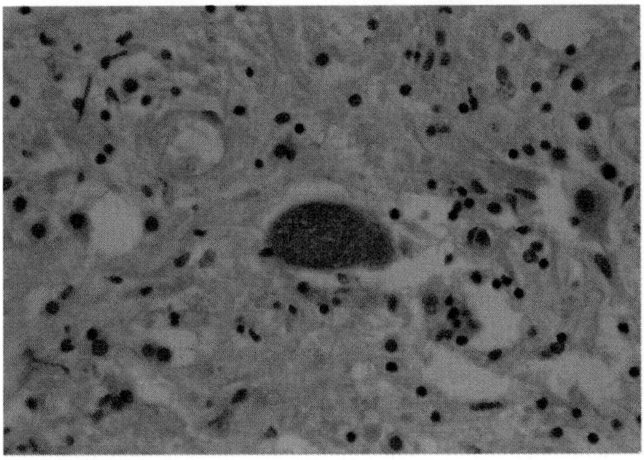

c.

Figure 14:4. ***Toxoplasma gondii*****. a. Tachyzoites in patient bone marrow—Giemsa stain, oil immersion. b. Cyst in mouse brain—Giemsa stain, oil immersion. c. Cyst in brain tissue of AIDS patient—Giemsa, oil immersion.**

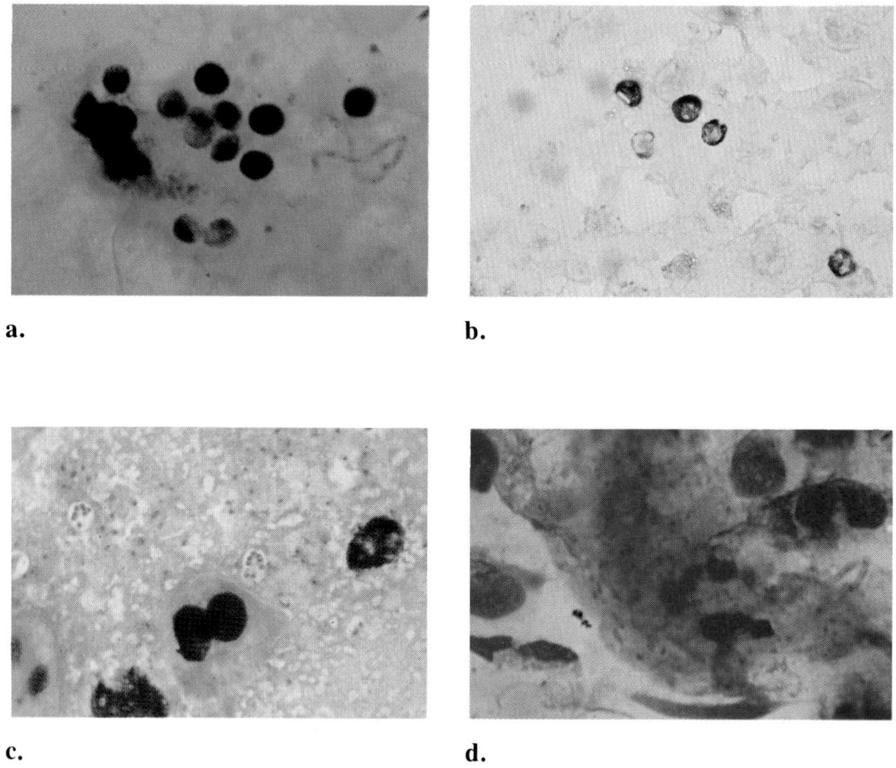

Figure 14:6. ***Pneumocystis carinii*****, oil immersion. a. Methenamine silver stain—note dark cup-shaped cysts. b. Gram Weigert stain—cysts with free trophozoites in surrounding tissue. c. Giemsa stain—note 8 intracystic organisms and halo due to unstained cyst wall. d. Giemsa stain—large clump of trophozoites characteristic of AIDS cases. Courtesy of M. Bartlett.**

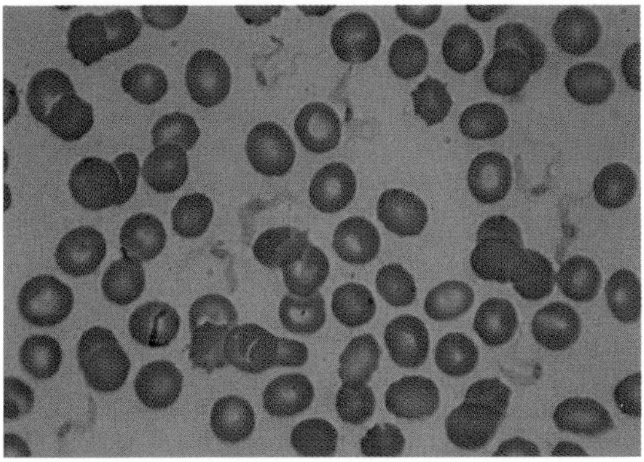

a.

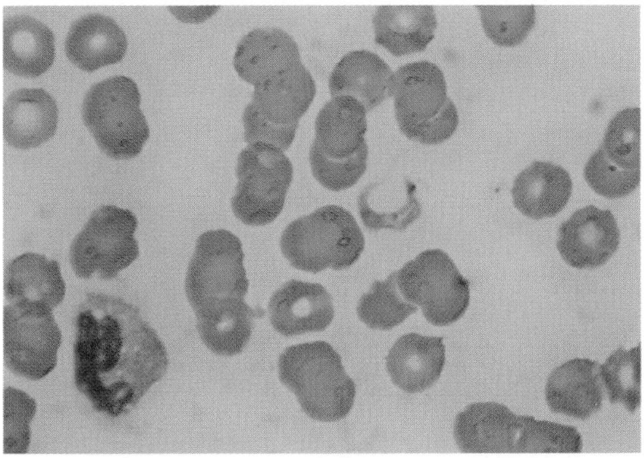

b.

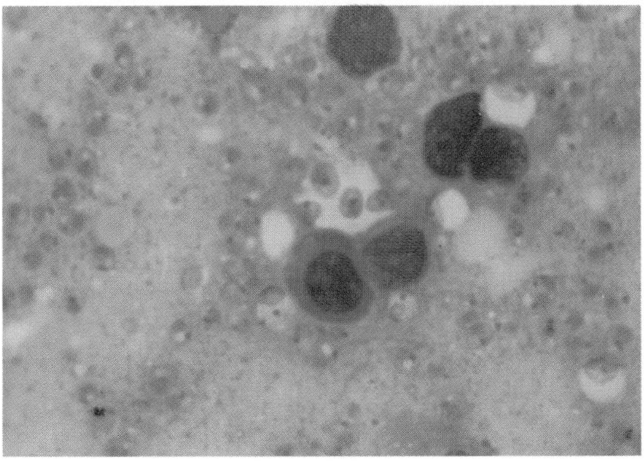

14:9

Figure 14:8. Trypanosomes, Giemsa stain, oil immersion. a. Trypomastigotes of the type seen in *Trypanosoma gambiense* or *T. rhodesiense*. b. Trypomastigotes of *T. cruzi*. Courtesy Dr. G.R. Healy.

Figure 14:9. *Leishmania donovani*, Giemsa stain, oil immersion. Numerous amastigotes with typical rod-shaped parabasal bodies. Courtesy Dr. G. R. Healy.

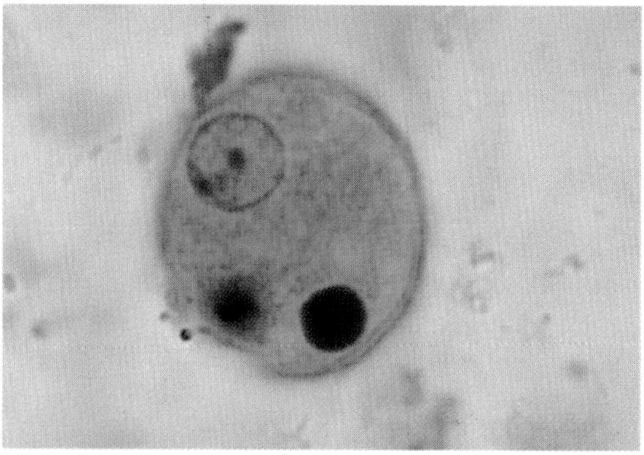

a.

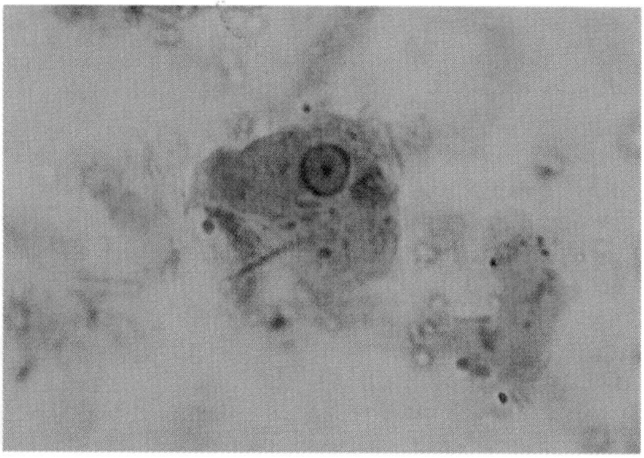

b.

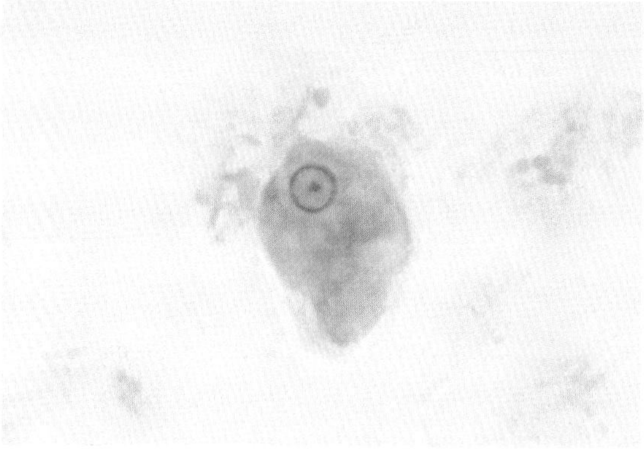

c.

Figure 15:4 Intestinal *Entamoeba* species in trichrome stain (1000X). a. This large *Entamoeba histolytica* trophozoite contains two ingested erythrocytes, one of which is slightly out of focus. The cytoplasm is delicate; the nucleus has evenly distributed peripheral chromatin and a small dot-like central karyosome. b. This *E. histolytica* trophozoite is smaller, but still has typical nuclear characteristics with evenly distributed peripheral chromatin and a dot-like central karyosome. c. This *E. histolytica* trophozoite has some vacuolization of the cytoplasm, probably a result of delayed fixation. The

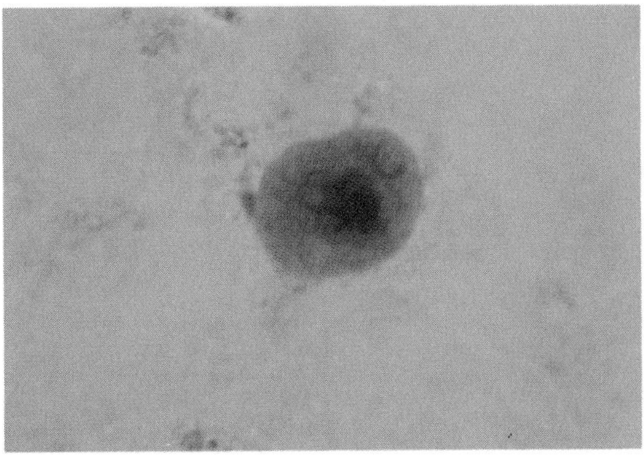

d.

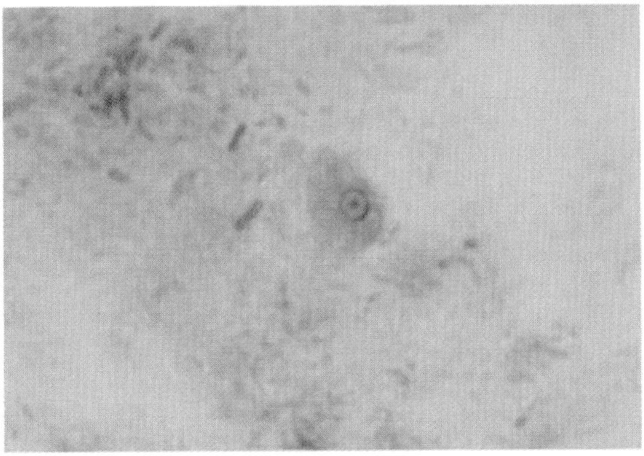

e.

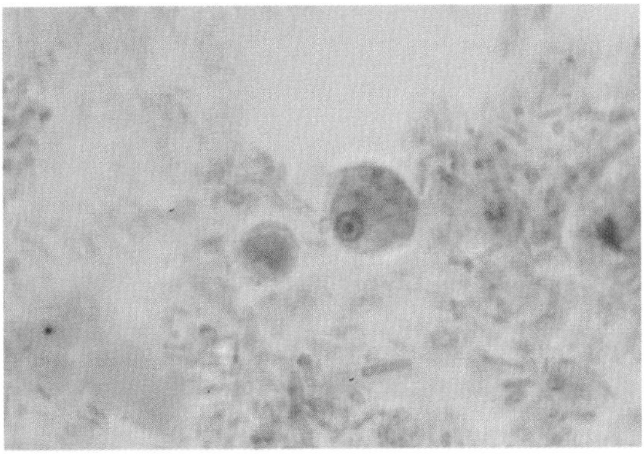

f.

nucleus has evenly distributed peripheral chromatin, but the karyosome is slightly eccentric. d. This *E. histolytica* cyst has four nuclei, but only three are in this focal plane. The red chromatoid body in the center is slightly out of focus. e. This *Entamoeba hartmanni* trophozoite has characteristics similar to those of *E. histolytica* but is smaller. Note that the cytoplasm is delicate and pale staining. f. This *E. hartmanni* trophozoite has typical nuclear and cytoplasmic characteristics. g. This binucleate *E. hartmanni* cyst has two red

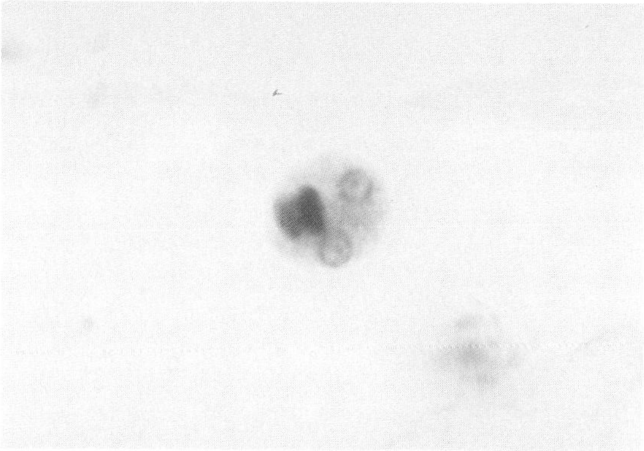
g.

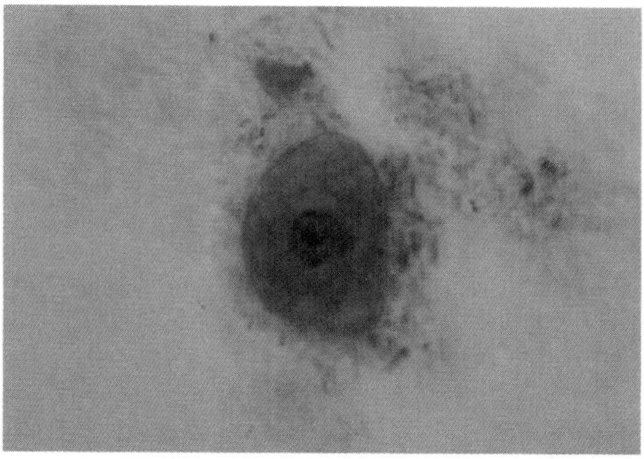
h.

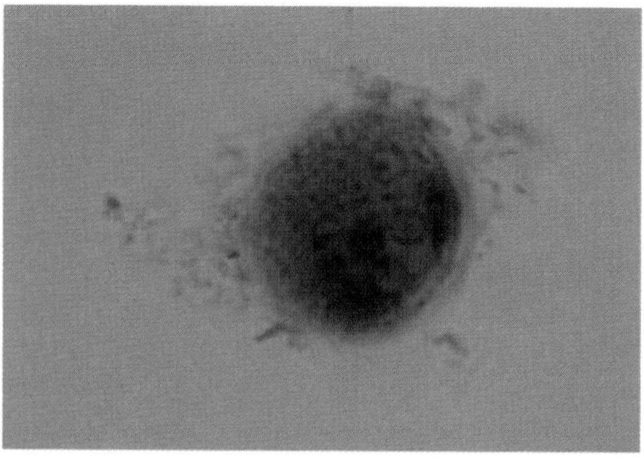
i.

chromatoid bodies with relatively even borders. Nuclear division is not yet complete. h. This *Entamoeba coli* trophozoite has a nucleus with irregularly distributed peripheral chromatin, a slightly eccentric large irregular karyosome, deeply and irregularly staining granules in the karyolymph space, cytoplasm has numerous vacuoles. i. This *E. coli* cyst has five of the eight nuclei in this focal plane and two reddish chromatoid bodies which are slightly out of focus. The upper nucleus has an eccentric karyosome, but nuclear details are not as distinct in the other nuclei.

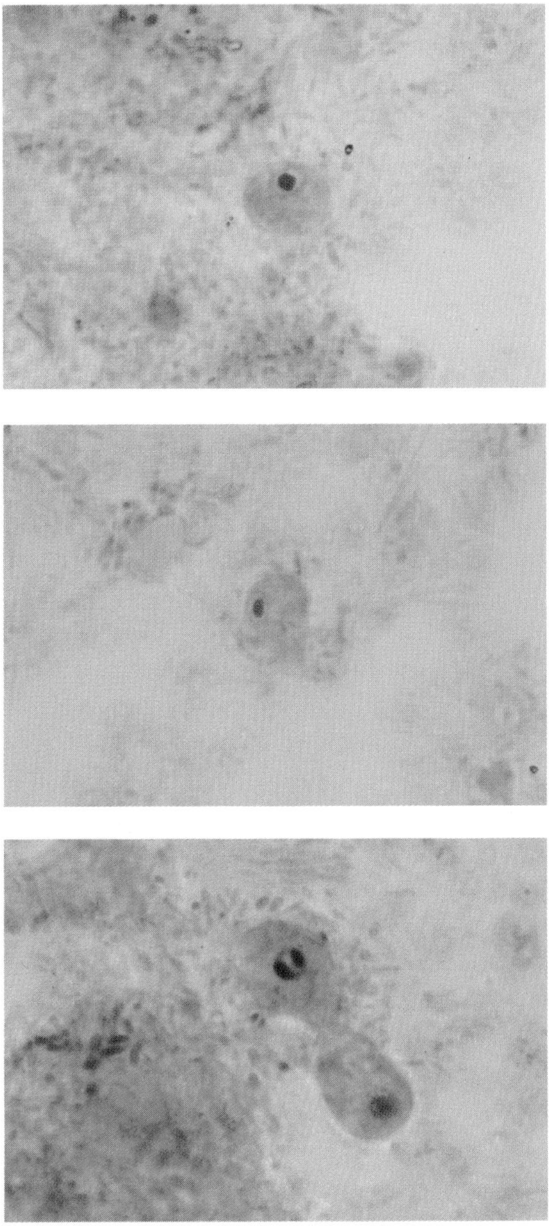

a.

b.

c.

Figure 15:5 Other intestinal amebae and *Dientamoeba fragilis* in trichrome stain (1000X). a. This *Endolimax nana* trophozoite has a nucleus with no peripheral chromatin and a large spherical karyosome surrounded by a clear karyolymph space. Cytoplasm is delicately stained. b. This trophozoite of *E. nana* has an elongated karyosome. c. This *E. nana* trophozoite has a split karyosome. Immediately below this *E. nana* trophozoite is an *Entamoeba*

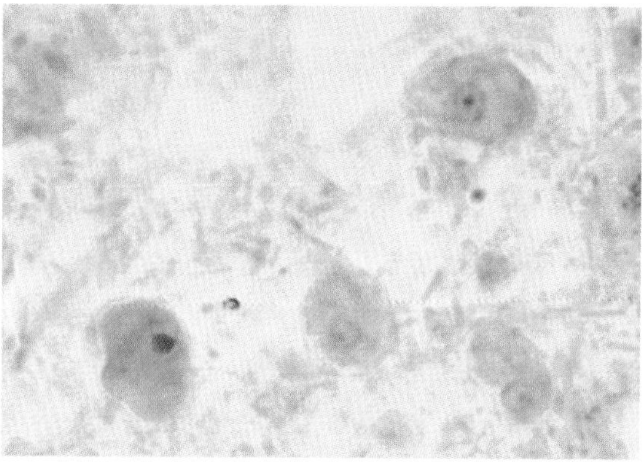

d.

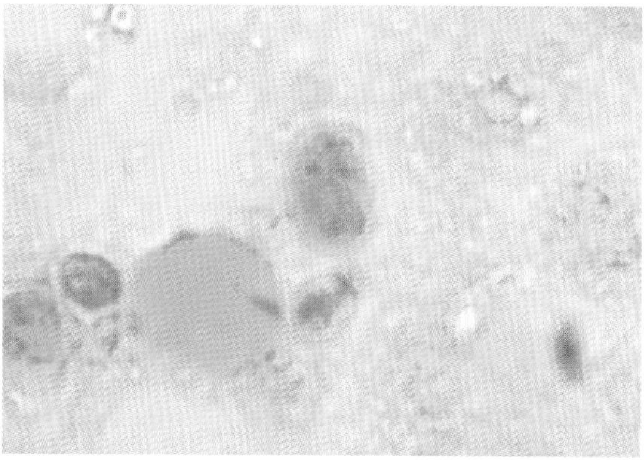

e.

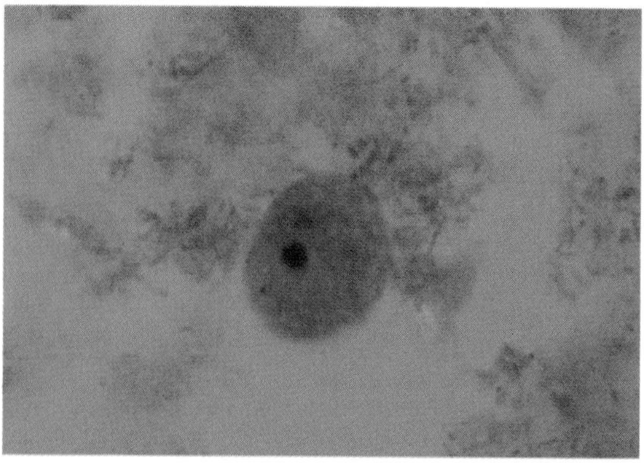

f.

hartmanni trophozoite. d. This *E. nana* trophozoite has a triangular karyosome. To the right are three delicately stained trophozoites of *E. hartmanni*. e. This cyst of *E. nana* has three nuclei at the upper pole, the fourth is in a different focal plane. f. This trophozoite of *Iodamoeba butschlii* has a spherical karyosome, the karyolymph space is muddy, and the nuclear membrane does not have peripheral chromatin. The cytoplasm is slightly vacuolated. g. This

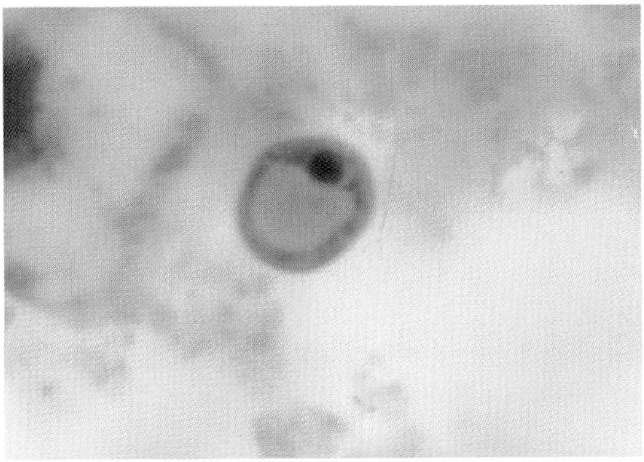

g.

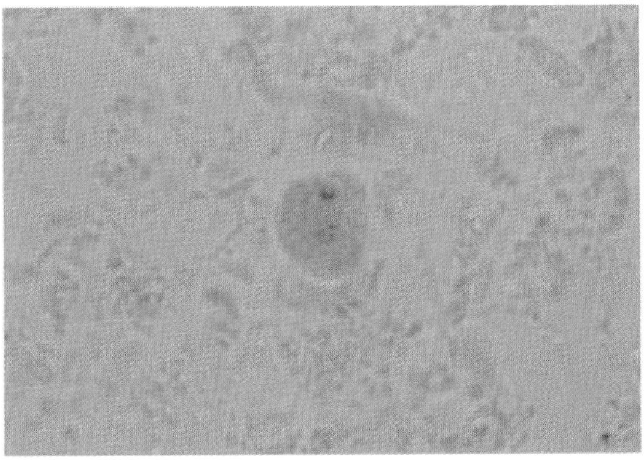

h.

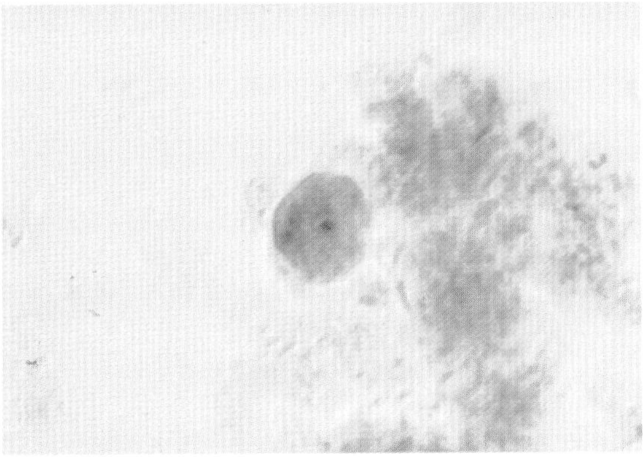

i.

cyst of *I. butschlii* contains a large vacuole which contained glycogen. The single nucleus has a prominent karyosome and muddy karyolymph space. h. This *D. fragilis* trophozoite has two nuclei which contain granular karyosomes. The cytoplasm is delicately stained. i. This *D. fragilis* trophozoite shows a tight clump of chromatin granules in the central nucleus. The more peripheral nucleus is slightly out of focus. The cytoplasm is delicately stained.

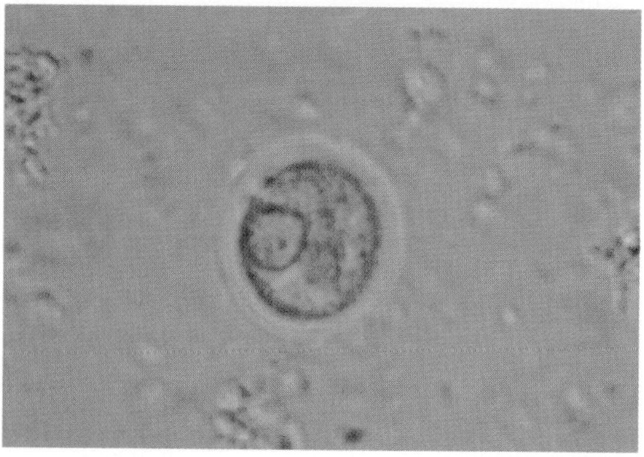

a.

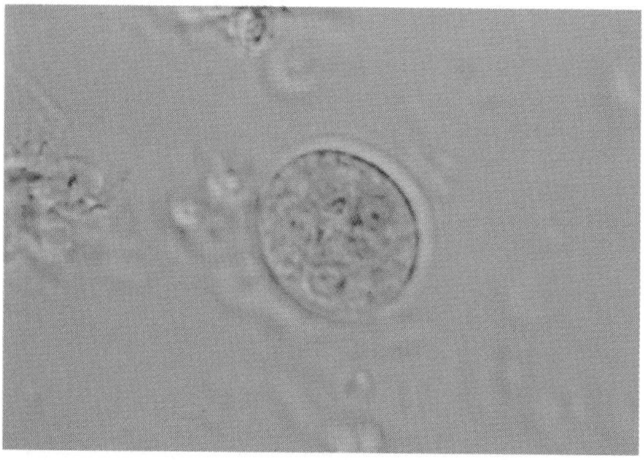

b.

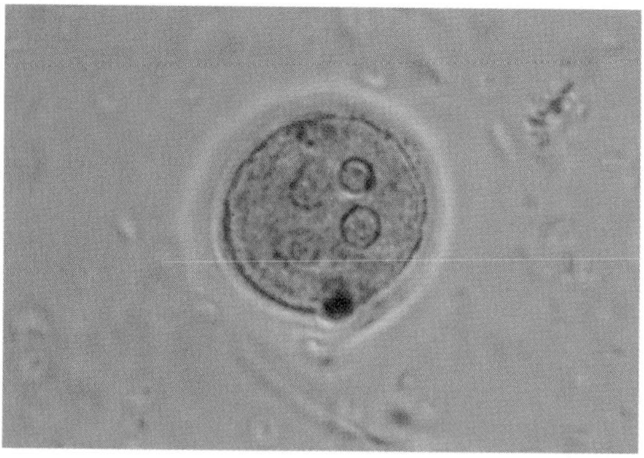

c.

Figure 15:6 Intestinal amebae and flagellates in wet mounts (1000X). a. This uninucleate cyst of *Entamoeba histolytica* has some pale areas in the cytoplasm which at one time contained glycogen masses which dissolved during storage in preservative. The nucleus is relatively large compared to the amount of cytoplasm. The dot-like karyosome is slightly eccentric. b. This mature cyst of *E. histolytica* shows two nuclei in this plane of focus. As nuclear division progresses the nuclei become smaller. c. This *Entamoeba coli* has two of the

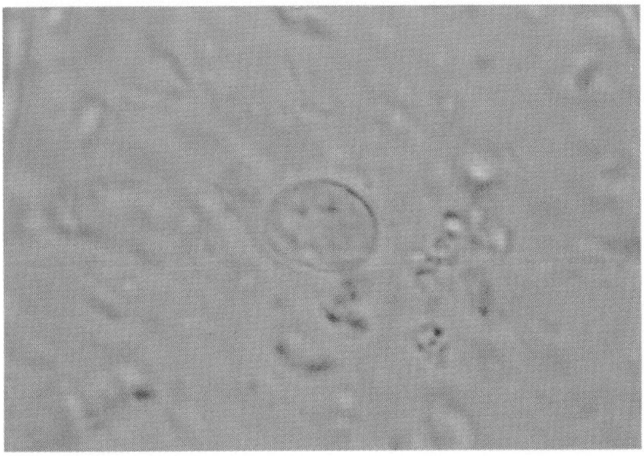

d.

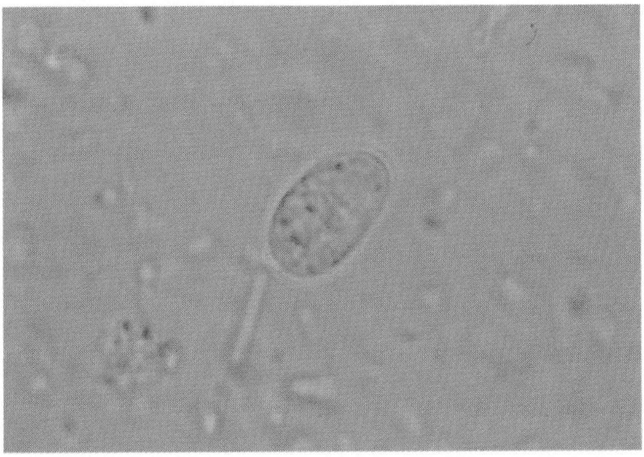

e.

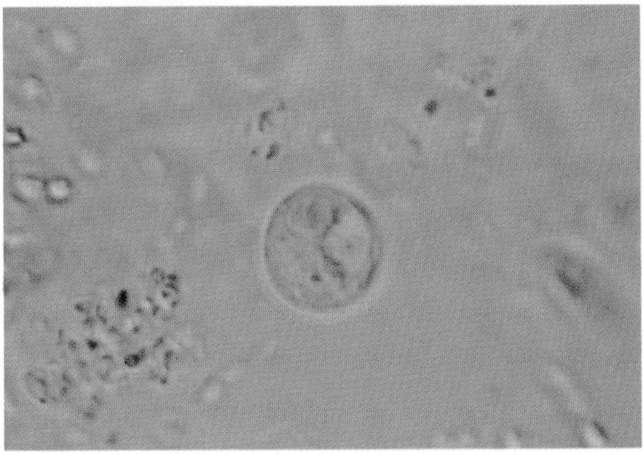

f.

eight nuclei in sharp focus at this focal plane. Nuclei in other focal planes are evident to the left. d. This cyst of *Endolimax nana* contains two nuclei in the plane of focus and two others in the lower portion of the organism which are slightly out of focus. e. This cyst of *E. nana* is oval and has all four nuclei visible at this plane of focus. f. This cyst of *Iodamoeba butschlii* has a faintly visible nucleus at the upper margin and a glycogen mass to the right. The glycogen mass does not stain with iodine because the glycogen has dissolved

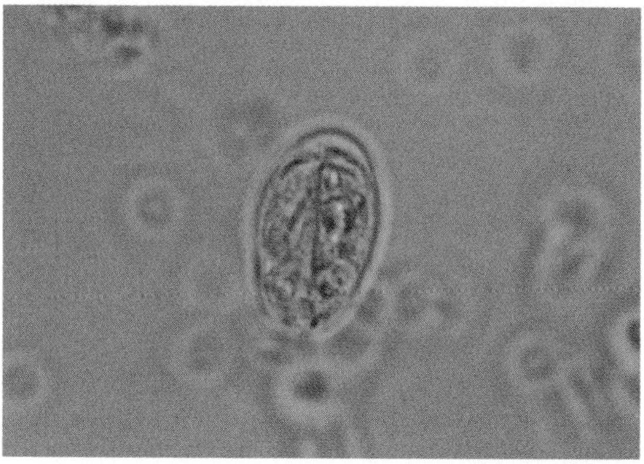

g.

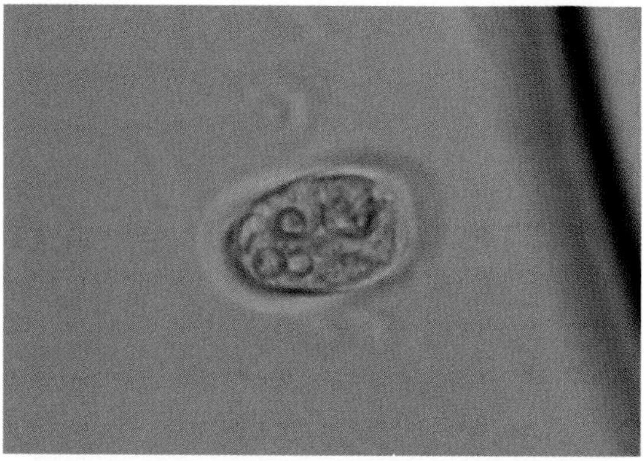

h.

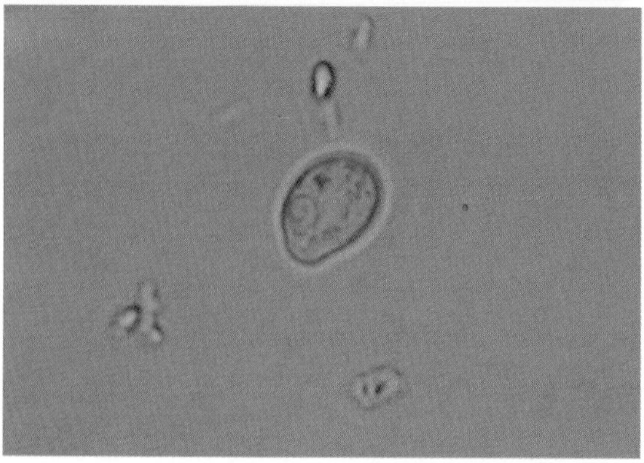

i.

during storage in formalin. g. This cyst of *Giardia lamblia* has two nuclei at the lower end and several fibrils within the cytoplasm. h. This cyst of *G. lamblia* has three nuclei in this plane of focus and one fibril is prominent. i. This cyst of *Chilomastix mesnili* is lemon-shaped with a nipple-like projection at the lower end and the nucleus above and to the left. Fibrils are evident in the cytoplasm.

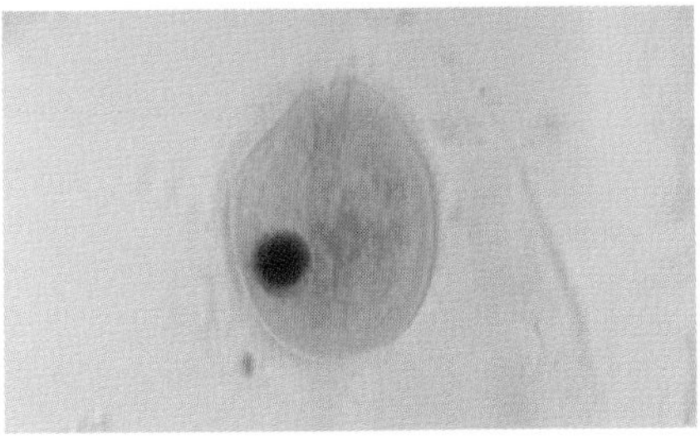

a.

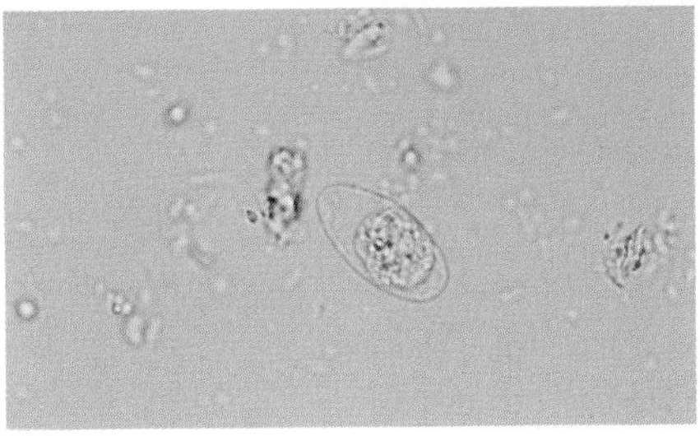

b.

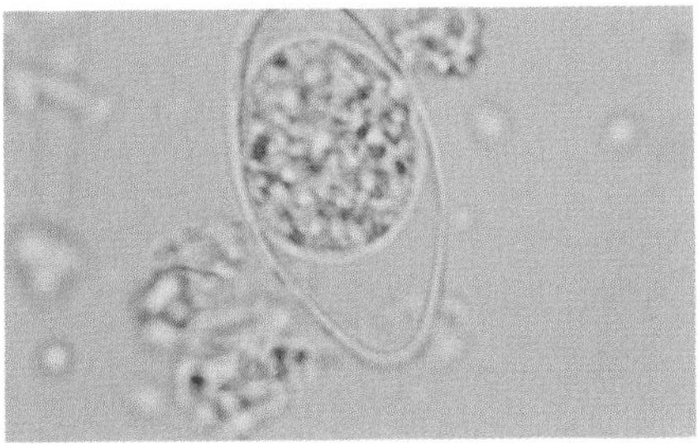

c.

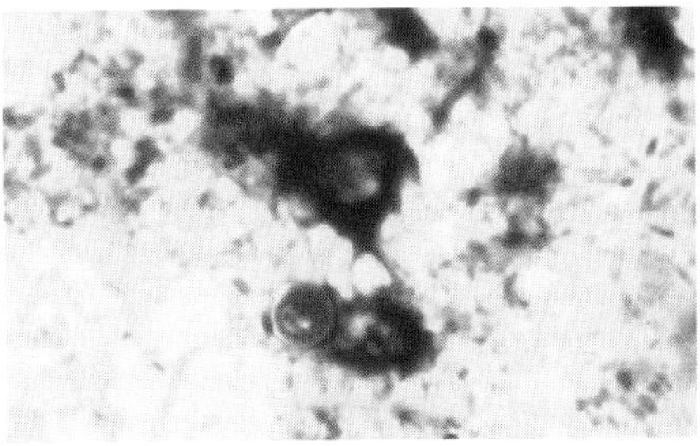

d.

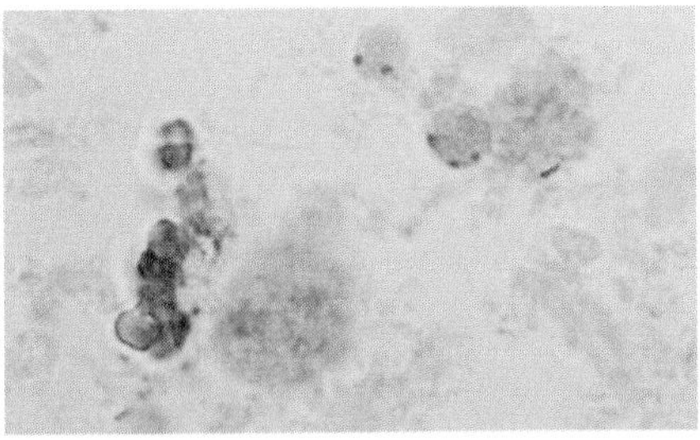

e.

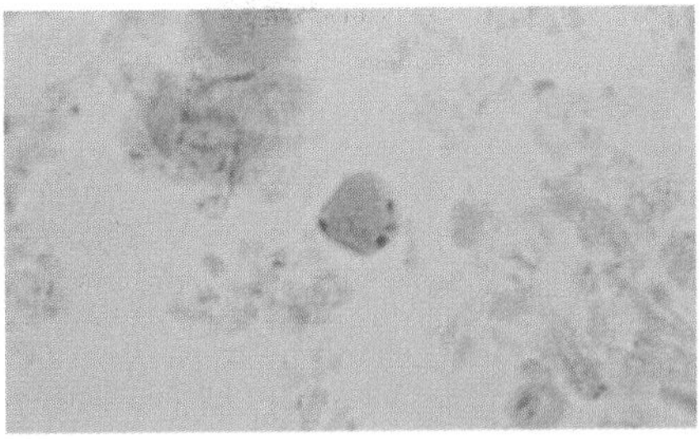

f.

Figure 15:8 Intestinal ciliate, coccidia and *Blastocystis*. a. This cyst of *Balantidium coli* stained with trichrome is photographed at low magnification (400X). There are numerous cilia over the surface and the cytostome is at the upper end. The large macronucleus is evident but the micronucleus is not seen. b. This oocyst of *Isospora* contains a sporoblast and would require several days of maturation before it was infective (unstained, 400X). c. This is the same oocyst of *Isospora* photographed at 1000X to show the undeveloped sporoblast. d. This acidfast stain of *Cryptosporidium* shows two bright red, thick-walled oocysts (1000X). e. There are two *Blastocystis* organisms in this field which have dark peripheral granules (Trichrome stain, 1000X). f. This *Blastocystis* organism has several peripheral granules which surround a central homogeneous blue area (Trichrome stain, 1000X).

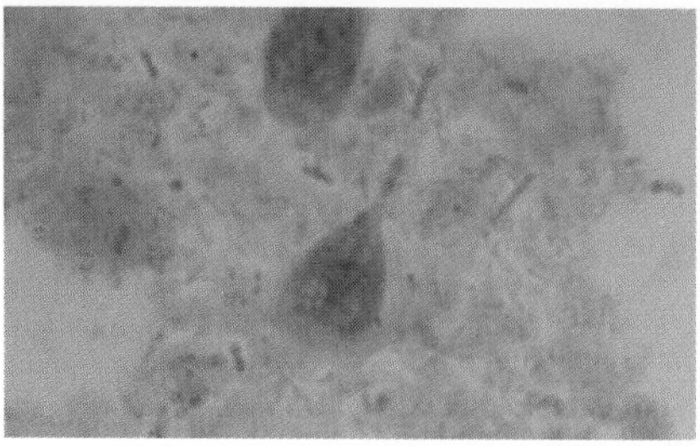

a.

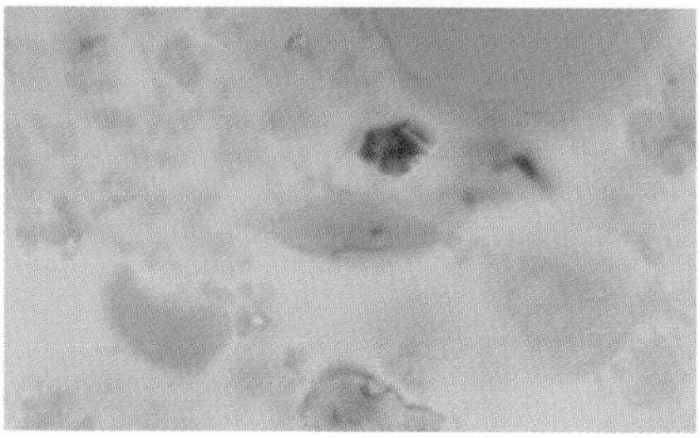

b.

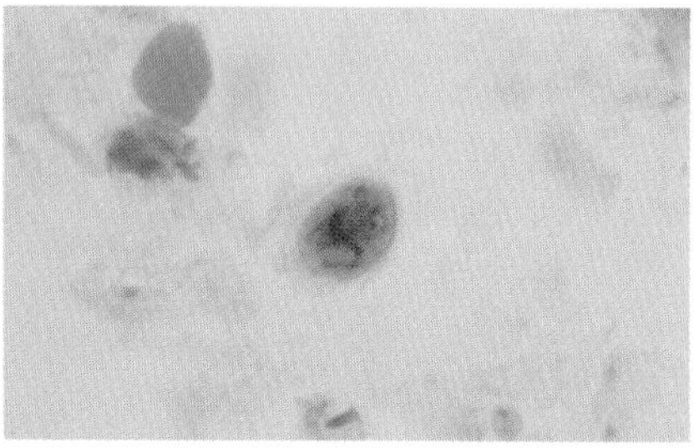

c.

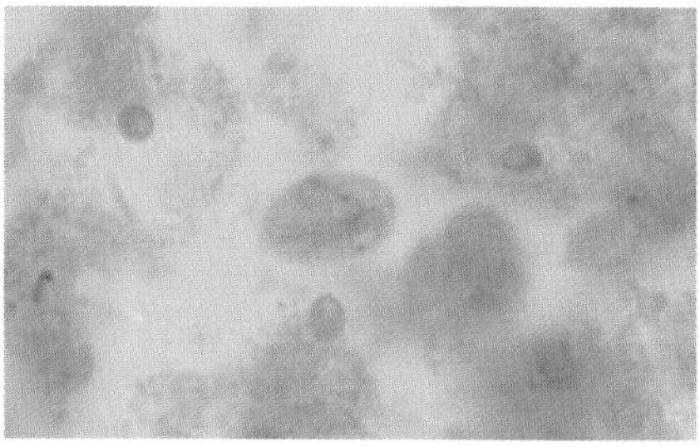

d.

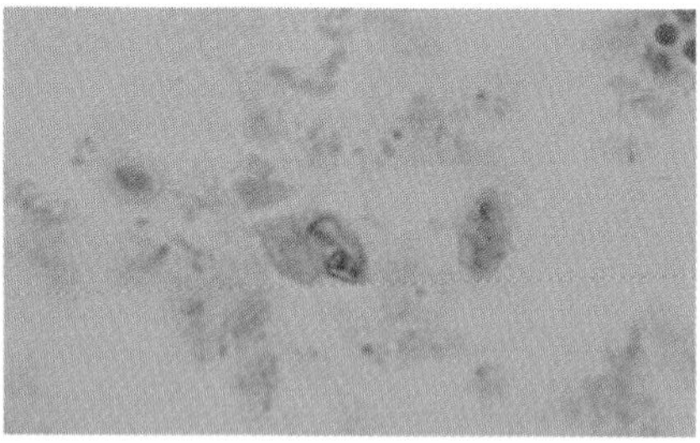

e.

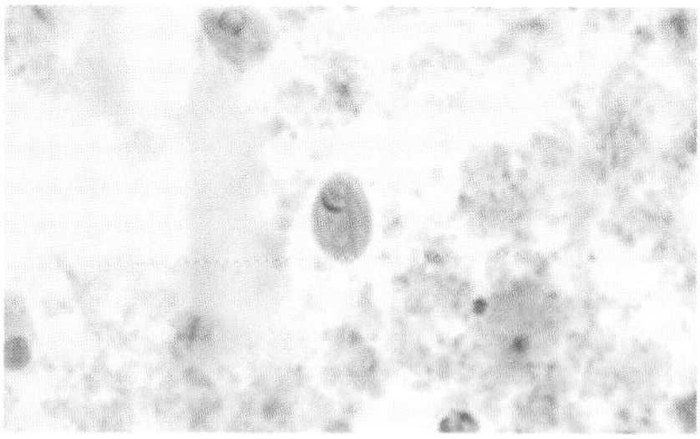

f.

Figure 15:10 Intestinal flagellates in trichrome stain (1000X). a. Two trophozoites of *Giardia lamblia* are evident in this field. The upper one shows greater detail with the nuclei and median body being prominent. b. This *G. lamblia* trophozoite is viewed from the side. Only one nucleus is evident at this plane of focus and it is immediately above the sucking disc which stains darker than the rest of the cell membrane. c. This *G. lamblia* cyst has two nuclei in this plane of focus plus fibrils in the cytoplasm. d. This *G. lamblia* cyst has three nuclei evident to the right with fibrils traversing the cytoplasm. e. This *Chilomastix mesnili* trophozoite has the nucleus to the right. Above it is the pale cytostome. Note the pointed posterior end to the left. f. This cyst of *C. mesnili* is lemon-shaped with the nucleus at the upper end and fibrils in the cytoplasm.

A

B

C

D

E

F

G

H

I

K

L

M

P

Q

R

S

U

V

W

X

Y

MANUFACTURERS CITED IN CHAPTERS

Abbott Laboratories
Diagnostic Division
P.O. Box 2020
Irving, TX 75061
1-800-323-9100

Aldrich Chemical Company, Inc.
940 West Saint Paul Avenue
Milwaukee, WI 53233
1-800-558-9160

Alcon Laboratories, Inc.
P.O. Box 1959
6201 South Freeway
Fort Worth, TX 76134
1-817-293-0450

American Scientific Products
1430 Waukegan Road
McGaw Park, IL 60085
1-312-689-8410

American Type Culture Collection
12301 Parklawn Drive
Rockville, MD 20852
1-800-638-6597

Amicon Division
W.R. Grace and Company
17 Cherry Hill Drive
Danvers, MA
617-777-3790

Analytab Products
Division of Sherwood Medical
200 Express Street
Plainville, NY 11803
1-800-645-0666

Baxter - Healthcare Corporation
Microscan Division
1584 Enterprise Blvd
West Sacramento, CA 95691
1-800-523-1186

BBL Microbiology Systems
Becton Dickinson and Company
P.O. Box 243
Cockeysville, MD 21030
1-800-638-8663

Becton Dickinson and Company
Diagnostic Instrument Systems
383 Hillen Road
P.O. Box 20086
Towson, MD 21204
1-800-638-8656

Calbiochem, Behring Diagnostics
P.O. Box 12087
San Diego, CA 92111
1-800-854-9256

Difco Laboratories, Inc.
P.O. Box 1058
Detroit, MI 48232
1-800-521-0851

Du Pont de Nemours
Clinical and Instrument Systems
Concord Plaza
Wilmington, DE 19898
1-800-435-7222

E.R. Squibb and Sons
P.O. Box 4000
Princeton, NJ 08540
1-609-921-4000

Evergreen Scientific
P.O. Box 58248
2300 East 49th Street
Los Angeles, CA 90058
1-800-421-6261

Fisher Scientific
711 Forbes Avenue
Pittsburgh, PA 15219
1-412-963-1664

Flow Laboratories, Inc.
7655 Old Springhouse Road
McLean, VA 22102
1-800-368-FLOW

Gibco Laboratories
P.O. Box 4385
2801 Industrial Drive
Madison, WI 53711
1-800-828-6686

Hedeco Corporation
2411 Pulgas Avenue
Palo Alto, CA 94303
1-415-325-7874

Hollister-Stier Laboratories
P.O. Box 3145, Terminal Annex
Spokane, WA 99220
1-509-489-5656

Immuno-Mycologics, Inc.
P.O. Box 1151
Norman, OK 73070
1-800-654-3639

International Biological
Laboratories, Inc.
Wampole Laboratories
Half Acre Road
Cranbury, NJ 08512
1-609-655-6000

Janssen Pharmaceutica
40 Kingsbridge Road
Piscataway, NJ 08554
1-800-624-0137

Johnston Laboratories
Diagnostic Instrument Systems
Becton Dickinson and Company
383 Hillen Road
Towson, MD 21204
1-301-377-8700

Otisville BioPharm, Inc.
P.O. Box 567
Otisville, NY 10963
914-386-2891

L.L. Pellet Company
8843 Larchwood Drive
Dallas, TX 75238
1-214-348-2967

M and Q Packaging Company
P.O. Box 180
Schuylkil Haven, PA
1-717-385-4991

M.A. Bioproducts
P.O. Box 127
Biggs Ford Road
Walkersville, MD 21793
1-800-638-8174

Meridian Diagnostics, Inc.
3471 River Hills Drive
Cincinnati, OH 45244
1-800-543-1980

Minnesota Mining and
Manufacturing Company
3M Center
St. Paul, MN 55144-1000
1-612-733-1110

Nolan-Scott Biological
Laboratories, Inc.
4958 Hammermill Road
Tucker, GA 30084
1-404-939-8284

Ortho Pharmaceutical Corporation
P.O. Box 202
Raritan, NJ 08869
1-800-631-5807

Pfizer Central Research,
Pfizer, Inc.
Eastern Point Road
Groton, CT 06340
1-203-441-4100

Polysciences, Inc.
400 Valley Road
Warrington, PA 18976
1-800-523-2575

Ramco Laboratories, Inc.
3801 Kirby Drive, Suite 170
Houston, TX 77098
1-800-231-6238

Remel Laboratories
12076 Santa Fe Drive
Lenexa, KS 66215
1-800-255-6730

Roche Laboratories
Division of Hoffman-LaRoche, Inc.
340 Kingsland Street
Nutley, NJ 07110
1-201-235-5082

Sigma Chemical Company
P.O. Box 14508
St. Louis, MO 63178
1-800-325-3010

Spectrum Diagnostics, Inc.
15413 Vantage Parkway East
Houston, TX 77032

Tekmar Company
P.O. Box 371856
Cincinnati, OH 45222
1-800-543-4461

Upjohn Company
301 Henrietta Street
Kalamazoo, MI 49007
1-800-253-9600

Vitek Systems
Subsidiary of McDonnell Douglas
595 Anglum Drive
Hazlewood, MO 63042
1-800-325-1977